Oxidative Stress in Cancer, AIDS, and Neurodegenerative Diseases

OXIDATIVE STRESS AND DISEASE

Series Editors

LESTER PACKER, PH.D.
University of California
Berkeley, California

ENRIQUE CADENAS, M.D., P.H.D.
University of Southern California School of Pharmacy
Los Angeles, California

1. Oxidative Stress in Cancer, AIDS, and Neurodegenerative Diseases, *edited by Luc Montagnier, René Olivier, and Catherine Pasquier*

Additional Volumes in Preparation

Understanding the Process of Aging: The Roles of Mitochondria, Free Radicals, and Antioxidants, *edited by Enrique Cadenas and Lester Packer*

Related Volumes

Vitamin E in Health and Disease: Biochemistry and Clinical Applications, *edited by Lester Packer and Jürgen Fuchs*

Vitamin A in Health and Disease, *edited by Rune Blomhoff*

Free Radicals and Oxidation Phenomena in Biological Systems, *edited by Marcel Roberfroid and Pedro Buc Calderon*

Biothiols in Health and Disease, *edited by Lester Packer and Enrique Cadenas*

Handbook of Antioxidants, *edited by Enrique Cadenas and Lester Packer*

Handbook of Synthetic Antioxidants, *edited by Lester Packer and Enrique Cadenas*

Vitamin C in Health and Disease, *edited by Lester Packer and Jürgen Fuchs*

Lipoic Acid in Health and Disease, *edited by Jürgen Fuchs, Lester Packer, and Guido Zimmer*

Flavonoids in Health and Disease, *edited by Catherine Rice-Evans and Lester Packer*

Oxidative Stress in Cancer, AIDS, and Neurodegenerative Diseases

edited by

LUC MONTAGNIER
Centre National de la Recherche Scientifique
Institut Pasteur
Paris, France

RENE OLIVIER
Institut Pasteur
Paris, France

CATHERINE PASQUIER
Centre National de la Recherche Scientifique
Faculté Xavier-Bichat
INSERM U294
Paris, France

CRC Press is an imprint of the
Taylor & Francis Group, an **informa** business

CRC Press
Taylor & Francis Group
6000 Broken Sound Parkway NW, Suite 300
Boca Raton, FL 33487-2742

First issued in paperback 2019

ISBN-13: 978-0-8247-9862-8 (hbk)
ISBN-13: 978-0-367-40082-8 (pbk)

Visit the Taylor & Francis Web site at
http://www.taylorandfrancis.com

and the CRC Press Web site at
http://www.crcpress.com

Series Introduction

Oxygen is a dangerous friend. Overwhelming evidence indicates that oxidative stress can lead to cell and tissue injury. However, the same free radicals that are generated during oxidative stress are produced during normal metabolism and thus are involved in both human health and disease.

Free radicals are molecules with an odd number of electrons. The odd, or unpaired, electron is highly reactive as it seeks to pair with another free electron.

Free radicals are generated during oxidative metabolism and energy production in the body.

Free radicals are involved in:

- Enzyme-catalyzed reactions
- Electron transport in mitochondria
- Signal transduction and gene expression
- Activation of nuclear transcription factors
- Oxidative damage to molecules, cells, and tissues
- Antimicrobicidal action of neutrophils and macrophages
- Aging and disease

Normal metabolism is dependent upon oxygen, a free radical. Through evolution, oxygen was chosen as the terminal electron acceptor for respiration. The two unpaired electrons of oxygen spin in the same direction; thus, oxygen is a biradical, but is not a very dangerous free radical. Other oxygen-derived free radical species, such as superoxide or hydroxyl radicals, formed during metabolism or by ionizing radiation are stronger oxidants and are therefore more dangerous.

In addition to research on the biological effects of these reactive oxygen species, research on reactive nitrogen species has been gathering momentum. NO, or nitrogen monoxide (nitric oxide), is a free radical generated by NO synthase (NOS). This enzyme modulates physiological responses such as vasodilation or signaling in the brain.

However, during inflammation, synthesis of NOS (iNOS) is induced. This iNOS can result in the overproduction of NO, causing damage. More worrisome, however, excess NO can react with superoxide to produce the very toxic product peroxynitrite. Oxidation of lipids, proteins, and DNA can result, thereby increasing the likelihood of tissue injury.

Both reactive oxygen and nitrogen species are involved in normal cell regulation in which oxidants and redox status are important in signal transduction. Oxidative stress is increasingly seen as a major upstream component in the signaling cascade involved in inflammatory responses, stimulating adhesion molecule and chemotractant production. Hydrogen peroxide, which breaks down to produce hydroxyl radicals, can also activate NFκB, a transcription factor involved in stimulating inflammatory responses. Excess production of these reactive species is toxic, exerting cytostatic effects, causing membrane damage, and activating pathways of cell death (apoptosis and/or necrosis).

Virtually all diseases thus far examined involve free radicals. In most cases, free radicals are secondary to the disease process, but in some instances free radicals are causal. Thus, there is a delicate balance between oxidants and antioxidants in health and disease. Their proper balance is essential for ensuring healthy aging.

The term *oxidative stress* indicates that the antioxidant status of cells and tissues is altered by exposure to oxidants. The redox status is thus dependent upon the degree to which a cell's components are in the oxidized state. In general, the reducing environment inside cells helps to prevent oxidative damage. In this reducing environment, disulfide bonds (S–S) do not spontaneously form because sulfhydryl groups kept in the reduced state (SH) prevent protein misfolding or aggregation. This reducing environment is maintained by oxidative metabolism and by the action of antioxidant enzymes and substances, such as glutathione, thioredoxin, vitamins E and C, and enzymes as superoxide dismutase (SOD), catalase, and the selenium-dependent glutathione and thioredoxin hydroperoxidases, which serve to remove reactive oxygen species.

Changes in the redox status and depletion of antioxidants occur during oxidative stress. The thiol redox status is a useful index of oxidative stress mainly because metabolism and NADPH-dependent enzymes maintain cell glutathione (GSH) almost completely in its reduced state. Oxidized glutathione (glutathione disulfide, GSSG) accumulates under conditions of oxidant exposure, and this changes the ratio of oxidized to reduced glutathione; and increased ratio indicates oxidative stress. Many tissues contain large amounts of glutathione 2–4 mM in erythrocytes or neural tissues and up to 8 mM in hepatic tissues. Reactive oxygen and nitrogen species can directly react with glutathione to lower the levels of this substance, the cell's primary preventative antioxidant.

Current hypotheses favor the idea that lowering oxidative stress can have a clinical benefit. Free radicals can be overproduced or the natural antioxidant system defenses weakened, first resulting in oxidative stress, and then leading to oxidative injury and disease. Examples of this process include heart disease and cancer. Oxidation of human low-density lipoproteins is considered the first step in the progression and eventual development of atherosclerosis, leading to cardiovascular disease. Oxidative DNA damage initiates carcinogenesis.

Compelling support for the involvement of free radicals in disease development comes from epidemiological studies showing that an enhanced antioxidant status is associated with reduced risk of several diseases. Vitamin E and prevention of cardiovascular disease is a notable example. Elevated antioxidant status is also associated with decreased incidence of cataracts and cancer, and some recent reports have suggested an inverse correlation between antioxidant status and occurrence of rheumatoid arthritis and diabetes mellitus. Indeed, the number of indications in which antioxidants may be useful in the prevention and/or the treatment of disease is increasing.

Oxidative stress, rather than being the primary cause of disease, is more often a secondary complication in many disorders. Oxidative stress diseases include inflammatory bowel disease, retinal ischemia, cardiovascular disease and restenosis, AIDS, ARDS, and neurodegenerative diseases such as stroke, Parkinson's disease, and Alzheimer's disease. Such indications may prove amenable to antioxidant treatment because there is a clear involvement of oxidative injury in these disorders.

In this new series of books, the importance of oxidative stress in diseases associated with organ systems of the body will be highlighted by exploring the scientific evidence and the medical applications of this knowledge. The series will also highlight the major natural antioxidant enzymes and antioxidant substances such as vitamins F, A, and C, flavonoids, polyphenols, carotenoids, lipoic acid, and other nutrients present in food and beverages.

Oxidative stress is an underlying factor in health and disease. More and more evidence is accumulating that a proper balance between oxidants and antioxidants is involved in maintaining health and longevity, and that altering this balance in favor of oxidants may result in pathological responses causing functional disorders and disease. This series is intended for researchers in the basic biomedical sciences and clinicians. The potential for healthy aging and disease prevention necessitates gaining further knowledge about how oxidants and antioxidants affect biological systems.

The first book in this series, *Oxidative Stress in Cancer, AIDS, and Neurodegenerative Diseases*, edited by Luc Montagnier, René Olivier, and Catherine Pasquier, includes contributions by leading researchers that provide unprecedented insight into the understanding of the role of oxidants and redox mechanisms in pathophysiological processes involved in various oxidative stress diseases. This book is timely and relevant to all those who work in biomedical sciences, and advances our understanding of oxidant and antioxidant actions in health and disease. This ultimately leads to new therapeutic strategies in the treatment and prevention of disease.

Lester Packer
Enrique Cadenas

Preface

Over the last 20 years, oxygen-free radicals have been of great interest as they have been found to be involved in many pathophysiological processes, including those of AIDS, cancer, inflammation, aging, and neurodegenerative diseases. Many symposia have addressed toxic effects on biological molecules such as lipids, proteins, sugars, and DNA, introducing the mechanisms of molecular, cellular, and tissue damage. At the same time, many antioxidants – vitamin E, β-carotene, lipoic acid, flavonoids, thiol derivatives, and precursors such as NAC – have been shown in vitro and in vivo to regulate the cellular redox status. Cell and tissue protection has been shown to occur with these molecules, even though some seem to also have antagonist pro-oxidant effects under specific conditions.

Cells are subjected to oxidative stress from a wide variety of sources, including environmental agents and oxidizing agents, generated within the cells or by other cells, causing damage to the components of biological systems. Much recent work has shown, however, that oxidative stress can also activate intracellular signal transduction. A symposium devoted to this subject was held in May 1996 at the Institut Pasteur in Paris. The aim of the symposium, "Oxidative Stress and Redox Regulation: Cellular Signaling, AIDS, Cancer and Other Disease," was to show a relationship between cytotoxic effects of reactive oxygen species (ROS) and regulation of intracellular signaling pathways by ROS. Since antioxidants have been shown to protect cells from cytotoxic effects, it appears that they also have a regulatory action on intracellular signaling.

This book presents different views of oxygen and nitrite-free radicals, from the chemical to the clinical standpoints, via biochemistry and cellular and molecular biology. These species are now turning up everywhere, and become more and more important with respect to their wide-ranging effects.

This book contains 50 chapters which can be divided into seven sections, beginning with chapters describing the importance of oxidant/antioxidant balance. Indeed, cellular well-being depends on achieving and maintaining a certain concentration of radicals, which is modulated by antioxidant production. The oxidative damage of biological molecules and of membranes and cells induced by free radicals, and which can be inhibited by various antioxidants, is described here as an introduction of the different functions that ROS induce from the outside to the inside of the cells, leading to cell injury or cell activation.

One current field of investigation in cellular activation is the intracellular signaling induced by ROS that controls the production of new proteins. The second section discusses different pathways leading to cell redox regulation through activation of the nuclear factor NFκB, demonstrating the active role that ROS play in inflammatory and infectious processes. A wide range of other intracellular signaling pathways are also reviewed, and it is shown that H_2O_2 as well as $NO^\bullet$ can induce protein synthesis and protein kinase activation by protein phosphorylation.

Apoptosis, presented in the third section, is the very important consequence of the action of ROS on several pathways that have not been completely elucidated. Apoptosis – physiological cell death – occurs in morphogenesis as well as in pathological processes such as neurodegenerative diseases. Hence, neuronal apoptosis may be a mechanism by which cells die in Alzheimer's disease and Parkinson's disease. Furthermore, in HIV-infected individuals, many of the CD4 lymphocytes that die are not infected by the virus but self-destruct through apoptosis. The external signals that lead to apoptosis are varied, and several are known to induce oxidative stress. It is now clear that increased generation of ROS and/or ROS effects participate in the degradation phase of apoptosis. The mechanisms by which ROS are involved in the induction of apoptosis could be the activation of a protease family related to interleukin-1 converting enzyme (ICE) and an alteration in thiol redox.

The fourth section presents the redox regulation by endogenous antioxidant molecules including thioredoxin and glutaredoxin; the exogenous antioxidant molecules, such as lipoic acid and flavonoids, have also been shown to be effective in therapeutics. However, it has been questioned whether some of them may sometimes be pro-oxidants instead of antioxidants.

The pathophysiological role of oxidative stress – a lack of balance between oxidants and reductants – is reviewed in the fifth section. It occurs in various cellular components, including DNA, mitochondria, and cytoskeleton, as well as during physiological processes such as perinatal development and during pathological processes such as malignant and inflammatory diseases.

The involvement of ROS in AIDS is then explored in the sixth section, in which glutathione is shown to be a key factor in the improvement of patient survival. Tests of antioxidant therapy in patients have shown efficacy of NAC in regenerating intracellular glutathione. In ex vivo cultures of AIDS patients, lymphocytes, antioxidant treatments (SOD, NAC, β-carotene, L-carnitine, etc.) have been tested on different deleterious oxidative mechanisms.

The last section discusses the importance of ROS in a wide array of pathologies, including neurodegenerative disease, disorders associated with aging, diabetes, cataracts, and inflammatory diseases such as rheumatoid arthritis.

This book brings together important work on the study of the intracellular pathways of ROS in cellular injury as well as in cellular signaling. It is addressed to fundamental researchers looking for intracellular mechanisms as well as to physicians working with diseases such as AIDS and neuronal or inflammatory diseases. Those valuable results bring us to a better understanding of the role of ROS in pathophysiology, and thus open a wide field of investigation connecting disciplines ranging from radical chemistry to antioxidant therapy.

Luc Montagnier
René Olivier
Catherine Pasquier

Contents

Series Introduction *Lester Packer and Enrique Cadenas* iii
Preface vii

Oxidative Stress in Cell Biology and Biochemistry

1. The Importance of Oxidant–Antioxidant Balance 1
   Joe M. McCord
2. The Effect of Oxidative Stress on Cells by Oxygen Radicals and Its Inhibition by Antioxidants 9
   Mareyuki Takahashi and Etsuo Niki
3. Cytotoxicity of Radiation-Derived Reactive Oxygen Species 15
   Manfred Saran and Wolf Bors

Oxidative Stress and Cellular Signaling

4. Regulation of the Bacterial Response to Hydrogen Peroxide 29
   Gisela Storz, Ming Zheng, Aixia Zhang, Ines Kullik, Michel Toledano, and Shoshy Altuvia
5. Activation of Lymphocyte Signal Pathways by Oxidative Stress: Role of Tyrosine Kinases and Phosphatases 35
   Gary L. Schieven
6. Oxidant-mediated Modulation of Phospholipases in Endothelial Cells: Possible Implication in Vascular Disorders 45
   V. Natarajan, Suryanarayana Vepa, and William M. Scribner

7. NFκB Transcription Factor Activation: Importance of the Redox Regulation 63
Bernard Piret, Giuseppina Bonizzi, Marie-Paule Merville, Vincent Bours, and Jacques Piette

8. Redox Regulation of the NFκB Signaling Pathway and Disease Control 75
Takashi Okamoto, Shinsaku Sakurada, Tetsuji Kato, Akihiko Yoneyama, Keiichi Tozawa, Jian-Ping Yang, and Naoko Takahashi

9. Mechanism of Activation of NFκB by Hydrogen Peroxide 83
Kirsty G. McPherson, John R. Babson, Anthony J. Kettle, and Christine C. Winterbourn

10. Self-regulation of the NFκB/IκBα System, and Its Positive and Negative Control by the Cellular Redox Status 89
Jean-Louis Virelizier, Fernando Arenzana-Seisdedos, Manuel Rodriguez, Catherine Dargemont, Ronald T. Hay, and Nicholas Hunt

11. The Role of Hydrogen Peroxide in Signal Transduction and the Involvement of Iron 97
B. S. van Asbeck, R. C. Sprong, T. van der Bruggen, J. F. L. M. van Oirschot, J. C. C. Borleffs

12. Redox Regulation of Heat Shock Protein Expression and Protective Effects Against Oxidative Stress 113
Abdelhamid El Yaagoubi, Ewa Mariéthoz, Muriel R. Jacquier-Sarlin, and Barbara S. Polla

13. Low-Density Lipoproteins Modulate Nitric Oxide and Heme Oxygenase Signal Transduction Pathways in Vascular Endothelium and Smooth Muscle 127
R. C. M. Siow, M. T. Jay, H. Sato, J. D. Pearson, and G. E. Mann

14. Nitric Oxide-stimulated Tyrosine Phosphorylation-dependent Signaling Pathways in Cultured Cells 139
Hugo P. Monteiro, Laura C. B. Oliveira, Tereza M. S. Peranovich, Renatta G. Penha, and Arnold Stern

15. Regulation of Cell Activation by Receptors for IgG 147
Marc Daëron and Wolf H. Fridman

16. Signal Transduction Pathways Activated by Mitogenic Neuropeptides 163
Enrique Rozengurt

Oxidative Stress and Apoptosis

17. Glutathione Metabolism During Apoptosis 179
Diels J. van den Dobbelsteen, C. Stefan I. Noble, Astrid Samuelsson, Sten Orrenius, and Andrew F. G. Slater

18. Endogenously Generated Superoxide and Hydrogen Peroxide in the Redox Modulation of Cell Proliferation, Apoptosis, and Virus Replication 191
Roy H. Burdon

19. Deoxy-D-Ribose-Induced GSH Depletion and Induction of Apoptosis in Human Quiescent Peripheral Blood Mononuclear Cells: Effects of *N*-Acetyl Cysteine and L-Buthionine-(*S,R*)-sulfoximine 199
B. Botti, D. Monti, S. Macchioni, S. Bergamini, F. Tropea, A. Tomasi, V. Vannini, and C. Franceschi

20. Nitric Oxide and Hydrogen Peroxide Production during Apoptosis of Human Neutrophils 205
Juanita Bustamante, Cecilia Carreras, Natalia Riobó, Alexis Tovar, Gonzalo Montero, Juan Jose Poderoso, and Alberto Boveris

21. The Mitochondrion as a Sensor/Effector of Oxidative Stress During Apoptosis 213
Philippe Marchetti, Santos A. Susin, Naoufal Zamzami, and Guido Kroemer

Oxidative Stress and Antioxidants

22. Antioxidants and the Regulation of Reactive Iron 223
John M. C. Gutteridge

23. Redox Regulation by the Thioredoxin and Glutaredoxin Systems 229
Arne Holmgren, Elias S. J. Arnér, Fredrik Åslund, Mikael Björnstedt, Zhong Liangwei, Johanna Ljung, Hajime Nakamura, and Dragana Nikitovic

24. Thioredoxin/Adult T Cell Leukemia-Derived Factor as the Key Redox Regulator of Signaling 247
Junji Yodoi, Yoshihisa Taniguchi, Tetsuro Sasada, and Kiichi Hirota

25. Therapeutic Potential of the Antioxidant and Redox Properties of α-Lipoic Acid 251
Chandan K. Sen, Sashwati Roy, and Lester Packer

26. Glutathione Compartmentation and Oxidative Stress 269
Melani K. Savage and Donald J. Reed

27. Flavonoid Functions In Vivo: Are They Predominantly Antioxidant? 285
Wolf Bors

28. Immunoregulatory Properties of Superoxide Dismutase 305
E. Postaire, B. Dugas, P. Debré, and C. Gudin

29. Pharmacological Inhibition of Endothelial Cytoskeleton Alterations Induced by Hydrogen Peroxide and TNF-α 313
P. d'Alessio, M. Moutet, C. Marsac, and J. Chaudière

30. Perinatal Development of Superoxide Dismutase and Other Antioxidant Enzyme Activities in Rat-brain Mitochondria 323
Ingrid Wiswedel, Sigrid Hoffmann, Heiko Noack, and Wolfgang Augustin

31. Coexistence of a "Reactive Oxygen Cycle" with "Q Cycle" and "H^+-Cycle" in the Respiratory Chain: A Hypothesis for the Generation, Partitioning, Targeting, and Functioning of Superoxide in Mitochondria 333
Shu-sen Liu

Oxidative Stress and Cancer

32. Endogenous Oxidative DNA Damage and Spontaneous Mutagenesis: Role of 8-Oxoguanine 351
Serge Boiteux

33. The Elevated Serum Level of Thioredoxin in Patients with Malignant Disease and Chronic Inflammatory Diseases 359
Hiro Wakasugi, Kunihisa Miyazaki, Michio Miyata, and Masaaki Terada

34. The Role of Free Radicals in Antitumor Effects of Various Cancer Treatments 369
Toshikazu Yoshikawa, Satoshi Kokura, and Motoharu Kondo

Oxidative Stress and AIDS

35. The Role of Cysteine and Glutathione in the Pathogenesis of HIV Infection: Effects of Treatment with N-Acetylcysteine 373
Wulf Dröge, Hans-Peter Eck, Heike Schenk, Klaus Schulze-Osthoff, Dagmar Galter, George Shubinsky, Sabine Mihm, Volker Hack, and Ralf Kinscherf

36. Low Glutathione Levels in CD4 T Cells Predict Poor Survival in AIDS; *N*-Acetylcysteine May Improve Survival 379
Leonard A. Herzenberg, Stephen C. De Rosa, and Leonore A. Herzenberg

37. Modulation of HIV-1 Long Terminal Repeat by Arachidonic Acid 389
Simonetta Camandola, Tiziana Musso, Gabriella Leonarduzzi, Rita Carini, Luigi Varesio, Patrick A. Baeuerle, and Giuseppe Poli

38. Protein Degradation in Lymphocytes as an Indicator of Oxidative Stress in HIV Infection 399
Giuseppe Piedimonte, Mauro Magnani, Dario Corsi, Javier F. Torres Roca, Denise Guetard, and Luc Montagnier

39. Oxidative Stress and AIDS: One-year Supplementation of HIV-positive Patients with Selenium or β-Carotene 409
C. Sergeant, C. Hamon, M. Simonoff, J. Constans, E. Peuchant, M. C. Delmas, J. L. Pellegrin, I. Pellegrin, M. Clerc, B. Leng, H. Fleury, and C. Conri

40. Glutathione Oxidation and Mitochondrial DNA Damage in AIDS: Effect of Zidovudine 429
José Viña, José García-de-la-Asunción, María L. Del Olmo, Arantxa Millán, Juan Sastre, José A. Martín, and Federico V. Pallardó

41. Effect of L-Carnitine Treatment In Vivo on Apoptosis and Ceramide Generation by Peripheral Blood Lymphocytes from AIDS Patients 437
M. G. Cifone, E. Alesse, L. Di Marzio, P. Roncaioli, F. Zazzeroni, S. Moretti, G. Famularo, S. Marcellini, G. Santini, V. Trinchieri, E. Nucera, and C. De Simone

42. Nutriceutical Modulation of Glutathione with a Humanized Native Milk Serum Protein Isolate, IMMUNOCAL™: Application in AIDS and Cancer 447
Sylvain Baruchel, Ginette Viau, René Olivier, Gustavo Bounos, and Mark A. Wainberg

43. Successful Antioxidant Therapy Including Superoxide Dismutase Associated with Antiretroviral Therapy in an HIV-Infected Patient with Hepatitis B-related Cirrhosis 463
J. Emerit, E. Postaire, D. Bonnefont-Rousselot, O. Lopez, and F. Bricaire

Oxidative Stress in Neurodegenerative Diseases and Other Pathologies

44. Parkinson's Disease, Apoptosis, and Oxidative Stress 469
Merle Ruberg, Valentine France-Lanord, Bernard Brugg, Stéphane Hunot, Philippe Anglade, Philippe Damier, Baptiste Faucheux, and Yves Agid

45. Protein Oxidation and Glycation in Neurodegenerative Diseases 485
Mark A. Smith, George Perry, and Lawrence M. Sayre

46. Glycation as Oxidative Stress and Redox Regulation: Implications for Aging, Diabetes, and Familial Amyotrophic Lateral Sclerosis 497
Naoyuki Taniguchi, Junichi Fujii, Hideaki Kaneto, Michio Asahi, Theingi Myint, Nobuko Miyazawa, Keiichiro Suzuki, and Kazi Nazrul Islam

47. Oxidative Stress in Ischemia – Reperfusion: Reilly's Irritation Syndrome Revisited 503
Makoto Suematsu and Masaharu Tsuchiya

48. Evaluation of the Antiperoxidase Defense System of the Lens Utilizing GSH Peroxidase Transgenics and Knockouts 507
Abraham Spector, Ye-Shih Ho, Ren-Rong Wang, Wanchao Ma, Yinqing Yang, and Wan-Cheng Li

49. The Rheumatoid Joint: Redox-Paradox? 517
Vanessa Gilston, David R. Blake, and Paul G. Winyard

50. Chronic Oxidative Stress in Rheumatoid Arthritis: Implications for T Cell Function 537
Madelon M. Maurice, Hajime Nakamura, Ellen A. M. van der Voort, Sussanne Thorell, Anita I. van Vliet, Paul-Peter Tak, Ferdinand C. Breedveld, and Cornelis L. Verweij

Index 547

1
The Importance of Oxidant–Antioxidant Balance

Joe M. McCord
Webb-Waring Institute
University of Colorado Health Sciences Center
Denver, Colorado

INTRODUCTION

For a number of years following the recognition that cellular metabolism can generate free radicals and other oxygen-derived oxidants, elevated oxidative status was viewed only as a potential cause of cellular damage. The antioxidant enzymes known as the superoxide dismutases (SOD), together with catalase and glutathione peroxidase, were viewed as essential, protective "housekeeping" enzymes, designed to effectively scavenge and eliminate their toxic substrates. One notable exception to this general concept was the realization that phagocytes produce superoxide radical not as a toxic by-product but rather as weapon in their war against invading microbes (1). Thus, in this limited context, superoxide production could be viewed as desirable or positive, even though its cellular action was still that of cytotoxicity directed against both the ingested microbe and the ingesting phagocyte. Extracellular superoxide dismutase, for example, was found capable of preventing the phagocytosis-induced death of the neutrophil (2).

We have gradually accumulated much evidence, both circumstantial and direct, suggesting that for an optimal state of health, whether cellular or organismal, there must be a balance between rate of production of superoxide radical and the amount of SOD present. Serious problems result from either the overproduction of superoxide or the underproduction of SOD. Ironically, similar problems result from the overproduction of SOD, and the presumed "overscavenging" of superoxide. A number of published reports document these concepts.

THE BELL-SHAPED CURVE

The toxicity of superoxide is readily demonstrated by such compounds as paraquat (1,1′-dimethyl-4,4′-bipyridinium dichloride) or MPTP (1-methyl-4-phenyl-1,2,3,6-tetrahydropyridine). These compounds are reduced within the cell and generate superoxide as they autoxidize (3). That their cytotoxicities are mediated primarily by superoxide generation is demonstrated by the fact that *Escherichia coli* that overexpress SOD are resistant to paraquat (4) and transgenic mice that overexpress SOD are resistant to the neurotoxicity of MPTP (5). Paraquat induces SOD synthesis in *E. coli* (6). Conversely, *E. coli* that are unable to make either Mn-SOD or Fe-SOD (*sodA sodB* mutants) are exquisitely sensitive to paraquat in rich medium, and are unable to grow aerobically in minimal medium (7). It was elegantly shown that wild-type phenotype could be restored in *sodA sodB* mutants by providing them with the ability to synthesize human Cu,Zn-SOD (8). Little controversy remains concerning the toxicity of superoxide radical and the physiological function of the SODs as scavengers of the superoxide radical.

The other side of the coin, that too much SOD may be toxic, is now the much more interesting observation. Twenty-five years ago when we began isolating and assaying SOD from a variety of organs and a variety of organisms, we were struck by the fact that this is a remarkably constant activity whether one examines *E. coli* or squash or human brain. This suggested that there might be some reason for keeping the radical at a relatively constant concentration. It would be very easy for an organism to produce more SOD if more were better, but for some reason that did not happen.

With the advent of molecular biology came the ability to do what nature did not. Cells of many types have been transformed or transfected to produce greater-than-normal amounts of SOD. Interestingly, these cultured cells often fare much better when challenged with an oxidative stress such as exposure to paraquat (9,10), but they are worse off under conditions of relatively normal oxidative stress. (It may be argued that all "normal" tissue culture conditions are oxidatively stressful.) Table 1 lists some of the problems associated with cells that express elevated amounts of SOD activity. It is of interest that mammalian cells or transgenic animals have not been produced that express SOD activity at more than 5 to 8 times the normal level. It has, in fact, been speculated that expression of SOD beyond this level would be lethal (11).

Transgenic mice have been endowed with the human gene for the Cu,Zn-SOD (*SOD1*) (16), the Mn-SOD (*SOD2*) (17), or the extracellular SOD (*SOD3*) (18). These animals express several times the normal amount of SOD activity in certain or all tissues, depending on the promoter used. As with the transfected cell lines, these mice often fair much better than wild-type when challenged with an oxidative stress such as pulmonary oxygen toxicity (17) or MPTP poisoning (5), but, once again, they are worse off under conditions of normal oxidative stress. Table 2 lists a number of morphological and functional abnormalities that have been observed in these animals. Note that many of the anomalies involve the nervous system. The first SOD transgenic mice were created as a

Table 1. Abnormalities Observed in Cells that Overexpress Human *SOD1*

Cell type	Observation	Ref.
Transformed *Escherichia coli*	Sensitivity to ionizing radiation	(12)
Down syndrome fibroblasts	Sensitivity to ionizing radiation	(13)
Transfected HeLa and mouse L cells	Increased lipid peroxidation	(9)
Transfected rat PC12 (neuronal) cells	Impaired neurotransmitter uptake	(14)
Platelets from transgenic mice and Down syndrome	Impaired serotonin uptake	(15)
Transfected bovine adrenocortical cells	Lethal cytotoxicity	(11)

Table 2. Abnormalities Observed in Mice Transgenic for Human *SOD1*

Observation	Ref.
Increased lipid peroxidation in the brain	(19,20)
Abnormal neuromuscular junction in the tongue	(21)
Premature aging of hindlimb neuromuscular junction	(22)
Prolonged expression of *hsp70* following cerebral ischemia	(23)
Prolonged expression of c-*fos* following cerebral ischemia	(24)
Increased apoptosis of thymocytes and bone marrow cells	(25)
Impaired uptake of serotonin by platelets	(15)
Vacuoles and mitochondrial degeneration in motor neurons	(26)

model for human trisomy-21 (Down syndrome) because the *SOD1* gene is located on this chromosome. Because *SOD1* is constitutively expressed (or nearly so), there is a gene dosage effect resulting in 1.5 times the normal amount of *SOD1* expression in individuals who possess three copies of chromosome 21. Many of the abnormalities seen in affected humans are also seen in *SOD1* transgenic mice.

WHY IS TOO MUCH SOD A PROBLEM?

We have observed anomalous, bell-shaped dose–response curves when using SOD to protect the postischemic heart from reperfusion injury (27,28). In several models, using several types of natural and recombinant SOD including both the Cu,Zn- and the Mn-containing enzymes, we saw substantial protection of myocardial structure and function when hearts were reperfused with SOD in the range of 1–10 μg/ml. When SOD concentrations increased toward 50 μg/ml, protection was always lost, and in some cases the injury was significantly exacerbated. Ischemia followed by reperfusion places tissue under oxidative stress, tipping the "oxidant/antioxidant balance" in the *oxidant* direction. As we treat the insulted organ with increasing concentrations of SOD, we bring the balance back toward the horizontal position. With higher and higher concentrations of SOD, the balance tips in the *antioxidant* direction, putting the organ once again at risk, presumably for some of the reasons illustrated above in Tables 1 and 2. An example of this dose–response behavior is shown in Figure 1. Isolated rabbit hearts were subjected to 60 min of no-flow ischemia followed by reperfusion for 15 min (29). Functional recovery was assessed by measuring developed pressure in the left ventricle and comparing it to the preischemic developed pressure measurement. Thiobarbituric acid-reactive substances (TBARS) were determined in heart homogenate supernate by the method of Ohkawa et al. (30). The results show that perfusing the postischemic heart with SOD at 5 mg/L was dramatically protective, allowing nearly complete recovery of function. At 0.5 mg/L there was no protection relative to buffer-perfused hearts (about 40% recovery in both groups), and at 50 mg/L there was no recovery at all–no heartbeat and no developed pressure. The reason for the paradoxical loss of function at the high dose of SOD is revealed when one examines the amount of lipid peroxidation taking place, also shown in Figure 1. Note that at the lower doses of SOD, lipid peroxidation is progressively inhibited. It becomes

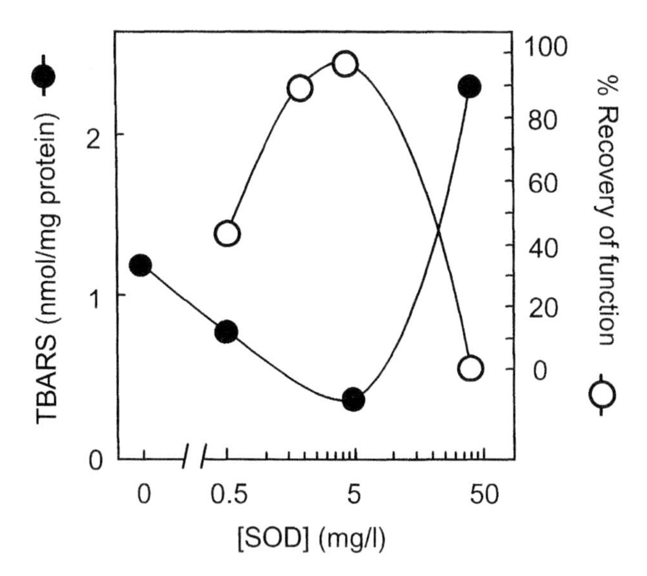

Figure 1. Relationships among SOD dose, lipid peroxidation, and functional recovery of postischemic isolated rabbit hearts. The optimally protective dose of SOD (5 mg/L) also produced the minimum amount of lipid peroxidation. (Data are replotted from Refs 28 and 29.)

minimal at 5 mg/L, where recovery of function is maximal. If SOD dose increases to 50 mg/L, however, there is a dramatic increase in lipid peroxidation with a concomitant complete loss of function.

The explanation for this odd behavior lies in the fact that lipid peroxidation is a free-radical chain process. A free radical is required to begin the chain reaction. Once begun, the chain reaction can only be terminated by the participation of another free radical. Superoxide radical (indirectly, at least) can play both roles. The overproduction of superoxide within a cell can initiate lipid peroxidation by mobilizing iron from the tissue storage protein ferritin (31,32). This Fe^{2+} can react with a pre-existing lipid hydroperoxide (LOOH) to produce a chain-initiating alkoxy radical ($LO^{\bullet}$). Superoxide can recycle the redox-active pool of iron, keeping it in the reduced state so that it can continue to initiate lipid peroxidation:

$$O_2^{\bullet -} + \text{ferritin-}Fe^{3+} \rightarrow O_2 + \text{ferritin} + Fe^{2+}$$

$$Fe^{2+} + LOOH \rightarrow LO^{\bullet} + OH^- + Fe^{3+}$$

$$LO^{\bullet} + LH \rightarrow L^{\bullet} + LOH$$

$$O_2^{\bullet -} + Fe^{3+} \rightarrow O_2 + Fe^{2+}$$

The chain propagating reactions ensue:

$$L^{\bullet} + O_2 \rightarrow LOO^{\bullet}$$

$$LOO^{\bullet} + LH \rightarrow LOOH + L^{\bullet}$$

Hence, it is well established that superoxide can indirectly *initiate* lipid peroxidation. Ironically, the only way to eliminate a radical is by reacting it with another radical. (Because radicals have an odd number of electrons, any reaction with a nonradical must yield a product that is another radical.) From a teleological perspective, it would be useful if a cell had available a continuous supply of a relatively unreactive free radical for the purpose of promoting a radical–radical annihilation to eliminate the very dangerous lipid dioxyl radical. Superoxide meets these criteria. It is produced as a by-product of oxygen metabolism, and is therefore continuously available in all cells. As free radicals go, it is relatively unreactive. Its concentration is maintained at a very low, *but nonzero* value by the cellular SOD content. If alkoxyl or dioxyl radicals are scavenged by $O_2^{\bullet -}$, then entire chains of reactions would be prevented or terminated:

$$LO^{\bullet} + O_2^{\bullet -} + H^+ \rightarrow LOH + O_2$$

$$LOO^{\bullet} + O_2^{\bullet -} + H^+ \rightarrow LOOH + O_2$$

Therefore, it was hypothesized that $O_2^{\bullet -}$, in addition to being able to liberate iron and to *initiate* lipid peroxidation, may also serve as a *terminator* of lipid peroxidation, such that overscavenging the radical may increase net lipid peroxidation (27,29). If this hypothesis is correct, then indices of lipid peroxidation should be high at both low concentrations of SOD (where initiation would be high) and at high concentrations of SOD (where termination would be low). At some intermediate SOD concentration, initiation by superoxide would be largely suppressed, but termination by superoxide would still be making a contribution, and net lipid peroxidation would be at a minimum. We found these predictions to be true (29), and experimental verification is shown in Figure 1. These results suggest that for any given level of oxidative stress, a single optimal concentration of SOD exists that will restore the superoxide level to yield the best combination of initiation and termination (i.e., the one producing minimal net lipid peroxidation). Any concentration of SOD other than the optimal leads to increased lipid peroxidation and therefore to increased oxidative stress.

DISCUSSION AND RAMIFICATIONS

The realization that cellular well-being depends on achieving and maintaining a certain concentration of superoxide radical, and that this is modulated by the concentration of SOD activity within that cell, provides us with a new perspective for the interpretation of past and future experiments. The increases in sensitivity to ionizing radiation noted in Table 1 now make sense. Ionizing radiation provides a superoxide-independent pathway for the initiation of lipid peroxidation. Without enough superoxide to act as terminator, longer chains of lipid peroxidation result. Similarly, all noted cases of increased lipid peroxidation in both Tables 1 and 2 can now be understood. The other functional aberrations cited (such as impaired neurotransmitter uptake, prolonged expression of *hsp70* and c-*fos*, increased rate of apoptosis, etc.) may or may not be reflections of lipid peroxidation rates, but further explorations of any such mechanistic associations should be readily testable.

REFERENCES

1. Babior BM. Oxygen-dependent microbial killing by phagocytes. N Engl J Med 1978; 298:659–668, 721–725.
2. Salin ML, McCord JM. Free radicals and inflammation: Protection of phagocytosing leukocytes by superoxide dismutase. J Clin Invest 1975; 56: 1319–1323.

3. Sandy MS, Di Monte D, Cohen P, Smith MT. Role of active oxygen in paraquat and 1-methyl-4-phenyl-1,2,3,6-tetrahydropyridine (MPTP) cytotoxicity. Basic Life Sci 1988; 49: 795–801.
4. Gruber MY, Glick BR, Thompson JE. Cloned manganese superoxide dismutase reduces oxidative stress in *Escherichia coli* and *Anacystis nidulans*. Proc Natl Acad Sci USA 1990; 87: 2608–2612.
5. Przedborski S, Kostic V, Jackson-Lewis V, Naini AB, Simonetti S, Fahn S, Carlson E, Epstein CJ, Cadet JL. Transgenic mice with increased Cu/Zn-superoxide dismutase activity are resistant to *N*-methyl-4-phenyl-1,2,3,6-tetrahydropyridine-induced neurotoxicity. J Neurosci 1992; 12: 1658–1667.
6. Hassan H, Fridovich I. Regulation of the synthesis of superoxide dismutase in *E. coli*: induction by methyl viologen. J Biol Chem 1977; 252: 7667–7672.
7. Carlioz A, Touati D. Isolation of superoxide dismutase mutants in *Escherichia coli*: is superoxide dismutase necessary for aerobic life? EMBO J 1986; 5: 623–630.
8. Natvig DO, Imlay K, Touati D, Hallewell RA. Human copper-zinc superoxide dismutase complements superoxide dismutase-deficient *Escherichia coli* mutants. J Biol Chem 1987; 262: 14697–14701.
9. Elroy-Stein O, Bernstein Y, Groner Y. Overproduction of human Cu/Zn-superoxide dismutase in transfected cells: extenuation of paraquat-mediated cytotoxicity and enhancement of lipid peroxidation. EMBO J 1986; 5: 615–622.
10. Krall J, Bagley AC, Mullenbach GT, Hallewell RA, Lynch RE. Superoxide mediates the toxicity of paraquat for cultured mammalian cells. J Biol Chem 1988; 263: 1910–1914.
11. Norris KH, Hornsby PJ. Cytotoxic effects of expression of human superoxide dismutase in bovine adrenocortical cells. Mutat Res 1990; 237: 95–106.
12. Scott MD, Meshnick SR, Eaton JW. Superoxide dismutase amplifies organismal sensitivity to ionizing radiation. J Biol Chem 1989; 264: 2498–2501.
13. Schwaiger H, Weirich HG, Brunner P, Rass C, Hirsch-Kauffmann M, Groner Y, Schweiger M. Radiation sensitivity of Down's syndrome fibroblasts might be due to overexpressed Cu/Zn-superoxide dismutase (EC 1.15.1.1). Eur J Cell Biol 1989; 48: 79–87.
14. Elroy-Stein O, Groner Y. Impaired neurotransmitter uptake in PC12 cells overexpressing human Cu/Zn-superoxide dismutase–implication for gene dosage effects in Down syndrome. Cell 1988; 52: 259–267.
15. Schickler M, Knobler H, Avraham KB, Elroy-Stein O, Groner Y. Diminished serotonin uptake in platelets of transgenic mice with increased Cu/Zn-superoxide dismutase activity. EMBO J 1989; 8: 1385–1392.
16. Epstein CJ, Avraham KB, Lovett M, Smith S, Elroy-Stein O, Rotman G, Bry C, Groner Y. Transgenic mice with increased Cu/Zn-superoxide dismutase activity: animal model of dosage effects in Down syndrome. Proc Natl Acad Sci USA 1987; 84: 8044–8048.
17. Wispe JR, Warner BB, Clark JC, Dey CR, Neuman J, Glasser SW, Crapo JD, Chang LY, Whitsett JA. Human Mn-superoxide dismutase in pulmonary epithelial cells of transgenic mice confers protection from oxygen injury. J Biol Chem 1992; 267: 23937–23941.
18. Oury TD, Ho YS, Piantadosi CA, Crapo JD. Extracellular superoxide dismutase, nitric oxide, and central nervous system O_2 toxicity. Proc Natl Acad Sci USA 1992; 89: 9715–9719.
19. Ceballos-Picot I, Nicole A, Briand P, Grimber G, Delacourte A, Defossez A, Javoyagid F, Lafon M, Blouin JL, Sinet PM. Neuronal-specific expression of human copper-zinc superoxide-dismutase gene in transgenic mice–animal-model of gene dosage effects in Down's syndrome. Brain Res 1991; 552: 198–214.
20. Ceballospicot I, Nicole A, Clement M, Bourre JM, Sinet PM. Age-related changes in antioxidant enzymes and lipid peroxidation in brains of control and transgenic mice overexpressing copper-zinc superoxide dismutase. Mutat Res 1992; 275: 281–293.
21. Avraham KB, Schickler M, Sapoznikov D, Yarom R, Groner Y. Down's syndrome: Abnormal neuromuscular junction in tongue of transgenic mice with elevated levels of human Cu/Zn-superoxide dismutase. Cell 1988; 54: 823–829.

22. Avraham KB, Sugarman H, Rotshenker S, Groner Y. Down's syndrome: Morphological remodelling and increased complexity in the neuromuscular junction of transgenic CuZn-superoxide dismutase mice. J Neurocytol 1991; 20: 208–215.
23. Kamii H, Kinouchi H, Sharp FR, Koistinaho J, Epstein CJ, Chan PH. Prolonged expression of hsp70 mrna following transient focal cerebral ischemia in transgenic mice overexpressing Cu, Zn-superoxide dismutase. J Cereb Blood Flow Metab 1994; 14: 478–486.
24. Kamii H, Kinouchi H, Sharp FR, Epstein CJ, Sagar SM, Chan PH. Expression of c-*fos* mRNA after a mild focal cerebral ischemia in SOD-1 transgenic mice. Brain Res 1994; 662: 240–244.
25. Peled-Kamar M, Lotem J, Okon E, Sachs L, Groner Y. Thymic abnormalities and enhanced apoptosis of thymocytes and bone marrow cells in transgenic mice overexpressing Cu/Zn-superoxide dismutase: implications for Down syndrome. EMBO J 1995;14: 4985–4993.
26. Dal Canto MC, Gurney ME. Neuropathological changes in two lines of mice carrying a transgene for mutant human Cu,Zn SOD, and in mice overexpressing wild type human SOD: A model of familial amyotrophic lateral sclerosis (FALS). Brain Res 1995; 676: 25–40.
27. Omar BA, Gad NM, Jordan MC, Striplin SP, Russell WJ, Downey JM, McCord JM. Cardioprotection by Cu,Zn-superoxide dismutase is lost at high doses in the reoxygenated heart. Free Radical Biol Med 1990; 9: 465–471.
28. Omar BA, McCord JM. The cardioprotective effect of Mn-superoxide dismutase is lost at high doses in the postischemic isolated rabbit heart. Free Radical Biol Med 1990; 9: 473–478.
29. Nelson SK, Bose SK, McCord JM. The toxicity of high-dose superoxide dismutase suggests that superoxide can both initiate and terminate lipid peroxidation in the reperfused heart. Free Radical Biol Med 1994; 16: 195–200.
30. Ohkawa H, Ohishi N, Yagi K. Assay for lipid peroxides in animal tissues by thiobarbituric acid reaction. Anal Biochem 1979; 95: 351–358.
31. Biemond P, van Eijk HG, Swaak AJG, Koster JF. Iron mobilization from ferritin by superoxide derived from stimulated polymorphonuclear leukocytes. Possible mechanism in inflammation diseases. J Clin Invest 1984; 73: 1576–1579.
32. Harris LR, Cake MH, Macey DJ. Iron release from ferritin and its sensitivity to superoxide ions differs among vertebrates. Biochem J 1994; 301: 385–389.

2
The Effect of Oxidative Stress on Cells by Oxygen Radicals and Its Inhibition by Antioxidants

Mareyuki Takahashi and Etsuo Niki
University of Tokyo
Tokyo, Japan

INTRODUCTION

Reactive oxygen species and free radicals attack cells and lipoproteins to induce the oxidation of lipids, proteins, sugars, and DNA, which results in membrane damage, protein modification, enzyme deactivation, DNA strand breaks, and base modification. Some proceed by nonradical mechanisms but many of them proceed by a free radical-mediated chain mechanism. Aerobic organisms are protected against such oxidative stress by an array of defense systems. We have been studying the oxidative damage of biological molecules, membranes, lipoproteins, and cells induced by reactive oxygen species and free radicals and its inhibition by antioxidants from basic chemical and biochemical points of view.

REACTIVE OXYGEN SPECIES AND FREE RADICALS

Various endogenous and exogenous reactive oxygen species and free radicals may be involved in oxidative stress in vivo. Table 1 summarizes the reactivities of several representative reactive oxygen species and free radicals in hydrogen atom abstraction from polyunsaturated lipids and addition to double bonds. It is seen that their reactivities vary extensively and also that they depend on the substrate. The reactivities of the free radicals toward hydrogens of lipids, proteins, sugars, and DNA bases can be estimated from the bond dissociation energy of X—H bonds, *D*(X—H). As summarized in Table 1, hydroxyl radical is extremely reactive, alkoxyl radical is also reactive, while peroxyl radical, the chain-carrying species in lipid peroxidation, is much less reactive and attacks only the reactive sites of lipids and proteins selectively. This also affects the efficacy of radical-scavenging antioxidants such as vitamin C and vitamin E; that is, these antioxidant vitamins act as potent scavengers of peroxyl radicals, but they may not scavenge hydroxyl nor alkoxyl radicals efficiently (1). Nitrogen dioxide, nitric oxide, and superoxide are not

Table 1. Reactivities of Free Radicals and Reactive Oxygen Species

Species X	Bond energy[a] D(X-H) (kcal/mol)	Rate constant k^b ($M^{-1}\ s^{-1}$)	
		k_{abst}	k_{add}
Hydroxyl radical, $HO^\bullet$	119	10^9	10^9
Alkoxyl radical, $RO^\bullet$	104	10^6	10^6
Peroxyl radical, $ROO^\bullet$	88	10^2	10^2
Nitrogen dioxide, $^\bullet NO_2$	?	small	small
Nitric oxide, $^\bullet NO$	50	~0	~0
Superoxide, $O_2^{\bullet -}$	?	~0	~0
Hydrogen peroxide, H_2O_2	–	0	0
Lipid hydroperoxide, LOOH	–	0	0
Singlet oxygen, $^1O_2$	–	0	10^6
Ozone, O_3	–	0	10^6

[a] Bond dissociation of X-H bond.
[b] The rate constants for the bis-allylic hydrogen atom abstraction from polyunsaturated lipids (k_{abst}) and for the addition to the double bond (k_{add}).

reactive per se and do not attack biological molecules at appreciable rate. It is known, however, that nitric oxide and superoxide react rapidly to give peroxynitrite, which induces the oxidation of various substrates (2). On the other hand, the nonradical reactive oxygen species such as hydrogen peroxide, hydroperoxides, and singlet oxygen do not abstract hydrogen atoms directly, but both singlet oxygen and ozone react rapidly with double bonds. This oxidation is stoichiometric and does not proceed by a chain mechanism; that is, one molecule of active species oxidizes only one molecule of substrate. Consequently, the total effect may be less significant than with free radicals. Ozone exerts protective effects in the stratosphere but it is toxic in vivo. Metal ions and complexes, hypochlorous acid, and oxygenases may also play an important role in oxidative stress.

It may be worthwhile to note that the sites of generation of reactive species and the permeability of these species to the membranes are also important factors. Nitric oxide and hydrogen peroxide are permeable, but superoxide is not.

OXIDATIVE HEMOLYSIS AND ITS INHIBITION BY ANTIOXIDANTS

The oxidation of erythrocytes serves as a model for the oxidative damage of biomembranes and cells (3). Oxidative damage in erythrocyte membranes during aging has been reported (4). It has been shown that the hemolysis of erythrocytes is induced by various oxidants such as hydrogen peroxide, dialuric acid, superoxide, and organic hydroperoxides. The peroxyl radicals generated by thermal decomposition of the water-soluble azo compound AAPH, 2,2′-azobis (2-amidinopropane) dihydrochloride, in air also induce hemolysis (5–8). Cigarette smoke also induces hemolysis (9). It has been observed that the free radicals generated outside erythrocytes attack the membrane lipids and proteins to induce their free radical-mediated chain oxidation (5). As the oxidation proceeds, polyunsaturated lipids, especially phosphatidylcholine, and proteins are oxidized (6) and the leakage of potassium ion is observed, followed by that of calcium ion, lactate dehydrogenase (LDH), glutamic-pyruvic transaminase (GPT), aspartate amino transferase (GOT), and hemoglobin (10). The erythrocyte membranes undergo hemolysis faster with increasing flux of free radicals but, interestingly, the extent of hemolysis was directly proportional to the total amount of free radicals formed and independent of their rate of formation (6). Radical-scavenging antioxidants such as vitamin E, vitamin C, and uric acid suppressed the hemolysis in a dose-dependent manner (6,7). Neither superoxide dismutase (SOD) nor catalase suppressed the free radical-induced hemolysis. Erythrocytes from vitamin E-deficient rats underwent hemolysis faster than those from control rats, but the extent of hemolysis was dependent on the extent of oxidation and independent of the presence or absence of vitamin E. Trolox, a water-soluble vitamin E analog, also suppressed the hemolysis, but its protective efficacy was only 1/50 of that of vitamin E in the membranes. All these results clearly show that the free radicals formed outside erythrocytes attack the membrane from outside to cause oxidative damage and that it is essential to break the chain propagation of the oxidation within the membranes.

Superoxide generated outside erythrocyte also induced hemolysis, which was suppressed by catalase and vitamin E, but not by SOD (11). The treatment of erythrocytes with carbon monoxide also suppressed hemolysis. In contrast to the oxidation induced by aqueous free radicals generated outside of erythrocytes from AAPH, phosphatidylethanolamine was oxidized more extensively than phosphatidylcholine in the superoxide-induced oxidation, suggesting that the inner membrane is oxidized preferentially to outer membrane.

FREE RADICAL-INDUCED CELL DAMAGE

The effects of AAPH upon cells were also studied. AAPH incubated with HeLa cells cultured in MEM Eagle's medium in humidified incubator under the atmosphere of air containing 5% CO_2 at 37°C suppressed the cell growth in a time- and dose-dependent manner. For example, Figure 1 shows the results of treatment of HeLa cells with AAPH on the cell growth. AAPH was added to the incubation medium and after some specific time it was removed from the medium. The figure shows that AAPH reduced the cell growth but that the cell number increased roughly at the same rate after removal of AAPH, implying that the damaging effect of AAPH in HeLa cells is not due to its incorporation and accumulation in the cells. Water-soluble antioxidants such as Trolox and uric acid added simultaneously with AAPH suppressed the cytotoxic effect of AAPH dose-dependently.

Oxidative damage of other cell lines was also studied. In THP-1 cells, AAPH caused time-dependent increase in phosphatidylcholine hydroperoxide content (PCOOH), indicating that lipid peroxidation was taking place in the cell (Figure 2). Free radical-induced lipid peroxidation of the cell was also confirmed in rat thymocytes, where peroxidation was monitored by cellular α-tocopherol content since α-tocopherol is supposed to be consumed only in the case of the peroxidation of the membranes in which it is located. As a result the α-tocopherol level decreased with time. Thus, free-radical species initiate lipid peroxidation in the cell as well as in erythrocyte membranes or artificial membranes as mentioned above. Cell damage brought about in this way usually results in cell death, mainly caused by the loss of membrane integrity.

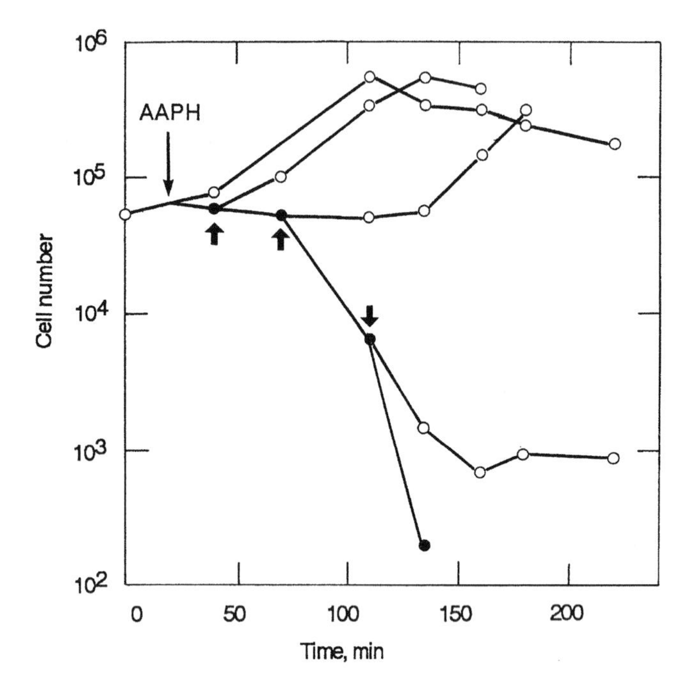

Figure 1. Effect of AAPH on the growth rate of HeLa cells. AAPH (2 mM) was added to the medium of cultured HeLa cells at the time 20 min, incubated, and removed at the time indicated by an arrow. The symbols (–●–) and (–○–) show the cell number in the presence of and after removal of AAPH, respectively.

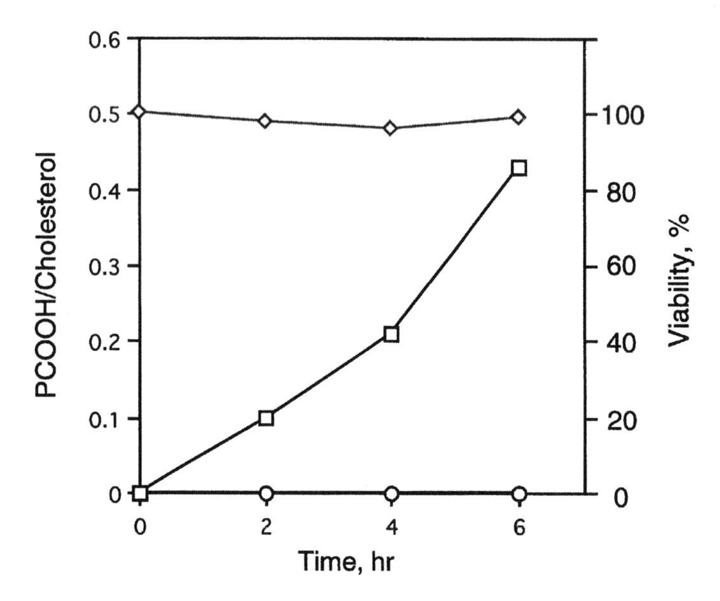

Figure 2. Peroxidation of THP-1 cells initiated with 10 mM AAPH at 37°C under air. Cells were incubated with (–□–) or without (–○–) AAPH. Phosphatidylcholine hydroperoxides formed in the cells were measured. Cell viability (–◇–) is also shown.

Another aspect of free radical-induced cell damage is the initiation of the signaling pathway for programmed cell death–apoptosis. In mouse early pre-B cell line GB11, AAPH induced cell death in a time- and dose-dependent manner (Figure 3). The cell death was apoptotic as judged by morphological alteration, (i.e., cell shrinkage), condensation of chromosomes and compartmentalization of the nucleus, and finally formation of apoptotic bodies. Apoptosis was also confirmed by the internucleosomal DNA fragmentation pattern. Reduction of cell viability and the progression of DNA fragmentation were found to be time and dose dependent. Trolox suppressed the apoptosis-inducing effect of AAPH effectively, indicating that apoptosis was due to free-radical species produced outside the cell.

The process of free-radical reaction in the plasma membranes was monitored using ESR spin probes. The spin probe doxylnitroxide stearic acid (NS) is known to react with reactive radical species, which are formed as intermediates in the radical chain reaction (12). The intensity of the ESR signal of 16-NS incorporated into the cell decreased time and dose-dependently. This reduction of signal intensity reflects the appearance of free-radical species within the lipid bilayers in the plasma membranes of the cell where 16-NS is incorporated. AAPH-induced oxidative stress is initiated with the oxidative damage of the components of the plasma membrane, unlike oxidative stress by hydrogen peroxide or irradiation where reactive oxygen species generated inside the cell are assumed to play a major role. Damaged membrane components may trigger the apoptotic signaling either by directly turning on the mechanism or by producing the secondary mediator such as lipid-derived hydroperoxides or aldehydes.

Other reactive oxygens species than AAPH-derived peroxyl radical were also found to induce apoptosis. Nitric oxide (NO) has been known to be cytotoxic to NO-producing cells such as macrophages. The effects of NO on the cell are generally attributed to the inhibition of the respiratory chain or glucose metabolism by NO itself or by peroxynitrite

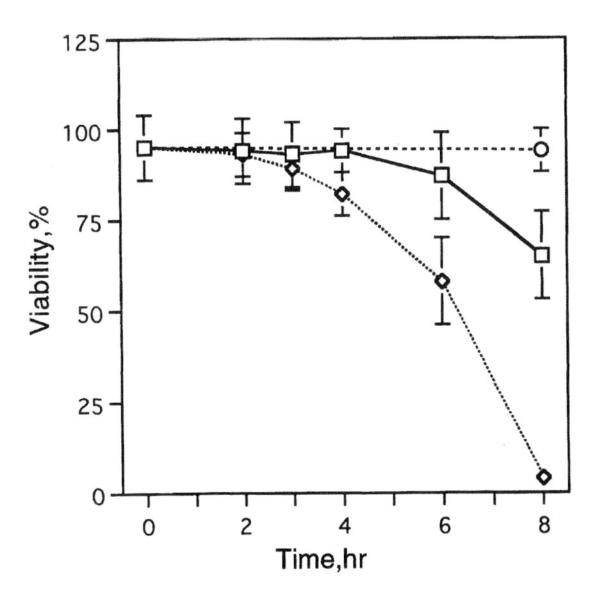

Figure 3. Cytotoxic effect of AAPH on GB11 cells. Cells were incubated with 10 mM (–□–), 30 mM (· · · ◇· · ·) or without (· · · ○· · ·) AAPH at 37°C in 5% CO_2–air. Viability was assessed by trypan blue exclusion.

formed by the reaction with superoxide anion. In GB11 cells NO applied extracellularly induced apoptosis in a dose-dependent manner. Apoptosis was suppressed by *N*-acetylcysteine as well as by hemoglobin, suggesting that reactive oxygen species including free radicals are involved in the pathway. Although NO has only a low oxidizing activity per se, peroxynitrite derived from NO may initiate peroxidation. In accord with this it was found that NO induced significant reduction of α-tocopherol content in the cell. It is possible that intracellular peroxynitrite induces apoptosis by damaging the membrane components or by altering the cellular conditions such as redox status.

The results shown here suggest that free radicals and reactive oxygen species damage the cell not only by destroying the cellular components but by mediating the intracellular signaling cascade.

ACKNOWLEDGMENT

The technical assistance by M. Oikawa and S. Kuroda is appreciated.

REFERENCES

1. Niki E, Noguchi N, Tsuchihashi H, Goto N. Interaction among vitamin C, vitamin E, and β-carotene. Am J Clin Nutr 1995; 62: 1322S–1326S.
2. Pryor WA, Squadrito GL. The chemistry of peroxynitrite: a product from the reaction of nitric oxide with superoxide. Am J Physiol 1995; 268: L699–L722.
3. Chiu D, Lubin B, Shohet SB. Peroxidative reactions in red cell biology. In Pryor WA, ed. Free Radicals in Biology, vol. 5. New York: Academic Press; 1982: 115–160.
4. Jain SK. Evidence for membrane lipid peroxidation during the in vivo aging of human erythrocytes. Biochim Biophys Acta 1988; 937: 205–210.
5. Yamamoto Y, Niki E, Kamiya Y, Miki M, Tamai H, Mino M. Free radical chain oxidation and hemolysis of erythrocytes by molecular oxygen and their inhibition by vitamin E. J Nutr Sci Vitaminol 1986; 32: 475–479.
6. Miki M, Tamai H, Mino M, Yamamoto Y, Niki E. Free-radical chain oxidation of rat red blood cells by molecular oxygen and its inhibition by α-tocopherols. Arch Biochem Biophys 1987; 258: 373–380.
7. Niki E, Komuro E, Takahashi M, Urano S, Ito E, Terao K. Oxidative hemolysis of erythrocytes and its inhibition by free radical scavengers. J Biol Chem 1988; 263: 19809–19814.
8. Sandhu IS, Ware K Grisham MB. Peroxyl radical-mediated hemolysis role of lipid, protein and sulfhydryl oxidation. Free Radical Res Commun 1992; 16: 111–122.
9. Minamisawa S, Komuro E, Niki E. Hemolysis of rabbit erythrocytes induced by cigarette smoke. Life Sci 1990; 47: 2207–2215.
10. Niki E, Minamisawa S, Oikawa M, Komuro E. Membrane damage from lipid oxidation induced by free radicals and cigarette smoke. Ann NY Acad Sci 1993; 686: 29–38.
11. Mino M, Miki M. Tamai H, Yasuda H, Maeda H. Membrane damage in erythrocytes induced by radical initiating reactions and the effect of tocopherol as a radical scavenger. In: Sevanian A, ed. Lipid Peroxidation in Biological Systems. Champain: American Oil Chemists Societies; 1993: 51–70.
12. Takahashi M, Tsuchiya J, Niki E. Scavenging of radicals by vitamin E in the membranes as studied by spin labeling. J Am Chem Soc 1989; 111: 6350–6353.

3

Cytotoxicity of Radiation-Derived Reactive Oxygen Species

Manfred Saran and Wolf Bors
Institut für Strahlenbiologie
GSF-Forschungszentrum Neuherberg
Oberschleissheim, Germany

According to current thinking, radiation-induced cell death may be caused by different primary events (cf. Figure 1):

1. Energy deposition in the nuclear compartment, either by direct interaction of radiation with the molecular target DNA or by "indirect hits," i.e., interactions with DNA of water radicals that are produced in close proximity and reach their target by diffusion.
2. The action (direct or indirect) of radiation on targets in the cytosolic compartment, where iron–sulfur clusters (aconitase) or heme-containing enzymes seem to be especially vulnerable entities.
3. Interactions of radicals with the membrane compartment, either producing "damage" or eliciting some membrane-derived signal that triggers physiological cascades that eventually lead to cell death.

Also amply discussed is the concept of "site specificity," which implies that radiation-derived molecular species (hydroperoxides, aldehydes) may react at specific cellular locations (e.g., metal binding sites), thus translating the effects of "indiscriminately" reacting radicals to the level of a defined chemical action. Experimental evidence has been presented for each of these aspects. It is very difficult, however, to assess which of the effects – for a given set of experimental conditions – is the decisive process.

One aspect that has not been much discussed in recent years is the extent to which radiation-derived chemicals produced in the extracellular space (E) might contribute to cell death. This deficit is easily understandable: In the early years of radiation biology many investigations were carried out to determine whether radiation-derived hydrogen peroxide (H_2O_2) could damage cells (1,2). It was soon realized, however, that H_2O_2 was not radiomimetic in a quantitative sense because addition to suspensions of cells of concentrations of H_2O_2 that corresponded to the amount produced by irradiation did not simulate the observed radiation effects. As no other long-lived product from the radiolysis

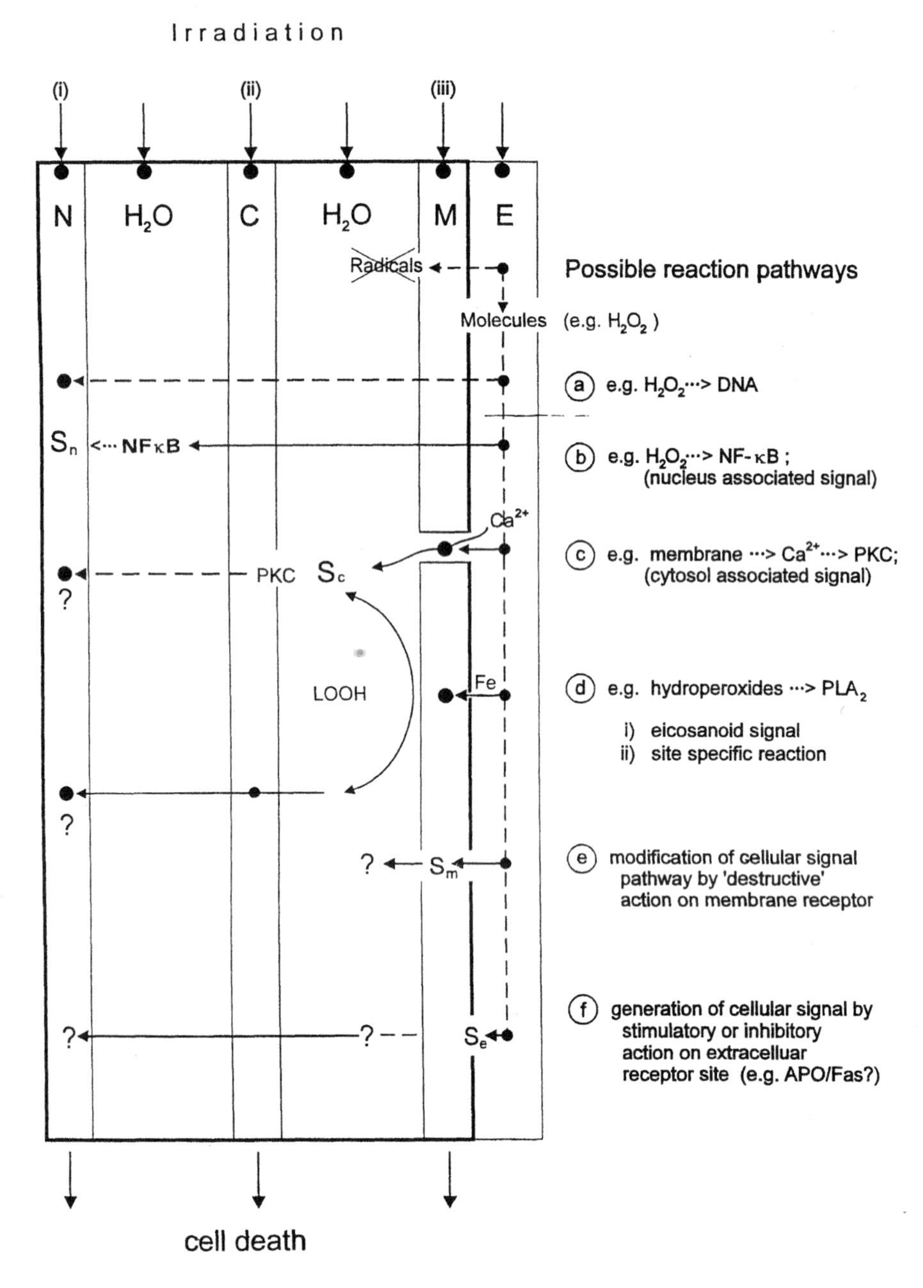

Figure 1. Different models have been developed to explain radiation-induced cell death: strand breaks induced in the compartment N (nucleus), metabolic disorders arising from the compartment C (cytosol), physiological disorders originating from the compartment M (membrane). How these disorders might be linked to signals (S) originating from the extracellular compartment (E) is indicated in the right-hand margin (see text for further explanation).

of water was known, the idea of extracellularly derived cell death was essentially abandoned in radiation biology.

With the knowledge that radiation can induce apoptosis and that oxygen radical-derived substances are involved in an intercellular pathway to induce it (3) the question should be addressed again: whether there is any correlation between radiation-derived substances and one of the cellular signaling routes that are decisive for cell survival.

POSSIBLE SIGNAL TRANSDUCTION PATHWAYS FOR EXTRACELLULARLY DETERMINED CELL DEATH

Accepting the fact that radicals per se are not able to cross membranes (apart from the messenger $NO^{\bullet}$, which certainly does not belong to the subset of radicals that derive from radiation), several aspects concerning stable molecules are worth discussing (cf. right margin of Figure 1): (a) Radiation-derived substances could diffuse to the nucleus and interact there with DNA. It has been shown repeatedly that H_2O_2 is able to do so (e.g. Ref. 4), but effects on DNA are only observable at such high levels of externally added H_2O_2 that this pathway does not seem to be of relevance in radiation biology. (b) H_2O_2 could activate some cytosolic entity, e.g., the transcription factor NFκB, which on translocation to the nucleus elicits some nucleus-associated signal there. Such mechanisms have been investigated intensively in recent years. However, the micromolar concentrations of H_2O_2 needed to evoke the response of the NFκB system seem too large for this pathway to be truly radiomimetic. Such concentration arguments would not hold, however, if an amplification step were involved in the chain of causality. (c) A rather unspecific way of inducing such an amplified cellular response would be alteration by radiation-derived substances of one of the components responsible for the maintenance of calcium homeostasis – e.g., disturbing the action of a channel protein or some other functional membrane structure. This would lead to a chain of calcium-dependent cytosolic events, including those of the protein kinase C cascade. (d) The generation of hydroperoxidic signals interfering with eicosanoid signaling pathways has often been proposed as a consequence of irradiation, but it is uncertain whether it is relevant in the lower dose range. The other two possibilities, (e) that radiation-derived substances directly interact with membrane receptors, thereby modifying the transmission or message content of a cellular signal, or (f) the interaction of a substance in the sense that it constitutes a specific "signal" by itself, have not yet been investigated.

It has been shown repeatedly that the presence of chloride during irradiation increases the efficiency of killing cells (4–9) to a considerable extent: hypochlorite (HOCl) has been proposed as the toxic species. In none of these cases, however, could the production of HOCl be verified experimentally under physiological conditions; the effects were only observable under artificial conditions, i.e., acidifying and saturating the solutions with N_2O during irradiation.

AMOUNTS OF HYDROGEN PEROXIDE AND/OR HYPOCHLORITE PRODUCED DURING IRRADIATION

The main problem in separating the contributions of H_2O_2 from those of HOCl is the fact that these substances interact with each other in a pH dependent way (10) and that at physiological pH or lower only chemically "inert" and physiologically harmless chloride ions and ground-state oxygen are detectable a few seconds after irradiation. (The well known reaction of H_2O_2 with HOCl producing singlet oxygen occurs only at alkaline pH.)

Therefore, experiments in which cells are incubated with both chemicals in order to compare the toxicity of the mixture to the processes that occur under irradiation meets with the difficulty that it is not known how many molecules of either substance have time to react with the cells before they deactivate each other. Analogously, experiments to determine the amount of either substance being produced during irradiation are hindered by the fact that determination of the substances *after* irradiation does not account for the transient concentrations of the substances *during* irradiation. Essential for the decision are the experimental means to distinguish H_2O_2 from HOCl. Figure 2 shows the development of the absorption at 351 nm of I_3^- when potassium iodide is added to solutions after irradiation under different conditions. The presence of formate in an oxygenated solution results in the greatest production of oxidant (light grey squares and solid triangles); the

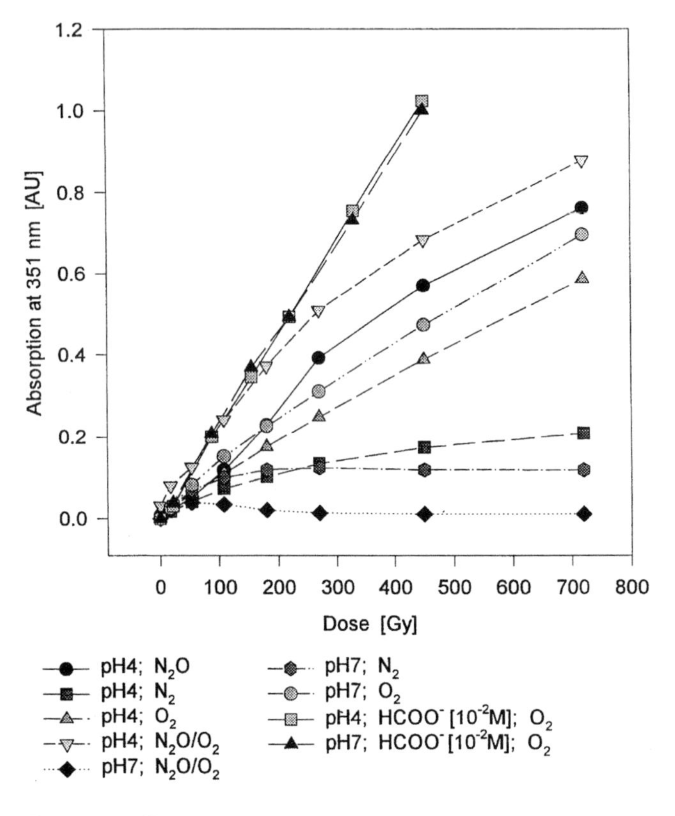

Figure 2. Generation of KI-reactive oxidant in irradiated PBS. The development of an absorption at 351 nm when potassium iodide solution is mixed with solutions that have been irradiated under different conditions of gas saturation and pH is used as an indication for the presence of oxidizing substances. As hydrogen peroxide is known to react stoichiometrically with KI, the curves suggest H_2O_2 as the responsible oxidant. [PBS = phosphate buffer (50 mM) + NaCl (140 mM); irradiation with ^{60}Co γ-rays at a dose rate of 17 Gy/min.]

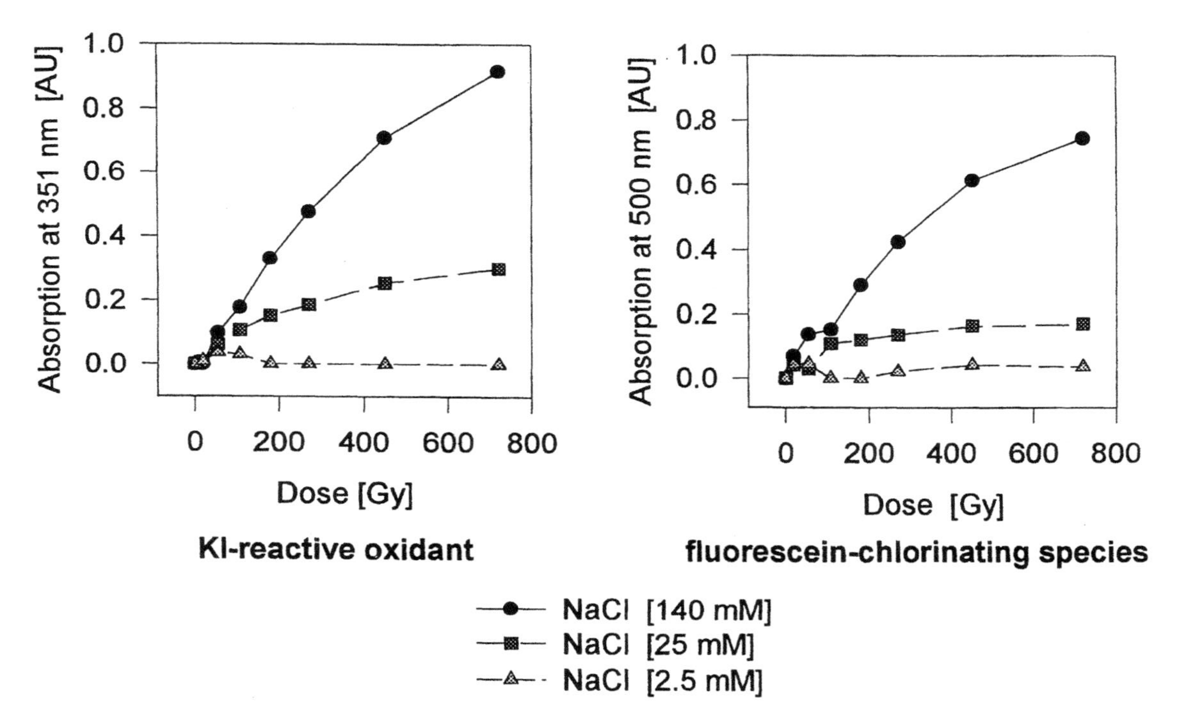

Figure 3. Comparison of different determination methods for reactive species in irradiated phosphate buffer. Left-hand panel, KI-oxidation; right-hand panel, chlorination of fluorescein. Whereas the left part of the figure might still suggest H_2O_2 as the responsible oxidant, the right panel proves that the oxidant is HOCl (as H_2O_2 is not able to chlorinate fluorescein). These curves were determined in 50 mM PB at pH 4, saturated with N_2O/O_2 7:1 (v/v), i.e., exactly the same conditions as for the inverted triangles in Figure 2. (This allows a reinterpretation of Figure 2 insofar as at some of the conditions presented there HOCl instead of H_2O_2 was the responsible oxidant.)

exclusion of oxygen (i.e., by purging the solutions with nitrogen) results in very little production of oxidant (solid squares and hexagons); and saturation with a mixture of N_2O and O_2 at pH 7 produces no oxidant at all (solid diamonds). Without going into the details of radiation chemistry, this means that large production rates for superoxide anions ($O_2^{\bullet -}$) (i.e., the formate experiment) leave large amounts of detectable oxidant; conditions of large production rates for OH radicals (i.e., N_2O saturation) leave no oxidant at all at pH 7. Cursory inspection of Figure 2 suggests H_2O_2 as the only oxidant. Figure 3, however, shows that this is not the case. The left panel depicts the chloride dependence of the development of KI-oxidant under N_2O/O_2 at pH 4, the right panel shows the absorption of fluorescein after addition to the irradiated solution. As the absorption at 500 nm displayed in the right panel is attributable to the chlorinated product of fluorescein (4′, 5′-dichlorofluorescein) (11), and keeping in mind the trivial statement that hydrogen peroxide cannot chlorinate fluorescein, we must conclude that the curves represented by the inverted triangles and the solid circles in Figure 2 (which refer to the same conditions of N_2O/O_2 saturation at acidic pH) result from the action of HOCl. Numerous control experiments we have carried out confirm that HOCl is the *only* species detectable at acidic pH under N_2O/O_2. Therefore, the enhanced killing of yeast in the presence of chloride (solid circles in Figure 4) clearly results from the action of hypochlorite. Figure 5 gives the corresponding cytotoxicities when commercial hydrogen peroxide or NaOCl are added to cell suspensions. Hypochlorite is about a thousandfold more toxic than hydrogen peroxide to the yeast cells that were used in this experiment. Figure 6 shows the fate of HOCl when it is irradiated (added at 50 μM concentration to phosphate-buffered saline (PBS) before irradiation): If no chloride is present, HOCl is depleted rather rapidly, at low

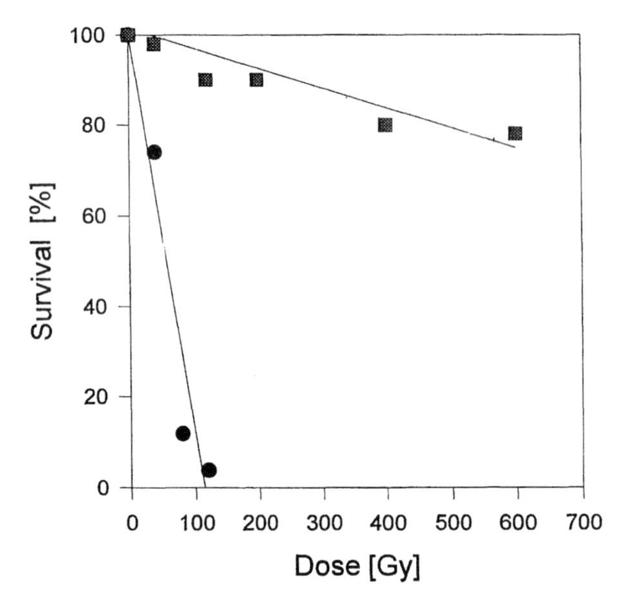

Figure 4. Chloride dependence of yeast cell survival after irradiation under N_2O/O_2 7:1 (v/v) at pH 4.4.

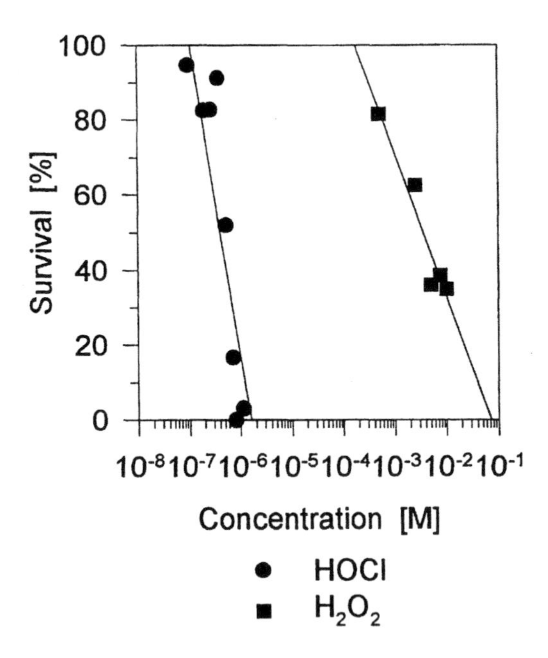

Figure 5. Cytotoxicity of HOCl and H_2O_2 towards *Saccharomyces cerevisiae*

chloride concentration the degradation is slightly inhibited, and at a chloride concentration greater than 50 mM the amount of HOCl exceeds the original level, indicating that HOCl is *generated* during irradiation. Figure 7 indicates even more complex processes: The left panel shows the decrease in I_3^- absorption that results from degradation of hydrogen peroxide which was added to PBS before irradiation at pH 7; the right panel shows the behavior of the same solution at pH 4. Whereas at pH 7 hydrogen peroxide is degraded linearly as a function of dose, at pH 4 I_3^- absorption starts to increase again after all the hydrogen peroxide is consumed. Comparison with parallel measurements of fluorescein chlorination showed that the rising portions of the curves in the right panel are attributable to the action of hypochlorite, which accumulates as its antagonist H_2O_2 is destroyed by irradiation. The picture is therefore kinetically complicated: Neither H_2O_2 nor HOCl is produced in a steady and independent manner during irradiation. They presumably react with each other during irradiation and seem to replace one another depending on the irradiation conditions.

Details of the underlying radiation chemistry need not to be presented here (see Ref. 12), but the consequences of the complex processes involved in the cytotoxicity of radiation-derived substances on cellular systems should be discussed briefly with the aid of the scheme shown in Figure 8, which roughly outlines the processes occurring in the aqueous phase during irradiation.

The primary products of water radiolysis are listed in the first line; line 2 gives the respective concentrations of radicals and molecular products which are produced by the absorption of 1 Gy (one gray) of radiation. Ignoring for the moment the particulate fraction of body fluids (i.e., the cells) and making the simplified assumption that the dissolved proteins are degraded into their constituent amino acids, we can calculate "molarities" of the different plasma constituents. An OH radical produced in the blood by

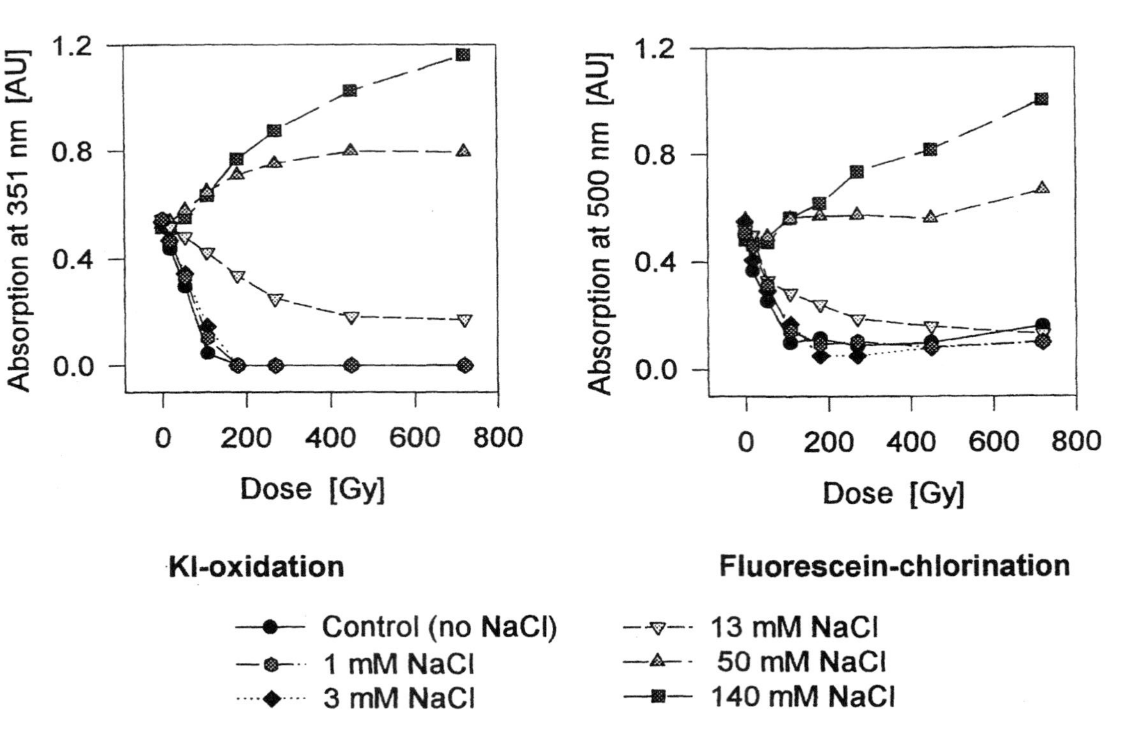

Figure 6. Generation/degradation of HOCl by irradiation. HOCl was added to solutions containing different concentrations of sodium chloride. At low NaCl concentration HOCl gradually vanishes during irradiation, at high NaCl concentration additional HOCl is formed. [PB (50 mM); pH 4.5; saturated with N_2O; 50 μM HOCl originally present.]

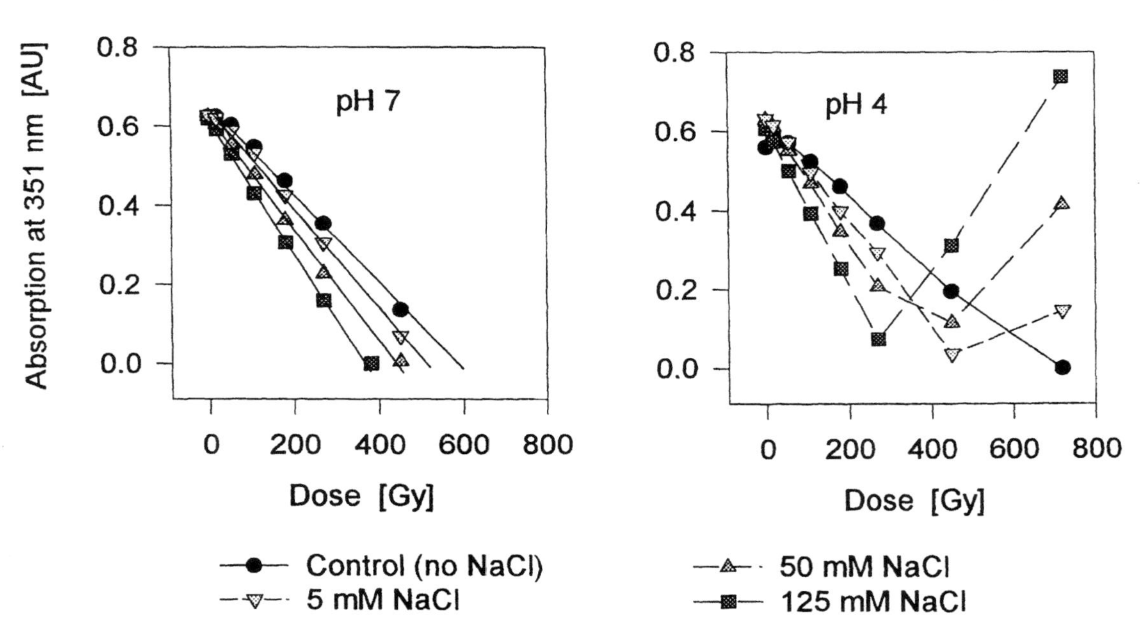

Figure 7. Degradation of H_2O_2 and generation of HOCl. Same experiment as for Figure 6, but H_2O_2 was added prior to irradiation instead of HOCl. The left-hand panel shows that H_2O_2 is gradually degraded at pH 7 (almost independently of NaCl concentration); the right-hand panel at acidic pH shows the onset of a steep rise in the concentration of KI-oxidant after the original H_2O_2 is consumed. (Parallel measurements using fluorescein chlorination proved that this rise is not due to H_2O_2 but rather to HOCl). [PB (50 mM); saturated with N_2O; 50 μM H_2O_2 initially present.]

1		$H_2O \xrightarrow{\text{ϟ}}$	$^{\bullet}OH$	e_{aq}^-	$H^{\bullet}$	H_2	H_2O_2
2	[µM/Gy]		.28	.28	.06	.05	.07

3		Protein	Cl^-	HCO_3^-	$H_2PO_4^-$	O_2
4	[mM]	600	140	25	1	<1

5 $$^{\bullet}OH + Cl^- \leftrightarrow HOCl^{\bullet -}$$

6 $$HOCl^{\bullet -} + H^+ \leftrightarrow H_2O + Cl^{\bullet}$$

7 $$Cl^{\bullet} + Cl^- \leftrightarrow Cl_2^{\bullet -}$$

8 $$2Cl_2^{\bullet -} \rightarrow Cl_2 + 2\,Cl^-$$

9 $$Cl_2 + H_2O \leftrightarrow HOCl + HCl$$

10 $$HOCl \leftrightarrow H^+ + OCl^-$$

Chemical halflife of H_2O_2 / HOCl:

11 $5 < pH < 7.5$: 6 sec

12 $pH < 5$: 90 sec

13 $$H_2O_2 + HOCl \rightarrow H_2O + H^+ + Cl^- + O_2$$

Figure 8. Radiation chemistry of "physiological aqueous solution."

radiolysis of water is confronted with the quantities of reactive targets indicated in line 4. (At a dose rate of 1 Gy per minute–producing 280 nmol of OH radicals–this means that any OH which comes into existence is exposed for 1 min to half a million chloride ions). Since the rate of the reaction $^{\bullet}OH + Cl^-$ is diffusion controlled, we must expect that a very large proportion of all OH radicals enter into the equilibrium given in line 5. Of course, seen on a statistical base, the protein fraction still exceeds chloride by a factor of 4, but considering that "protein," in the simplification made above, represents a great variety of *different* molecules, we must conclude that $HOCl^{\bullet -}$ is the most abundant *single molecular* "sink" of OH radicals in the solution. (Analogous arguments must also hold for OH radicals produced in the blood by other processes, e.g., during reperfusion or by Fenton catalysis or the like.) As $HOCl^{\bullet-}$ can liberate OH radicals (the equilibrium lies on the left side), chloride actually acts like a sponge taking up OH radicals and, depending on the environmental conditions, releases them for reaction with a suitable target. Lines 6 to 10 represent, in an abbreviated form, the pathway to formation of hypochlorous acid (HOCl) which becomes observable at pH values less than 6.5 and reaches its maximum yield at pH 4.2. With increasing proton concentration the $HOCl^{\bullet-}$ equilibrium is pulled to the right and a reaction sequence with intermediate formation of chlorine radicals leads to molecular chlorine, which in aqueous solution is known to exist in equilibrium with the bleaching agent and microbicide HOCl. The further fate of HOCl depends on the pH and the presence of H_2O_2: if the latter is also present in relevant quantities, the half-lives of both species range from a few seconds to some tens of seconds before they react to indiscernible reaction products (as indicated in the last three lines of the scheme).

EXTENT OF HYDROGEN PEROXIDE AND HYPOCHLORITE CONTRIBUTION TO CELL DEATH

The time scale of the interaction of hydrogen peroxide and hypochlorite is the key to understanding the processes which occur during irradiation. When mixed together in a rapid mixing device, these substances react with each other rather rapidly; at neutral and acidic pH the reaction products are chloride and ground-state oxygen. (It should be noted that only at alkaline pH is some of the generated oxygen in a reactive singlet configuration.) During irradiation the situation is different, however: The substances are not mixed instantaneously but are formed by different processes with different buildup kinetics. The kinetics depend in a complicated way on the dose rate, i.e., the generation rate of OH radicals, the availability of oxygen, and the presence of catalytic metals. From detailed studies of the underlying processes (12,13) we know that the net formation of hydrogen peroxide is more efficient than that of HOCl: mainly because processes that destroy H_2O_2 (e.g., the reaction $^{\bullet}OH + H_2O_2$) are slower than the processes that destroy HOCl. Thus the faster depletion of HOCl keeps its concentration low. The difference in the mechanism of the generation processes is also important. The hydrogen peroxide originates partially from processes within the primary ionization clusters called "spurs." From these local accumulations, H_2O_2 diffuses out into the bulk of the solution. The fraction that is generated in the spurs is not accessible to scavenging by constituents of the solution, i.e., does not depend on the irradiation conditions. Hypochlorite, in contrast, derives from OH radicals in the bulk solution. This is one of the reasons why the buildup of both species during irradiation responds differently to changes in solution composition. During the first period of irradiation, when the concentrations of both species are low, they do not have a great chance to meet (whereby they would deactivate their cytotoxic potential) and we assume that they have ample time to react with cells which are suspended in the solution. These arguments are supported by control experiments which we carried out with cell suspensions (10^4–10^9 cells per ml) irradiated in a buffer of "physiological" composition. According to these experiments the development of measurable HOCl and/or H_2O_2 under different irradiation conditions was essentially unaltered whether cells were present or not. This may be explained by the fact that even the highest achievable cell density in a liquid culture corresponds to only some picomoles of cells per liter; therefore, at a dose rate of 10 Gy, which continuously produces micromolar quantities of HOCl and H_2O_2, the chemistry of these two substances proceeds as if no cells were present. However, as a substantial part of the cells had lost their colony forming ability during that irradiation, we must conclude that the amount of cytotoxic material that was sufficient to kill them represented only such a small fraction of the total chemical turnover that the material balance of the overall chemistry was not altered to any measurable extent.

In summary, our experiments lead us to assume that under the conditions of in vitro irradiation of cell suspensions the predominant part of cell death (i.e., loss of reproductive ability) derives from chloride-derived substances produced in the extracellular medium rather than from direct hits of radiation on DNA or other cellular targets. The main difficulty is to extrapolate from these findings to the in vivo condition of whole-body irradiation. Here the ratio of cell mass to liquid phase (interstitial fluid or blood) is quite different from the in vitro conditions. Therefore, an answer to the question how cells in a coherent tissue respond to radiation-derived molecules can only be found if progress is made in explaining the participation of the oxygen radical-derived compounds H_2O_2 and HOCl in those pathways that control intercellular communication and intercellular induction of apoptosis. Final answers to these interesting questions await further elucidation.

ACKNOWLEDGMENT

The help of Dr. David Tait in preparing this manuscript and in making numerous scientific contributions is gratefully appreciated.

REFERENCES

1. Alper T, Ebert M. Influence of dissolved gases on H_2O_2-formation and bacteriophage inactivation by radiation. Nature 1954; 173: 987–989.
2. Frey HE, Pollard EC. Ionizing radiation and bacteria: nature of the effect of irradiated medium. Radiat Res 1966; 28: 668–676.
3. Bauer G. Elimination of transformed cells by normal cells: a novel concept for the control of carcinogenesis. Histol Histopathol 1996; 11: 237–255.
4. Schraufstätter IU, Cochrane CG. H_2O_2 induced cellular DNA damage and its consequences. In: Davies KJA, ed. Oxidative Damage & Repair: Chemical, Biological and Medical Aspects. Oxford: Pergamon Press; 1991: 481–486.
5. Matsuyama A, Namiki M, Okazawa Y. Alkali halides as agents enhancing the lethal effect of ionizing radiations on microorganisms. Radiat Res 1967; 30: 687–701.
6. Brustad T, Wold E. Long-lived species in irradiated N_2O-flushed saline phosphate buffer, with toxic effect upon *E. coli* K-12. Radiat Res 1976; 66: 215–230.
7. Czapski G, Goldstein S, Andorn N, Aronovitch J. Radiation-induced generation of chlorine derivatives in N_2O-saturated phosphate buffered saline: Toxic effects on *Escherichia coli* cells. Free Rad Biol Med 1992; 12: 353–364.
8. Saran M, Bertram H, Bors W, Czapski G. On the cytotoxicity of irradiated media. To what extent are stable products of radical chain reactions in physiological saline responsible for cell death? Int J Radiat Biol 1993; 64: 311–318.

9. Wolcott RG, Franks BS, Hannum DM, Hurst JK. Bactericidal potency of hydroxyl radical in physiological environments. J Biol Chem 1994; 269: 9721–9728.
10. Connick RE. The interaction of hydrogen peroxide and hypochlorous acid in acidic solutions containing chloride ion. J Am Chem Soc 1947; 69: 1509–1514.
11. Hurst JK, Albrich JM, Green TR, Rosen H, Klebanoff S. Myeloperoxidase-dependent fluorescein chlorination by stimulated neutrophils. J Biol Chem 1984; 259: 4812–4821.
12. Saran M, Bors W. Radiation chemistry of physiological saline re-investigated: Evidence that chloride derived intermediates play a key role in cytotoxicity. Radiat Res 1997; 147: 70–77.
13. Saran M, Hamm U, Friedl AA, Bors W. Radiation-induced cell killing is highly dependent upon buffer treatment (filtration compared to autoclaving) due to metal-catalyzed formation of hypochlorite: A cautionary note. Radiat Res 1996; 146: 232–235.

4

Regulation of the Bacterial Response to Hydrogen Peroxide

Gisela Storz, Ming Zheng, Aixia Zhang, Ines Kullik,* Michel Toledano,† and Shoshy Altuvia‡
National Institute of Child Health and Human Development
Bethesda, Maryland

Reactive oxygen species (superoxide anion, hydrogen peroxide, and hydroxyl radical) can lead to the damage of almost all cell components (DNA, lipid membranes, and proteins) and have been implicated as causative agents in many degenerative diseases. The bacterial responses to superoxide anion and hydrogen peroxide have proven to be ideal models for studying how cells sense and adapt to oxidative stress.

SoxR, SoxS

Bacterial cells are capable of adapting to superoxide-generating compounds, and treatment of bacteria with compounds such as paraquat, menadione, and plumbagin leads to the induction of a distinct group of proteins (see Ref. 1 and 2 for comprehensive reviews). Key regulators of the adaptive response to superoxide in *E. coli* are the SoxR and SoxS proteins. Genes under the control of these regulators include *sodA* (manganese superoxide dismutase), *nfo* (the endonuclease IV DNA repair enzyme), *zwf* (glucose-6-phosphate dehydrogenase, which can increase the reducting power of the cell), *fumC* and *acnA* (superoxide-resistant isozymes of fumarase and aconitase, respectively), *fpr* (NADPH:ferredoxin oxidoreductase), *acrAB* (a transporter), and the *micF* RNA (interferes with the translation of the *ompF*-encoded outer membrane porin) (1–7).

Regulation of the *soxRS* regulon occurs by a two-step process: SoxR is first converted to an active form which enhances *soxS* transcription (8,9). The increased levels of SoxS in turn activate expression of the regulon. The genes encoding the two regulators

---

Current affiliations:
* University of Zurich, Zurich, Switzerland.
† Rutgers College of Pharmacy, Piscataway, New Jersey.
‡ The Hebrew University–Hadassah Medical School, Jerusalem, Israel

overlap each other, with the *soxR* promoter embedded in the *soxS* structural gene and transcribed in the opposite direction. The constitutively expressed SoxR protein resembles MerR, a regulator of mercury resistance. Like MerR, SoxR is a dimer and has four C-terminal cysteines that are critical for activity. SoxR can be isolated as Fe-free or Fe-containing forms. Both forms can bind the *soxS* promoter, but only the Fe-form, with two [2Fe:2S] centers per dimer, is able to activate transcription in vitro (10–12). The mechanism of SoxR activation and the nature of the signaling molecule is still under debate. Possibly, SoxR exists as an apoprotein and the full [2Fe:2S] clusters in SoxR are assembled with the iron released from superoxide-sensitive enzymes in the cell (11). Alternatively, the SoxR protein is normally in a reduced $[2Fe:2S]^{+}$ state and is activated by oxidation to a $[2Fe:2S]^{2+}$ state (12). This oxidation may occur through direct exposure to superoxide anion or through changes in the levels of NADPH or reduced flavodoxins or ferredoxins (6). Any model for SoxR activation must take into account the observation that SoxR is also activated by nitric oxide in vivo (13). Once SoxR is activated, the regulator appears to increase *soxS* transcription by distorting the *soxS* promoter (11).

The SoxS protein activates the promoters of the *soxRS* regulon by mechanisms which involve binding near or at the –35 hexamer. SoxS and a MalE–SoxS fusion protein have been purifed and shown to bind to several SoxS-regulated promoters, and the core sequence of a proposed SoxS box is AnnGCAY (14,15). For some promoters such as *zwf* and *fpr*, the sequences bound by SoxS do not overlap the –35 promoter sequence, and in vitro transcription experiments have shown that activation of these promoters requires the C-terminal domain of the α subunit of RNA polymerase, indicating direct contact between SoxS and RNA polymerase (16). In contrast, at other promoters such as *micF* and *fumC*, SoxS binds at sites overlapping the –35 hexamer, and the C-terminal domain of the α subunit is not required for activation. SoxS also binds to its own promoter where it appears to repress its own transcription (17).

OXYR

Treatment of bacterial cells with low doses of hydrogen peroxide results in the induction of at least 30 proteins and resistance to killing by higher doses of hydrogen peroxide (1,2). The expression of at least nine of the hydrogen peroxide-inducible proteins is controlled by OxyR in *E. coli* and *Salmonella typhimurium*. Several of the genes whose expression is activated by OxyR have been identified and include *katG* (hydroperoxidase I), *ahpCF* (an alkyl hydroperoxide reductase), *dps* (a nonspecific DNA binding protein that may protect against oxidative DNA damage), and *gorA* (glutathione reductase), all encoding activities which have understandable roles in protecting the cell against oxidative damage (1,2,18). OxyR also activates the expression of a small untranslated regulatory RNA denoted OxyS, which protects against mutagenesis (S. Altuvia, D. Weinstein, A. Zhang, L. Postow and G. Storz, unpublished). OxyR has also been shown to be a repressor and negatively autoregulates its own expression so that a constant level of OxyR is maintained in the cell (1,2). In addition, OxyR represses the expression of the Mu phage *mom* gene and the *E. coli flu* gene, neither of which has an understandable role in the oxidative stress response (1,2,19). An interesting direction for future studies will be to elucidate all of the roles of OxyR within the cell.

The tetrameric OxyR protein is a member of the LysR family of transcriptional activators and has been characterized extensively (20,21). The protein appears to exist in two forms, reduced and oxidized, but only the oxidized form is able to activate transcription. Direct oxidation of OxyR is therefore the likely mechanism whereby the cells sense hydrogen peroxide and induce the OxyR regulon. The redox-active center in OxyR has been proposed to be a single cysteine, but this hypothesis needs to be tested with additional biochemical experiments (21).

OxyR has been shown to bind to promoters by contacting four major grooves on one face of the DNA, and a consensus of four ATAGnt repeats was defined for OxyR (22). Interestingly, the two forms of OxyR appear to have different binding specificities. The reduced form is able to bind the *oxyR* and *mom* promoters, but not the *katG* and *ahpC* promoters, and contacts ATAGnt repeats in two pairs of adjacent major grooves separated by one helical turn. In contrast, oxidized OxyR has been found to bind all OxyR-regulated promoters that have been tested and binds in four adjacent major grooves. The differences in binding may allow OxyR to carry out different functions under different conditions. Therefore OxyR can repress the *oxyR* and *mom* promoters during normal growth and activate *katG* and *ahpC* in response to oxidative stress. OxyR activates transcription by increasing the binding of RNA polymerase to the promoters and has recently been shown to require specific surfaces on the C-terminal domain of the α subunit of RNA polymerase to activate transcription (23,24).

σ^S SUBUNIT

An additional regulator that is important for survival against oxidative stress in *E. coli* is the *rpoS*-encoded σ^S subunit of RNA polymerase (see Ref. 25 for a comprehensive review). This sigma factor is important for the expression of a large group of genes that are induced when cells encounter a number of different stresses, including starvation, osmotic stress, and acid stress as well as upon entry into stationary phase. Starved and stationary phase cells are intrinsically resistant to a variety of stress conditions, including high levels of hydrogen peroxide, and RpoS has been shown to regulated the expression of *katG* (hydroperoxidase I), *katE* (hydroperoxidase II), *dps* (nonspecific DNA binding protein), *xthA* (exonuclease III), and *gorA* (glutathione reductase) (18, 25, 26). The *katG*, *dps*, and *gorA* genes are also activated by OxyR, suggesting that *E. coli* cells have two regulons that can protect against exposure to hydrogen peroxide; the OxyR regulon during exponential growth and the σ^S regulon in stationary phase. It seems likely some of the SoxS-regulated genes are also regulated by σ^S, and one SoxS target, *acrAB*, has recently been shown to be induced in stationary phase (7).

The regulation of σ^S levels occurs at multiple levels and much remains to be learned (25). The transcription, translation, and stability of σ^S are all modulated in response to different signals, including the starvation signal ppGpp, a cell density signal homoserine lactone, cAMP, and UDP-glucose (25, 27–29). It is not yet known whether oxidative stress has a direct impact on σ^S levels.

SUMMARY

The regulation of the *E. coli* response to oxidative stress involves several key regulators: SoxR, SoxS, OxyR, and σ^S. Two of these transcription factors, SoxR and OxyR, appear to be direct sensors of oxidative stress. An exciting challenge for the future will be to elucidate the nature of the chemical reactions leading to the activation of the two regulators.

REFERENCES

1. Demple B. Regulation of bacterial oxidative stress genes. Annu Rev Genet 1991; 25: 315–317.
2. Farr SB, Kogoma T. Oxidative stress responses in *Escherichia coli* and *Salmonella typhimurium.* Microbiol Rev 1991; 55: 561–585.
3. Liochev SI, Fridovich I. Fumarase C, the stable fumarase of *Escherichia coli*, is controlled by the *soxRS* regulon. Proc Natl Acad Sci USA 1992; 89: 5892–5896.
4. Chou JH, Greenberg JT, Demple B. Posttranscriptional repression of *Escherichia coli* OmpF protein in response to redox stress: Positive control of the *micF* antisense RNA by the *soxRS* locus. J Bacteriol 1993; 175: 1026–1031.
5. Gruer MJ Guest JR. Two genetically-distinct and differentially-regulated aconitases (AcnA and AcnB) in *Escherichia coli.* Microbiology 1994; 140: 2531–2541.
6. Liochev SI, Hausladen A, Beyer WF Jr, Fridovich I. NADPH: ferredoxin oxidoreductase acts as a paraquat diaphorase and is a member of the *soxRS* regulon. Proc Natl Acad Sci USA 1994; 91: 1328–1331.
7. Ma D, Alberti M, Lynch C, Nikaido H, Hearst JE. The local repressor AcrR plays a modulating role in the regulation of *acrAB* genes of *Escherichia coli* by global stress signals. Mol Microbiol 1994; 19: 101–112.
8. Nunoshiba T, Hidalgo E, Amábile-Cuevas CF, Demple B. Two-stage control of an oxidative stress regulon: The *Escherichia coli* SoxR protein triggers redox-inducible expression of the *soxS* regulatory gene. J Bacteriol 1992; 174: 6054–6060.
9. Wu J, Weiss B. Two-stage induction of the *soxRS* (superoxide response) regulon of *Escherichia coli.* J Bacteriol 1992; 174: 3915–3920.
10. Hidalgo E, Demple B. An iron–sulfur center essential for transcriptional activation by the redox-sensing SoxR protein. EMBO J 1994; 13: 138–146.
11. Hildalgo E, Bollinger JM Jr, Bradley TM, Walsh CT, Demple, B. Binuclear [2Fe-2S] clusters in the *Escherichia coli* SoxR protein and role of the metal centers in transcription. J Biol Chem 1995; 270: 20908–20914.
12. Wu J, Dunham WR, Weiss B. Overproduction and physical characterization of SoxR, a [2Fe-2S] protein that governs an oxidative response regulon in *Escherichia coli.* J Biol Chem 1995; 270: 10323–10327.
13. Nunoshiba T, DeRojas-Walker T, Wishnok JS, Tannenbaum SR, Demple B. Activation by nitric oxide of an oxidative-stress response that defends *Escherichia coli* against activated macrophages. Proc Natl Acad Sci USA 1993; 90: 9993–9997.
14. Fawcett WP, Wolf RE Jr. Purification of a MalE–SoxS fusion protein and identification of the control sites of *Escherichia coli* superoxide-inducible genes. Mol Microbiol 1994; 14: 669–679.
15. Li Z, Demple B. SoxS, an activator of superoxide stress genes in *Escherichia coli*: Purification and interaction with DNA. J Biol Chem 1994; 269: 18371–18377.
16. Jair K-W, Fawcett WP, Fujita N, Ishihama A, Wolf RE Jr. Ambidextrous transcriptional activation by SoxS: Requirement for the C-terminal domain of the RNA polymerase alpha subunit in a subset of *Escherichia coli* superoxide-inducible genes. Mol Microbiol 1996; 19: 307–317.
17. Nunoshiba T, Hidalgo E, Li Z, Demple B. Negative autoregulation by the *Escherichia coli* SoxS protein: A dampening mechanism for the *soxRS* redox stress response. J Bacteriol 1993; 175: 7492–7494.

18. Altuvia S, Almirón M, Huisman G, Kolter R, Storz G. The *dps* promoter is activated by OxyR during growth and by IHF and σ^s in stationary phase. Mol Microbiol 1994; 13: 265–272.
19. Henderson IR., Meehan M, Owen P. A novel regulatory mechanism for a novel phase-variable outer membrane protein of *Escherichia coli.* In Paul, PS, Francis, DH, Benfield, D, eds Mechanisms in the Pathogenesis of Enteric Diseases. New York: Plenum; 1996.
20. Kullik I, Stevens J, Toledano MB, Storz G. Mutational analysis of the redox-sensitive transcriptional regulator OxyR: Regions important for DNA binding and multimerization. J Bacteriol 1995; 177: 1285–1291.
21. Kullik I, Toledano MB, Tartaglia LA, Storz G. Mutational analysis of the redox-sensitive transcriptional regulator OxyR: Regions important for oxidation and transcriptional activation. J Bacteriol 1995; 177: 1275–1284.
22. Toledano MB, Kullik I, Trinh F, Baird PT, Schneider TD, Storz, G. Redox-dependent shift of OxyR-DNA contacts along an extended DNA-binding site: A mechanism for differential promoter selection. Cell 1994; 78: 897–909.
23. Tao K, Fujita N, Ishihama A. Involvement of the RNA polymerase α subunit C-terminal region in co-operative interaction and transcriptional activation with OxyR protein. Mol Microbiol 1993; 7: 859–864.
24. Tao K, Zou C, Fujita N, Ishihama A. Mapping of the OxyR protein contact site in the C-terminal region of RNA polymerase α subunit. J Bacteriol 1995; 177: 6740–6744.
25. Loewen PC, Hengge-Aronis R. The role of the sigma factor σ^s (KatF) in bacterial global regulation. Annu Rev Microbiol 1994; 48: 53–80.
26. Becker-Hapak M, Eisenstark A. Role of *rpoS* in the regulation of glutathione oxidoreductase (gor) in *Escherichia coli.* FEMS Microbiol Lett 1995; 134: 39–44.
27. Böhringer J, Fischer D, Mosler G, Hengge-Aronis R. UDP-glucose is a potential intracellular signal molecule in the control of expression of σ^s and σ^s-dependent genes in *Escherichia coli.* J Bacteriol 1995; 177: 413–422.
28. Huisman GW, Kolter R. Sensing starvation: A homoserine lactone-dependent signaling pathway in *Escherichia coli.* Science 1994; 265: 537–539.
29. Lange R, Hengge-Aronis R. The cellular concentration of the σ^s subunit of RNA polymerase in *Escherichia coli* is controlled at the levels of transcription, translation, and protein stability. Genes Dev 1994; 8: 1600–1612.

5

Activation of Lymphocyte Signal Pathways by Oxidative Stress: Role of Tyrosine Kinases and Phosphatases

Gary L. Schieven
Bristol–Myers Squibb Pharmaceutical Research Institute
Seattle, Washington

INTRODUCTION

Cells may be subjected to oxidative stress from a wide variety of sources, including environmental agents, oxidizing agents generated within the cell or by other cells of the organism, and by radiation. Although these agents are well known to cause damage to the components of biological systems, recent work has shown that oxidative stress can activate tyrosine phosphorylation signal pathways (see Ref. 1 for review). We have employed lymphocytes as a model system to investigate the effects of oxidizing agents, ionizing radiation, and UV radiation on cellular tyrosine phosphorylation signaling. Lymphocytes are well suited for this purpose for two reasons. First, tyrosine phosphorylation is a central regulatory pathway in lymphocytes. Tyrosine phosphorylation is essential for lymphocyte activation, with both the antigen receptor complex and numerous accessory signaling molecules employing this pathway (2–4). Second, lymphocytes are highly sensitive to oxidative stress. These cells are frequently exposed to oxidative stress during the course of their normal function at sites of inflammation. Lymphocytes are also deliberately subjected to oxidative stress during the course of medical therapy. For example, ionizing radiation followed by bone marrow transplantation is employed for the treatment of T and B cell malignancies. Lymphocytes, particularly T cells, are also very sensitive to UV radiation (5). UV radiation can inhibit graft versus host disease and improve graft survival in transplants; and has also been used in the treatment of cutaneous T cell lymphoma (6).

ANTIGEN RECEPTOR SIGNALING IN LYMPHOCYTES

Tyrosine phosphorylation is essential for lymphocyte activation (2,3). Cellular tyrosine phosphorylation is induced by the concerted action of specific protein tyrosine kinases

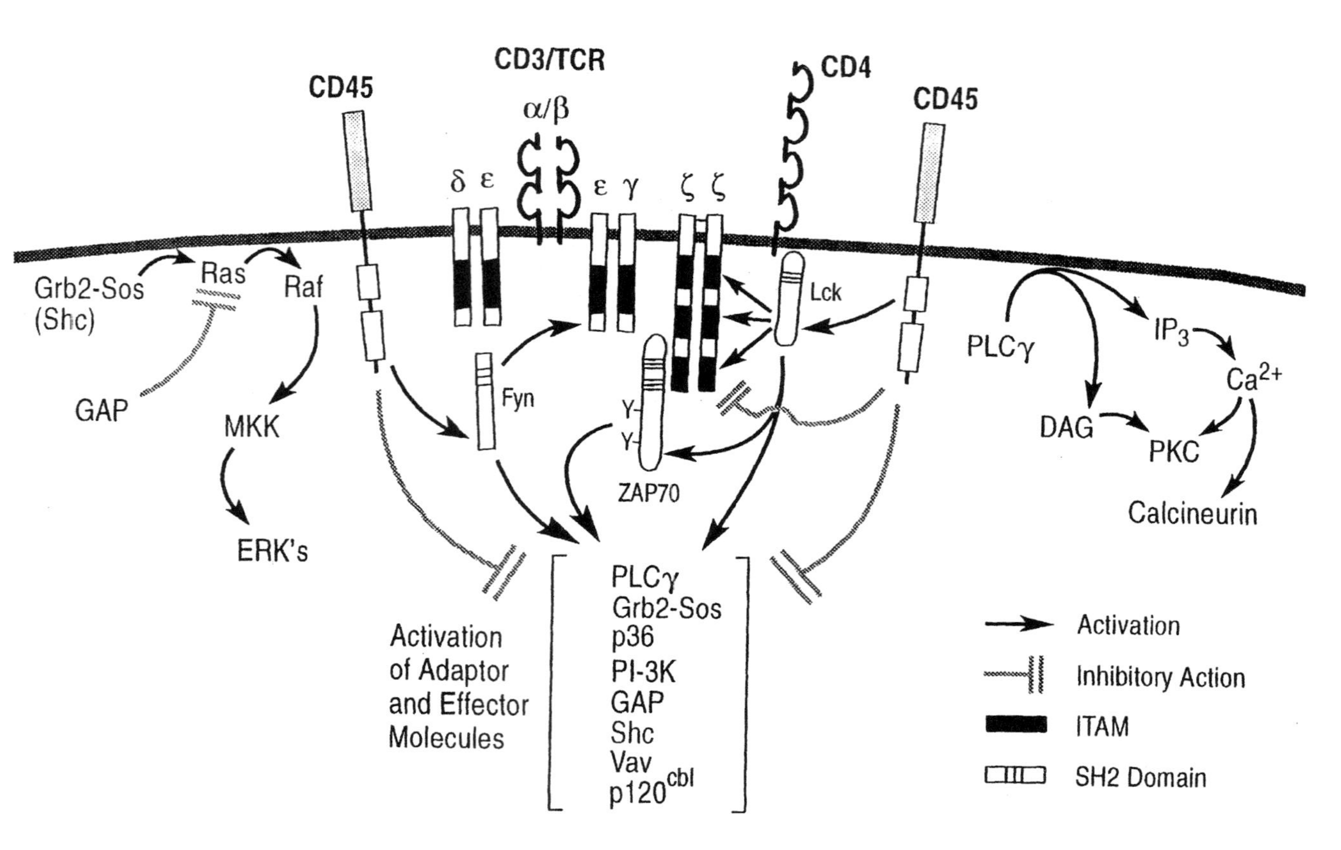

Figure 1. Antigen receptor signaling in lymphocytes. (From Schieven GL, in Oxidative Stress and Signal Transduction, Cadenas E and Forman HJ, eds. New York: Chapman & Hall.)

(PTK) and phosphotyrosine phosphatases (PTP). The Src-family tyrosine kinases are activated first, followed by the Syk-family kinases (see Ref. 7 for review). In T cells, the T cell receptor (TCR) binds a peptide antigen presented by major histocompatibility complex (MHC). CD4 binds to MHC II, stabilizing the interaction between the TCR and MHC II, whereas CD8 performs a similar function in binding to MHC I. Lck is associated with CD4 and CD8. These binding events lead to the activation of the Src-family kinases Lck and Fyn (Figure 1). These kinases then act to phosphorylate the ζ, δ, ϵ, and γ chains of the TCR at dual tyrosines present in ITAM (immunoreceptor tyrosine activation motif) sequences. The Syk-family kinases Syk and ZAP-70 then bind to the phosphorylated ITAM sequences via their SH2 domains (Figure 1). SH2 domains bind phosphotyrosine in the context of specific amino acid sequences and are responsible for the phosphotyrosine-dependent assembly of multiprotein signaling complexes (see Ref. 8 for review). Although both Syk and ZAP-70 are present in T cells; ZAP-70 is essential for productive T cell receptor signaling (9). Syk is primarily involved in B cell and myeloid cell signaling. Lck activates ZAP-70 by phosphorylating it on Tyr-493 after ZAP-70 has bound via its SH2 domains to tyrosine phosphorylated ζ chain (7). When ZAP-70 and Syk are activated, they phosphorylate themselves at multiple additional sites, so that they become scaffolds for the assembly of signaling complexes of proteins through SH2 interactions (10,11).

The activated Src-family and Syk-family kinases phosphorylate downstream signaling molecules (Figure 1). One of the most important of these is PLCγ, which associates with ZAP-70, the 36 kDa adapter protein Lnk, and other signaling proteins via phosphotyrosine – SH2 complexes (12–14). PLCγ hydrolyzes phosphatidylinositol 4,5-bisphosphate to generate diacylglycerol and inositol 1,4,5-trisphosphate (IP_3) (Figure 1). IP_3 binding to its receptor results in the mobilization of Ca^{2+} ions from intracellular stores. Ca^{2+} activates the calcium-dependent phosphatase calcineurin in an essential step of T cell activation. Ca^{2+} also acts with diacylglycerol to activate protein kinase C.

B cell antigen receptor signaling is similar to the process in T cells, except that an important difference between Syk and ZAP-70 can lead to differing sensitivity to oxidative stress and PTP inhibition. In B cells, surface immunoglobulin (sIg) binds antigen directly. This induces activation of Src-family kinases such as Lyn, Fyn, and Blk, leading to tyrosine phosphorylation of the ITAM sequences of the Igα and β chains of the B cell antigen receptor. Syk is directly activated by binding tyrosine-phosphorylated ITAM sequences without the need for further activation by a Src kinase. This can make Syk much more susceptible to activation by PTP inhibition than ZAP-70, and indeed the PTP inhibitor BMOV activates Syk in B cells but not ZAP-70 in T cells (15).

Phosphotyrosine phosphatases (PTP) both positively and negatively regulate lymphocyte tyrosine phosphorylation. Src-family kinases are inhibited by phosphorylation of a conserved C-terminal tyrosine (see Figure 2 for Lck). When this tyrosine is phosphorylated, it binds intramolecularly to the SH2 domain of the kinase, holding it in an inactive conformation. The transmembrane PTP CD45 is essential for lymphocyte signal transduction because it is responsible for the dephosphorylation of this C-terminal site on Src-family kinases such as Lck and Fyn. In addition to this positive regulation of signaling, CD45 can also negatively regulate signaling by dephosphorylating substrates such as the ζ chain of the TCR and the adapter protein $p36^{lnk}$ (16,17).

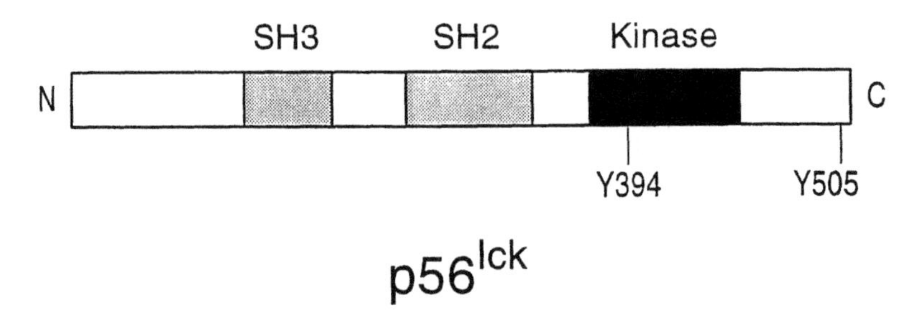

Figure 2. Lck tyrosine kinase. (From Schieven GL, in Oxidative Stress and Signal Transduction, Cadenas E and Forman HJ, eds. New York: Chapman & Hall.)

IONIZING AND ULTRAVIOLET RADIATION

B cell malignancies derived from B cell precursors are highly sensitive to ionizing radiation, leading to the use of radiation therapy followed by bone marrow transplantation for the treatment of this disease. Although sufficient doses of ionizing radiation are capable of killing any cell by the well-known mechanism of DNA damage, recent work has shown that ionizing radiation acts on the signal pathways of these cells. Ionizing radiation strongly induces tyrosine phosphorylation in cell lines derived from B cell lymphomas such as Ramos and Daudi (13). Tyrosine kinase inhibitors such as herbimycin and genistein block this phosphorylation, whereas the PTP inhibitor vanadate greatly increased the response to radiation (13). These results suggest that radiation activates tyrosine kinases, whereas PTP act to limit the response. Further studies have shown that the Src-family kinases Lck (18) and Lyn (19) can be activated by ionizing radiation. Tyrosine kinase inhibitors also inhibit the radiation-induced activation of protein kinase C (20), NFκB (21), and the induction of c-*jun* (22). Most importantly, the tyrosine kinase inhibitors blocked radiation-induced apoptosis and clonogenic cell death, whereas the PTP inhibitor vanadate greatly augmented radiation-induced cell death (13). These results indicate that the activation of tyrosine kinases by radiation strongly contributes to lymphocyte cell death, whereas the PTP serve to protect the cells.

Lymphocytes are known to be highly sensitive to UVB (290–320 nm) and UVC (200–290 nm) but not to UVA (320–400 nm) radiation (23). UVB and UVC, but not UVA, radiation strongly induce tyrosine phosphorylation in T and B lymphocytes (24). In contrast to ionizing radiation, UV radiation can also induce Ca^{2+} signals via the tyrosine phosphorylation of PLCγ, but only in T cells (24) (Figure 3). These effects of UV irradiation occur in normal human peripheral blood lymphocytes as well as transformed cell lines (25). The pattern of tyrosine phosphorylation induced by UV is extremely similar to that induced by antigen receptor signaling (24). It is likely that PTP are responsible for the lack of Ca^{2+} signaling in response to UV in B cells, because treatment of the cells with the PTP inhibitor vanadate resulted in strong Ca^{2+} signals with UV irradiation (Figure 3A). The vanadate treatment also increased the T cell Ca^{2+} response to UV (Figure 3B). These results suggest that PTP act to limit UV-induced signaling, and that this regulation differs between T and B cells.

Why would UV, commonly thought of as only inducing DNA damage, also induce such a similar pattern of tyrosine phosphorylation as antigen receptor-induced signaling?

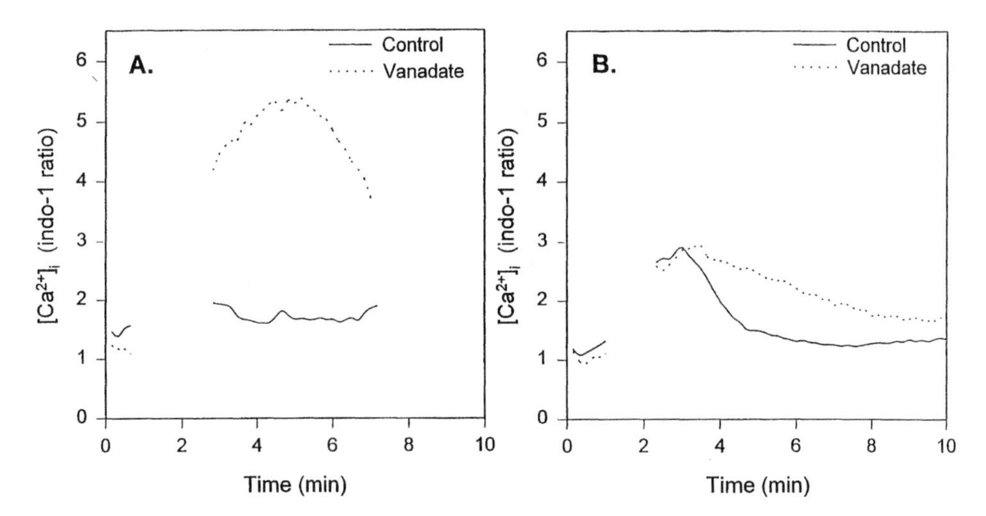

Figure 3. UV-induced Ca^{2+} signaling. Cells were cultured for 16 h in the presence or absence of 50 μM sodium orthovanadate. The baseline level of intracellular free Ca^{2+} $[Ca^{2+}]_i$ was measured by flow cytometry using the the Ca^{2+} as previously described (24). The cells were then irradiated with 1200 J/m^2 UVC (gap in the trace) and the $[Ca^{2+}]_i$ levels in the cells were immediately measured. A. B cells; B. T cells.

UV activates the same key kinases as do the receptors, namely, Syk in B cells (26) and ZAP-70 in T cells (27). Furthermore, T cell receptor expression is required for UV-induced tyrosine phosphorylation, ZAP-70 kinase activation, Ca^{2+} signaling, and the activation of NFκB (27), indicating that UV acts at the receptor level to initiate these signals. UV has also been found to act at the receptor level in the Src-and Ras-dependent UV activation of gene transcription in HeLa cells (28,29). UV thus activates receptors in a ligand-independent manner, inducing downstream signaling and gene transcription.

H_2O_2 BYPASSES RECEPTORS TO ACTIVATE SIGNAL PATHWAYS

H_2O_2 can also activate tyrosine kinase signal pathways in lymphocytes. In B cells, H_2O_2 induces cellular tyrosine phosphorylation and Ca^{2+} signaling (26). Antigen receptor Ca^{2+} signals are dependent on IP_3 production (Figure 1), and H_2O_2 treatment also induced inositol 1,4,5-trisphosphate (IP_3) production within 10 seconds of exposure (21), suggesting a common mechanism. Tyrosine kinase inhibitors block both the IP_3 production and the calcium signals (21,26). In T cells, H_2O_2 also induces Ca^{2+} signals and cellular tyrosine phosphorylation in a pattern very similar to that induced by antigen receptor stimulation (27). These similar patterns of tyrosine phosphorylation and Ca^{2+} signaling suggest that both antigen receptor signaling and H_2O_2-induced signaling share a common pathway. Examination of the Syk-family and Src-family kinases revealed that although Src-family kinases are not activated by H_2O_2 (26), Syk is activated by H_2O_2 in B cells (26) and ZAP-70 is activated by H_2O_2 in T cells (27). Since these kinases are normally activated by antigen receptor signaling, it is not surprising that their activation by H_2O_2 gives a similar pattern of tyrosine phosphorylation as does antigen receptor stimulation.

H_2O_2 stimulation of these kinases differs from that by antigen receptor or UV radiation in that H_2O_2 is capable of bypassing the receptor. H_2O_2 can induce Ca^{2+} signals and cellular tyrosine phosphorylation, and activate ZAP-70 in a T cell line that does not express the antigen receptor (27). However, the H_2O_2-induced signals are less rapid and less intense in these cells, suggesting that H_2O_2 can act via receptor-dependent as well as receptor-independent pathways.

ROLE OF PHOSPHOTYROSINE PHOSPHATASES

Syk is not activated directly by H_2O_2 (26), suggesting that such oxidizing agents act via other cellular components. One likely pathway is via inhibition of PTP. PTP are very sensitive to oxidative stress because they contain in their active site an essential cysteine residue that is highly reactive (30). As a result, oxidizing agents such as H_2O_2 are potent PTP inhibitors. Recently, we have found that bis(maltolato)oxovanadium(IV) (BMOV), previously reported to be an insulin-mimetic agent (31), is a PTP inhibitor with strong activity toward CD45 and PTP1B (15). This compound selectively induces cellular tyrosine phosphorylation in B cells relative to T cells and activates Syk, demonstrating that PTP inhibition can induce Syk-family kinase activation (15).

We have investigated the effects of oxidative stress in PTP inhibition by comparing two PTP inhibitors, BMOV and the peroxovanadium compound sodium oxodiperoxo(1,10-phenanthroline)vanadate(V) [pV(phen)] (Figure 4), also reported to be an insulin-mimetic agent (32). Pervanadate, a peroxovanadium compound generated by the reaction of H_2O_2 plus vanadate, has been reported to strongly induce tyrosine phosphorylation in lymphocytes (21,33,34) through its ability to inhibit PTP while activating tyrosine kinases such as Lck and Fyn (21,34). Pervanadate is highly unstable, whereas the greater stability of pV(phen) permits further investigation into its mechanism of action.

BMOV induced little cellular oxidative stress in lymphocytes as measured by change in DCF-DA fluorescence, whereas pV(phen) strongly induced oxidative stress (35). Whereas BMOV selectively induced tyrosine phosphorylation in B cells but not T cells (15), pV(phen) rapidly induced massive cellular tyrosine phosphorylation in both T and B

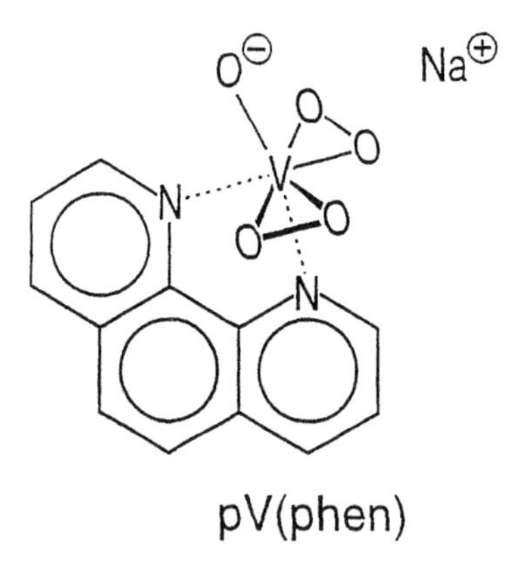

Figure 4. Structure of sodium oxodiperoxo(1,10-phenanthroline)vanadate(V) (pV(phen)).

lymphocytes, and also activated Syk and ZAP-70. The antioxidant pyrrolidine dithiocarbamate (PDTC) blocked intracellular oxidation by pV(phen) and inhibited pV(phen)-induced cellular tyrosine phosphorylation and activation of Syk and ZAP-70, while not blocking the ability of pV(phen) to inhibit PTP1B in direct enzyme assays (35). PDTC also inhibits the activation of MAP kinase by pV(phen). Taken together, these results indicate that the oxidative stress induced by peroxovanadium compounds plays a key role in their activation of tyrosine kinases, which, when combined with PTP inhibition, leads to massive accumulation of tyrosine phosphorylation and the activation of downstream signaling molecules such as MAP kinase.

However, studies with these PTP inhibitors suggest that some of their effects on downstream signaling molecules can be separable from the induction of oxidative stress. The transcription factor NFκB plays an important role in the expression of a variety of genes, particularly those involved in inflammatory and immune responses. Extensive studies have indicated that reactive oxygen species are involved in the activation of NFκB by many stimuli (36). Both BMOV and pV(phen) can activate NFκB in Jurkat T cells (35), suggesting that PTP can play an important role in the regulation of NFκB. Strikingly, BMOV is able to activate NFκB in the presence of high doses of *N*-acetylcysteine. Although *N*-acetylcysteine cannot be used in combination with pV(phen) owing to the direct chemical reaction of these molecules, the antioxidant PDTC that inhibits the induction of oxidative stress by pV(phen) actually increased the activation of NFκB (35). These results suggest that PTP inhibitors may be able, at least in part, to bypass the need for reactive oxygen species in the activation of NFκB. Part of the mechanism of action of reactive oxygen species in NFκB activation may be to inhibit phosphatases, so phosphatase inhibitors could also fill this role.

CONCLUSION

These findings indicate that in addition to damage of cellular components by oxidative stress leading to loss of function, oxidative stress can also activate receptor-linked signaling pathways independent of normal ligand control by acting on tyrosine kinases and phosphotyrosine phosphatases. The directed targeting of oxidative stress via molecules such as pV(phen) can help elucidate the pathways affected by oxidative stress and may prove useful in the development of drugs capable of oxidizing specific molecular targets.

REFERENCES

1. Schieven GL, Ledbetter JA. Activation of tyrosine kinase signal pathways by radiation and oxidative stress. Trends Endocrinol Metab 1994; 5: 383–388.
2. June CH, Fletcher MC, Ledbetter JA, Schieven GL, Siegel JN, Phillips AF, Samelson LE. Inhibition of tyrosine phosphorylation prevents T-cell receptor-mediated signal transduction. Proc Natl Acad Sci USA 1990; 87: 7722–7726.
3. Lane PJL, Ledbetter JA, McConnell FM, Draves K, Deans J, Schieven GL, Clark EA. The role of tyrosine phosphorylation in signal transduction through surface Ig in human B cells. J Immunol 1991; 146: 715–722.
4. Mustelin T, Coggeshall M, Isakov N, Altman A. T cell antigen receptor-mediated activation of phospholipase C requires tyrosine phosphorylation. Science 1990; 247: 1584–1587.

5. Pamphilon DH, Alnaqdy AA, Wallington TB. Immunomodulation by ultraviolet light: clinical studies and biological effects. Immunol Today 1991; 12: 119–123.
6. Ramsay DL, Lish KM, Yalowitz CB, Soter NA. Ultraviolet-B phototherapy for early-stage cutaneous T-cell lymphoma. Arch Dermatol 1992; 128: 931–933.
7. Weiss A, Littman DR. Signal transduction by lymphocyte antigen receptors. Cell 1994; 76: 263–274.
8. Pawson T. Protein modules and signalling networks. Nature 1995; 373: 573–579.
9. Arpaia E, Shahar M, Dadi H, Cohen A, Roifman CM. Defective T cell receptor signaling and $CD8^+$ thymic selection in humans lacking ZAP-70 kinase. Cell 1994; 76: 947–958.
10. Law CL, Chandran KA, Sidorenko SP, Clark EA. Phospholipase C-γ1 interacts with conserved phosphotyrosyl residues in the linker region of Syk and is a substrate of Syk. Mol Cell Biol 1996; 16: 1305–1315.
11. van Oers NSC, Weiss A. The Syk/ZAP-70 protein tyrosine kinase connection to antigen receptor signalling processes. Semin Immunol 1995; 7: 227–236.
12. Huang X, Li Y, Tanaka K, Moore KG, Hayashi JI. Cloning and characterization of Lnk, a signal transduction protein that links T-cell receptor activation signal to phospholipase $C\gamma_1$, Grb2, and phosphatidylinositol 3-kinase. Proc Natl Acad Sci USA 1995; 92: 11618–11622.
13. Gilliland LK, Schieven GL, Norris N, Kanner SB, Aruffo A, Ledbetter JA. Lymphocyte lineage-restricted tyrosine-phosphorylated proteins that bind PLCγ1 SH2 domains. J Biol Chem 1992; 267: 13610–13616.
14. Nel AE, Gupta S, Lee L, Ledbetter JA, Kanner SB. Ligation of the T-cell antigen receptor (TCR) induces association of hSos1, ZAP-70, phospholipase Cγ 1, and other phosphoproteins with Grb2 and the zeta-chain of the TCR. J Biol Chem 1995; 270: 18428–18436.
15. Schieven GL, Wahl AF, Myrdal S, Grosmaire L, Ledbetter JA. Lineage-specific induction of B cell apoptosis and altered signal transduction by the phosphotyrosine phosphatase inhibitor bis(maltolato)oxovanadium(IV). J Biol Chem 1995; 270: 20824–20831.
16. Furukawa T, Itoh M, Krueger NX, Streuli M, Saito H. Specific interaction of the CD45 protein-tyrosine phosphatase with tyrosine-phosphorylated CD3 zeta chain. Proc Natl Acad Sci USA 1994; 91: 10928–10932.
17. Ledbetter JA, Schieven GL, Uckun FM, Imboden JB. CD45 cross-linking regulates phospholipase C activation and tyrosine phosphorylation of specific substrates in CD3/Ti stimulated cells. J Immunol 1991; 146: 1577–1583.
18. Waddick KG, Chae HP, Tuel-Ahlgren L, Jarvis LJ, Dibirdik I, Myers DE, Uckun FM. Engagement of the CD19 receptor on human B-lineage leukemia cells activates LCK tyrosine kinase and facilitates radiation-induced apoptosis. Radiat Res 1993; 136: 313–319.
19. Kharbanda S, Yuan ZM, Rubin E, Weichselbaum R, Kufe D. Activation of Src-like p56/p53*lyn* tyrosine kinase by ionizing radiation. J Biol Chem 1994; 269: 20739–20743.
20. Uckun FM, Schieven GL, Tuel-Ahlgren LM, Dibirdik I, Myers DE, Ledbetter JA, Song CW. Tyrosine phosphorylation is a mandatory proximal step in radiation-induced activation of the protein kinase C signaling pathway in human B-lymphocyte precursors. Proc Natl Acad Sci USA 1993; 90: 252–256.
21. Schieven GL, Kirihara JM, Myers DE, Ledbetter JA, Uckun FM. Reactive oxygen intermediates activate NF-KB in a tyrosine kinase dependent mechanism and in combination with vanadate activate the $p56^{lck}$ and $p59^{fyn}$ tyrosine kinases in human lymphocytes. Blood 1993; 82: 1212–1220.
22. Chae HP, Jarvis LJ, Uckun FM. Role of tyrosine phosphorylation in radiation-induced activation of c-*jun* protooncogene in human lymphohematopoietic precursor cells. Cancer Res 1993; 53: 447–451.
23. Kripke ML. Immunological unresponsiveness induced by ultraviolet radiation. Immunol Rev 1984; 80: 87–102.
24. Schieven GL, Kirihara JM, Gilliland LK, Uckun FM, Ledbetter JA. Ultraviolet radiation rapidly induces tyrosine phosphorylation and calcium signaling in lymphocytes. Mol Biol Cell

1993; 4: 523–530.
25. Schieven GL, Ledbetter JA. Ultraviolet radiation induces differential calcium signals in human peripheral blood lymphocyte subsets. J Immunother 1993; 14: 221–225.
26. Schieven GL, Kirihara JM, Burg DL, Geahlen RL, Ledbetter JA. p72^{syk} tyrosine kinase is activated by oxidizing conditions which induce lymphocyte tyrosine phosphorylation and Ca^{2+} signals. J Biol Chem 1993; 268: 16688–16692.
27. Schieven GL, Mittler RS, Nadler SG, Kirihara JM, Bolen JB, Kanner SB, Ledbetter JA. ZAP-70 tyrosine kinase, CD45 and T cell receptor involvement in UV and H_2O_2 induced T cell signal transduction. J Biol Chem 1994; 269: 20718–20726.
28. Devary Y, Rosette C, Didonato JA, Karin M. NF-κB activation by ultraviolet light not dependent on a nuclear signal. Science 1993; 261: 1442–1445.
29. Devary Y, Gottlieb RA, Smeal T, Karin M. The mammalian ultraviolet response is triggered by activation of src tyrosine kinases. Cell 1992; 71: 1081–1091.
30. Barford D. Protein phosphatases. Curr Opin Struct Biol 1995; 5: 728–734.
31. McNeill JH, Yuen VG, Hoveyda HR, Orvig C. Bis(maltolato)oxovanadium(IV) is a potent insulin mimic. J Med Chem 1992; 35: 1489–1491.
32. Posner BI, Faure R, Burgess JW, Bevan AP, Lachance D, Zhang-Sun G, Fantus IG, Ng JB, Hall DA, Soo Lum B, Shaver A. Peroxovanadium compounds. J Biol Chem 1994; 269: 4596–4604.
33. O'Shea JJ, McVicar DW, Bailey TL, Burns C, Smyth MJ. Activation of human peripheral blood T lymphocytes by pharmacological induction of protein-tyrosine phosphorylation. Proc Natl Acad Sci USA 1992; 89: 10306–10310.
34. Secrist JP, Burns LA, Karnitz L, Koretzky GA, Abrahams RT. Stimulatory effects of the protein tyrosine phosphatase inhibitor, pervanadate, on T-cell activation events. J Biol Chem 1993; 268: 5886–5893.
35. Krejsa CM, Nadler SG, Esselstyn JM, Kavanagh TJ, Ledbetter JA, Schieven GL. Role of oxidative stress in the action of vanadium phosphotyrosine phosphatase inhbitors: redox independent activation of NF-κB. J Biol Chem 1997; in press.
36. Schreck R, Reiber P, Baeuerle PA. Reactive oxygen intermediates as apparently widely used messengers in the activation of the NF-κB transcription factor and HIV-1. EMBO J 1991; 10: 2247.

6

Oxidant-mediated Modulation of Phospholipases in Endothelial Cells: Possible Implication in Vascular Disorders

V. Natarajan, Suryanarayana Vepa, and William M. Scribner
Indiana University School of Medicine Indianapolis, Indiana

INTRODUCTION

The vascular endothelium plays a critical role in maintenance of normal lung function. Perturbation of vascular endothelial cells (ECs) results in barrier dysfunction (1,2), a hallmark of vascular disorders including adult respiratory distress syndrome (ARDS), sepsis, tumor metastasis, and atherosclerosis (3–7). A variety of inflammatory mediators and reactive oxygen species (H_2O_2, superoxide anion, hydroxyl radical, nitric oxide, peroxynitrite, and oxidized LDL) alter EC barrier function, which has been implicated in the pathogenesis of vascular disorders (8–10). Earlier studies suggest that oxidant stress alters EC high-energy compounds, glutathione levels, and intracellular free Ca^{2+} (11,12); however, the mechanisms of oxidant-induced EC barrier dysfunction leading to vascular disorders are not well understood (13). It is hypothesized that oxidant-mediated alterations in edemogenic protein kinases and phosphatases regulate EC barrier function via direct modulation of lipid-derived second messengers. In this study, we have investigated the effect of oxidants in the modulation of phospholipases A_2, C, and D in vascular endothelial cells and possible implication in vascular disorders.

EXPERIMENTAL PROCEDURES

Bovine pulmonary artery endothelial cells (BPAECs) and rabbit femoral artery smooth-muscle cells were cultured as described earlier (14). Phospholipase A_2 (PLA_2) activity was assayed in intact cells by measuring [^{3}H]arachidonic acid and the metabolites released in response to vehicle or oxidants (15), while phospholipase D (PLD) was quantified as [^{32}P]phosphatidylethanol (PEt) or phosphatidylbutanol (PBt) formed in the presence of ethanol or butanol, respectively (14,16), which served as an index of PLD activation (17). Oxidant-mediated changes in protein tyrosine phosphorylation were measured by western blot analysis with antiphosphotyrosine antibodies and enhanced chemiluminescence

(ECL) (18). Diacyglogycenol (DAG) and protein kinase C (PKC) activities were determined in total cell lysates by enzymatic assay and western blot analysis, respectively (19). Immunoprecipitation of focal adhesion kinase and mitogen-activated protein kinase (MAPK) were carried out in total cell lysates with specific monoclonal antibodies (20).

RESULTS AND DISCUSSION

Oxidants Activate Phospholipase A_2 and Release Arachidonic Acid Metabolites

Stimulation of BPAECs by H_2O_2, 4-hydroxynonenal (4-HNE), or pervanadate (V^{4+}-OOH) resulted in accumulation of arachidonic acid metabolites in the medium (Figure 1). The oxygenated derivative of arachidonic acid, derived from arachidonic acid

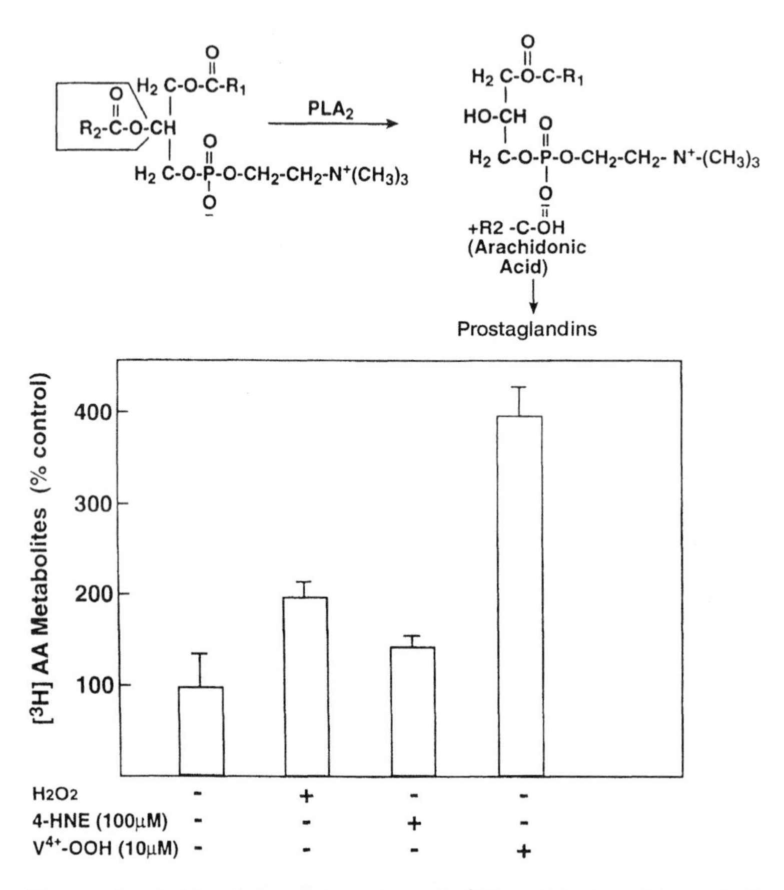

Figure 1. Oxidant-induced formation of [^{3}H]arachidonic acid metabolites. BPAECs (5×10^5/dish) were labeled with [^{3}H]arachidonic acid (0.5 μCi/dish) in MEM for 24 h. The cells were washed and challenged with medium (control) or medium containing H_2O_2 (100 μM) or 4-HNE (100 μM) or V^{4+}-OOH (10 μM) for 30 min. Radioactivity in the medium was counted and expressed as % Control. Greater than 80% of the counts in the medium was associated with 6-keto-$PGF_{1\alpha}$ a stable metabolite of prostacyclin.

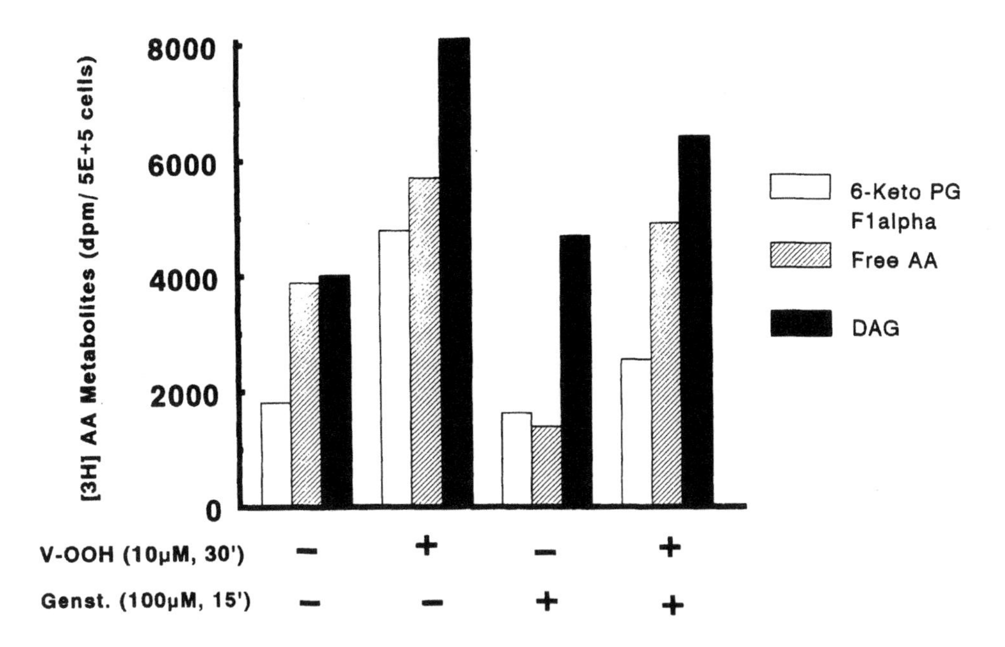

Figure 2. Effect of genistein on vanadyl hydroperoxide-mediated release of [^{3}H]arachidonic acid and DAG. BPAECs (5×10^5 in 35 mm dishes) were labeled with [^{3}H]arachidonic acid (0.5 μCi/dish) for 24 h. Labeled cells were preincubated with genistein (100 μM) for 30 min before challenging with medium (control) or medium containing V^{4+}-OOH for 30 min. At the end of the incubation, the medium was aspirated, centrifuged, and counted for radioactivity. The cells were treated with methanol under acidic conditions and lipids were extracted. 3H-labeled free fatty acids and DAG were separated on TLC using hexane–diethylether–glacid acetic acid (50:50:1) and areas corresponding to fatty acid and DAG were counted. Values are expressed as dpm/5×10^5 cells. Counts in the medium represent 6-keto-$PGF_{1\alpha}$ (>85% of the counts in medium).

released by phospholipase A_2 activation was characterized as 6-keto-$PGF_{1\alpha}$ (a stable derivative of prostacyclin). Prostacyclin (PGI_2) is a potent vasodilator and an inhibitor of platelet activation. A similar increase in PGI_2 synthesis after exposure to H_2O_2 was observed in ECs derived from aorta and human umbilical cord (15,21). The regulation of oxidant-induced PGI_2 release in vascular endothelium may involve: PKC, tyrosine kinases (tyrks), and GTP-binding proteins increase in intracellular Ca^{2+}. The vanadyl hydroperoxide-induced formation of 6-keto-$PGF_{1\alpha}$ and DAG was attenuated by genistein, a tyrk inhibitor (Figure 2). These data suggest that H_2O_2-induced formation of PGI_2 involves the activation of tyrks. At present, it is not clear whether the activation of PLA_2 or cyclooxygenase or both is regulated by tyrosine kinases/protein tyrosine phosphatases in vascular endothelium (22).

Oxidants and Activation of Phospholipase C

Activation of phosphatidyinositol 4,5-bisphosphate (PIP_2)-specific PLC generates inositol 1,4,5-trisphosphate (IP_3) and DAG (23,24). IP_3 stimulates the release of intracellular free Ca^{2+} from the endoplasmic reticulum (25) while DAG activates PKC (23). PKC is a serine/threonine kinase and requires phosphatidylserine and Ca^{2+} for activation (23). At least eleven PKC isotypes have been reported and classified into (i) conventional (α, βI, βII and γ), (ii) novel (δ,ε,η, and θ), and (iii) atypical (ζ,λ,ι, and μ) isoenzymes based on structural and cofactor regulation (26). In BPAECs, PKCα and PKCε have been identified as the two major PKC isotypes present in the cytosol and membrane fractions, respectively (27). Treatment of BPAECs with H_2O_2 (1 mM) caused partial translocation of PKCα from the cytosol to the membrane fraction as determined by immunoblot analysis with PKCα antibody. Under similar experimental conditions, diperoxovanadate showed no effect, while 12-)-tetradecanoylphorbol 13-acetate (TPA), an activator of PKC, completely translocated. PKCα to the membrane (Figure 3). The H_2O_2-induced translocation of PKCα was dose-dependent and was seen even at 100 μM H_2O_2 (19). Activation of PKCα by H_2O_2 was further confirmed by increases in intracellular free Ca^{2+} and generation of $[^3H]IP_3$ (Figure 4). Interestingly, pervanadate-induced accumulation of (^3H) IP_3 was greater than that observed with H_2O_2 (Fig. 4), but pervanadate failed to activate PKCα (Figure 3). The reason for this discrepancy is not clear. In addition to changes in Ca^{2+} and IP_3, H_2O_2 and pervanadate treatment increased DAG levels (data not shown). These data clearly indicate that H_2O_2-induced activation of PKCα is due to PIP_2 hydrolysis by PLC resulting in IP_3 and DAG production (Figure 5). In addition to IP_3-induced Ca^{2+} release, oxidants can alter Ca^{2+} homeostasis through modulation of Ca^{2+} flux (28).

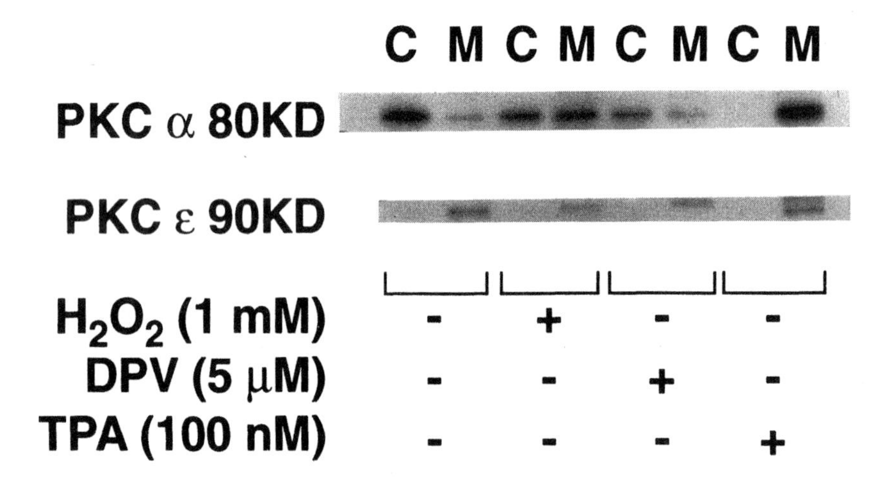

Figure 3. Effect of oxidants and TPA on PKCα translocation to the membrane. BPAECs (5×10^6 cells/T-75 flask) were challenged with medium or medium containing H_2O_2 (1 mM) or diperoxovandate (5 μM) or TPA (100 nM) for 30 min at 37°C. Cells were washed in ice-cold phosphate-buffered saline, scraped, and sonicated in 20 mM Hepes, 1 mM EGTA, and 150 mM NaCl, pH 7.4. Cell lysates were subjected to high-speed differential centrifugation to obtain 105,000 *g* cytosol and membrane fractions. Protein (10 gmg) from the cytosol and membrane fractions was subjected to SDS-PAGE/western blotting with PKCα (1:5000dilution) or PKCε (1:2500 dilution) and detected using enhanced chemiluminescence (ECL). C = Cytosol, M = Membrane.

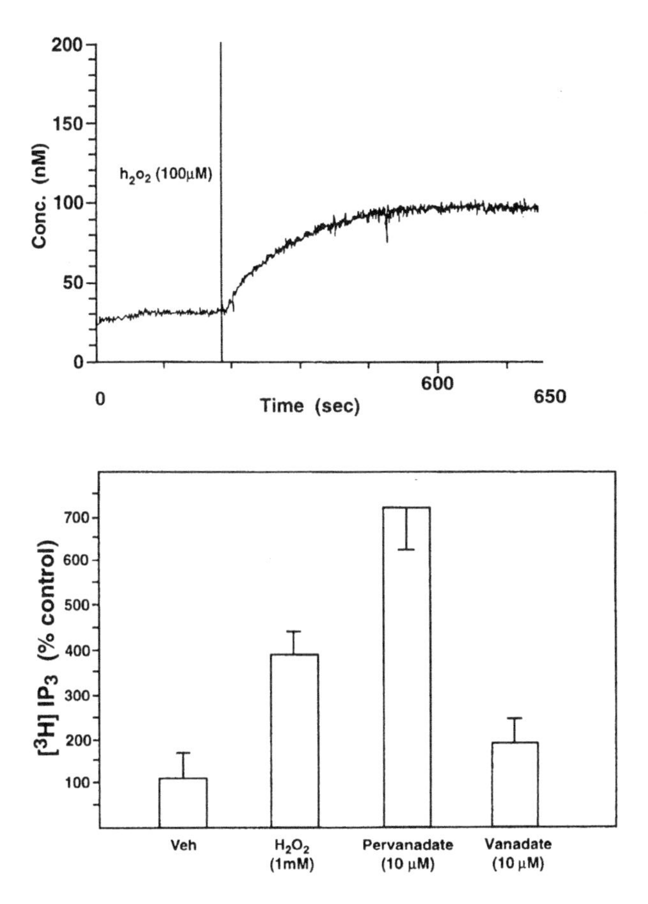

Figure 4. Oxidant-induced alterations in intracellular Ca^{2+} and IP_3. H_2O_2-induced changes in intracellular Ca^{2+} were measured in BPAEC monolayers grown on cover slips using Fura-2/AM and SLM/AMINCO spectroflurometer (ratio 340/380). For [^{3}H] IP_3 level, cells were labeled with [^{3}H]inositol (5 μCi/T-25 flasks) and challenged with vehicle (control) or H_2O_2 (1 mM) or pervanadate (10 μM) for 30 min. The aqueous methanol phases of the lipid extracts were subjected Dowex AG 1X8 formate anion exchange chromatography to separate IP_3, IP_2, and IP from inositol. In vehicle-treated cells, count in [^{3}H]IP_3 was 1795 ± 179 dpm/T25 flask.

Pathways of DAG Production by H_2O_2

The H_2O_2-induced DAG generation was dose- and time-dependent (19) and interestingly exhibited a biphasic response (Figure 6). Phase I of H_2O_2-induced DAG formation occurred within 5 min (140% over control) while phase II was observed after 10 min. Unlike phase I, the phase II accumulation was sustained for a longer period of time (Figure 6). It is tempting to speculate that phase I represents the PLC-catalyzed hydrolysis of PIP_2, while phase II may be due to the activation of PLD, thus generating phosphatidic acid (PA). The PA thus generated can be converted to DAG by PA Pase present in ECs (29). Thus, the DAG generated under phase I and phase II periods of H_2O_2 treatment can activate PKCα and cause translocation. Also, the DAG generated by the PLC pathway can be converted to PA by DAG kinase (30). These data indicate

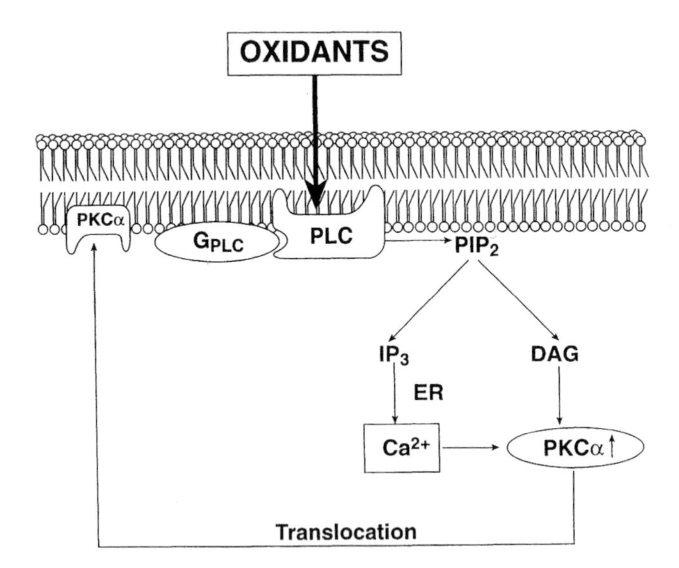

Figure 5. Schematic illustration of oxidant-induced PLC activation and translocation of PKCα.

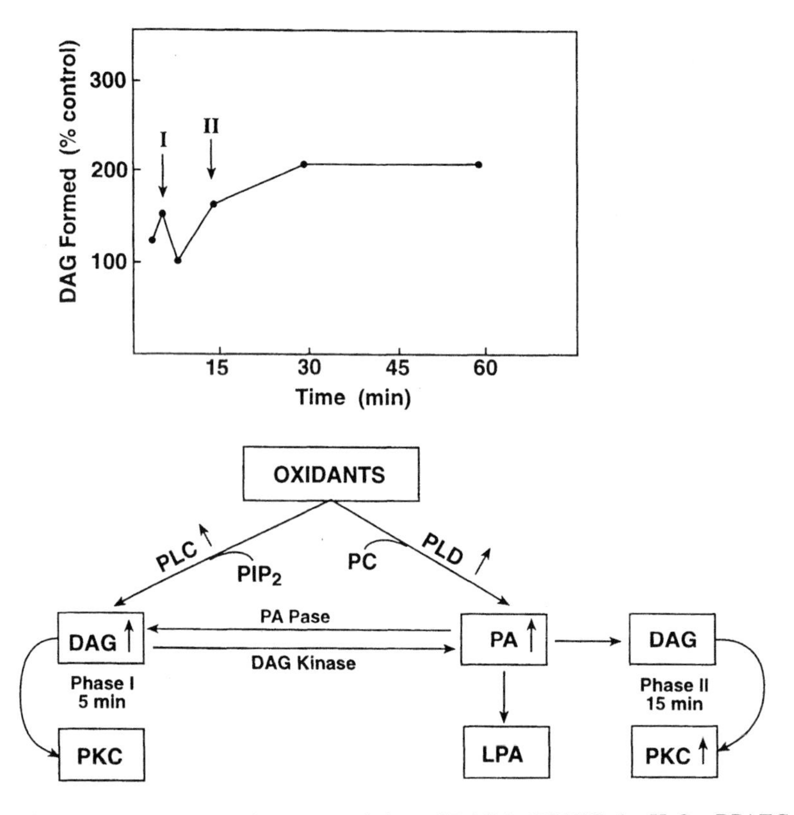

Figure 6. Time-dependent accumulation of DAG in BPAECs by H_2O_2. BPAECs (1×10^6 cells/ T25 flasks) were challenged with vehicle (control) or vehicle containing H_2O_2 (1 mM) for varying time periods. Lipids were extracted under nonacidic conditions and DAG levels in the extract were determined (19) using commercial DAG kinase from *E. coli*. Values are expressed as % of control.

that oxidant-induced DAG generation in ECs may involve the PLC and/or PLD pathways. The role of phosphatidylcholine (PC)-specific PLC in oxidant-induced DAG production is not clear.

Oxidants and Activation of Phospholipase D

The biphasic nature of DAG accumulation in BPAECs in response to H_2O_2 lead us to investigate the activation of PLD by oxidants. Phospholipase D, a predominantly membrane-associated enzyme, catalyzes the hydrolysis of PC or phosphatidylethanolamine to PA and free choline or ethanolamine, respectively (31). In addition to hydrolytic activity, the PLD also exhibits phosphatidyltransferase activity whereby it catalyzes the transphosphatidylation of PA to primary alcohols (C–2–C–6), thereby generating the corresponding phosphatidylalcohol, an unnatural acidic phospholipid (32). As formation of phosphatidylalcohol is directly coupled to PLD activation, this reaction serves as an index for PLD activation in mammalian cells (33). An outline of PLD activation and phosphatidylalcohol formation is given in Figure 7.

In BPAECs, exogenous addition of H_2O_2 or linoleic acid hydroperoxide enhanced [^{32}P]phosphatidylethanol (PEt) formation (Figure 8). The oxidant-induced PLD activation was not associated with cytotoxicity as determined by [^{3}H]deoxyglucose or LDH release (34). The H_2O_2-induced PLD activation was enhanced by ferrous chloride (50 μM) and attenuated by deferoxamine, a chelator of iron, suggesting a role for hydroxyl radicals in PLD activation (Figure 9). In addition to H_2O_2 and linoleic acid hydroperoxide, 4-HNE,

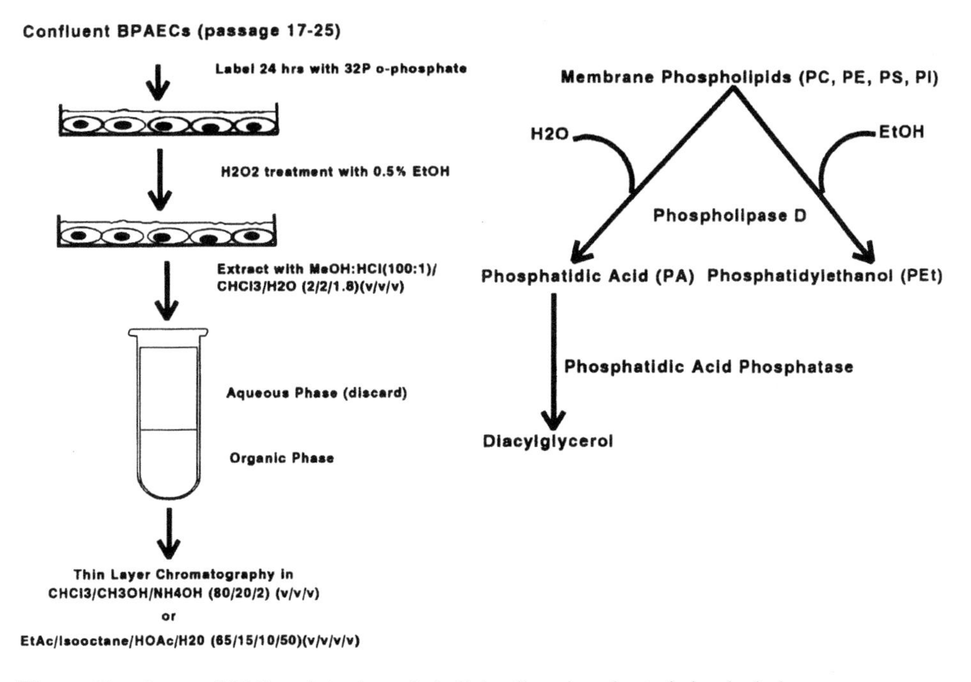

Figure 7. Assay of PLD activity in endothelial cells using short-chain alcohol.

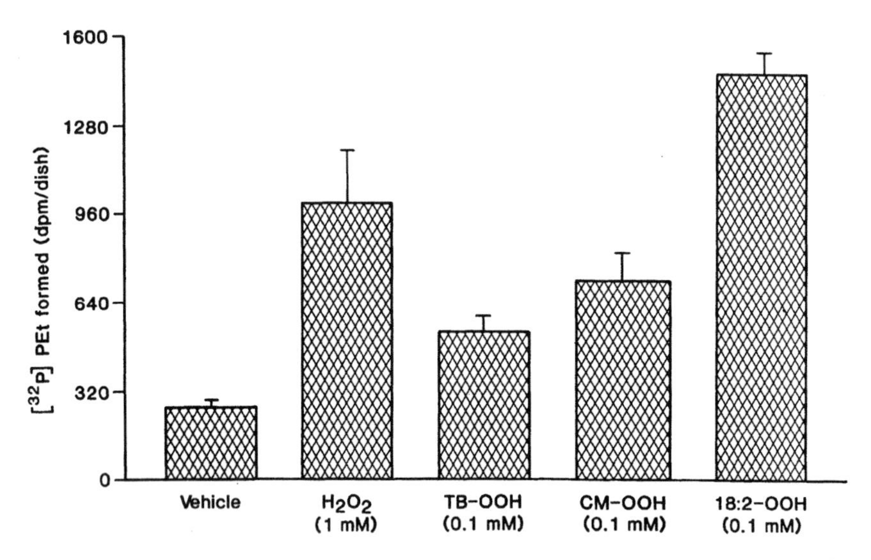

Figure 8. Oxidant-induced activation of PLD in endothelial cells. BPAECs (5×10^5 cells/35 mm dish) were labeled with [^{32}P] orthophosphate (5 μCi/dish) in DMEM phosphate-free medium for 18–24 h. Cells were washed in medium and challenged with medium (control) or medium containing H_2O_2 (1 mM) or *t*-butylhydroperoxide (TB-OOH) (0.1 mM) or cumene hydroperoxide (CM-OOH) (0.1 mM) or linoleic acid hydroperoxide (18:2-OOH) (0.1 mM) for 60 min in the presence of 0.5% ethanol. Lipids were extracted under acidic conditions and [^{32}P]PEt was separated by TLC (see Figure. 7) and quantified by scintillation counting. Values are mean ± SD of triplicate determination.

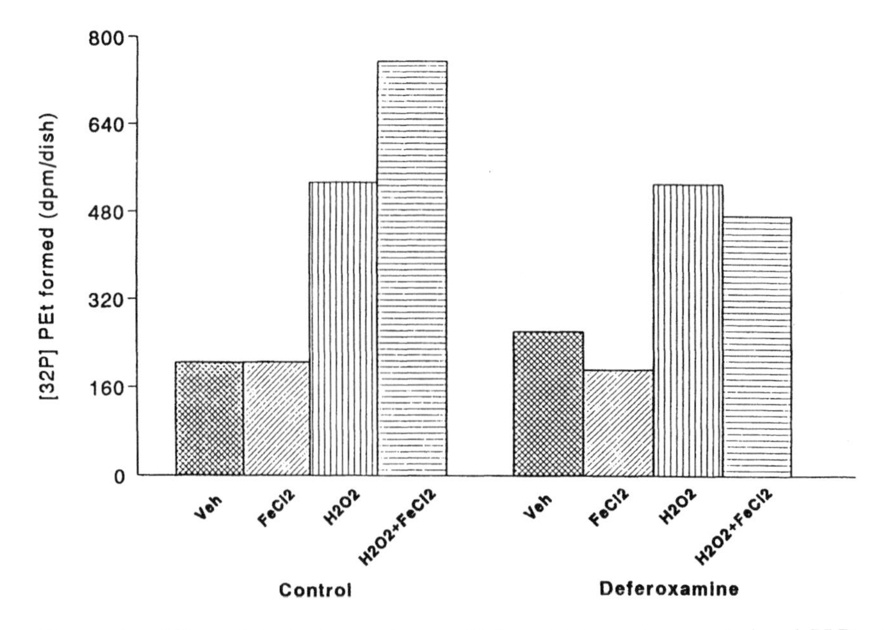

Figure 9. Effect of desferricoxamine on H_2O_2-and H_2O_2 plus iron-induced PLD activation. [^{32}P]Orthosphosphate-labeled BPAECs (as described in Figure 8) were pretreated with decferrioxamine (1 mM) for 30 min before challenging with H_2O_2 (1 mM) or $FeCl_2$ (100 μM) or H_2O_2 + $FeCl_2$ (1 mM + 100 μM) for 60 min in the presence of 0.5% ethanol. Lipids were extracted under acidic condition and [^{32}P]PEt formed was quantified after separation by TLC.

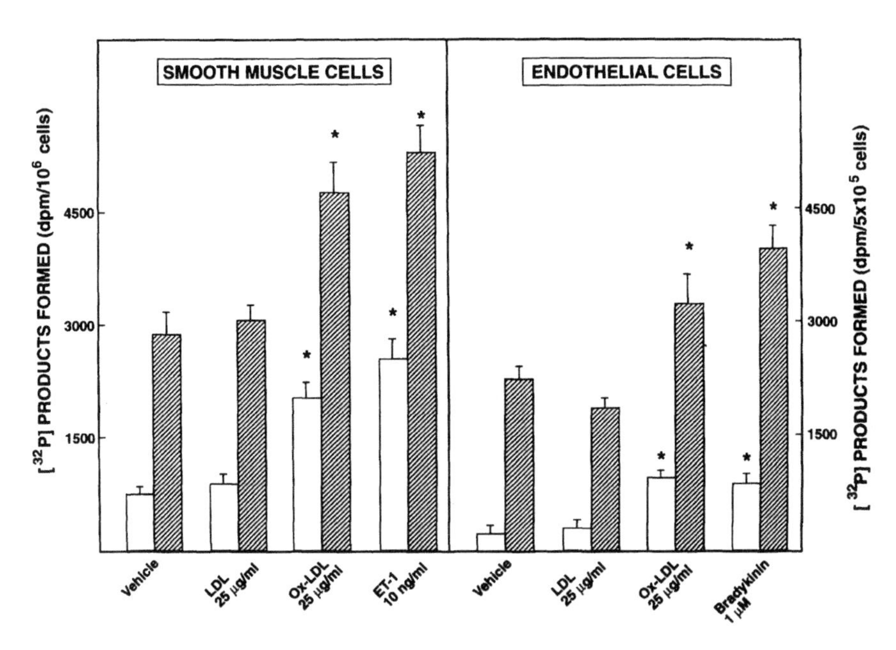

Figure 10. Effect of native and Ox-LDL on PLD activation in smooth-muscle and endothelial cells. Rabbit femoral artery smooth-muscle cells and BPAECs were labeled with [^{32}P]orthophosphate as described in Figure 8. Cells were washed and treated with LDL (25 µg/mL) or Ox-LDL (25 µg/mL) or bradykinin (1 µM) for 30 min in the presence of 0.5% ethanol. Lipids were extracted under acidic condition and [^{32}P]PEt was quantified by TLC. Values are mean ± SD of triplicate determinations (open bars, PEt; hatched bars, PA).

a major metabolite of lipid peroxidation, also activated BPAEC PLD (35). Among the various short-chain aldehydes tested, only *trans*-2-nonenal and *trans*-2-*cis*-nonadienal activated PLD, suggesting a specific requirement of a *trans* double bond at carbon 2 and a hydroxyl group at carbon 4. Oxidative modification of low-density lipoproteins (Ox-LDL), similar to H_2O_2 or fatty acid hydroperoxide, also increased [^{32}P]PEt formation in smooth-muscle cells and endothelial cells (Figure 10). This activation of PLD by Ox-LDL was specific as n-LDL or acetylated LDL had no effect (36).

Mechanism(s) of Oxidant-induced PLD Activation

Previous studies from my laboratory have established that agonist-induced activation of PLD in ECs is downstream to PLC stimulation and is PKC dependent (16,17). Furthermore, the activation of PLD by H_2O_2, 4-HNE or Ox-LDL was not abolished by chelation of either extracellular Ca^{2+} with EGTA or intracellular Ca^{2+} with BAPTA-AM, suggesting that oxidant-induced PLD activation was Ca^{2+}-independent. The oxidant-induced PLD activation was not attenuated by staurosporine or calphostin C or bisindolylmaleimide, indicating a PKC-independent pathway in PLD activation. The noninvolvement of PKC was also confirmed by the inability to attenuate oxidant (H_2O_2, 4-HNE, or linoleic acid hydroxyperoxide) induced PLD activation after the down-regulation of PKC with prolonged exposure to TPA. These data suggest that oxidant-induced PLD activation in ECs is PKC and Ca^{2+} independent (34,37).

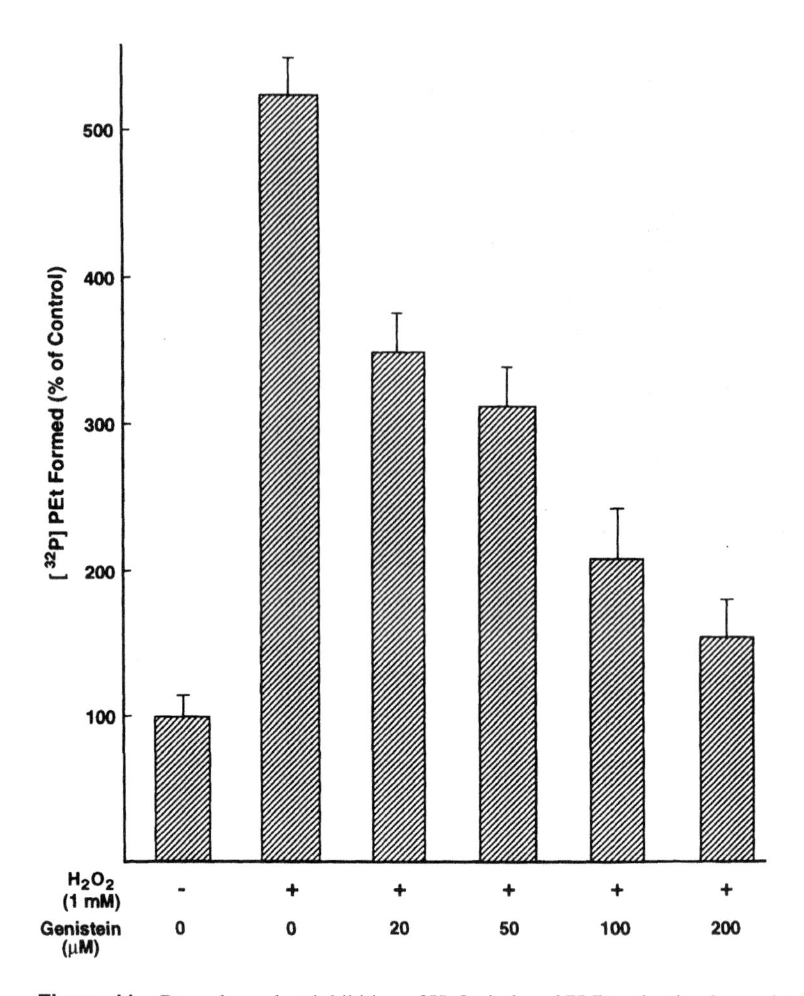

Figure 11. Dose-dependent inhibition of H_2O_2-induced PLD activation by genistein. BPAECs were labeled with [^{32}P]orthophosphate as described in Figure 8. Cells were pretreated with varying concentrations of genistein for 60 min. Cells were washed and challenged with medium (control) or medium containing H_2O_2 (1 mM) for 30 min in the presence of 0.5% ethanol. Lipids were extracted under acidic conditions and [^{32}P]PEt was quantified after separation by TLC. Data are expressed as % of control.

It was hypothesized that oxidant-induced PLD activation may involve tyrosine kinases (tyrks) and protein tyrosine phosphatases. To characterize the role of protein tyrosine phosphorylation in H_2O_2-induced PLD activation, the effect of genistein, a putative tyrk inhibitor, was examined. Genistein, in a dose-dependent manner, inhibited the H_2O_2-induced [^{32}P]PEt formation (Figure 11) with an IC_{50} of about $\approx 100\ \mu M$. Genistein by itself, at concentrations tested, had no effect on basal PLD activity (18). The inhibitory effect of genistein was specific for H_2O_2-induced PLD activation as it did not modulate the TPA- or bradykinin-mediated PLD stimulation (Figure 12).

To investigate whether H_2O_2 treatment modulated protein tyrosine phosphorylation, cell lysates from control and H_2O_2-treated BPAECs were subjected to SDS-PAGE and western blot analysis with antiphosphotyrosine antibodies. As seen in Figure 13, exposure to H_2O_2 (1–60 min) enhanced tyrosine phosphorylation of several proteins with apparent

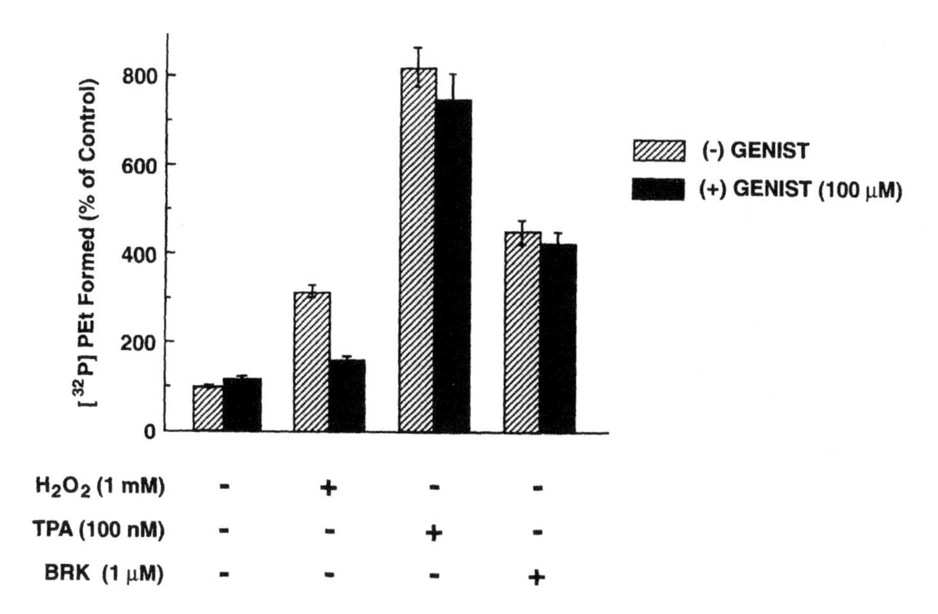

Figure 12. Specificity of genistein towards H_2O_2-induced PLD activation. BPAECs were labelled with [^{32}P]orthophosphate as described in Figure 8. Cells were pretreated with genistein (100 μM) for 60 min before challenging with medium (control) or medium containing H_2O_2 (1 mM) or TPA (100 nM) or bradykinin (1 μM) for 30 min in the presence of 0.5% ethanol. [^{32}P]PEt was quantified in the lipid extracts after separation by TLC.

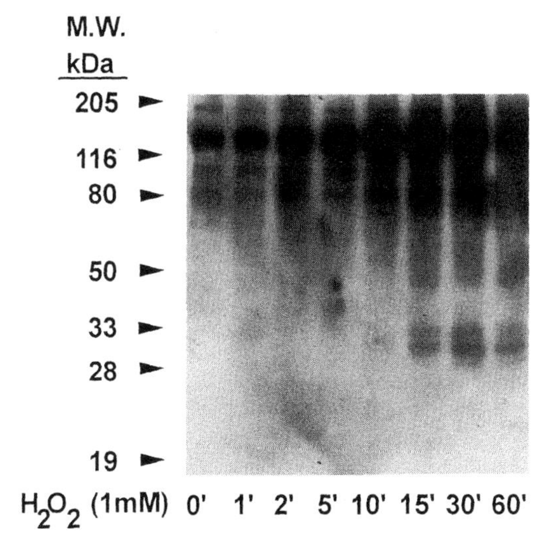

Figure 13. Effect of H_2O_2 on endothelial cells protein tyrosine phosphorylation. BPAECs (1 × 10^6 cells) were challenged with H_2O_2 (1 mM) for varying time periods. Cell lysates were subjected to SDS-PAGE (10% gel) followed by transfer to membranes. Membranes were immunoblotted using antiphosphotyrosine antibody (1:2500 dilution) and bands were detected by ECL.

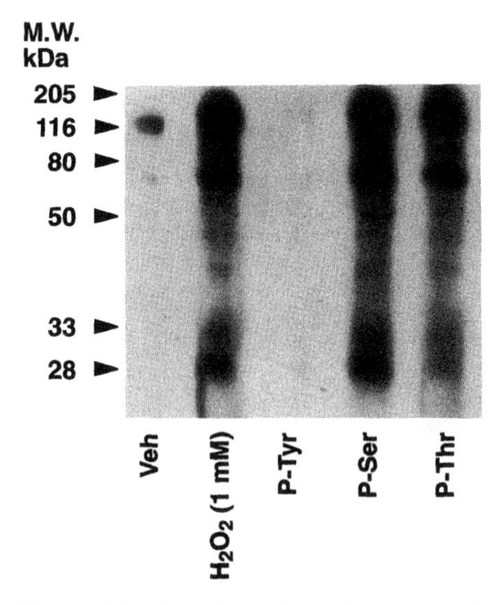

Figure 14. Specificity of antiphosphotyrosine antibody immunodetection. Lysates from vehicle or H_2O_2 (1 mM) treated cells were subjected to SDS-PAGE and membrane transfer. As indicated, during western blotting with antiphosphotyrosine antibody, phosphotyrosine (2 mM) or phosphoserine (2 mM) or phosphothreonine (2 mM) was added to compete with the antibody. ECL was used to detect the tyrosine phosphorylated protein.

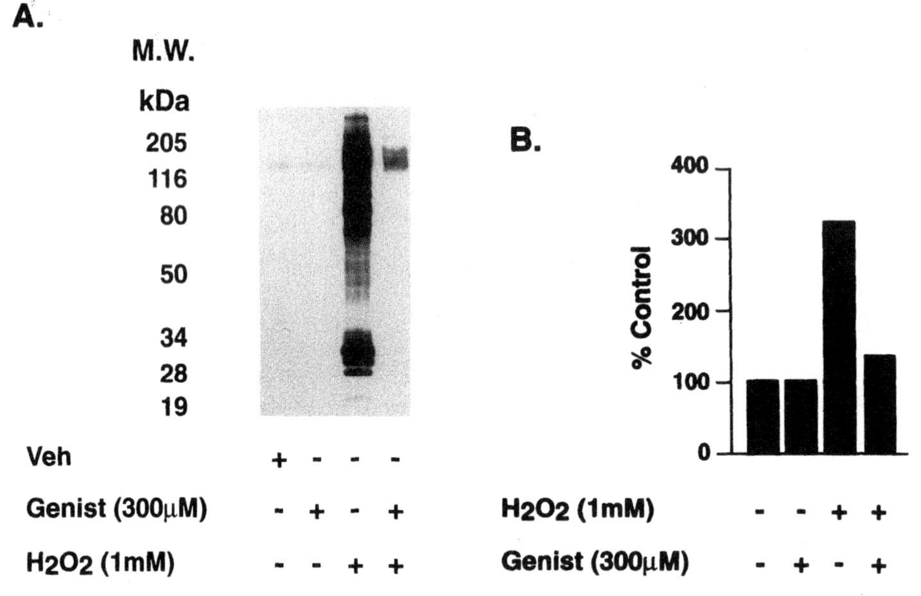

Figure 15. Effect of genistein on H_2O_2-induced protein tyrosine phosphorylation. Panel A, BPAEC (starved in RPMI-1640 serum-free medium for 3 h) were incubated with either medium alone (control) or medium containing genistein (300 μM) for 30 min at 37°C. The cells were then washed and treated with H_2O_2 (1 mM) for 60 min at 37°C. Cell lysates were subjected to 10% SDS-PAGE and immunoblotting as described under "Experimental Procedures". Panel B, after ECL detection, the lanes containing the tyrosine phosphorylated proteins were scanned with a densitometer and the relative intensities are presented.

molecular masses of 20–205 kDa. The enhanced protein tyrosine phosphorylation by H_2O_2 was detectable as early as 2 min and declined after 30 min (Figure 13). The immunodetection of tyrosine phosphorylated proteins was blocked by coincubating the blots with excess phosphotyrosine (2 mM) and antiphosphotyrosine antibody (Figure 14). To further confirm that H_2O_2-induced protein phosphorylation was tyrosine kinase mediated, the effect of the tyrk inhibitor genistein was investigated. Genistein, at a concentration of 300 μM, completely blocked H_2O_2-induced tyrosine phosphorylation (Figure 15), and a correlation between H_2O_2-induced PLD activation and protein tyrosine phosphorylation was observed (18).

Effect of Phosphatase Inhibitors on PLD Activation and Protein Tyrosine Phosphorylation

Protein tyrosine phosphorylation is a balance between protein tyrosine phosphorylation and dephosphorylation mediated by tyrks and protein tyrosine phosphatases (PTPases), respectively (38,39). Inhibition of PTPases should upregulate tyrk-mediated protein tyrosine phosphorylation. To test whether PTPase inhibitors modulate H_2O_2-induced PLD activation and 4-HNE-mediated protein tyrosine phosphorylation, BPAECs were treated with sodium orthovanadate. As shown in Figures 16 and 17, vanadate potentiated

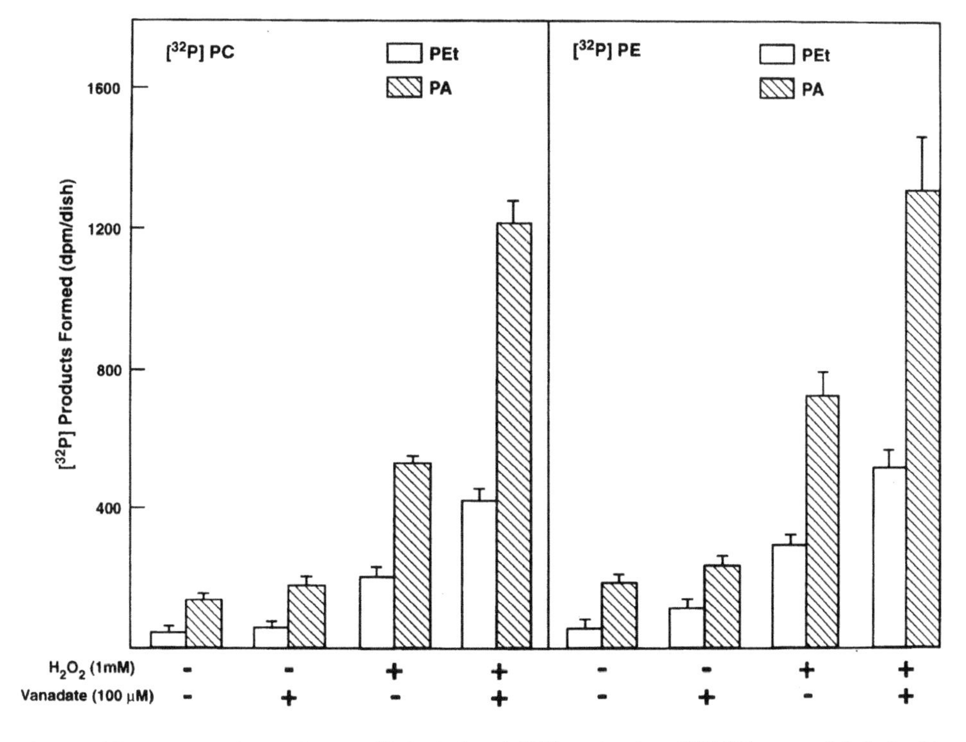

Figure 16. Effect of vanadate on H_2O_2-induced PLD activation. BPAECs were labeled with [^{32}P]Lyso-PC (2 μCi/dish) or [^{32}P]lyso-PE (2 μCi/dish) for 48 h. Cells were pretreated with vanadate (100 μM) for 15 min before challenging with vehicle (control) or H_2O_2 (1 mM) for 30 min, in the presence of 0.5% ethanol. Lipids were extracted under acidic condition and [^{32}P]PEt and PA were separated by TLC.

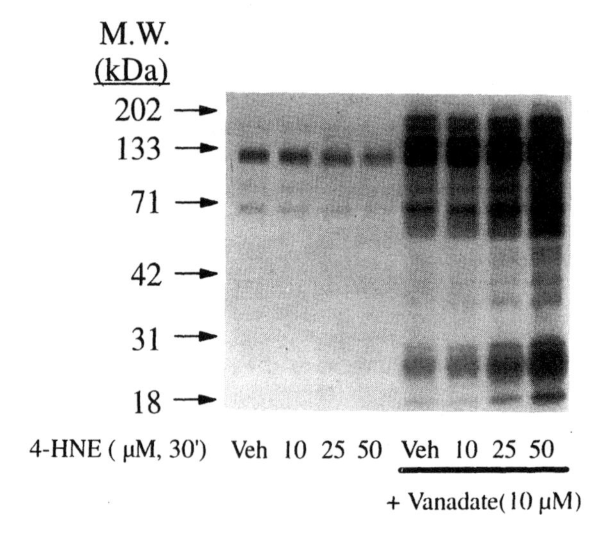

Figure 17. Effect of vanadate on 4-HNE-induced protein tyrosine phosphorylation. BPAECs (5 $\times 10^5$ cells/dish) were pretreated with medium (control) or medium containing vanadate (10 μM) for 30 min. Cells were challenged with varying concentrations of 4-HNE. Cell lysates were subjected to SDS-PAGE/membrane transfer and tyrosine phosphorylated proteins were detected with antiphosphotyrosine antibody and ECL.

H_2O_2-induced PLD activation and protein tyrosine phosphorylation, respectively. Similar modulation in PLD activation and protein tyrosine phosphorylation was observed upon treatment with the thiol oxidizing agent phenylarsineoxide or diamide (data not shown). However, calyculin, a specific inhibitor of PTPase 1 and PTPase 2A, had no effect on H_2O_2-induced PLD activation. These data point out that tyrk/PTPase regulate oxidant-mediated PLD activation in ECs (18,40).

Identification of H_2O_2-induced Tyrosine Phosphorylated Proteins in ECs

We have identified the 22, 42–44, 68, and 125 kDa tyrosine phosphorylated proteins as caveolin (41), ERK-1/ERK-2 (MAP kinase) (20), paxillin, and focal adhesion kinase (42), respectively. A plausible link between these H_2O_2-induced tyrosine phosphorylated proteins and PLD activation in ECs is currently under investigation.

PHYSIOLOGICAL SIGNIFICANCE OF OXIDANT-INDUCED PLD ACTIVATION

Oxidants have been implicated in the pathophysiology of a number of vascular disorders including ARDS, vasculitis, pulmonary hypertension, and atherosclerosis. Alterations in the activities of protein kinases and phosphatases may affect endothelial cell permeability to circulating cells and molecules, causing pulmonary edema and alveolar flooding. Oxidants also modulate the generation of second-messengers like Ca^{2+}, diacylglycerol, phosphatidic acid, and metabolites of arachidonic acid which in turn affect cellular responses and functions. For example, PA generated by the PLD pathway can activate PKC (43) and tyrosine kinases (44) and serve as a source of arachidonic acid for the generation of prostanoids (45). Recent studies suggest that PA and lyso-PA have mitogenic properties by increasing DNA synthesis and cell proliferation (46). Proliferation of smooth-muscle cells in the endothelium may be important in the formation of atherosclerotic plaques (3,4). Our studies show that Ox-LDL is an activator of PLD in smooth-muscle cells (36). As Ox-LDL is a risk factor in atherosclerosis, it was hypothesized that Ox-LDL-mediated PLD activation represents an initial signal for smooth-muscle cell proliferation. Exogenous addition of Ox-LDL, or PA or LPA to quiescent smooth-muscle cells increased [^{3}H]thymidine incorporation into total cellular DNA, suggesting increased cell proliferation. These studies point out a possible link between Ox-LDL-induced PLD activation and smooth-muscle cell proliferation (36).

In summary, the data reported here suggest that oxidants modulate endothelial and smooth-muscle cell signaling pathways and bring about the generation of second-messengers (Figure 18). Alternations in oxidant-induced signal transduction may explain the pathophysiology of vascular disorders including atherosclerosis.

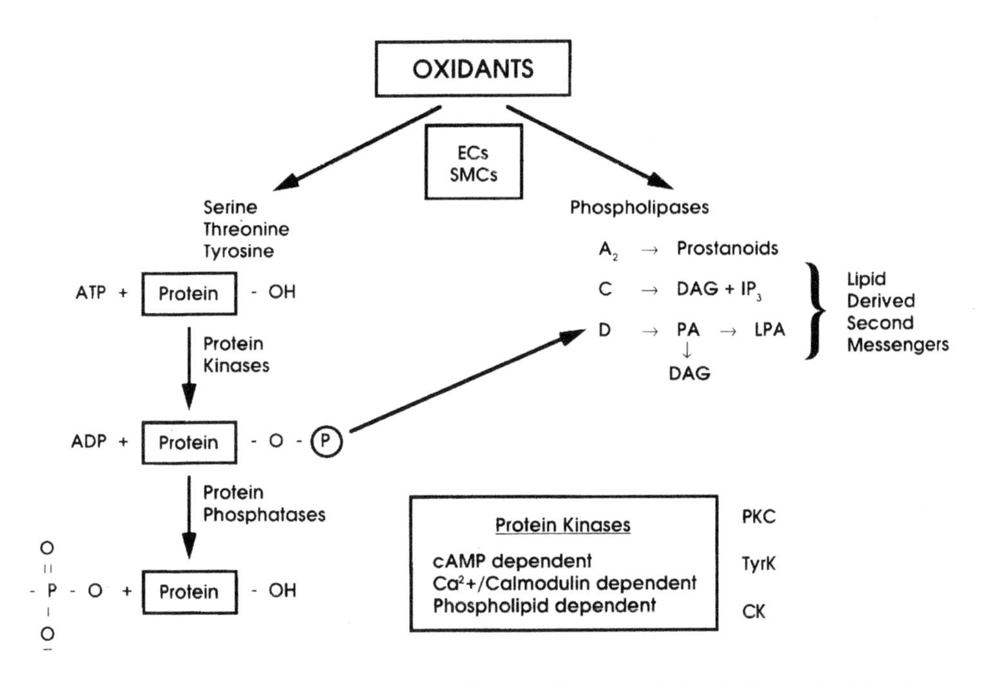

Figure 18. Role of oxidants in modulation of protein kinases and phospholipases. In this scheme it is shown that treatment of smooth-muscle or endothelial cells with oxidants results in modulation of protein kinases and phosphatases. Specifically, PKC and tyrosine kinase are activated, which in turn modulates phospholipases resulting in generation of lipid-derived second-messengers like DAG, PA, and LPA.

ACKNOWLEDGMENTS

This work was supported by grants from National Institutes of Health (HLBI47671 and K04HL03095) and the American Lung Association Career Investigator Award (V.N.) and an American Heart-Indiana Affiliate Post-Doctoral Fellowship (S.V.).

REFERENCES

1. Brigham KL. Role of free radicals in lung injury. [Review]. Chest 1986; 89(6): 859–863.
2. Garcia JG, Davis HW, Patterson CE. Regulation of endothelial cell gap formation and barrier dysfunction: role of myosin light chain phosphorylation. J Cell Physiol 1995; 163(3): 510–522.
3. Ross R, Glomset JA. The pathogenesis of atherosclerosis (second of two parts). [Review]. N Engl J Med 1976; 295(8): 420–425.
4. Ross R, Glomset JA. The pathogenesis of atherosclerosis (first of two parts). [Review]. N Engl J Med 1976; 295(7): 369–377.
5. Cross CE, Halliwell B, Borish ET, et al. Oxygen radicals and human disease [clinical conference]. [Review]. Ann Intern Med 1987; 107(4): 526–545.
6. Bishop CT, Mirza Z, Crapo JD, Freeman BA. Free radical damage to cultured porcine aortic endothelial cells and lung fibroblasts: modulation by culture conditions. In Vitro Cell Dev Biol 1985; 21(4): 229–236.
7. Brigham KL. Oxidant stress and adult respiratory distress syndrome. [Review]. Eur Respir J Suppl 1990; 11: 482s–484s.
8. Killackey JJ, Johnston MG, Movat HZ. Increased permeability of microcarrier-cultured endothelial monolayers in response to histamine and thrombin. A model for the in vitro study of increased vasopermeability. Am J Pathol 1986; 122(1): 50–61.
9. Lum H, Malik AB. Regulation of vascular endothelial barrier function. [Review]. Am J Physiol 1994; 267(3 Pt 1): L223–L241.
10. Garcia JGN, Pavalko FM, Patterson CE. Vascular endothelial cell activation and permeability responses to thrombin. Blood 1995; 6: 609–626.
11. Suttorp N, Toepfer W, Roka L. Antioxidant defense mechanisms of endothelial cells: glutathione redox cycle versus catalase. Am J Physiol 1986; 251(5 Pt 1): C671–C680.
12. Elliott SJ, Meszaros JG, Schilling WP. Effect of oxidant stress on calcium signaling in vascular endothelial cells. [Review]. Free Radical Biol Med 1992; 13(6): 635–650.
13. Ward PA. Mechanisms of endothelial cell injury. [Review]. J Lab Clin Med 1991; 118(5): 421–426.
14. Garcia JG, Fenton JW 2d, Natarajan V. Thrombin stimulation of human endothelial cell phospholipase D activity. Regulation by phospholipase C, protein kinase C, and cyclic adenosine 3′5′-monophosphate. Blood 1992; 79(8): 2056–2067.
15. Chen SH, Wu HL, Lin MT, et al. Cytoprotective effect of reduced glutathione in hydrogen peroxide-induced endothelial cell injury. Prostaglandins Leukotrienes & Essential Fatty Acids 1992; 45(4): 299–305.
16. Natarajan V, Garcia JG. Agonist-induced activation of phospholipase D in bovine pulmonary artery endothelial cells: regulation by protein kinase C and calcium. J Lab Clin Med 1993; 121(2): 337–347.
17. Natarajan V. Oxidants and signal transduction in vascular endothelium [see comments]. [Review]. J Lab Clin Med 1995; 125(1): 26–37.
18. Natarajan V, Vepa S, Verma RS, Scribner WM. Role of tyrosine phosphorylation in H_2O_2-induced activation of endothelial cell phospholipase D. Am J Physiol 1996; 271(15): L400–408.

19. Taher MM, Garcia JG, Natarajan V. Hydroperoxide-induced diacylglycerol formation and protein kinase C activation in vascular endothelial cells. Arch Biochem Biophys 1993; 303(2): 260–266.
20. Scribner WM, Vepa S, Natarajan V. Diperoxovanadate-induced activation of MAP kinase (abstract). 9th International Conference on Second Messengers & Phosphoproteins 1993: 182.
21. Lewis MS, Whatley RE, Cain P, McIntyre TM, Prescott SM, Zimmerman GA. Hydrogen peroxide stimulates the synthesis of platelet-activating factor by endothelium and induces endothelial cell-dependent neutrophil adhesion. J Clin Invest 1988; 82(6): 2045–2055.
22. Glenney JR Jr. Tyrosine-phosphorylated proteins: mediators of signal transduction from the tyrosine kinases. [Review]. Biochim Biophys Acta 1992; 1134(2): 113–127.
23. Nishizuka Y. Intracellular signaling by hydrolysis of phospholipids and activation of protein kinase C. [Review]. Science 1992; 258(5082): 607–614.
24. Garcia JG, Natarajan V. Signal transduction in pulmonary endothelium. Implications for lung vascular dysfunction. [Review]. Chest 1992; 102(2): 592–607.
25. Berridge MJ, Irvine RF. Inositol trisphosphate, a novel second messenger in cellular signal transduction. [Review]. Nature 1984; 312(5992): 315–321.
26. Newton AC. Protein kinase C: structure, function, and regulation. [Review]. J Biol Chem 1995; 270(48): 28495–28498.
27. Pottratz ST, Hall TD, Scribner WM, Jayaram HN, Natarajan V. Selectin-mediated attachment of small cell lung carcinoma to endothelial cells. Am J Physiol 1996; 271(15): L918–L923.
28. Schilling WP, Elliott SJ. Ca2+ signaling mechanisms of vascular endothelial cells and their role in oxidant-induced endothelial cell dysfunction. [Review]. Am J Physiol 1992; 262(6 Pt 2): H1617–H1630.
29. Natarajan V, Jayaram HN, Scribner WM, Garcia JG. Activation of endothelial cell phospholipase D by sphingosine and sphingosine-1-phosphate. Am J Respir Cell Mol Biol 1994; 11(2): 221–229.
30. Shukla SD, Halenda SP. Phospholipase D in cell signalling and its relationship to phospholipase C. [Review]. Life Sci 1991; 48(9): 851–866.
31. Liscovitch M, Chalifa V. Signal-Activated Phospholipase D. In Liscovitch M, ed. Signal-Activated Phospholipases. Austin TX: Landes Co; 1994: 31–63.
32. Gustavsson L, Alling C. Formation of phosphatidylethanol in rat brain by phospholipase D. Biochem Biophys Res Commun 1987; 142(3): 958–963.
33. Kobayashi M, Kanfer JN. Phosphatidylethanol formation via transphosphatidylation by rat brain synaptosomal phospholipase D. J Neurochem 1987; 48(5): 1597–1603.
34. Natarajan V, Taher MM, Roehm B, et al. Activation of endothelial cell phospholipase D by hydrogen peroxide and fatty acid hydroperoxide. J Biol Chem 1993; 268(2): 930–937.
35. Natarajan V, Scribner WM, Taher MM. 4-Hydroxynonenal, a metabolite of lipid peroxidation, activates phospholipase D in vascular endothelial cells. Free Radic Biol Med 1993; 15(4): 365–375. [Published erratum appears in Free Radical Biol Med 1994 Feb; 16(2): 295].
36. Natarajan V, Scribner WM, Hart CM, Parthasarathy S. Oxidized low density lipoprotein-mediated activation of phospholipase D in smooth muscle cells: a possible role in cell proliferation and atherogenesis. J Lipid Res 1995; 36: 2005–2016.
37. Natarajan V, Scribner WM, Suryanarayana V. Regulation of phospholipase D by tyrosine kinases. Chem Phys Lipids 1996; 80: 133–142.
38. Hunter T, Cooper JA. Protein-tyrosine kinases. [Review]. Annu Rev Biochem 1985; 54: 897–930.
39. Fischer EH, Charbonneau H, Tonks NK. Protein tyrosine phosphatases: a diverse family of intracellular and transmembrane enzymes. [Reviews]. Science 1991; 253(5018): 401–406.
40. Natarajan V, Scribner WM, Vepa S. Phosphatase inhibitors potentiate 4-hydroxynonenal-induced phospholipase D activation in vascular endothelial cells. Am J Respir Cell Mol Biol 1997; in press.

41. Vepa S, Scribner-WM, Natarajan V. Activation of protein phosphorylation by oxidants in vascular endothelial cells: Identification of tyrosine phosphorylation of caveolin. Free Radic Biol Med 1996; 22: 25–35.
42. Vepa S, Scribner WM, Natarajan V. Hydrogen peroxide induces tyrosine phosphorylation of FAK in endothelial-cells (abstract). 9th International Conference on Second Messengers & Phosphoproteins 1995: 181.
43. Stasek JE Jr, Natarajan V, Garcia JG. Phosphatidic acid directly activates endothelial cell protein kinase C. Biochem Biophys Res Commun 1993; 191(1): 134–141.
44. Moolenaar WH, van der Bend RL, van Corven EJ, Jalink K, Eichholtz T, van Blitterswijk WJ. Lysophosphatidic acid: a novel phospholipid with hormone- and growth factor-like activities. [Review]. Cold Spring Harbor Symp Quant Biol 1992; 57: 163–167.
45. Fernandez B, Balboa MA, Solis-Herruzo JA, Balsinde J. Phosphatidate-induced arachidonic acid mobilization in mouse peritoneal macrophages. J Biol Chem 1994; 269(43): 26711–26716.
46. Moolenaar WH. Lysophosphatidic acid, a multifunctional phospholipid messenger. [Review]. J Biol Chem 1995; 270(22): 12949–12952.

7

NFκB Transcription Factor Activation: Importance of the Redox Regulation

Bernard Piret, Giuseppina Bonizzi, Marie-Paule Merville, Vincent Bours, and Jacques Piette
University of Liège
Liège, Belgium

INTRODUCTION

One of the most intensive investigative efforts in cellular and molecular biology is devoted to the elucidation of molecular switches modulating gene expression. The transcriptional regulation is used as a predominant strategy to control the production of new proteins in response to extracellular stimuli. At the level of gene transcription, the initiation of mRNA synthesis is used most frequently to govern gene expression and is controlled by activator proteins binding in a sequence-specific manner to responsive elements situated in gene promoters. The activity of these transcription factors can be controlled by signaling pathways through a cascade of events involving kinases. Usually transcription factors form families, whereas individual members perform specific, distinct, or similar tasks. One such family includes the Rel/NFκB protein, which has the unique property of being (i) sequestered in the cytoplasm in association with inhibitory proteins and (ii) translocated into the nucleus when cells are subjected to a large number of stimuli.

NFκB TRANSCRIPTION FACTOR

NFκB and the other members of the Rel family of transcriptional activator proteins are a focal point for understanding how extracellular signals induce the expression of specific sets of genes in higher eukaryotes [for reviews, see Refs. 1 to 4]. Unlike most transcriptional factors, this family of proteins resides in the cytoplasm and must therefore translocate into the nucleus to function. The nuclear translocation of Rel proteins is induced by an extraordinarily large number of agents ranging from bacterial and viral pathogens to immune and inflammatory cytokines to a variety of agents that damage cells. Remarkably, an even larger number of genes appear to be targets for the activation by Rel proteins.

The Rel and IκB Families

The Rel protein family has been divided into two groups based on differences in their structures, functions, and modes of synthesis. The first group consists of p50 (NF-κB1) and p52 (NF-κB2), which are synthesized as precursor proteins of 105 and 100 kDa, respectively (Figure 1A). The mature proteins, which are generated by proteolytic processing, have a so-called Rel homology domain that includes DNA-binding, dimerization motives, and a nuclear localization signal (Figure 1A). The mature proteins form functional Rel dimers with other members of the family, while dimers containing the unprocessed proteins remain sequestered in the cytoplasm. The second group of Rel proteins, which includes p65 (RelA), Rel (c-Rel), RelB, and the drosophila Rel proteins dorsal and Dif, are not synthesized as precursors (Figure 1A). In addition to the Rel homology domain, they possess one or more transcriptional activation domains. Members of both groups of Rel proteins can form homo- or heterodimers; e.g., NFκB is the classical p50-p65 heterodimer which binds to the 5′-GGGANNYYCC-3′ consensus sequence.

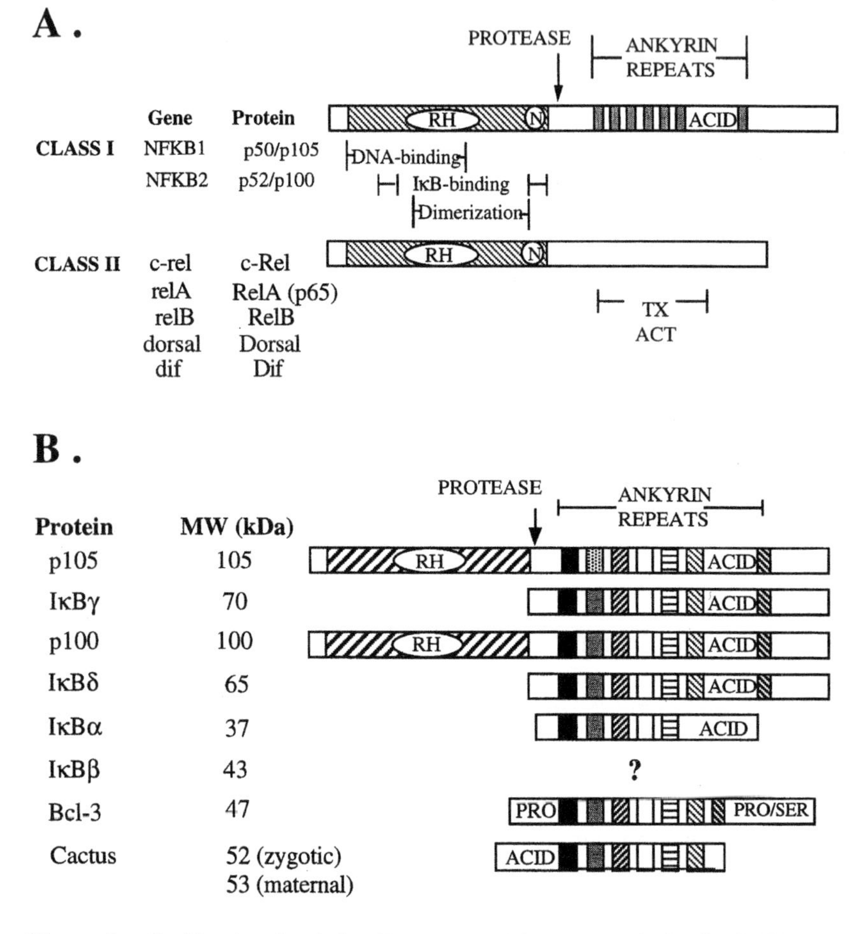

Figure 1. (A) The class I and class II members of the Rel protein family (B) The members of the IκB family.

Two types of Rel protein complexes are found in the cytoplasm prior to induction. The first type consists of Rel homo- or heterodimers (e.g., p50 and p65) bound to a member of the IκB family of inhibitor proteins (IκBα, IκBβ, IκBγ, Bcl-3, and the drosophila protein cactus; Figure 1B). Members of this family share a characteristic ankyrin repeat motif that is required for their interactions with Rel proteins and a C-terminal PEST sequence thought to be involved in protein degradation (5) (Figure 2). The second type of complex consists of a heterodimer formed between a mature Rel protein (e.g., p65) and an unprocessed Rel protein precursor (e.g., p105).

Induction signals lead to the phosphorylation of both inhibitors, IκB and p105 (6) and these phosphorylations are the signals for IκB degradation and p105 processing, both of which generate active Rel dimeric complexes that translocate to the nucleus and activate genes containing Rel protein-binding sites (κB sites). Although phosphorylation of p105 or of p65 by an associated serine kinase has been implicated in the activation of NFκB (7), in most cases the ultimate target of these signaling molecules is IκB (8). From many observations, a model was proposed in which phosphorylation of IκB results in its degradation, thereby freeing NFκB to translocate to the nucleus and bind κB sites (9). Although recent experiments demonstrate that this model needs further improvement to explain in vivo activation of NFκB, it is certain that IκB-α phosphorylation is an integral and obligatory step in NFκB translocation into the nucleus. The induced IκBα phosphorylation sites have been identified by mutagenesis (10), resulting in the identification of amino-terminal deletion mutants that did not undergo phosphorylation when stably expressed in reporter cells. This observation led to the discovery of the potential key phospho-acceptor residues (Ser-32 and Ser-36, Figure 2) when mutated individually or in combination failed to yield phosphorylated IκBα following stimulation (11). Since the S32/S36A IκBα mutant is resistant to phosphorylation induced by various stimuli (interleukins, tumor necrosis factor, phorbol esters, ...) different signaling pathways seem to result in the phosphorylation of these amino acids. Recently, it has been shown that the C-terminal 40 residues of IκBα (amino-acids 277 to 317), which include a PEST-like domain, are entirely dispensable for tumor necrosis factor-α-induced degradation but a glutamine- and leucine-rich (QL) region of IκBα located between residues 263 and 277 and overlapping with the sixth ankyrin repeat seems to be required

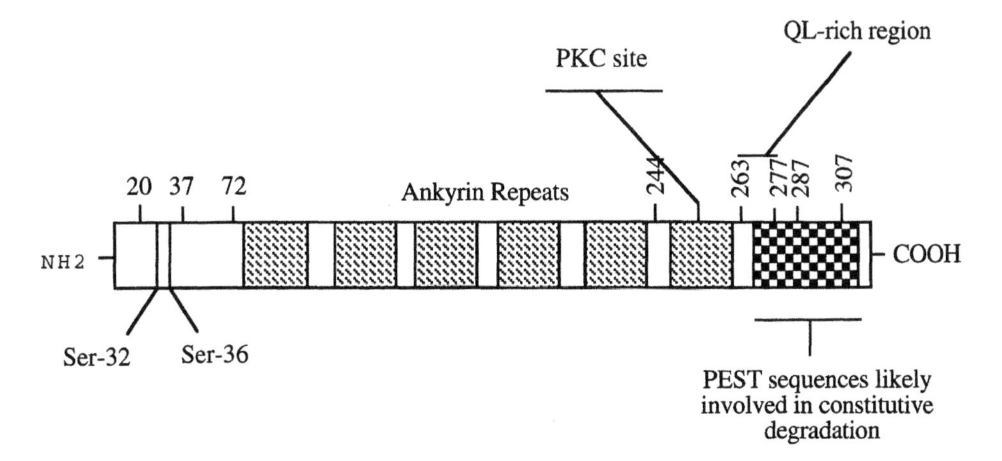

Figure 2. Schematic summary of various functional domains identified within IκBα.

for inducible degradation (12) (Figure 2). The regulation of IκBα degradation by this QL-rich region appears to occur independently of phosphorylation at Ser-32 and Ser-36, indicating that IκBα is organized within distinct modular domains displaying different functional and regulatory properties (Figure 2).

IκBα undergoes complete degradation following stimulation but prior to NFκB activation (10,13,14). This induced degradation is rapid and in some cell types it is complete within 10 minutes. It is also extensive and produces no obvious intermediates. Thus, the induced IκBα degradation is a remarkably efficient process which is an obligatory step in the activation of NFκB. The phosphorylation event on Ser-32 and Ser-36 is required for an additional modification of IκBα, namely, multi-ubiquitination (15). The ubiquitinated IκBα remains associated with NFκB and is specifically degradated by the 26S proteasome. Ubiquitin is conjugated at Lys-21 and Lys-22 in the amino-terminal part of IκBα (16,17). In summary, the sequence of events leading to NFκB activation may require phosphorylation of IκBα at several important residues such as Ser-32 and Ser-36, followed by phosphorylation-dependent multi-ubiquitination of IκBα at Lys-21 and Lys-22 and degradation of IκBα by ubiquitin-dependent proteasome and finally the release of free NFκB transcription factor.

NFκB Responds to Oxidative Stress

The common step in all these induction pathways is the involvement of reactive oxygen species (ROS) (18,19). This conclusion is based largely on the inhibition of NFκB activation by a series of antioxidants. Antioxidants have been reported to block NFκB activation in many instances, although the extent of this block appears to vary depending on cell and signal. Inhibitory antioxidants with diverse chemical properties include *N*-acetyl-L-cysteine, dithiocarbamates, vitamin E derivatives, glutathione peroxidase activation, and various metal chelators. Support for the involvement of ROS as common messenger also derives from evidence showing elevated cellular level of ROS in response to TNF-α, IL-1, phorbol myristate acetate, lipopolysaccharide and UV light, which all are known to be very potent NFκB inducers. Further evidence for an essential role of ROS came from experiments using exogenously added pro-oxidants. In Jurkat and other T cell lines, addition of 100–300 μM H_2O_2 or peroxide-containing molecules resulted in a rapid activation of NFκB. Incubation with various superoxide-, hydroxyl radical-, or NO-generating compounds, however, failed to induce activation, suggesting that, like bacterial transcription factor OxyR, NFκB activation is selectively mediated by peroxides. Compelling genetic evidence for an involvement of H_2O_2 came also from a recent study with catalase- and Cu/Zn superoxide dismutase-overexpressing cells. In a catalase-overexpressing cell line, NFκB activation was substantially suppressed compared with the parental cells, whereas it was superinduced in a Cu/Zn–SOD-overexpressing cell line (20). Likewise, overexpression of thioredoxin or addition of recombinant thioredoxin to the cell culture medium prevented NFκB activation by phorbol ester (21).

Among the ROS determining the redox status of the cell, the major species are superoxide anion, H_2O_2, and singlet oxygen. When singlet oxygen ($^1O_2$) is generated extracellularly by the thermal decomposition of endoperoxide, it cannot act as a NFκB inducer nor it does not lead to transcriptional activation of a reporter gene driven by the HIV-1 Long Terminal Repeat (LTR) (22). To determine whether an intracellular generation of $^1O_2$ would be a signal triggering NFκB activation in T cells, we have subjected these cells to photosensitization mediated by either methylene blue (MB), rose bengal (RB), or proflavin (PF). These molecules exhibit rather important singlet oxygen quantum yield and can be taken up by cells either before or during the photosensitization reaction (23,24). To determine whether MB photosensitization-mediated oxidative damage could lead to a similar induction of NFκB, EMSA experiments have been performed with nuclear extracts from photosensitized cells. In the hours following the photooxidative stress mediated by MB, we detected in the nuclear extracts of phototreated T cells a factor capable of binding to the κB sites of the HIV-1 enhancer (Figure 3) and this activity did not appear in nonilluminated cells. The specificity for these κB sites was shown by the fact that the labeled probe was efficiently competed out by an excess of unlabeled probe but not by the same excess of unlabeled probe mutated on the κB sites. Another complex, showing a lower electrophoretic mobility, seemed to be induced with the same kinetic characteristics, but disappeared when competition experiments were done with both wild-type and mutated κB probes. Thus, this complex did not seem to involve a nuclear factor which binds specifically to the κB sites. The intensity of the specific complex was not affected when cycloheximide (CHX), an inhibitor of protein synthesis, was added to the cells immediately after illumination, demonstrating that the induction of this complex occurred through a posttranscriptional mechanism. However, nonilluminated cells cultivated in the presence of CHX showed a slight NFκB induction. The induction of the κB-binding complex was directly dependent upon the visible light illumination in

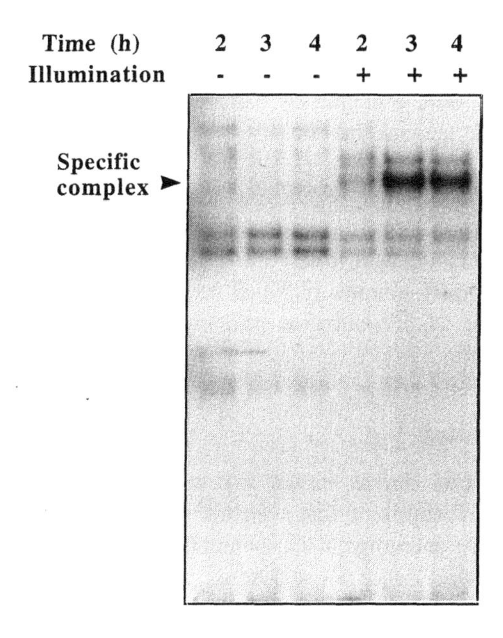

Figure 3. Effect of a photosensitization mediated by MB on κB-binding activities in the nucleus of ACH-2 cells. Rapid induction of a nuclear κB enhancer DNA-binding protein by treatment of ACH-2 cells with 4 μM methylene blue and illuminated for 6 min with visible light. Nuclear extracts were prepared 120, 180, and 240 min after the photoreaction with equal amounts of proteins extracted from nucleus of cells treated and mixed with a ^{32}P-labeled probe encompassing the κB elements of the HIV-1 enhancer. Samples were loaded on 6% native polyacrylamide gels and electrophoresed at 180 V. Autoradiogram of the gel is shown and the arrow indicates the position of the specific complex.

the presence of MB. Illumination without MB led only to a slight induction, which was not higher than in control cells.

Using antibodies directed against proteins of the NFκB family, we determined which subunits composed the κB-binding factor that was induced by the MB-mediated photooxidative stress. The supershift experiments showed that the addition of antibodies recognizing p50 (NFκB1) or p65 (RelA), respectively, induced a further shift of the specific complex, while an antibody raised against c-Rel, another member of the NFκB family, did not modify the mobility of the κB-binding complex. Thus, the NFκB factor induced by MB photosensitization, which binds to the κB sites of the HIV-1 enhancer, is the heterodimeric form p50/p65 of NFκB.

Photosensitization carried out in the presence of rose bengal, another potent singlet oxygen producer which is known to penetrate inside both the cytoplasm and the nucleus of cells (24), led to induction of the same κB-binding complex with similar kinetic characteristics. Proflavin is a cationic molecule having a high affinity for DNA capable of intercalating between DNA base pairs. Upon photosensittization, intercalated PF molecules oxidize guanine residues and generate DNA single-strand breaks. In lymphocytes or monocytes latently infected with HIV-1 (ACH-2 or U1, respectively), this photosensitizing treatment induced cytotoxicity, induction of NFκB, and reactivation of HIV-1 in cells surviving the treatment (23). Unexpectedly, NFκB induction by PF-mediated photosensitization was not affected by the presence of an antioxidant like *N*-acetyl-L-cysteine. Another transcription factor like AP-1 is less activated by this photosensitizing treatment. In comparison with other inducing treatments such as PMA or tumor necrosis factor-α (TNF-α), the activation of NF-κB is slow, being optimal 120 minutes after treatment. These kinetic data were obtained by following, on the same samples, both the appearance of NFκB in the nucleus and the disappearance of IκBα in cytoplasmic extracts.

All the data obtained with photosensitizing drugs which generate $^1O_2$ inside cells indicate that this ROS is capable of triggering signaling events initiated by oxidated targets. The signal is then transmitted into the cytoplasm, where the inactive NFκB factor is resident, allowing the translocation of p50/p65 subunits of NFκB to the nucleus. The nature of the oxidized targets, and the nature of the oxidation products (lipoperoxides, hydroperoxides, ...) remains to be discovered.

DNA Damage Can Induce NFκB Activation

Treatment of T-leukemic cells (ACH-2, CEM, Jurkat, Jurkat JR) with some DNA-damaging drugs leads to an activation of NFκB similar to that observed after an oxidative stress. The topoisomerase poisons amsacrine, actinomycin D, camptothecin, etoposide, and daunomycin at concentrations of 1–10 μM cause an activation of NFκB with a maximum after 2 or 3 hours of drug treatment (Figure 4). The NFκB complex activated by these agents, detected by electrophoretic mobility shift assay, had the same mobility and reacted in competition and supershift experiments in the same way as the complex induced by hydrogen peroxide or PF + light. Hence it is identified as a p50/p65 complex. Strikingly, neither drugs causing adducts on DNA (4-nitroquinoline *N*-oxide, *trans*-platin) and/or DNA–DNA cross-links (mitomycin C, *cis*-platin), nor a DNA polymerase inhibitor (cytosine arabinoside) activate NFκB efficiently in T cells. Topoisomerase poisons react with open complexes formed between a topoisomerase enzyme and DNA (nicked on one or both strands by the enzyme), and stabilize them by preventing the religation of the

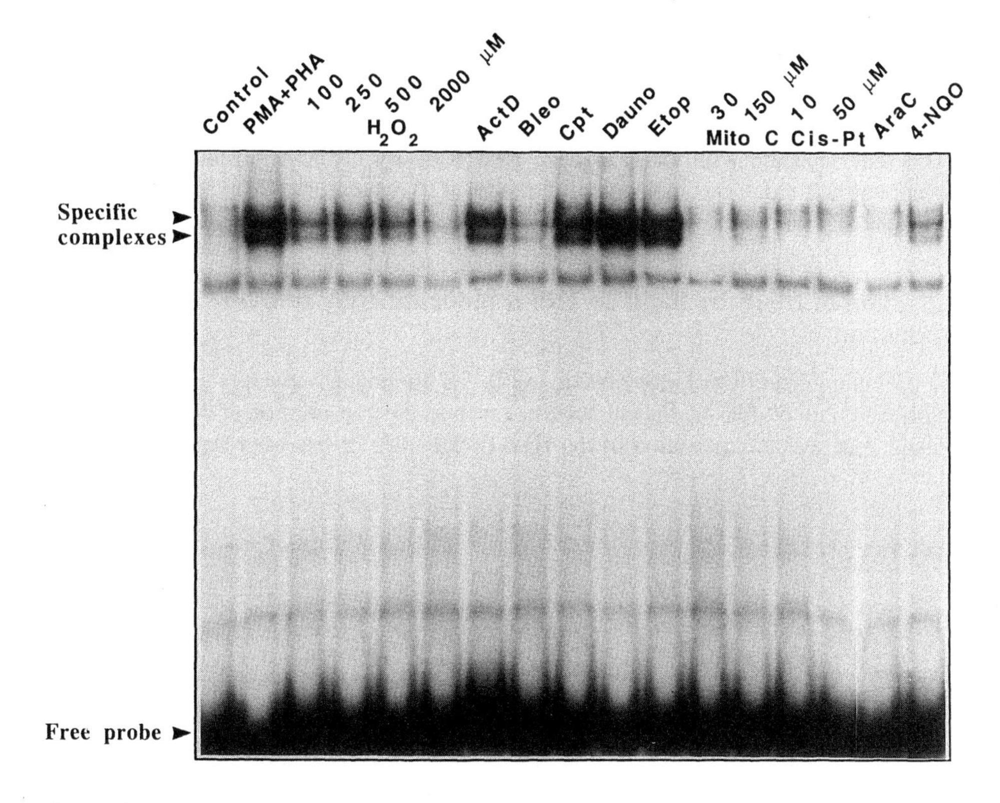

Figure 4. Exponentially growing CEM cells were exposed to the following treatments: Control; no treatment; PMA+PHA, 100 nM phorbol myristate acetate and 10 μg/ml phytohemagglutinin for 40 min; H_2O_2, Hydrogen peroxide at 100, 250, 500, or 2000 μM for 25 h; ActD, actinomycin D at 2 μM for 3 h; Bleo, bleomycin 100 μg/ml for 3 h; Cpt, camptothecin at 10 μM for 2 h; Dauno, daunomycin at 4 μM for 3 h; Etop, etoposide 10μM for 3 h; Mito C, mitomycin C at 30 or 150 μM for 3 h; Cis-Pt, cis-platine at 10 or 50 μM for 3 h; AraC, cytosine arabinoside 10 μM for 2 h; 4-NQO, 4-nitroquinoline *N*-oxide 10 μM for 3 h. The nuclear proteins were then analysed by EMSA with a ^{32}P-labeled probe containing the HIV LTR κB site.

DNA nicks by the topoisomerase. Interaction of such a complex with the DNA replication or transcription machineries causes its disruption with generation of double-strand breaks and release of the DNA ends (25). Since treatment with PF + light causes DNA strand breaks too, this suggests that these are the type of DNA lesions that can trigger the induction of NFκB. Another drug, bleomycin, which causes strand breaks by a radical-generating process, activates NFκB in ACH-2 cells but moderately, probably owing to an inefficient incorporation of the drug by the cells (26). NFκB activation by daunomycin and etoposide (drugs that react with type II topoisomerase–DNA complexes) does not seem to require interaction with the DNA replication machinery, while activation by camptothecin (which which reacts with type I topoisomerase–DNA complexes) does. The activation triggered by this latter drug only is inhibited by DNA polymerase inhibitors (aphidicolin and cytosine arabinoside). Pyrrolidine dithiocarbamate (PDTC), a metal chelator and free radical scavenger, inhibits NFκB activation by the topoisomerase poisons tested (camptothecin, daunomycin, and etoposide). This shows that the pathway of NFκB activation by topoisomerase poisons converges with pathways of NFκB activation by other agents (PMA, TNF,...) which are also inhibited by PDTC (19), probably at a step which involves free radicals as intracellular messengers. However, PDTC may interfere with DNA damage generation by daunomycin and etoposide by preventing their redox metabolization (at the level of cytochrome P450 monooxygenase, for instance).

Activation of NFκB by IL-1β or TNF-α Independently of ROS Production

The proinflammatory cytokines TNF-α and IL-1β are potent activators of NFκB, which mediates several of their biological activities such as the transcription of the IL-2 receptor α-chain gene and the activation of the HIV-1 LTR (1,2). It has been suggested that the

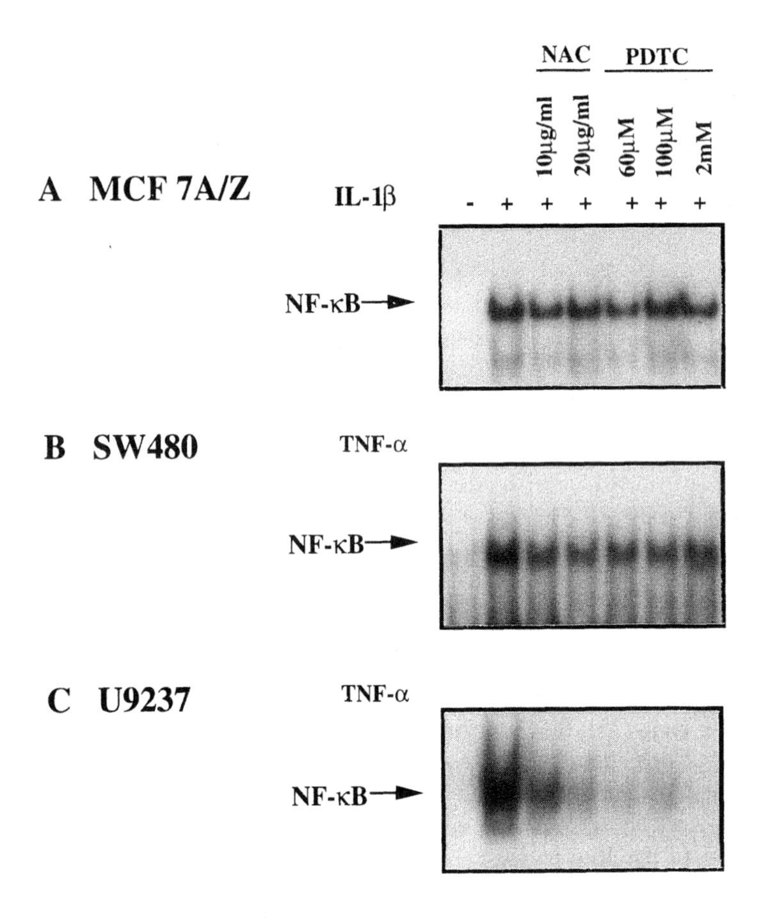

Figure 5. SW480 cells (colon adenocarcinoma cell line) and U937 cells (monocytic cell line) were incubated with 100 U/ml of TNF-α, and MCF7 A/Z cells (breast adenocarcinoma cell line) were treated with 50 U/ml of IL-1β for 15 min. The cells were preincubated for 2 h with *N*-acetyl-L-cysteine (NAC) or pyrrolidine dithiocarbamate (PDTC) at the indicated concentrations priors to cytokine stimulation. Nuclear extracts of these cells were analyzed by electrophoretic mobility shift assay with a specific κB probe. The arrows indicate the specific NFκB complexes.

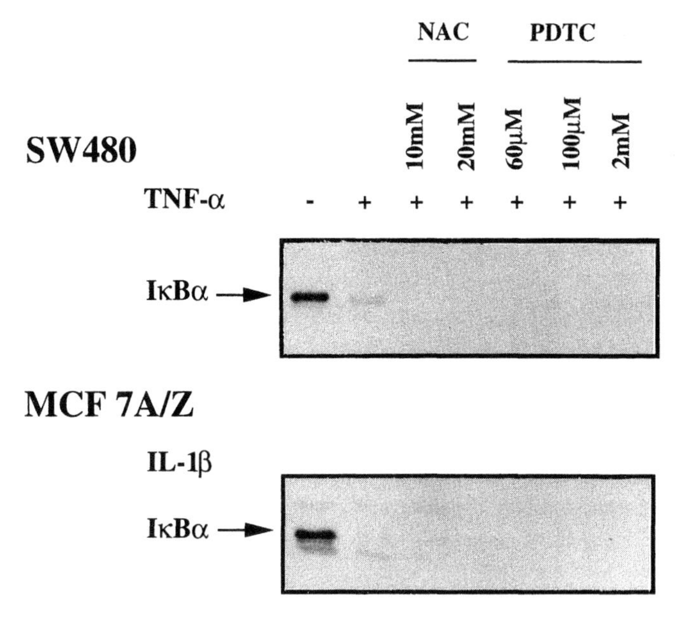

Figure 6. SW480 and MCF7 A/Z were treated as described in Figure 5 and cytoplasmic extracts were analyzed on an immunoblot revealed with a specific IκBα antibody. The arrows indicate the specific IκBα band.

TNF-α- or IL-1β-mediated induction of NFκB in lymphoid cells proceeds through intracellular production of ROS as it is inhibited by antioxidants (18,19,27).

We stimulated various adenocarcinoma cell lines with TNF-α or IL-1β and demonstrated by electrophoretic mobility shift assay a rapid and strong induction of endonuclear NFκB DNA-binding activity (Figure 5 and Ref. 28). This activation of NFκB is not inhibited by preincubation of the cells with antioxidants such as *N*-acetyl-L-cysteine (NAC) or PDTC before TNF-α or IL-1β stimulation (Figure 5) suggesting that proinflammatory cytokines induce NFκB in adenocarcinoma cells through a signaling pathway independent of the production of ROS. However, NFκB induction by TNF-α or IL-1β in lymphoid or monocytic cells (U937) is completely blocked by antioxidants (Figure 4 and Ref. 27).

We also showed that IL-1β stimulation of MCF7 A/Z cells or TNF-α treatment of SW480 cells induced a rapid and complete degradation of the IκBα protein (Figure 6). Again this degradation, which is responsible for NFκB nuclear translocation, is not affected by preincubation with antioxidants.

CONCLUSIONS

A large variety of signals can activate the transcription factor NFκB. These signals include cytokines, infectious agents, and chemical and physical stresses. Following its activation, NFκB induces the transcription of effector genes and thus plays a major role in various biological and pathological functions such as the immune response, inflammation, and HIV-1 replication (2). A large body of work has been devoted to the study of the signaling pathways leading to NFκB activation and IκBα degradation, but many steps of these pathways remain unknown.

Some studies have claimed that any signal activating NFκB proceeds through the intracellular production of ROS (18). Indeed, at least in lymphoid or monocytic cells, antioxidants such as NAC and PDTC are able to inhibit NFκB activation by all the compounds investigated (18). Moreover, NFκB activation by ROS is thought to play an active role in processes such as inflammation and HIV-1 replication. We confirmed that an oxidative stress can induce NFκB in various cell types. Indeed, methylene blue, rose bengal and proflavin activate NFκB, presumably though the production of singlet oxygen, but the targets of singlet oxygen and the pathway to IκB degradation remain to be discovered. DNA-damaging agents such as UV-irradiation or X-rays can also induce NFκB (29,30). We demonstrated that several DNA-damaging agents used as cytotoxic drugs activate NFκB. Our data suggest that the generation of DNA double-strand breaks could trigger the activation of NFκB, possibly again through the production of free radicals as this induction is inhibited by PDTC. However, we cannot exclude that PDTC interferes with DNA damage generated by the drugs. We will have therefore to explore how the signal is transmitted from the damaged DNA to the cytoplasm, leading to IκBα degradation and NFκB nuclear translocation.

However, we were able to demonstrate that the induction of NFκB following TNF-α or IL-1β stimulation of epithelial transformed cell lines (MCF7 A/Z, SW480, but also OVCAR-3 and SKOV-3) occurs independently of ROS production as it is not inhibited by antioxidants. Interestingly, NFκB activation by the same stimuli in lymphoid or monocytic cells is completely inhibited by antioxidants, indicating that a single stimulus induces a transcription factor through clearly distinct pathways in various cell types.

Our data thus show clearly that the redox regulation plays a major role in NFκB activation in a number of circumstances. However, some signals can induce NFκB in a redox-independent way.

ACKNOWLEDGMENTS

J.P., M.P.M. and V.B. are Research Director and Research Associates of the Belgian National Fund for Scientific Research (NFSR, Brussels, Belgium). B.P. is a grantee from the FRIA (Brussels, Belgium). G.B. is a fellow of Anti-Cancer Research Fund (Liège, Belgium). These research programs were funded by the NFSR, a concerted action program (Communauté Française de Belgique), and Télévie (NFSR, Brussels, Belgium).

REFERENCES

1. Baeuerle PA, Henkel T. Function and activation of NFκB in the immune system. Annu Rev Immunol 1994; 12: 141–179.
2. Siebelnist U, Franzoso G, Brown K. Structure, regulation and function of NFκB. Annu Rev Cell Biol 1994; 10: 405–455.
3. Grilli M, Chiu JJ-S, Lenardo MJ. NFκB and rel-participants in a multiform transcriptional regulatory system. Int Rev Cytol 1993; 143: 1–62.
4. Verma IM, Stevenson JK, Schwarz EM, Van Antwerpen D, Miyamoto S. Rel/NFκB/IκB family: intimate tales of association and dissociation. Genes Dev 1995; 9: 2723–2735.

5. Beg AA, Baldwin AS. The IκB proteins: multifunctional regulators of Rel/NFκB transcription factors. Genes Dev 1993; 7: 2064–2070.
6. Miyamoto S, Verma IM. Rel/NF-κB/IκB story. Adv Cancer Res 1995; 66: 255–292.
7. Hayashi T, Sekine T, Okamoto T. Identification of a new serine kinase that activates NFκB by direct phosphorylation. J Biol Chem 1993; 268: 26790–26795.
8. Israel A. A role for phosphorylation and degradation in the control of NFκB activity. Trends Genet 1995; 11: 203–205.
9. Baeuerle PA. The inducible transcription factor NF-κB-regulation by distinct protein subunits. Biochim Biophys Acta 1991; 1072: 63–80.
10. Brown K, Gersterberger S, Carlson L, Franzoso G, Siebelnist U. Control of IκBα proteolysis by site-specific, signal induced phosphorylation. Science 1995; 267: 1485–1488.
11. Traenckner EBM, Pahl HL, Henkel T, Schmidt S, Wilk S, Baeuerle PA. Phosphorylation of human IκBα on serines 32 and 36 controls IκBα protreolysis and NF-κB activation in response to diverse stimuli. EMBO J 1995; 14: 5433–5441.
12. Sun S-C, Elwood J, Greene WC. Both amino- and carboxy-terminal sequences within IκBα regulate its inducible degradation. Mol Cell Biol 1996; 16: 1058–1065.
13. Beg AA, Finco TS, Nantermet PV, Baldwin AS. Tumor necrosis factor and interleukin 1 lead to phosphorylation and loss of IκBα–a mechanism for NF-κB activation. Mol Cell Biol 1993; 13: 3301–3310.
14. Cordle SR, Donald R, Read MA, Hawiger J. Liposaccharide induces phosphorylation of MAD3 and activation of C-rel and related NF-κB proteins in human monocytic THP-1 cells. J Biol 1993; 268: 11803–11810.
15. Chen ZJ, Hagler J, Palombella J, Melandri F, Scherer D, Ballard D, Maniatis T. Signal-induced site-specific phosphorylation targets IκBα to the ubiquitin-proteasome pathways. Genes Dev 1995; 9: 1586–1597.
16. Rodriguez MS, Michapoulos I, Arenzana-Seisdedos F, Hay RT. Inducible degradation of IκBα in vitro and in vivo requires the acidic C-terminal domain of the protein. Mol Cell Biol 1995; 15: 2413–2419.
17. Scherer DC, Brockman JA, Chen Z, Maniatis T, Ballard DW. Signal-induced degradation of IκBα requires site-specific ubiquitination. Proc Natl Acad Sci USA 1995; 92: 11259–11263.
18. Schreck R, Rieber P, Baeuerle PA. Reactive oxygen intermediates as apparently widely used messengers in the activation of the NFκB transcription factor and HIV-1. EMBO J 1991; 10: 2247–2258.
19. Schreck R, Meier B, Mannel DN, Droge W, Baeuerle PA. Dithiocarbamates as potent inhibitors of nuclear factor kappa B activation in intact cells. J Exp Med 1992; 175: 1181–1194.
20. Schmidt KN, Amstad P, Cerutti P, Baeuerle PA. The roles of hydrogen peroxide and superoxide as messengers in the activation of transcription factor NF-κB. Chem Biol 1995; 2: 13–22.
21. Schenk H, Klein M, Erdbürger W, Dröge W, Schultze-Osthoff K. Distinct effects of thioredoxin and other antioxidants on the activation of NF-κB and AP-1. Proc Natl Acad Sci USA 1994; 91: 1672–1676.
22. Legrand-Poels S, Hoebeke M, Vaira D, Rentier B, Piette J. HIV-1 promoter activation following an oxidative stress mediated by singlet oxygen. J Photochem Photobiol B 1993; 17: 229–237.
23. Legrand-Poels S, Bours V, Piret B, Pflaum M, Epe B, Rentier B, Piette J. Transcription factor NF-κB is activated by photosensitization generating oxidative DNA damages. J Biol Chem 1995; 270: 6925–6934.
24. Piret B, Legrand-Poels S, Sappey C, Piette J. NF-κB transcription factor and human immunodeficiency virus type 1 (HIV-1) activation by methylene blue photosensitization. Eur J Biochem 1995; 228: 447–455.

25. Froelich-Ammon SJ, Osheroff N. Topoisomerase poisons: harnessing the dark side of enzyme mechanism. J Biol Chem 1995; 270: 21429–21432.
26. Povirk LF, Austin MJF. Genotoxicity of bleomycin. Mutat Res 1991; 257: 127–143.
27. Bonizzi G, Dejardin E, Piret B, Piette J, Merville MP, Bours V. Interleukin-1β induces nuclear factor κB in epithelial cells independently of the production of reactive oxygen intermediates. Eur J Biochem 1996; 242: 544–549.
28. Schreck R, Albermann K, Baeuerle PA. Nuclear factor kappa B: an oxidative stress-responsive transcription factor of eukaryotic cells (a review). Free Radical Res Commun 1992; 17: 221–227.
29. Brach MA, Haas R, Sherman ML, Gunji H, Weichselbaum R, Kufe D Ionizing radiation induces expression and binding activity of the nuclear factor kappa B. J Clin Invest 1991; 88: 691–695.
30. Stein B, Krämer M, Rahmsdorf HJ, Ponta H, Herrlich P. UV-induced transcription from the human immunodeficiency virus type 1 (HIV-1) long terminal repeat and UV-induced secretion of an extracellular factor that induces HIV-1 transcription in nonirradiated cells. J Virol 1989; 63: 4540–4544.

8

Redox Regulation of the NFκB Signaling Pathway and Disease Control

Takashi Okamoto, Shinsaku Sakurada, Tetsuji Kato, Akihiko Yoneyama, Keiichi Tozawa, Jian-Ping Yang, and Naoko Takahashi
Nagoya City University Medical School Nagoya, Aichi, Japan

INTRODUCTION

Oxygen plays complexed roles in the cell. While oxygen is indispensable for an aerobic cell to obtain the essential chemical energy in the form of ATP, it is often transformed into highly reactive forms, radical oxygen intermediates (ROI), which are often toxic for the cell. In order to defend against the toxic actions of ROI, cells have acquired multiple endogenous antioxidant systems. These defense mechanisms include redox enzyme systems such as glutaredoxin and thioredoxin (Holmgren, 1985; Holmgren, 1989). Recent studies of cell biology and biochemistry have revealed the involvement of these molecules in the intracellular signal transduction pathways (Holmgren, 1985; Ziegler, 1985; Holmgren, 1989; Allen, 1993). The term "redox regulation" has thus been proposed to indicate the role of oxidoreductive modifications of proteins in regulating their functions. Oxidoreductive reaction of biomolecules, mostly proteins, which used to be considered as oxidative "stress," is now considered as a "signal" that contains biological information that is necessary for the maintenance of cellular homeostasis in response to the extracellular environment. The nature of redox regulation of one of the transcription factors, nuclear factor kappa B (NFκB) will be discussed as an example.

TRANSCRIPTION FACTOR NFκB AS A PATHOPHYSIOLOGIC DETERMINANT

NFκB regulates expression of a wide variety of cellular and viral genes (for reviews, see Gilmore, 1990; Baeuerle, 1991; Baeuerle and Henkel, 1994; Thanos and Maniatis, 1995). These genes include cytokines such as IL-2, IL-6, IL-8, GM-CSF and TNF; cell adhesion molecules such as ICAM-1 and E-selectin; inducible nitric-oxidase synthase (iNOS); and viruses such as human immunodeficiency virus (HIV) and cytomegalovirus (Maekawa et al., 1989; Okamoto et al., 1989; Mukaida et al., 1990; Schreck and Baeuerle, 1990; Stade et al., 1990; Whelan et al., 1991; Donnelly et al., 1993; Schindler and Baichwal, 1994; Xie et al., 1994; Roebuck et al., 1995; Staynov et al., 1995). Through the causal relationship with these genes, NFκB is considered to be involved in the currently intractable diseases such as acquired immunodeficiency syndrome (AIDS), hematogenic cancer cell metastasis, and rheumatoid arthritis (RA). Although the genes induced by NFκB are also under the control of other transcription factors and the levels of expression are variable according to the context of cell lineage, NFκB plays a major role in the transcriptional regulation and thus contributes a great deal to the pathogenesis. Therefore, biochemical intervention of NFκB should conceivably interfere with the pathogenic processes and be effective in treatment.

NFκB consists of two subunit proteins, p65 and p50, and usually exists as a molecular complex with an inhibitory protein, IκB, in the cytosol (Sen and Baltimore, 1986; Nabel and Baltimore, 1987; Baeuerle and Baltimore, 1988a,b; Ghosh et al., 1990; Ghosh and Baltimore, 1990; Read et al., 1994). Upon stimulation of the cells, for example, by proinflammatory cytokines IL-1 and TNF, IκB is dissociated and NFκB is translocated to the nucleus and activates expression of target genes. Thus the activity of NFκB itself is regulated by the upstream regulatory mechanism. There are at least two independent steps in the NFκB activation cascade: kinase pathways and redox signaling pathway. These two distinct pathways are involved in the NFκB activation cascade in a coordinated fashion, which may contribute to a fine tuning as well as fail-safe regulation of NFκB activity.

ACTIVATION OF NFκB BY THE KINASE AND REDOX CASCADES

At least two distinct kinase pathways are known to be involved in NFκB activation: NFκB kinase and IκB kinase (Figure 1). We found a 43 kDa serine kinase, named NFκB kinase, that is associated with NFκB (Hayashi et al., 1993a). This kinase phosphorylates both subunits of NFκB and dissociates it from IκB. There is another kinase (or kinases) known to phosphorylate IκB (Shirakawa and Mizel, 1989; Meichle et al., 1990; Feuillard et al., 1991; Ostrowski et al., 1991; Schutze et al., 1992; Brown et al., 1995; Cao et al., 1996; Chen et al., 1996). As a candidate for IκB kinases, IRAK (IL-1 receptor-associated kinase) was recently cloned and found to share structural similarity to Pelle, a protein kinase known to be involved in the activation of a NFκB homologue in *Drosophila* (Cao et al., 1996). Consistent with these findings, NFκB was shown to be phosphorylated in some cell lines and IκB was phosphorylated in others in response to stimulation with TNF or IL-1 (Li et al., 1994; Naumann and Scheidereit, 1994). In most of cases, NFκB dissociation by a kinase cascade(s) is considered to be a primary step of NFκB activation.

After dissociation from IκB, however, NFκB must go through the redox regulation by a cellular reducing catalyst, thioredoxin (TRX) (Okamoto et al., 1992; Hayashi et al., 1993b). TRX is a cellular reducing catalyst and is known to participate in redox reactions through reversible oxidation of its active-center dithiol to a disulfide (Figure 2). Interestingly, human TRX has been initially identified as a factor responsible for induction of the α subunit of interleukin-2 receptor, which is now known to be under the control of NFκB (Tagaya et al., 1989; Wakasugi et al., 1990). It is known that NFκB does not bind to the κB DNA sequence of the target genes until it is reduced (Molitor et al., 1991; Schreck et al., 1991; Toledano and Leonard, 1991; Matthews et al., 1992; Okamoto et al., 1992). In our previous paper we have assigned the cysteine residue at the amino acid position 62 of p50 subunit as a target of the redox regulation based on the high localized

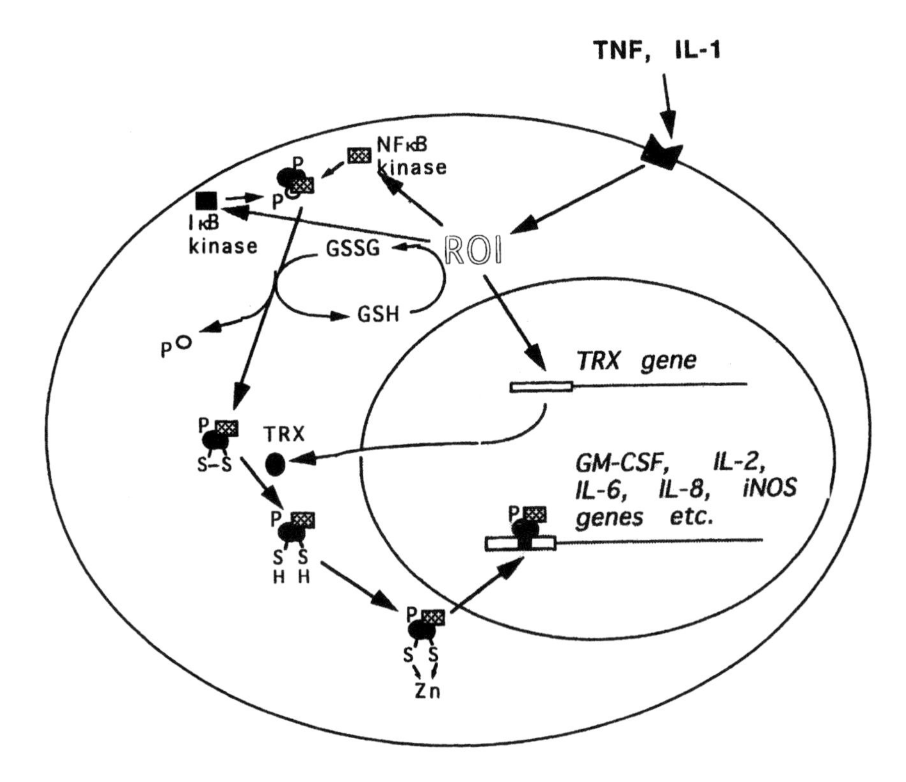

Figure 1. Signal transduction pathways for NFκB activation. The first step involves kinase pathways such as by NFκB and IκB kinases. The second step involves "redox regulation" by thioredoxin (TRX). See text for the details. Extracellular signals such as those elicited by IL-1 and TNF are mediated by production of radical oxygen intermediates (ROI) which then stimulate the kinase and TRX-mediated NFκB activation.

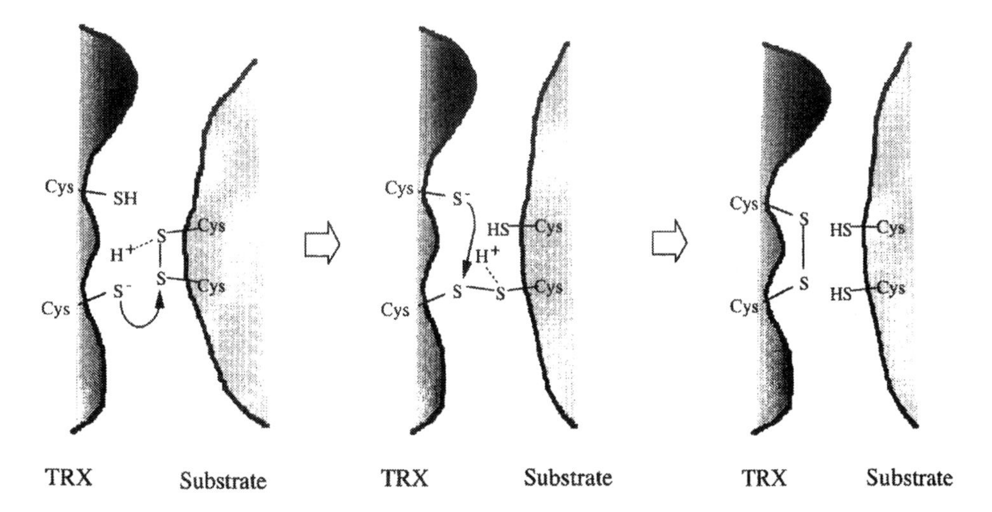

Figure 2. Action of TRX. Reduction of the oxidized protein by TRX is carried out by successive donation of protons from the cysteines on TRX. Those redox-active cysteines susceptible for dithiol–disulfide exchange may be determined by the positions and the adjacent amino acid residues.

p*I* values near one of the conserved cysteine residues conserved among the NFκB/Rel family members (Hayashi et al., 1993b).

Structural biological approaches have provided physicochemical background for the redox regulation of NFκB by thioredoxin. First, in early 1995 two US groups independently demonstrated the three-dimensional (3D) structure of the NFκB subunit p50 homodimer cocrystalized with the target DNA (Ghosh et al., 1995; Müller et al., 1995). NFκB appears to have a novel DNA-binding structure called the β-barrel, a group of β-sheets stretching toward the target DNA. There is a loop peptide extending from one of the tips of the β that intercalates with the nucleotide bases and makes a direct contact with the κB DNA. This DNA-binding loop contains the cysteine-62 that we predicted to be the target of redox regulation by thioredoxin (Hayashi et al 1993b). Although in both studies this cysteine was replaced by alanine, presumably for technical reasons in crystalization, these observations nevertheless confirm our earlier speculations (Hayashi et al., 1993b). Second, using NMR a group in NIH recently solved the 3D structure of the TRX molecule that is associated with the DNA-binding loop of p50 (Qin et al., 1995). A boot-shaped hollow on the surface of TRX containing the redox-active cysteines could stably recognize the DNA-binding loop of p50 and is likely to reduce the oxidized cysteine by donating protons in a structure-dependent way. Therefore, the reduction of NFκB by TRX is considered to be specific. TRX was shown to be translocated to the nucleus simultaneously with NFκB when the cells were stimulated with TNF-α (Sakurada et al., in press). Interestingly, this comigration of TRX with NFκB was only transient; after 30 minutes of stimulation, TRX was localized only in the cytoplasm while NFκB was still predominantly detected in the nucleus. Although we do not have direct evidence of TRX–NFκB cross-linking, it is assumed that TRX makes direct contact with NFκB by forming a disulfide bridge during the reducing reaction as suggested by the action of TRX (Figure 2).

INVOLVEMENT OF RADICAL OXYGEN INTERMEDIATES (ROI)

What triggers these NFκB signaling cascades? Not much is known about what happens immediately downstream of the cell surface receptor. However, involvement of ROI is suggested upstream of the NFκB activation pathway since the signaling was efficiently blocked by pretreatment of the cells with antioxidants such as *N*-acetyl-cysteine (NAC) or α-lipoic acid (Roederer et al., 1990; Schreck et al., 1991; Suzuki et al., 1992; Biswas et al., 1993; Meyer et al 1993; Suzuki and Packer, 1994; Packer et al., 1995). Therefore, antioxidants are now considered to be effective NFκB inhibitors.

Figure 3 illustrates the intracellular redox cascade involving successive reduction of oxygen and redox regulation of a target protein. Among these ROI, hydrogen peroxide has the longest half-life and is considered to be a major mediator of the oxidative signal. On the other hand, cellular reducing systems such as TRX counteract the action of hydrogen peroxide. The intensity of the oxidative signal may be modulated by the internal GSH level. Similarly, total GSH/GSSG content may influence the responsivenes of the cellular redox signaling. Therefore, intracellular thiol redox status appears to be a critical determinant of NFκB activation. NAC is considered to inhibit NFκB activation by replenishing intracellular cysteine required to produce GSH. We found that NAC could also block the induction of TRX (Sachi et al., 1995; Okamoto et al., in preparation). Therefore, anti-NFκB actions of antioxidants are considered to be twofold: (1) blocking

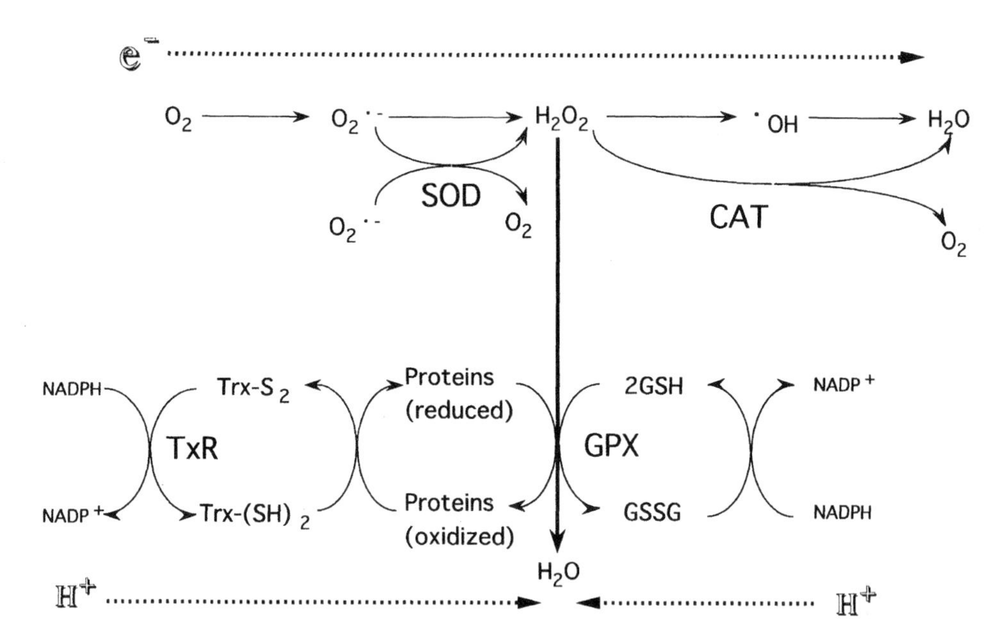

Figure 3. Intracellular redox system. ROI are generated during the process of oxygen reduction. Electrons are most likely provided from mitochondrial electron transport system. Among ROI, hydrogen peroxide may play a principal role as a mediator of oxidative signal. The cellular reducing system, on the other hand, particularly TRX counteracts the oxidative signal by subsequent reduction of the oxidized target molecule. SOD, superoxide dismutase; CAT, catalase; TxR, thioredoxin reductase; TRX, thioredoxin; GPX, glutathione peroxidase.

the signaling immediately downstream of the elicitation of signal, and (2) suppressing induction of the redox effector TRX.

Interestingly, we recently demonstrated that even the fully activated NFκB could still be blocked by gold ion by a redox mechanism (Yang et al., 1995). We found that the zinc ion is a necessary component of the active NFκB and that addition of monovalent gold ion could efficiently block its activity by oxidizing the redox-active cysteines on NFκB. Since gold did not appear to replace zinc (Yang et al., 1995), it is likely that gold ion oxidizes these thiolate anions on NFκB into disulfides and thus aborogates the DNA-binding activity because of its higher oxidation potential over zinc ion. It is notable that gold compounds have been successfully used for the treatment of RA (Skosey, 1993; Insel, 1990). Our finding could explain why gold is effective in RA and suggests that NFκB might have a crucial role in the disease process (Handel et al., 1996; Sakurada et al., in press; Kato et al., submitted). It may be that gold compounds are potentially effective in other diseases where NFκB plays a pathological role.

INTERVENTION IN NFκB ASSOCIATED DISEASES

The pivotal role of NFκB in the life cycle of HIV, especially in the virus reactivation process within latently infected cells, has been widely accepted. After activation through intracellular signaling pathways such as those elicited by the T cell receptor–antigen complex or by receptors for IL-1 or TNF, NFκB initiates HIV gene expression by binding to the target DNA element within the promoter region of HIV LTR (Nabel and Baltimore, 1987; Bohnlein et al., 1988; Okamoto et al., 1989, 1990). Then, the virus-encoded *trans*-activator Tat is produced and triggers explosive viral replication (Arya et al., 1985; Sodroski et al., 1985; Okamoto and Wong-Staal, 1986; Peterlin et al., 1986). Since the activation pathway of HIV gene expression by cellular transcription factor NFκB conceptually precedes activation by viral *trans*-activators, NFκB is considered to be a determinant of the maintenance and breakdown of the viral latency. Attempts to control the NFκB activation pathway will give us a clue to biochemical intervention in the clinical development of AIDS. In recent years, cumulative evidence has indicated that antioxidants could be effective in treating AIDS by blocking HIV replication (Roederer et al., 1990; Suzuki et al., 1992).

Another situation where NFκB plays a role is hematogenic cancer cell metastasis (Tozawa et al., 1995). NFκB induces E-selectin (also known as ELAM-1) on the surface of vascular endothelial cells (Montgomery et al., 1991; Whelan et al., 1991). Since some cancer cells constitutively express a ligand for E-selectin the sialyl-Lewis^X glycoprotein molecule, expressed on their cell surface, induction of E-selectin is considered to be a rate-determining step in cancer cell–endothelial cell interaction (Dejana et al., 1988; Takada et al., 1993). We examined the role of NFκB in induction of E-selectin (Tozawa et al., 1995). When primary human umbilical venous endothelial cells (HUVEC) were treated with IL-1 or TNF, nuclear translocation of NFκB was observed, followed by the augmented expression of E-selectin. We examined the cell-to-cell interaction between HUVEC and QG90 cells, a tumor cell line derived from human small-cell carcinoma of the lung expressing sialyl-Lewis^X antigen, and found that IL-1 was able to induce the attachment of cancer cells to HUVEC. However, pretreatment of HUVEC with *N*-acetylcysteine, aspirin, or pentoxyphillin efficiently blocked the cell-to-cell attachment in a dose-dependent manner.

SCREENING STRATEGY FOR ANTI-NFκB COMPOUNDS

As an application of basis studies on transcriptional regulation of HIV, we have constructed a plasmid to genetically transduce the regulatory cascade of virus replication (Kira et al., 1995). The pKO plasmid has been created containing Tat gene under the control of cytomegalovirus promoter which is positively controlled by NFκB. This plasmid also contains hygromycin B resistance gene (hygromycin B phosphotransferase gene) under the control of HIV LTR which is induced by Tat and NFκB. The cells stably transfected with the pKO plasmid could survive in the presence of the antibiotic hygromycin B as long as the hygromycin B resistance gene was expressed by the combined actions of NFκB and Tat (kira et al., 1995). This system could be used not only for anti-Tat but also for anti NFκB compounds (Kira et al., in press; Merin et al., in press). In the use of pKO for anti-NFκB screening, we assume that Tat is acting simply as an amplifier of the action of NFκB as in the case of HIV replication. Since the pKO plasmid can endow any cell with drug resistance to hygromycin B, irrespective of the NFκB activation cascade which is now known to be variable in different cell lineages,this strategy would be feasible for the screening of anti-NFκB drug in various disease conditions by use of the relevant cells.

SUMMARY

Recognition of ROI and redox-mediated protein modifications as signals has opened up a new field of cell regulation and provided a novel means of controlling disease processes. One such approach has been shown to be feasible for gene expression governed by transcription factor NFκB. We have identified novel signal transduction cascades in the activation of NFκB involving redox control effector TRX. By controlling redox cascades by using antioxidants, for example, currently incurable diseases such as hematogenic cancer cell metastasis and AIDS are now within the scope of effective therapy.

REFERENCES

Allen JF, FEBS Lett 1993; 332: 203–207.
Arya SK, Guo C, Josephs SF, Wong-Staal F. Science 1995; 229: 69–73.
Baeuerle PA, Biochim Biophys Acta 1991; 1072: 63–80.
Baeuerle PA, Baltimore D Cell 1988a; 53: 211–217.
Baeuerle PA, Baltimore D. Science 1988b; 242: 540–546.
Baeuerle PA, Henkel T. Annu Rev Immunol 1994; 12: 141–179.
Biswas DK, Dezube BJ, Ahlers CM, Pardee AB J AIDS 1993; 6: 778–786.
Bohnlein E, Lowenthal JW, Siekevitz M, Ballard DW, Franza BR, Greene WC, 1988C; Cell 53: 827–836.
Brown K, Gerstberger S, Carlson L, Franzoso G, Siebenlist U. Science 1995; 267: 1485–1488.
Cao Z, Henzel WJ, Gao X. Science 1996; 271: 1128–1131.
Chen ZJ, Parent L, Maniatis T. Cell 1996; 84: 853–862.
Dejana E, Bertocci F, Bortolami MC, Regonesi A, Tonta A, Breviario F, Giavazzi R. J Clin Invest 1988; 82: 1466–1470.
Donnelly RP, Crofford LJ, Freeman SL, Buras J, Remmers E, Wilder RL, Fenton MJ. J Immunol 1993; 151: 5603–5612.
Feuillard J, Gouy H, Bismuth G, Lee LM, Debre P, Korner M. Cytokine 1991; 3: 257–265.
Ghosh S, Baltimore D. Nature 1990; 344: 678–682.
Ghosh S, Gifford AM, Riviere LR, Tempst P, Nolan GP, Baltimore D. Cell 1990; 62: 1019–1029.
Ghosh G, Van Duyne G, Ghosh S, Sigler PB. Nature 1995; 373: 303–310.
Gilmore TD Cell 1990; 62: 841–843.
Handel ML, McMorrow LB, Gravallese EM. Arthritis Rheum 1996; 38: 1762–1770.
Hayashi T, Sekine T, Okamoto T. J Biol Chem 1993a; 826: 26790–26795.
Hayashi T, Ueno Y, Okamoto T. J Biol Chem 1993b; 268: 11380–11388.
Holmgren A. Annu Rev Biochem 1985; 54: 237–271.
Holmgren A. J Biol Chem 1989; 264: 13963–13966.
Insel PA. Gilman G et al., eds. Autacoids: Drug Therapy of Inflammation. New York: Macmillan; 1990; 670–681.
Kira T, Merin JP, Baba M, Shigeta S, Okamoto T. AIDS Res Hum Retrovir 1995; 11: 1359–1366.
Li C-C H, Dai R-M, Chen E, Longo DL. J Biol Chem 1994; 269: 30089–30092.
Maekawa T, Itoh F, Okamoto T, Kurimoto M, Imamoto F, Ishii S. J Biol Chem 1989; 264: 2826–2831.
Matthews JR, Wakasugi N, Virelizier JL, Yodoi J, Hay RT. Nucleic Acids Res 1992; 20: 3821–3830.
Meichle A, Schutze S, Hensel G, Brunsing D, and Kronke M. J Biol Chem 265, 8339–8343 (1990).
Meyer M, Schreck R, Baeuerle PA. EMBO J 1993; 12: 2005–2015.
Molitor JA, Ballard DW, Greene WC. New Biol 1991; 3: 987–996.

Montgomery KF, Osborn L, Hession C, Tizard R, Goff D, Vassallo C, Tarr PI, Bomsztyk K, Lobb R, Harlan JM, Pohlman TH. Proc Natl Acad Sci USA 1991; 88: 6523–6527.
Mukaida NY, Mahey Matsushima K. J Biol Chem 1990; 265: 21128–21133.
MΓler CW, Rey FA, Sodeoka M, Verdine GL, Harrison SC. Nature 1995; 373: 311–317
Nabel G, Baltimore D. Nature 1987; 326: 711–713.
Naumann M, Scheidereit C. EMBO J 1994; 13: 4597–4607.
Okamoto T, Wong-Staal F. Cell 1986; 47: 29–35.
Okamoto T, Matsuyama T, Mori S, Hamamoto Y, Kobayashi N, Yamamoto N, Josephs SF, Wong-Staal F, Shimotohno K. AIDS Res Hum Retrovir 1989; 5: 131–138.
Okamoto T, Benter T, Josephs SF, Sadaie MR, Wong-Staal F. Virology 1990; 177: 606–614.
Okamoto T, Ogiwara H, Hayashi T, Mitsui A, Kawabe T, Yodoi J. Int Immunol 1992; 4: 811–819.
Okstrowski J, Sims JE, Sibley CH, Valentine MA, Dower SK, Meier KE, Bomsztyk K. J Biol Chem 1991; 266: 12722–12733.
Packer L, Witt EH, Tritschler HJ. Free Radical Biol Med 1995; 19: 227–250.
Peterlin BM, Luciw PA, Barr PJ, Walker MD. Proc Natl Acad Sci USA 1986; 83: 9734–9738.
Qin J, Clore GM, Kennedy WMP, Huth JR, Gronenborn AM Structure 1995; 3: 289–297.
Read MA, Whitley MZ, Williams AJ, Collins T. J Exp Med 1994; 179: 503–512.
Roebuck KA, Rahman A, Lakshminarayanan V, Janakidevi K, Malik AB. J Biol Chem 1995; 270: 18966–18974.
Roederer M, Staal FJT, Raju PA, Ela SW, Herzenberg LA, Herzenberg LA, Proc Natl Acad Sci USA 1990; 87: 4884–4888.
Sachi Y, Hirota K, Masutani H, Toda K, Okamoto T, Takigawa M, Yodoi J. Immunol Lett 1995; 44: 189–193.
Schindler U, Baichwal VR. Mol Cell Biol 1994; 14: 5820–5831.
Schreck R, Baeuerle PA. Mol Cell Biol 1990; 10: 1281–1286.
Schreck RP, Rieber P. Baeuerle PA. EMBO J 1991; 10: 2247–2258.
Schutze S, Potthoff K, Machleidt T, Berkovic D, Wiegmann K, Kronke M. Cell 1992; 71: 765–776.
Sen R, D Baltimore D. Cell 1986; 46: 705–716.
Shirakawa F, Mizel SB. Mol Cell Biol 1989; 9: 2424–2430.
Skosey JL. In McCarty DJ, Koopman WJ, eds. Arthritis and Allied Conditions Philadelphia: Lea & Febiger; 1993: 603–614.
Sodroski J, Patarca R, Rosen C. Science 1985; 229: 74–77.
Stade BG, Messer G, Riethmuller G, Johnson JP. Immunobiology 1990; 182: 79–87.
Staynov DZ, Cousins DJ, Lee TH. Proc Natl Acad Sci USA 1995; 92: 3606–3610.
Suzuki YJ, Aggarwal BB, Packer L. Biochem Biophys Res Commun 1992; 189: 1709–1715.
Suzuki YJ, Packer L. J Immunol 1994; 153: 5008–5015.
Tagaya Y, Maeda Y, Mitsui A, Kondo N, Matsui H, Hamuro J, Brown J, Arai KI, Yokota T, Wakasugi H, Yodoi J. EMBO J 1989; 8: 757–764.
Takada A, Ohmori K, Yoneda T, Tsuyuoka K, Hasegawa A, Kiso M, Kannagi R. Cancer Res 1993; 53: 354–361.
Thanos D, Maniatis T. Cell 1995; 80: 529–532.
Toledano MB, Leonard WJ. Proc Natl Acad Sci USA 1991; 88: 4328–4332.
Tozawa KS, Sakurada S, Kohri K, Okamoto T. Cancer Res 1995; 55: 4162–4167.
Wakasugi N, Tagaya Y, Wakasugi H, Mitsui A, Maeda M, Yodoi J, Tursz T. Proc Natl Acad Sci USA 1990; 87: 8282–8286.
Whelan J, Ghersa P, Huijsduijnen RH, Gray J, Chandra G, Talabot F, DeLamarter JF. Nucleic Acids Res 1991; 19: 2645–2653.
Xie Q-W, Kashiwabara Y, Nathan C. J Biol Chem 1994; 269: 4705–4708.
Yang JP, Merin JP, Nakano T, Kato T, Kitade Y, Okamoto T. FEBS Lett 1995; 361: 89–96.
Ziegler DM. Annu Rev Biochem 1985; 54: 305–329.

9

Mechanism of Activation of NFκB by Hydrogen Peroxide

Kirsty G. McPherson, John R. Babson,* Anthony J. Kettle, and Christine C. Winterbourn
Christchurch School of Medicine
Christchurch, New Zealand

Nuclear transcription factor κB (NFκB) regulates the expression of genes involved in the immune response, acute-phase reaction, and inflammation (1,2). It also induces expression of AIDS virus genes in infected cells. It is normally present in the cytoplasm in an inactive form bound to its inhibitor, IκB. Activation involves phosphorylation and proteolytic degradation of the inhibitor and translocation to the nucleus (1–4). Diverse agents, including tumor necrosis factor-α (TNF-α) and other cytokines, lipopolysaccharide, phorbol esters, and H_2O_2 can activate NFκB.

A large body of evidence supports the contention that gene expression controlled by NFκB is redox regulated. At one level, its binding to DNA requires a key cysteine residue that must be reduced (5). On the other hand, reactive oxygen species appear to serve as common second-messenger-like molecules in the various pathways leading to NFκB activation (2). There are, however, conflicting reports on how this occurs. Some investigators have found that activation by TNF-α and other cytokines depends on the redox (thiol) status of the cell and can be inhibited by numerous antioxidants (6–10). Others, though, report activation that is refractory to antioxidants (11,12). Furthermore, although exogenous H_2O_2 can trigger NFκB activation, this is restricted to a few cell types (11,13). Since antioxidants inhibit more broadly, it has been suggested that there may be more than one oxidant-sensitive step, one possibly involving a redox-sensitive kinase or phosphatase controlling IκB phosphorylation (13,14).

Various mechanisms have been proposed for activation of NFκB. These include transformation of the cell to a more oxidized redox state, either by decreasing the concentration of reduced thiols or increasing the ratio of glutathione disulfide to reduced glutathione (GSSG:GSH), or by generation of the hydroxyl radical or another reactive oxidant. It is generally assumed that this would result in oxidation of some critical protein thiol. Antioxidants are variously proposed to work by increasing intracellular GSH or total

* *Current affiliation*: University of Rhode Island, Kingston, Rhode Island

thiols, scavenging free radicals or iron chelation. There is, however, no direct evidence for any of these proposed reactions.

An alternative explanation that has received little attention is that the oxidative step is more specific, perhaps involving an enzymatic reaction of H_2O_2. We have investigated this possibility for H_2O_2-mediated activation of NFκB in Jurkat JR (Wurzburg) cells. As reported by others (15), we observed NFκB activation by submillimolar concentrations of H_2O_2 in this T cell line. This contrasts with other Jurkat cells that respond to TNF-α and other stimuli but not H_2O_2 (13). In our hands, activation by H_2O_2 (100 μM) as detected with an electrophoretic mobility shift assay was evident at 30 min and maximum at about 2 h, whereas activation by TNF-α was maximal within about 15 min.

To determine how NFκB activation relates to consumption of H_2O_2 added to the cells, we measured H_2O_2 in the medium (Figure 1). Most was consumed by the cells within 15 min, as compared with a slow disappearance in medium alone. Thus an initial reaction of H_2O_2 sets in motion events that subsequently lead to NFκB activation. The heme poisons azide and cyanide, and the catalase inhibitor aminotriazole, were added to determine how they affected H_2O_2 removal. They slowed the rate substantially, implying that catalase and other heme enzymes are responsible for most of the H_2O_2 consumption by the cells.

We expected that prolonging the existence of H_2O_2 would enhance NFκB activation. Contrary to this, however, azide and cyanide were inhibitory (Figure 2). Aminotriazole (not shown) had no effect. The cells remained viable with these treatments, as assessed by trypan blue exclusion. Activation by TNF-α was not affected by either azide or cyanide at these concentrations, indicating that there are differences in the activation pathways for TNF-α

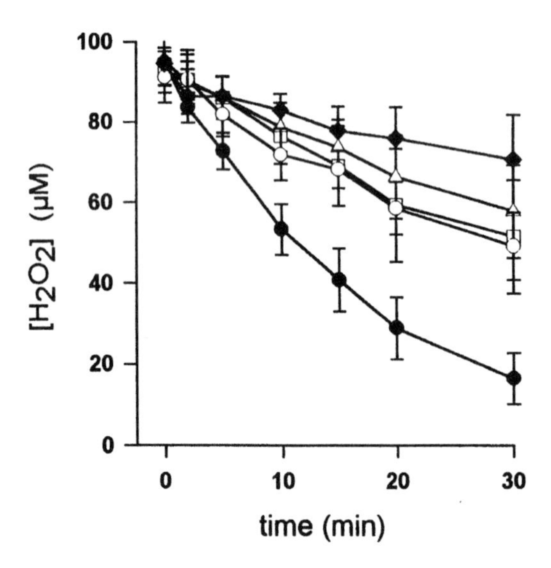

Figure 1. Consumption of H_2O_2 (100 μM) added to Wurzburg cells (10^6/ml) in conditioned RPMI medium. H_2O_2 was added to the cell suspensions, then at designated times the cells were spun down for 10 s and the supernatants were immediately analyzed for H_2O_2 using the FOX1 assay (20). ● Cells in medium; ◆ medium separated from the cells immediately before H_2O_2 addition; □ cells plus azide (100 μM); △ cells plus cyanide (1 mM); ○ cells plus aminotriazole (25 mM). Each curve is the mean (±SD) for three experiments.

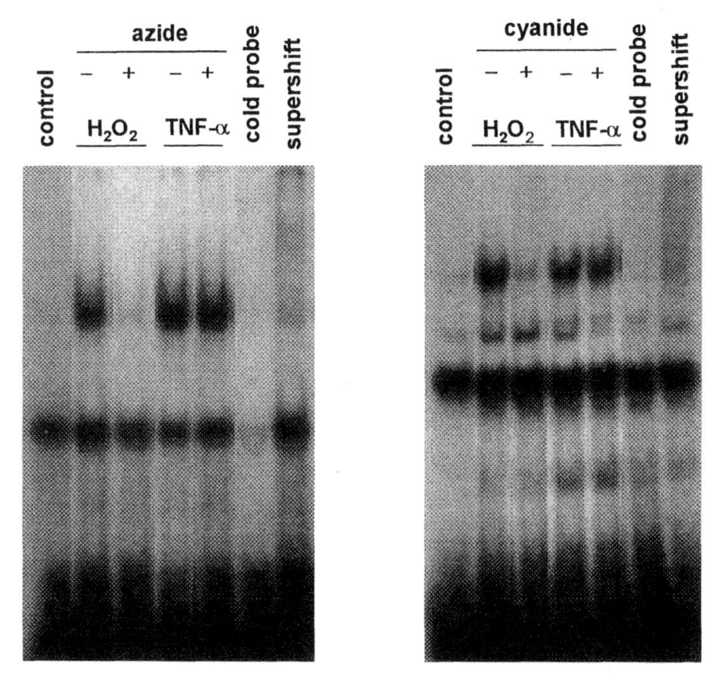

Figure 2. Electrophoretic mobility shift assay (EMSA) of cell extracts from Wurzburg cells treated with H_2O_2 (100 μM) or TNF-α (50 ng/ml), plus azide (100 μM) or cyanide (1 mM) as indicated. Extracts were prepared after 2 h incubation in RPMI plus fetal calf serum and the EMSA was performed using ^{32}P-labeled DNA probe containing the NFκB consensus binding element using standard methods (21). Lanes 7 and 8 represent the addition of cold probe and a supershift with an antibody against the p65 subunit of NFκB, respectively.

and H_2O_2. This result also provides a control for the heme inhibitors and shows that they did not act by compromising cell metabolism. We conclude, therefore, that the action of H_2O_2 cannot be simply to oxidize GSH or other molecules directly, but must involve a sequence with an azide- or cyanide-sensitive step. The difference in H_2O_2 consumption between cells treated with these inhibitors and those treated with aminotriazole was slight. Since aminotriazole inhibits catalase but did not prevent NFκB activation, it appears that the activation process consumes only a small amount of added H_2O_2.

Azide and cyanide are classic heme poisons. They would be expected to inhibit peroxidases and pseudoperoxidases including prostaglandin H synthase (16,17), cytochrome P450 (18) and possibly nitric-oxide synthase and heme oxygenase. They may also interact with other enzymes with a redox-active metal at their active site. Lipoxygenase is an iron protein that shows pseudoperoxidase activity (17). Further support for peroxidase-like activity involvement in NFκB activation could come from the inhibitor profile of the process. Many of the compounds classified as "antioxidants" that have been shown to inhibit NFκB activation are phenolic compounds that are good peroxidase inhibitors or substrates. Many are effective at concentrations that may be too low to be compatible with a radical-scavenging role and their effects might be better explained if they were acting on a more specific enzymatic process.

Studies using enzyme inhibitors may be useful for distinguishing between potential candidates for the activity we propose. However, compounds that inhibit lipoxygenase are

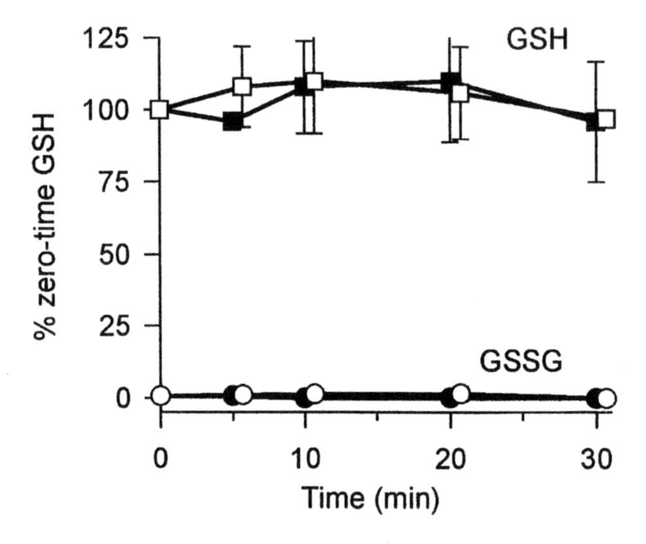

Figure 3. Reduced glutathione (GSH; ■, □) and glutathione disulfide (GSSG; ●, ○) content of Wurzburg cells incubated in the absence (filled symbols) and presence (open symbols) of 100 μM H_2O_2. Cells were incubated in RPMI and pelleted as in Figure 1. Pellets were processed immediately after separation from supernatants, and analyzed as dansyl derivatives by HPLC (22). Results are means of three experiments.

also likely to act as substrates/inhibitors of prostaglandin H synthase and other peroxidase-like enzymes (17,19) as well as being radical scavengers, and to lack sufficient specificity. Much more rigorous examination of inhibitor profiles is required before definitive conclusions can be drawn.

Our results in Figure 1 imply that glutathione peroxidase does not play a major role in H_2O_2 catabolism by Wurzburg cells. They also suggest that direct or glutathione peroxidase-catalyzed GSH oxidation by H_2O_2 is unlikely to be responsible for activating NFκB. In spite of prevailing hypotheses favoring such a mechanism, there have been few studies documenting changes in GSH and GSSG during activation of NFκB. When we made such measurements during the period of H_2O_2 consumption by Wurzburg cells (Figure 3), we observed no detectable loss of GSH or gain in GSSG. Therefore, unless changes small enough to be within our margin of error are sufficient, we must conclude that H_2O_2-mediated activation of NFκB cannot simply be explained by alteration of the cell's redox state brought about by loss of GSH or an increase in GSSG:GSH ratio.

Overall, our findings with Wurzburg cells suggest that activation of NFκB by H_2O_2 is due to a more specific oxidative reaction rather than a general alteration in thiol redox state, and that it is brought about by the peroxidative action of a heme protein or perhaps other metalloenzyme. This conclusion differs from commonly held views of redox regulation, and highlights the need to test more rigorously the favored hypotheses.

ACKNOWLEDGMENTS

This work was supported by grants from the Health Research Council and Lottery Health of New Zealand.

REFERENCES

1. Baeuerle PA, Henkel T. Function and activation of NFκB in the immune system. Annu Rev Immunol 1994; 12: 141–179.
2. Siebenlist U, Franzoso G, Brown K. Structure, regulation and function of NFκB. Annu Rev Cell Biol 1994; 10: 405–455.
3. Palombella VJ, Rando OJ, Goldberg AL, Maniatis T. The ubiquitin-proteasome pathway is required for processing the NFκB1 precursor protein and the activation of NFκB. Cell 1994; 78: 773–785.
4. Traenckner EBM, Wilk S, Baeuerle PA. A proteasome inhibitor prevents activation of NFκB and stabilizes a newly phosphorylated form of IκB-α that is still bound to NFκB. EMBO J 1994; 13: 5433–5441.
5. Matthews JR, Wakasugi N, Virelizier JL, Yodoi J, Hay RT. Thioredoxin regulates the DNA binding activity of NFκB by reduction of a disulphide bond involving cysteine 62. Nucleic Acids Res 1992; 20: 3821–3830.
6. Staal FJT, Roederer M, Herzenberg LA. Intracellular thiols regulate activation of nuclear factor κB and transcription of human immunodeficiency virus. Proc Natl Acad Sci USA 1991; 87: 9943–9947.
7. Simon G, Moog C, Obert G. Valproic acid reduces the intracellular level of glutathione and stimulates human immunodeficiency virus. Chem Biol Interact 1994; 91: 111–121.
8. Mihm S, Galter D, Dröge W. Modulation of transciption factor NFκB activity by intracellular glutathione levels and by variations of the extracellular cysteine supply. FASEB J 1995; 9: 246–252.
9. Israël N, Gougerot-Pocidalo MA, Aillet F, Virelizier JL. Redox status of cells influences constitutive or induced NFκB translocation and HIV long terminal repeat activity in human T and monocytic cell lines. J Immunol 1992; 149: 3386–3393.
10. Galter D, Mihm S, Dröge W. Distinct effects of glutathione disulphide on the nuclear transcription factors κB and the activator protein-1. Eur J Biochem 1994; 221: 639–648.
11. Brennan P, O'Neill LAJ. Effects of oxidants and antioxidants on nuclear factor κB activation in three different cell lines: evidence against a universal hypothesis involving oxygen radicals. Biochim Biophys Acta 1995; 1260: 167–175.
12. Suzuki YJ, Mizuno M, Packer L. Signal transduction for nuclear factor κB activation. Proposed location of antioxidant-inhibitable step. J Immunol 1994; 5008–5015.
13. Anderson MT, Staal FJT, Gitler C, Herzenberg LA. Separation of oxidant-initiated and redox-regulated steps in the NFκB signal transduction pathway. Proc Natl Acad Sci USA 1994; 91: 11527–11531.
14. Schmidt KN, Traencker EB-M, Meier B, Baeuerle PA. Induction of oxidative stress by okadaic acid is required for activation of transcription factor NFκB. J Biol Chem 1995; 270: 27136–27142.
15. Schreck R, Meier B, Männel DN, Dröge W, Baeuerle PA. Dithiocarbamates as potent inhibitors of nuclear factor κB activation in intact cells. J Exp Med 1992; 175: 1181–1194.
16. Eling TE, Thompson DC, Foureman GL, Curtis JF, Hughes MF. Prostaglandin H synthase and xenobiotic oxidation. Annu Rev Pharmacol Toxicol 1990; 30: 1–45.
17. Smith WL, DeWitt DL. Biochemistry of prostaglandin endoperoxide H synthase-1 and synthase-2 and their differential susceptibility to nonsteroidal anti-inflammatory drugs. Semin Nephrol 1995; 15: 179–194.
18. Hollenberg PF. Mechanisms of cytochrome P450 and peroxidase-catalysed xenobiotic metabolism. FASEB J 1992; 6: 686–694.
19. Ford-Hutchinson AW, Gresser M, Young RN. 5-Lipoxygenase. Annu Rev Biochem 1994; 63: 383–417.

20. Wolff SP. Ferrous ion oxidation in presence of ferric ion indicator xylenol orange for measurement of hydroperoxides. Methods Enzymol 1994; 233: 182–189.
21. Zabel U, Schreck R, Baeuerle PA. DNA binding of purified transcription factor NFκB. J Biol Chem 1991; 266: 252–260.
22. Martin J, White INH. Fluorimetric determination of oxidised and reduced glutathione in cells and tissues by high-performance liquid chromatography following derivatisation with dansyl chloride. J Chromatogr 1991; 568: 219–225.

10

Self-regulation of the NFκB/IκBα System, and Its Positive and Negative Control by the Cellular Redox Status

Jean-Louis Virelizier, Fernando Arenzana-Seisdedos, and Manuel Rodriguez
Institut Pasteur
Paris, France

Catherine Dargemont
Institut Curie
Paris, France

Ronald T. Hay
University of St. Andrews
St. Andrews, Scotland

Nicholas Hunt
University of Sydney
Sydney, Australia

The aim of this brief overview is not to discuss exhaustively a very large research area but rather to re-examine the intriguing influence of the cellular redox status on transcription of some genes, in light of recent findings indicating that regulation of NFκB function occurs in both cytoplasm and nucleus, and in both a positive and a negative manner.

THE NFκB/IκBα SYSTEM

Many cellular and viral promoters contain DNA sequences able to bind the transcription factor NFκB. Such binding is triggered by cellular activation and results, in conjunction with that of other transcription factors on their respective responsive DNA elements, in initiation or amplification of gene expression. The transcriptional role of NFκB is particularly critical for the expression of the HIV provirus integrated in host cell DNA, and also for that of many cytokine genes. Two of these cytokines, tumor necrosis factor (TNF) and interleukin 1 (IL-1), are depending on NFκB for their transcription, and are capable of activating NFκB function through their respective transmembrane receptors. In addition, two genes of the c-*rel* family are themselves NFκB-dependent. One is the *p105* gene, which codes for a precursor protein giving rise after proteolysis, to the p50 subunit of NFκB. The other is *IκBα*, the product of which is a main inhibitor of NFκB, in that it associates with p50/p65 NFκB heterodimers and retains them in the cytoplasm. The p65 subunit appears to be constitutively expressed, and does not depend on NFκB for transcription. The trimeric complex containing p50, p65, and IκBα is retained in the cytoplasm of unstimulated cells, since the nuclear localization sequences (NLS) of NFκB are not exposed (for review see Ref. 1).

The IκBα protein, associated with NFκB in these trimeric complexes in the cytoplasm, is the main target for the transduction signal activating NFκB in cells stimulated with TNF or IL-1. A complex cascade of transduction events eventually results in phosphorylation of IκBα on two critical serine residues (S32 and S36) by an as yet uncharacterized kinase (2). This phosphorylation event is necessary for TNF- or IL-1-induced IκBα ubiquitination on the four lysine residues K21, k22, K38, and k47 (3), a phenomenon which in turn is essential for the 26S proteasome to degrade IκBα (4). The degradation of IκBα within the trimer leaves the NFκB NLS exposed, thus permitting the transport of p50/p65 heterodimers into the nucleus, where transcription of NFκB-dependent genes is initiated.

TERMINATION OF NFκB FUNCTION BY NUCLEAR TRANSLOCATION OF IκBα PROTEINS

Once translocated into the nucleus, the fate of NFκB heterodimers is not well known, and has indeed not been much studied. The observations briefly reviewed above all bear on NFκB activation, and it remains to be ascertained whether NFκB function could also be subjected to active control mechanisms in the nucleus. Such control would terminate the rapid and intense cellular response to immune and inflammatory stimulation, which if prolonged might be deleterious for cells.

After cytoplasmic degradation of IκBα, the IκBα pool is replenished by rapid production of the protein, induced transcriptionally (5) through interaction of NFκB with DNA-binding sites in the IκBα promoter (6). This led us to hypothesize that large amounts of newly synthesized IκBα could exceed the capacity of cytoplasmic NFκB proteins to associate with the inhibitor, and would be free to translocate into the nucleus and associate there with DNA-bound NFκB heterodimers. This was indeed shown to occur in Hela cells stimulated with IL-1, or submitted to a short pulse (2 h) of TNF followed by washing (7). Western blot analysis showed that IκBα is degraded in the cytoplasm within minutes of cell stimulation, and can be detected in the nucleus as soon as 40 min after stimulation, at a time when cytoplasmic IκBα remains undetectable. Nuclear IκBα expression persists up to 200 min and this correlates with a complete disappearance of nuclear NFκB binding activity in the nucleus. At this stage, coimmunoprecipitation assays clearly showed that IκBα–NFκB complexes are detected in the nuclei of TNF-pulsed or IL-1-stimulated cells. Finally, it was shown that nuclear expression of newly synthesized IκBα correlates with inhibition of NFκB-dependent transcription, as shown by the suppression of HIV LTR activity in transient transfection assays (7).

These results clearly indicate that TNF and IL-1 affect NFκB function successively in a positive and a negative manner. Figure 1 schematically represents these successive events. Early after stimulation, NFκB is activated and rapidly initiates transcription of many genes, including that of IκBα. If cells are permanently exposed to TNF, as is done usually in vitro, newly synthesized IκBα is permanently degraded. Thus, NFκB remains bound to DNA, and

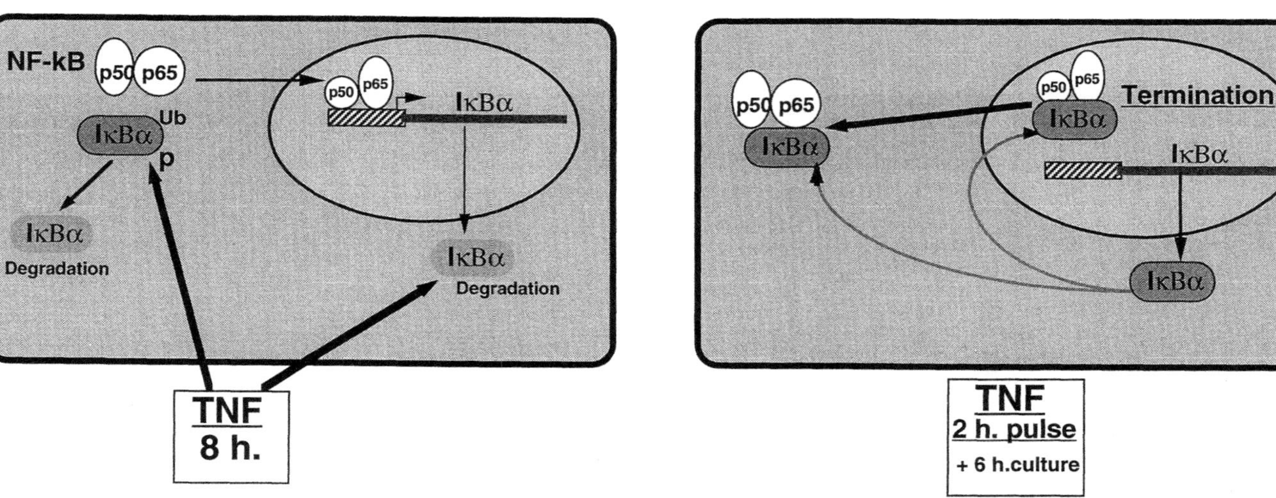

Figure 1. The NFκB/IκBα system: termination of NFκB function by nuclear IκBα.

can induce gene transcription for long periods of time. In contrast, a short TNF pulse, or stimulation with IL-1 (which soon downregulates its receptor and blocks its own activating effects), result in immediate IκBα degradation, followed by accumulation of newly synthesized IκBα in the nucleus, and termination of NFκB function. The precise mechanism by which IκBα releases p50/p65 NFκB from DNA is not known. In any case, the newly formed IκBα/p50/p65 trimer is indeed dissociated from DNA, since our recent results, obtained in both *Xenopus* oocytes and mammalian cells, indicate that IκBα transports NFκB back to the cytoplasm and identify the nuclear export sequence in IκBα responsible for that retrograde transport (8).

POSITIVE AND NEGATIVE CONTROL OF NFκB FUNCTION BY CELLULAR REDOX STATUS

The cellular redox status influences NFκB function in at least three ways, as schematically represented in Figure 2.

Enhancement of NFκB DNA Binding and Transcriptional Activity by Thioredoxin, and Its Decrease by Nitric Oxide

In contrast to the constitutive expression of glutathione, the expression of the thioredoxin (TRX) gene appears to be inducible, at least in lymphocytes.

We have observed (N. Wakasugi, unpublished results) that resting lymphocytes do not detectably express TRX transcripts or proteins, but that TRX expression is triggered by T cell activation. Others (J. Yodoi) have shown that the TRX protein can be found in the cell nucleus. In its reduced form, human TRX is a potent reducing protein, and the inducibility of the TRX gene makes it likely that it controls the redox status of nuclei in a manner that may not be redundant to that of glutathione. With this concept in mind, we tested the DNA-binding activity of recombinant p50 protein after addition of recombinant human TRX in gel retardation assays. TRX clearly enhanced the DNA-binding activity of wild-type, but not cysteine-62-mutated, p50. The reduction by TRX of a disulfide bond involving cysteine 62 may thus increase NFκB binding activity in the nucleus. Consistent with this concept, we observed that cotransfection of a TRX-expression vector increases TNF-induced HIV LTR activity in an NFκB-dependent manner (9).

Recent evidence indicates that nitric oxide (NO) inactivates NFκB DNA-binding activity, as shown in vitro with NO-donor compounds. Binding inhibition was observed with both homodimers and heterodimers of the p50 and p65 NFκB subunits, and was dependent on the redox-sensitive cysteine-62 residue (10). Cellular activation induces expression of the inducible NO synthase (reviewed in Ref. 10) and NO production. If, as is very likely, the highly diffusible NO molecule is present in the nucleus, it might exert a negative effect on transcriptional NFκB function. Thus both a positive (TRX-mediated) and negative (NO-mediated) control of NFκB activity might take place in the nuclear environment.

Cellular Reduction by Antioxidants Blocks IκBα Degradation and NFκB Activation

It is well known that incubation of cells with antioxidants suppresses NFκB activation. This has been interpreted as indicating that reactive oxygen intermediates (ROI) mediate

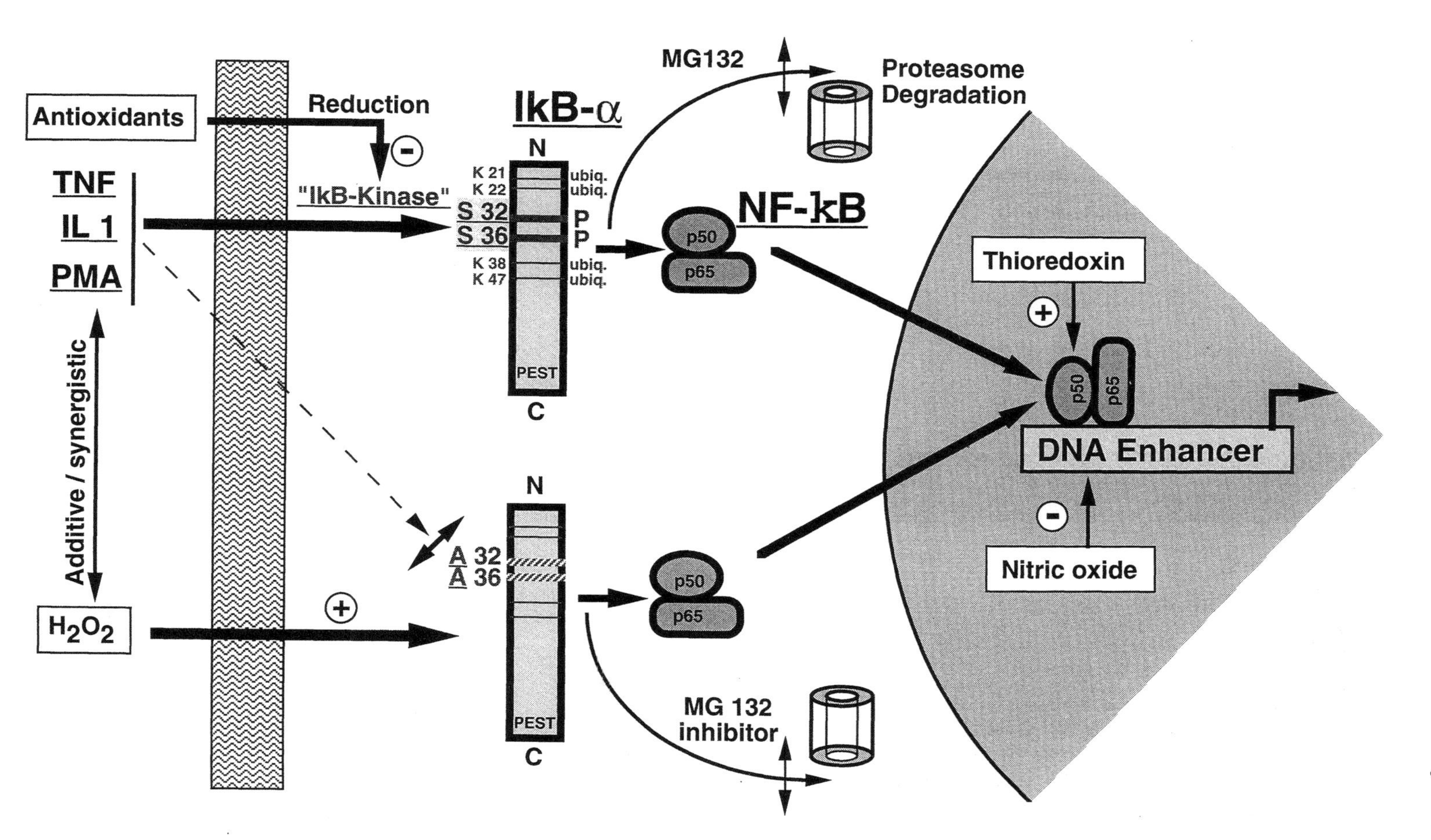

Figure 2. Redox control of the NFκB/IκBα system.

NFκB activation, thus behaving as second-messengers in the cascade of transduction events leading to IκBα degradation when inducers such as TNF or IL-1 are used (11). Our own results, however, are not compatible with that interpretation. We found that phorbol esters (PMA) and H_2O_2 synergize in NFκB activation, which should not happen if ROI were induced by PMA and were mediating PMA-induced NFκB activation. Furthermore, we found that incubation with antioxidants alone, but not TNF, increases the total thiol content of the lymphoblastoid T cell lines tested (12). This led us to suggest that there is no need to postulate a role of ROI in the mechanism of action of NFκB activators such as PMA or TNF. It rather appears that an imbalance toward oxidation facilitates, but does not directly mediate, NFκB activation. The simplest explanation is that the reduction of cells induced by antioxidants suppresses one or more steps leading to activation of the elusive "IkB kinase." Indeed, our more recent results indicate that the antioxidant 3-aminothiolphenol blocks TNF-induced degradation of endogenous IκBα, but not that of an expressed IκBα protein mutated in the two critical serine residues S32 and S36.

H_2O_2 Induces a Slow, Phosphorylation-Independent, Proteasome-Mediated IκBα Degradation

Further evidence against the concept that ROI directly mediate NFκB activation comes from our recent observations in HeLa or U937 monocytic cells transfected with IκBα mutants with specific mutations in residues 32 and 36 (S→A). As pointed out above, such mutants cannot be degraded by stimulation with TNF. In contrast, we found that mutated IκBα proteins, lacking the two critical phosphorylation sites, are still degraded upon cell treatment with H_2O_2. This clearly indicates that ROI and TNF do not induce IκBα degradation through identical mechanisms. Moreover, the kinetics of the events were quite different, since TNF induced IκBα degradation in 5–10 min, whereas H_2O_2-induced degradation was only observed after 1 h. If ROI were mediating TNF-induced IκBα proteolysis, they should act more rapidly, and in a phosphorylation-dependent manner. Both types of inducers, however, degraded IκBα in a proteasome-dependent manner, as shown by the use of MG132, a peptide-aldehyde inhibitor specific for the proteasome. Our conclusion that production of ROI does not mediate NFκB activation after physiological stimulation is consistent with results showing that overexpression of catalase, a specific scavenger of H_2O_2, does not block TNF-induced NFκB activation (13).

CONCLUSION

Much remains to be understood about the redox control of the NFκB/IκBα system. This system is regulated in both a positive and negative way, and each step of that regulation may be subjected to control by oxidation and reduction. The effects of cellular reduction appear to differ in the cellular compartment considered, since reduction suppresses NFκB translocation from the cytoplasm, but may enhance NFκB function in the nucleus. Since newly synthesized IκBα is expressed in the nucleus and terminates NFκB function, it will be important to know whether the nuclear redox status also controls this novel, negative function of IκBα. The intimate mechanisms underlying the intriguing effects of the redox status on NFκB activation and function are not yet ascertained, but the molecules involved are likely to be protein thiols such as thioredoxin and glutathione, rather than ROI. Similarly to what is now believed in the field of apoptosis and ROI (14), it may be suggested that ROI are capable of degrading IκBα and activating NFκB, but are not required to execute the NFκB-activating program triggered by physiological inducers. This area of research is likely to provide us with more surprises, hopefully providing better clues for the design of redox-based future therapeutic interventions.

ACKNOWLEDGMENTS

The support of a Concerted Action of the EU (project ROCIO II) is gratefully acknowledged.

REFERENCES

1. Liou HC, Baltimore D. Regulation of the NF-κB/rel transcription factor and IκB inhibitor system. Curr Opin Cell Biol 1993; 5: 477–487.
2. Chen ZJ, Parent L, Maniatis T. Site-specific phosphorylation of the IκBα by a novel ubiquitination-dependent protein kinase activity. Cell 1996; 84: 853–862.
3. Rodriguez MS, Wright J, Thompson J, Thomas D, Baleux F, Virelizier J-L, Hay RT, Arenzana-Seisdedos F. Identification of lysine residues required for signal-induced ubiquitination and degradation of IκBα in vivo oncogene 1996; 12, 2425–2435.
4. Palombella V, Rando OJ, Goldberg AL, Maniatis T. The ubiquitin-proteasome pathway is required for processing the NF-κB1 precursor protein and the activation of NF-κB. Cell 1994; 78: 773–785.
5. Sun SC, Ganchi PA, Ballard DW, Greene WC. NF-κB controls expression of inhibitor IκBα: evidence for an inducible autoregulation. Science 1993; 259: 1912–1915.
6. Le Bail O, Schmidt-Ulrich R, Israël A. Promoter analysis of the gene encoding the IκBα/MAD-3 inhibitor of NF-κB: positive regulation by members of the rel/NF-κB family. EMBO J 1993; 11: 5043–5049.
7. Arenzana-Seisdedos F, Thompson J, Rodriguez MS, Bachelerie F, Thomas D, Hay RT. Inducible nuclear expression of newly synthesized IκBα negatively regulates DNA-binding and transcriptional activities of NF-κB. Mol Cell Biol 1995; 15: 2689–2696.
8. Arenzana-Seisdedos F, et al. J Cell Sci 1997; 110: 369–378.
9. Matthews JR, Wakasugi N, Virelizier J-L, Yodoi J, Hay RT. Thioredoxin regulates the DNA-binding activity of NF-κB by reduction of a disulphide bond involving cysteine 62. Nucleic Acids Res 1992; 20: 3821–3830.
10. Matthews JR, Botting CH, Panico M, Morris HR, Hay RT. Inhibition of NF-κB DNA-binding activity by nitric oxide. Nucleic Acids Res 1996; 24: 2236–2243.
11. Schreck R, Rieber P, Bauerle PA. Reactive oxygen species intermediates as apparently widely used messengers in the activation of the NF-κB transcription factor and HIV. EMBO J 1991; 10: 2247–2252.
12. Israël N, Gougerot-Pocidalo M-A, Aillet F, Virelizier J-L. The redox status of cells influences constitutive or induced NF-κB translocation and HIV-LTR activity in human T and monocytic cell lines. J Immunol 1992; 149: 3386–3393.
13. Suzuki YJ, Mizumo M, Packer L. Transient overexpression of catalase does not inhibit TNF- or PMA-induced NF-κB activation. Biochem Biophys Res Commun 1995; 210: 537–541.
14. Jacobson MD. Reactive oxygen species and programmed cell death. TIBS 1996; 21: 83–86.

11

The Role of Hydrogen Peroxide in Signal Transduction and the Involvement of Iron

B. S. van Asbeck, R. C. Sprong, T. van der Bruggen, J. F. L. M. van Oirschot, and J. C. C. Borleffs
University Hospital Utrecht
Utrecht, The Netherlands

INTRODUCTION

Ever since the presence of molecular oxygen in the earth's early atmosphere, living cells have had to adapt to the potential hazards of oxygen, mediated by its metabolites such as superoxide, which may be formed when oxygen is reduced to water. Thus, oxygen toxicity necessitated aerobic as well as anaerobic life forms to develop systems whose function was to rapidly inactivate reactive oxygen species. This might be the reason that the rate of evolution of superoxide dismutase, which catalyzes the reduction of superoxide to hydrogen peroxide (H_2O_2) is among the fastest observed for any protein (1). There may, however, also be another evolutionary reason for effective H_2O_2 formation. That is the notion that H_2O_2 seems to function in signal transduction. For example, the observation that H_2O_2 can induce viral replication (2) supports a role for H_2O_2 in basic biochemical processes that control protein synthesis. An important mechanism by which H_2O_2 acts as a messenger molecule in mammalian cells is through the activation of the transcription factor nuclear factor kappa B (NFκB) (2). Via this pathway H_2O_2 can induce early gene expression of cytokines (3). In addition, H_2O_2 may play a prominent role in programmed cell death (4). Since bacterial endotoxin (lipopolysaccharide; LPS) as well as cytokines, such as tumor necrosis factor-α (TNF-α), may stimulate leukocytes and endothelial cells to produce reactive oxygen species, including H_2O_2, it is obvious that there is an increasing interest in understanding the role of H_2O_2 in the pathogenesis of inflammatory diseases. In this overview, the role of H_2O_2 in NFκB-mediated TNF-α synthesis and replication of human immunodeficiency virus (HIV) will be discussed.

MESSENGER FUNCTIONS OF H_2O_2

Because H_2O_2 is relatively nontoxic and readily diffuses through cell membranes, this molecule is pre-eminently suitable to act as a diffusible signal for the induction of biochemical processes. Hydrogen peroxide mediates the expression of genes associated with immune responses (3) and cell adaptation to adverse conditions (4). This has been observed not only in animals but also in plants, in which H_2O_2 induced the activation of defense genes encoding cellular protectants during the development of plant disease resistance (5,6). A similar mechanism has been reported in bacteria, where low concentrations of H_2O_2 function as an oxidative stress signal for the induction of the synthesis of antioxidant proteins (7,8). It has also been shown that H_2O_2 is required for platelet-derived growth factor signal transduction in vascular smooth-muscle cells, including tyrosine phosphorylation, protein kinase stimulation, DNA synthesis, and chemotaxis (9). Furthermore, it has been shown that H_2O_2 stimulates the synthesis of platelet-activating factor (10,11) and the expression of the endothelial cell adhesion molecules granule

Table 1. Second-messenger-like Functions of Oxidants in Inflammation

Inflammatory process	Oxidant	Stimulation/ inhibition	Model
Lymphocyte proliferation	H_2O_2	−	Human mixed lymphocytes (68)
	H_2O_2	−	Human T-cells (69)
	$^{\bullet}OH$	+	Human mixed lymphocytes (68)
Arachidonic acid metabolism			
lipoxygenase			
5HETE	H_2O_2	+	Perfused rat lung (70)
	H_2O_2	−	Alveolar macrophages (71)
cycloxygenase			
TxA_2, prostacyclin	Oxidants	+	Purified cyclooxygenase (72–74)
	H_2O_2	+	Perfused lungs (70,75,76)
	H_2O_2	+	Rat alveolar macrophages (77)
	H_2O_2, oxidants	+	Human platelets (78)
	H_2O_2	+	Endothelial cells (79,80)
	H_2O_2	−	Purified cycloxygenase (73,81)
	H_2O_2	−	Endothelial cells (82)
Platelet activating factor synthesis	*t*-BuOOH, H_2O_2	+	Endothelial cells (10,11)
Adhesion molecule gene expression			
VCAM-1	Oxidants	+	Endothelial cells (83)
GMP-140	H_2O_2	+	Endothelial cells (12)
ICAM-1	H_2O_2	+	Endothelial cells (13,14)
Chemoattractant gene expression			
JE/MCP-1	Oxidants	+	Murine mesangial cells (84)
α_1-Antiprotease activity	HOCl, $^{\bullet}OH$	−	Purified human α_1-proteinase-inhibitor (85,86)
Cytokine production			
CSF-1	Oxidants	+	Murine mesangial cells (84)
IL-8	H_2O_2, $^{\bullet}OH$	+	Human whole blood (16)
TNF-α	Oxidants	+	Serum of mice and dogs (18,17)
	Oxidants	+	Mono Mac 6 cells (19)
	Oxidants	+	Murine macrophages (20)

membrane protein-140 (12) and intracellular adhesion molecule 1 (13,14). The effects of oxidants on inflammatory mediators are summarized in Table 1. Table 2 summarizes the effects of H_2O_2 on cellular messenger systems. The effect of oxidants on second-messengers of endothelial cells has been reviewed extensively by Natarajan (15).

Table 2. Intracellular Actions of Oxidants

Cellular regulation mechanism	Oxidant	Stimulation/ inhibition	Model
Guanylate cyclase	H_2O_2	+	Rat lung homogenate (87)
	H_2O_2	+	Perfused rabbit lung (88)
	H_2O_2	+	Isolated bovine arterial rings (89)
Proteine kinase C	H_2O_2	+	Boviner endothelial cells (41)
	H_2O_2	+/–	C6 glioma cells,
	H_2O_2	+/–	B16 melanoma cells (15)
	H_2O_2	+	Perfused guinea-pig lungs (90)
Tyrosine kinase	H_2O_2	+	Rat hepatoma cells (42)
	H_2O_2	+	B cell line (43,91)
	H_2O_2-derived species	+	Insulin receptors of adipocytes (92)
Tyrosine phosphatase	H_2O_2	–	Rat hepatoma cells (93)
	H_2O_2	+	Human skin fibroblasts (94)
Phospholipases			
PLD	H_2O_2, oxidants	+	Bovine endothelial cells (95)
		+	Human neutrophils (96)
PLC	H_2O_2	+	Rat alveolar type II cells (97)
PLA_2	*t*-BuOOH	+	Bovine endothelial cells (98)
Cytosolic Ca^{2+}	H_2O_2	+	Human endothelial cells (99)
	H_2O_2	+	Canine endothelial cells (100)
	H_2O_2	+	Renal tubular cells (101)
	H_2O_2	+	Human B cells (43)
	H_2O_2	+	Rat alveolar type II cells (97)
	H_2O_2	+	Sheep sarcoplasmic reticulum (102)
Inositol phosphate	*t*-BuOOH, H_2O_2	–	Hepatocytes (103)
	t-BuOOH, H_2O_2	–	Endothelial cells (82)
	H_2O_2	–	Rat alveolar type II cells (104)
	$O_2^{-\bullet}$	+	Human B cells (91)
	$O_2^{-\bullet}$	+	Sarcoplasmic reticulum (105)
cAMP	*t*-BuOOH	–	Rat alveolar type II cells (104)
Na/K pump activity	H_2O_2	+	Bovine endothelial cells (106)
ATP	H_2O_2	–	Rat endothelial cells (107)
	H_2O_2	–	Human platelets (108)
Transcription factors			
NFκB	H_2O_2	+	Jurkat T cells (2)
	H_2O_2	+	Hela cells (22)
Platelet-derived growth factor signal transduction	H_2O_2	+	Vascular smooth-muscle cells (9)

OXIDANT EFFECTS ON TNF-α PRODUCTION

When LPS is added to human whole blood, peak levels of TNF-α can be measured after 4 h of incubation (Figure 1). The involvement of oxidants in cytokine production was obtained from a study in which H_2O_2 dose-dependently stimulated interleukin-8 (IL-8) release in human whole blood (16). In addition, the oxidant scavengers dimethyl sulfoxide and dimethylthiourea (DMTU) inhibited the LPS-stimulated release of IL-8 (16). Production of TNF-α is probably also regulated by redox-dependent mechanisms, since treatment with the oxidant scavenger *N*-acetylcysteine (NAC) reduced TNF-α activity in endotoxemic mice (17) and dogs (18). In addition, the oxidant scavengers pyrrolidine dithiocarbamate and dimethylsulfoxide reduced TNF-α production in LPS-activated Mono Mac 6 cells, a human monocytic cell line (19), and murine peritoneal macrophages (20). *N*-acetylcysteine (Figure 2) and the hydroxyl radical (•OH) scavenger DMTU (Figure 3) also inhibited the release of TNF-α by LPS-stimulated human monocytes. The mechanism by which antioxidants inhibit the expression of TNF-α is not yet clear, but there is increasing evidence that the redox status of the cell plays an important role in signal transduction events (21) and the activation of transcription factors (2,22). Since the iron chelator deferoxamine reduces plasma TNF-α levels after LPS stimulation of human whole blood (Figure 4), it is suggested that not H_2O_2 itself but the •OH, which can be generated from H_2O_2 by iron catalysis is the final responsible oxidant. The promoter sequences of the human TNF-α gene contain four binding sites for the NFκB, with the strongest binding at the –605 motif (19). Expression of TNF-α is probably not due solely to NFκB activation, since TNF-α production has also been reported to be regulated by the transcription factors c/EBP (23) and SP-1 (19). These factors, however, are not affected by the oxidant pyrrolidine dithiocarbamate (19).

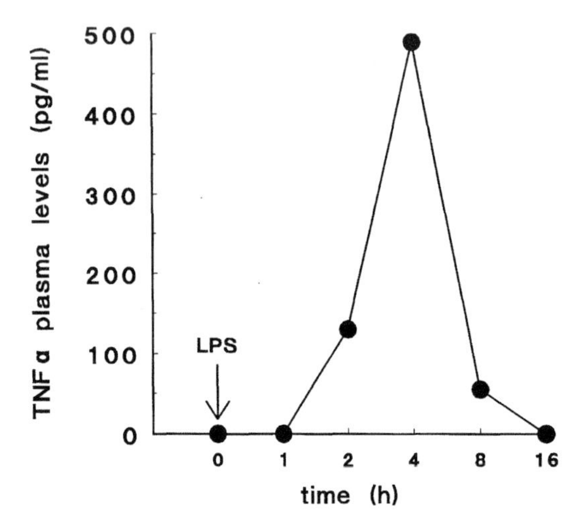

Figure 1. LPS enhances TNF-α levels in human whole blood. EDTA whole blood was incubated (37°C, 5% CO_2 in air) on a 48-well plate (1 ml/well) in the absence (control) or presence of LPS (10 ng/ml). Plasma samples were collected at the indicated times and used for the determination of TNF-α with an ELISA.

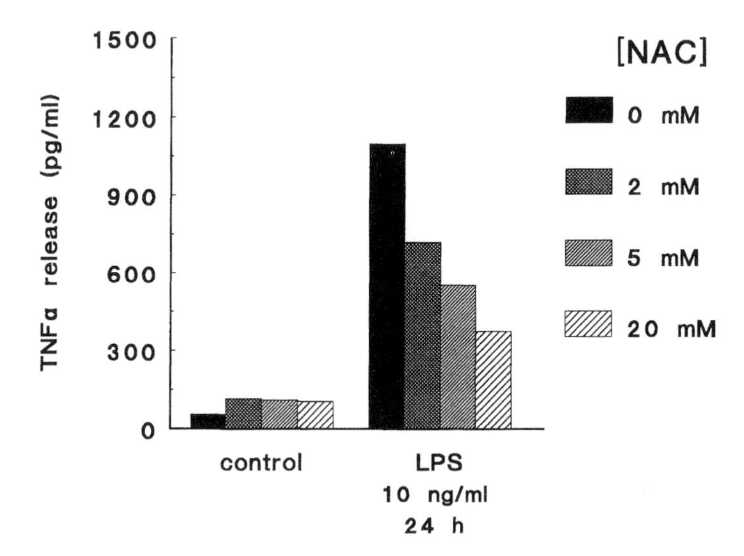

Figure 2. *N*-Acetylcysteine (NAC) inhibits LPS-induced TNF-α release by human monocytes. Elutriated monocytes (10^6 cells/ml) were incubated (37°C, 5% CO_2 in air) on a 96-well polypropylene plate (100 μl/well) with the indicated concentrations of NAC in Iscove's Modified Dubelco's Medium (IMDM) medium supplemented with 10% fetal calf serum. Supernatants were collected 20 h after addition of LPS (10 ng/ml) and used for determination of TNF-α levels (pg/ml) with an ELISA.

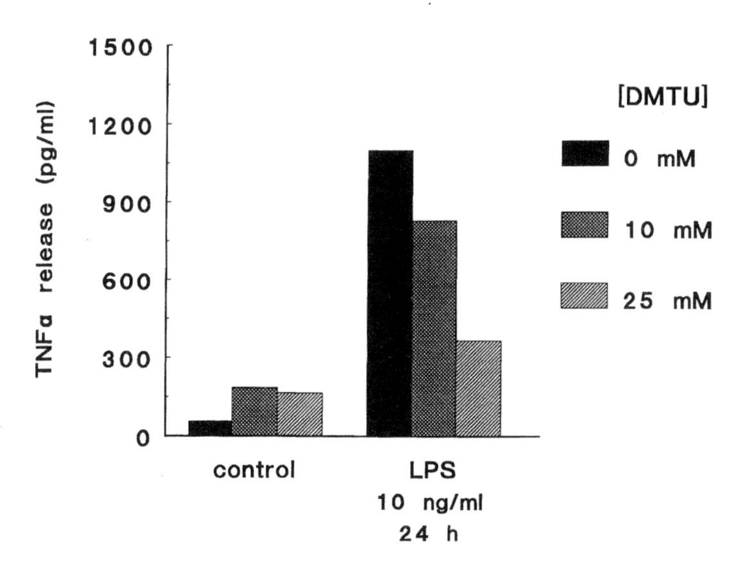

Figure 3. Dimethylthiourea (DMTU) inhibits LPS-induced TNF-α release by human monocytes. Monocytes were isolated by elutriation from healthy volunteers. Cells (10^6/ml) were incubated in a 96-well polypropylene plate (100 μl/well) with indicated concentrations of DMTU in culture medium (IMDM/10% fetal calf serum) at 37°C and 5% CO_2 in air. Supernatants were collected 20 h after addition of LPS (10 ng/ml). TNF-α was quantified in the supernatants with an ELISA.

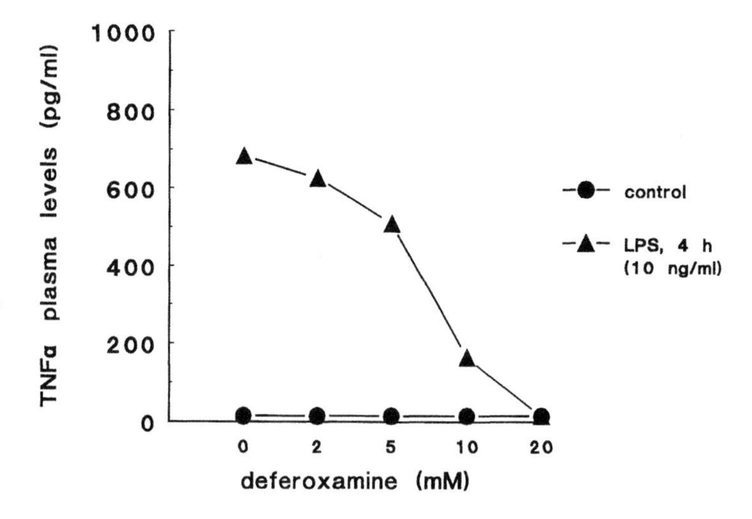

Figure 4. Deferoxamine inhibits LPS-induced TNF-α release in human whole blood. EDTA whole blood was incubated (37°C, 5% CO_2 in air) with indicated concentrations of deferoxamine (Desferal™) on a 48-well plate (0.5 ml/well). Plasma samples were collected 4 h after addition of LPS (10 ng/ml) and used for quantification of TNF-α with an ELISA.

OXIDANT-INDUCED NFκB ACTIVITY

The transcription factor NFκB consists of two polypeptides with apparent molecular masses of 50 and 65 kDa, are referred to as p50 and p65 (see Figure 7). Both p50 and p65 bind specifically to the NFκB consensus locus of genes, but only the p65 subunit can initiate transcription (24). The two subunits form dimers in vivo. Both heterodimers of p50–p65 and homodimers of p50–p50 and p65–p65 are formed in vitro when isolated subunits are mixed (25). Based on their migration differences, they can be distinguished in electrophoretic mobility shift assays (EMSAs). The mobility of the p50–p65-DNA complex was intermediate to that of the faster-migrating complex consisting of the NFκB p50 homodimer bound to DNA and of the p65 homodimer–DNA complex, which migrates more slowly. Because of the trans-activation capacity of p65, binding of the p50–p65 dimer to DNA may result in gene activation, while the p50–p50 dimer, which lacks the transactivation capacity, will not initiate transcription. It has been proposed that the NFκB p50 homodimer may act as a regulatory protein, inhibiting NFκB-mediated gene expression (26).

In unstimulated cells, much of the NFκB is present in the cytoplasm, complexed with an inhibitory subunit, IκB, that binds preferentially to the p65 rather than to the p50 subunit (24). After activation with an appropriate stimulus such as LPS, TNF-α, IL-1, or phorbol 12-myristate-13-acetate (PMA), the IκB unit is released from the NFκB complex (25,27,28). The active NFκB is subsequently transported to the nucleus where it can bind to DNA (28). The mechanism by which NFκB is activated is not completely clear, but evidence for the involvement of protein kinases has been obtained from studies which showed that the protein kinase C activator PMA induces NFκB activation (29). In addition, NFκB activation was inhibited by herbimycin A (30), providing evidence for tyrosine kinases in NFκB activation.

A number of studies performed with HIV-infected cells provided evidence that suggests that NFκB activity is controlled by a redox-dependent mechanism. Stimulation of Jurkat T cells with PMA and TNF-α activates NFκB and decreases intracellular thiols (2,31). *N*-Acetylcysteine prevented both the increase in NFκB and the decrease in thiols, suggesting oxidant involvement (18,27). Furthermore, the chemically different anti-oxidants NAC and pyrrolidine dithiocarbamate inhibited NFκB activation in T cell lines (2,32), B-cell lines (32), and HeLa cells (22) which had been activated by several unrelated agonists such as LPS, PMA, IL-1, and TNF-α. Other antioxidants have also been shown to inhibit NFκB activation, including ebselen, 2-mercaptoethanol, glutathione, vitamin C (33), L-cysteine (34), butylated hydroxyanisole (35,36), α-lipoic acid (37), and vitamin E derivates (38). More direct evidence for oxidant involvement in NFκB activation was obtained from studies in which H_2O_2 induced NFκB activity in Jurkat T cells (2) and HeLa cells (23). NAC and reduced glutathione (GSH) can inhibit the activation of NFκB by H_2O_2 (2) as well as its activation by TNF-α and PMA (31), suggesting that H_2O_2 can serve as a messenger that directly and indirectly activates the NFκB transcription factor. As iron chelators that withhold iron from catalyzing $^{\bullet}OH$ generation also inhibited the activation of NFκB by H_2O_2 (32), it is suggested that H_2O_2 acts via $^{\bullet}OH$ generation. It has indeed been shown that the iron chelator deferoxamine inhibits NFκB activation by H_2O_2 (32,39)

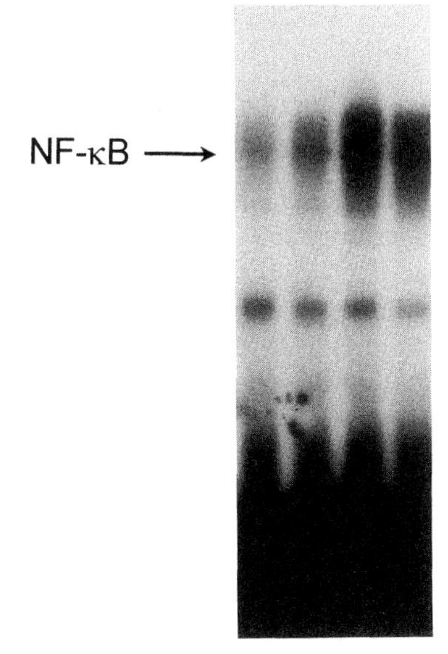

Figure 5. NFκB DNA-binding activity is increased in Mono Mac 6 cells after administration of LPS or H_2O_2. Mono Mac 6 cells (2.5×10^6 cells/ml) were incubated (37°C, 5% CO_2 in air) on a 24-well plate with buffer only, or with LPS (10 ng/ml), H_2O_2 (40 μM), or both for 1 h. Nuclear extracts were isolated (5×10^6 cells/sample) and used (3 μg protein/lane) for electrophonetic mobility shift assay (EMSA) with a ^{32}P-labeled oligonucleotide encoding an NFκB binding site from the HIV promoter (5′-tcgacagagGGGACTTTCCgagaggc-3′).

and PMA (39) and protects against the H_2O_2- and TNF-α-induced activation of HIV-1 (39). Furthermore it was found that the rate of progression was significantly smaller in thalassemia major patients – who have iron overload – receiving a higher dose of deferoxamine (40). These observations are consistent with the above-mentioned inhibition of TNF-α production by deferoxamine (see Figure 4). Evidence for a redox-dependent NFκB-mediated expression of the TNF gene has been obtained from a study in which the oxidant scavengers pyrrolidine dithiocarbamate and dimethylsulfoxide suppressed NFκB mobilization, TNF-α transcripts, and TNF-α protein in LPS-activated Mono Mac 6 cells (19,20). Consistently with this, similarly as the effect of LPS (10 ng/ml) incubation of Mono Mac 6 cells with H_2O_2 (40 μM) results in activation of NFκB (Figure 5).

It is not known whether oxidants directly activate NFκB by oxidative modification of the inhibitory subunit IκB. Oxidation as a mechanism for regulation has also been studied using *Salmonella typhimurium* and *Escherichia coli* (7). It was demonstrated that the prokaryotic factor OxyR, a transcriptional activator of genes induced by H_2O_2, is activated directly by oxidation and is therefore both the sensor and the transducer of an oxidative stress signal (7). However, H_2O_2 was not able to activate isolated NFκB complexed with IκB in vitro in a cell-free system (2), suggesting involvement of other intermediaries, such as second-messenger systems. Phosphorylating enzymes are likely candidates, since oxidants have been shown to activate protein kinase C (41) and tyrosine kinases (42,43). Prolonged action of tyrosine-phosphorylated proteins due to oxidant-mediated inactivation of tyrosine phosphatases has also been described. Recently, it was shown that inhibitors of the pp1 and pp2a phosphatases, which are important in the regulation of serine and threonine protein phosphorylation, activate NFκB in an antioxidant-independent way, suggesting that oxidants are involved in the upstream signal-transducing pathway (44).

OXIDANT-INDUCED HIV REPLICATION

Antioxidants (2,32,45–47) and iron chelators (3,39,40) inhibit HIV replication in vitro. This suggests that reactive oxygen species play an important role in the expression of HIV and in the infection process. It has been shown that H_2O_2 can cause the production of newly generated HIV-1 virions in latently infected T cells (2). TNF-α, the synthesis of which can be induced by H_2O_2, also stimulates the proliferation of HIV by inducing transcription from the HIV long terminal repeat (LTR), which contains the viral enhancer and promoter sequences (48–50). The mechanism underlying these effects involves the cytoplasmic multisubunit transcription factor NFκB that can rapidly stimulate HIV enhancer in activated T cells (51) and monocytic cells (50) as well as the expression of genes encoding cytokines such as IL-1, TNF-α (52), IL-6 (53), and IL-8 (16). Activation of NFκB and synthesis of TNF-α is inhibited by IL-10 (54). Activation of HIV-1 LTR (45) and reverse transcriptase (46) by phorbol myristate acetate, TNF-α, and IL-6 can be inhibited by NAC, a scavenger of H_2O_2 and prodrug of glutathione synthesis (55).

With regard to the role of TNF-α and H_2O_2 in HIV disease, it is remarkable that TNF-α has a positive influence on its own synthesis (56,57) and that increased serum levels of TNF-α (58) and decreased plasma concentrations of GSH (59,60) are found in asyptomatic HIV seropositives and patients with AIDS (58,59,61). A decreased glutathione concentration was also found in peripheral blood mononuclear cells and monocytes of HIV-seropositive patients with lymphadenopathy and AIDS (59).

Furthermore, the HIV-induced GSH deficiency is present in lung epithelial lining fluid (60) and in both CD4 and CD8 T-cells (62,63). Thus antioxidants should be able to inhibit HIV replication. This has indeed been shown in vitro using NAC (45,46,64). Virus production by HIV-infected lymphocytes was almost completely inhibited by 300–500 μM NAC (45). HIV production by the chronically infected promonocytic cell line U-1, which serves as model for clinical latency, was completely inhibited by 1 mM NAC (64). Furthermore, NAC also attenuated the TNF-α-mediated decrease of GSH concentrations in HIV-infected Molt-4 T lymphocytes and an increase in the production of the viral core protein p24 (45). These results suggest that TNF-α-mediated generation of H_2O_2, which may lead to GSH consumption and stimulation of HIV replication, is inhibited by NAC by either a direct action of NAC on TNF-α synthesis or as a result of H_2O_2 inactivation, thereby blocking its role in signal transduction. On the other hand, we recently showed that NAC at a concentration of 2.5 mM enhanced the self-sustained HIV-1 replication in monocyte-derived macrophages (65). The effect of NAC was inhibited by the $^{\bullet}OH$ scavenger dimethylthiourea, which agrees with the observation that NAC can reduce Fe^{3+} to Fe^{2+}, thereby catalysing $^{\bullet}OH$ generation from H_2O_2, following the Fenton reaction: $H_2O_2 + Fe^{2+} \rightarrow {}^{\bullet}OH + OH^- + Fe^{3+}$. This observation, which suggests a role for $^{\bullet}OH$ in the pathogenesis of HIV disease, is in accordance with the inhibitory effect of deferoxamine on HIV replication (39). The HIV-stimulating effect of thiols has also been observed by other investigators. It was recently shown that homocysteine stimulates the PMA-mediated HIV-LTR transactivation in Jurkat T cells (66). Furthermore, it has been shown that HIV and its target cell engage in a thiol–disulfide interchange mediated by protein disulfide-isomerase and it has been suggested that the reduction of critical disulfides in viral envelope glycoproteins may be the initial event that triggers conformational changes required for HIV entry and cell infection (67).

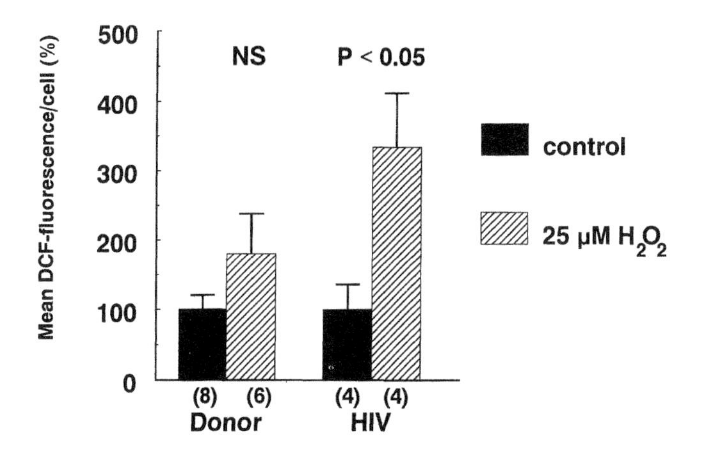

Figure 6. Increased peroxide retention in peripheral blood lymphocytes (PBL) from HIV seropositive patients. Isolated PBL (5×10^6 cells/ml) were incubated (37°C, 5% CO_2 in air) for 18 h with 5 μM 2′, 7′-dichlorofluorescein diacetate (DCFH-DA), followed by an incubation period of 30 min with 25 μM H_2O_2. Thereafter cells were fixed with 2% paraformaldehyde. Fluorescence of the oxidized DCFH-DA (DCF) was then measured by flow cytometry. For each analysis 10^4 events were accumulated. Results are expressed as percentage increase of control PBL (set to 100%) of mean fluorecsence intensity per cell. Values are mean±SEM and the number of subjects is presented in parentheses. Statistical analysis was performed using Student's *t*-test.

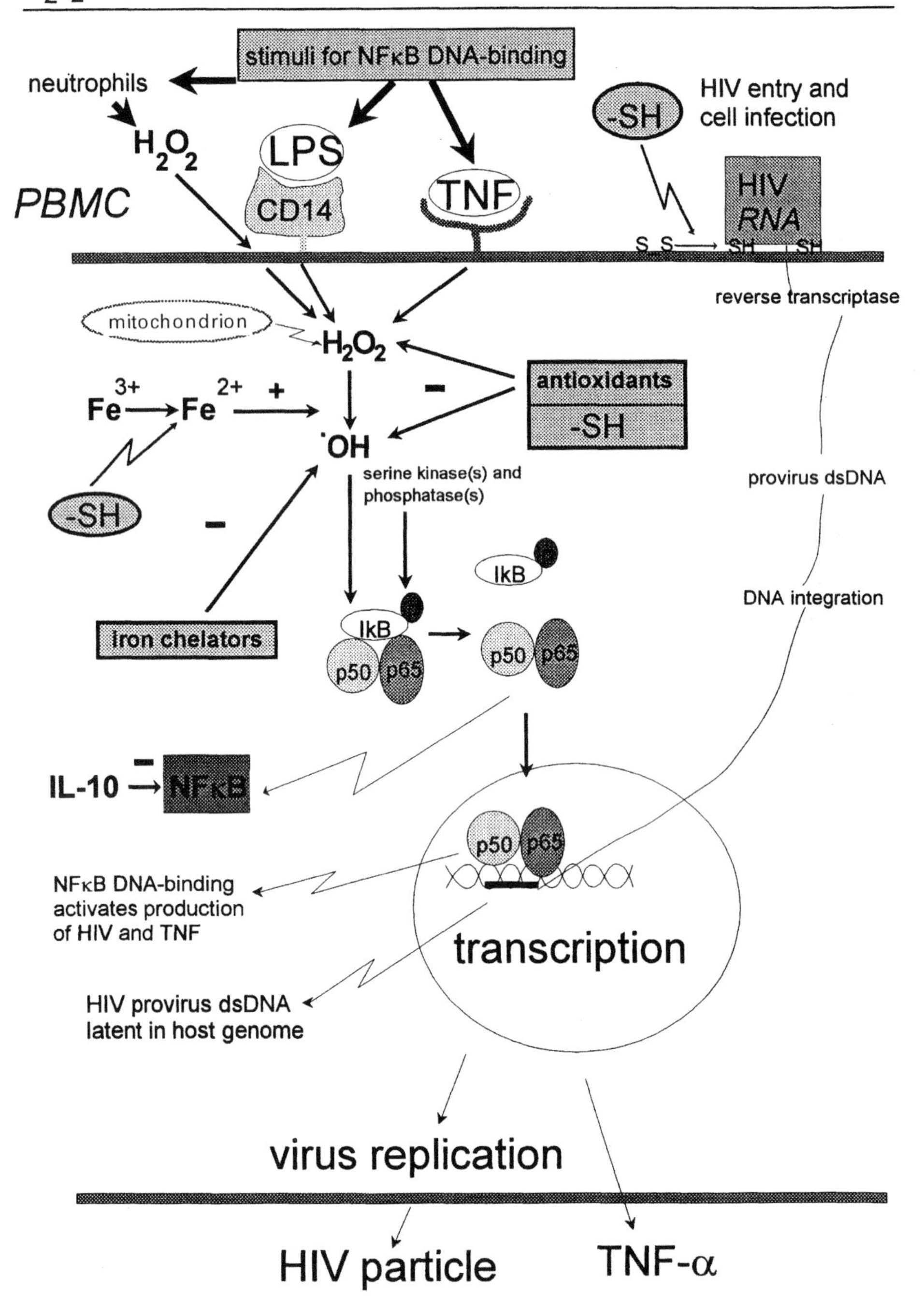

Figure 7. H_2O_2-mediated HIV and TNF production: role of Fe and thiols.

Another aspect, which also suggest that the redox state of the cellular milieu plays an important role in the ability of the virus to multiply is our preliminary observation that the capacity of peripheral blood lymphocytes (PBL) from HIV-seropositive patients to inactivate H_2O_2 is significantly less than the antioxidant activity of PBL obtained from healthy donors (Figure 6). This is in agreement with the fact that catalase concentrations in patients' PBL were lower ($p < 0.05$) than in donor PBL (0.05 ± 0.028 pg/cell, $n = 8$, vs. 0.12 ± 0.08 pg/cell, $n = 18$, mean ± SD). Thus it seems that the intracellular milieu of lymphocytes from patients infected with HIV favors the generation of $^{\bullet}OH$. This may lead to NFκB activation and consequently to HIV-1 replication and TNF-α production, which in turn may activate NFκB. Therefore, it makes sense to develop therapies for the treatment of AIDS that increase the cellular antioxidant capacity and withhold iron from being catalytically active for the generation of $^{\bullet}OH$ (see Figure 7).

REFERENCES

1. Lee YM, Friedman DJ, Ayala FJ. Superoxide dismutase: an evolutionary puzzle. Proc Natl Acad Sci 1985; 82: 824–828.
2. Schreck R, Rieber P, Bauerle PA. Reactive oxygen intermediates as apparently widely used messengers in the activation of the NFκB transcription factor and HIV-1. EMBO J 1991; 10: 2247–2258.
3. Los M, Dröge W, Stricker K, Baeuerle PA, Schulze-Oshoff K. Hydrogen peroxide as a potent activator of T lymphocyte functions. Eur J Immunol 1995; 25: 159–165.
4. Hockenbery DM, Oltvai ZN, Yin XM, Milliman CL, Korsmeyer SJ. Bcl-2 Functions in an antioxidant pathway to prevent apoptosis. Cell 1993; 75: 241–251.
5. Chen Z, Silva H, Klessig D. Active oxygen species in the induction of plant systemic acquired resistance by salicylic acid. Science 1993; 262: 1883–1886.
6. Livine A, Tenhaken R, Dixon R, Lamb C. H_2O_2 from the oxidative burst orchestrates the plant hypersensitive disease resistance response. Cell 1994; 79: 583–593.
7. Storz G, Tartaglia LA, Ames BN. Transcriptional regulator of oxidative stress-inducible genes: direct activation by oxidation. Science 1990; 248: 189–194.
8. Zheng H, Hassett DJ, Bean K, Cohen MS. Regulation of catalase in *Neisseria gonorrhoeae* Effects of oxidant stress and exposure to human neutrophils. J Clin Invest 1992; 90: 1000–1009.
9. Sundaresan M, Yu Z-X, Ferrans VJ, Irani K, Finkel T. Requirement for generation of H_2O_2 for platelet-derived growth factor signal transduction. Science 1995; 270: 296–299.
10. Lewis MS, Whatley RE, Cain P, McIntyre TM, Prescott SM, Zimmerman GA. Hydrogen peroxide stimulates the synthesis of platelet-activating factor by endothelium and induces endothelial cell-dependent neutrophil adhesion. J Clin Invest 1988; 82: 2045–2055.
11. Patel KD, Zimmerman GA, Prescott SM, McIntyre TM. Novel leukocyte antagonists are released by endothelial cells exposed to peroxide. J Biol Chem 1992; 266: 15168–15175.
12. Patel KD, Zimmerman GA, Prescott SM, McEver RP, McIntyre TM. Oxygen radicals induce human endothelial cells to express GMP-140 and bind neutrophils. J Cell Biol 1991; 112: 749–759.
13. Lo SK, Janakidevi K, Lai L, Malik AB. Hydrogen peroxide-induced increase in endothelial adhesiveness is dependent on ICAM-1 activation. Am J Physiol 1993; 264: L406–L412.
14. Bradley JR, Johnson DR, Pober JS. Endothelial activation by hydrogen peroxide. Selective increases of intercellular adhesion molecule-1 and major histocompatibility complex class 1. Am J Pathol 1993; 142: 1598–1609.
15. Natarajan V. Oxidants and signal transduction in vascular endothelium. J Lab Clin Med 1995; 125: 26–37.

16. DeForge LE, Fantone JC, Kenney JS, Remick DG. Oxygen radical scavengers selectively inhibit interleukin 8 production in human whole blood. J Clin Invest 1992; 90: 2123–2129.
17. Peristeris P, Clark BD, Gatti S, Raffaella R, Mantovani A, Mengozzi M, Orencole SF, Sironi M, Ghezzi P. *N*-Acetylcysteine and glutathione as inhibitors of tumor necrosis factor production. Cell Immunol 1992; 140: 390–399.
18. Zhang H, Spapen H, Ngyen DN, Benlabed M, Buurman WA, Vincent JL. Protective effects of *N*-acetyl-L-cysteine in endotoxemia. Am J Physiol 1994; 266: H1746–H1754.
19. Ziegler-Heitbrock HWM, Stensdorf T, Liese J, Belohradsky B, Weber C, Wedel A, Schreck R, Bauerle P, Strobel M. Pyrrolidine dithiocarbamate inhibits NF-κB mobilization and TNF production in human monocytes. J Immunol 1993; 151: 6986–6993.
20. Kelly-KA, Hill MR, Yukohana K, Wanker F, Gimble JF. Dimethyl sulfoxide modulates NF-κB and cytokine activation in lipopolysaccharide-treated murine macrophages. Infect Immun 1994; 62: 3122–3128.
21. Staal FJT, Anderson MT, Staal GEJ, Herzenberg LA, Gitlers CG, Herzenberg LA. Redox regulation of signal transduction: tyrosine phosphorylation and calcium influx. Proc Natl Acad Sci USA 1994; 91: 3619–3622.
22. Meyer M, Pahl HL, Bauerle PA. Regulation of the transcription factors NF-κB and AP-1 by redox changes. Chem Biol Interact 1994; 91: 91–100.
23. Pope RM, Leutz A, Ness SA. C/EBPß regulation of the tumor necrosis factor α gene. J Clin Invest 1994; 94: 1449–1455.
24. Urban MB, Schreck R, Bauerle PA. NF-κB contacts DNA by a heterodimer of the p50 and the p65 subunit. EMBO J 1991; 10: 1817–1825.
25. Urban MB, Bauerle PA. The role of p50 and p65 subunits of NF-κB in the recognition of cognate sequences. New Biologist 1991; 3: 279–288.
26. Kang SM, Tran AC, Grilli M, Leonardo MJ. NF-κB subunit regulation in non-transformed CD4⁺ T lymphocytes. Science 1992; 256: 1452–1456.
27. Urban MB, Bauerle PA. The 65-kD subunit of NF-κB is a receptor for IκB and a modulator of DNA-binding specificity. Genes Dev 1990; 4: 1975–1984.
28. Bauerle PA, Baltimore D. Activation of DNA-binding activity in an apparently cytoplasmic precursor of the NF-κB transcription factor. Cell 1988; 53: 211–217.
29. Sen R, Baltimore B. Inducibility of κ immunoglobulin enhancer-binding protein NF-κ by a posttranslational mechanism. Cell 1986; 47: 921–928.
30. Iwasaki T, Uehara Y, Graves L, Rachie N, Bomsztyk K. Herbimycin A blocks IL-1-induced NF-κB DNA-binding activity in lymphoid cell lines. FEBS Lett 1992; 298: 240–244.
31. Staal FJT, Roederer M, Herzenberg LA, Herzenberg LA. Intracellular thiols regulate activation of nuclear factor κB and transcription of human immunodeficiency virus. Proc Natl Acad Sci USA 1990; 87: 9943–9947.
32. Schreck R, Meier B, Männel DN, Dröge W, Bauerle PA. Dithiocarbamates as potent inhibitors of nuclear factor κB activation in intact cells. J Exp Med 1992; 175: 1181–1194.
33. Staal FJT, Roederer M, Raju PA, Anderson MT, Ela SW, Herzenberg LA, Herzenberg LA. Antioxidants inhibit stimulation of HIV transcription. AIDS Res Hum Retroviruses 1993; 9: 299–306.
34. Mihm S, Ennen J, Pessara U, Kurth R, Dröge W. Inhibition of HIV-1 replication and NF-κB activity by cysteine and cysteine derivates AIDS 1991; 5: 497–503.
35. Israel N, Gougerot-Pocidalo MA, Aillet F, Virelizier JL. Redox status of cells influences constitutive or induced nuclear factor-κB translocation and HIV long terminal repeat activity in human T and monocytic cell lines. J Immunol 1992; 149: 3386–3393.
36. Schulze-Osthoff K, Beyaert R, Vandevoorde V, Haegeman G, Fiers W. Depletion of the mitochondrial electron transport abrogates the cytotoxic and gene-inductive effects of TNF. EMBO J 1993; 12: 3095–3104.

37. Suzuki YJ, Aggarwal BB, Packer L. α-Lipoic acid is a potent inhibitor of NF-κB activation in human T-cells. Biochem Biophys Res Commun 1992; 189: 1709–1715.
38. Suzuki YJ, Packer L. Inhibition of NF-κB activation by vitamin E derivatives. Biochem Biophys Res Commun 1993; 193: 277–283.
39. Sappey C, Boelaert JR, Legrand-Poels S, Forceille C, Favier A, Piette J. Iron chelation decreases NF-κB and HIV type 1 activation due to oxidative stress. Aids Res Human Retroviruses 1995; 11: 1049–1061.
40. Costagliola DG, De Montalembert M, Lefrère J, Briand C, Rebulla P, Baruchel S, Dessi C, Fondu P, Karagiorga M, Perrimond H, Girot R. Dose of desferrioxamine and evolution of HIV-1 infection in thalassaemic patients. Br J Haem 1994; 87: 849–952.
41. Siflinger-Birnboim A, Goligorsky MS, Del Vecchio PJ, Malik AB. Activation of protein kinase C pathway contributes to hydrogen peroxide-induced increase in endothelial permeability. Lab Invest 1992; 67: 24–32.
42. Koshio O, Akanuma Y, Kasuga M. Hydrogen peroxide stimulates tyrosine phosphorylation of the insulin receptor and its tyrosine kinase activity in intact cells. Biochem J 1988; 250: 95–101.
43. Schieven GL, Kirihara JM, Myers DE, Ledbetter JA, Uckun FM. Reactive oxygen intermediates activate NF-κB in a tyrosine kinase-dependent mechanism and in combination with vanadate activate the $p56^{lck}$ and $p59^{fyn}$ tyrosine kinases in human lymphocytes. Blood 1993; 82: 1212–1220.
44. Suzuki YJ, Mizuni M, Packer L. Signal transduction for nuclear factor κB activation. Proposed location of antioxidant-inhibitable step. J Immunol 1994; 153: 5008–5015.
45. Roederer M, Staal FJT, Raju PA, Ela SW, Herzenberg LA, Herzenberg LA. Cytokine-stimulated HIV replication is inhibited by *N*-acetylcysteine. Proc Natl Acad Sci USA 1990; 87: 4884–4888.
46. Kalebic T, Kinter A, Poli G, Anderson ME, Meister A, Fauci AS. Suppression of HIV expression in chronically infected monocytic cells by glutathione, glutathione ester, and *N*-acetylcysteine. Proc Natl Acad Sci USA 1991; 88: 986–990.
47. Lioy J, When-Zhe Ho, Cutilli JR, Polin RA, Douglas SD. Thiol suppression of human immunodeficiency virus type 1 replication in primary cord blood monocyte-derived macrophages in vitro. J Clin Invest 1993; 91: 495–498.
48. Ito M, Baba M, Sato A, Hirabayashi K, Tanabe F, Shigeta S, de Clercq E. Tumor necrosis factor enhances replication of human immunodeficiency virus (HIV) in vitro. Biochem Biophys Res Commun 1989; 158: 307–312.
49. Duh EJ, Maury WJ, Folks TM, Fauci AS, Rabson AB. Tumor necrosis factor ß activates human deficiency virus type I through induction of nuclear factor binding to the HF-B sites in the long terminal repeat. Proc Natl Acad Sci USA 1989; 86: 5974–5978.
50. Griffin GE, Leung K, Folks TM, Kunkel S, Nabel GJ. Activation of HIV gene expression during monocyte differentiation by induction of NF-κB. Nature 1989; 339: 70–73.
51. Nabel G, Baltimore D. An inducible transcription factor activates expression of human immunodeficiency virus in T cells. Nature 1987; 326: 711–713.
52. Anisowicz A, Messineo M, Lee LW, Sager R. An NF-κB like transcription factor mediates IL-1/TNF-alpha induction of *gro* in human fibroblasts. J Immunol 1991; 147: 520–527.
53. Zhang Y, Lin JX, Vilcek J. Interleukin-6 induction by tumor necrosis factor and interleukin-1 in human fibroblasts involves activation of nuclear factor binding to a κB like sequence. Mol Cell Biol 1990; 10: 3818–3823.
54. Wang P, Wu P, Siegel MI, Egan RW, Billah MM. Interleukin IL-10 inhibits nuclear factor kB (NF-κB) activation in human monocytes. J Biol Chem 1995; 270: 9558–9563.
55. Meister A, Anderson ME. Glutathione. Annu Rev Biochem 1983; 52: 711–760.
56. Philip R, Epstein LB. Tumour necrosis factor as immunomodulator and mediator of monocyte cytotoxicity induced by itself, γ-interferon and interleukin-1. Nature 1986; 323: 86–89.

57. Cordingley FT, Hoffbrand AV, Heslop HE, Turner M, Bianchi A, Reittie JE, Vyakarnam A, Meager A. Tumour necrosis factor as an autocrine tumour growth factor for chronic B-cell malignancies. Lancet 1988; i: 969–971.
58. Wright SC, Jewett A, Mitsuyasu R, Bonavida B. Spontaneous cytotoxicity and tumor necrosis factor production by peripheral blood monocytes from AIDS patients. J Immunol 1988; 141: 99–104.
59. Eck H-P, Gmunder H, Hartmann M, Petzoldt D, Daniel V, Dröge W. Low concentrations of acid-soluble thiol(cysteine) in the blood plasma of HIV-1-infected patients. Biol Chem Hoppe-Seyler 1989; 370: 101–108.
60. Buhl R, Holroyd KJ, Mastrangeli A, Cantin AM, Jaffe HA, Wells FB, Saltini C, Crystal RG. Systematic glutathione deficiency in asymptomatic HIV-seropositive individuals. Lancet 1989; ii: 1294–1298.
61. Lahdevirta J. Maury CPG, Teppo AM, Repo H. Elevated levels of circulating cachetin tumor necrosis factor in patients with acquired immune deficiency syndrome. Am J Med 1988; 85: 289–291.
62. Staal FJT, Ela SW, Roederer M, Anderson MT, Herzenberg LA, Herzenberg LA. Glutathione deficiency and human immunodeficiency virus infection. Lancet 1992; 339: 909–912.
63. Staal FJT, Roederer M, Israelski DM, Bubp J, Mole LA, McShane D, Peresinski S, Ross W, Sussman H, Raju PA, Anderson MT, Moore W, Ela S, Herzenberg LA, Herzenberg LA. Intracellular glutathione levels in T cell subsets decrease in HIV infected individuals. AIDS Res Hum Retroviruses 1992; 8: 305–311.
64. Roederer M, Raju PA, Staal FJT, Herzenberg LA, Herzenberg LA. *N*-Acetylcysteine inhibits latent HIV expression in chronically infected cells. AIDS Res Hum Retroviruses 1991; 7: 563–567.
65. Nottet HSLM, Asbeck BS van, Graaf L de, Vos NM de, Visser MR, Verhoef, J. Role for oxygen radicals in the self-sustained HIV-1 replication in monocyte-derived macrophages: enhanced HIV-1 replication by *N*-acetyl-L-cysteine. J Leukocyte Biol 1994; 56: 702–707.
66. Simon S, Moog C, Obert G. Effects of glutathione precursors on human immunodeficiency virus replication. Chem Biol Interact 1994; 91: 217–224.
67. Ryser HJP, Levy EM, Mandel R, DiSciullo GJ. Inhibition of human immuno-deficiency virus infection by agents that interfere with thiol–disulfide interchange upon virus–receptor interaction. Med Sci 1994; 91: 4559–4563.
68. Rush DN, McKenna RM, Walker SM, Bakkestad-Legare P, Jeffrey JR. Catalase increases lymphocyte proliferation in mixed lymphocyte culture. Transplant Proc 1988; 20: 1271–1273.
69. Patterson DA, Rapoport R, Patterson MAK, Freed BM, Lempert N. Hydrogen peroxide-mediated inhibition of T-cell response to mitogens is a result of direct action on T cells. Arch Surg 1988; 123: 300–304.
70. Burghuber OC, Strife RJ, Zirrolli J, Henson PM, Henson JE, Mathias MM, Reeves JT, Murphy RC, Voelkel NF. Leukotriene inhibitors attenuate rat lung injury induced by hydrogen peroxide. Am Rev Respir Dis 1985; 131: 778–785.
71. Sporn PHS, Peters-Golden M. Hydrogen peroxide inhibits alveolar macrophage 5-lipoxygenase metabolism in association with depletion of ATP. J Biol Chem 1988; 29: 14776–14783.
72. Polgar P, Taylor L. Stimulation of prostaglandin synthesis by ascorbic acid via hydrogen peroxide formation. Prostaglandins 1980; 19: 693–700.
73. Hemmler ME, Cook HW, Lands WE. Prostaglandin synthesis can be triggered by lipid peroxides. Arch Biochem Biophys 1979; 193: 340–345.
74. Hemmler ME, Lands WE. Evidence for a peroxide-mediated free radical mechanisms of prostaglandin biosynthesis. J Biol Chem 1980; 255: 6253–6261.
75. McDonald RJ, Berger EM, Repine JE. Neutrophil-derived oxygen metabolites stimulate thromboxane release, pulmonary artery pressure increases and weight gains in isolated perfused rat lungs. Am Rev Respir Dis 1987; 135: 957–959.

76. Tate RM, Morris HG, Schroeder WR, Repine JE. Oxygen metabolites stimulate thromboxane production and vasoconstriction in isolated saline-perfused rabbit lungs. J Clin Invest 1984; 74: 608–613.
77. Sporn PHS, Peters-Golden M, Simon RH. Hydrogen peroxide-induced arachidonic acid metabolism in rat alveolar macrophage. Am Rev Respir Dis 1988; 137: 49–56.
78. Practico D, Iuliano L, Pulcinelli FM, Bonavita MS, Gazzaniga P, Violi F. Hydrogen peroxide triggers activation of human platelets selectively exposed to nonaggregating concentrations of arachidonic acid and collagen. J Lab Clin Med 1992; 119: 364–370.
79. Miller DK, Sadowski S, Soderman DD, Kuehl FA. Endothelial cell prostacyclin production induced by activated neutrophils. J Biol Chem 1985; 260: 1006–1014.
80. Harlan JM, Callahan KS. Role of hydrogen peroxide in the neutrophil-mediated release of prostacyclin from cultured endothelial cells. J Clin Invest 1984; 74: 442–448.
81. Egan RW, Gale PH, Kuehl FA. Reduction of hydroperoxides in the prostaglandin biosynthetic pathway by a microsomal peroxidase. J Biol Chem 1979; 254: 3295–3302.
82. Vercellotti GM, Severson SP, Duane P, Moldow CF. Hydrogen peroxide alters signal transduction in human endothelial cells. J Lab Clin Med 1991; 117: 15–24.
83. Marui N, Offerman MK, Swerlick R, Kunsch C, Rosen CA, Ahmad M, Alexander RW, Medford RM. Vascular cell adhesion molecule-1 (VCAM-1) gene transcription and expression are regulated through an antioxidant-sensitive mechanism in human vascular endothelial cells. J Clin Invest 1993; 92: 1866–1874.
84. Satriano JA, Shuldiner M, Hora K, Xing Y, Shan Z, Schlondorff D. Oxygen radicals as second messengers for expression of the monocyte chemoattractant protein, JE/MCP-1, and the monocyte colony-stimulating factor CSF-1, in response to tumor necrosis factor-α and immunoglobulin G. J Clin Invest 1993; 92: 1564–1571.
85. Ossanna PJ, Test ST, Matheson NR, Regioni S, Weiss SJ. Oxidative regulation of neutrophil elastase–alpha-1-proteinase inhibitor interactions. J Clin Invest 1986; 77: 1939–1951.
86. Stroncek DF, Vercellotti GM, Woh Huh P, Jacob HS. Neutrophil oxidants inactivate alpha-1-protease inhibitor and promote PMN-mediated detachment of cultured endothelium. Protection by methioninc. Arteriosclerosis 1986; 6: 332–340.
87. White AA, Crawford KM, Patt CS, Lad PJ. Activation of soluble guanylate cyclase from rat lung by incubation or by hydrogen peroxide. J Biol Chem 1976; 251: 7304–7312.
88. Burke-Wolin T, Abate CJ, Wolin MS, Gurtner GH. Hydrogen peroxide-induced pulmonary vasodilation: role of guanosine 3′,5′-cyclic monophospate. Am J Physiol 1991; 261: L393–L398.
89. Burke TM, Wolin MS. Hydrogen peroxide elicits pulmonary arterial relaxation and guanylate cyclase activation. Am J Physiol 1987; 252: H721–H732.
90. Johnson A, Phillips P, Hocking D, Tsan MF, Ferro T. Protein kinase inhibitor prevents pulmonary edema in response to H_2O_2. Am J Physiol 1989; 256: H1012–H1022.
91. Schieven GL, Kirihara JM, Burg DL, Geahlen RL, Ledbetter JA. $p72^{syk}$ Tyrosine kinase is activated by oxidizing conditions that induce lymphocyte tyrosine phosphorylation and Ca^{2+} signals. J Biol Chem 1993; 268: 16688–16692.
92. Hayes GR, Lockwood DH. Role of insulin receptor phosphorylation in the insulinomimetic effects of hydrogen peroxide. Proc Natl Acad Sci USA 1987; 84: 8115–8119.
93. Hecht D, Zick Y. Selective inhibition of protein tyrosine phosphatase activities by H_2O_2 and vanadate in vitro. Biochem Biophys Res Commun 1992; 188: 773–779.
94. Keyse SM, Emslie EA. Oxidative stress and heat shock induce a human gene encoding a protein-tyrosine phosphatase. Nature 1992; 359: 644–647.
95. Natarajan V, Taher MM, Roehm B, Parinandi NL, Schmid HHO, Kiss Z, Garcia JGN. Activation of endothelial cell phospholipase D by hydrogen peroxide and fatty acid hydroperoxide. J Biol Chem 1993; 268: 930–937.
96. Uings IJ, Thompson NT, Randall RW. Tyrosine phosphorylation is involved in receptor

coupling to phospholipase D but not phospholipase C in the human neutrophil. Biochem J 1992; 281: 597–600.

97. Rice KL, Duane PG, Archer SL, Gilboe DP, Niewoehner DE. H_2O_2 injury causes Ca^{2+}-dependent and -independent hydrolysis of phosphatidylcholine in alveolar epithelial cells. Am J Physiol 1992; 263: L430–L438.
98. Chakraborti S, Michael JR, Gurtner GH, Ghosh SS, Dutta G, Merker A. Role of a membrane-associated serine esterase in the oxidant activation of phospholipase A_2 by t-butyl hydroperoxide. Biochem J 1993; 292: 585–589.
99. David M, Horvath GY, Schimke I, Nagy I, Mueller MM. Comparative drug influence on peroxide-mediated increase of cytosolic calcium in human endothelial cells. Clin Chim Acta 1993; 223: 1–7.
100. Doan TN, Gentry DL. Taylor AA, Elliot SJ. Hydrogen peroxide activates agonist-sensitive Ca^{2+}-flux pathways in canine endothelial cells. Biochem J 1994; 297: 209–215.
101. Ueda N, Shah SV, Role of intracellular calcium in hydrogen peroxide-induced renal tubular cell injury. Am J Physiol 1992; 263: F214–F221.
102. Boraso A, Williams AJ. Modification of the gating of the cardiac sarcoplasmic reticulum Ca^{2+}-release channel by H_2O_2 and dithiothreitol. Am J Physiol 1994; 267: H1010–H1016.
103. Bellomo G, Thor H, Orrenius S. Alterations in inositol phosphate production during oxidative stress in isolated hepatocytes. J Biol Chem 1987; 262: 1530–1534.
104. Brown LAS. Glutathione protects signal transduction in type II cells under oxidant stress. Am J Physiol 1994; 266: L172–L177.
105. Suzuki YJ, Ford GD. Superoxide stimulates IP_3-induced Ca^{2+} release from vascular smooth muscle sarcoplasmic reticulum. Am J Physiol 1992; 262: H114–H116.
106. Meharg JV, McGowan-Jordan J, Charles A, Parmelee JT, Cutaia MV, Rounds S. Hydrogen peroxide stimulates sodium–potassium pump activity in cultured pulmonary arterial endothelial cells. Am J Physiol 1993; 265: L613–L1621.
107. Varani J, Phan SH, Gibbs DF, Ryan US, Ward PA. H_2O_2-mediated cytotoxicity of rat pulmonary endothelial cells. Lab Invest 1990; 63: 683–689.
108. Holmsen H, Robkin L. Hydrogen peroxide lowers ATP levels in platelets without altering adenylate energy charge and platelet function. J Biol Chem 1977; 252: 1752–1757.

12

Redox Regulation of Heat Shock Protein Expression and Protective Effects Against Oxidative Stress

Abdelhamid El Yaagoubi, Ewa Mariéthoz, Muriel R. Jacquier-Sarlin, and Barbara S. Polla
Université Paris V
Paris, France

INTRODUCTION

Reactive oxygen species (ROS) are generated from molecular oxygen and include the free radicals superoxide ($^{\bullet}O_2-$), hydroxyl ($^{\bullet}OH$), and nitric oxide ($NO^{\bullet}$), as well as nonradical intermediates such as hydrogen peroxide (H_2O_2) and singlet oxygen ($^1O_2$). During normal cellular respiration, ROS are constantly produced at low rate, in particular by mitochondria. At these low concentrations, ROS can act as second-messengers and as mediators for cell activation (1). However, during infection or inflammation, or upon exposure to environmental stresses such as ionizing or nonionizing radiation, metals, or various toxic compounds, ROS can accumulate to deleterious levels and damage almost all cellular components (1,2). Endogenous mechanisms to detoxify the oxidants and to repair the damage caused by ROS include, among others, the heat shock/stress proteins (HSP).

The HSP are a set of extremely conserved proteins induced in prokaryotes and eukaryotes by elevated temperatures and a variety of physical and chemical injuries, including oxidative stress (3). Human cells in particular respond to cellular injury by inducing the binding of pre-existing transcriptional activators, the heat shock factor(s) (HSF) and subsequently synthesizing the HSP (reviewed in Ref. 4). HSP are generally classified into families according to their apparent molecular masses, their respective inducers, and their functions (Table 1).

HSP expression is not limited to cells undergoing acute stress, and several members of HSP families are constitutively expressed (HS cognates). HSP function, at least in part, to maintain cellular homeostasis by acting as molecular chaperones: they participate in the folding and assembly of nascent and unfolded polypeptides, and facilitate protein transport and degradation. HSP thus represent an essential and adaptive response to cellular stress in general and to oxidative stress in particular.

There is a dual reason to include HSP in the redox responses to oxidative stress: first, they are induced by oxidants, indicating that the cells have developed mechanisms to sense the ROS and to transduce this signal into increased HSP expression; and second, they contribute to the cells' antioxidant defenses. In this chapter, we review recent work on the redox regulation of HSP expression and on the protective effects of HSP. With respect to protection, we will essentially concentrate on the inducible hsp70, although other members of the HSP/stress protein families probably contribute, together with hsp70, to protection against oxidative stress, at least in vitro (5,6) (Table 1).

Table 1. Main Stress Proteins Involved in Antioxidant Protection

HSP	Inducers	Protective functions
hsp110	HS, EP, cadmium	Thermotolerance
hsp90	HS, EP, cadmium, ROS, PMA	Chaperoning and regulation of steroid receptor activity Protection of transduction signal pathways (kinases)
hsp70	HS, EP, cadmium, ROS, PMA, UV radiation, arsenite, ischemia/reperfusion	Thermotolerance Protection against ROS Mitochondrial protection Protection from apoptosis
hsp 60	HS, EP, cadmium, ROS, PMA	Chaperoning of mitochondrial proteins
hsp47	HS, abnormal collagen	Collagen chaperone
TRX	HS, ROS	PDI and molecular chaperone Regenerates proteins inactivated by ROS Keeps cytoplasmic proteins reduced Antioxidant scavenger
SOD	HS, ROS, TNF-α	Dismutation of O_2^- to H_2O_2 Antioxidant
HO-1	HS, EP, ROS,	Antioxidant Ferritin inducer
Ferritin	Iron	Iron chelator Antioxidant (inhibitor of Fenton reaction)
Metallothioneins	Metals, cytokines	Antioxidant Protection against metal toxicity
bcl-2	HS, ROS, radiation	Protection from apoptosis Antioxidant
Ubiquitine		Degradation of irreversibly damaged proteins

HSP and other stress proteins (TRX; thioredoxin; SOD, superoxide dismutase; HO-1, heme oxygenase-1; metallothioneins; *bcl-2*; and ubiquitin) are all involved in protection against oxidative injury. Some of them contribute to correct folding of unfolded or misfolded polypeptides in specific subcellular compartments (hsp70 in cytoplasm; hsp60 in mitochondria) while others are "private chaperones" (hsp90 for steroid receptor or hsp47 for collagen) or play important roles in redox reactions (TRX, SOD, ferritin, metallothioneins, *bcl-2*).

Main inducers and major protective functions are listed.

HS, heat shock; EP, erythrophagocytosis; ROS, reactive oxygen species; PMA, phorbol myristate acetate.

REDOX REGULATION OF HSP EXPRESSION: PROPOSED ROLE OF THIOREDOXIN

Transcriptional induction of HS genes requires the activation of the HS factor(s) (HSF), which bind to the consensus HS element (HSE) located in the promoter region of the HS genes (5′-n-GAA-nn-TTC-n-3′). Although, to date, three HSF (HSF-1, HSF-2, HSF-4) have been characterized in mammalian cells and a fourth (HSF-3) has been identified in avian cells (7–9), the data reviewed here apply essentially to HSF-1.

In unstressed cells, HSF exists in a cytoplasmic, non-DNA-binding form; activation leads to oligomerization and nuclear translocation of HSF (10). Heat shock causes two major changes in HSF. First, HSF multimerizes to form a trimer and acquires the ability to bind HSEs (11). Second, it undergoes posttranslational modifications and phosphorylation at multiple positions (12). Activated HSF may interact directly with the polymerase to stimulate elongation (possibly by removing or altering nucleosomes or by antagonizing histone H1, which can act as repressor of transcription (13)), and increase the rates of initiation and elongation.

Some experiments indicate that the activity of HSF in vivo cannot be a simple function of the absolute environmental temperature but is influenced by cellular factors whose ability to act on HSF(s) from different species is not necessarily conserved. Intracellular accumulation of abnormal or degraded proteins has been proposed as a common signal for the induction of stress protein synthesis (14), while HSP (most likely hsp70) appear to act as homeostatic suppressors of HSF activity (15).

We propose that ROS represent ubiquitous second messengers for HSP induction. The effects of oxidants on the HS response have been studied by a number of groups, including ours, often with conflicting results, suggesting that there is a multistep and complex regulation of this response according to the cell type, the particular oxidant used, the subcellular location of its production, and the time allowed for recovery from stress. ROS lead to HSF activation in both *Drosophilia* and human cells, but oxidative stress induces significantly less HSP expression than heat shock (16–19). We compared the regulation of HSF activation by heat shock and by oxidants by investigating the effect of the membrane-permeant H_2O_2 on HSF activation and binding and on HSP synthesis. We confirmed that H_2O_2 leads to the activation of HSF-1 in human cells and showed a dual regulation of HSF by oxidants (Figure 1).

This dual regulation appears to be under the control of thioredoxin (TRX), a small ubiquitous protein containing two redox-active cysteine residues in the active site (-Trp-**Cys**-Gly-Pro-**Cys**-), whose reduced form functions as a hydrogen donor for the reduction of protein disulfides (20). TRX is a molecular chaperone which, depending on the redox environment and on the nature of the polypeptide substrate, might catalyze the isomerization, oxidation, or reduction of protein thiols/disulfides, and, as protein disulfide isomerase (PDI), favors disulfide bond formation and has been involved in the regeneration of cellular proteins inactivated by oxidative stress in vivo (20–22).

TRX is itself an enzyme which is also regulated by oxidoreduction of SH groups, and in the absence of other reductants TRX may be inactivated by SH group oxidation. In vivo, TRX is maintained in its reduced form by TRX reductase and NADPH, and functions not only as a proton donor for numerous proteins but also scavenges ROS (in particular H_2O_2 and hydroxyl radicals) and, as mentionned above, reactivates denatured proteins that contain mispaired disulfide bonds. However, while the refolding activity of TRX has been clearly established, this is not the case for its scavenging activity. TRX may

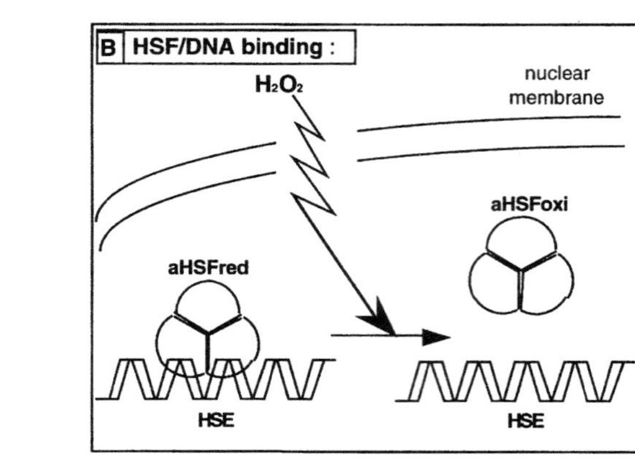

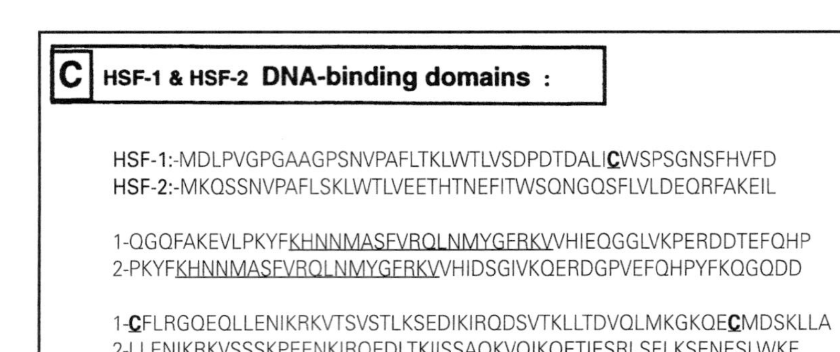

Figure 1. Dual regulation of HSF activation and DNA binding by H_2O_2. (A): H_2O_2 induces activation and DNA binding of HSF. H_2O_2 releases HSFi (inactivated HSF) from HSF–hsp70 complex, and induces trimerization and nuclear translocation of HSF. aHSF (activated HSF) interacts with HSE to activate HSP gene transcription. (B): While aHSFred efficiently interacts with HSE, H_2O_2 oxidizes HSF (aHSFoxi) and inhibits HSF-DNA binding activity. (C) Conserved DNA binding domains of HSF-1 and HSF-2 (underlined). There are three cysteine residues around this domain in HSF-1 (bold).

regulate the scavenging activity of other antioxidants by facilating proton transfer, or exert a direct effect as a free radical quencher.

In order to test the involvement of sulfhydryl groups on HSF–HSE complex formation, we examined the effects of a number of redox- and thiol-reactive agents including TRX on HSF-DNA-binding activity, Oxidation-mediated inhibition of HSF-DNA-binding was reversed by exogenous reducing agents including TRX, while the alkylation-mediated inhibition was irreversible, suggesting that one or more reduced cysteine residues within the DNA-binding domain of HSF are critical for binding. Cysteine residue(s) within and around the conserved DNA-binding domain of HSF may thus be a substrate for ROS, leading to oxidation of HSF, which is then unable to bind HSE. Interestingly, only some HSF contain cysteine residues within their DNA-binding domains (for example, human HSF-2 and yeast HSF do not). The selective activation of HSF-1 by ROS might thus depend upon its molecular structure.

Furthermore, we found that TRX expression was increased during heat shock and oxidative stress (19). This inducible expression of TRX leads us to suggest that TRX might be a new member of the stress protein families. Interestingly, the recently published sequence of the promoter region of human TRX reveals the presence of many regulatory elements compatible with HSF-responsive elements (23). We propose that the time required for TRX induction provides an explanation for the lack of HSP synthesis upon exposure of cells to ROS, despite the activation of HSF, as reported in a number of studies (17). Indeed, the time-dependent induction of TRX by H_2O_2 correlates well with the increase in HSF-DNA-binding activity (19). Previous studies were usually limited to 4h of recovery after stress, a period which in U937 cells, for example, was insufficient to allow the H_2O_2-mediated induction of HSP synthesis.

Figure 2 schematically summarizes our current view on the role of TRX in the redox regulation of HSF: upon exposure to oxidative stress, hsp70 dissociates from HSF, favoring its trimerization and nuclear translocation (activated HSF, aHSF). However, HSF

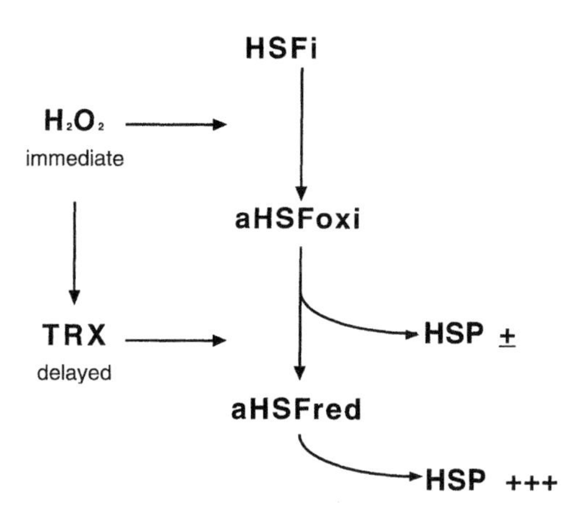

Figure 2. Role of TRX in the redox regulation of HSF. Upon exposure to H_2O_2, hsp70 dissociates from HSF, favoring its activation (activated HSF, aHSF). Simultaneously, aHSF is oxidized, thus becoming unable to bind HSE efficiently (see also Figure 1), leading to low levels of HSP synthesis (±). In parallel, with a certain delay, TRX is induced, catalyzing the transformation of aHSFoxi into aHSFred, which, in contrast to aHSFoxi, interacts efficiently with HSE, leading to active HSP synthesis (HSP+++).

becomes oxidized simultaneously with its activation (aHSFoxi), is thus unable to efficiently bind HSE (see also Figure 1) and subsequently induces only low levels of HSP (±). In parallel, soon after oxidative stress, TRX is induced, aiming at radical scavenging by direct interaction with ROS or regulation of other antioxidants and protection of HSF. After a prolonged recovery time, upregulation of TRX catalyses the transformation of aHSFoxi into the reduced form (aHSFred), which, in contrast to aHSFoxi, now efficiently interacts with HSE, leading to high transcriptional activation of HS genes and HSP synthesis (HSP+++), until sufficient amounts of HSP are synthesized to again chaperone and deactivate HSF.

In conclusion, HSF belongs to the ROS-modulated transcription factors and is regulated at two distinct levels: activation and DNA binding. These observations indicate interesting similarities between HSF and NFκB.

A COMPARISON BETWEEN HSF AND NFκB REDOX REGULATION

The nuclear transcription factor NFκB (discussed in more detail in numerous other chapters) was initially described as an activity that specifically bound DNA fragments containing the decameric DNA sequence motif 5′-GGGACTTTCC-3′ (reviewed in Refs. 24,25). This motif was first identified as a B cell-specific element, but it soon became evident that it was also functional in many other cells. NFκB is now recognized as a ubiquitous factor whose activation is independent of protein synthesis. Interestingly, the HSE and the NFκB-binding element are quite similar:

5′-GG**GA**CT**TTCC**-3′ (NFκB-binding element) and

5′-**GAA**nn**TTCC**-3′ (HSE)

Furthermore, NFκB and other related factors of the v- and c-Rel oncoprotein family share a characteristic sequence motif with a cysteine and three arginine residues within the DNA-binding region. Interestingly, there also are two arginine residues within the conserved DNA binding domains of HSF1 and of HSF2:

HSF1-HVFDQGQFAKEVLPKYFKHNNMASFV**R**OLNMYGF**R**KVVHIEQGGLVKPERDDTEFQHP

HSF2-AKEILPKYFKHNNMASFV**R**OLNMYGF**R**KVVHIDSGIVKQERDGPVEFQHPYFKQGQDD

A great variety of agents can activate NFκB in vivo, including phorbol esters, antigens, superantigens and antibodies, pro-inflammatory cytokines, viral (HIV) and bacterial proteins, lipopolysaccharides, and ionizing and nonionizing radiations (26–28). Interestingly, all these factors share the potential to induce the production of low, nontoxic, cell-activating concentrations of ROS. In vivo ROS activate NFκB by stimulating release from IκB, while in vitro ROS have also been shown to inhibit NFκB binding to DNA (29).

These are not the only similarities in the redox regulation of HSF and other transcription factors. As for HSF, the activation of NFκB by oxidants is selective for the type of oxidant, H_2O_2 being more effective than $^{\bullet}O_2^-$ in activating both transcriptional regulators. Along these lines, cell lines overexpressing catalase, but not superoxide dismutase (SOD), are deficient in activating NFκB (30).

Also similar to what has been observed for HSF is the likely mode of action of ROS upon NFκB activation: as does HSF, NFκB and AP-1 harbor conserved cysteines within their DNA-binding site, which has previously been proposed as the target for the modulatory effects of oxidants and reducing agents on those transcription factors (31). Furthermore, TRX may also play similar regulatory roles for NFκB as for HSF. While glutathione was the first physiologically relevant modulator of the redox activation of NFκB to be identified, other endogenous or exogenous antioxidants, including TRX and *N*-acetylcysteine, also decrease NFκB activation, perhaps in conjunction with glutathione (32,33). TRX, owing to its catalytic mechanism and ubiquitous distribution, may restore the activity of various oxidatively damaged redox-regulated transcription factors, including HSF and NFκB (32,34).

HSP-MEDIATED PROTECTION AGAINST OXIDATIVE STRESS

The induction of HSP has been observed as an adaptive response to oxidative stress in both prokaryotes and eukaryotes and the gene products induced after HSF activation play an established role in survival upon exposure to ROS. HSP induce thermotolerance and protection against a number of stresses distinct from heat shock, including tumor necrosis factor-α (TNFα) and, in vivo, ischemia and reperfusion, sepsis, and acute inflammation (5,35–37), conditions which are all characterized by an increased oxidative burden (2,38).

HSP-induced eukaryotic cellular protection against ROS may be targeted to any of the following: membranes (lipid peroxidation), proteins, DNA, and mitochondria. The protective effects of hsp70 toward lipid peroxidation and DNA damage have been reviewed (39). Protection in terms of membrane alteration (measured by lipid peroxidation) ranges from 20% to 25% (39) and protection at the level of DNA (measured by ethidium bromide fluorescence) varies according to in vitro cell aging (40; M. Perin-Minisini and B.S. Polla, unpublished). Protection was highest for mitochondria (60–100%) (41), indicating that these organelles might represent a primary target for HS-induced cellular protection, as also suggested by others (42,43).

We found that the induction of heat shock response prevented alterations in mitochondrial membrane potential upon exposure to H_2O_2 (41). The kinetics for protection correlated with those of hsp70 accumulation (and not for other HSP tested) within the heat-shocked cells, while protection was abolished when heat shock was performed in the presence of the transcriptional inhibitor actinomycin D. These results are consistent with the fact that hsp70 likely represents the major protective stress protein, as also proposed by a number of investigators (36,37).

Hsp70-mediated chaperoning and protection may be exerted at the mitochondrial membrane and/or within the mitochondria as well as within the cytoplasm, since among the members of the hsp70 family, there are members locating themselves specifically to these as well as other subcellular compartments (44). Although the precise mitochondrial function that is protected remains to be determined, several arguments favor some component of the respiratory chain. First, ATPase activity undergoes thermotolerance in *Saccharomyces cerevisiae* (42) and heat shock prevents the inhibition of oxidative phosphorylation induced by the ATPase inhibitor oligomycin in cells strictly dependent upon oxidative metabolism (45). Second, induction of a HS response in vivo in rats prevents alterations in mitochondrial respiration upon subsequent exposure of isolated perfused hearts to H_2O_2 and, more specifically, in state III respiration (46; L. Bornman et al., unpublished). These data are in agreement with those of Borkan and colleagues who demonstrated thermotolerance in state III mitochondrial respiration in inner-medullary collecting duct cells from rat kidneys (47). While the target for mitochondrial oxidative damage does not appear to be mitochondrial DNA (41), an attractive hypothesis is that hsp70 in some way stabilizes the mitochondrial membrane and prevents the release from the mitochondria of the "death factor" recently described by Kroemer (48,49). Thus, as proposed by Samali (50), heat shock might protect cells against apoptosis by inhibiting or modifying one or more of the death proteins.*

A number of similarities between heat shock and the *bcl-2*-induced antioxidant effects suggest the possibility that overexpression of *bcl-2* may contribute to the mechanisms by which heat shock prevents oxidative damage (51,52). Indeed, both heat shock and/or HSP, and *bcl-2*, protect cells from oxidative stress; both decrease ROS production; and in both cases, protection is distal to the Ca^{2+} entry that is induced by oxidative stress (52,53). Furthermore, heat shock prevents apoptosis, as does *bcl-2* (50,54,55). We found that heat shock increased the expression of *bcl-2* in a number of human cell lines (41). The promoter region of *bcl-2* contains a number of sequences resembling the HS element (HSE) (including n-AGAA-n), as well as a classical HSE (n-GAA-nn-TTC-n) at position -622/-630, which may explain the increased expression of *bcl-2* protein after heat shock (Figure 3) (56,57). As proposed for TRX, *bcl-2* might thus also proove to belong to the stress protein families.

Apoptosis, or programmed cell death, is used by almost all living organisms to self-destruct cells by an active "suicide" program when they are no longer needed, or have become seriously damaged (reviewed in Refs. 58,59). The morphological characteristics of apoptosis include nuclear and cytosolic condensation, and the formation of membrane-bound apoptotic bodies that are rapidly engulfed and digested by neighboring cells, in particular by macrophages. Apoptotic cells do not release their potentially noxious contents and phagocytosis of apoptotic bodies occurs without activation of the macrophages. In contrast, cells undergoing necrosis (as a consequence of overwhelming injuries or loss in their capacity to adequately cope with stress) swell, lyse, and release cytoplasmic material which can trigger an inflammatory response. Phagocytosis of necrotic cells or debris activates the phagocytes, thus leading to amplification of damage. As a consequence, whether a given cell undergoes apoptosis or necrosis is crucial in the control of inflammation.

Mitochondria have been shown to affect profoundly the cellular "choice" between necrosis and apoptosis, and have been proposed as a "switchboard for apoptosis" or "masters of cell death" (60,61). Mitochondrial functional integrity and maintenance of ATP levels are important during the early phases of apoptosis, when a variety of energy-requiring intracellular processes occur (60,61). In contrast, if intracellular ATP levels are very low, and if mitochondrial function is impaired or damaged, necrosis eventually occurs. Given the differences described between apoptosis and necrosis in terms of amplification or limitation of inflammation, the maintenance of mitochondrial function may protect cells and tissues from inflammatory damage, as a result of the cells' "choice" to undergo the active apoptotic program rather than passive necrosis. Under the control of HSP (hsp70), mitochondria might thus also represent the switchboard defining the outcome of acute inflammation, whether complete healing or chronic inflammation,

---

*Note added in proof: see p. 125.

GCGCCCGCCCCTCCGCGCCGCCTGCCCGCCCGCCCGCCGCGCTCCCGCCCGCCGCTCTCGGTGGCCCC

GCCGCGCTGCCGCCGCCGCCGCTGCCAGCGAAGGTGCCGGGGCTCCGGGCCCTCCCTGCCGGCGGCC

-1300
GTCAGCGCTCGGAGCGAACTGCGCGACGGGAGGTCCGGGAGGCGACCGTAGTCGCGCCGCCGCGCA

-1200
GGACCAGGAGGAGGAGAAAGGGTGCGCAGCCCGGAGGCGGGGTGCGCCGGTGGGGTGCAGCGGAA

GAGGGGGTCCAGGGGGCAGAACTTCGTAGCAGTCATCCTTTTTAGGAAAAGAGGGAAAAAATAAAA

-1100
CCCTCCCCCACCACCTCCTTCTCCCCACCCCTCGCCGGACCACACACAGCGCGGGCTTCTAGCGCTCG

-1000
GCACCGGCGGGCCAGGCGCGTCCTGCCTTCATTTATCCAGCAGCTTTTCGGAAAATGCATTTGCTGTT

CGGAGTTTAATCAGAAGACGATTCCTGCCTCCGTCCCCGGCTCCTTCATCGTCCCATCTCCCCTGTCT

-900
CTCTCCTGGGGAGGCGTGAAGCGGTCCCGTGGATAGAGATTCATGCCTGTGTCCGCGCGTGTGTGCGC

-800
GCGTATAAATTGCCGAGAAGGGGAAAACATCACAGGACTTCTGCGAATACCGGACTGAAAATTGTA

ATTCATCTGCCGCCGCCGCTGCCAAAAAAAAACTCGAGCTCTTGAGATCTCCGGTTGGGATTCCTGCG

-700
GATTGACATTTCTGTGAAGCAGAAGTCTGGGAATCGATCTGGAAATCCTCCTAATTTTTACTCCCTCT

-600
CCCCCCGACTCCTGATTCATTGGGAAGTTTCAAATCAGCTATAACTGGAGAGTGCTGAAGATTGATGG

CATCGTTGCCTTATGCATTTGTTTTGGTTTTACAAAAAGGAAACTTGACAGAGGATCATGCTGTACTT

-500
AAAAAATACAAGTAAGTCTCGCACAGGAAATTGGTTTAATGTAACTTTCAATGGAAACCTTTGAGAT

-400
TTTTTACTTAAAGTGCATTCGAGTAAATTTAATTTCCAGGCAGCTTAATACATTGTTTTTAGCCGTGTT

ACTTGTAGTGTGTATGCCCTGCTTTCACTCAGTGTGTACAGGGAAACGCACCTGATTTTTTACTTATTA

-300
GTTTGTTTTTTCTTTAACCTTTCAGCATCACAGAGGAAGTAGACTGATATTAACAATACTTACTAATAA

-200
TAACGTGCCTCATGAAATAAAGATCCGAAAGGATTTGGAATAAAAATTTCCTGCCTCTCATGCCAAG

AGGGAAACACCAGAATCAAGTCTTCCGCGTGATTGAAGACACCCCCTCGTCCAAGAATGCAAAGCA

-100
CATCCAATAAAATAGCTGGATTATAACTCCTCTTCTTTCTCTGGGGGCCGTGGGGTGGGAGCTGGGGC

+1
GAGAGGTCCCCTTGGCCCCCCTTGCTTTTCCTCTGGGAAGG MET
ATG

Figure 3. The *bcl-2* promoter contains HSE-like sequences (boxes). The promoter region of *bcl-2* contains a number of sequences resembling the HS element (HSE) (including n-AGAA-n), as well as a classical HSE (n-GAA-nn-TTC-n) at position -622/-630. The sequence data is from Refs. 56 and 57.

amplification of inflammation being associated with or secondary to necrosis, and limitation of inflammation with apoptosis (62).

FURTHER IMPLICATIONS OF REDOX REGULATION OF HSF AND POTENTIAL APPLICATIONS OF THE PROTECTIVE EFFECTS OF HSP AGAINST OXIDATIVE INJURY

One situation of interest where regulation of oxidative stress-inducible genes modulates cellular responses is bacterial infection and phagocytosis. When infected, human phagocytes activate the complex respiratory-burst enzyme NADHP oxidase, leading to production, by the host cell, of $^{\bullet}O_2^-$ generated from molecular oxygen at the expense of NADPH. The ROS produced during phagocytosis and infection are usually considered as aiming to kill the responsible microorganism. However, infection-associated ROS also activate a number of redox-regulated transcription factors and play a role in the inducible protective responses such as HSP and SOD, in both host cells and microorganisms (63–65). The respective induction of host and pathogen HSP is tightly regulated by the type and the subcellular location of ROS production, $^{\bullet}O_2^-$ in itself being insufficient for host HSF activation. Inducible protective responses such as HSP might determine the outcome of infections, and further studies on the redox regulation of inducible genes during infection should be of interest.

Protection from oxidative injury has wide medical implications, including inflammation, HIV infection, bacterial infection, and other diseases or physiological conditions such as cancer and aging in which an imbalance between oxidants and antioxidants plays a key role (2,66,67). Therapeutic applications of HSP are being considered in all of these and other clinical conditions. Recently, BRLP-42 (Bioclomol®) (BIOREX R&D), the first nontoxic inducer of molecular chaperones, has been developed (68). Administration of BRLP-42 in vivo diplays striking effects on a number of conditions associated with an increased oxidative burden, further supporting the potential of HSP as therapeutic agents.

REFERENCES

1. Fridovich I. The biology of oxygen radicals. Science 1978; 201:75–879.
2. Halliwell B, Gutteridge JMC. Role of free radicals and catalytic metal ions in human disease: an overview. In Packer L, Glazer AN, eds. Methods in Enzymology. New York: Academic Press; 1990; 86:1–85.
3. Feige U, Polla BS. Heat shock proteins: the hsp 70 family. Experientia 1994; 50:979–986.
4. Wu C. Heat shock transcription factors: structure and regulation. Annu Rev Cell Dev Biol 1995; 11:441–469.
5. Melhen P, Preville X, Chareyron P, Briolay J, Klemenz R, Arrigo PA. Constitutive expression of human hsp27, Drosophila hsp27, or human αβ-crystallin confers resistance to TNF- and oxidative stress-induced cytotoxicity in stable transfected murine L929 fibroblasts. J Immunol 1995; 154:363–374.
6. Melhen P, Kretz-Remy C; Preville X, Arrigo PA. Human hsp27, Drosophila hsp27 and human αβ-crystallin expression-mediated increase in glutathione is essential for the protective activity of these proteins against TNFα induced cell death. EMBO J 1996; 15:2695–2706.
7. Rabindran SK, Giorgi G, Clos J, Wu C. Molecular cloning and expression of human heat shock factor, HSF1. Proc Natl Acad Sci USA 1991; 88:6906–6910.
8. Schuetz TJ, Gallo GJ, Sheldon L, Tempst P, Kingston RE. Isolation of a cDNA for HSF2: evidence for two heat shock factor genes in humans. Proc Natl Acad Sci USA 1991; 88:6911–6915.
9. Sistonen L, Sarge KD, Morimoto RI. Human heat shock factors 1 and 2 are differentially activated and can synergistically induce hsp70 gene transcription. Mol Cell Biol 1994; 14:2087–2099.
10. Sarge KD, Zimarino V, Holm K, Wu C, Morimoto RI. Activation of heat shock gene transcription by heat shock factor 1 involves oligomerization, acquisition of DNA-binding activity, and nuclear localization and can occur in the absence of stress. Genes Dev 1991; 5:1902–1911.

11. Westwood JT, Clos J, Wu C. Stress-induced oligomerisation and chromosomal relocalisation of heat-shock factor. Nature 1991; 353:822–827.
12. Sorger PK. Yeast heat shock factor contains separable transient and sustained response transcriptional activators. Cell 1990; 62:793–805.
13. Croston GE, Kerrigan LA, Lira LM, Marshak DR, Kadonaga JT. Sequence-specific antirepression of histone H1-mediated inhibition of basal RNA polymerase II transcription. Science 1991; 251:643–649.
14. Ananthan J, Golderg AL, Voellmy R. Abnormal proteins serve as eukaryotic stress signals and trigger the activation of heat shock genes. Science 1986; 232:522–524.
15. Clos J, Westwood T, Becker PB, Wilson S, Lambert K, Wu C. 1990. Molecular cloning and expression of a hexameric Drosophila heat shock factor subject to negative regulation. Cell 1990; 63:1085–1097.
16. Polla BS, Healy AM, Wojno WC, Krane SM. Hormone 1α, 25-dihydroxyvitamin D_3 modulates heat shock response in human monocytes. Am J Physiol 1987; 252:C640–C649.
17. Becker J, Metzger V, Courgeon AM, Best-Belpomme M. Hydrogen peroxide activates immediate binding of a Drosophila factor to DNA heat-shock regulatory element in vivo and in vitro. J Biochem 1990; 189:553–558.
18. Bruce JL, Price BD, Calderwood SK. Oxidative injury rapidly activates the heat shock transcription factor but fails to increase levels of heat shock proteins. Cancer Res 1993; 53:12–15.
19. Jacquier-Sarlin MR, Polla BS. Dual regulation of heat-shock transcription factor (HSF) activation and DNA-binding activity by H_2O_2: role of thioredoxin. Biochem J 1996; 318: 187–193.
20. Holmgren A. Thioredoxin. Annu Rev Biochem 1985; 54:237–271.
21. Fernando MR, Nanri, Yoshitake S, Nagata-Kuno K, Minakami S. Thioredoxin regenerates proteins inactivated by oxidative stress in endothelial cells. Eur J Biochem 1992; 209:917–922.
22. Bardwell JC, McGovern K, Beckwith J. Identification of a protein required for disulfide bond formation in vivo. Cell 1991; 67:581–589.
23. Kaghad M, Dessarps F, Jacquemin-Sablon H, Caput D, Fradelizi D, Wollman EE. Genomic cloning of human thioredoxin-encoding gene: mapping of the transcription start point and analysis of the promoter. Genes 1994; 140:273–278.
24. Grimm S, Baeuerle PA. The inducible transcription factor NFκB: structure–function relationship of its protein subunits. Biochem J 1993; 290:297–308.
25. Verma IM, Stevenson JK, Schwarz EM, Van Antwerp D, Miyamoto S. Rel/Nf-κB/IκB family: intimate tales of association and dissociation. Genes Dev 1995; 9:2723–2735.
26. Staal FJT, Roederer M, Herzenberg LA, Herzenberg LA. Intracellular thiols regulate activation of nuclear factor κB and transcription of human immunodeficiency virus. Proc Natl Acad Sci USA 1990; 87:9943–9947.
27. Schreck R, Rieber P, Baeuerle PA. Reactive oxygen intermediates as apparently widely used messengers in the activation of the NF-κB transcription factor and HIV-1. EMBO J 1991; 10:2247–2258.
28. Schreck R, Albermann K, Baeuerle PA. Nuclear factor κB: an oxidative stress responsive transcription factor of eukaryotic cells. Free Radical Res Commun 1992; 17:221–237.
29. Toledano MB, Leonard WJ. Modulation of transcription factor NF-κB binding activity by oxidation–reduction in vitro. Proc Natl Acad Sci USA 1991; 88:4328–4332.
30. Schmidt KN, Amstad P, Cerutti P, Baeuerle PA. The roles of hydrogen peroxide and superoxide as messengers in the activation of transcription factor NF-κB. Chem Biol 1995; 2:13–22.
31. Sharif M, Worrall JG, Singh B, Gupta RS, Lydyard PM, Lambert C, McCulloch J, Rook GA. The development of monoclonal antibodies to the human mitochondrial 60-kd heat shock protein, and their use in studying the expression of the protein in rheumatoid arthritis. Arthritis Rheum 1992; 35:1427–1433.

32. Hayashi T, Ueno Y, Okamoto T. Oxidative regulation of nuclear factor κB. Involvement of a cellular reducing catalyst thioredoxin. J Biol Chem 1993; 268:11380–11388.
33. Schenk H, Klein M, Erdbrügger W, Dröge W, Schulze-Osthoff K. Distinct effects of thioredoxin and antioxidants on the activation of transcription factors NF-κB and AP-1. Proc Natl Acad Sci USA 1994; 91:1672–1676.
34. Matthews JR, Wakasugi N, Virelizier JL, Yodoi J, Hay RT. Thioredoxin regulates the DNA binding activity of NF-κB by reduction of a disulphide bond involving cysteine 62. Nucleic Acids Res 1992; 20:2005–2015.
35. Jäättelä M. Overexpression of major heat shock protein hsp70 inhibits tumor necrosis factor-induced activation of phospholipase A2. J Immunol 1993; 151:4286–4294.
36. Plumier JC, Ross BM, Currie RW, Angelidis CE, Kazlaris H, Kollias G, Pagoulatos GN. Transgenic mice expressing the human heat shock protein 70 have improved post-ischemic myocardial recovery. J Clin Invest 1995; 95:1854–1860.
37. Villar J, Ribeiro SP, Brendan J, Mullen M, Kuliszewski M, Post M, Slutsky AS. Induction of the heat shock response reduces mortality rate and organ damage in a sepsis-induced acute lung injury model. Crit Care Med 1994; 22:914–921.
38. McCord JM. Oxygen-derived free radicals in postischemic tissue injury. N Engl J Med 1985; 312:159–163.
39. Jacquier-Sarlin MR, Fuller K, Dinh-Xuan AT, Richard MJ, Polla BS. Protective effects of hsp70 in inflammation. Experientia 1994; 50:1031–1038.
40. Perin-Minisini M, Kantengwa S, Polla BS. DNA damage and stress protein synthesis induced by oxidative stress proceed independently in the human premonocytic line U937. Mutat Res 1994; 315:169–179.
41. Polla BS, Kantengwa S, Salvioli FS, Franceschi C, Marsac C, Cossarizza A. Mitochondria are selective targets for the protective effects of heat shock against oxidative injury. Proc Natl Acad Sci USA 1996; 93:6458–6463.
42. Patriarca EJ, Maresca B. Acquired thermotolerance following heat shock protein synthesis prevents impairment of mitochondrial ATPase activity at elevated temperatures in *Saccharomyces cerevisiae*. Exp Cell Res 1990; 190:57–64.
43. Gabai VL, Kabakov AE. Rise in heat-shock protein level confers tolerance to energy deprivation. FEBS Lett 1993; 327:247–250.
44. Stuart RA, Cyr DM, Neupert W. Hsp70 in mitochondrial biogenesis: from chaperoning nascent polypeptide chains to facilitation of protein degradation. Experientia 1994; 50:1002–1011.
45. Polla BS, Bonventre JV. Heat shock protects cells dependent on oxidative metabolism from inhibition of oxidative phosphorylation. Clin Res 1987; 35:555A (abstract).
46. Bornman L, Gericke GS, Steinmann CML, Polla BS. State 3 respiration: thermotolerant protection against oxidative injury. J Free Radical Biol Med 1994; 2:R6 (abstract).
47. Borkan SC, Emami A, Schwartz, JH. Heat stress protein-associated cytoprotection of inner medullary collecting duct cells from rat kidney. Am J Physiol 1993; 265:F333–F341.
48. Zamzami N, Susin SA, Marchetti, Hirsch T, Gòmez-Monterrey I, Castedo, M, Kroemer G. Mitochondrial control of nuclear apoptosis. J Exp Med 1996; 183:1533–1544.
49. Henkart PA, Grinstein S. Apoptosis: mitochondria resurrected? J Exp Med 1996; 183:1293–1295.
50. Samali A, Cotter TG. Heat shock proteins increase rsistance to apoptosis. Exp Cell Res 1996; 223:163–170.
51. Kane DJ, Sarafian TA, Anton R, Hahn H, Gralla E B, Selverstone Valentine J, Ord T, Bredesen DE. Bcl-2 inhibition of neural death: decreased generation of reactive species. Science 1993; 262:1274–1276.
52. Hockenbery DM, Oltvai ZN, Yin XM, Milliman CL, Korsmeyer SJ. Bcl-2 functions in an antioxidant pathway to prevent apoptosis. Cell 1993; 75:241–251.
53. Polla BS, Bonventre JV, Krane SM. 1,25-Dihydroxyvitamin D_3 increases the toxicity of hydrogen peroxide in the human monocytic line U937: the role of calcium and heat shock. J

Cell Biol 1988; 107:373–380.
54. Driggers WJ, LeDoux SP and Wilson GL. Repair of the oxidative damage within the mitochondrial DNA of RINr 38 cells. J Biol Chem 1993; 268:22042–22045.
55. Mailhos C, Howard M K, Latchman DS. Heat shock protects neuronal cells from programmed cell death by apoptosis. Neuroscience 1993; 55:621–627.
56. Cleary ML, Sklar J. Nucleotide sequence of a t(14;18) chromosomal breakpoint in follicular lymphoma and demonstration of a break point-cluster region near transcriptionally active locus on chromosome 18. Proc Natl Acad Sci USA 1985; 82:7439–7443.
57. Cleary ML, Smith SD, Sklar J. Cloning and structural analysis of cDNAs for bcl-2/immunoglobulin transcript resulting from the t(14;18) translocation. Cell 1986; 47:19–28.
58. Steller H. Mechanisms and genes of cellular suicide. Science 1995; 267:1445–1449.
59. Thompson CB. Apoptosis in the pathogenesis and treatment of disease. Sciences 1995; 267:1456–1462.
60. Richter C. On the role of energy decline in muscle cell apoptosis. Basic Appl Myol 1996; 6:215–220.
61. Richter C, Schweizer M, Cossarizza A, Franceschi C. Control of apoptosis by the cellular ATP level. FEBS Lett 1996; 378:107–110.
62. Mangan DF, Mergenhagen SE, Wahl SM. Apoptosis in human monocytes: possible role in chronic inflammatory diseases. J Periodontol 1993; 64:461–466.
63. Kantengwa S, Polla BS. Phagocytosis of *Staphylococcus aureus* induces a selective stress response in human Monocytes-Macrophages (Mϕ): modulation by Mϕ differentiation and by iron. Infect Immun 1993; 61:1281–1287.
64. Kantengwa S, Müller I, Louis J, Polla BS. Infection of human and murine macrophages with *Leishmania* major is associated with early parasite heat shock protein synthesis but fails to induce a host cell stress response. Immunol Cell Biol 1995; 73:73–80.
65. Polla BS, Mariéthoz E, Hubert D, Barazzone C. Heat shock proteins in host–pathogen interactions: implications for cystic fibrosis. Trends Microbiol 1995; 3:392–396.
66. Cross CE, Halliwell B, Borish ET, Pryor WA, Ames BN, Saul R, McCord JM, Harman D. Oxygen radicals and human disease. Ann Intern Med 1987; 107: 526–545.
67. Ames BN, Shigenaga MK. Oxidants are a major contributor to aging. In Franceschi C, Crepaldi G, Cristofalo VJ, Vijg J, eds. Aging and Cellular Defense Mechanisms. New York Academy of Sciences; 1992:85–96.
68. Vigh L, Litérati PN, Horvath I, Török Z, Balogh G, Glatz A, Kovacs, Duda E, Boros I, Ferdinandy P, Multhoff G, Farkas B, Jaslitz L, Koranyi L, Maresca B. BRLP-42, a new generation of drugs acting as a chaperone inducer. In: Molecular Chaperones and the Heat Shock Response. Cold Spring Harbor, NY: Cold Spring Harbor Laboratory Press; 1996; 345 (abstract).

Note added in proof

One of those "death proteins" might be cytochrome *c*, and the release of cytochrome *c* from mitochondria, a primary site for regulation of apoptosis, whether by *bcl-2* or by hsp70.

Yang J, Liu X, Bhalla K, Kim CN, Ibrado AM, Cai J, Oeng TI, Jones DP, Wang X. Prevention of apoptosis by *Bcl-2*: release of cytochrome *c* from mitochondria blocked. Science 1997; 275: 1129–1132.
Kluck RM, Bossy-Wetzel E, Green DR, Newmeyere DD. The release of cytochrome *c* from mitochondria: a primary site for *Bcl-2* regulation of apoptosis. Science 1997; 275: 1132–1136.

13

Low-Density Lipoproteins Modulate Nitric Oxide and Heme Oxygenase Signal Transduction Pathways in Vascular Endothelium and Smooth Muscle

R. C. M. Siow, M. T. Jay, H. Sato, J. D. Pearson, and G. E. Mann
King's College
University of London
London, England

INTRODUCTION

The development of atherosclerotic lesions is thought to result from endothelial dysfunction, involving alterations in adhesion molecule expression and diminished release of prostacyclin (PGI_2) and nitric oxide (NO), which normally inhibit smooth muscle contraction and proliferation, platelet aggregation, and monocyte adhesion to the endothelial surface (1,2). Focal accumulation within the arterial intima of excess amounts of cholesterol-rich low-density lipoprotein (LDL) leads to the migration and recruitment of monocytes into these areas. Monocytes then differentiate into macrophages after taking up large amounts of oxidized LDL via their scavenger receptor to become lipid-laden "foam cells" within the subendothelial space, an early characteristic of the atherosclerotic lesion (3–6). Concomitant release of cytokines, such as interleukin-1, tumor necrosis factor-α, and transforming growth factor-β, by activated lesion macrophages and smooth muscle cells exposed to oxidized LDL modulates smooth muscle cell phenotype and proliferation (7).

Endothelial cells synthesize NO from L-arginine via a constitutive Ca^{2+}/calmodulin-sensitive NO synthase (eNOS), and relaxation of vascular smooth muscle by endothelium-derived NO is mediated through the activation of soluble guanylate cyclase and elevation of intracellular cGMP (8). In addition to eNOS, an inducible Ca^{2+}-insensitive NO synthase (iNOS) has been identified in endothelial cells, smooth muscle cells and macrophages activated with proinflammatory cytokines or bacterial lipopolysaccharide (9). Generation of NO by iNOS occurs with a lag period of 2 h, reaching a maximum between 6 and 12 h, and can be inhibited by glucocorticoids and inhibitors of protein synthesis. It is generally accepted that oxidized LDL and hyperlipidemia impair endothelium-dependent relaxation (10–13), yet the existing literature on the effects of oxidized LDL on eNOS and iNOS expression remains inconclusive, since oxidized LDL has been reported to either enhance (14,15) or reduce (16,17) NOS expression.

Accumulating evidence now implicates carbon monoxide (CO), generated from heme metabolism via heme oxygenases, as another regulator of soluble guanylate cyclase and cGMP levels in the brain and vasculature (18–21). Heme oxygenase is the most efficient heme-degrading system and produces almost exclusively biliverdin and carbon monoxide as heme degradation products (18). The cytosolic enzyme biliverdin reductase then catalyzes the two-electron reduction of biliverdin to bilirubin, a powerful chain-breaking antioxidant (22). Two forms of heme oxygenase have been identified, a constitutive from HO-2 (~34 kDa) and an inducible form HO-1 (~32 kDa), which is expressed at a low basal level in vascular smooth muscle cells and induced by heavy metals, oxidative stress, inflammatory mediators, and oxidatively modified LDL (23–25). The only major chemical constraint on CO as a cellular messenger, activator of soluble guanylate cyclase, and vasodilator is the lower affinity of CO for the heme in soluble guanylate cyclase. Nevertheless, it is conceivable that, as in the brain (26), diminished production of or sensitivity to NO in atherogenesis may be compensated for by an induction of heme oxygenase and generation of CO.

In the present work we review our studies of the effects of native, minimally modified, and highly oxidized human LDL on (i) initial rates of L-arginine transport in human endothelial and smooth muscle cells, (ii) basal and histamine-stimulated NO and PGI_2 release in human endothelial cells, and (iii) expression of heme oxygenase-1 in smooth muscle cells.

METHODS

Endothelial and Smooth Muscle Cell Culture

Human umbilical vein endothelial cells were isolated by collagenase (0.5 mg/ml) digestion and cultured in medium 199 containing 5 mM D-glucose and supplemented with 10% fetal calf serum, 10% newborn calf serum, 5 mM glutamine, 100 IU/ml penicillin–streptomycin, and 0.03 mg/ml gentamycin at 37°C in a 5% CO_2 atmosphere (27). Confluent second-passage cells were used 48–60 h after subculture into microtiter plates. Endothelial cells were identified by their cobblestone morphology, ability to take up acetylated low-density lipoprotein, and the absence of immunostaining for α-actin. Smooth muscle cells from medial explants of human umbilical artery or porcine aorta were cultured in Dulbecco's modified Eagle's medium (DMEM) supplemented with 10% fetal calf serum, 2 mM L-glutamine, 30 mM $NaHCO_3$, 100 U/ml penicillin, 100 μg/ml streptomycin at 37 °C in a 5% CO_2 atmosphere. Passages 3–6 cells were used and confirmed as smooth muscle cells by their "hill and valley" morphology and positive immunofluorescent staining for α-actin using mouse antibody against α-actin and FITC-conjugated antimouse IgG (23).

Preparation of Native, Minimally Modified, and Oxidized Low-density Lipoproteins

As described previously (28), LDL (density 1.019–1.063 g/ml) was isolated from normal human blood by ultracentrifugation in the presence of EDTA, followed by dialysis against (mM): NaCl 154, NaH_2PO_4 16.7, Na_2HPO_4 21.1, EDTA 0.1; pH 7.4. Mildly oxidized LDL was formed by incubating LDL at 100 μg protein/ml at 37°C in (mM): NaCl 137, KCl 2.68, NaH_2PO_4 8.1, KH_2PO_4 1.47, pH 7.4 with $CuSO_4$ (net concentration 5 μM above the EDTA present). The absorbance at 234 nm was monitored to measure the conjugated dienes produced until the absorbance had increased by 0.2. 1 mM EDTA was then added to stop the oxidation. Highly oxidized LDL was formed by incubating the LDL in the above way but for 24 h before adding 1 mM EDTA. The density of the oxidized LDL was then raised to 1.2 g/ml with KBr (in the presence of Chelex-100) and then concentrated by ultracentrifugation and dialyzed several times (28). LDLs were sterilized by membrane filtration (0.2 μm) and their protein and lipid hydroperoxide contents and mobility in agarose gels were determined (28,29). Lipid hydroperoxides in the native (nLDL), mildly oxidized (mmLDL), and highly oxidised (oxLDL) LDL were 40, 64, and 80 nmol/mg protein, respectively, and the relative electrophoretic mobilities of mmLDL and oxLDL (compared to nLDL) were 1.3 and 4.6, respectively.

Effects of LDL on L-Arginine Transport in Endothelial and Smooth Muscle Cells

Confluent cell monolayers were incubated for 0–24 h in the absence or presence of 100 μg protein/ml native or highly oxidized LDL. Monolayers were then rinsed with warmed (37°C) Krebs solution [(mM): NaCl 131, KCl 5.6, $NaHCO_3$ 25, NaH_2PO_4 1, D-glucose 5, Hepes 20, $CaCl_2$ 2.5, $MgCl_2$ 1; pH 7.4), and initial rates of L-[^{3}H]arginine influx (100 μM) were then measured over 1 min in Krebs. Tracer uptake was terminated by removal of the medium 1 s before rinsing the monolayer three times with 200 μl ice-cold stop solution (Krebs buffer containing 10 mM unlabeled substrate) (27). Radioactivity in formic acid cell digests was determined by liquid scintillation counting.

Endothelial Cell Nitric Oxide (NO) and Prostacyclin (PGI_2) Synthesis

Endothelial cell monolayers were treated for 1–24 h with 100 μg protein/ml native, minimally modified, or highly oxidized LDL. Cells were then incubated for 15 min with Krebs solution (37°C) containing 100 μML-arginine and the phosphodiesterase inhibitor 3-isobutyl-1-methylxanthine (IBMX, 0.5 mM). The preincubation medium was removed and 500 μl Krebs containing IBMX (with or without 10 μM histamine) was added to the wells for a further 5 min (37°C). The supernatant was removed for analysis of PGI_2 release by radioimmunoassay of its stable metabolite 6-keto-$PGF_{1\alpha}$ (27). Cells were then placed on ice and incubated with 0.1M HCl (1 ml/well, 60 min) and HCl cell extracts were taken for radioimmunoassay of cGMP (assay for NO) (27). The effect of the NOS inhibitor N^G-nitro-L-arginine methyl ester (L-NAME) on cGMP levels was assessed by preincubating cells for 15 min with Krebs containing 100 μM L-NAME, and then for a further 5 min with these agents in the absence or presence of 10 μM histamine.

Western Blot Analysis for Heme Oxygenase-1 in Vascular Smooth Muscle Cells

Porcine aortic smooth muscle cells were incubated in the absence or presence of 100 μg protein/ml LDL for 6–48 h, and cells were then solubilized (2% sodium dodecyl sulphate (SDS), 10% glycerol, 50 mM Tris-HCl, pH 6.8) and boiled for 5 min. Equal protein concentrations from each sample were then boiled in a mixture of 1% 2-mercaptoethanol and 0.05% bromophenol blue for 3 min. As described previously (23), proteins were separated by SDS–polyacrylamide gel electrophoresis using a 13% acrylamide resolving gel, transferred to a polyvinylidine difluoride membrane and probed with a polyclonal rabbit anti-rat heme oxygenase antibody (30). A horseradish peroxidase-conjugated goat anti-rabbit secondary antibody was used in conjunction with enhanced chemiluminescence to detect the HO-1 bands (23).

Materials

Newborn and fetal calf serum and all other reagents were purchased from Sigma. Collagenase type II from *Clostridium histolyticum* was from Boehringer Mannheim, Germany, and Bradford protein reagent from BioRad Laboratories, UK. L-[2,3,-^{3}H]arginine (36.1 Ci/mmol) was from New England Nuclear, Dreieich, Germany, 3′,5′-cyclic GMP-TME, [tyrosine-^{125}I] from ICN, UK, and [^{125}I]6-keto-PGF$_{1\alpha}$ from Advanced Magnetics Inc., UK. Horseradish peroxidase-conjugated goat anti-rabbit immunoglobuin and BCA protein assay reagents were purchased from Pierce, Chester, UK. Polyvinylidine difluoride membrane (Immobilon-P) was purchased from Millipore, Watford, UK, and enhanced chemiluminesence western blotting detection reagents and Hyperfilm-MP autoradiography film from Amersham International, UK.

Statistics

Data are expressed as mean ± S.E. of measurements in 3–6 different cell cultures. Statistical analyses were performed using a Student's paired *t*-test, with $P<0.05$ considered significant.

RESULTS

Effects of Oxidatively Modified LDL on L-Arginine Transport in Endothelial and Smooth Muscle Cells

Initial rates of L-arginine transport were measured in human endothelial and smooth muscle cell monolayers preincubated for 3–24 h with complete medium in the absence or presence of either 100 μg protein/ml native or highly oxidized LDL (Figure 1). Rates of L-arginine transport were unaffected after 3–24 h pretreatment with either native or highly oxidized LDL. Furthermore, incubation of cells for 1 h in Krebs medium containing 1–100 μg protein/ml native, minimally modified, or highly oxidized LDL had no effect on L-arginine transport (data not shown). These LDL concentrations were not cytotoxic, as neither lactate dehydrogenase release nor the activity of active mitochondrial dehydrogenases was altered.

Effects of Native and Oxidatively Modified LDL on Endothelial Cell cGMP and PGI$_2$ Production

As shown in Figure 2, preincubation of human endothelial cells for 1 h (data not shown) or 24 h in the absence or presence of native, minimally modified, or highly oxidized

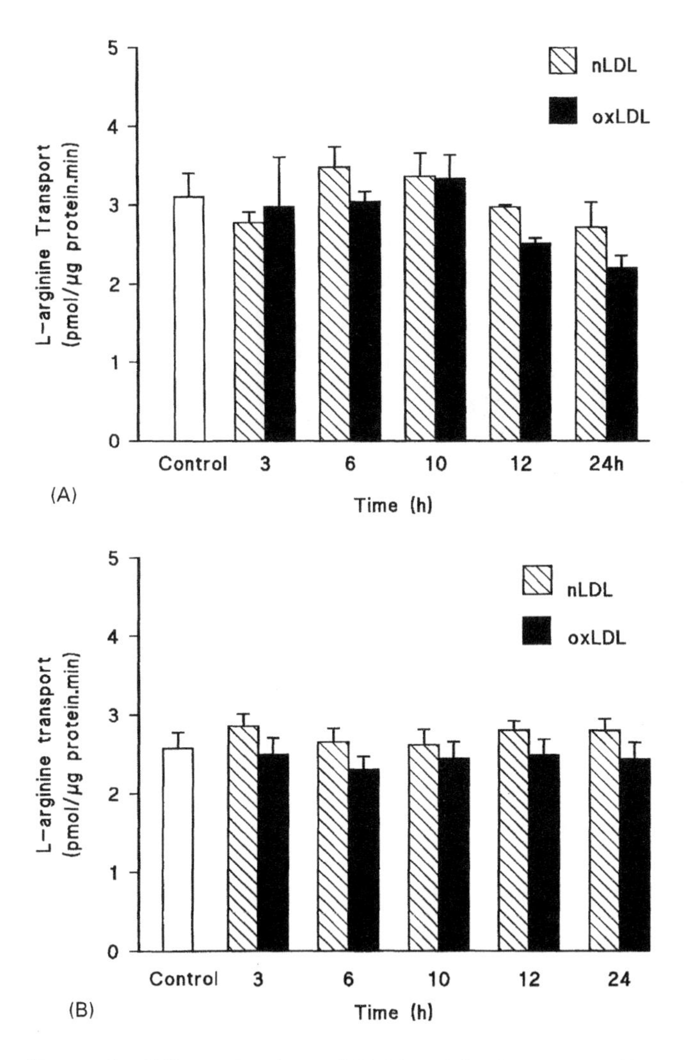

Figure 1. Effects of native and highly oxidized LDL on initial rates of L-arginine transport in human umbilical vein endothelial cells (A) and umbilical artery smooth muscle cells (B). Confluent cell monolayers were cultured in complete medium in the absence (control) or presence of either 100 μg protein/ml native LDL (nLDL) or highly oxidized LDL (oxLDL) for 3–24 h. Transport of L-[^{3}H]arginine (100 μM, 1 min) was then measured in the same cell monolayers incubated in Krebs–Henseleit medium. Values denote the mean ± S.E. of measurements in three different cell cultures. (Data from Jay et al. (31) and unpublished findings from R.C.M. Siow and G.E. Mann.)

LDL (100 μg protein/ml) had no significant effect on basal rates of cGMP production. Activation of endothelial cells with 10 μM histamine elevated cGMP levels from 1.3 ± 0.2 to 8.5 ± 1.9 cells per pmol/10^6 cells per 5 min (n=6). Although exposure of endothelial cells for 1 h to native, minimally modified or highly oxidized LDL had no effect on histamine-stimulated cGMP levels, pretreatment of cells for 24 h with highly oxidized LDL reduced agonist-stimulated cGMP production (Figure 2). Histamine-mediated changes in endothelial cell cGMP accumulation were mediated by NO, since

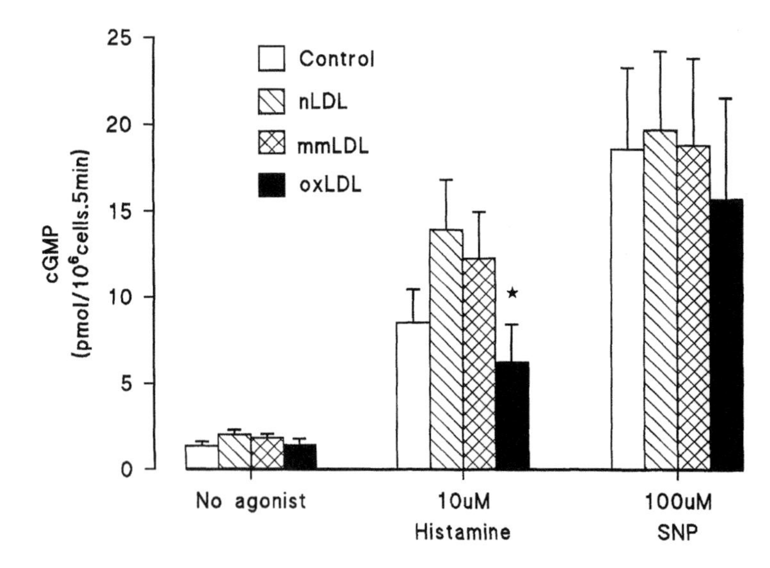

Figure 2. Effects of native and oxidatively modified LDL on intracellular cGMP accumulation in human umbilical vein endothelial cells. Confluent cell monolayers were cultured in medium 199 for 24 h in the absence (control, open column) or presence of either 100 μg protein/ml native (nLDL), minimally modified (mmLDL), or highly oxidized (oxLDL) LDL. Cell monolayers were rinsed with Krebs medium, and basal (no agonist), histamine-stimulated, and sodium nitroprusside-stimulated (SNP) cGMP production was monitored incubated in Krebs medium containing 0.5 mM isobutylmethylxanthine. Values denote the mean ± S.E. of measurements in 5–6 different cell cultures. *$P<0.05$ compared to native LDL. (Data from Jay et al. (31).)

the NO synthase inhibitor L-NAME (100 μM) abolished agonist-induced increases in cGMP (Figure 3).

When endothelial cells were challenged with sodium nitroprusside (100 μM), which liberates NO in aqueous solution, intracellular cGMP accumulation increased 19-fold but was unaffected by pretreatment of cells with native, minimally modified or highly oxidized LDL (Figure 2), indicating that the activity of soluble guanylate cyclase was not impaired by LDL.

In the same experiments, basal release of prostacyclin (PGI_2) was unaffected after pretreatment of endothelial cells for 1 h or 24 h with native, minimally modified, or highly oxidized LDL (100 μg protein/ml). Histamine increased PGI_2 production from 14 ± 6 to 102 ± 22 pmol/10^6 cells per min (n=6), and treatment of endothelial cells for 1 h or 24 h with native or minimally modified LDL (100 μg protein/ml) had no effect on rates of PGI_2 production (data not shown). In contrast, pretreatment of cells for 24 h with highly oxidized LDL reduced histamine-stimulated PGI_2 release by 50%.

Oxidatively Modified LDL Enhances Expression of Heme Oxygenase-1 in Vascular Smooth Muscle Cells

When smooth muscle cells were incubated for 48 h with native, minimally modified, or highly oxidized LDL (100 μg protein/ml), expression of heme oxygenase (HO-1) increased with the degree of LDL oxidation (Figure 4). In subsequent experiments, we examined time-dependent changes in HO-1 expression in response to 100 μg protein/ml

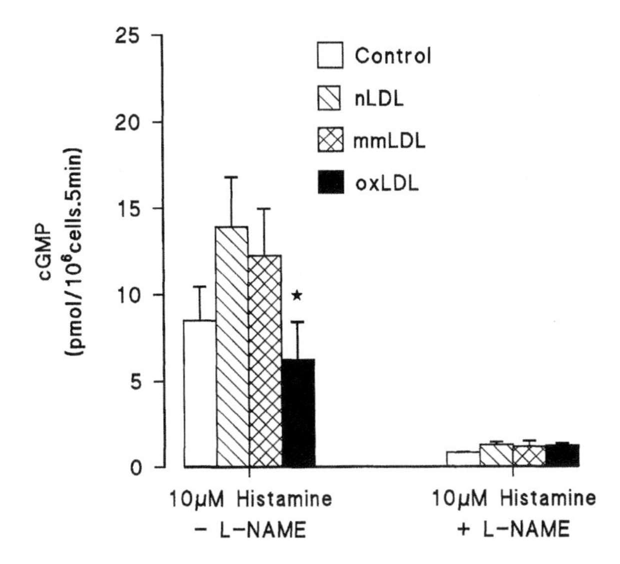

Figure 3. Inhibition of nitric oxide synthase by L-NAME abolishes histamine-stimulated cGMP production in human umbilical vein endothelial cells. Incubation of cell monolayers with the NO synthase inhibitor L-NAME prevented histamine-mediated increases in cGMP accumulation in cells preincubated for 24 h in the absence (control) or presence of 100 μg protein/ml nLDL, mmLDL, or oxLDL. Histamine data replotted from Figure 2. Values denote the mean ± S.E. of measurements in 3–6 different cell cultures. (Data from Jay et al. (31.).)

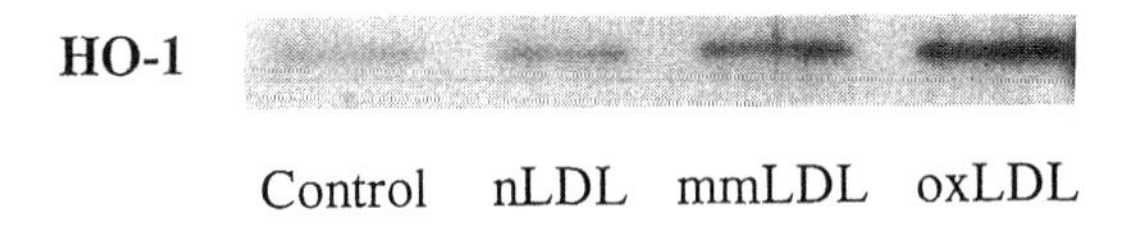

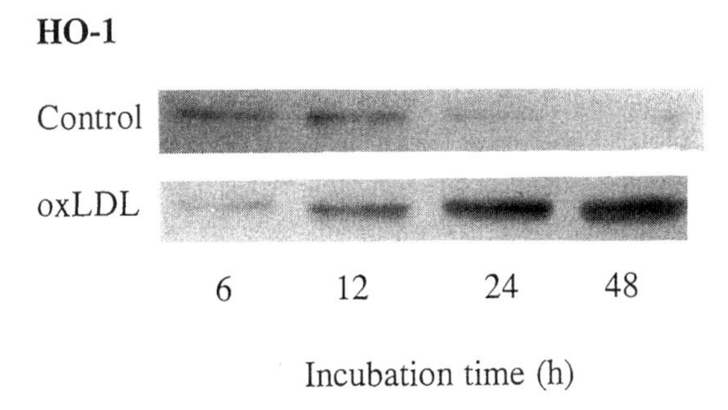

Figure 4. Western blot analyses of heme oxygenase-1 expression in porcine aortic smooth muscle cells exposed to native, minimally modified, or highly oxidized LDL. Confluent cell monolayers were incubated for 6–48 h in the absence (control) or presence of 100 μg protein/ml nLDL, mmLDL, and oxLDL. HO-1 expression was assayed by western blot analysis (see Methods). Data are representative of experiments in three different cell cultures. (Data replotted from Figure 3 and 4 of Siow et al. (23).)

highly oxidized LDL, a dose reported to produce maximal expression of HO-1 in peritoneal macrophages (32). Although HO-1 expression was not enhanced within 6 h, levels were increased maximally after 24 h and remained elevated for 48 h (Figure 4).

DISCUSSION

The present study has established that native and oxidatively modified human LDL does not inhibit L-arginine transport or basal synthesis of NO and PGI_2 in cultured vascular endothelial and smooth muscle cells. Moreover, stimulation of NO and PGI_2 production by human endothelial cells in response to histamine was insensitive to pretreatment with native or minimally modified LDL, whereas prolonged exposure (24 h) to highly oxidized LDL inhibited agonist-induced release of these vasodilators. Although oxidized LDL has been reported to inactivate soluble guanylate cyclase (sGC) (33), we found that direct activation of sGC by the NO donor sodium nitroprusside was unaffected by either native or oxidatively modified LDL. Thus, over the time course of our experiments, using carefully defined LDL preparations (29), it seems unlikely that LDL inhibited sGC activity.

The conflicting reports on the effects of oxidized LDL on eNOS expression in endothelial cells (14–17) may well reflect differences in the isolation, preparation, and definition of oxidatively modified LDL. In rabbit aortic endothelial cells, exposure for 16 h to oxidized LDL, but not native LDL, increased Ca^{2+} uptake and NO release but inhibited protein tyrosine phosphatase, suggesting that LDL may modulate both NO synthesis and tryosine dephosphorylation (14). In contrast, native LDL has been reported to reduce basal, bradykinin- (~10%) and A23187- (80%) stimulated cGMP levels in porcine aortic endothelial cells, while sodium nitroprusside (SNP) induced increases in cGMP were unaffected (34). Unlike our present findings in human endothelial cells, oxidized LDL augmented bradykinin-mediated increases in intracellular Ca^{2+} without altering agonist-induced increases in cGMP in porcine aortic endothelial cells (34). Paradoxically, oxidized LDL attenuated SNP induced cGMP increases in these same experiments (34). These discrepancies cannot be attributed to differences in experimental technique or cell type, since other studies report that oxidized LDL either decreases eNOS mRNA levels and conversion of radiolabeled L-arginine to L-citrulline (17) or increases eNOS expression (15). The literature on effects of LDL is further confounded by the observation that butylated native LDL perturbs endothelial cell oxidative metabolism and uncouples L-arginine metabolism from NO, resulting in an increased generation of superoxide anions via eNOS (35).

Inactivation of endothelium-derived NO by superoxide anions or oxidatively modified LDL itself will reduce delivery of NO to target smooth muscle cells. Recent studies have shown that oxidized LDL attenuates the antiaggregating properties of endothelium-derived NO apparently by scavenging NO rather than inhibiting its synthesis (36). Studies in human platelets have also reported that oxidized LDL causes a concentration-dependent increase in agonist-induced platelet aggregration which is reduced by pretreatment of platelets with L-arginine (37). Unlike the present findings, oxidized LDL was also found to inhibit uptake of L-arginine by platelets, which the authors concluded may account for the reduced activity and expression of NO synthase (37).

Prolonged exposure of bovine aortic endothelial cells to oxidized LDL (~3 days) decreases basal and bradykinin-stimulated PGI_2 production (38). In another study, using the same endothelial cell type, native LDL, as well as oxidized glycated LDL, was found to increase basal PGI_2 production (39). Unlike the latter findings, we were unable to detect LDL-induced changes in basal PGI_2 release from human endothelial cells. The inhibition of histamine-stimulated PGI_2 generation in our study is consistent with the observations reported for bradykinin in bovine aortic endothelial cells (38), and may reflect inhibition of the cyclooxygenase pathway and/or receptor activation by histamine.

Oxidized LDL and reactive oxygen species influence gene expression and activate nuclear transcription factors involved in the expression of iNOS and heme oxygenase-1 (40–42). Our findings of enhanced expression of HO-1 in vascular smooth muscle cells in response to either minimally modified or highly oxidized LDL confirms previous findings in cultured macrophages (32). Since oxidized LDL inhibits iNOS expression in macrophages (16), this suggests that, as in the brain (26), induction of HO-1 may compensate for diminished NOS activity in atherosclerotic lesions. The findings that brain levels of glutathione (GSH) are significantly reduced following inhibition of heme oxygenase activity with Sn-protoprophyrin (19) may have similar implications for the sensitivity of vascular endothelial and smooth muscle cells to oxidative stress agents including oxidized LDL (43).

Thus, if oxidized LDL inhibits eNOS activity in endothelial cells and iNOS is not expressed in smooth muscle cells in response to oxidized LDL, we hypothesize that the parallel induction of heme oxygenase-1 in both endothelial and smooth muscle cells will generate CO to elevate intracellular cGMP levels (24,25) to sustain normal vascular tone, whilst bilirubin would contribute to cellular antioxidant defenses (22).

ACKNOWLEDGMENTS

This work was supported in part by the Ministry of Agriculture, Fisheries and Food (ANO4/13) (UK), King's College University of London and a British Council Collaborative Research Link between Dr. G.E. Mann, King's College University of London, and Professer S. Bannai, University of Tsukuba, Japan. We thank Dr. David Leake and Mr. Justin Richards (University of Reading) for supplying us with oxidatively modified low-density lipoproteins and Dr. S. Taketani (Kansai Medical University, Japan) for providing us with the antibody for HO-1.

REFERENCES

1. Flavahan NA. Atherosclerosis or lipoprotein-induced endothelial dysfunction. Circulation 1992; 85: 1927–1938.
2. O'Brien KD, Allen MD, McDonald TO, Chait A, Harlan JM, Fishbein D, McCarty J, Fergason M, Hudkins K, Benjamin CD, Lobb R, Alpers CE. Vascular cell adhesion molecule-1 is expressed in human coronary atherosclerotic plaques. J Clin Invest 1993; 92: 945–951.
3. Witztum JL, Steinberg D. Role of oxidised low density lipoprotein in atherogenesis. J Clin Invest 1991; 88: 1785–1792.
4. Ross R. The pathogenesis of atherosclerosis: a perspective for the 1990s. Nature 1993; 362: 801–809.
5. Esterbauer H, Wag G, Puhl H. Lipid peroxidation and its role in atherosclerosis. Br Med Bull 1993; 49: 566–576.
6. Reidy MA, Boyer DE. Control of arterial smooth muscle cell proliferation. Curr Opin Lipidol 1993; 4: 349–354.

7. Campbell JH, Campbell GR. The role of smooth muscle cells in atherosclerosis. Curr Opin Lipidol 1994; 5: 323–330.
8. Moncada S, Palmer RMJ, Higgs EA. Nitric oxide: physiology, pathophysiology and pharmacology. Pharmacol Rev 1991; 43: 109–142.
9. Knowles RG, Moncada S. Nitric oxide synthases in mammals. Biochem J 1994; 298: 249–258.
10. Creager MA, Cooke JP, Mendelsohn ME, Gallagher SJ, Coleman SM, Loscalzo J, Dzau VJ. Impaired vasodilatation of forearm resistance vessels in hypercholesterolemic humans. J Clin Invest 1990; 86: 228–234.
11. Plane F, Bruckdorfer KR, Kerr P, Steuer A, Jacobs M. Oxidative modification of low density lipoproteins and the inactivation of relaxations mediated by endothelium-derived nitric oxide in rabbit aorta. Br J Pharmacol 1992; 105: 216–222.
12. Chin JH, Azhar S, Hoffman BB. Inactivation of endothelial derived relaxing factor by oxidized lipoproteins. J Clin Invest 1992; 89: 10–18.
13. Galle J, Ochslen M, Schollmeyer P, Wanner C. Oxidized lipoproteins inhibit endothelium-dependent vasodilation. Hypertension 1994; 23: 556–564.
14. Fries DM, Penha RG, D'Amico EA, Abdalla DSP, Monteiro HP. Oxidized low-density lipoprotein stimulates nitric oxide release by rabbit aortic endothelial cells. Biochem Biophys Res Commun 1995; 207: 231–237.
15. Hirata K, Miki N, Kuroda Y, Sakoda T, Kawashima S, Yokoyama M. Low concentration of oxidized low-density lipoprotein and lysophosphatidylcholine upregulate constitutive nitric oxide synthase mRNA expression in bovine aortic endothelial cells. Circ Res 1995; 76: 958–962.
16. Yang X, Cai B, Sciacca RR, Cannon PJ. Inhibition of inducible nitric oxide synthase in macrophages by oxidised low density lipoproteins. Circ Res 1994; 74: 318–328.
17. Liao JK, Shin WS, Clark SL. Oxidized low-density lipoprotein decreases the expression of endothelial nitric oxide synthase. J Biol Chem 1995; 270: 319–324.
18. Maines MD. Heme oxygenase: function, multiplicity, regulatory mechanisms and clinical applications. FASEB J 1988; 2: 2557–2568.
19. Ramos KS, Lin H, McGrath JJ. Modulation of cyclic guanosine monophosphate levels in cultured aortic smooth muscle cells by carbon monoxide. Biochem Pharmacol 1989; 38: 1368–1370.
20. Furchgott RF, Jothianandan D. Endothelium-dependent and independent vasodilation involving cyclic GMP: relaxation induced by nitric oxide, carbon monoxide and light. Blood Vessels 1991; 28: 52–61.
21. Schmidt HHHW, Lohmann SM, Walter U. The nitric oxide and cGMP signal transduction system: regulation and mechanism of action. Biochim Biophys Acta 1993; 1178: 153–175.
22. Stocker R, Yamamoto Y, McDonagh AF, Glazer AN, Ames BN. Bilirubin is an antioxidant of possible physiological significance. Science 1987; 235: 1043–1046.
23. Siow RCM, Ishii T, Sato H, Taketani S, Leake DS, Sweiry JH, Pearson JD, Bannai S, Mann GE. Induction of the antioxidant stress proteins heme oxygenase-1 and MSP23 by stress agents and oxidized LDL in cultured vascular smooth muscle cells. FEBS Lett 1995; 368: 239–242.
24. Morita, T, Perrella MA, Lee M-E, Kourembanas S. Smooth muscle cell-derived carbon monoxide is a regulator of vascular cGMP. Proc Natl Acad Sci USA 1995; 92: 1475–1479.
25. Christodoulides N, Durante W, Kroll MH, Schafer AI. Vascular smooth muscle cell heme oxygenases generate guanylyl cyclase stimulatory carbon monoxide. Circulation 1995; 91: 2306–2309.
26. Maines MD, Mark JA, Ewing JF. Heme oxygenase, a likely regulator of cGMP production in the brain: induction in vivo of HO-1 compensates for depression in NO synthase activity. Mol Cell Neurosci 1993; 4: 398–405.
27. Sobrevia L, Yudilevich DL, Mann GE. Diabetes-induced activation of system y^+ and nitric

oxide synthase in human endothelial cells: association with membrane hyperpolarization. J Physiol 1995; 489: 183–192.
28. Wilkins GM, Leake DS. The effect of inhibitors of free radical generating-enzymes on low density lipoprotein oxidation by macrophages. Biochim Biophys Acta 1994; 1211: 69–78.
29. Rice-Evans C, Leake DS, Bruckdorfer KR, Diplock, AT. Practical approaches to low density lipoprotein oxidation: whys, wherefores and pitfalls. Free Radical Res 1996; 25: 285–311.
30. Taketani S, Sato H, Yoshinaga T, Tokunaga R, Ishii T, Bannai S. Induction in mouse peritioneal macrophages of 34 kDa stress protein and heme oxygenase by sulfhydryl reactive agents. J Biochem 1990; 108: 28–32.
31. Jay MT, Chirico S, Bruckdorfer KR, Jacobs M, Leake DS, Pearson JD, Mann GE. Effects of native and oxidatively modified low density lipoproteins on L-arginine transport and nitric oxide and prostacyclin production in human endothelial cells. Exp Physiol 1996; 1997; 82: 349–360.
32. Yamaguchi M, Sato H, Bannai S. Induction of stress proteins in mouse peritoneal macrophages by oxidized low density lipoprotein. Biochem Biophys Res Commun 1993; 193: 1198–1201.
33. Schmidt K, Klatt P, Graier WF, Kostner GM, Kukovetz WR. High density lipoprotein antagonises the inhibitory effect of oxidised lipoprotein and lysolecithin on soluble guanylate cyclase. Biochem Biophys Res Commun 1992; 182: 302–308.
34. Pohl U, Heydari N, Galle J. Effects of LDL on intracellular free calcium and nitric oxide dependent cGMP formation in porcine endothelial cells. Atheriosclerosis 1995; 117: 169–178.
35. Pritchard KA, Groszek L, Smalley DM, Sessa WC, Wu M, Villalon P, Wolin MS, Stemerman MB. Native low density lipoprotein increases endothelial cell nitric oxide synthase generation of superoxide anion. Circ Res 1995; 77: 510–518.
36. Minuz P, Lechi C, Gaino S, Bonapace S, Fontana L, Gargin U, Paluani F, Cominacini L, Zatti M, Lechi A. Oxidized LDL and reduction of the antiaggregating activity of nitric oxide derived from endothelial cells. Thromb Haemostasis 1995; 74: 1175–1179.
37. Chen LY, Mehta P, Mehta JL. Oxidized LDL decreases L-arginine uptake and nitric oxide synthase protein expression in human platelets: relevance of the effect of oxidized LDL on platelet function. Circulation 1996; 93: 1740–1746.
38. Thorin E, Hamilton CA, Cominiczak MH, Reid JL. Chronic exposure of cultured bovine endothelial cells to oxidized LDL abolishes prostacyclin release. Arteriosclerosis Thromb 1994; 14: 453–459.
39. Kobayashi K, Watanabe J, Umeda F, Taniguchi S, Masakado M, Yamauchi T, Nawata H. Enhancement of prostacyclin production in cultured bovine aortic endothelial cells by oxidized glycated low density lipoprotein. Prostaglandins Leukotrienes Essential Fatty Acids 1995; 52: 263–270.
40. Andalibi A, Liao F, Imes S, Fogelman AM, Lusis AJ. Oxidized lipoproteins influence gene expression by causing oxidative stress and activating the transcription factor NF-κB. Biochem Soc Trans 1993; 21: 651–655.
41. Adcock IM, Brown CR, Kwon O, Barnes PJ. Oxidative stress induces NFκB DNA binding and inducible NOS mRNA in human epithelial cells. Biochem Biophys Res 1994; Commun 199; 1518–1524.
42. Lavrovsky Y, Schwartzman ML, Levere RD, Kappas A, Abraham NG. Identification of binding sites for transcription factor NFκB and AP-2 in the promoter region of the human heme oxygenase-1 gene. Proc Natl Acad Sci USA 1994; 91: 5987–5991.
43. Jornot L, Junod AF. Variable glutathione levels and expression of antioxidant enzymes in human endothelial cells. Am J Physiol 1993; 264: L482–L489.

14

Nitric Oxide–stimulated Tyrosine Phosphorylation-dependent Signaling Pathways in Cultured Cells

Hugo P. Monteiro, Laura C.B. Oliveira, Tereza M.S. Peranovich, and Renatta G. Penha
Fundação Pró-Sangue Hemocentro de São Paulo
São Paulo, Brazil

Arnold Stern
New York University Medical Center
New York, New York

INTRODUCTION

Reactive oxygen species, sulfhydryl oxidants, and the reactive nitrogen species nitric oxide (NO) are thought to mediate signal transduction events in a number of cell lines (1). In particular, NO plays a major role in the vascular homeostasis controlling the enzyme guanylyl cyclase and modulating the activities of protein tyrosine kinases (PTK) and protein tyrosine phosphatases (PTP) (2,3). Tyrosine phosphorylation is associated with mitogenesis, cell transformation, and cell death. It is a dynamic and reversible process controlled by the opposing actions of PTK and PTP (4). Binding of a polypeptide growth factor (e.g., the epidermal growth factor, EGF; platelet-derived growth factor, PDGF; or fibroblast growth factor, FGF) to its cognate receptor triggers the receptor tyrosine kinase activity, leading to a cascade of phosphorylation reactions. In this report we will focus on the NO-modulated tyrosine phosphorylation/dephosphorylation signaling pathways in cultured cells. NO generated through autoxidation of sodium nitroprusside (SNP) and *S*-nitroso-*N*-acetylpenicillamine stimulated tyrosine phosphorylation of a group of proteins (p126, p56, and p43) in HER14 cells (murine fibroblasts expressing the human EGF receptor (5)). Furthermore, NO also potentiated EGF-stimulated PTK activity in these cells, promoting tyrosine phosphorylation of the same set of proteins (3,6). NO-dependent inhibition of PTP activities in HER14 cells was also reported (3,6). Using monoclonal antibodies against proteins of the growth factor-mediated signaling cascade, we obtained evidences indicating that the p126 and p43 proteins are focal adhesion kinase (FAK) and mitogen-activated protein (MAP) kinase, respectively. These results suggest that the NO-dependent tyrosine phosphorylation signaling pathway in murine fibroblasts associates FAK and MAP kinase; in this aspect it is similar to the integrin-mediated cell adhesion process in these cells (7). Further work using cultured smooth-muscle cells from rabbit aorta (RASMC) stimulated with lipopolysaccharide (LPS) was centered on the endogenous production of NO and its implications on tyrosine dephosphorylation signaling processes. RASMC at different densities under LPS stimulation produced different amounts of NO. Nonconfluent RASMC stimulated with the endotoxin released significant amounts of NO. Conversely, cells in multilayers were much less effective. Additionally, PTP activities from RASMC stimulated with LPS were also affected according to the cell density.

MATERIALS AND METHODS

Treatment of Intact Cells

HER14 cells (murine fibroblasts expressing the human EGF receptor) were cultivated in Dulbecco's modified Eagle's medium (DMEM) supplemented with 10% FBS. For experiments, cells were starved in DMEM supplemented with 0.5% FBS for 48 h. They were incubated, with or without agents, in DMEM supplemented with 20 mM Hepes buffer, pH 7.5, and 1 mg/ml bovine serum albumin (BSA) for 10 min or as otherwise indicated, at 25°C. RASMC were grown in Ham's nutrient mixture F12 (F12 medium) supplemented with 10% FBS. For determinations of NO release and PTP activities, RASMC were starved during 24 h in F12 medium supplemented with 1 mg/ml BSA and stimulated with LPS at different concentrations (10–100 ng/ml) during 24 and 48 h.

Assay for Nitric Oxide

NO released by RASMC incubated in the presence and absence of LPS was estimated by determining the concentrations of nitrite in the incubation medium using the Griess reagent as previously described (8)

Assay of PTP Activities

PTP activities were determined in RASMC lysates as the EGF receptor tyrosine phosphatase activity (9). After incubations, cells were lysed in lysis buffer A (20 mM Hepes, pH 7.5; 150 mM NaCl; 10% glycerol; 1% Triton X-100; 1.5 mM $MgCl_2$; 1 mM EGTA; 1 μg/ml aprotinin; 1 μg/ml leupeptin, and 1 mM phenylmethylsulfonylfluoride). After 30 min incubation on ice, lysates were clarified by centrifugation at 12,000 *g* for 10 min at 4°C. The supernatants were collected, their protein content was determined (10), and they served as the source of PTP activity. PTP activity was assayed by mixing 100 μg protein from lysates with a suspension containing immunoprecipitated ^{32}P-labeled EGF receptors, prepared as described by Honegger et al. (5). Reactions were carried out for 30 min at 37°C with agitation, and terminated by aspiration of the supernatants. The radiolabeled immunoprecipitates were collected and mixed with 50 μl of 2-fold concentrated Laemmli's sample buffer and resolved on 7.5% SDS-PAGE. After drying, the gels were exposed overnight to autoradiographic films. PTP activity was estimated as a decrease of the intensity of the corresponding band to the EGF receptor in the autoradiographs. Band intensities in the autoradiographs were subjected to densitometric analysis using an LKB Ultroscan Laser Enhanced Densitometer.

Immunoblotting

For immunoblotting experiments, HER14 cells were lysed in lysis buffer A containing phosphatases inhibitors. Total cell lysates had their protein content determined (10) and equal amounts of proteins (75 μg) were resolved in 10% SDS–polyacrylamide gels. Gels were blotted onto nitrocellulose sheets and blots were probed with monoclonal antibodies against tyrosine phosphorylated proteins (anti-PY; Upstate Biotechnology), MAP kinase (anti-MAPK; Gibco-BRL), and FAK (kindly provided by Dr. J. Schlessinger, NYU Medical Center). After incubation with horseradish peroxidase-conjugated-secondary antibody, the blots were developed using the enhanced chemiluminescence system ECL from Amersham.

RESULTS AND DISCUSSION

Effects of Exogenous NO

Incubation of HER14 cells with the NO donor SNP resulted in stimulation of tyrosine phosphorylation of a set of proteins: p126, p56, and p43 (Figure 1) (3). Because MAP kinases (40–44 kDa proteins) are major routes of intracellular signaling pathways (11), and their activities could be stimulated under oxidizing conditions (12), we examined

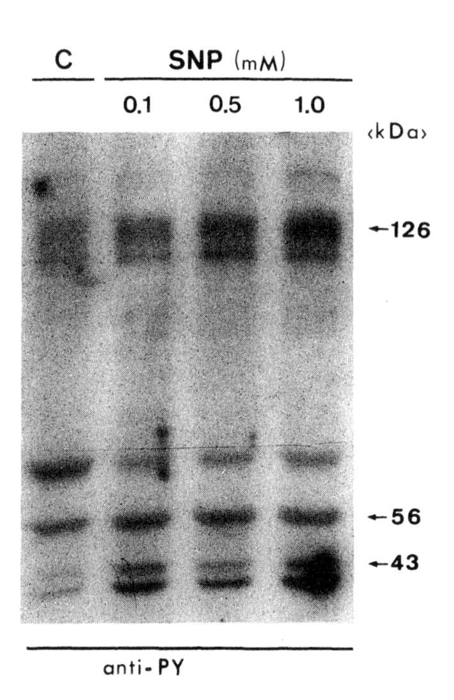

Figure 1. Effects of SNP on tyrosine phosphorylation in HER14 cells starved for 48 h. Cells were starved during 48 h and after this period were treated with increasing concentrations of SNP (0–1.0 mM). After 10 min incubation, cells were lysed. Lysate protein content was determined and equal amounts of protein were resolved by SDS-PAGE (7.5% gel) and transferred to nitrocellulose membranes. Immunoblotting was performed with anti-PY antibodies as described in Materials and Methods. Lane 1, control; lane 2, 0.1 mM SNP; lane 3, 0.5 mM; lane 4, 1.0 mM.

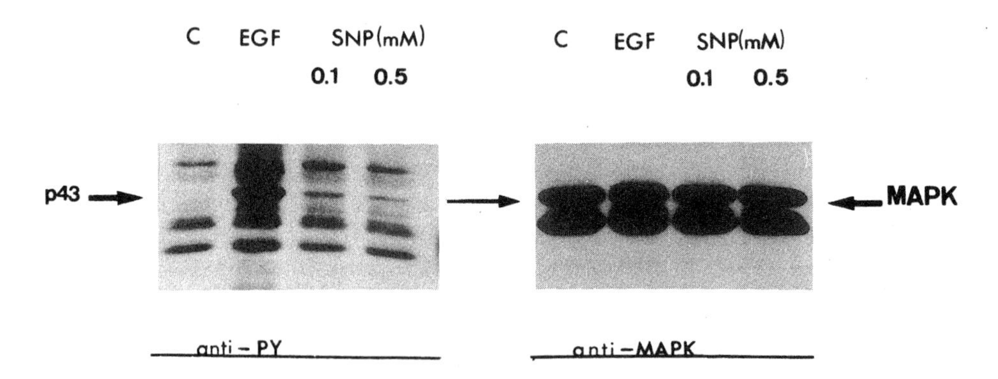

Figure 2. Comigration of tyrosine phosphorylated p43 with a MAP kinase immunoreactive protein. Cells were starved during 48 h and after this period were treated with EGF (83 nM) or SNP (0.1 and 0.5 mM). After 5 min (EGF) or 10 min (SNP) incubation, cells were lysed. Their protein content was determined and equal amounts of protein were resolved by SDS-PAGE (10% gel) and transferred to nitrocellulose membranes. Immunoblotting was performed with anti-PY (autoradiogram on the left) and anti-MAPK (autoradiogram on the right) antibodies, as described in Materials and Methods.

whether p43 was related to these kinases. Furthermore, p21ras, an activator of the MAP kinases (13), is stimulated by NO (14). We showed that the p43 protein phosphorylated on tyrosine after incubation with the NO donor comigrates with the 44 kDa protein band of a doublet (42 and 44 kDa) that is immunoreactive with the anti-MAP2 kinase mouse monoclonal antibody (Materials and Methods) (Figure 2). Integrins, growth factors, and hormones have been reported to induce the tyrosine phosphorylation of a 125 kDa protein called focal adhesion kinase (FAK) (15). Furthermore, integrin-mediated cell adhesion causes the activation of FAK and MAP kinases. Although not proven, both kinases could be part of the same signaling pathway (7). Thus, on the basis of these observations and the suggested activation of MAP kinases by NO (this work), we investigated whether NO-dependent tyrosine phosphorylated p126 was related to FAK. The p126 comigrates with the 125 kDa protein band that is immunoreactive with the mouse monoclonal antibody anti-FAK (Figure 3).

Taken together, the results described above suggest that the NO-mediated signaling pathway in fibroblasts apparently involves the activation of MAP kinases and FAK. An initiating event in this signaling cascade seems to be the activation of p21ras by NO (14), followed by phosphorylation of the MAP kinases (this work). Furthermore, FAK could undergo tyrosine phosphorylation caused by NO (this work). Putatively, phosphorylated FAK would recruit adapter proteins such as GRB2 and possibly Src-like kinases, linking tyrosine phosphorylated FAK to its downstream targets such as MAP kinases. We summarize the proposed pathway in Figure 4.

Effects of Endogenous NO

Bacterial LPS is an inducer of NO synthase in vascular smooth-muscle cells (16). Taking advantage of this system, we investigated the effects of the endogenously produced NO on tyrosine phosphorylation and proliferation of RASMC stimulated with LPS from *E. coli*.

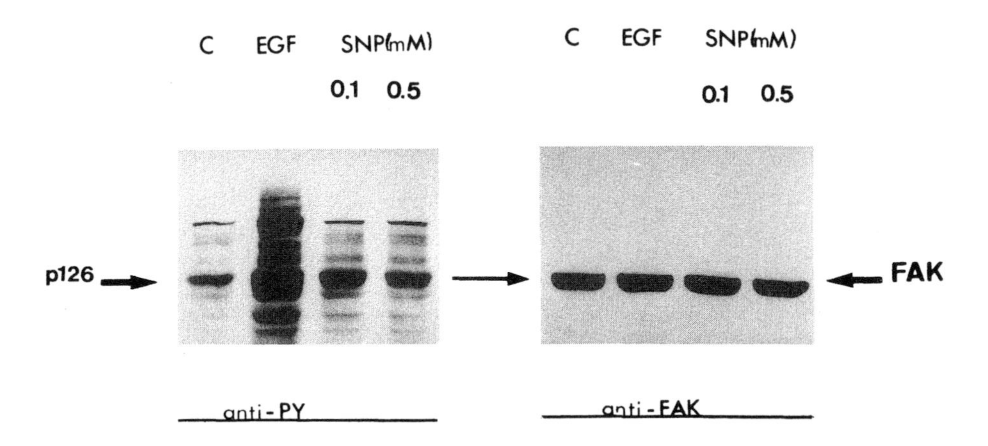

Figure 3. Comigration of tyrosine phosphorylated p43 with a focal adhesion kinase (FAK) immunoreactive protein. Cells were starved during 48 h and after this period were treated with EGF (83 nM) or SNP (0.1 and 0.5 mM). After 5 min (EGF) or 10 min (SNP) incubation, cells were lysed. Their protein content was determined and equal amounts of protein were resolved by SDS-PAGE (10% gel) and transferred to nitrocellulose membranes. Immunoblotting was performed with anti-PY (autoradiogram on the left) and anti-FAK (autoradiogram on the right) antibodies as described in Materials and Methods.

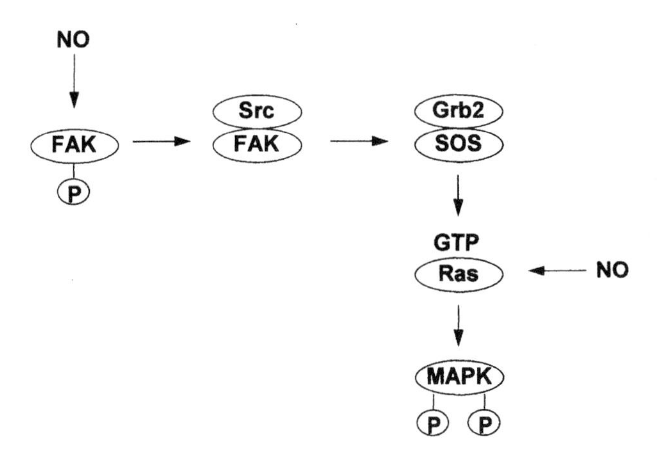

Figure 4. Schematic view of a putative NO-mediated signaling pathway.

Initially, a time course for stimulation of RASMC with different doses of the endotoxin was established. If cells were allowed to grow to multilayers (seeded at 6.5×10^5 cells/ $56\,cm^2$ dish), presenting their typical "hill and valley" morphology, after stimulation with LPS (0, 10 or 100 ng/ml) during 24 h there was no significant production of NO above basal levels (Figure 5A). However, nonconfluent cultures of RASMC (seeded at 2.5×10^5 cells/$56\,cm^2$ dish) responded positively to 10 ng/ml LPS stimulation after 24 h incubation (Figure 5B). It is noteworthy that nonconfluent cultures of RASMC in the absence of stimulus released greater amounts of NO in comparison to cells in multilayers.

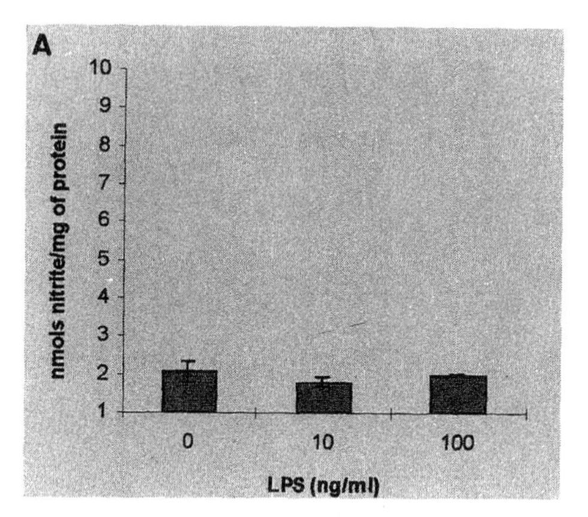

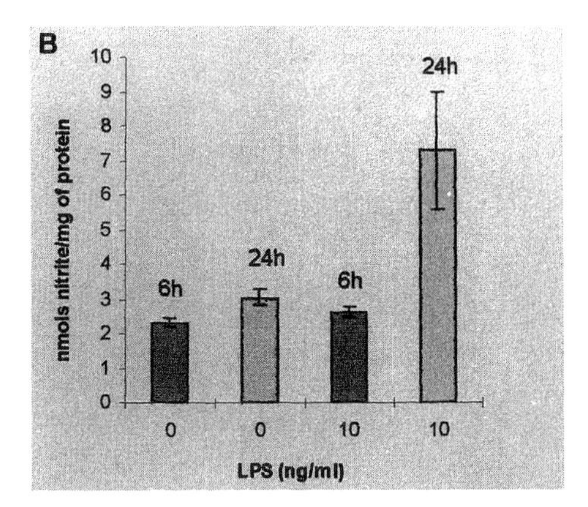

Figure 5. Effects of LPS on NO release by RASMC. RASMC were serum-starved during 24 h before incubation with LPS. (A) Cells in multilayers were incubated with LPS at the indicated concentrations during 24h. (B) Nonconfluent cells were incubated with LPS (10 ng/ml) during 6 and 24 h. After incubation, NO release was determined as accumulated nitrite in the supernatants of the cells cultures as described in Materials and Methods.

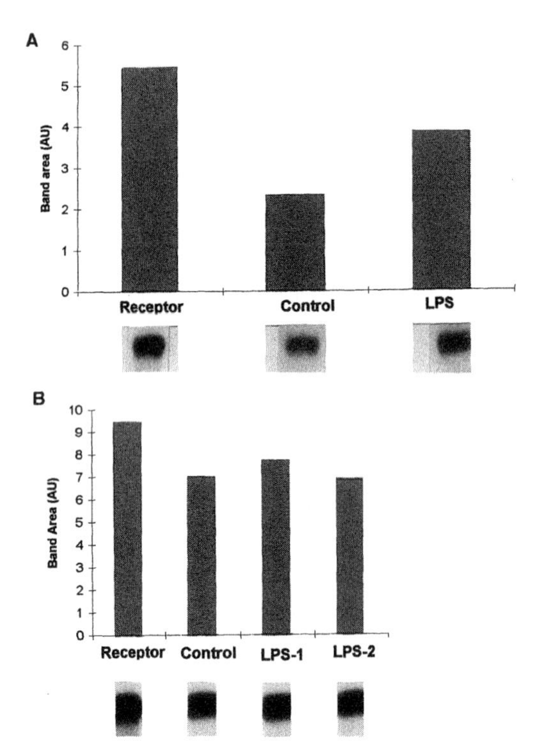

Figure 6. Modulation of PTP activities by LPS in nonconfluent RASMC cultures and in RASMC cultivated in multilayers. (A) Serum-starved nonconfluent cultures of RASMC were incubated during 24 h with LPS (10 ng/ml). (B) Serum-starved cultures of RASMC in multilayers were incubated with two concentrations of LPS, LPS-1 (10 ng/ml) and LPS-2 (100 ng/ml). Control cells were maintained in the absence of LPS for the incubation period. Reaction was terminated and homogenates were prepared to measure PTP activity using ^{32}P-labeled EGF receptors as substrate, as described in Materials and Methods. The corresponding area for the EGF receptor band was determined by laser densitometric analysis.

Concomitantly with the measurements of NO production in the supernatants, we also investigated the consequences of LPS stimulation on PTP activities in RASMC. RASMC-associated PTP activities, determined as the EGF receptor tyrosine phosphatase activity, were inhibited when cells from nonconfluent cultures were incubated with 10 ng/ml LPS during 24 h (Figure 6A). Conversely, under the same experimental conditions, no inhibition of PTP activities was detected in cells in multilayers (Figure 6B). Vanadate, an inhibitor of PTP activities, was shown to be growth stimulatory (17,18). Consequently, the inhibition of PTP activities in LPS-treated nonconfluent RASMC could promote cell growth. Furthermore, Clementi et al. (19) have shown an NO-mediated production of cyclic GMP by the growth factors, EGF, PDGF, and FGF in cultured fibroblasts. Production of the cyclic nucleotide was intrinsically associated with the growth factor-mediated signaling. In conclusion, these observations taken together suggest that NO could be a modulator of tyrosine phosphorylation/dephosphorylation-dependent signaling processes in living cells.

ACKNOWLEDGMENTS

The authors acknowledge the financial support of FAPESP-Brazil (Proc. no. 94/1418–6) to H.P.M. and of NIH-USA (ESO3425) to A.S.

REFERENCES

1. Monteiro HP, Stern A. Redox modulation of tyrosine phosphorylation/dependent signal transduction pathways. Free Radical Biol Med; 1996; 21: 323–333.
2. Ignarro LJ. Endothelium-derived nitric oxide: actions and properties. FASEB J 1989; 3: 31–36.
3. Peranovich, TMS, da Silva AM, Fries DM, Stern A, Monteiro HP. Nitric oxide stimulates tyrosine phosphorylation in murine fibroblasts in the absence and presence of epidermal growth factor. Biochem J 1995; 305:613–619.
4. Ullrich A, Schlessinger J. Signal transduction by receptors with tyrosine kinase activity. Cell 1990; 61:203–212.
5. Honegger AM, Dull JS, Felder S, van Obberghen E, Bellot F, Szapary D, Schmidt A, Ullrich A, Schlessinger, J. Point mutation at the ATP binding site of EGF receptor abolishes protein-tyrosine kinase activity and alters cellular routing. Cell 1987; 51: 199–209.
6. Monteiro HP, Peranovich TMS, Fries DM, Stern A, Silva AM. Nitric oxide potentiates EGF-stimulated tyrosine kinase activity in 3T3 cells expressing human EGF receptors. In Asada K, Yoshikawa T, eds. Frontiers of Reactive Oxygen Species in Biology and Medicine. Amsterdam: Elsevier Science, 1994:215–218.
7. Chen Q, Kinch MS, Lin TH, Burridge K, Juliano RL. Integrin-mediated cell adhesion activates mitogen-activated protein kinases. J Biol Chem 1994; 269: 26602–26605.
8. Stuehr DJ, Nathan CF. Nitric oxide: a macrophage product responsible for cytostasis and respiratory inhibition in tumor target cells. J Exp Med 1989; 169: 1543–1545
9. Monteiro HP, Ivaschenko Y, Fischer R, Stern A. Inhibition of protein tyrosine phosphatase activity by diamide is reversed by epidermal growth factor in fibroblasts. FEBS Lett 1991; 295:146–148.
10. Bradford MM. A rapid and sensitive method for the quantitation of microgram quantities of protein utilizing the principle of protein–dye binding. Anal Biochem 1976; 72:248–254.
11. Seger R, Krebs EG. The MAPK signaling cascade. FASEB J 1995; 9:726–735.

12. Stevenson MA, Pollock SS, Coleman CN, Calderwood SK. X-Irradiation, phorbol esters and H_2O_2 stimulate mitogen-activated protein kinase activity in NIH-3T3 cells through the formation of reactive oxygen intermediates. Cancer Res 1994; 54:12–15.
13. Leevers SJ, Marshall CJ. Activation of extracellular signal-regulated kinase, ERK2, by p21 ras oncoprotein. EMBO J 1992; 11:569–574.
14. Lander HM, Ogiste JS, Pearce SFA, Levi R, Novogrodsky A. Nitric oxide-stimulated guanine nucleotide exchange on p21 ras J Biol Chem 1995; 270:7017–7020.
15. Schaller MD, Parsons JT. Focal adhesion kinase and associated proteins. Curr Opin Cell Biol 1994; 6:705–710.
16. Mckenna TM. Prolonged exposure of rat aorta to low levels of endotoxin in vitro results in impaired contractility. J Clin Invest 1990; 86:160–168.
17. Klarlund JK. Transformation of cells by an inhibitor of phosphatases acting on phosphotyrosine in proteins. Cell 1985; 41:707–717.
18. Itkes AV, Imamova LR, Alexandrova NM, Favorova OO, Kisselev LL. Expression of c-*myc* gene in human ovary carcinoma cells treated with vanadate. Exp Cell Res 1993; 188:169–171.
19. Clementi E, Sciorati C, Riccio M, Miloso M, Meldolesi, J, Nistico G. Nitric oxide action on growth factor-elicited signals. J Biol Chem 1995; 270:22277–22282.

15

Regulation of Cell Activation by Receptors for IgG

Marc Daëron and Wolf H. Fridman
Laboratoire d'Immunologie Cellulaire et Clinique
INSERM U255
Institut Curie
Paris, France

THE IMMUNORECEPTOR FAMILY

Receptors involved in the recognition of antigens by cells of the immune system are of three types: B cell receptors (BCR), T cell receptors (TCR), and Fc receptors (FcR). They are now understood as structurally and functionally related members of the same family, referred to as the immunoreceptor family.

BCR are composed of an antigen recognition subunit, consisting of a membrane immunoglobulin molecule, and of heterodimeric Igα-Igβ signal transduction subunits (1). TCR are composed of a peptide + MHC recognition subunit, consisting of an α-β or γ-δ heterodimer, and of three signal transduction subunits: two heterodimers, CD3ε-δ and CD3ε-γ, and one homodimer, TCRζ (2). FcR are, for most of them, composed of an immunoglobulin-binding α subunit, associated with a homodimeric FcRγ subunit (3). In mast cells, FcR that possess FcRγ also associate with a single-chain FcRβ subunit (4) having four transmembrane domains and whose amino- and carboxy-terminal ends are both intracytoplasmic (5). FcRγ and TCRζ are functionally and structurally related (6). They are both hardly exposed to the outside and they possess a transmembrane domain having a charged residue, involved in the association with the other subunits. They can each associate either with FcR or TCR and, in some cases, they can form γ-ζ heterodimers (6).

BCR, TCR, and FcR possess a common motif, in the intracytoplasmic domains of the signal transduction subunits Igα, Igβ, CD3ε, CD3δ, CD3γ, TCRζ, FcRγ, and FcRβ (7). This motif, composed by a twice-repeated YxxL sequence flanking seven variable residues, was understood to account for the cell-triggering properties of receptors possessing these subunits and designated as immunoreceptor Tyrosine-based activation motif (ITAM) (8). An ITAM was also found in the intracytoplasmic domain of a single-chain low-affinity receptor for IgG unique to humans and referred to as FcγRIIA. The FcγRIIA ITAM, however, possess 12 residues, instead of seven, between the two YxxL sequences (7). Upon receptor aggregation, ITAMs become rapidly tyrosine-phosphorylated by one or several protein tyrosine kinases of the *Src* family. Phospho-ITAMs then provide docking sites for cytoplasmic protein tyrosine kinases of the *Syk* family whose two SH2 domains bind each to one phosphorylated YxxL motif of ITAMs. Phospho-ITAM-bound *Syk* or ZAP70 become phosphorylated by *Src* kinases and activated. They then phosphorylate other substrates, among which is phospholipase C_γ, which initiates the turnover of phosphatidylinositides, itself resulting in calcium mobilization and, ultimately, in cell activation (9).

Although FcR that possess ITAMs can trigger cell activation, not all FcR possess ITAMs. FcγRIIB are single-chain low-affinity IgG receptors that have no ITAM (10–15) and cannot activate cells (16,17). They are encoded by a single gene in which a separate exon encodes the transmembrane domain and three other exons the intracytoplasmic domain (18). This enables several isoforms to be generated by alternative splicing of corresponding sequences. In humans and mice, FcγRIIB1 retain sequences encoded by all four exons, whereas FcγRIIB2 lack sequences encoded by the first intracytoplasmic exon. The FcγRIIB1-specific insertion encoded by the first IC exon is longer, however, in the murine receptor (47 amino acids) than in the human receptor (19 amino acids). An additional murine isoform, resulting from the use of a cryptic splice donor site, in the first intracytoplasmic murine exon, has been described by us (19). Rather than FcγRIIB1, it is the actually the murine homologue of human FcγRIIB1 and, for this reason, we have named it FcγRIIB1′. FcγRIIB are widely expressed by cells of hematopoietic origin. FcγRIIB1 are preferentially expressed by cells of the lymphoid lineage, FcγRIIB2 by cells of the myeloid lineage and murine FcγRIIB1′ by cells of both lineages.

When aggregated by multivalent ligands, FcγRIIB are involved in capping, endocytosis, and phagocytosis. A tyrosine-containing 13-amino-acid sequence, encoded by the third intracytoplasmic exon, determines the ability of both murine and human FcγRIIB2 to mediate the endocytosis of soluble immune complexes via clathrin-coated pits (20). Another tyrosine-containing sequence, nearer to the carboxy-terminus, accounts for the ability of murine FcγRIIB2 to trigger the phagocytosis of particulate immune complexes (21). Human FcγRIIB2, in which the carboxy-terminal tyrosine involved in phagocytosis by murine FcγRIIB2 is not conserved, do not trigger phagocytosis (22). Sequences encoded by the first intracytoplasmic exon inhibit the internalization properties of carboxy-terminal sequences (23). They probably account for the inability of murine FcγRIIB1 and FcγRIIB1′ to mediate endocytosis and phagocytosis and for the inability of human FcγRIIB1 to mediate endocytosis. Finally, sequences encoded by the first intracytoplasmic exon are necessary and sufficient for capping (24), a property of FcγRIIB1 and FcγRIIB1′ (19), when aggregated at 37 °C. FcγRIIB therefore appear as a subfamily of receptors, with a wide tissue distribution, unable to activate cells, and whose biological properties depend on the various combinations of effector and regulatory sequences in the intracytoplasmic domain which characterizes each isoform.

FcγRIIB exert another major biological function when they are coaggregated to other immunoreceptors, instead of being simply aggregated to themselves. By contrast with antigen receptors that are clonally expressed on lymphocytes, FcR are not clonally distributed, and a single cell usually expresses more than one type of FcR. On many cells, FcγRIIB are coexpressed with immunoreceptors that can trigger cell activation. FcγRIIB are expressed together with BCR on B cells (25), together with TCR on activated T cells (26,27), together with FcεRI and FcγRIIIA on mast cells (28) and

Langerhans cells (29), together with FcγRI and FcγRIIIA on macrophages (30). This enables FcγRIIB to become coaggregated with adjacent immunoreceptors. Under these conditions, FcγRIIB negatively regulate cell activation triggered by ITAM-containing immunoreceptors.

FcγRIIB INHIBIT BCR-MEDIATED B CELL ACTIVATION

Passively administered antigen-specific IgG antibodies were shown, thirty years ago, to inhibit primary responses to that antigen in vivo (31) and in vitro (32). These early experiments (reviewed in Ref. 33) prompted immunologists to adopt the endocrinology-derived concept of negative feedback regulation for antibody production. The latter was extended to other antibody-dependent immunoregulatory phenomena such as allotype suppression, induced by anti-allotype antibodies (34), and idiotype suppression, induced by anti-idiotype antibodies (35).

The regulatory effects of IgG antibodies were demonstrated to depend on the Fc portion of the suppressive antibodies. In order to inhibit a primary anti-sheep red blood cell (SRBC) antibody response in vivo, anti-SRBC IgG must indeed possess an intact Fc portion (36). Likewise, allotype suppression and idiotype suppression were not induced by $F(ab')_2$ fragments of anti-allotype (34) or anti-idiotype (37) antibodies, respectively. Finally, anti-IgM IgG antibodies were found to be unable to activate B cells, whereas $F(ab')_2$ fragments of the same antibodies triggered polyclonal B cell activation (38). These observations together supported the concept first proposed by Sinclair as the *tripartite model of inactivation*, according to which the aggregation of membrane immunoglobulins on B cells, by multivalent antigens, anti-allotype, anti-idiotype, or anti-isotype $F(ab')_2$, would lead to B cell activation, whereas the coaggregation of membrane immunoglobulins to FcγR by corresponding intact antibodies, on the same cell, would induce a state of B cell inactivation (36).

A direct demonstration of the role of B cell FcγR in antibody-mediated inhibition was made possible when the rat anti-mouse FcγR monoclonal antibody 2.4G2 became available (39). Indeed, intact IgG anti-mouse immunoglobulin antibodies could stimulate B cells if FcγR were made inaccessible to the Fc portion of antibodies by preincubating cells with 2.4G2 (40). After cDNAs encoding murine FcγR were cloned and the distinction between three low-affinity FcγR was made (10–12), mouse B cells were found to express only FcγRIIB1 (25). Definitive evidence for the role of this FcγR isoform in inhibition was obtained by reconstitution experiments in the FcγR-negative murine lymphoma B cell IIA1.6 (41). These cells have a functional BCR whose aggregation either by $F(ab')_2$ fragments of rabbit anti-mouse IgG (RAM) or by intact RAM IgG triggers an increase of the concentration of intracellular Ca^{2+} and the secretion of IL-2. IIA1.6 cells were stably transfected with cDNA encoding mouse FcγRIIB1, FcγRIIB1′, FcγRIIB2, or FcγRIIB whose intracytoplamsic domain had been deleted, and stimulated with either RAM $F(ab')_2$ or RAM IgG. RAM $F(ab')_2$ triggered the secretion of IL-2 in all three transfectants whereas RAM IgG, which stimulated tail-less FcγRIIB-expressing transfectants, failed to stimulate cells expressing FcγRIIB1, FcγRIIB1′, or FcγRIIB2 (19,24). The same results were obtained with human FcγRIIB1, FcγRIIB2, or tail-less FcγRIIB, respectively, stably expressed in IIA1.6 cells (42). Taken together these experiments showed that the coaggregation of murine BCR to murine or human FcγRIIB of either isoform inhibited cell activation provided they had an intact intracytoplasmic domain. That all FcγRIIB isoforms could inhibit B cell activation equally well was somewhat surprising for a regulatory process restricted so far to B cells, which do not express FcγRIIB2.

FcγRIIB INHIBIT FcR-MEDIATED MAST CELL ACTIVATION

If one now accepts BCR as being one among other ITAM-possessing immunoreceptors, one may speculate that the regulatory properties of FcγRIIB might not be restricted to B cells.

We and others reported previously that mouse mast cells coexpress FcεRI, FcγRIIIA, and FcγRIIB (28,43). In order to examine whether FcγRIIB might inhibit mast cell activation induced by IgE antibodies, we constructed a transfectant model in the rat mast cell line RBL-2H3 (44). These cells express functional FcεRI whose aggregation triggers the release of inflammatory mediators and cytokines. They were stably transfected with cDNAs encoding mouse FcγRIIB1, FcγRIIB2, or FcγRIIB whose intracytoplasmic domain had either been deleted or replaced by that of the α subunit of mouse FcγRIIIA. Transfectants released mediators upon FcεRI aggrgegation, but not upon murine FcγRIIB aggregation (16). Mediator release, triggered by FcεRI aggregation, was inhibited when FcεRI were coaggregated to FcγRIIB. Both FcγRIIB1 and FcγRIIB2 were inhibitory, but not FcγRIIB without intracytoplasmic domain or with that of FcγRIIIA (45).

Likewise, IgE-induced mediator release, triggered by FcεRI aggregation, was inhibited when FcεRI were coaggregated to human FcγRIIB stably expressed in RBL cells. Both human FcγRIIB1 and FcγRIIB2 were inhibitory, but not FcγRIIB without intracytoplasmic domain (27).

Since human FcγRIIA also possess an ITAM, RBL transfectants expressing human FcγRIIB1 or FcγRIIB without intracytoplasmic domain were retransfected with cDNA encoding human FcγRIIA. Both transfectants released serotonin in response to human FcγRIIA aggregation. Serotonin release, however, was inhibited when FcγRIIA were coaggregated with human FcγRIIB1, but not when they were coaggregated with FcγRIIB without intracytoplasmic domain (27).

FcγRIIB INHIBIT TCR-MEDIATED T CELL ACTIVATION

Since, as anticipated, the inhitory properties of FcγRIIB are not restricted to BCR-mediated B cell activation, but affect also FcR-mediated mast cell activation, one can expect that FcγRIIB might inhibit TCR-mediated T cell activation as well. This was investigated using, as a model system, the murine BW5147 thymoma cells whose TCR had previously been reconstituted by the stable transfection of cDNA encoding mouse CD3ε and TCRζ (46). These cells were stably transfected with cDNA encoding murine FcγRIIB1, FcγRIIB2, or FcγRIIB without intracytoplasmic domain. Resulting transfectants had a functional TCR whose aggregation triggered cells to secrete IL-2. IL-2 secretion was inhibited if BW5147 TCR were coaggregated to FcγRIIB1 or FcγRIIB2. Under the same conditions, no inhibition was observed when TCR was coaggregated to FcγRIIB without intracytoplasmic domain (27). Like BCR-mediated B cell responses and FcR-mediated mast cells responses, TCR-mediated T cell responses were therefore inhibited when TCR was coaggregated to FcγRIIB having an intact intracytoplasmic domain.

FcγRIIB INHIBIT CELL ACTIVATION BY ALL ITAM-BASED IMMUNORECEPTORS

To determine whether negative regulation exerted by FcγRIIB requires associated chains of multisubunit receptors, chimeric molecules were constructed by joining cDNA sequences encoding the ITAM-containing intracytoplasmic domain of an immunoreceptor and cDNA sequences encoding the transmembrane and extracellular domains of other molecules. Thus, chimeras made of the extracellular and transmembrane domains of human IgM and of the intracytoplasmic domain of murine Igα (IgM-Igα) or Igβ (IgM-Igβ) were expressed in IIA1.6 cells (47); and chimeras made of the extracellular and transmembrane domains of the human IL-2 receptor α-chain and of the intracytoplasmic domain of either FcRγ (IL2Rα-FcRγ) or TCRζ (IL2Rα-TCRζ) were expressed in RBL cells (27). Calcium mobilization triggered in B cells by the aggregation of IgM-Igα or of IgM-Igβ was inhibited when these chimeras were coaggregated to murine FcγRIIB (47). Likewise, serotonin release triggered in mast cells by the aggregation of IL2Rα-FcRγ or IL2Rα-TCRζ chimeras was inhibited when these chimeras were coaggregated to murine FcγRIIB (27). Because TCRζ has three ITAMs an IL2Rα-TCRζ chimera whose intracytoplasmic domain was restricted to the 42 amino-terminal residues, i.e. contained the first TCRζ ITAM only, was stably expressed in RBL cells. When aggregated at the cell surface, this chimera triggered the release of serotonin. Serotonin release was inhibited when it was coaggregated to murine FcγRIIB (27). It follows that chimeric molecules having a single ITAM are sufficient targets to enable FcγRIIB to inhibit B cell or mast cell activation.

This finding extends FcγRIIB-dependent inhibition to all immunoreceptors associated with ITAM-containing subunits, besides BCR and TCR. Since FcRγ associates to the ligand-binding α subunit of FcεRI, FcγRI, FcγRIIIA, and FcαRI, FcγRIIB may regulate various IgE- IgG-, and IgA-induced responses of cells that coexpress FcγRIIB and corresponding receptors. In mice these include mast cells, macrophages, monocytes, and Langerhans cells, and in humans they include macrophages and monocytes, Langerhans cells, polymorphonuclear neutrophils, basophils, and eosinophils. NK cells are a notable exception. These cells express FcγRIIIA (48,49), but do not express FcγRIIB, which might negatively regulate antibody-dependent cell-mediated cytotoxicity (ADCC). Interestingly, however, murine and human NK cells express inhibitory receptors that recognize MHC class I molecules on target cells and whose coaggregation with FcγRIIIA inhibits ADCC (50–52). In humans, the same molecules are also expressed by some T cells and they can inhibit TCR-dependent T cell functions (53). Murine and human inhibitory receptors belong to structurally unrelated families of molecules. They nevertheless possess a motif which is reminiscent of the motif that accounts for the inhibitory properties of FcγRIIB.

THE ITIM MOTIF

An inhibitory sequence in murine FcγRIIB was first identified by Amigorena and colleagues, who examined the ability of murine FcγRIIB2 bearing intracytoplasmic deletions of increasing length to inhibit B cell activation in IIA1.6 transfectants (24). The deletion of the 16 carboxy-terminal residues of the 47-amino-acid intracytoplasmic domain of FcγRIIB2 had no effect on inhibition, whereas deletion of the next 13 residues abolished inhibition. The inhibitory properties of this 13-amino-acid sequence were confirmed by Muta and co-workers, who showed that murine FcγRIIB having an intracytoplasmic domain whose 16 carboxy-terminal residues were deleted and whose 18-amino-acid terminal residues were replaced by those of TCRζ retained an ability to inhibit B cell activation (47). Combined, these results indicate that a 13-amino-acid intracytoplasmic sequence, common to murine FcγRIIB1 and FcγRIIB2, is both necessary and sufficient to inhibit BCR-dependent B cell activation.

The same FcγRIIB2 deletants were expressed in RBL cells and their effects on serotonin release were examined when they were coaggregated to FcεRI. The same deletions which abrogated inhibition in B cells also abrogated inhibition in mast cells (27). In the 13-amino-acid inhibitory sequence, one finds a tyrosine residue (Y26). There is another tyrosine (Y43), in the 16 carboxy-terminal residues that are not required for inhibition. FcγRIIB2 in the intracytoplasmic domain of which either Y26 or Y43 was mutated were expresssed in RBL cells (27). When coaggregated to FcεRI, FcγRIIB2 with a mutated Y43 inhibited serotonin release, but not FcγRIIB2 with a mutated Y26.

To determine the sequence that accounts for inhibition of TCR-dependent T cell activation, the same FcγRIIB2 mutants were coexpressed in RBL cells, together with IL2Rα-TCRζ chimeric molecules, and serotonin release was examined when the two molecules were coaggregated at the cell surface. The deletion of the same 13-amino-acid sequence, but not that of the 16 carboxy-terminal residues, abolished the inhibition of IL2Rα-TCRζ-induced mediator release, as did the point mutation of Y26, but not that of Y43 (27).

Taken together, these data indicate that the same tyrosine-containing 13-amino-acid murine sequence accounts for inhibition of BCR-dependent B cell activation, of FcR-dependent mast cell activation, and of TCR-dependent T cell activation.

The same approach applied to human FcγRIIB yielded the same results and led to the conclusion that the same 13-amino-acid sequence accounts for inhibition of FcεRI-mediated mast cell activation (27). As stressed earlier, human and murine FcγRIIB are highly homologous. Residues are particularly conserved in an intracytoplasmic stretch that corresponds to the inhibitory sequence. Among conserved residues, one notices the tyrosine whose mutation is sufficient to abrogate the inhibitory properties of the receptor. One also notices that it is followed, at the position Y+3, by a leucine residue. This YslL motif is reminiscent of the double YxxL motif that constitutes ITAMs and determines the cell-triggering ability of immunoreceptors. The comparison of human and murine sequences is not sufficient to define a consensus inhibitory motif. In view of the critical role played by each YxxL of ITAMs in immunoreceptors, the structural analogy with ITAMs, however, was sufficiently suggestive to prompt several groups, including us, to adopt the acronym ITIM, for immunoreceptor tyrosine-based inhibition motif, which designates a yet to be defined inhibitory motif in the FcγRIIB inhibitory sequence. Interestingly, inhibitory receptors expressed by NK cells, in which they inhibit FcγRIIIA-dependent ADCC as well as NK cytotoxicity, and by some T cells, in which they inhibit TCR-dependent cytotoxicity, possess an intracytoplasmic sequence composed of two YxxL motifs separated by 26 residues (50–52). In view of their inhibitory properties, on the one hand, and of the structural constraints imposed by the interaction of phospho-ITAMs with kinases of the Syk family on the other, this sequence was thought to correspond to two ITIMs, rather than to an ITAM with an usually long intervening sequence between the two YxxL motifs (54,55). The comparison of sequences flanking these motifs and the FcγRIIB motif, made apparent the conservation of a valine or an isoleucine residue at position Y–2, which is not conserved in ITAMs (54). If this residue proves to be functionally significant, it should be taken into account in a preliminary structural definition of ITIM which might thus be built on the V/Ix YxxL backbone.

BIOCHEMICAL MECHANISMS OF NEGATIVE REGULATION BY FcγRIIB

The mechanism of FcγRIIB-mediated inhibition has been so far investigated only in B cells. A critical event following BCR aggregation is a rapid and transient efflux of Ca^{2+} from intracellular stores, followed by a sustained influx of extracellular Ca^{2+} across the plasma membrane. The coaggregation of BCR to FcγRIIB did not affect the initial increase in Ca^{2+} concentration resulting from the mobilization of intracellular Ca^{2+} stores. No subsequent influx of extracellular Ca^{2+}, however, was observed when the two receptors were coaggregated (56,57). When studied in IIA1.6 transfectants expressing the various FcγRIIB mutants described earlier, a correlation was found between inhibition of cytokine secretion and inhibition of extracellular Ca^{2+} influx (47,58).

The opening of plasma membrane Ca^{2+} channels is believed to depend on the increase in the intracellular Ca^{2+} concentration resulting from the release of Ca^{2+} from the endoplasmic reticulum, which results from the generation of inositol phosphates, itself resulting from the activation of phospholipase C_γ (PLC_γ) (59,60). Events involved in this intracellular cascade were examined in A20 lymphoma B cells and in IIA1.6 transfectants. No significant change was observed either in total inositol phosphates generated during the first few minutes following the coaggregation of BCR and FcγRIIB, or in the phosphorylation of PLC_γ (47,57).

Finally, when early phosphorylation events were examined in B cells whose BCR had been coaggregated to FcγRIIB, no obvious inhibition was observed. The phosphorylation of intracellular substrates was not affected in whole cell lysates, and neither the phosphorylation of BCR ITAMs nor that of Syk seemed to be significantly inhibited (47,57). These results indicate that inhibition does not result from FcγRIIB preventing BCR aggregation at the cell surface.

That inhibition is apparently not a passive phenomenon was further suggested by the observation that, when coaggregated to BCR, FcγRIIB became tyrosine phosphorylated (47). FcγRIIB whose ITIM had the tyrosine mutated failed to become phosphorylated (58). Since such FcγRIIB mutants are no longer capable of inhibiting B cell activation, one may assume that the phosphorylation of the tyrosine residue, in the ITIM, is critical for inhibition. The kinase(s) responsible for this phosphorylation has not yet been identified. By analogy with ITAMs, one anticipates that a phospho-ITIM could be recognized by the SH2 domains of tyrosine kinases, tyrosine phosphatases, phospholipases, and other molecules (61). To examine this question, synthetic peptides corresponding to nonphosphorylated and phosphorylated ITIM were used as immunoadsorbents to precipitate molecules in cell lysates from B cells. Peptides corresponding to phospho-ITIM, but not peptides corresponding to nonphosphorylated ITIM, precipitated three proteins, two of which could be identified as the cytoplasmic phosphatases PTPIC and PTPID. These phosphatases were recently renamed as SHP-1 and SHP-2, respectively (62). SHP-1 and SHP-2 have the distinctive characteristic of possessing two tandem SH2 domains (63), reminiscent of the two tandem SH2 domains of the cytoplasmic tyrosine kinases of the *Syk* family. Neither Syk nor ZAP70, however, bound to the phospho-ITIM peptide. Conversely, FcγRIIB that had been phosphorylated upon coaggregation with BCR were precipitated by GST fusion proteins corresponding to the SH2 domains of SHP-1 incubated with lysates from B cells. Nonphosphorylated FcγRIIB were not precipitated. Finally, small amounts of SHP-1 and of SHP-2 were coprecipitated with FcγRIIB in lysates of A20 B cells whose BCR were coaggregated with FcγRIIB (58). These data led to the conclusion that FcγRIIB-mediated inhibition of BCR-dependent B cell activation may involve the recruitment of cytoplasmic phosphatases having SH2 domains capable of binding to the phospho-ITIM in FcγRIIB. Interestingly, phosphorylated peptides corresponding to the ITIM-like motifs of MHC receptors expressed by NK cells also bound SHP-1 and SHP-2 (54,55). One may hypothesize that, if they are indeed recruited in vivo by the phospho-ITIM, these phosphatases might be functional, as d'Ambrosio and colleagues found that the phosphatase activity of SHP-1 increased severalfold in the presence of phospho-ITIM, but not in the presence of phospho-ITAM (58).

BIOLOGICAL SIGNIFICANCE OF NEGATIVE REGULATION BY FcγRIIB

Evidence that FcγRIIB can negatively regulate all ITAM-based immunoreceptors was obtained under rather artifactual conditions. One may therefore wonder how biologically significant this conclusion is. The situation is not comparable for BCR, TCR, and FcR for the following two reasons. First, conditions under which coaggregation can take place are obvious for BCR and FcR since BCR and antibodies can recognize epitopes borne by the same antigen molecule; this is not so for TCR, which recognizes a peptide + MHC complex that antibodies do not see. Second, more or less physiological conditions were explored for the three receptors. These were both in vitro and in vivo

In vitro evidence that IgG immune complexes can inhibit B cell activation was documented extensively in early works. These were reviewed above.

In vitro evidence that IgG antibodies can negatively regulate IgE-dependent mast cell activation was obtained using bone marrow-derived mast cells (BMMC). These cells consist of a homogeneous population of muscosal-type mast cells which express FcεRI and FcγRIIB1 (28). BMMC were sentitized with anti-ovalbumin IgE antibodies and challenged with DNP-ovalbumin. DNP-ovalbumin triggered IgE-sensitized BMMC cells to release serotonin. Serotonin release was dose-dependently inhibited if DNP-ovalbumin was complexed to increasing concentrations of mouse monoclonal IgG anti-DNP antibodies (45). Under conditions in which no competition for antigen could take place between IgE and IgG antibodies, FcεRI and FcγRIIB1, constitutively expressed by nontransformed mouse mast cells, could therefore be coaggregated by immune complexes the antigen moiety of which bound to the Fab portions of receptor-bound IgE and the antibody moiety of which bound to adjacent FcγRIIB1 via their Fc portion. In vitro evidence that constitutive human FcγRIIB can inhibit IgE-dependent basophil activation was obtained with human basophils purified by elutriation from peripheral blood. Again, the coaggregation of FcεRI to FcγRII constitutively expressed by human blood basophils inhibited IgE-induced histamine release (27). Likewise, the coaggregation of FcγRIIB to FcγRIIA prevented the latter triggering mediator release (27). This explains why human basophils are usually unable to respond to IgG immune complexes (64) although they express FcγRIIA. Indeed, human basophils react with the FcγRIIA-specific mouse monoclonal antibody VI.3 (65).

Finally, in vitro evidence that TCR-dependent T cell activation can be negatively regulated by constitutively expressed FcγRIIB1 has so far been obtained using T cells lines only. A murine hybridoma T cell, 2B4 (66), and a murine lymphoma T cell, RMA (67), which express a functional TCR, were both found to react with 2.4G2 in immunofluorescence. They contained only FcγRIIB1 transcripts when analyzed by RT-PCR using FcγRIIB-specific oligonucleotide probes. When plated onto anti-CD3 $F(ab')_2$-coated dishes, the two cells were triggered to secrete IL-2. When plated onto dishes coated with a mixture of anti-CD3 $F(ab')_2$ and of increasing concentrations of 2.4G2 $F(ab')_2$, IL-2 secretion was dose-dependently inhibited (27). This indicates that the coaggregation of TCR to FcγRIIB1 constitutively expressed by murine T cell lines could inhibit TCR-dependent T cell activation leading to cytokine response.

In vivo evidence of the biological significance of the negative regulation exerted by FcγRIIB is more indirect, especially for FcR and TCR. There are, however, a number of in vivo observations which have not been given satisfactory explanations and which could be readily accounted for by a negative effect of FcγRIIB.

Historical in vivo experiments, which seeded the concept of a negative feedback regulation of antibody production by IgG, amply validate the physiological relevance of the negative effects of FcγRIIB expressed by B cells. Recent experiments showing that antibody responses are enhanced in FcγRIIB-knockout mice further confirmed the phenomenon (68). Likewise, *motheaten* mice (69), which bear a genetic defect in the SHP-1 gene and which lack this phosphatase (70), show polyclonal B cell activation, have a wide array of autoantibodies, and possess B cells that can be activated by intact anti-immunoglobulin IgG antibodies (58).

Mast cells from FcγRIIB-knockout mice also provided evidence that the prediction that all ITAM-based immunoreceptors should be susceptible to the inhibitory effects of FcγRIIB might be correct. Mast cell degranulation triggered by IgG immune complexes or anti-receptor monoclonal antibody was increased in FcγRIIB-deficient mice, and IgG-induced passive cutaneous anaphylactic reactions were augmented in FcγRIIB-knockout mice compared to normal mice (68).

The most suggestive in vivo evidence of the effectiveness of IgG immune complexes in controlling IgE-induced mast cell/basophil activation may be immunotherapy performed in allergic patients, the beneficial effects of which remain unexplained. This treatment consists of injecting increasing doses of allergen to patients, starting with doses that are too low to trigger anaphylactic reactions. As injections are repeated with higher doses, an anti-allergen IgG response develops, and positive responses to treatment may be correlated with the concentration of these antibodies in the serum (71). The possible role of IgG immune complexes, generated in allergic patients treated by immunotherapy, is supported by reports showing that the injection of preformed immune complexes, made of specific allergen and IgG antibodies to the same allergen, into patients with allergic asthma significantly decreased allergic symptoms (72,73). If such mechanisms operate in pathological situations, they are likely to operate also under physiological conditions. One may thus speculate that, in normal individuals, mast cells and basophils are constantly stimulated by IgE antibodies generated in response to allergens to which nonallergic individuals are exposed as well as allergic individuals. No allergic symptom appears in normal subjects, possibly because allergens become complexed to specific IgG antibodies before they come into contact with mast cells or basophils and keep cell responses under control. Whether such control mechanisms are overridden in allergic patients is a hypothesis that deserves to be examined.

Evidence supporting the possibility that FcγRIIB might negatively regulate TCR-mediated T cell activation in vivo is scarce. FcγRIIB, expressed on T cells, can conceivably be coaggregated to TCR when T cells recognize peptide + MHC on antigen-presenting cells or target cells in the presence of IgG antibodies directed to epitopes expressed on the same cell. This is likely to happen in several situations. The enhancement of allogeneic tumors (74) or the immunological facilitation of normal tissues (75) by passive antibody are two such situations. They require the Fc portion of antibodies (76), and alloactivated cytotoxic T cells express FcγR (77). Another situation might be viral infections in which cytotoxic T cells kill infected cells in a MHC-restricted way and can keep infection asymptomatic for long periods. Viruses also induce antibodies against antigens expressed at the surface of infected cells. Finally, a proportion of serum immunoglobulins were found to be autoantibodies in normal individuals (78). One may speculate that such autoantibodies, directed to peptide-presenting cells or to invariant epitopes of the TCR itself, could prevent autoreactive T cell clones which escaped from negative selection in the thymus from being harmful. The beneficial effects of the intravenous injections of polyclonal immunoglobulins into patients suffering of autimmune diseases (79) might be partly accounted for by such a mechanism.

CONCLUSION

FcγRIIB are the most ubiquitous of all FcR (3). The finding that FcγRIIB can negatively regulate all ITAM-based immunoreceptors endows IgG antibodies with a wide spectrum of immunoregulatory properties. IgG antibodies might possibly affect the onset of an immune response by inhibiting the activation of helper T cells by antigen-presenting cells in the presence of IgG to structures expressed by the latter and by exerting a negative feedback regulation of antibody production. They might affect also cell-mediated immune responses by inhibiting cytolytic functions of CTL recognizing target cells in the presence of antibodies to molecules expressed by the latter. They might affect inflammatory responses resulting from the interaction of IgE, IgG, or IgA with corresponding FcR expressed by mast cells, macrophages, or polymorphonuclear cells of the three types. They might affect antibody-dependent cell-mediated cytotoxicity exerted by FcγRIIIA-expressing effector cells other than NK cells.

Like most other receptors, FcR have been understood not to be functional when expressed on a plasma membrane. They become functional after they have been aggregated by antibodies and multivalent antigens, and the molecular significance of FcR aggregation has only recently been clarified by the concept of transphosphorylation (80). Under physiological conditions, however, FcR aggregation is probably a rare event for the following two reasons. First, FcR of several types, with identical or different isotypic specificities, are coexpressed on most cells of hematopoietic origin. Second, immune complexes have no reason to be made of a single class or subclass of antibodies. As a consequence, when immune complexes interact with FcR on a single cell, they coaggregate different adjacent FcR on the same membrane. The various cell activation-triggering FcR and/or cell activation-regulating FcR that are coexpressed on a single cell thus function as the subunits of multichain receptors which form when they become coaggregated by immune complexes. The qualitative and quantitative composition of multichain receptors of that kind is not predetermined. It depends on the cell type, on cytokines which differentially regulate the expression of the various FcR and on the composition of immune complexes with which they interact. Resulting receptor complexes will transduce signals which, instead of being simply "on" or "off," might establish sophisticated intracellular communication networks through which versatile messages can be elaborated. Cell metabolism may be finely tuned by such messages in response to stimuli delivered by environmental factors. Under these conditions, FcR with ITAMs would function as positive coreceptors for each other, FcγRIIB as negative coreceptors for ITAM-based immunoreceptors. It remains to be determined whether their regulatory effect is restricted to receptors with ITAMs.

ACKNOWLEDGMENTS

Experimental data from our laboratory discussed in this review were from the works of Odile Malbec, Sylvain Latour, Sebastian Amigorena, Christian Bonnerot, Eric Espinosa, and Patrick Pina.

REFERENCES

1. Pleinman C, D'Ambrosio D, Cambier J. The B-cell antigen receptor complex: structure and signal transduction. Immunol Today 1994; 15: 393–399.
2. Weiss A. T cell antigen receptor signal transduction: a tale of tails and cytoplasmic protein-tyrosine kinases. Cell 1993; 73: 209–212.
3. Hulett MD, Hogarth PM. Molecular basis of Fc Receptor function. Adv Immunol 1994; 57: 1–127.
4. Kurosaki T, Gander I, Wirthmueller U, Ravetch JV. The β subunit of the FcεRI is associated with the FcγRIII on mast cells. J Exp Med 1992; 175: 447–451.
5. Kinet JP, Blank U, Ra C, White K, Metzger H, Kochan J. Isolation and characterization of cDNAs coding for the β subunit of the high-affinity receptor for immunoglobulin E. Proc Natl Acad Sci USA 1988; 85: 6483–6487.
6. Orloff DG, Ra C, Frank SJ, Klausner RD, Kinet JP. The zeta and eta chains of the T cell receptor and the gamma chain of Fc receptors form a family of disulfide-linked dimers. Nature 1990; 347: 189–191.
7. Reth MG. Antigen receptor tail clue. Nature 1989; 338: 383–384.
8. Cambier JC. New nomenclature for the Reth motif (or ARH1/TAM/ARAM/YXXL). Immunol Today 1994; 16: 110.
9. Heldin CH. Dimerization of cell surface receptors in signal transduction. Cell 1995; 80: 213–223.
10. Hibbs ML, Walker ID, Kirszbaum L, Pietersz GA, Deacon NJ, Chambers GW, McKenzie IFC, Hogarth PM. The murine Fc receptor for immunoglobulin: purification, partial amino acid sequence, and isolation of cDNA clones. Proc Natl Acad Sci USA 1986; 83: 6980–6984.
11. Lewis VA, Koch T, Plutner H, Mellman I. A complementary DNA clone for a macrophage-lymphocyte Fc receptor. Nature 1986; 324: 372.
12. Ravetch JV, Luster AD, Weinshank R, Kochan J, Pavlovec A, Portnoy DA, Hulmes J, Pan YCE, Unkeless JC. Structural heterogeneity and functional domains of murine immunoglobulin G Fc receptors. Science 1986; 234: 718–725.
13. Stuart SG, Simister NE, Clarkson SB, Kacinski BM, Shapiro M, Mellman I. Human IgG Fc receptor (hFcRII; CD32) exists as multiple isoforms in macrophages, lymphocytes, and IgG-transporting placental epithelium. EMBO J. 1989; 8: 3657–3666.
14. Hibbs ML, Bonadonna L, Scott BM, McKenzie IFC, Hogarth PM. Molecular cloning of a human immunoglobulin G Fc receptor. Proc Natl Acad Sci USA 1988; 85: 2240–2244.

15. Brooks DG, Qiu WQ, Luster AD, Ravetch JV. Structure and expression of human IgG FcRII (CD32). Functional heterogeneity is encoded by the alternatively spliced products of multiple genes. J Exp Med 1989; 170: 1369–1386.
16. Daëron M, Bonnerot C, Latour S, Fridman WH. Murine recombinant FcγRIII, but not FcγRII, trigger serotonin release in rat basophilic leukemia cells. J Immunol 1992; 149: 1365–1373.
17. Bonnerot C, Amigorena S, Choquet D, Pavlovich R, Choukroun V, Fridman WH. Role of associated γ chain in tyrosine kinase activation via murine FcγRIII. EMBO J 1992; 11: 2747–2757.
18. Hogarth PM, Witort E, Hulett MD, Bonnerot C, Even J, Fridman WH, McKenzie IFC. Structure of the mouse βFcγ receptor II gene. J Immunol 1991; 146: 369–376.
19. Latour S, Fridman WH, Daëron M. Identification, molecular cloning, biological properties and tissue distribution of a novel isoform of murine low-affinity IgG receptor homologous to human FcγRIIB1. J Immunol 1996; 157: 189–197.
20. Miettinen HM, Rose JK, Mellman I. Fc receptor isoforms exhibit distinct abilities for coated pit localization as a result of cytoplasmic domain heterogeneity. Cell 1989; 58: 317–327.
21. Daëron M, Malbec O, Latour S, Bonnerot S, Segal DM, Fridman WH. Distinct intracytoplasmic sequences are required for endocytosis and phagocytosis via murine FcγRII in mast cells. Int Immunol 1993; 5: 1393–1401.
22. Van den Herik-Oudijk IE, Capel PJA, Van der Bruggen T, Van de Winkel JGJ. Identification of signalling motifs within human FcγIIA and FcγRIIB isoforms. Blood 1995; 85: 2202–2211.
23. Miettinen HM, Matter K, Hunziker W, Rose JK, Mellman I. Fc receptor endocytosis is controlled by a cytoplasmic domain determinant that actively prevents coated pit localization. J Cell Biol 1992; 116: 875.
24. Amigorena S, Bonnerot C, Drake J, Choquet D, Hunziker W, Guillet JG, Webster P, Sautès C, Mellman I, Fridman WH. Cytoplasmic domain heterogeneity and functions of IgG Fc receptors in B-lymphocytes. Science 1992; 256: 1808–1812.
25. Amigorena S, Bonnerot C, Choquet D, Fridman WH, Teillaud JL. FcγRII expression in resting and activated B lymphocytes. Eur J Immunol 1989; 19: 1379–1385.
26. Néauport-Sautès C, Dupuis D, and Fridman WH. Specificity of Fc receptors of activated T cells. Relation with released immunoglobulin-binding factor. Eur J Immunol 1975; 5: 849–854.
27. Daëron M, Latour S, Malbec O, Espinosa E, Pina P, Pasmans S, Fridman WH. The same tyrosine-based inhibition motif, in the intracytoplasmic domain of FcγRIIB, regulates negatively BCR-, TCR-, and FcR-dependent cell activation. Immunity 1995; 3: 635–646.
28. Benhamou M, Bonnerot C, Fridman WH, Daëron M. Molecular heterogeneity of murine mast cell Fcγ receptors. J Immunol 1990; 144: 3071–3077.
29. Esposito-Farese M-E, Sautès C, de la Salle H, Latour S, Bieber T, de la Salle C, Ohlmann P, Fridman WH, Cazenave J-P, Teillaud J-L, Daëron M, Bonnerot C, Hanau D. Membrane and soluble FcγRII/III modulate the antigen-presenting capacity of murine dendritic epidermal Langerhans cells for IgG-complexed antigens. J Immunol 1995; 154: 1725–1736.
30. Daëron M, Bonnerot C, Latour S, Benhamou M, Fridman WH. The murine αFcγR gene product: identification, expression and regulation. Mol Immunol 1990; 27: 1181–1188.
31. Möller G, Wigzell H. Antibody synthesis at the cellular level. Antibody-induced suppression of 19S and 7S antibody response. J Exp Med 1965; 121: 969.
32. Henry C, Jerne NK. Competition of 19S and 7S antigen receptors in the regulation of the primary immune response. J Exp Med 1968; 128: 133–145.
33. Uhr JW, Möller G. Regulatory effect of antibody on the immune response. Adv Immunol 1968; 8: 81.
34. Shek PN, Dubiski S. Allotypic suppression in rabbits: competition for target cell receptors between isologous and heterologous antibody and between native antibody and antibody fragments. J Immunol 1975; 114: 621.

35. Kohler H, Richardson B, Rowley DA, Smyk S. Immune response to phosphorylcholine. III. Requirement of the Fc portion and equal effectiveness of IgG subclasses in anti-receptor antibody-induced suppression. J Immunol 1977; 119: 1979–1986.
36. Sinclair NRS, Chan PL. Regulation of the immune response. IV. The role of the Fc-fragment in feedback inhibition by antibody. Adv Exp Med Biol 1971; 12: 609–615.
37. Kohler H, Richardson BC, Smyk S. Immune response to phosphorylcholine. IV. Comparison of homologous and isologous anti-idiotypic antibody. J Immunol 1978; 120: 233–238.
38. Sidman CL, Unanue ER. Requirements for mitogenic stimulation of murine B cells by soluble anti-IgM antibodies. J Immunol 1979; 122: 406–413.
39. Unkeless JC. Characterization of monoclonal antibody directed against mouse macrophage and lymphocyte Fc receptors. J Exp Med 1979; 150: 580–596.
40. Phillips NE, Parker DC. Cross-linking of B lymphocyte Fcγ receptors and membrane immunoglobulin inhibits anti-immunoglobulin-induced blastogenesis. J Immunol 1984; 132: 627–632.
41. Jones B, Tite JP, Janeway CA Jr. Different phenotypic variants of the mouse B cell tumor A20/2J are selected by antigen- and mitogen-triggered cytotoxicity of L3T4-positive, I-A-restricted T cell clones. J Immunol 1986; 136: 348–356.
42. Van den Herik-Oudijk IE, Westerdaal NAC, Henriquez NV, Capel PJA, Van de Winkel JGJ. Functional analysis of human FcγRII (CD32) isoforms expressed in B lymphocytes. J Immunol 1994; 152: 574–585.
43. Katz HR, Arm JP, Benson AC, Austen KF. Maturation-related changes in the expression of FcγRII and FcγRIII on mouse mast cells derived in vitro and in vivo. J Immunol 1990; 145: 3412–3417.
44. Barsumian EL, Isersky C, Petrino MG, Siraganian RP. IgE-induced histamine release from rat basophilic leukemia cell lines: isolation of releasing and nonreleasing clones. Eur J Immunol 1981; 11: 317.
45. Daëron M, Malbec O, Latour S, Arock M, Fridman WH. Regulation of high-affinity IgE receptor-mediated mast cell activation by murine low-affinity IgG receptors. J Clin Invest 1995; 95: 577–585.
46. Wegener A-M, Letourneur F, Hoeveler A, Brocker T, Luton F, Malissen B. The T cell receptor/CD3 complex is composed of at least two autonomous transduction modules. Cell 1992; 68: 83–95.
47. Muta T, Kurosaki T, Misulovin Z, Sanchez M, Nussenzweig MC, Ravetch JV. A 13-amino-acid motif in the cytoplasmic domain of FcγRIIB modulates B-cell receptor signalling. Nature 1994; 368: 70–73.
48. Perussia B, Tutt MM, Qui WQ, Kuziel WA, Tucker PW, Trinchieri G, Bennett M, Ravetch JV, Kumar V. Murine natural killer cells express functional Fcγ receptor II encoded by the FcγRα gene. J Exp Med 1989; 170: 73–86.
49. Bonnerot C, Amigorena S, Fridman WH, Even J, Daëron M. Unmethylation of specific sites in the 5′ region is critical for the expression of murine αFcγR gene. J Immunol 1990; 144: 323–328.
50. Colonna M, Samaridis J. Cloning of immunoglobulin-superfamily members associated with HLA-C and HLA-B recognition by human natural killer cells. Science 1995; 268: 405–408.
51. Wagtmann N, Biassoni R, Cantoni C, Verdiani S, Malnati MS, Vitale M, Bottino C, Moretta L, Moretta A, Long E. Molecular clones of the p58 NK cell receptor reveal immunoglobulin-related molecules with diversity in both the extra- and intracellular domains. Immunity 1995; 2: 439–449.
52. d'Andrea A, Chang C, Franz-Bacon K, McClanahan T, Phillips JH, Lanier LL. Molecular cloning of NKB1, a natural killer cell receptor for HLA-B allotypes. J Immunol 1995; 155: 2306–2310.
53. Nakajima H, Tomiyama H, Takiguchi M. Inhibition of γδ T cell recognition by receptors for MHC class I molecules. J Immunol 1995; 155: 4139–4142.

54. Burshtyn DN, Scharenberg AM, Wagtmann N, Rajogopalan S, Berrada K, Yi T, Kinet J-P, Long EO. Recruitment of tyrosine phosphatase HCP by the killer cell inhibitory receptor. Immunity 1996; 4: 77–85.
55. Olcese L, Lang P, Vély F, Cambiaggi A, Marguet D, Bléry M, Hippen KL, Biassoni R, Moretta A, Moretta L Cambier, JC, Vivier E. Human and mouse killer-cell inhibitory receptors recruit PTP1C and PTP1D protein tyrosine phosphatases. J Immunol 1996; 156: 4531–4534.
56. Choquet D, Partiseti M, Amigorena S, Bonnerot C, Fridman WH, Korn H. Cross-linking of IgG receptors inhibits membrane immunoglobulin-stimulated calcium influx in B lymphocytes. J Cell Biol 1993; 121: 355–363.
57. Diegel ML, Rankin BM, Bolen JB, Dubois PM, Kiener PA. Cross-linking of Fcγ receptor to surface immunoglobulin on B cells provides an inhibitory signal that closes the plasma membrane calcium channel. J Biol Chem 1994; 15: 11407–11416.
58. D'Ambrosio D, Hippen KH, Minskoff SA, Mellman I, Pani G, Siminovitch KA, Cambier JC. Recruitment and activation of PTP1C in negative regulation of antigen receptor signaling by FcgRIIB1. Science 1995; 268: 293–296.
59. Claphman DE. Calcium signaling. Cell 1995; 80: 259–268.
60. Divecha N, Irvine RF. Phospholipid signaling. Cell 1995; 80: 269–278.
61. Cohen GB, Ren R, Baltimore D. Modular binding domains in signal transduction proteins. Cell 1995; 80: 237–248.
62. Adachi M, Fischer EH, Ihle J, Imai K, Jirik F, Neel B, Pawson T, Shen S-H, Thomas M, Ullrich A, Zhao Z. Mammalian SH2-containing protein tyrosine phosphatases. Cell 1996; 85: 15.
63. Yi T, Cleveland JL, Ihle JN. Protein tyrosine phosphatase containing SH2 domains: characterization, preferential expression in hematopoietic cells, and localization to human chromosome 12p12-p13. Mol Cell Biol 1992; 12: 836–846.
64. Van Toorenenbergen AW, Aalberse RC. IgG4 and passive sensitization of basophil leukocytes. Int Arch Allergy Appl Immunol 1981; 65: 432–440.
65. Anselmino LM, Perussia B, Thomas LL. Human basophils selectively express the FcγRII (CDw32) subtype of IgG receptor. J Allergy Clin Immunol 1989; 84: 907–914.
66. Frank SJ, Niklinska BB, Orloff DG, Mercép M, Ashwell JD, Klausner RD. Structural mutations of the T cell receptor ζ chain and its role in T cell activation. Science 1990; 249: 174–177.
67. Ljunggren HG, Kärre K. Host resistance directed selectively against H-2-deficient lymphoma variants: analysis of the mechanism. J Exp Med 1985; 162: 1745–1759.
68. Takai T, Ono M, Hikida M, Ohmori H, Ravetch JV. Augmented humoral and anaphylactic responses in FcγRII-deficient mice. Nature 1996; 379: 346–349.
69. Greene MC, Shultz LD. Motheaten, an immunodeficient mutant of the mouse. Genetics and pathology. J Hered 1975; 66: 250–258.
70. Shultz LD, Schweitzer PA, Rajan TV, Yi T, Ihle JN, Mattews RJ, Thomas ML, Beier DR. Mutations at the murine motheaten locus are within the hematopoietic cell protein-tyrosine phosphatase (Hcph) gene. Cell 1993; 73: 1445–1454.
71. Gleich GJ, Zimmermann EM, Henderson LL, Yunginger JW. Effect of immunotherapy on immunoglobulin E and immunoglobulin G antibodies to ragweed antigens: a six-year prospective study. J Allergy Clin Immunol 1978; 62: 261.
72. Machiels JJ, Somville MA, Jacquemin MG, Saint-Rémy JMR. Allergen-antibody complexes can efficiently prevent seasonal rhinitis and asthma in grass pollen hypersensitive patients. Allergy 1991; 46: 335–348.
73. Machiels JJ, Lebrun PM, Jacquemin MG, Saint-Rémy JMR. Significant reduction of nonspecific bronchial reactivity in patients with *Dermatophagides pteronyssinus*-sensitive allergic asthma under therapy with allergen–antibody complexes. Am Rev Respin Dis 1993; 147: 1407–1412.
74. Kaliss N. Immunological enhancement of tumor homografts in mice. A review. Cancer Res 1958; 18: 992–1035.

75. Voisin GA. Immunological facilitation, a broadening of the concept of the enhancement phenomenon. Prog Allergy 1971; 15: 328–375.
76. Capel PJA, Tamboer WPM, De Waal RMW, Jansen JLJ, Koene RAP. Passive enhancement of skin grafts by alloantibodies is Fc dependent. J Immunol 1979; 122: 421–429.
77. Leclerc JC, Plater C, Fridman WH. The role of Fc receptors (FcR) on thymus-derived lymphocytes. I. Presence of FcR on cytotoxic lymphocytes and absence of direct role in cytotoxicity. Eur J Immunol 1977; 7: 543–548.
78. Avrameas S, Guilbert B, Dighiero G. Natural antibodies against actin, tubulin, myoglobin, thyroglobulin, fetuin, albumin and transferin are present in normal human sera and monoclonal immunoglobulins from multiple myeloma and Waldeström macroglobulinemia. Ann Immunol (Inst Pasteur) 1981; 132C: 231–240.
79. Hall PD. Immunomodulation with intravenous immunoglobulin. Pharmacotherapy 1993; 13: 564–573.
80. Pribluda VS, Pribluda C, Metzger H. Transphophorylation as the mechanism by which the high affinity receptor for IgE is phosphorylated upon aggregation. Proc Natl Acad Sci USA 1994; 91: 11246–11250.

16

Signal Transduction Pathways Activated by Mitogenic Neuropeptides

Enrique Rozengurt
Imperial Cancer Research Fund
London, England

INTRODUCTION

Multicellular organisms have developed highly efficient regulatory networks to control cell proliferation. These involve cellular interactions with positive and negative diffusible modulators as well as with the extracellular matrix proteins. In fully mature organisms, the cells of many tissues and organs are maintained in a nonproliferating state, but can be stimulated to resume DNA synthesis and cell division in response to external stimuli such as hormones, antigens, or growth factors. In this manner the growth of individual cells is regulated according to the requirements of the whole organism. The regulation of normal cell proliferation is, therefore, central to many physiological processes, including embryogenesis, growth and development, selective cell survival, tissue repair, and immune responses.

It has become evident that cultured cancer cells, which are characterized by unrestrained proliferation, acquire complete or partial independence of mitogenic signals in the extracellular environment through different mechanisms (1,2). These include production of growth factors that act on the same cells that produced them or on adjacent cells, alterations in the number or structure of cellular receptors, and changes in the activity of postreceptor signaling pathways that either stimulate or suppress cell growth (3,4). For these reasons the identification of the extracellular factors that modulate cell proliferation and the elucidation of the molecular mechanisms involved have emerged as fundamental problems in cancer biology.

In recent years an increasing number of small regulatory peptides or neuropeptides, including bombesin/gastrin-releasing peptides (GRP) have been discovered in the neural and neuroendocrine cells of the gastrointestinal tract and central nervous system (5). Defining the physiological functions of these peptides is complicated by the different modes of action of these molecules. Some are localized in neurons and act as neurotransmitters in the central or peripheral nervous system, while others are released by endocrine cells and have effects both as systemic hormones circulating through the bloodstream and by acting in a paracrine or autocrine fashion. Moreover, a number of peptides are found in both neuronal and endocrine cells, and a major effect of some regulatory peptides in vivo (e.g., bombesin/GRP) is to stimulate the release of other biologically active peptides. The role of these peptides as fast-acting neurohumoral signalers has recently been expanded by the discovery that they also stimulate cell proliferation (6–8). Furthermore, indirect evidence is accumulating that the mitogenic effects of neuropeptides such as bombesin may be relevant for a number of normal and abnormal biological processes, and in particular tumorigenesis. Consequently, it is very important to understand in detail the receptors and signal transduction pathways that mediate the mitogenic action of bombesin and other neuropeptides because they may provide potential targets for therapeutic intervention.

Many studies to identify the molecular pathways by which neuropeptide mitogens elicit cellular growth have exploited cultured murine 3T3 cells as a model system (7,9,10). The list of neuropeptides that can act as mitogens in these cells has now grown to a considerable size and includes bombesin, bradykinin, endothelin, vasopressin, vasoactive intestinal peptide, and adrenomedullin (11–20). Evidence for direct growth-promoting activities of several other peptides including cholecystokinin, galanin, gastrin, and neurotensin has come from work using another cultured cell model system, namely, cell lines established from small-cell lung carcinoma (15–19,21). Some fundamental features of the mechanism of action of bombesin as a growth factor in 3T3 cells will be discussed in the following review.

EARLY SIGNALING EVENTS INDUCED BY BOMBESIN

The early cellular and molecular responses elicited by bombesin and structurally related peptides in 3T3 cells have been elucidated in detail (Figure 1). Bombesin is a 14-amino-acid peptide first isolated from the skin of the frog *Bombina bombina* (22). Many bombesin-related peptides have subsequently been isolated from various species and classified into the three subfamilies bombesin, ranatensin, and litorin according to their C-terminal hexapeptide sequence homology. The principal mammalian counterparts are GRP and neuromedin B, members of the bombesin and ranatensin subfamilies, respectively.

There are a number of considerations that make bombesin attractive as a model peptide with which to investigate the mechanisms underlying peptidergic regulation of cell growth. In serum-free medium it stimulates DNA synthesis and cell division in the absence of other growth-promoting agents (12). Bombesin also promotes the expression of the genes encoding ribonucleotide reductase subunits, the enzyme responsible for the in vivo production of deoxyribonucleotides for DNA synthesis (23). The ability of bombesin, like platelet-derived growth factor (PDGF), to act as a sole mitogen for these cells contrasts with other peptide growth factors that are active only in synergistic combinations (10). The mitogenic effects of bombesin are markedly potentiated by insulin, which both increases the maximal response and reduces the bombesin concentration required for half-maximal effect (12). Furthermore, receptors for bombesin-like peptides have been well characterized at the molecular level (see below). The cause–effect relationships and temporal organization of the early signals and molecular events induced by bombesin provide a paradigm for the study of other growth factors and mitogenic neuropeptides and illustrate the activation and interaction of a variety of signaling pathways.

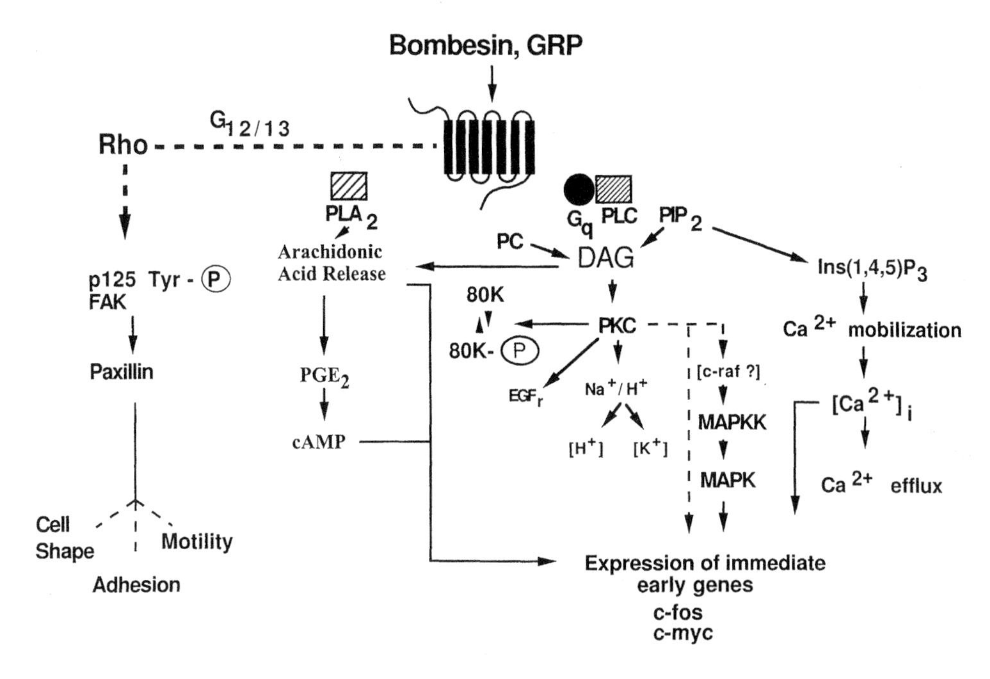

Figure 1. Signal transduction pathways activated by bombesin/GRP. Detailed explanation can be found in the text.

Specific Receptors

Bombesin and GRP bind to a single class of high-affinity receptors in Swiss 3T3 cells. The receptors are transmembrane glycoproteins of M_r = 75,000–85,000 with a core of M_r = 43,000 (24). The receptor is coupled to one or more G-proteins as judged by the modulation of ligand binding either in membrane preparations or in receptor solubilized preparations and of signal transduction in permeabilized cells (25). The bombesin/GRP receptor has been cloned and sequenced (26,27). The deduced amino acid sequence predicts a polypeptide core of M_r = 43,000 and demonstrates that it belongs to the superfamily of heterodimeric G-protein coupled receptors. These receptors are characterized by seven hydrophobic domains thought to traverse the cytoplasmic membrane and cluster to form a ligand-binding pocket (26,27).

The neuromedin-B receptor has also been cloned and shown to be another member of the heterodimeric G-protein receptor superfamily that induces cell proliferation (28,29). The predicted amino acid sequence has 56% homology with the GRP receptor, and in stably transfected cells the neuromedin-B receptor binds GRP with low affinity and neuromedin-B with high affinity (28,29).

Mutational analysis of these receptors is beginning to reveal which parts of the receptors are important for ligand binding, G-protein interaction, and receptor internalization. Thus, the fifth transmembrane domain of the neuromedin-B receptor has been shown to reduce or abolish neuromedin-B binding, suggesting that this part of the receptor is critical for ligand binding (30). In other studies the carboxyl cytoplasmic tail of the GRP receptor has been truncated, resulting in diminished rates of receptor internalization following ligand binding, without affecting the affinity of the receptor for ligand or activation of the heterodimeric G-protein (31). In a number of other members of the seven transmembrane domain receptor superfamily, the third cytoplasmic loop appears to be particularly critical in determining G-protein coupling to the activated receptor–ligand complex (32). Other neuropeptide mitogens with seven transmembrane domain receptors include endothelin, vasopressin, bradykinin, substance K, and substance P (33).

Inositol Phosphatidyl Turnover, Ca^{2+} Mobilization and Activation of Protein Kinase C

Binding of bombesin/GRP to its receptor initiates a cascade of intracellular signals culminating in DNA synthesis 10 to 15 hours later. One of the earliest events to occur after the binding of bombesin to its specific receptor is the activation of the heterodimeric G-protein, which in turn stimulates phospholipase C (PLC)-catalyzed hydrolysis of phosphatidyl inositol 4,5-bisphosphate (PIP_2) in the plasma membrane. This reaction produces inositol 1,4,5-trisphosphate ($Ins[1,4,5]P_3$), which, as a second-messenger binds to an intracellular receptor and induces the release of Ca^{2+} from internal stores (34). Bombesin causes a rapid increase in $Ins[1,4,5]P_3$, which coincides with a transient increase in the intracellular concentration of Ca^{2+} ($[Ca^{2+}]_i$) and with Ca^{2+} efflux from the cells (35–37). Depletion of Ca^{2+} from internal stores, as induced by bombesin, other mitogenic neuropeptides, and growth factors, could play a role as one of the synergistic signals that contribute to stimulating the transition from G_0 to DNA synthesis (38).

Our understanding of the molecular identity of the G-proteins and phospholipases involved has been greatly enhanced by the cloning, expression and in vitro reconstruction of activity of these proteins. The binding of bombesin to its receptor causes activation of pertussis toxin-insensitive G-proteins (39–43), probably of the G_q subfamily (44,45). Recently, neuropeptide activated $G_{\alpha q}$ has been shown to stimulate the PLC-β isoform of PLC (46–50). This, however, has not been specifically shown for bombesin-activated receptors. Furthermore, while there is direct evidence indicating that the γ isoform of PLC is a target for growth factor receptors with intrinsic tyrosine kinase activity, it is less clear which of the four PLC-β isoforms plays the most important role in G-protein-linked bombesin-stimulated signal transduction. Likewise, the role of the βγ subunits of the heterodimeric G-protein remains poorly defined (51). In summary, a role of PLC-β and G_q in the mechanism of action of bombesin and other growth-promoting peptides is plausible, but definitive evidence is not yet available.

PLC-mediated hydrolysis of PIP_2 also generates 1,2-diacylglycerol (DAG). DAG can also be generated from other sources, such as phosphatidylcholine (PC) hydrolysis, and acts as a second messenger in the activation of protein kinase C (PKC) by multiple extracellular stimuli (52) including bombesin (40). In accord with this, bombesin strikingly increases the phosphorylation of a major protein kinase C substrate that migrates with an apparent molecular mass of 80 kDa termed "80 K" (53,54). Recently the cDNA encoding this substrate from Swiss 3T3 cells has been cloned (55). 80 K has been shown to have 66–74% homology with a myristolated alanine-rich C-kinase substrate (MARCKS) cloned from human and bovine brain, but only one 80 K/MARCKS gene exists in each species. This implies that rodent 80 K and bovine and human MARCKS are not distinct members of a gene family but equivalent genes in different species (56). Interestingly, PKC activation induced by bombesin causes a striking translocation of 80 K/MARCKS from the membrane to the cytosolic fraction (57) and a dramatic downregulation of the expression of mRNA and protein of the 80 K/MARCKS substrate in Swiss 3T3 cells through a posttranscriptional mechanism (55,58,59). This novel result suggests that this PKC substrate, which appears to be a calmodulin- and actin-binding protein (60), may play a suppressor role in the control of cell proliferation.

As depicted in Figure 1, bombesin/GRP also stimulates a rapid exchange of Na^+, H^+, and K^+ ions across the cell membrane, leading to cytoplasmic alkalinization and increased intracellular [K^+], and induces a striking PKC-dependent transmodulation of the epidermal growth factor (EGF) receptor (7,24).

Role of Mitogen-activated Protein Kinase

Mitogen-activated protein (MAP) kinases (ERKs) are a family of highly conserved serine/threonine kinases that are activated in response to a wide range of extracellular signals. The two best-characterized isoforms, $p42^{MAPK}$ (ERK2) and $p44^{MAPK}$ (ERK2) can be stimulated by both heterotrimeric G-protein-linked receptors and receptors that posses intrinsic tyrosine kinase activity (61–63). The signaling pathways that lead from the activated receptor to MAP kinase are different for these two classes of receptors. Thus PDGF or EGF induce tyrosine phosphorylation of their respective receptors, which permits the association of the two linking molecules Grb2/SEM 5 and Sos. This complex facilitates the activation of *Ras*, by exchanging GDP for GTP on this protein. The active GTP-bound *Ras* then directly associates with the serine/threonine kinase, *Raf*, which in turn stimulates MAP kinase kinase (MEK). Subsequently, MEK activates MAP kinase by tyrosine and serine phosphorylation (64).

The pathways leading from the activated heterotrimeric G-protein-linked receptors to MAP kinase activation are less clear. Bombesin has recently been shown to stimulate MAP kinase via PKC activation, while EGF uses an alternative pathway that involves *Ras* activation in Swiss 3T3 cells (65). Interestingly, bombesin did not measurably activate *Ras* (i.e., loading of *Ras* with GTP) or stimulate *Raf*-1 above basal levels (66) (Seufferlein, Withers and Rozengurt, unpublished results). However, in Rat-1 fibroblasts, EGF and 1-oleoyl-lysophosphatidic acid (LPA; like bombesin signals via heterotrimeric G-proteins) have both been shown to activate MAP kinase via a *Ras*-dependent pathway (67). These data indicate that different G-protein-linked receptors may activate MAP kinase via separate pathways in different cells (68,69).

In summary, it is probable that bombesin stimulates MAP kinase via PKC in Swiss 3T3 cells (see Figure 1). The events that take place between PCK and MAP kinase activation following bombesin stimulation are under investigation. MAP kinase is itself activated by MEK. This in turn has a number of activators termed "MAP kinase kinase kinases" which include *Raf.* Interestingly, PKC translocates to the plasma membrane for activation by DAG, while activated MAP kinase subsequently translocates to the nucleus to phosphorylate a number of substrates that directly cause induction of early response genes such as c-*fos* (62).

Arachidonic Acid Release and Prostaglandin Synthesis: Differential Effects of Bombesin and Vasopressin

While bombesin and structurally related mammalian peptides stimulate DNA synthesis in the absence of other factors, vasopressin is mitogenic for Swiss 3T3 cells only in synergistic combination with other factors (11,12). Binding of vasopressin to its distinct receptor on quiescent cultures of Swiss 3T3 cells causes a rapid production of

$Ins(1,4,5)P_3$, mobilization of Ca^{2+} from intracellular stores, and sustained activation of PKC via G-protein-linked transduction pathway (7). Since the initiation of DNA synthesis is triggered by independent signal transduction pathways that act synergistically in mitogenic stimulation, the ability of bombesin to act as a sole mitogen could be due to activation of a signaling pathway not stimulated by vasopressin.

Bombesin, but not vasopressin, has been shown to induce a marked, biphasic release of arachidonic acid into the extracellular medium (70,71). A first phase involves rapid activation of phospholipase A_2 (PLA_2). These results showed a clear difference in the pattern of early signals induced by the neuropeptides bombesin and vasopressin in Swiss 3T3 cells. The stimulation of arachidonic acid release by bombesin is likely to contribute to bombesin-induced mitogenesis because externally applied arachidonic acid potentiates mitogenesis induced by agents that stimulate polyphosphoinositide breakdown but not arachidonic acid release (e.g., vasopressin) (70).

Arachidonic acid released by bombesin is converted into E-type prostaglandins which, acting in an autocrine and paracrine manner, enhance cyclic adenosine monophosphate (cAMP) accumulation in the cell (Figure 1). Since elevated cAMP levels constitute a mitogenic signal for Swiss 3T3 cells (21,72), at least one consequence of arachidonic acid release may be the modulation of intracellular cAMP levels. However, other arachidonic acid metabolites may also play a role in mitogenic signal transduction by bombesin.

Bombesin Induction of the Proto-oncogenes c-*fos* and c-*myc*

In addition to the events in the membrane and cytosol described above, bombesin rapidly and transiently induces the expression of the cellular oncogenes c-*fos* and c-*myc* in quiescent fibroblasts (10). Since these cellular oncogenes encode nuclear proteins, it is plausible that their transient expression may play a role in the transduction of the mitogenic signal in the nucleus (73). The demonstration that the product of the proto-oncogene c-*jun*, identified as a major component of the *trans*-acting factor AP-1, forms a tight complex with *fos* protein is consistent with a role for c-*fos* in the regulation of gene transcription (73).

There has been considerable interest in elucidating the signal transduction pathways involved in c-*fos* induction. There is evidence implicating PKC activation in the sequence of events linking receptor occupancy and proto-oncogene induction (74). Accordingly, bombesin-induced oncogene expression is markedly reduced by downregulation of PKC. As mentioned above, PKC activation leads to the activation of MAP kinase, which directly phosphorylates transcription factor regulators, resulting in the increased expression of c-*fos* (62,75). However, neither direct activation of PKC by phorbol esters nor addition of vasopressin evoke a maximal increase in c-*fos* mRNA levels. It is likely that the induction of c-*fos* by bombesin is mediated by the coordinated effects of PKC activation, Ca^{2+} mobilization, and an additional pathway dependent on arachidonic acid release (7).

Bombesin Stimulation of Tyrosine Phosphorylation: Focal Adhesion Kinase ($p125^{FAK}$) and Paxillin; Cytoskeletal Link

It has been established unequivocally that the receptor for peptides of the bombesin family is coupled to G-proteins and does not possess intrinsic tyrosine kinase activity. However, bombesin has been shown to increase rapidly tyrosine phosphorylation of multiple substrates in intact quiescent Swiss 3T3 cells (76). Vasopressin and endothelin elicit a similar response. The substrates for neuropeptide tyrosine phosphorylation in these cells appear to be unrelated to known targets for the PDGF receptor including the GTPase-activating protein (GAP), the PLC-γ, and the phosphatidylinositol 3′-kinase (P13 kinase). Thus, bombesin and other neuropeptides that act through receptors linked to G-proteins can increase tyrosine phosphorylation of protein substrates in intact cells (Figure 1). Neuropeptide stimulation of Swiss 3T3 cells increases tyrosine phosphorylation in cell-free preparations of these cells of both a major endogenous p115 substrate and exogenously added substrates (77). These results suggested that neuropeptide stimulation of tyrosine phosphorylation reflects activation of a tyrosine kinase but provided no clue to the identity of either kinase or substrate.

A novel cytosolic protein tyrosine kinase was recently identified in the search for substrates of the oncogene pp60$^{v\text{-}src}$ tyrosine kinase (78). This new kinase colocalized with several components of cellular focal adhesions, such as tensin, vinculin, and talin, which are important in regulating cytoskeletal structure and probably signal transduction (78). Hence, it has been named focal adhesion kinase (p125^{FAK}). Although p125^{FAK} contains a central catalytic domain that exhibits all the conserved motifs characteristic of the catalytic domains of other protein tyrosine kinases (PTKs), this region is flanked by large N-terminal and C-terminal domains lacking any significant homology with other PTKs, such as Src homology 2 (SH2) and SH3 domains (78,79). A stretch of 159 amino acids within the C-terminal region of p125^{FAK}, named focal adhesion targeting (FAT) sequence, is essential for focal adhesion localization. p125^{FAK} has been shown to associate with other signaling proteins, including pp60^{src} and paxillin (80–82).

It is now recognized that p125^{FAK} is a major substrate for bombesin-stimulated tyrosine phosphorylation in Swiss 3T3 cells (83). It has also become apparent that p125^{FAK} is tyrosine phosphorylated in response to many agents that regulate cell growth and differentiation, including neuropeptides, bioactive lipids, growth factors, and ligands of the adhesion receptors of the integrin family (84–89). Thus, p125^{FAK} appears to be a point of convergence in the action of multiple extracellular signals and could be involved in the regulation of cell shape, adhesion, and motility (88,89). Indeed, the recent demonstration that p125^{FAK}-deficient cells do not display polar migrating shape and exhibit a striking reduction in motility (90) points to a key role of p125^{FAK} in cell locomotion.

At present, it is not known how the bombesin receptor is linked to p125^{FAK} phosphorylation. Stimulation of p125^{FAK} tyrosine phosphorylation by bombesin is not dependent on either PKC activation or Ca^{2+} mobilization or on the activity of P13 kinase (91). There is evidence from studies of cytoskeletal changes induced by bombesin in Swiss 3T3 cells, that the small GTP-binding Ras-related protein, p21^{Rho}, may be associated with p125^{FAK} phosphorylation (92). Thus, bombesin promotes a rapid increase in actin stress fibers and focal adhesions that can be blocked by inhibitors of p21^{Rho} (92). Cytochalasin D, an agent that selectively disrupts the network of actin microfilaments, completely inhibits bombesin induced p125^{FAK} tyrosine phosphorylation (93). Furthermore, bombesin is unable to stimulate tyrosine phosphorylation of p125^{FAK} when p21^{Rho} is inactivated by ADP-ribosylation with botulinum C3 exoenzyme (94). A functional connection between p21^{Rho} and p125^{FAK} has also been established using permeabilized Swiss 3T3 cells stimulated with nonhydrolyzable GTP analogs (95). Thus it is attractive to speculate that p125^{FAK} tyrosine phosphorylation induced by bombesin is mediated by p21^{Rho} activation (89).

The focal adhesion-associated protein paxillin has also been identified as a prominent tyrosine phosphorylated protein in Swiss 3T3 cells following stimulation with bombesin and other neuropeptides (84). Recent molecular cloning of paxillin revealed a multidomain protein that may function as an adaptor capable of associating with $p125^{FAK}$, Crk, Src, and vinculin (96,97). As with $p125^{FAK}$, the pathway leading from the bombesin receptor to tyrosine phosphorylation of paxillin does not involve PKC activation or Ca^{2+} mobilization, but it is critically dependent on the integrity of the actin filament network (84). The coordinated regulation of tyrosine phosphorylation of $p125^{FAK}$ and paxillin may reflect the fact that paxillin could be a direct substrate of $p125^{FAK}$ (98).

Interestingly, high concentrations of PDGF inhibit tyrosine phosphorylation of $p125^{FAK}$ and paxillin and cytoskeletal changes, which are normally seen when Swiss 3T3 cells are stimulated with low concentrations of PDGF (87). Furthermore, high concentrations of PDGF abolish the ability of bombesin to stimulate both cytoskeletal changes and tyrosine phosphorylation of $p125^{FAK}$ and paxillin (87). This suggests that there is a novel crosstalk between PDGF and bombesin on the tyrosine phosphorylation of $p125^{FAK}$ and paxillin that occurs as a result of their opposing effects on the integrity of focal adhesions and the actin cytoskeleton.

Given the importance of tyrosine phosphorylation in the action of growth factors and nonreceptor oncogenes, it is plausible that this novel event plays a role in neuropeptide mitogenic signaling. The tyrphostins are a series of cell-permeable molecules that inhibit the tyrosine kinase activity of various receptors and block cell proliferation (99). They provide potential antiproliferative agents that could inhibit cell proliferation through a novel mechanism. Studies have shown that tyrphostin selectively inhibits bombesin stimulation of tyrosine phosphorylation including $p125^{FAK}$ in Swiss 3T3 cells (100). Since tyrphostin also inhibited bombesin-induced DNA synthesis over a similar concentration range and to a similar degree, it is tempting to speculate that tyrosine phosphorylation plays a role in bombesin-mediated mitogenesis (100).

THE BOMBESIN/GRP RECEPTOR TRANSFECTED INTO Rat-1 CELLS COUPLES TO MULTIPLE PATHWAYS

As summarized in a previous section, several members of the bombesin-like receptor family have been cloned and shown to belong to the G-protein-linked receptor superfamily. They are the bombesin/GRP-preferring, the neuromedin B (NmB)-preferring, and BSR-3 (26–28,30), a bombesin-like receptor whose physiological ligand is not yet known. It has generally been assumed that all the early signals and the subsequent mitogenic response elicited by bombesin are mediated by occupancy of the bombesin/GRP receptor. This assumption has recently been questioned. Bold and colleagues reported that antagonists that inhibit bombesin-induced Ca^{2+} mobilization do not prevent bombesin-induced cell growth in human gastric adenocarcinoma cells (101). However, tyrphostin blocked bombesin-induced proliferation of cells. Bold et al (101) suggested that a bombesin receptor subtype couples to Ca^{2+} mobilization and a different bombesin receptor subtype, not yet identified, couples to the protein tyrosine phosphorylation pathway and cell growth. However, no direct evidence supporting this hypothesis has been provided.

To examine whether the bombesin/GRP-preferring receptor couples to both PLC activation and tyrosine kinase pathway, the cDNA encoding the bombesin/GRP receptor has been expressed in Rat-1 fibroblasts, a cell line that can be reversibly arrested in the G_0/G_1 phase of the cell cycle (102). The relative abilities of bombesin-related peptides to displace ^{125}I-Labeled GRP in these transfected Rat-1 cells are consistent with the binding properties of endogenous bombesin/GRP receptors found in Swiss 3T3 fibroblasts and other cell types. Furthermore, the apparent affinity of GRP for the transfected receptor is similar to that for the endogenous receptor.

Agonist binding to the transfected bombesin/GRP receptor elicits multiple responses in Rat-1 cells: (1) mobilization of Ca^{2+} from internal stores, leading to a rapid increase in $[Ca^{2+}]_i$ and increased phosphorylation of the 80 K/MARCKS protein mediated by PKC; (2) reinitiation of DNA synthesis and cell proliferation in serum-free medium; and (3) tyrosine phosphorylation of multiple proteins including broad bands of M_r = 110,000–130,000 and 70,000–90,000 (102). We identified two major substrates in the transfected cells as $p125^{FAK}$ and paxillin. None of these responses were elicited by bombesin in the parental Rat-1 cells.

The characteristics of protein tyrosine phosphorylation induced by agonist binding to the bombesin/GRP receptor transfected into Rat-1 cells were examined in detail in two clones, BOR 5 and BOR 15 (102). An increase in the tyrosine phosphorylation of multiple proteins, including $p125^{FAK}$ and paxillin, occurred at low concentrations of bombesin, the half-maximum being 0.3 nM. This is similar to that observed in Swiss 3T3 cells (half-maximum 0.08–0.3 nM) (83,93). Both Swiss 3T3 cells and transfected Rat-1 cells show similar time dependence of tyrosine phosphorylation: increases are detected within seconds of bombesin addition, reach a maximum at 1 minute, and remain elevated for hours. Bombesin stimulation of tyrosine phosphorylation in bombesin/GRP receptor-transfected Rat-1 cells, as in Swiss 3T3 cells, is not downstream of PKC. $p125^{FAK}$ and paxillin are localized in the focal adhesions which form at the termini of actin stress fibers and disruption of the actin cytoskeleton by cytochalasin D prevents tyrosine phosphorylation of these proteins in Swiss 3T3 cells. In BOR 5 and BOR 15 cells, bombesin-induced tyrosine phosphorylation of multiple proteins was also prevented by cytochalasin D. Therefore, the protein tyrosine phosphorylation pathway activated by the transfected bombesin/GRP receptor in Rat-1 cells shows the same characteristics as that induced by bombesin stimulation of Swiss 3T3 cells (102).

In the transfected Rat-1 cells the increases in Ca^{2+} mobilization, protein tyrosine phosphorylation, and DNA synthesis were inhibited by the specific bombesin/GRP receptor antagonist, D-F_5-Phe^6, D-Ala^{11} Bombesin (6–13) OMe, further substantiating that all these responses emanate from the same receptor. These results demonstrate that ligand activation of a transfected bombesin/GRP receptor elicits multiple signaling pathways and leads to cell proliferation. The conclusion that a single receptor is linked to multiple intracellular pathways is further substantiated by experiments showing that the transfected neuromedin-B receptor also mediates Ca^{2+} mobilization, 80 K/MARCKS phosphorylation, reinitiation of DNA synthesis, and tyrosine phosphorylation of $p125^{FAK}$ and paxillin (29).

These findings raise important questions regarding the mechanism(s) by which a single seven transmembrane receptor subtype can couple to multiple signaling pathways. While the mechanism by which bombesin/GRP receptor and other seven transmembrane domain receptors couple to PLC through heterodimeric G-proteins of the G_q subfamily is increasingly understood, little is known about the pathways coupling the bombesin/GRP receptor to tyrosine phosphorylation of $p125^{FAK}$ and paxillin. As mentioned before, recent work has implicated activation of the monomeric G-protein of the $p21^{Rho}$ subfamily as one step in the signaling pathway leading to tyrosine phosphorylation of these proteins. Thus, our results suggest that the bombesin/GRP receptor couples to both heterotrimeric (G_q) and monomeric ($p21^{Rho}$) G proteins.

The demonstration that certain seven transmembrane domain receptors couple to $p21^{Ras}$ via $\beta\gamma$ subunits of the heterotrimeric G-protein of the G_i subfamily (103) suggests a possible coupling mechanism between the bombesin/GRP receptor and $p21^{Rho}$. By analogy, the $\beta\gamma$ generated by activation of G_q/G_{11} could be preferentially coupled to activation of $p21^{Rho}$ and thereby to the tyrosine phosphorylation pathway, whereas the α subunit of G_q stimulates PLC (104). Alternatively, one domain of the bombesin/GRP receptor may couple to G_q and thereby to PLC, whereas a separate domain could lead to activation of $p125^{FAK}$ and paxillin through interaction with a different heterotrimeric G-protein. In this context the recent demonstration that $G\alpha_{12}$ and $G\alpha_{13}$, but not $G\alpha_q$, regulate *Rho*-dependent actin polymerization is highly relevant (105). This suggests that the bombesin/GRP receptor could couple to *Rho* and tyrosine phosphorylation of $p125^{FAK}$ and paxillin via different heterotrimeric G-proteins than those which couple it to PLC (Figure 1). Further experimental work will be required to distinguish between these models of signal transduction.

REFERENCES

1. Cross M, Dexter TM. Growth factors in development, transformation, and tumorigenesis. Cell 1991; 64: 271–280.
2. Westermark B, Heldin CH. Platelet-derived growth factor in autocrine transformation. Cancer Res 1991; 51: 5087–5092.
3. Bishop JM. Molecular themes in oncogenesis. Cell 1991; 64: 235–248.
4. Sager R. Tumor suppressor genes: the puzzle and the promise. Science 1989; 246: 1406–1412.
5. Walsh JH. Gastrointestinal hormones. In Johnson LR, ed. Physiology of the Gastrointestinal Tract. Raven Press; New York: 1987; 181.
6. Zachary I, Woll PJ, Rozengurt E. A role for neuropeptides in the control of cell proliferation. Dev Biol 1987; 124: 295–308.
7. Rozengurt E. Neuropeptides as cellular growth factors: role of multiple signalling pathways. Eur J Clin Invest 1991; 21: 123–134.
8. Rozengurt E. Polypeptide and neuropeptide growth factors: signalling pathways and role in cancer. In Peckham M, Pinedo H, Veronesi U, eds. Oxford Textbook of Oncology. Oxford: Oxford University Press; 1995: 12–20.
9. Rozengurt E. The mitogenic response of cultures 3T3 cells: integration of early signals and synergistic effects in a unified framework. In Cohen P, Housley M, eds. Molecular Mechanisms of Transmembrane Signalling. Elsevier; 1985; 429–452.
10. Rozengurt E. Early signals in the mitogenic response. Science 1986; 234: 161–166.
11. Rozengurt E, Legg A, Pettican P. Vasopressin stimulation of mouse 3T3 cell growth. Proc Natl Acad Sci USA 1979; 76: 1284–1287.
12. Rozengurt E, Sinnett-Smith J. Bombesin stimulation of DNA synthesis and cell division in cultures of Swiss 3T3 cells. Proc Natl Acad Sci USA 1983; 80: 2936–2940.
13. Woll PJ, Rozengurt E. Two classes of antagonist interact with receptors for the mitogenic neuropeptides bombesin, bradykinin, and vasopressin. Growth Factors 1988; 1: 75–83.
14. Zurier RB, Kozma M, Sinnett Smith J, Rozengurt E. Vasoactive intestinal peptide synergistically stimulates DNA synthesis in mouse 3T3 cells: role of cAMP, Ca^{2+}, and protein kinase C. Exp Cell Res 1988; 176: 155–161.

15. Sethi T, Rozengurt E. Galanin stimulates Ca^{2+} mobilization, inositol phosphate accumulation, and clonal growth in small cell lung cancer cells. Cancer Res 1991; 51: 1674–1679.
16. Sethi T, Rozengurt E. Multiple neuropeptides stimulate clonal growth of small cell lung cancer: effects of bradykinin, vasopressin, cholecystokinin, galanin, and neurotensin. Cancer Res 1991; 51: 3621–3623.
17. Sethi T, Rozengurt E. Gastrin stimulates Ca^{2+} mobilization and clonal growth in small cell lung cancer cells. Cancer Res 1992; 52: 6031–6035.
18. Sethi T, Rozengurt E. Gastrin as a growth factor for small cell lung cancer cells in vitro. In Walsh JH, ed. Gastrin. New York: Raven Press; 1993: 395–406.
19. Sethi T, Herget T, Wu SV, Walsh JH, Rozengurt E. CCKA and CCKB receptors are expressed in small cell lung cancer lines and mediate Ca^{2+} mobilization and clonal growth. Cancer Res 1993; 53: 5208–5213.
20. Withers DJ, Coppock HA, Seufferlein T, Smith DM, Bloom SR. Adrenomedullin stimulates DNA synthesis and cell proliferation via elevation of cAMP in Swiss 3T3 cells. FEBS Lett 1996; 378: 83–87.
21. Rozengurt E, Neuropeptide growth factors: mechanisms of action and role in small cell lung carcinoma. In Carneg DN, ed. Lung Cancer. London: Arnold; 1995: 180–190.
22. Anastasi A, Erspamer V, Bucci M. Isolation and structure of bombesin and alytesin, 2 analogous active peptides from the skin of the European amphibians *Bombina* and *Alytes*. Experientia 1971; 27: 166–167.
23. Albert DA, Rozengurt E. Synergistic and coordinate expression of the genes encoding ribonucleotide reductase subunits in Swiss 3T3 cells: effects of multiple signal-transduction pathways. Proc Natl Acad Sci USA 1992; 89: 1597–1601.
24. Rozengurt E. Bombesin stimulation of mitogenesis. Specific receptors, signal transduction, and early events. Am Rev Respir Dis 1990; 142: 511–514.
25. Rozengurt E, Fabregat I, Coffer A, Gil J, Sinnett Smith J. Mitogenic signalling through the bombesin receptor: role of a guanine nucleotide regulatory protein. J Cell Sci Suppl 1990; 13: 43–56.
26. Battey JF, Way JM, Corjay MH, Shapira H, Kusano K. Molecular cloning of the bombesin/gastrin-releasing peptide receptor from Swiss 3T3 cells. Proc Natl Acad Sci USA 1991; 88: 395–399.
27. Spindel ER, Giladi E, Brehm P, Goodman RH, Segerson TP. Cloning and functional characterization of a complementary DNA encoding the murine fibroblast bombesin/gastrin-releasing peptide receptor. Mol Endocrinol 1990; 4: 1956–1963.
28. Wada E, Way J, Shapira H, Kusano K, Lebacq-Verheyden AM, Coy D, Jensen R, and Battey J. cDNA cloning, characterization, and brain region-specific expression of a neuromedin-B-preferring bombesin receptor. Neuron 1991; 6: 421–430.
29. Lach EB, Broad S, Rozengurt E. Mitogenic signaling by transfected neuromedin B receptors in Rat-1 cells. Cell Growth Differ 1995; 6: 1427–1435.
30. Fathi Z, Benyas RV, Shapira H, Jenson RT, Battey JF. The fifth transmembrane segment of the neuromedin B receptor is critical for high affinity neuromedin B binding. J Biol Chem 1993; 268: 14622–14626.
31. Benya RV, Fathi Z, Battey JF, Jensen RT. Serines and threonines in the gastrin-releasing peptide receptor carboxyl terminus mediate internalisation. J Biol Chem 1993; 268: 20285–20290.
32. Probst WC, Snyder LA, Schuster DI, Brosius J, Sealfon SC. Sequence alignment of the G-protein coupled receptor superfamily. DNA Cell Biol 1992; 11: 1–20.
33. Strosberg AD. Structure/function relationship of proteins belonging to the family of receptors coupled to GTP-binding proteins. Eur J Biochem 1991; 196: 1–10.
34. Berridge MJ. Inositol trisphosphate and calcium signaling. Ann NY Acad Sci 1995; 766: 31–43.
35. Mendoza SA, Schneider JA, Lopez-Rivas A, Sinnett-Smith J, Rozengurt E. Early events

elicited by bombesin and structurally related peptides in quiescent Swiss 3T3 cells. II. Changes in Na^+ and Ca^{2+} fluxes, Na^+/K^+ pump activity, and intracellular pH. J Cell Biol 1986; 102: 2223–2233.

36. Lopez-Rivas A, Mendoza SA, Nanberg E, Sinnett-Smith J, Rozengurt E. Ca^{2+}-mobilizing actions of platelet-derived growth factor differ from those of bombesin and vasopressin in Swiss 3T3 mouse cells. Proc Natl Acad Sci USA 1987; 84: 5768–5772.
37. Nanberg E Rozengurt E. Temporal relationship between inositol polyphosphate formation and increases in cytosolic Ca^{2+} in quiescent 3T3 cells stimulated by platelet-derived growth factor, bombesin and vasopressin. EMBO J 1988; 7: 2741–2747.
38. Charlesworth A, Rozengurt E. Thapsigargin and di-tert-butylhydroquinone induce synergistic stimulation of DNA synthesis with phorbol ester and bombesin in Swiss 3T3 cells. J Biol Chem 1994; 269: 32528–32535.
39. Zachary I, Millar J, Nanberg E, Higgins T, Rozengurt E. Inhibition of bombesin-induced mitogenesis by pertussis toxin: dissociation from phospholipase C pathway. Biochem Biophys Res Commun 1987; 146: 456–463.
40. Erusalimsky JD, Friedberg I, Rozengurt E. Bombesin, diacylglycerols, and phorbol esters rapidly stimulate the phosphorylation of an M_r = 80,000 protein kinase C substrate in permeabilized 3T3 cells. Effect of guanine nucleotides. J Biol Chem 1988; 263: 19188–19194.
41. Erusalimsky JD, Rozengurt E. Vasopressin rapidly stimulates protein kinase C in digitonin-permeabilized Swiss 3T3 cells: involvement of a pertussis toxin-insensitive guanine nucleotide binding protein. J Cell Physiol 1989; 141: 253–261.
42. Coffer A, Fabregat I, Sinnett-Smith J, Rozengurt E. Solubilisation of the bombesin receptor from Swiss 3T3 cell membranes: functional association to a guanine nucleotide regulatory protein. FEBS Lett 1990; 263: 80–84.
43. Murphy AC, Rozengurt E. Pasteurella multocida toxin selectively facilitates phosphatidylinositol 4,5-bisphosphate hydrolysis by bombesin, vasopressin, and endothelin. Requirement for a functional G protein. J Biol Chem 1992; 267: 25296–25303.
44. Strathmann M, Simon MI. G protein diversity: A distinct class of α subunits is present in vertebrates and invertebrates. Proc Natl Acad Sci USA 1990; 87: 9113–9117.
45. Wilkie TM, Scherle PA, Strathmann MP, Slepak VZ. Characterization of G-protein α subunits in the Gq class: expression in murine tissues and in stromal and hematopoietic cell lines. Proc Natl Acad Sci USA 1991; 88: 10049–10053.
46. Blank JL, Ross AH, Exton JH. Purification and characterization of two G-proteins that activate the beta 1 isozyme of phosphoinositide-specific phospholipase C. Identification as members of the Gq class. J Biol Chem 1991; 266: 18206–18216.
47. Gutowski S, Smrcka A, Nowak L, Wu DG, Simon M. Antibodies to the alpha q subfamily of guanine nucleotide-binding regulatory protein alpha subunits attenuate activation of phosphatidylinositol 4,5-bisphosphate hydrolysis by hormones. J Biol Chem 1991; 266: 20519–20524.
48. Shenker A, Goldsmith P, Unson CG, Spiegel AM. The G protein coupled to the thromboxane A2 receptor in human platelet is a member of the novel Gq family. Science 1991; 266: 9309–9313.
49. Smrcka AV, Hepler JR, Brown KO, Sternweis PC. Regulation of polyphosphoinositide-specific phospholipase C activity by purified Gq. Science 1991; 251: 804–807.
50. Berstein G, Blank JL, Smrka AV, Higashijima T, Sternweis PC. Reconstitution of agonist-stimulated phosphatidylinositol 4,5-bisphosphate hydrolysis using purified m1 muscarinic receptor, Gq/11, and phospholipase C-β1. J Biol Chem 1992; 267: 8081–8088.
51. Sternweis PC, Smrcka AV. Regulation of phospholipase C by G proteins. Trends Biochem Sci 1992; 17: 502–506.
52. Nishizuka Y. The molecular heterogeneity of protein kinase C and its implications for cellular regulation. Nature 1988; 334: 661–665.

53. Rozengurt E, Rodriguez-Pena A, Smith KA. Phorbol esters, phospholipase C, and growth factors rapidly stimulate the phosphorylation of a M_r 80,000 protein in intact quiescent 3T3 cells. Proc Natl Acad Sci USA 1983; 80: 7224–7248.
54. Erusalimsky JD, Brooks SF, Herget T, Morris C, Rozengurt E. Molecular cloning and characterization of the acidic 80-kDa protein kinase C substrate from rat brain. Identification as a glycoprotein. J Biol Chem 1991; 266: 7073–7080.
55. Brooks SF, Herget T, Erusalimsky JD, Rozengurt E. Protein kinase C activation potently down-regulates the expression of its major substrate, 80 K, in Swiss 3T3 cells. Embo J 1991; 10: 2497–2505.
56. Herget T, Brooks SF, Broad S, Rozengurt E. Relationship between the major protein kinase C substrates acidic 80-kDa protein-kinase-C substrate (80 K) and myristoylated alanine-rich C-kinase substrate (MARCKS). Members of a gene family or equivalent genes in different species. Eur J Biochem 1992; 209: 7–14.
57. Herget T, Rozengurt E. Bombesin, endothelin and platelet-derived growth factor induce rapid translocation of the myristoylated alanine-rich C-kinase substrate in Swiss 3T3 cells. Eur J Biochem 1994; 225: 539–548.
58. Brooks SF, Herget T, Broad S, Rozengurt E. The expression of 80 K/MARCKS, a major substrate of protein kinase C (PKC), is down-regulated through both PKC-dependent and -independent pathways. Effects of bombesin, platelet-derived growth factor, and cAMP. J Biol Chem 1992; 267: 14212–14218.
59. Herget T, Brooks SF, Broad S, Rozengurt E. Expression of the major protein kinase C substrate, the acidic 80-kilodalton myristoylated alanine-rich C kinase substrate, increases sharply when Swiss 3T3 cells move out of cycle and enter G0. Proc Natl Acad Sci USA 1993; 90: 2945–2949.
60. Blackshear PJ. The MARCKS family of cellular protein kinase C substrates. J Biol Chem 1993; 268: 1501–1504.
61. Crews CM, Erikson RL. Extracellular signals and reversible protein phosphorylation: what to Mek of it all. Cell 1993; 74: 215–217.
62. Davis RJ. The mitogen-activated protein kinase signal transduction pathway. J Biol Chem 1993; 268: 14553–14556.
63. Marshall CJ. Specificity of receptor tyrosine kinase signaling: transient versus sustained extracellular signal-regulated kinase activation. Cell 1995; 80: 179–185.
64. Egan SE, Weinberg RA. The pathway to signal achievement [news]. Nature 1993; 365: 781–783.
65. Pang L, Decker SJ, Saltiel AR. Bombesin and epidermal growth factor stimulate the mitogen-activated protein kinase through different pathways in Swiss 3T3 cells. Biochem J 1993; 289: 283–287.
66. Mitchell FM, Heasley LE, Qian NX, Zamarripa J, Johnson GL. Differential modulation of bombesin-stimulated phospholipase C beta and mitogen-activated protein kinase activity by [D-Arg1, D-Phe5, D-Trp7,9,Leu11] substance P. J Biol Chem 1995; 270: 8623–8628.
67. Cook SJ, Rubinfeld B, Albert I, McCormick F. RapV12 antagonizes Ras-dependent activation of ERK1 and ERK2 by LPA and EGF in Rat-1 fibroblasts. Embo J 1993; 12: 3475–3485.
68. Seufferlein T, Withers DJ, Broad S, Herget T, Walsh JH. The human CCKB/gastrin receptor transfected into rat1 fibroblasts mediates activation of MAP kinase, p74raf-1 kinase, and mitogenesis. Cell Growth Differ 1995; 6: 383–393.
69. Seufferlein T, Rozengurt E. Sphingosylphosphorylcholine activation of mitogen-activated protein kinase in Swiss 3T3 cells requires protein kinase C and a pertussis toxin-sensitive G protein. J Biol Chem 1995; 270: 24334–24342.
70. Millar JB, Rozengurt E. Arachidonic acid release by bombesin. A novel postreceptor target for heterologous mitogenic desensitization. J Biol Chem 1990; 265: 19973–19979.
71. Domin J, Rozengurt E. Platelet-derived growth factor stimulates a biphasic mobilization of

arachidonic acid in Swiss 3T3 cells. The role of phospholipase A2. J Biol Chem 1993; 268: 8927–8934.

72. Withers DJ, Bloom SR, Rozengurt E. Dissociation of cAMP-stimulated mitogenesis from activation of the mitogen-activated protein kinase cascade in Swiss 3T3 cells. J Biol Chem 1995; 270: 21411–21419.
73. Lewin B. Oncogenic conversion by regulatory changes in transcription factors. Cell 1991; 64: 303–312.
74. Rozengurt E, Sinnett-Smith J. Early signals underlying the induction of the c-*fos* and c-*myc* genes in quiescent fibroblasts: studies with bombesin and other growth factors. Prog Nucleic Acid Res Mol Biol 1988; 35: 261–295.
75. Treisman R. The serum response element. Trends Biochem Sci 1992; 17: 423–426.
76. Zachary I, Gil J, Lehmann W, Sinnett Smith J, Rozengurt E. Bombesin, vasopressin, and endothelin rapidly stimulate tyrosine phosphorylation in intact Swiss 3T3 cells. Proc Natl Acad Sci USA 1991; 88: 4577–4581.
77. Zachary I, Sinnett Smith J, Rozengurt E. Stimulation of tyrosine kinase activity in anti-phosphotyrosine immune complexes of Swiss 3T3 cell lysates occurs rapidly after addition of bombesin, vasopressin, and endothelin to intact cells. J Biol Chem 1991; 266: 24126–24133.
78. Schaller MD, Borgman CA, Cobb BS, Vines RR, Reynolds AB. pp125FAK a structurally distinctive protein-tyrosine kinase associated with focal adhesions. Proc Natl Acad Sci USA 1992; 89: 5192–5196.
79. Hanks SK, Calalb MB, Harper MC, Patel SK. Focal adhesion protein-tyrosine kinase phosphorylated in response to cell attachment to fibronectin. Proc Natl Acad Sci USA 1992; 89: 8487–8491.
80. Cobb BS, Schaller MD, Leu TH, Parsons JT. Stable association of pp60src and pp59fyn with the focal adhesion-associated protein tyrosine kinase, pp125FAK. Mol Cell Biol 1994; 14: 147–155.
81. Hildebrand JD, Schaller MD, Parsons JT. Paxillin, a tyrosine phosphorylated focal adhesion-associated protein binds to the carboxyl terminal domain of focal adhesion kinase. Mol Biol Cell 1995; 6: 637–647.
82. Parsons JT. Integrin-mediated signalling: regulation by protein tyrosine kinases and small GTP-binding proteins. Curr Opin Cell Biol 1996; 8: 146–152.
83. Zachary I, Sinnett-Smith J, Rozengurt E. Bombesin, vasopressin and endothelin stimulation of tyrosine phosphorylation in Swiss 3T3 cells: Identification of a novel tyrosine kinase as a major substrate. J Biol Chem 1992; 267: 19031–19034.
84. Zachary I, Sinnett Smith J, Turner CE, Rozengurt E. Bombesin, vasopressin, and endothelin rapidly stimulate tyrosine phosphorylation of the focal adhesion-associated protein paxillin in Swiss 3T3 cells. J Biol Chem 1993; 268: 22060–22065.
85. Seufferlein T, Rozengurt E. Lysophosphatidic acid stimulates tyrosine phosphorylation of focal adhesion kinase, paxillin, and p130. Signaling pathways and cross-talk with platelet-derived growth factor. J Biol Chem 1994; 269: 9345–9351.
86. Seufferlein T, Rozengurt E. Sphingosine induces p125FAK and paxillin tyrosine phosphorylation, actin stress fiber formation, and focal contact assembly in Swiss 3T3 cells. J Biol Chem 1994; 269: 27610–27617.
87. Rankin S Rozengurt E. Platelet-derived growth factor modulation of focal adhesion kinase (p125FAK) and paxillin tyrosine phosphorylation in Swiss 3T3 cells. Bell-shaped dose response and cross-talk with bombesin. J Biol Chem 1994; 269: 704–710.
88. Zachary I, Rozengurt E. Focal adhesion kinase (p125FAK): a point of convergence in the action of neuropeptides, integrins, and oncogenes. Cell 1992; 71: 891–894.
89. Rozengurt E. Convergent signalling in the action of integrins, neuropeptides, growth factors and oncogenes. Cancer Surv 1995; 24: 81–96.

90. Llic D, Furuta Y, Kanazawa S, Takeda N, Sobue K. Enhanced focal adhesion contact formation in cells from FAK-deficient mice. Nature 1995; 377: 539–544.
91. Rankin S, Hooshmand-Rad R, Claesson-Welsh L, Rozengurt E. Requirement for phosphatidylinositol 3′-kinase activity in platelet-derived growth factor-stimulated tyrosine phosphorylation of p125 focal adhesion kinase and paxillin. J Biol Chem 1996; 271: 7829–7834.
92. Ridley AJ, Paterson HF, Johnston CL, Diekmann D, Hall A. The small GTP-binding protein rac regulates growth factor-induced membrane ruffling. Cell 1992; 70: 401–410.
93. Sinnett-Smith J, Zachary I, Valverde AM, Rozengurt E. Bombesin stimulation of p125 focal adhesion kinase tyrosine phosphorylation: role of protein kinase C, Ca^{2+} mobilization and the actin cytoskeleton. J Biol Chem 1993; 268: 14261–14268.
94. Rankin S, Morii N, Narumiya S, Rozengurt E. Botulinum C3 exoenzyme blocks the tyrosine phosphorylation of p125FAK and paxillin induced by bombesin and endothelin. FEBS Lett 1994; 354: 315–319.
95. Seckl MJ, Morii N, Narumiya S, Rozengurt E. Guanosine 5′-3-O-(thio) triphosphate stimulates tyrosine phosphorylation of p125FAK and paxillin in permeabilized Swiss 3T3 cells. Role of p21rho. J Biol Chem 1995; 270: 6984–6990.
96. Bellis SL, Miller JT, Turner CE. Characterization of tyrosine phosphorylation of paxillin in vitro by focal adhesion kinase. J Biol Chem 1995; 270: 17437–17341.
97. Salgia R, Li JL, Lo SH, Brunkhorst B, Kansas GS. Molecular cloning of human paxillin, a focal adhesion protein phosphorylated by P210BCR/ABL. J Biol Chem 1995; 270: 5039–5047.
98. Turner CE, Schaller MD, Parsons JT. Tyrosine phosphorylation of the focal adhesion kinase pp125FAK during development: relation to paxillin. J Cell Sci 1993; 105: 637–645.
99. Levitzki A. Tyrphostins: tyrosine kinase blockers as novel antiproliferative agents and dissectors of signal transduction. FASEB J 1992; 6: 3275–3282.
100. Seckl M, Rozengurt E. Tyrphostin inhibits bombesin stimulation of tyrosine phosphorylation, *c-fos* expression, and DNA synthesis in Swiss 3T3 cells. J Biol Chem 1993; 268: 9548–9554.
101. Bold RJ, Lowry PS, Ishizuka J, Battey JF, Townsend CM Jr. Bombesin stimulates the in vitro growth of a human gastric cancer cell line. J Cell Physiol 1994; 161: 519–525.
102. Charlesworth A, Broad S, Rozengurt E. The bombesin/GRP receptor transfected into Rat-1 fibroblasts couples to phospholipase C activation, tyrosine phosphorylation of $p125^{FAK}$ and paxillin and cell proliferation. Oncogene 1996; 12: 1337–1345.
103. Koch WJ, Hawes BE, Allen LF, Lefkowitz RJ. Direct evidence that G_i-coupled receptor stimulation of mitogen-activated protein kinase is mediated by $G\beta\gamma$ activation of $p21^{ras}$. Proc Natl Acad Sci USA 1994; 91: 12706–12710.
104. Lee SB, Rhee SG. Significance of PIP_2 hydrolysis and regulation of phospholipase C isozymes. Curr Opin Cell Biol 1995; 7: 183–189.
105. Buhl AM, Johnson NL, Dhanasekaran N, Johnson GL. G alpha 12 and G alpha 13 stimulate Rho-dependent stress fiber formation and focal adhesion assembly. J Biol Chem 1995; 270: 24631–24634.

17
Glutathione Metabolism During Apoptosis

Diels J. van den Dobbelsteen, C. Stefan I. Nobel, Astrid Samuelsson, Sten Orrenius, and Andrew F. G. Slater
Karolinska Institutet, Stockholm, Sweden

INTRODUCTION

In apoptosis an individual cell undergoes an internally controlled transition from an intact metabolically active state into a number of shrunken remnants more or less retaining their membrane integrity (1,2). Lysis of cellular organelles does not occur until late in this process and little leakage of the contents of the dying cell can be detected. As a consequence, apoptotic cells do not induce an inflammatory response in vivo. Instead the shrunken apoptotic bodies are specifically recognized and phagocytosed by other cells, allowing their contents to be recycled (3). Therefore, apoptosis provides an organism with both a safe and efficient way to continuously turn over cells in any tissue, and the capability to remove specific cells during development.

Since the beginning of the 1990s an increasing number of studies have shown that oxidants and other agents promoting a pro-oxidant state are potent triggers of apoptotic cell death. The list of agents of this type comprises not only classical oxidants such as hydrogen peroxide (4,5), quinones (6), and diamide (7), but also oxidized low-density lipoproteins (8), lipid hydroperoxides (9), and agents increasing the cellular content of redox-active copper (e.g., pyrrolidine dithiocarbamate as shown by Nobel and colleagues (10)). From most of these studies it became clear that these agents should be used at the optimum concentration to induce the appropriate amount of oxidative stress able to trigger apoptotic cell death. Whereas serious damage to proteins, lipids, and DNA leads to a dramatic disturbance in the cellular homeostasis directly culminating in necrosis, only mild oxidative stress is able to induce apoptosis (5–7).

In addition to this list of papers reporting oxidants to be potent inducers of cell death, several studies clearly indicate the presence of oxidative changes in cells undergoing apoptosis induced by nonoxidative stimuli. The oxidative changes observed range from decreased levels of reduced glutathione (11–15), α-tocopherol (11), and protein thiols (11) to downregulation of primary antioxidant defense enzymes such as catalase, manganese superoxide dismutase (Mn-SOD), Cu/Zn superoxide dismutase (Cu/Zn-SOD), DT-diaphorase, and thioredoxin (16). Furthermore, some of these studies show that antioxidants can protect against apoptotic cell death induced by nonoxidative stimuli (12,17,18). For example, cultured primary sympathetic neurons die by apoptosis when deprived of nerve growth factor (NGF) and an increased peroxide tone can be detected several hours after the withdrawal. The apoptotic cell death can be totally inhibited by injection of these cells with either Cu/Zn-SOD protein or cDNA if they are injected within the first few hours of removal of NGF (18). (For a more detailed review on redox changes during apoptosis, see Slater et al. (19,20).)

Although it is now widely accepted that oxidative stress is one of the many stimuli capable of promoting apoptosis and that oxidative changes occur during this process, it is much less clear whether any common oxidative changes universally accompany this type of cell death. Several experiments performed in a low-oxygen environment (21–24) suggest that oxygen is not required for the execution of apoptosis. Furthermore, Hug and colleagues (25) suggested that in Fas-mediated apoptosis reactive oxygen species were not required because they observed no inhibition of apoptosis by antioxidants. In contrast, Stefanelli and colleagues have reported that glucocorticoid-induced thymocyte apoptosis is greatly reduced when oxygen tension is lowered below 5% (26). Thus, the issue of redox changes being required for the execution of apoptotic cell death is still unresolved, especially when it is considered that cellular redox changes can occur independently of ambient oxygen tension. We believe that there are ways to generate an oxidative tone in an apoptotic cell without the involvement of reactive oxygen species.

EFFLUX OF REDUCED GLUTATHIONE DURING APOPTOSIS

To further investigate possible mechanisms behind the cell's shift toward a more oxidized state during apoptosis, we have studied the metabolism of the major low-molecular-mass antioxidant in cells, namely reduced glutathione (GSH). Human Jurkat T lymphocytes are rapidly induced to undergo apoptosis after exposure to anti-Fas/APO-1 antibody (27,28). Most characteristics of apoptosis such as activation of the interleukin-1β-converting enzyme-like protease Yama/CPP-32/apopain, plasma membrane blebbing, degradation of the cytoskeleton, phosphatidylserine exposure on the cell surface nuclear fragmentation, and DNA degradation into high-molecular-mass and oligonucleosomal fragments are observed (29–32). We have demonstrated that the intracellular GSH content of these cells dropped approximately 30 min after exposure to anti-Fas/APO-1 antibody (33). Surprisingly, this decrease in intracellular GSH level was not due to an inhibition of synthesis, oxidation to GSSG, or a deficit in the GSH salvage pathway (34–36), but to an increased rate of efflux of GSH (Figure 1). The reduced form of the tripeptide was quantitatively recovered from the medium when the cells were incubated in a cystine-free medium (necessary to prevent fast transhydrogenation reaction between GSH and cystine) (Figure 1). In other cell types, such as rat thymocytes, monocytic U937 cells, and mouse fibroblast L929 cell line, a similar GSH extrusion has also been detected using other apoptosis-inducing stimuli such as methylprednisolone, tumor necrosis factor, or etoposide (our unpublished results). Thus, the efflux of GSH during apoptosis seems to be a general phenomenon occurring irrespective of cell type or apoptotic stimulus applied. The fact that all the cell types studied maintained their membrane integrity for at least another 2 h after the onset of GSH efflux suggested that a transmembrane channel, rather than nonspecific leakiness, could be responsible for the GSH extrusion.

Previously, Kaplowitz and coworkers (37–40) have described different GSH-specific transporters located in both plasma and mitochondrial membranes of cells. Rat

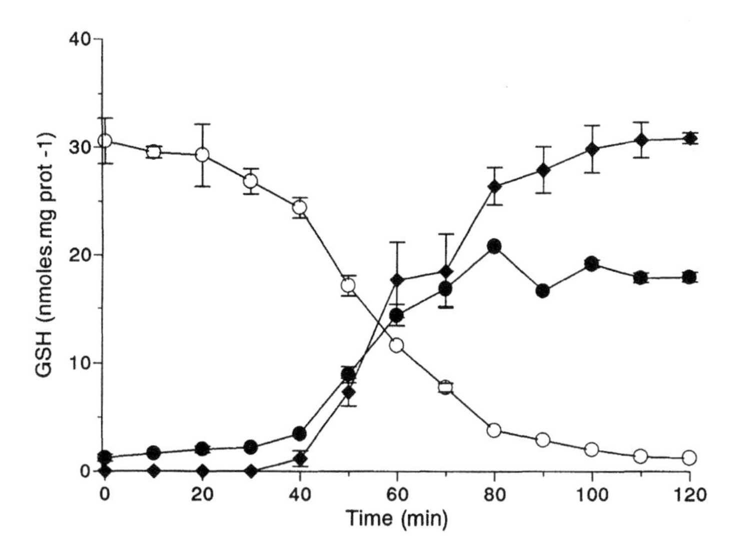

Figure 1. Glutathione is rapidly extruded in its reduced form after Jurkat T cells are exposed to anti-Fas/APO-1 antibody. Jurkat T cells (4×10^6/ml) were incubated in cystine-free RPMI 1640 medium in the absence (not shown, see Ref. 33) or presence of 250 ng/ml anti-Fas/APO-1 antibody. At the indicated times 150 μl of the cell suspension was taken in triplicate and centrifuged, and GSH was determined both in the medium (● without, and ◆ with dithiothreitol reduction prior to monobromobimane derivatization) and the cell pellet (○). Data points show mean ± S.D. of the anti-Fas/APO-1-treated cells. Control cells are not shown but demonstrate a small decrease in intracellular GSH (due to GSH depletion in cystine-free medium) and no significant increase in GSH or total glutathione in the medium. For more experimental details see Ref. 33.

Table 1. Known Transmembrane GSH Transporters

Transmembrane GSH transporters	Characteristics	Inhibitors
GSH specific transporters		
Canalicular GSH transporter	In most cell types	Bromosulfophthalein Dibromosulfophthalein
Sinusoidal GSH transporter	Liver specific	L-Methionine, L-cystathionine, BSP-GSH conjugate
Mitochondrial GSH transporter	Proton gradient driven	Glutamate
Other transporters		
MRP1	Transports GS-X conjugates and GSH	GS-alkyl conjugates *s*-nonyl GSH

(a)

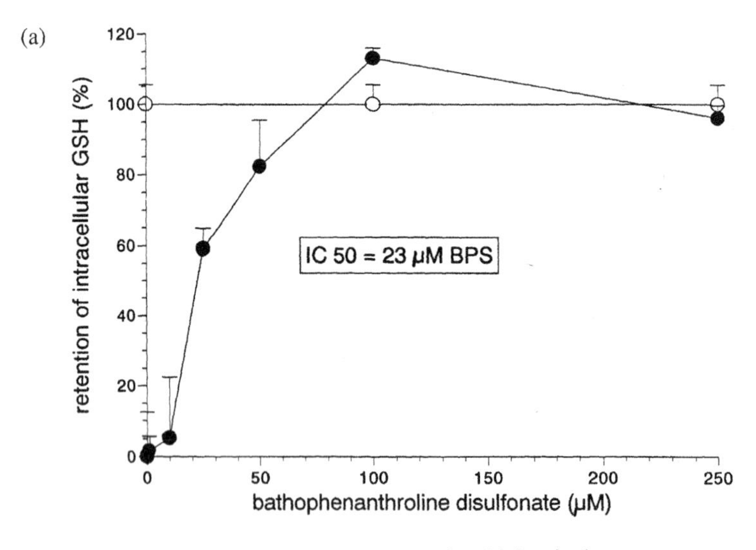

(b)

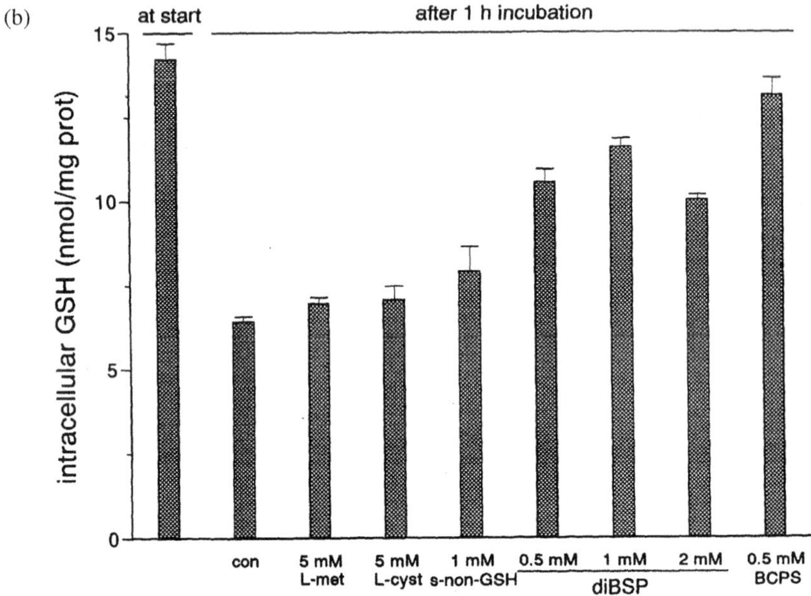

Figure 2. Bathophenanthroline disulfonate and bathocuproine disulfonate are potent inhibitors of GSH efflux in anti-Fas/APO-1 antibody-treated Jurkat T cells and apoptotic thymocytes. (a) Jurkat T cells (4×10^6) were pretreated for 15 min with different concentrations of BPS before adding control medium (○) or 250 ng/ml anti-Fas/APO-1 antibody (●). Intracellular GSH was determined 2 h after addition of the antibody as described (33). Results are expressed as percentage retention of the intracellular GSH. The intracellular GSH values of control and anti-Fas/APO-1-treated cells were 32.3 ± 1.1 and 18.6 ± 1.8 nmol/mg protein, respectively. (b)Rat thymocytes were treated with 25 µM etoposide for 2 h in RPMI 1640 with 2% FCS and fractionated on a Percoll gradient as described (11). Thymocytes (2.5×10^7 cells/ml) of the apoptotic fraction (V) were incubated in cystine-free RPMI for 1 h in the absence or presence of different efflux inhibitors. Intracellular GSH was determined at the beginning and end this incubation as described in Ref. 33. BPS, bathophenanthroline disulfonate; BCPS, bathocuprione disulfonate; L-met, L-methionine; L-cyst, L-cystathionine; s-non-GSH, *s*-nonyl-GSH;diBSP, dibromosulfophthalein.

hepatocytes are the best-characterized cell system with respect to these GSH transporters and have been shown to contain two different transporters in their plasma membrane which can be distinguished by their different sensitivity to inhibitors (Table 1; based on Refs. 34–49). Furthermore, a mitochondrial GSH transporter has been described which is thought to be responsible for the active import of GSH to build up the antioxidant defense of this organelle (39,46,47). In addition, the transmembrane multidrug resistance-associated protein 1 (MRP1) has been shown to excrete GSH when cells are treated with chemotherapeutic agents (48,49)

The anti-Fas/APO-1 antibody-induced GSH efflux can be blocked by both bromosulfophthalein (BSP) and phenol-3,6-dibromosulfophthalein disulfonate (diBSP) suggesting the involvement of a transporter with characteristics resembling the hepatocyte canalicular membrane GSH transporter. Inhibitors selective for the other known GSH transporters listed in Table 1 were ineffective in blocking this apoptosis-associated GSH excretion (33), excluding their involvement in the event (see also Figure 2b). Unfortunately, only short-term incubations (30–60 min) in the presence of the known inhibitors of the canalicular GSH transporter could be performed, as BSP and diBSP lowered the GSH content of the control cells by direct conjugation (50) and thereby negatively affected cell viability in the long-term experiments. Both compounds also interfered with the reduction of 3-(4,5-dimethylthiazol-2-yl)-2,5-diphenyltetrazolium (MTT) in cell viability assays. Thus, it is difficult to use these compounds to study the role of inhibition of GSH efflux and maintenance of GSH levels on the other biochemical changes occurring during anti-Fas/APO-1 antibody-induced apoptosis.

In the search for other agents that could inhibit the hepatocyte canalicular membrane GSH transporter without conjugating to GSH, we have found two copper-chelating compounds, bathophenanthroline disulfonate (BPS) and bathocuproine disulfonate (BCPS), to be highly efficient in inhibiting GSH efflux from anti-Fas/APO-1 antibody-treated Jurkat T lymphocytes or etoposide-treated rat thymocytes (Figure 2a,b). Major advantages of the use of these disulfonates were their relative efficiency (BPS has an IC_{50} of about 25 μM), their inability to conjugate to GSH, and the low toxicity associated with their use. The copper-chelating, nonsulfonated analogs bathophenanthroline (BP) and bathocuproine (BCP) were completely without effect, demonstrating that the copper-chelating properties were not required for the inhibitory effect on the GSH transporter (Table 2).

Table 2. GSH Efflux from Anti-Fas/APO-1-treated Jurkat T Cells is Inhibited by the Sulfonated Analog of Bathophenanthroline

	Intracellular GSH (nmol/mg protein)	
	Untreated	+Anti-Fas/APO-1 antibody
Control	34.1±1.1	18.6±1.8
Bathophenanthroline (50 μM)	32.6±1.8	19.2±2.4
Bathophenanthroline disulfonate (50 μM)	32.2±1.8	30.1±1.9

Jurkat T lymphocytes (4×10^6) were preincubated with the bathophenanthroline analog 15 min before adding 250 ng/ml anti-Fas/APO-1 antibody and 2 h after this addition the cells were harvested for intracellular GSH determination as described (33). Results are expressed as mean ± S.D. of triplicates.

THE EFFECT OF GSH EFFLUX ON CELLULAR CHANGES DURING APOPTOSIS

The discovery of the sulfonated phenanthroline derivatives as potent nontoxic inhibitors of GSH efflux in cells undergoing apoptosis enabled us to study the potential role of intracellular GSH in the development/generation of other apoptotic changes. From Table 3 it is clear that the inhibition of GSH extrusion from the apoptotic cell does not lead to increased viability as assessed by reductive MTT metabolism. Morphological changes such as plasma membrane blebbing and nuclear fragmentation were similarly unaffected by inhibition of GSH efflux (Table 3). In addition, the redistribution of phosphatidylserine phospholipid from the inner leaflet of the plasma membrane to the surface of the cell could not be prevented by BCPS in anti-Fas/APO-1-treated Jurkat T cells (Figure 3a). Finally, the agarose gel shown in Figure 3b demonstrates that neither BPS nor BCPS could prevent the formation of oligonucleosomal DNA fragments in the same experimental system. Thus, neither DNA degradation and nuclear fragmentation nor plasma membrane blebbing and phosphatidylserine exposure seem to be dependent on GSH efflux in this apoptotic model. However, there is a potential link between GSH efflux and cell shrinkage, another hallmark for apoptosis, as suggested by preliminary experiments performed with etoposide-treated rat thymocytes. After a 5 h incubation with 25 μM etoposide, the mean volume of the thymocytes decreased to 88 fl (98 fl for controls) and this loss of volume was almost completely blocked by the addition of 250 μM BCPS (96 fl after 5 h treatment with etoposide and BCPS). The findings of Häussinger and colleagues are interesting in this regard; they observed GSH extrusion from hepatocytes during a volume regulatory K^+ efflux induced by perfusion with hypotonic medium (51). As ion fluxes have been implicated previously in cell shrinkage during apoptosis, additional experiments are currently being performed to further investigate this potential link between GSH efflux and cell volume regulation.

Table 3. Inhibition of GSH Efflux Has No Effect on Survival in Anti-Fas/APO-1-treated Jurkat T Cells

Parameter	Control	100 μM BPS
MTT metabolism (% of control)		
Untreated	100 ± 9	112 ± 9
Anti-Fas/APO-1 antibody	55 ± 4	41 ± 2
Plasma membrane blebbing (%)		
Untreated	3 ± 1	4 ± 2
Anti-Fas/APO-1 antibody	67 ± 4	71 ± 5
Nuclear fragmentation (%)		
Untreated	2 ± 1	3 ± 1
Anti-Fas/APO-1 antibody	40 ± 6	39 ± 6

Jurkat T cells were left untreated or incubated with BPS or BCPS for 15 min before the addition of 250 ng/ml anti-Fas/APO-1 antibody or control medium. MTT metabolism was determined for 60 min 3 h after the addition of the antibody as in Ref. 33. At 2 h cells were examined for plasma membrane blebbing by light microscopy after mixing 1:1 in trypan blue (0.16%) or nuclear fragmentation by fluorescence microscopy after paraformaldehyde (4%) fixation and staining with Hoechst 33342 dye (10 μg/ml).

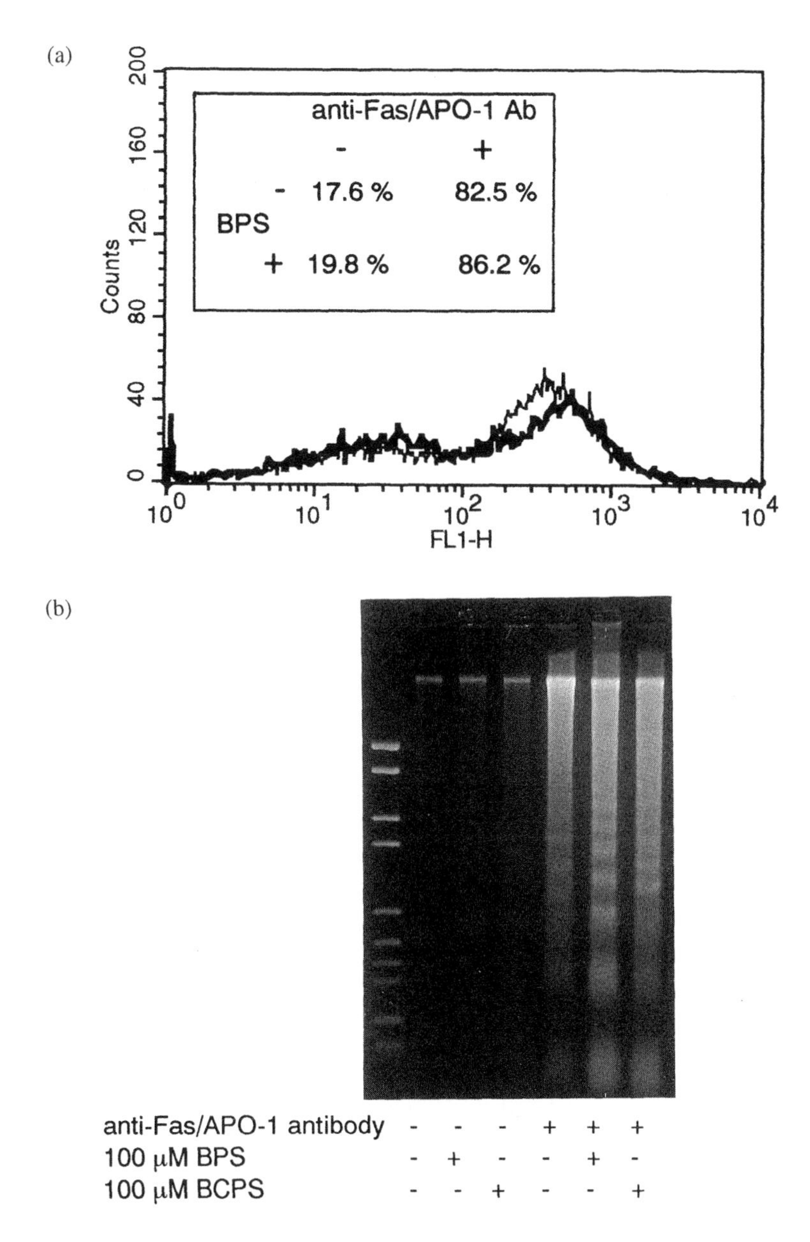

Figure 3. Inhibition of GSH efflux has no effect on phosphatidylserine exposure and DNA fragmentation induced by anti-Fas/APO-1 antibody treatment of Jurkat T cells. (a) Jurkat T cells (4×10^6) were incubated for 15 min with 250 μM BPS prior to addition of anti-Fas/APO-1 antibody. After 2 h cells were harvested, washed, and labeled with FITC-labeled annexin V in order to detect phosphatidylserine exposure on the cells' outer surface. FACS analysis shows the fluorescence distribution of anti-Fas/APO-1-treated Jurkat T cells in the absence (bold line) or presence (thin line) of 250 μM BPS. The inset shows the percentage of cells exposing phosphatidylserine. (b) Jurkat T cells (4×10^6) were incubated for 15 min in the absence (lanes 1 and 4) or presence of 100 μM BPS (lanes 2 and 5) or 100 μM BCPS (lanes 3 and 6) prior to addition of 250 ng/ml anti-Fas/APO-1 antibody. After 2 h, 1.2×10^6 cells were harvested for agarose gel electrophoresis essentially as described by Wyllie (2).

DISCUSSION

GSH efflux occurs during apoptosis in all cell types we have studied (Jurkat T cells, thymocytes, U937 monocytic cells, and L929 fibroblasts). In addition to chromatin fragmentation, nuclear condensation, plasma membrane blebbing, phosphatidylserine exposure, and cell shrinkage, GSH efflux thus appears to be an integral part of the apoptotic process. All these events are probably initiated by the activation of members of the family of interleukin-1β-converting enzyme (ICE)-like proteases. The GSH transporter, with characteristics similar to the hepatocyte canalicular membrane GSH transporter, is the most likely candidate for the efflux of the reduced tripeptide. This fits very well with the ubiquitous presence of this protein in a variety of cell types (35,38,39). We have studied the consequences of inhibition of GSH efflux on DNA degradation, nuclear fragmentation, membrane blebbing, and phosphatidylserine exposure, but all of these biochemical changes developed in a similar way irrespective of the absence or presence of GSH efflux inhibitors. Thus, it appears that these events evolve independently of the GSH efflux. Preliminary experiments reveal, however, that the export of GSH may be linked to the transmembrane fluxes of ions and water involved in the shrinkage of the cell during apoptosis.

Besides the role GSH efflux could play in cell volume regulation, other possibilities for a physiological function of this phenomenon can be envisaged. First, it may be necessary for a cell undergoing apoptosis to lose reducing capacity. By exporting its major intracellular antioxidant, the cell does so without the need for the generation of oxidants with their attendant nonspecific reactivities. As many biochemical reactions are dependent on adequate GSH levels, the low intracellular GSH content will eventually lead to downregulation of cellular metabolism and thereby limit persistence of apoptotic bodies. Second, GSH release may be of nutritional significance for normal cells surrounding an apoptotic neighbor. The expelled GSH could become substrate for the plasma membrane-associated γ-glutamyl transpeptidase (and dipeptidase) and thereby rapidly become available for re-uptake by surrounding cells. As reducing agents and other compounds increasing the intracellular GSH level are well known to enhance the function of lymphoid cells in vitro (52–54), it is possible that GSH efflux contributes to the local stimulation of phagocytic cells.

In previous experiments we showed that increasing the intracellular GSH content by preloading the cells with GSH-esters delayed the decrease in MTT metabolism and the formation of high-molecular-mass DNA fragments caused by anti-Fas/APO-1 antibody treatment of Jurkat T cells (33). Presumably, the apoptotic events in the preloaded cells were inhibited by the supraphysiological levels of reducing equivalents, but this protective effect was counteracted by a greatly increased GSH efflux. Numerous other studies recording protective effects of thiol reductants against apoptosis have been published. For example, Fernandez and colleagues (14) showed that DNA fragmentation in glucocorticoid-treated thymocytes could be inhibited by preincubating the cells with high concentrations (25 mM) of the GSH precursor *N*-acetylcysteine (NAC). In addition, the antioxidant WR-1065 (10 mM) protects thymocytes from DNA fragmentation induced by dexamethasone, γ-irradiation, or calcium ionophore A23187 (55). It is important to emphasize that the effect of inhibition of the GSH efflux by BPS or BCPS is completely different from attempts to increase intracellular GSH levels by means of GSH-esters, NAC, or other antioxidants. Whereas inhibition of GSH efflux does not perturb the normal redox state in the cell, allowing its effect on other apoptotic processes to be studied under normal redox conditions, high intracellular concentrations of GSH or other antioxidants will disrupt the cell's delicate redox balance and are thus of limited physiological relevance.

ACKNOWLEDGMENTS

We are grateful to Anna Carin Hellerqvist for excellent secretarial assistance and Ian Cotgreave for helpful discussions. This work was supported by a fellowship from the European Science Foundation (to D.v.d.D.) and a grant from the Swedish Medical Research Council.

REFERENCES

1. Kerr JFR, Wyllie AH, Currie AR. Apoptosis: a basic biological phenomenon with wide ranging implications in tissue kinetics. Br J Cancer 1972; 26: 239–257.
2. Wyllie AH, Kerr JFR, Currie AR. Cell death: the significance of apoptosis. Int Rev Cytol 1980; 68: 251–306.
3. Savill JS, Fadok V, Henson P, Haslett C. Phagocyte recognition of cells undergoing apoptosis. Immunol Today 1993; 14: 131–136.
4. Lennon SV, Martin SJ, Cotter TG. Dose-dependent induction of apoptosis in human tumour cell lines by widely diverging stimuli. Cell Proliferation 1991; 24: 203–204.
5. Ueda N, Shah SV. Endonuclease-induced DNA damage and cell death in oxidant injury to renal tubular epithelial cells. J Clin Invest 1991; 90: 2593–2597.
6. Dypbukt JM, Ankarcrona M, Burkitt M, Sjöholm Å, Ström K, Orrenius S, Nicotera P. Different prooxidant levels stimulate cell growth, activate apoptosis, or produce necrosis in insulin-secreting RINm5F cells. J Biol Chem 1994; 269: 30553–30560.
7. Sato N, Iwata S, Nakamura K, Hori T, Mori K, Yodoi J. Thiol-mediated redox regulation of apoptosis. J Immunol 154; 1995: 3194–3203.
8. Escargueil I, Nègre-Salvayre A, Pieraggi M-T, Salvayre R. Oxidized low density lipoproteins elicit DNA fragmentation of cultured lymphoblastoid cells. FEBS Lett 1992; 305: 155–159.
9. Sandstrom PA, Tebbey PW, Van Cleave S, Buttke TM. Lipid hydroperoxides induce apoptosis in T cells displaying a HIV-associated glutathione peroxidase deficiency. J Biol Chem 1994; 269: 798–801.
10. Nobel CSI, Kimland M, Lind B, Orrenius S, Slater AFG. Dithiocarbamates induce apoptosis in thymocytes by raising the intracellular level of redox-active copper. J Biol Chem 1995; 270: 26202–26208.
11. Slater AFG, Nobel CSI, Maellaro E, Bustamante J, Kimland M, Orrenius S. Nitrone spin traps and a nitroxide antioxidant inhibit a common pathway of thymocyte apoptosis. Biochem J 1995; 306: 771–778.
12. Bustamante J, Slater AFG, Orrenius S. Antioxidant inhibition of thymocyte apoptosis by dihydrolipoic acid. Free Radical Biol Med 1995; 19: 339–347.
13. Beaver JP, Waring P. A decrease in intracellular glutathione concentration precedes the onset of apoptosis in murine thymocytes. Eur J Cell Biol 1995; 68: 47–54.
14. Fernandez A, Kiefer J, Fosdick L, McConkey DJ. Oxygen radical production and thiol depletion are required for Ca^{2+}-mediated endogenous endonuclease activation in apoptotic thymocytes. J Immunol 1995; 155: 5133–5139.
15. Ghibelli L, Coppola S, Rotilio G, Lafavia E, Maresca V, Ciriolo MR. Nonoxidative loss of glutathione in apoptosis via GSH extrusion. Biochem Biophys Res Commun 1995; 216: 313–320.

16. Briehl MM, Cotgreave IA, Powis G. Downregulation of the antioxidant defense during glucocorticoid-mediated apoptosis. Cell Death Differentiation 1995; 2: 41–46.
17. Sandstrom PA, Mannie MD, Buttke TM. Inhibition of activation-induced death in T cell hybrydomas by thiol antioxidants: oxidative stress as a mediator of apoptosis. J Leukocyte Biol 1994; 55: 221–226.
18. Greenlund LJS, Deckwerth TL, Johnson EM Jr. Superoxide dismutase delays neuronal apoptosis: a role for reactive oxygen species in programmed neuronal death. Neuron 1995; 14: 303–314.
19. Slater AFG, Nobel CSI, Van den Dobbelsteen DJ, Orrenius S. Signalling mechanisms and oxidative stress in apoptosis. Toxicol Lett 1995; 82/83: 149–153.
20. Slater AFG, Nobel CSI, Van den Dobbelsteen DJ, Orrenius S. Intracellular redox changes during apoptosis. Cell Death Differentiation 1995; 3: 57–62.
21. Muschel RJ, Bernhard EJ, Garza L, McKenna WG, Koch C. Induction of apoptosis at different oxygen tensions: evidence that oxygen radicals do not mediate apoptotic signaling. Cancer Res 1995; 55: 995–998.
22. Jacobson MD, Raff MC. Programmed cell death and Bcl-2 protection in very low oxygen. Nature 1995; 374: 814–816.
23. Shimizu S, Eguchi Y, Kosaka H, Kamiike W, Matsuda H, Tsujimoto Y. Prevention of hypoxia-induced cell death by Bcl-2 and Bcl-xL. Nature 1995; 374: 811–813.
24. Jacobson MD. Reactive oxygen species and programmed cell death. Trends Biochem Sci 1996; 21: 83–86.
25. Hug H, Enari M, Nagata S. No requirement of reactive oxygen intermediates in Fas-mediated apoptosis. FEBS Lett 1994; 351: 311–313.
26. Stefanelli C, Stanic I, Bonavita F, Muscari C, Pignatti C, Rossoni C, Caldera CM. Oxygen tension influences DNA fragmentation and cell death in glucocorticoid-treated thymocytes. Biochem Biophys Res Commun 1995; 212: 300–306.
27. Eischen CM, Dick CJ, Leibson PJ. Tyrosine kinase activation provides an early and requisite signal for Fas-induced apoptosis. J Immunol 1994; 153: 1947–1954.
28. Enari M, Hug H, Nagata S. Involvement of an ICE-like protease in Fas-mediated apoptosis. Nature 1995; 375: 78–81.
29. Tewari M, Quan LT, O'Rourke K, Desnoyers S, Zeng Z, Beidler DR, Poirier GG, Salvesen GS, Dixit VM. Yama/CPP32 β, a mammalian homolog of CED-3 is a CrmA-inhibitable protease that cleaves the death substrat poly(ADP-ribose) polymerase. Cell 1995; 18: 801–809.
30. Schlegel J, Peters I, Orrenius S, Miller DK, Thornberry NA, Yamin TT, Nicholson DW. CPP32 is the ICE-like protease in Fas-mediated apoptosis. J Biol Chem 1996; 271: 1841–1844.
31. Weiss M, Schlegel J, Kass GEN, Holmström TH, Peters I, Eriksson J, Orrenius S, Chow SC. Cellular events in Fas/APO-1-mediated apoptosis in JURKAT T lymphocytes. Exp Cell Res 1995; 219: 699–708.
32. Martin SJ, Reutelingsberger CPM, McGahon AJ, Rader JA, Van Schie RCAA, LaFace DM, Green DR. Early redistribution of plasma membrane phosphatidylserine is a general feature of apoptosis regardless of the initiating stimulus: inhibition by overexpression of Bcl-2 and Abl. J Exp Med 1995; 182: 1545–1556.
33. Van den Dobbelsteen DJ, Nobel CSI, Schlegel J, Cotgreave IA, Orrenius S, Slater AFG. Rapid and specific efflux of reduced glutathione during apoptosis induced by anti-Fas/APO-1 antibody. J Biol Chem 1996; 271: 15420–15427.
34. Meister A. Glutathione, ascorbate, and cellular protection. Cancer Res 1994; [suppl] 54: 1969s–1975s.
35. Deneke SM, Fanburg BL. Regulation of cellular glutathione. Am J Physiol 1989; 257: L163–L173.
36. Bannai S, Tatishi N. Role of membrane transport in metabolism and function of glutathione in mammals. J Membrane Biol 1986; 89: 1–8.
37. Fernández-Checa J, Yi J-R, Garcia-Ruiz C, Knezic Z, Tahara SM, Kaplowitz, N. Expression

of rat liver reduced glutathione transport in *Xenopus laevis* oocytes. J Biol Chem 1993; 268: 2324–2328.
38. Yi J-R, Lu S, Fernández-Checa J, Kaplowitz N. Expression cloning of a rat hepatic reduced glutathione transporter with canalicular characteristics. J Clin Invest 1993; 93: 1841–1845.
39. García-Ruiz C, Morales A, Colell A, Rodés J, Yi J-R, Kaplowitz N, Fernández-Checa J. Evidence that the rat hepatic mitochondrial carrier is distinct from the sinusoidal and canalicular transporters for reduced glutathione. J Biol Chem 1995; 270: 15946–15949.
40. Yi J-R, Lu S, Fernández-Checa J, Kaplowitz N. Expression cloning of the cDNA for a polypeptide associated with rat hepatic sinusoidal reduced glutathione transport: characteristics and comparison with the canalicular transporter. Proc Natl Acad Sci USA 1995; 92: 1495–1499.
41. Kaplowitz N, Fernández-Checa J, Kannan R, Garcia-Ruiz C, Ookhtens M, Yi J-R. GSH transporters: molecular characterization and role in GSH homeostasis. Biol Chem Hoppe-Seylers vol. 20 1997; in press.
42. Lu SC, Sun W-M, Yi J, Ookhtens M, Sze G, Kaplowitz N. Role of two recently cloned rat liver GSH transporters in the ubiquitous transport of GSH in mammalian cells. J Clin Invest 1997; in press.
43. Sze G, Kaplowitz N, Ookhtens M, Lu S. Bidirectional membrane transport of intact glutathione in Hep G2 cells. Am J Physiol 1993; 265: G1128–G1134.
44. Garcia-Ruiz C, Fernández-Checa J, Kaplowitz N. Bidirectional mechanism of plasma membrane transport of reduced glutathione in intact rat hepatocytes and membrane vesicles. J Biol Chem 1992; 267: 22256–22264.
45. Ballatori N, Dutczak WJ. Identification and characterization of high and low affinity transport systems for reduced glutathione in liver cell canalicular membranes. J Biol Chem 1994; 269: 19731–19737.
46. Mårtensson J, Lai JCK, Meister A. High-affinity transport of glutathione is part of a multicomponent system essential for mitochondrial function. Proc Natl Acad Sci USA 1990; 87: 7185–7189.
47. Kurosawa K, Hayashi N, Sato N, Kamada T, Tagawa K. Transport of glutathione across the mitochondrial membranes. Biochem Biophys Res Commun 1990; 167: 367–372.
48. Müller M, Meijer C, Zaman GJR, Borst P, Scheper RJ, Mulder NH, De Vries EGE, Jansen PLM. Overexpression of the gene encoding the multidrug resistance-associated protein results in increased ATP-dependent glutathione S-conjugate transport. Proc Natl Acad Sci USA 1994; 91: 13033–13037.
49. Zaman GJR, Lankelma J, Van Tellingen O, Beijnen J, Dekker H, Paulusma C, Oude Elferink RPJ, Baas F, Borst P. Role of glutathione in the export of compounds from cells by the multidrug resistance-associated protein. Proc Natl Acad Sci USA 1995; 92: 7690–7694.
50. Habig WH, Pabst MJ, Jakoby WB. Glutathione S-transferases. The first enzymatic step in mercapturic acid formation. J Biol Chem 1974; 249: 7130–7139.
51. Häussinger D, Lang F, Bauers K, Gerok W. Control of hepatic nitrogen metabolism and glutathione release by cell volume regulatory mechanisms. Eur J Biochem 1990; 193: 891–898.
52. Fidelius RK, Ginouves P, Lawrence D, Tsan MF. Modulation of intracelluar glutathione concentrations alters lymphocyte activation and proliferation. Exp Cell Res 1987; 170: 269–278.
53. Gmunder H, Eck P, Benninghoff B, Roth S, Dröge W. Macrophages regulate intracellular glutathione levels of lymphocytes: evidence for an immunoregulatory role of cysteine. Cell Immunol 1990; 129: 32–46.
54. Fay M, Jampy-Fay M, Akarid K, Gougerot-Pocidalo M-A. Protective effect of LPS and poly A:U against immune oxidative injury: role of thiols released by activated macrophages. Free Radical Biol Med 1995; 18: 649–654.
55. Ramakrishan N, Catravas GN. *N*-(2-Mercaptoethyl)-1,3-propanediamine (WR-1065) protects thymocytes from programmed cell death. J Immunol 1992; 148: 1817–1821.

18

Endogenously Generated Superoxide and Hydrogen Peroxide in the Redox Modulation of Cell Proliferation, Apoptosis, and Virus Replication

Roy H. Burdon
University of Strathclyde
Glasgow, Scotland

EFFECTS OF EXOGENOUS SUPEROXIDE AND HYDROGEN PEROXIDE

Some years ago, we reported that low concentrations of superoxide or hydrogen peroxide were growth stimulatory when added to cultures of BHK-21 hamster fibroblasts (1). This now appears to be a wide-ranging phenomenon. A considerable variety of non-inflammatory mammalian cells exhibit positive growth responses when exposed to low levels of exogenous superoxide or hydrogen peroxide (2). In our studies, while hydrogen peroxide stimulated the growth of both nontransformed and polyoma virus-transformed BHK cells, superoxide was notably stimulatory toward the tumor virus-transformed cells. Differences in growth stimulatory responses elicited by exogenously added superoxide and hydrogen peroxide have been observed by others (3). In the case of superoxide, certain of its effects, such as increasing intracellular pH and calcium, are extremely rapid and can be inhibited by anion-channel blockers. Hydrogen peroxide, which can permeate cell membranes, does not elicit these early growth-related changes but, in common with superoxide, can upregulate the expression of "early growth response genes."

We have recently shown that if the concentration of exogenously added hydrogen peroxide is increased, proliferation of BHK-21 cells is progressively depressed. Higher levels also increase the rate of apoptosis (4). Additionally under these conditions, we have observed excessive DNA repair (4) as well as gene sequence-specific DNA damage. In the case of BHK cells exposed to 1 mM hydrogen peroxide, around four times the normal cellular amount of DNA accumulated (4). In HeLa cells, we observed that sequences associated with genes encoding catalase, α_1-antitrypsin, and β-actin are more susceptible to damage than sequences associated with genes for heat shock protein 60, H-Ras, or p53 (5).

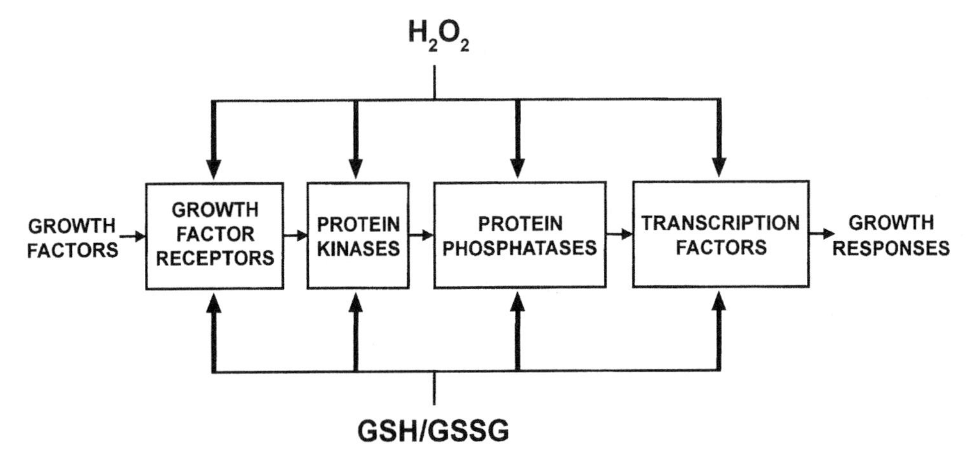

Figure 1. An overview of the multiplicity of cellular targets for hydrogen peroxide, GSH, or GSSG that may be relevant to proliferation responses.

MOLECULAR MECHANISMS OF GROWTH CONTROL BY OXIDANTS

In mammalian cells there is as yet no evidence for proteins similar to the bacterial SoxR or OxyR capable of specifically "sensing" superoxide (6) or hydrogen peroxide (7), which might be an initial step in switching on genes relevant to mammalian cell proliferation responses. In contrast there has been an impressive accumulation of experimental data to indicate that the functioning of many growth signal transducing proteins is significantly dependent on their redox state. Such proteins include growth factor receptors, protein kinases, and protein phosphatases as well as a number of important transcription factors, including AP-1 and NFκB (2). It has been speculated by us that the adjustment of thc redox states of individual signal transduction proteins within cells is a prerequisite for their optimal functioning in the transmission of growth responses (2). Such protein redox regulation could be achieved through the direct oxidative interaction of hydrogen peroxide (or superoxide), or indirectly through changes in cellular levels of GSH and GSSG, for example involving the participation of glutathione peroxidase (Figure 1).

Another type of observation which suggests that the growth modulatory effects of hydrogen peroxide are generally through redox modulation of regulatory proteins, rather than as a result of direct interaction with specific "receptor" proteins, is that *t*-butylhydroperoxide will also elicit similar concentration-dependent positive and negative growth regulatory effects (2).

ENDOGENOUS GENERATION OF SUPEROXIDE AND HYDROGEN PEROXIDE AND REGULATORY MECHANISMS

Whereas both superoxide and hydrogen peroxide are established products of the "respiratory burst" when the plasma membrane NADPH-oxidase of neutrophils and macrophages is activated, it is now becoming clear (Figure 2) that superoxide and hydrogen peroxide are released by a range of noninflammatory cells (2). For instance, primary human skin fibroblasts stimulated with cytokines such as interleukin-1, or tumor

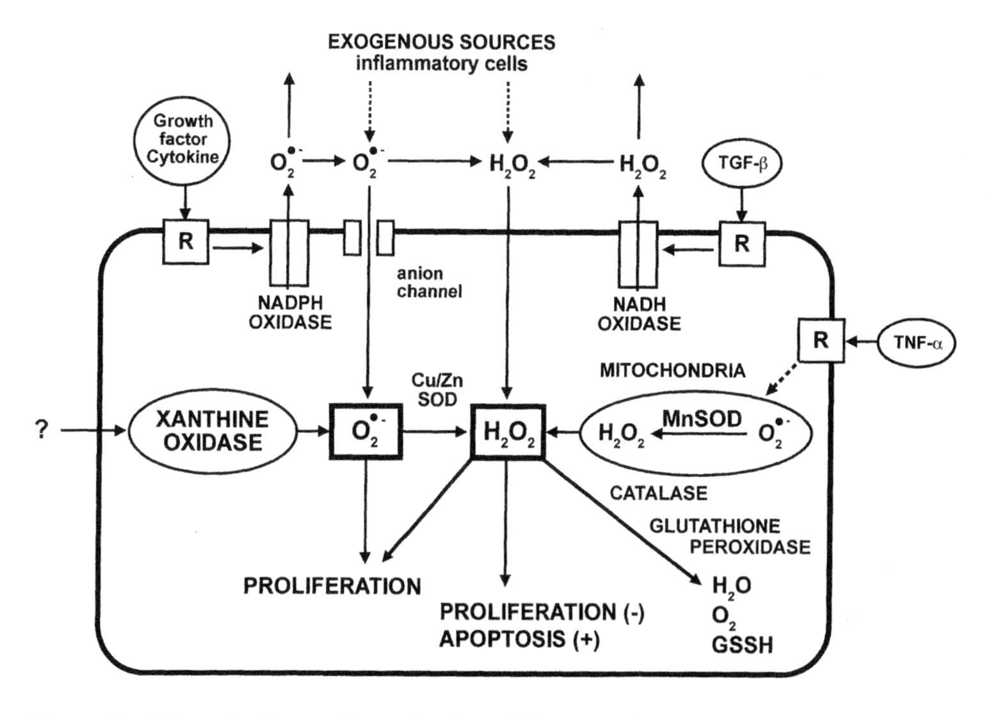

Figure 2. Schematic diagram illustrating the cellular generation and release of superoxide and hydrogen peroxide in relation to their possible role as cell signals in the redox regulation of proliferation or apoptosis. TGF-β, transforming growth factor-β; TNF-α, tumor necrosis factor-α; SOD, superoxide dismutase.

necrosis factor-α, release significant levels of superoxide (8). Endothelial cells also release superoxide, but again this release is greatly stimulated by cytokines, in this case interferon and interleukin-1 (9). In the case of Balb/3T3 cells, platelet-derived growth factor is required (10). Recently it has been reported that transforming growth factor-β1 is required to activate a plasma membrane hydrogen peroxide-generating NADH oxidase in human lung fibroblasts (11). On the other hand, *constitutive* release of hydrogen peroxide has been detected in a wide range of human tumor cells (12).

Within mammalian cells it has also been possible to detect superoxide generation (13). In BHK and HeLa cells possible sources include xanthine oxidase and mitochondria (13). Although it is difficult to evaluate the comparative rate of superoxide generation from mitochondria *in intact cells*, mitochondrial generation is nevertheless clear and very greatly stimulated in mouse L-929 cells exposed to the cytokine tumor necrosis factor-α (14).

While the activity of xanthine oxidase in mouse epidermal and endothelial cells can be stimulated by phorbol esters and interferon, respectively (15, 16), we found that overall intracellular generation of superoxide in BHK and HeLa cells (from xanthine oxidase and possibly mitochondria) was maintained by the presence of serum antioxidants (13). Deprivation of such serum antioxidants, or the addition to the cultures of certain polyunsaturated fatty acids (PUFAs), caused increased rates of intracellular superoxide generation. Normal rates of generation could be restored, however, by the addition of α-tocopherol to the cells (13).

INTRACELLULAR LEVELS OF HYDROGEN PEROXIDE IN RELATION TO PROLIFERATION AND APOPTOSIS

The previous section illustrated the variety of systems available to noninflammatory cells for the endogenous generation of superoxide and hydrogen peroxide, and how these systems can be regulated by important growth factors, cytokines, and other serum molecules. Within cells, further systems that can regulate levels of superoxide and hydrogen peroxide include the superoxide dismutases, catalase, and glutathione peroxidase. In BHK cells we found that the activities of both catalase and glutathione peroxidase were influenced by cell growth state. In early stages of growth, the activity of glutathione peroxidase was high but declined during growth. The opposite was true of catalase activity.

When intracellular levels of hydrogen peroxide were determined in BHK cells, these levels were not static but declined progressively with growth (17). Exposure of these cells to inhibitors of glutathione peroxidase (mercaptosuccinate, mercaptovaline) or catalase (aminotriazole) not only reduced this growth-associated decline but also depressed rates of cell proliferation and increased rates of apoptosis. This suggests that these antioxidant enzymes may have a novel, but critical, role as regulatory molecules. Another type of enzyme-related agent which also reduces the growth-associated decline of intracellular hydrogen peroxide comprises cell-permeable, low-molecular-mass mimics of superoxide dismutase. One of these, copper(II)-(3,5-diisopropylsalicylate)$_2$, also brought about reduced BHK (4) and HeLa (18) cell proliferation and increased apoptosis.

As mentioned previously, exposure of BHK cells to certain PUFAs results in increased rates of intracellular superoxide generation. In the case of docosahexaenoic acid-treated cells there are also elevated levels of intracellular hydrogen peroxide, reduced rates of cell proliferation, and increased occurrence of apoptosis. In this particular case these negative effects could be reversed by the addition of α-tocopherol (19).

ABNORMALLY HIGH OR LOW INTRACELLULAR LEVELS OF HYDROGEN PEROXIDE

Although the previous section would suggest that procedures resulting in increases in intracellular hydrogen peroxide usually lead to reduced rates of cell proliferation and increased apoptosis, early experiments of ours suggested that exposure of BHK or HeLa cell cultures to exogenously added superoxide dismutase or catalase also produced the same outcomes (18,20). In addition, exposure of BHK cells to *N*-acetyl-L-cysteine, although resulting in elevated cellular GSH levels, nevertheless caused a dramatic reduction in cellular levels of hydrogen peroxide (17) and after 48 hours resulted in both reduced proliferation and some increase in apoptosis. In summary, for BHK or HeLa cells, it appears that procedures that lead to either abnormally higher or abnormally lower intracellular levels of hydrogen peroxide can result in reduced rates of proliferation and increased apoptosis.

GROWTH STIMULATORY EFFECTS OF HYDROGEN PEROXIDE

Because intracellular levels that are either abnormally high or abnormally low can result in reduced proliferation and increased apoptosis, it may be asked why exposure of fibroblasts such as BHK cells to low concentrations of exogenous hydrogen peroxide (for example, 1 μM) will stimulate proliferative responses (17). In fact, hydrogen peroxide addition at this low concentration does not actually alter intracellular hydrogen peroxide levels. Rather it brings about a small decline in cellular GSH levels (21). Thus, growth stimulation may be achieved not by direct oxidative interaction of hydrogen peroxide with signal transduction proteins, but more likely indirectly through changes in cellular GSH (or GSSG) levels that would affect the proteins' redox status

A REDOX REGULATORY PARADIGM

An Overview

Cell-specific signaling is currently dominated by a single paradigm whereby this is accomplished by molecules that bind noncovalently to specific receptors through complementarity of shape. We have proposed a novel growth regulatory paradigm (2) that is superimposed on established cell signal transduction pathways. This may involve the direct oxidative modification of growth signal transduction proteins such as receptors, protein kinases, protein phosphatases, and transcription factors by hydrogen peroxide or superoxide. Alternatively, hydrogen peroxide may modulate the redox state and activity of these important signal transduction proteins through changes in cellular levels of GSH and GSSG.

Redox Regulation of Proliferative Responses

In this proposal the optimal intracellular level of hydrogen peroxide and superoxide would be "set" through the modulation of NADPH- and NADH-oxidases, xanthine oxidase, and mitochondrial activities by the appropriate growth factors or cytokines, and by the relative activities of cellular superoxide dismutases, catalase, and glutathione peroxidase (see Figure 2). However, abnormally elevated or lowered levels of hydrogen peroxide can prejudice optimal growth responses and favor reduced proliferation, as well as apoptosis. It is now clear from the work of other groups that the endogenous generation of such active oxygen species is a necessary prerequisite for cytokine and growth factor responses. Examples include the induction of c-*fos* expression in bovine chondrocytes by tumor necrosis factor-α and basic fibroblast growth factor (22), the angiotensin II-induced AP-1 DNA-binding activity and proliferative hypertrophic responses in mouse myogenic cells (23), the stimulation of rat vascular smooth-muscle cells by platelet-derived growth factor (24), and the downregulation of DNA synthesis in mouse osteoblastic cells exposed to transforming growth factor-β1 (25).

Redox Regulation of Apoptosis

Although the generation of reactive oxygen species appears necessary for the transduction of growth signals, it is not yet clear whether it is actually necessary for the signaling of apoptosis. While higher than normal levels of hydrogen peroxide can induce apoptosis, apoptosis can also be brought about in conditions of very low intracellular hydrogen peroxide, suggesting that the involvement of reactive oxygen species in apopotosis may not be obligatory. Rather, normal apoptosis pathways may simply be favored under cellular conditions which are either abnormally "oxidatively" or "reductively" stressful. In situations where increased intracellular generation of superoxide has been provoked in BHK cells by acute serum withdrawal or exposure to the PUFA docosahexaenoic acid, the ensuing "oxidative" stress that can lead to apoptosis can nevertheless be blocked by α-tocopherol, which can restore intracellular levels of superoxide generation to normal. α-Tocopherol, on the other hand, does not block apoptosis in these cells induced by direct exposure to exogenous hydrogen peroxide.

The Redox Regulatory Paradigm and Virus Replication

While abnormal intracellular levels of hydrogen peroxide can reduce proliferation of BHK cells, we find that they can also reduce the replication of herpes simplex virus in these cells as host. Previously, reactive oxygen intermediates have been reported to promote the replication of HIV-1 in a human T-cell line (26).

Intercellular Communication

In physiological terms, relatively high concentrations of exogenous hydrogen peroxide and superoxide could be available at sites of inflammation. In principle this could provide a pathologically important means of influencing the redox status of cell proteins. On the other hand, if noninflammatory cells can also generate and release these reactive oxygen species, then it is also possible to envisage a means whereby these could influence the redox status of proteins in adjacent cells of a tissue. Indeed, the effects of growth factor or cytokine interaction with only a single cell could be transmitted to many surrounding cells in a tissue. While there would likely be a gradient of diminishing influence within the tissue, this could have relevance to the development of tissue morphology. Developmental biologists have previously hypothesized tissue gradients of diffusive developmental "morphogens" (27).

REFERENCES

1. Burdon RH, Rice-Evans C. Free radicals and the regulation of mammalian cell proliferation. Free Radical Res Commun 1989; 6: 345–358.
2. Burdon RH Superoxide and hydrogen peroxide in relation to mammalian cell proliferation. Free Radical Biol Med 1995; 18: 775–794.
3. Ikebuchi Y, Masumoto K, Tasaka K, Koike K, Kasahara K, Miyake A, Tanizawa O. Superoxide anion increases intracellular pH, intracellular free calcium and arachidonate release in human amnion cells. J Biol Chem 1991; 266: 13233–13237.
4. Burdon RH, Hill V, Alliangana D. Hydrogen peroxide in relation to proliferation and apoptosis in BHK-21 hamster fibroblasts. Free Radical Res 1996; 24: 81–93.
5. Burdon RH, Gill V, Boyd PA, Rahim RA. Hydrogen peroxide and sequence-specific DNA damage in human cells. FEBS Lett 1996; 383: 150–154.
6. Hildago E, Demple B. An iron–sulphur centre essential for transcription activation by redox sending Sax R protein. EMBO J 1994; 13: 138–146.
7. Stortz G, Tartaglia LA. OxyR: a regulator of antioxidant genes. J Nutr 1992; 122: 627–630.
8. Meier B, Radeke HH, Selle S, Younes M, Seis H, Resch K, Habermehl GG. Human fibroblasts release active oxygen species in response to interleukin-1 or tumour necrosis factor-α. Biochem J 1989; 263: 539–545.

9. Matsuyama T, Ziff M. Increased superoxide anion release from endlothelial cells in response to cytokines. J Immunol 1986; 137: 3295–3304.
10. Shibanuma M, Kuroki T, Nose K. Stimulation by hydrogen peroxide of DNA synthesis competence family gene expression and the phosphorylation of a specific protein in quiescent Balb/3T3 cells. Oncogene 1990; 3: 27–32.
11. Thannickal VJ, Fanbury BL. Activation of an H_2O_2-generating NADH oxidase in human lung fibroblasts by transforming growth factor β1. J Biol Chem 1995; 51: 30334–30338.
12. Szatrowski TP, Nathan CF. Production of large amounts of hydrogen peroxide by human tumour cells. Cancer Res 1991; 51: 794–798.
13. Burdon RH, Gill V, Rice-Evans C. Reduction of a tetrazolium salt and superoxide generation in human tumour cells (HeLa). Free Radical Res Commun 1993; 18: 369–388.
14. Hennet T, Richter C, Peterhaus C. Tumour necrosis-α induces superoxide generation in mitochondria of L-929 cells. Biochem J 1993; 289: 587–592.
15. Reiners JJ, Pence BC, Barcus MCS, Cantu AR. 12-*O*-tetradecanoylphorbol-13-acetate dependent induction of xanthine dehydrogenase and conversion to xanthine oxidase in murime epidermis. Cancer Res 1987; 47: 1775–1779.
16. Dupont GP, Huecksteadt TP, Marshall BC, Ryan US, Michael JR, Hoidal JR. Regulation of xanthine dehydrogenase and xanthine oxidase activity and gene expression in cultured rat pulmonary endolethial cells. J Clin Invest 1992; 89: 197–202.
17. Burdon RH, Alliangana D, Gill V. Hydrogen peroxide and the proliferation of BHK-21 cells. Free Radical Res, 1995; 23: 471–486.
18. Burdon RH, Gill V. Cellularly generated active oxygen species and HeLa cell proliferation. Free Radical Res Commun 1993; 19: 203–213.
19. Burdon RH, Gill V, Rice-Evans C. Active oxygen species in the promotion and suppression of tumour cell growth. In Davies KJA, ed. Oxidative Damage and Repair. Oxford: Pergamon Press; 1991: 791–795.
20. Burdon RH. Cell proliferation and oxidative stress: basis for anticancer drugs. Proc R Soc Edin 1992; 99B: 169–176.
21. Burdon RH, Alliangana D, Gill V. Endogenously generated active oxygen species and cellular glutathione levels in relation to BHK-21 cell proliferation. Free Radical Res 1994; 21: 121–133.
22. Lo YYC, Cruz TF. Involvement of reactive oxygen species in cytokine and growth factor induction of c-fos expression in chondrocytes. J Biol Chem 1995; 270: 11727–11730.
23. Puri PL, Avantaggiati ML, Burgio VL, Chirillo P, Collepando D, Natoli G, Balsano, C, Levero M. Reactive oxygen intermediates mediate angiotension II-induced c-Jun.c-Fos heterodimer DNA binding activity and proliferative responses in myogenic cells. J Biol Chem 1995; 270: 22129–22134.
24. Sundaresan M, Yu Z-X, Ferrans VJ, Irani K, Finkel T. Requirement for the generation of H_2O_2 for platelet-derived growth factor signal transduction. Science 1995; 270: 296–299.
25. Nose K, Ohba M, Shibanuma M, Kuroki T. Involvement of hydrogen peroxide in the actions of TGF-β1. In Pasquier C, Oliver RY, Auclair C, Packer L, eds. Oxidative Stress, Cell Activation and Viral Infection Basel: Birkauser Verlag; 1994: 21–34.
26. Schreck R, Rieber P, Baeverle PA. Reactive oxygen intermediates as apparently widely used messengers in the activation of NF-kappa B transcription factor and HIV-1. EMBO J 1991; 10: 2247–2258.
27. Wolpert L. Positional information and pattern formation. Curr Topics Dev Biol 1971; 6: 183–224.

19

Deoxy-D-Ribose-Induced GSH Depletion and Induction of Apoptosis in Human Quiescent Peripheral Blood Mononuclear Cells: Effects of *N*-Acetylcysteine and L-Buthionine-(*S*,*R*)-sulfoximine

B. Botti and V. Vannini
Università di Pavia Pavia, Italy

D. Monti, S. Macchioni, S. Bergamini, F. Tropea, A. Tomasi, and C. Franceschi
Università di Modena Modena, Italy

INTRODUCTION

It has previously been shown that 2-deoxy-D-ribose (dRib), a reducing sugar, induced apoptosis in human quiescent peripheral blood mononuclear cells (PBMC), which are relatively insensitive to apoptosis, while *N*-acetylcysteine (NAC), at the concentration of 10 mM, completely protected from dRib-induced apoptosis (1).

In recent experiments, cultured PBMC from centenarian donors appeared more resistant to dRib-induced apoptosis in comparison with those from young donors (Monti D, et al, unpublished). Similarly, cultured PBMC from uremic patients undergoing hemodialysis have been shown to exhibit higher resistance to dRib-induced apoptosis (2). PBMC from uremic patients also exhibited higher content of intracellular reduced glutathione (GSH), in comparison with normal subjects. Several studies have reported that reactive oxygen species are produced in chronic uremic patients undergoing hemodialisis, and the extensive radical formation in vivo has been related to an increased risk of malignancies and atherosclerosis.

The relationship between oxidative stress and apoptosis has attracted increased interest. Oxidant agents have been indicated to induce apoptosis; in contrast, many inhibitors of apoptosis have antioxidant activities or enhance cellular antioxidant defenses.

We have studied the mechanism of action of dRib-induced apoptosis in cultured PBMC from young and healthy donors. In particular, we investigated on the possibility of the involvement of an alteration of the redox status of the cell and the role of GSH.

The mechanism of dRib-induced GSH depletion and apoptosis was investigated by following the effects of NAC, a precursor of GSH synthesis, and of L-buthionine-(*S*,*R*)-sulfoximine (BSO), which induces GSH depletion through inhibition of γ-glutamylcysteine synthetase.

The effects of desferal, a compound having excellent copper- and iron-chelating properties, on dRib-induced GSH depletion and apoptosis in PBMC were also examined in order to investigate on the involvement of "free" iron in the oxidative stress induced by dRib.

MATERIALS AND METHODS

Cell Cultures

Human PBMC were obtained from whole blood of young and healthy donors (mean age 25 ± 3 years) anticoagulated with EDTA. PBMC were obtained by sedimentation over Lymphoprep gradient, according to the method of Boyum (3) PBMC were resuspended at a density of 10^6 cells/ml in a RPMI 1640 culture medium containing 2 mM L-glutamine, penicillin (100 units/ml) streptomycin (100 μg/ml), and 10% heat-inactivated fetal calf serum. Cell suspensions were seeded into culture plates in the absence or presence (10 mM) of dRib and then incubated for 2, 24, 48, and 72 h at 37°C in a humidified atmosphere of 5% CO_2 in air. Apoptosis and GSH content were analyzed at the same times of incubation.

Flow Cytometry

Apoptosis was detected in PBMC by reduced fluorescence of propidium iodide (PI), a DNA-binding dye, in the apoptotic nuclei, according to the method of Nicoletti et al. (4). The cellular pellet (10^6 cells) was resuspended in a hypotonic fluorochrome solution (PI 50 μg/ml in 0.1% sodium citrate plus 0.1% Triton X-100 in double-distilled water). Cells were analyzed after a minimum of 20 min of incubation in this solution. Analysis was performed using a FACScan flow cytometer (Becton Dickinson) equipped with a single 488 nm argon laser. Orange fluorescence due to PI staining of DNA, detected in fluorescence 2 (FL2), was registered on a logarithmic scale. A minimum of 10,000 cells were acquired in list mode and analyzed by the LYSYS II software program. Debris was gated out on light scatter measurements before the single-parameter histograms were drawn.

HPLC Analysis

GSH was measured by HPLC as described by Reed et al. (5). PBMC (6×10^6 cells) were collected at 0, 2, 24, 48, 72 h of incubation in the absence or presence of dRib (10 mM). The cellular pellet was deproteinized by mixing with 1% picric acid containing 0.5 mM desferal. After centrifugation at 1500 rpm, the acidic supernatant was derivatized for HPLC measurements. The sample was mixed with 100 mM iodoacetic acid solution containing excess sodium bicarbonate. The reaction was allowed to proceed for 1 h in the dark; then 0.75% dinitrofluorobenzene in HPLC-grade absolute ethanol (Sanger's reagent) was added and allowed to react overnight in the dark at room temperature. Aliquots of

100 μl were analyzed using a 5 μm Spherisorb amino column (Phase Separations, 25 cm × 4.6 mm) in a Hewlett Packard 1090 Liquid Chromatograph equipped with a diode-array detector. The column was eluted isocratically with solvent A (80% methanol–20% water) for 10 min, followed by a 20 min gradient to 90% solvent B (80% ammonium acetate buffer–20% solvent A). GSH was calculated at 357 nm against an external standard.

RESULTS AND CONCLUSIONS

dRib (10 mM) induced apoptosis and GSH depletion in cultured PBMC. GSH depletion in cultured PBMC was evident at 2 h of incubation with 10 mM dRib and the effect was maximum at 24 h (Figure 1A), while the onset of apoptosis began at 48 h of exposure (Figure 1B).

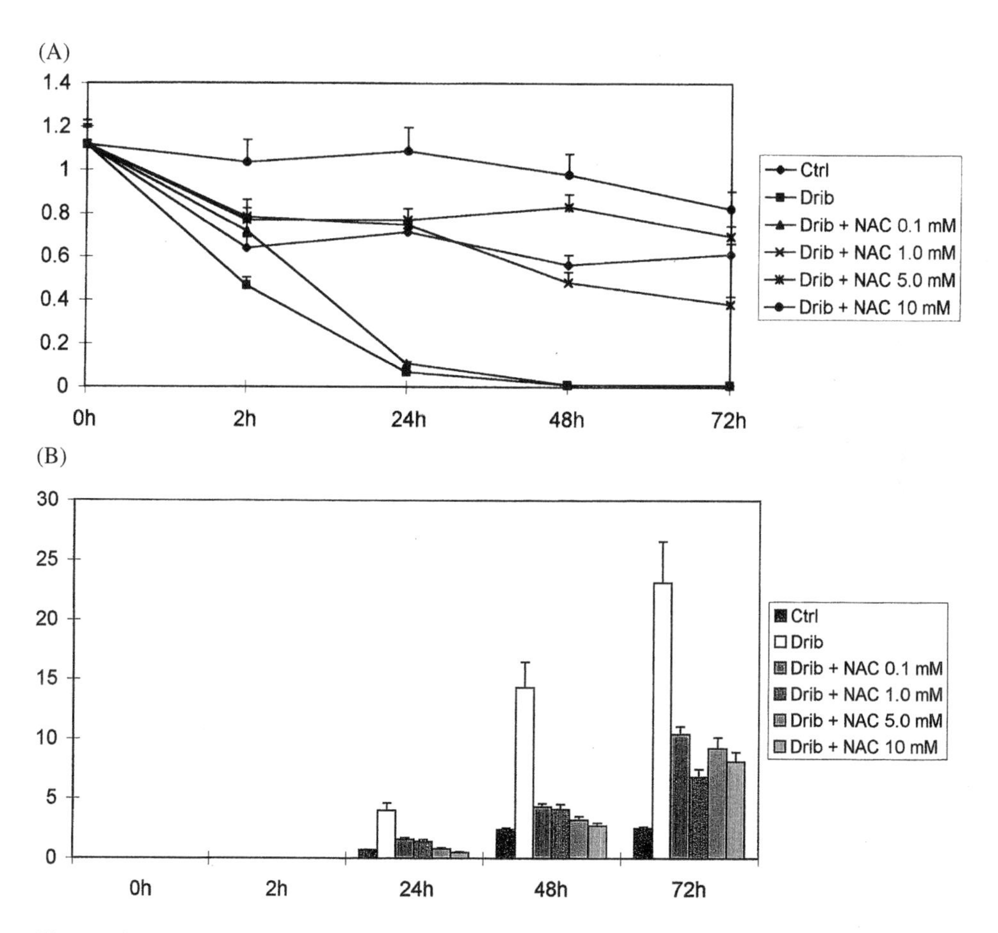

Figure 1. *N*-Acetylcysteine (NAC)-induced prevention of dRib-induced GSH depletion and apoptosis in PBMC from young donors. NAC (0.1–10 mM) prevented dRib-induced GSH depletion in a dose-dependent manner (A), and inhibited dRib-induced apoptosis even at the lower concentration used (0.1 mM) (B). GSH content is expressed as nmol/10^6 cells. Apoptosis is expressed as percentage of apoptotic nuclei.

NAC (0.1–10 mM) prevented dRib-induced apoptosis and GSH depletion in cultured PBMC. The protective effect of NAC on dRib-induced GSH depletion was dose dependent, resulting in an increase of GSH content over control levels at the concentration of 10 mM (Figure 1A,) while the protective effect of NAC on dRib-induced apoptosis was observable even at the lower concentration used (0.1 mM NAC) (Figure 1B).

BSO alone (500 μM) induced GSH depletion while it did not alter the kinetics of dRib-induced GSH depletion (Figure 2A). BSO potentiated dRib-induced apoptosis. The effect was observed at 48 h of exposure (Figure 2B).

Interestingly, even in the conditions where GSH resynthesis was inhibited by BSO, NAC (10 mM) could partially prevent BSO-induced (Figure 3A) and BSO+dRib-induced GSH depletion (Figure 3B), while it completely prevented BSO+dRib-induced apoptosis (Figure 3C).

Desferal, when added to the incubation medium at a final concentration of 0.5 mM, did not show any appreciable effect on dRib-induced GSH depletion and apoptosis (data not shown).

The results obtained are consistent with the hypothesis of the existence of a threshold value for GSH, below which the apoptotic process may be initiated. In addition, the NAC-dependent prevention of dRib-induced apoptosis in PBMC, which was evident also when GSH resynthesis was inhibited by BSO, might well be related to the ability of NAC to maintain the cellular -SH redox equilibrium, instead of GSH, and/or preventing the utilization of GSH in cultured PBMC.

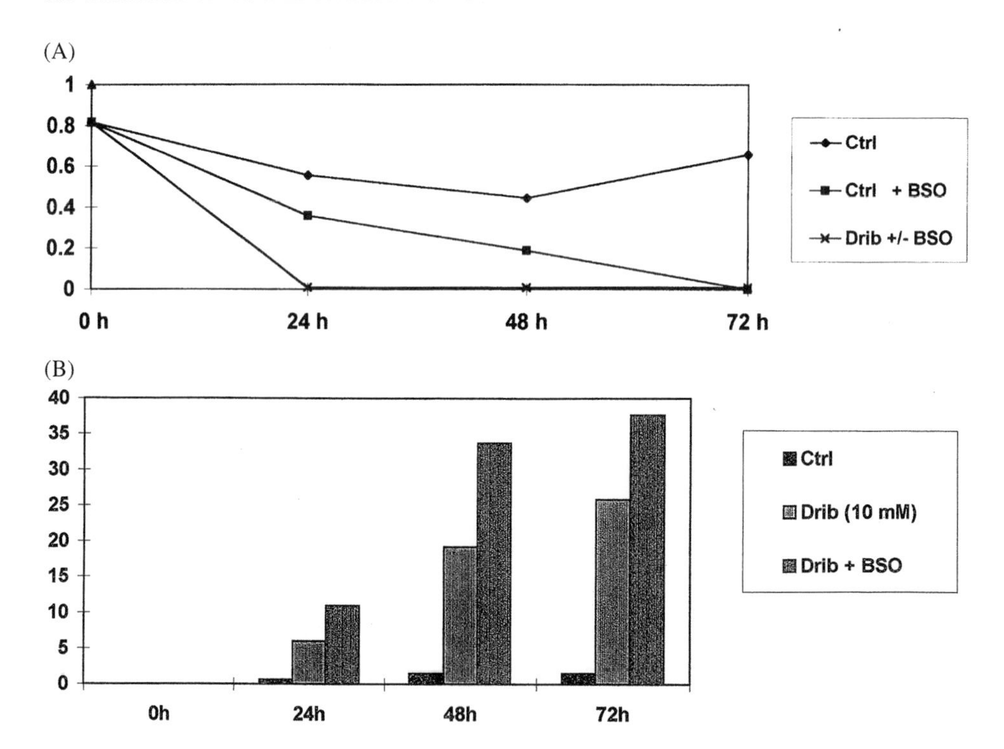

Figure 2. L-Buthionine-(*S*,*R*)-sulfoxine (BSO) alone (500 μM) induced GSH depletion in PBMC while it did not alter the kinetics of dRib-induced GSH depletion (A). BSO potentiated dRib induced apoptosis at 48 h of exposure (B). GSH content is expressed as nmol/10^6 cells. Apoptosis is expressed as percentage of apoptotic nuclei.

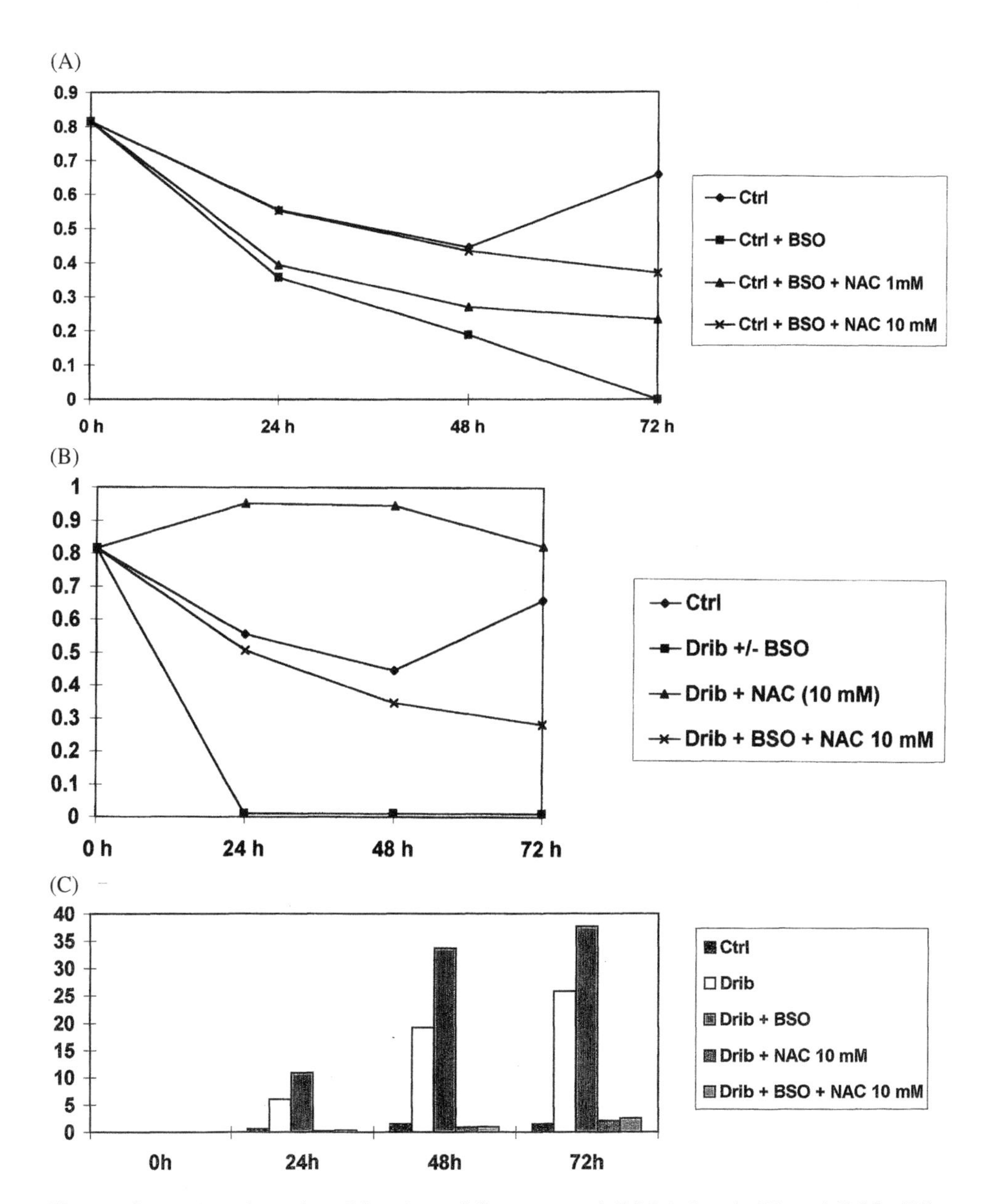

Figure 3. *N*-Acetylcysteine (10 mM) partially prevented BSO-induced (A) and BSO+dRib-induced GSH depletion (B), while it completely prevented BSO+dRib-induced apoptosis (C) in PBMC. GSH content is expressed as nmol/10^6 cells. Apoptosis is expressed as percentage of apoptotic nuclei.

REFERENCES

1. Barbieri D, Grassilli E, Monti D, Salvioli S, Franceschini MG, Franchini A, Bellesia E, Salomoni P, Negro P, Capri M, Troiano L, Cossarizza A, Franceschi C. D-Ribose and deoxy-D-ribose induce apoptosis in human quiescent peripheral blood mononuclear cells. Biochem Biophys Res Commun. 1994; 201: 1109–1116.

2. Tropea F, Tetta C, Wratten ML, Botti B, Baraldi A, Rapanà R, Monti D, Macchioni S, Kalasnikova G, Petruzzi E, Troiano L, Franceschi C. Peripheral blood mononuclear cells from uremic patients are resistant to induced apoptosis (manuscript submitted for publication).
3. Boyum A. Separation of lymphocytes, granulocytes and monocytes from human blood using iodinated density gradient. Methods Enzymol. 1984; 108: 88–102.
4. Nicoletti I, Migliorati G, Pagliacci MC, Grignani F, Riccardi C. A rapid and simple method for measuring thymocyte apoptosis by propidium iodide staining and flow cytometry. J Immunol Methods 1997; 139: 271–279.
5. Reed DJ, Babson RJ, Beatty PW, Brodie AE, Ellis WW, Potter DW. High-performance liquid chromatography analysis of nanomole levels of glutathione, glutathione disulfide, and related thiols and disulfides. Anal Biochem. 1980; 106: 55–62.

20

Nitric Oxide and Hydrogen Peroxide Production during Apoptosis of Human Neutrophils

Juanita Bustamante, Cecilia Carreras, Natalia Riobó, Alexis Tovar, Gonzalo Montero, Juan Jose Poderoso, and Alberto Boveris
University of Buenos Aires
Buenos Aires, Argentina

INTRODUCTION

Polymorphonuclear leukocytes (PMN) play determinant roles in immunity and inflammation. These processes are mediated by the activation of the NADPH oxidase, which generates superoxide anion (O_2^-), and by the activation of a cytoplasmic, constitutive and inducible nitric oxide synthase (NOS and iNOS), which produces nitric oxide ($NO^•$) (1–5). Superoxide anion, H_2O_2, and $NO^•$ are not only efficient antimicrobial effector molecules, they are key mediators in inflammatory reactions in which they participate, altering the expression of endothelial adhesion molecules and inducing the production of different cytokines (6–8). Together with these molecules and proteolytic enzymes, PMN cells can mediate intra- and extracellular cytotoxicity and modulate different cellular functions in adjacent cells. After degradation of the ingested material, polymorphonuclear cells die. These terminally differentiated cells are not able to proliferate and in vitro they demise rapidly by an apoptotic mechanism; PMN senecent cells are then recognized by macrophages by a cell-surface mechanism mediated by vitronectin receptor (9).

The molecular mechanisms by which spontaneous apoptosis of human PMN occurs are not known. The present study deals with the role of O_2^-, H_2O_2, and $NO^•$ as inducers of spontaneous apoptosis in human neutrophils and with the inhibition of this physiological process by the antioxidant compound lipoic acid.

RESULTS

Unstimulated PMN produce detectable amounts of superoxide, hydrogen peroxide, and nitric oxide (10), presumably owing to activation during their procurement and isolation. Table 1 shows that unstimulated neutrophils consume 0.56±0.03 nmol O_2 and produce

Table 1. Production of Superoxide Anion, Hydrogen Peroxide, and Nitric Oxide by Human Neutrophils

	Oxygen uptake	Superoxide anion	Hydrogen peroxide	Nitric oxide
Unstimulated PMN	0.56 ± 0.03	0.35 ± 0.03	0.20 ± 0.05	0.08 ± 0.01
FMLP-treated PMN	3.0 ± 0.18	3.8 ± 0.25	1.6 ± 0.18	0.21 ± 0.01
PMA-treated PMN	4.0 ± 0.27	5.0 ± 0.5	2.0 ± 0.2	0.58 ± 0.04

Stoichiometric ratios of the rate of oxygen uptake, and superoxide anion, hydrogen peroxide, and nitric oxide production by unstimulated human polymorphonuclear cells and treated with PMA (0.1 μ/ml) and FMLP (1 μM).

Values are expressed in nmol/min per 10^6 cells and are mean ± S.E.M. of five experiments.

0.35±0.03 nmol O_2^-, 0.20±0.05 nmol H_2O_2, and 0.08±0.008 nmol $NO^{\bullet}$ per min per 10^6 cells. These values increases severafold after cell stimulation with FMLP (1 μM) and phorbol myristate acetate (1 μg/ml) (Table 1).

Unstimulated PMN, as well as rat thymocytes, incubated in RPMI 1640 plus 5% FCS at 37°C, clearly showed spontaneous apoptosis as observed in Figure 1. Kinetic studies showed low internucleosomal DNA fragmentation at 6 h, and a visible polynucleosome ladder at 12 h of incubation, the DNA fragmentation being stronger after 24 h of incubation. DNA laddering was determined by gel electrophoresis as described previously (11). Cell viability of human PMN cells assayed by trypan blue exclusion after 72 h of incubation was 15%, 39%, 21%, 32% for control, catalase-, Superoxide dismutase

Figure 1. Kinetics of spontaneous apoptosis in human neutrophils. Human circulating PMN were obtained from healthy donors by Ficoll–Hypaque gradient centrifugation. Cells were incubated in RPMI 1640 supplemented with 5% FCS, streptomycin (100 μg/ml), penicillin (100 U/ml), and 0.2% bicarbonate. At 6, 12, and 24 h the cells were collected and gel electrophoresis was carried out in 1.8% agarose.

(SOD)-, and *N*-monomethyl-L-arginine (L-NMMA) treated cells, respectively (Figure 2). After neutrophil activation, NADPH oxidase generates O_2^- via the univalent reduction of O_2. The O_2^- then undergoes rapid spontaneous or enzymatic dismutation to produce H_2O_2 (2). As the neutrophil membrane starts to engulf the particle within a phagocytic vacuole, O_2^- and H_2O_2 are released directly into the extracellular medium. Therefore, the H_2O_2 generated should be in a concentration equilibrium between vacuolar, cytosolic, and extracellular pools owing to the high diffusion coefficient of H_2O_2. Although in unstimulated PMN cells the vacuolar pool may not exist, extracellular and cytosolic pools are in equilibrium. The H_2O_2 in the cytoplasm is primarily utilized by either catalase or glutathione peroxidase. The intracellular H_2O_2 steady-state concentration in unstimulated PMN can be estimated by the assumption that the rate of H_2O_2 production equals the rate of H_2O_2 consumption by catalase as indicated in equations (1) and (2)

$$\frac{d[H_2O_2]}{dt} = \frac{-d[H_2O_2]}{dt} = k \cdot [H_2O_2] \cdot [\text{catalase}]$$

$$[H_2O_2] = \frac{d[H_2O_2]/dt}{k \cdot [\text{catalase}]} = 2\,\mu\text{M}$$

where $k = 3.4 \times 10^7\ \text{M}^{-1}\text{s}^{-1}$ (12); catalase = 0.7×10^{-13} mol $\times$ 10^6 cells was determined in human PMN unstimulated cells by the decrease in absorption at 240 nm in 50 mM phosphate buffer and 1.5 mM H_2O_2, using the pseudo-first order reaction constant ($k' = k \cdot [\text{catalase}]$ for

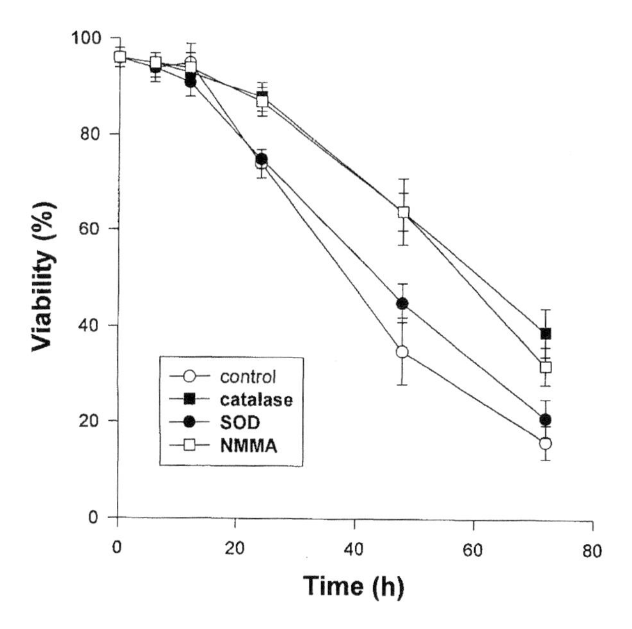

Figure 2. Effect of SOD catalase and L-NMMA on human PMN cell viability. Trypan blue exclusion of PMN cells was performed after isolation (time 0), or after incubation [help]far[/help] 6, 12, 24, 48, and 72 h in RPMI 1640 with 5% FCS as described previously. Percentage viability untreated PMN cells (○), and PMN treated with SOD (50 μg/ml) ((●), catalase (50 μg/ml) (■), and L-NMMA (1 mM) (□) are shown. Results are expressed as mean ± S.E.M. of three different experiments.

the decrease in H_2O_2 absortion. Total catalase content is expressed in mol/10^6 cells. The hydrogen peroxide production $d[H_2O_2]/dt$ measured in unstimulated cells was 0.20 nmol/minper/10^6 cells (Table 1). With these data we calculate a steady-state intracellular hydrogen peroxide concentration $[H_2O_2]$ for unstimulated human PMN cells of 2 μM, increasing to 12 μM after stimulation with PMA (0.1 μg/ml). As well as O_2^- and H_2O_2,

Figure 3. SOD stimulates spontaneous apoptosis in human neutrophils. Human neutrophils were incubated in the presence of SOD (50 μg/ml) in RPMI 1640 with 5% FCS as described previously. After 6 h the cells were collected and gel electrophoresis was performed in 1.8% agarose as described previously. The gels were representative of three individual experiments.

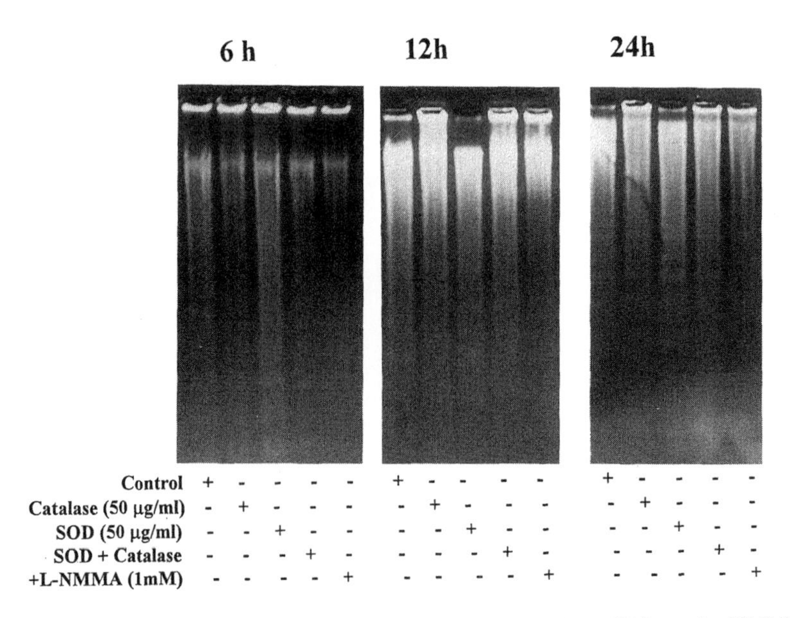

Figure 4. Modulation of spontaneous apoptosis by catalase, SOD, and L-NMMA in PMN cells. Cells were exposed to catalase (50 μg/ml), SOD (50 μg/ml), and L-NMMA (1 mM) and incubated for 6, 12, and 24 h. Agarose gel DNA electrophoresis was carried out. This figure is representative of three different experiments.

Figure 5. Lipoic acid inhibits spontaneous apoptosis in human neutrophils. Human neutrophils were subjected to gel electrophoresis after 12 h incubation in the presence of α-lipoic acid (4 mM) and lipoamide (2 mM).

unstimulated PMN cells produce $NO^{\bullet}$; this molecule can react with superoxide to generate ONOO–, NO_2, and $HO^{\bullet}$, which are highly detrimental to PMN cells. It has been reported that $NO^{\bullet}$ can react with O_2^- under certain conditions, neutralizing its cytotoxicity and its metabolic effects (13, 14).

In our studies SOD (50 μ/ml) was able to positively modulate spontaneous apoptosis in human neutrophils, as observed in Figure 2. This increase mediated by SOD was slightly diminished by the addition of catalase (50 μ/ml). When PMN cells were incubated in the presence of catalase (50 μg/ml) spontaneous apoptosis was inhibited, since no laddering was observed after 6, 12, or 24 h (Figure 3). iNOS can be inhibited by the addition of L-NMMA. As observed in Figure 4, no laddering was formed after treatment with L-NMMA (1 mM), indicating that spontaneous apoptosis of human PMN cells was inhibited. The antioxidant treatment of PMN cells with the α-lipoic acid (4 mM) and its metabolite lipoamide (2 mM) showed a complete inhibition of the apoptotic process after 12 h as shown in Figure 5.

DISCUSSION

Activation of the oxidase system and nitric oxide synthase (NOS) in human PMN cells must be independently and rigorously controlled (15). Our results show that unstimulated human neutrophils are able to produce reactive oxygen and nitrogen species such as superoxide (O_2^-), hydrogen peroxide (H_2O_2), and nitric oxide ($NO^{\bullet}$) at low but measurable rates (Table 1). The production of these molecules could be closely related to the in vitro spontaneous apoptosis founded in unstimulated human PMN cells.

This spontaneous process analyzed by internucleosomal DNA fragmentation in agarose gel electrophoresis was evident at 12 h of cell incubation, reaching the maximum at 24 h. In turn, these cells showed 75% viability at 24 h, as has been observed by other laboratories (16). Although the intensity of the internucleosomal DNA fragmentation exhibits some variation between the different samples, it was never observed before 6 h of cell incubation.

The signals that induce human PMN cell apoptosis are not known, although our results indicate that a slight activation of the respiratory burst and the NOS occurs during their procurement and isolation, and the reactive oxygen and nitrogen species produced could be the signals for the induction of apoptosis in these cells. Activation of human PMN cells with PMA and FMLP increases superoxide, hydrogen peroxide, and nitric oxide production severalfold (Table 1). These results were not correlated with a major increase in internucleosomal DNA fragmentation (results not shown).

Treatment of these cells with SOD did not show the expected SOD protection; this enzyme clearly stimulated spontaneous apoptosis in human neutrophils, this being more evident at earlier stages of the process. Moreover, catalase together with SOD decreased the intensity of SOD-mediated apoptosis, and was able to inhibit spontaneous apoptosis in PMN cells; no DNA laddering was observed after catalase cell treatment.

Important inhibition of internucleosomal DNA fragmentation in human neutrophils was observed after L-NMMA treatment, indicating that $NO^{\bullet}$, rather than NO_2^- and H_2O_2, play a determinant role in human PMN apoptosis. Superoxide anion and hydrogen peroxide production should not be affected in presence of L-NMMA; however, no laddering after 24 h of PMN incubation was observed in these conditions. Our results are reinforced by the fact that SOD can extend the effects of $NO^{\bullet}$ by decreasing O_2^- concentration and the reaction of $NO^{\bullet}$ with O_2^- (k=6.7$\times 10^9$ $M^{-1}s^{-1}$) (13) and by a second mechanism (reversible conversion of $NO^{\bullet}$ to NO^-) (14). Furthermore, catalase also can interfere with the availiability of $NO^{\bullet}$, reacting with $NO^{\bullet}$ and forming a CAT–NO complex through to its native [(FeIII)] form (14). $NO^{\bullet}$ production has been linked with apoptosis by its effects on peritoneal macrophage cell death (17, 18) and also by the ability of different $NO^{\bullet}$ donors to induce apoptosis in RAW 264.7 macrophages (19).

These results agree with our previous results on the inhibition of thymocyte apoptosis induced by lipoamide (11, 20) as well as other antioxidants (21). However, in rat thymocytes α-lipoic acid was not able to inhibit the apoptotic process, indicating important differences in the metabolism of these compounds in human PMN cells and in rat thymocytes.

The results obtained in this study with SOD, catalase, and L-NMMA on human PMN point to nitric oxide as the molecule that could modulate the biochemical pathways of human PMN apoptosis.

REFERENCES

1. Segal AW. Electron transport chain of the microbicidal oxidase of phagocytic cells and its involvement in the molecular pathology of chronic granulomatous dssease. J Clin Invest 1989; 83: 1785–1793.
2. Fridovich Y. Biological effects of superoxide radical. Arch Biochem Biophys 1986; 247: 1–11.
3. Hibbs JB Jr, Vavrin Z, Taintor RR. L-Arginine is required for expression of the activated macrophage effector mechanism causing selective metabolic inhibition in target cells. J Immunol 1987; 138: 550–565.

4. Hibbs JB Jr, Vavrin Z, Taintor RR. Macrophage cytotoxicity: role for arginine deiminase activity and imino nitrogen oxidation to nitrite. Science 1987; 235: 473–476.
5. Marletta MA, Yoon PS, Iyengar R, Leaf CD, Wishnor JS. Macrophage oxidation of L-arginine to nitrite and nitrate: nitric oxide is an intermediate. Biochemistry 1988; 27: 8706–8711.
6. Winrow VR, Winyard PG, Morris CJ, Blade DR. Free radicals in inflammation: second messengers and mediators of tissue destruction. Br Med Bull 1993; 49: 506.
7. Janssen YMW, Van Houten B, Borm PJA, Mossman BT. Cell and tissue responses to oxidative stress. Lab Invest 1993; 69: 261.
8. Weiss SJ. Tissue destruction by neutrophils. N Engl J Med 1989; 320: 365.
9. Savill J, Dransfield I, Hogg N, Haslett Ch. Vitronectine receptor-mediated phagocytosis of cells undergoing apoptosis. Nature 1990; 343: 170–173.
10. Carreras MC, Pargament G, Catz S, Poderoso J, Boveris A. Kinetics of nitric oxide and hydrogen peroxide production and formation of peroxinitrite during the respiratory burst of human neutrophils. 1994; 341: 65–68.
11. Bustamante J, Slater A, Orrenius S. Antioxidant inhibition of thymocytes apoptosis by dihydrolipoic acid. Free Radicals Biol Med. 1995; 19: 339–347.
12. Aebi H. Catalase in vitro. Methods Enzymol 1984; 105: 121–126.
13. Beckman JS, Beckman TW, Chen J, Marshall P, Freeman B. Apparent hydroxyl radical production by peroxinitrite: implication for endothelial injury from nitric oxide and superoxide. Proc Natl Acad Sci USA 1990; 87: 1620–1624.
14. Murphy M, Sies H. Reversible convertion of nitroxyl anion to nitric oxide by superoxide dismutase. Proc Natl Acad Sci USA 1991; 88: 10860–10864.
15. Bastian N, Hibbs JB Jr. Assembly and regulation of NADPH oxidase and nitric oxide synthase. Curr Opin Immunol 1994; 6: 131–139.
16. Colotta F, Re F, Polentarutti N, Sozanni S, Mantovani A. Modulation of granulocyte survival and programmend cell death by cytokines and bacterial products. Blood 1992; 80: 2012–2020.
17. Albina JE, Cui S, Mateo RB, Reichner JS. Nitric oxide-mediated apoptosis in murine peritoneal macrophages. J Immunol 1993; 150: 5080–5085.
18. Sarih M, Souvannavong V, Adam A. Nitric oxide induces macrophage death by apoptosis. Biochem Biophys Res Commun 1993; 191: 503–508.
19. Mebmer K, Lapetina EG, Brune B. Nitric oxide-induced apoptosis in Raw 264.7 macrophages is antagonized by protein kinase C- and protein kinase A-activating compounds. Mol Pharmacol 1995; 47: 757–765.
20. Bustamante J, Slater A, Orrenius S. Lipoic acid and the redox regulation of apoptosis In Fuch J, Zimmer G, Packer L, eds. Lipoic Acid in Health and Disease; in press.
21. Slater A, Nobel S, Maellaro E, Bustamante J, Kimland M, Orrenius S. Nitrone spin traps and a nitroxide antioxidant inhibit a common pathway of thymocyte apoptosis. Biochem J 1995; 306: 771–778.

21

The Mitochondrion as a Sensor/Effector of Oxidative Stress During Apoptosis

Philippe Marchetti, Santos A. Susin, Naoufal Zamzami, and Guido Kroemer
CNRS-UPR420
Villejuif, France

INTRODUCTION

The process of apoptosis can be subdivided into at least three different phases. During the *initiation* phase cells either suffer subnecrotic damage or receive "death signals" via specific receptors (or are deprived of "survival signals") (1–5). Nonspecific or receptor-mediated death induction involves a stimulus-dependent ("private") biochemical pathway, and it is only after this initiation phase that common pathways come into action. It is generally assumed that the *execution* phase of apoptosis defines the "decision to die" at the "point of no return" of the apoptotic cascade. It is at this level that the different private pathways converge into one (or few) common pathway(s) and that cellular processes (redox potentials, expression levels of oncogene products, etc.) still have a decisive regulatory function. Once the cell has been irreversibly committed to death, the different manifestations classically associated with apoptosis such as DNA fragmentation become manifest. This *degradation* phase is the same in all cell types and involves the action of catabolic enzymes including nucleases and proteases.

It has become increasingly clear that increased generation of reactive oxygen species (ROS) and/or ROS effects participate in the degradation phase of apoptosis (6–14). Moreover, it appears that ROS may be involved in some private pathways of apoptosis induction. Here we attempt to integrate these observations into a novel concept of apoptosis control. This chapter is designed to convey our current working hypothesis that mitochondria control the execution phase of apoptosis and that mitochondria can function both as sensors of redox potentials and as as producers of reactive oxygen species.

MITOCHONDRIAL IMPLICATION IN THE EXECUTION PHASE OF APOPTOSIS

Although a lot of information is available on the biochemistry of DNA and protein degradation, little is known about the precise nature of the execution phase of apoptosis. We and others have observed that cells that are on the way to undergoing apoptosis but still lack ultrastructural signs of apoptosis or DNA fragmentation do manifest a disruption of the mitochondrial inner transmembrane potential ($\Delta\Psi_m$). This $\Delta\Psi_m$ collapse appears to be a constant feature of early apoptosis in the sense that it is found in very different cell types (neurons, fibroblasts, T and B lymphocytes, thymocytes, pre-B cells, cells from the myelomonocytic lineage), in response to very different receptor-mediated induction protocols of apoptosis (cross-linking of Fas/Apo-1/CD95; ligation of the tumor necrosis factor receptor; activation of T and B cells via their respective antigen receptors; ligation of glucocorticoid receptors; etc.), as well as in nonspecific protocols of apoptosis induction (γ-irradiation, drugs used in chemotherapy such as etoposide, exogenous sources of ROS such as *t*-butylhydroperoxide, H_2O_2, etc.) (5,14–21). The $\Delta\Psi_m$ dissipation thus appears a constant and early feature of the apoptotic process, indicating that it reflects a common event of the apoptotic execution phase.

That mitochondria are indeed involved in the regulation of apoptosis is suggested by a number of additional observations. (i) Inhibition of the respiratory chain induces apoptosis (22,23). (ii) Isolated mitochondria can provoke nuclear apoptosis in cell-free systems of apoptosis (24,25). (iii) Apoptosis-inhibitory oncoproteins of the Bcl-2 family must be expressed at the level of the mitochondrion to inhibit apoptosis (26–28). (iv) The Bcl-2 antagonist Bax only kills cells possessing functional, respiring mitochondria (28). On the basis of these observations, we set out to characterize the molecular mechanisms of the preapoptotic $\Delta\Psi_m$ disruption.

PERMEABILITY TRANSITION AS A CENTRAL COORDINATING EVENT IN APOPTOSIS

To understand the mechanism by which cells undergoing apoptosis lose their $\Delta\Psi_m$, we performed a series of experiments in which cells were first labeled with $\Delta\Psi_m$-sensitive fluorochromes and then purified in a fluorocytometer, on the basis of their $\Delta\Psi_m$. In appropriate conditions, this procedure allows for the purification of cells with low $\Delta\Psi_m$ values and a still normal DNA content and morphology (preapoptotic cells) or, alternatively, of cells with a still high $\Delta\Psi_m$ that will lose their $\Delta\Psi_m$ upon short-term (30–120 min) culture (14,17,29). We have used this system to show that $\Delta\Psi_m^{low}$ (but not $\Delta\Psi_m^{high}$) cells will undergo oligonucleosomal DNA fragmentation upon short-term culture at 37°C. Moreover, we have found that some drugs inhibit the $\Delta\Psi_m$ loss of $\Delta\Psi_m^{high}$ cells, namely cyclosporin A (CsA) and bongkrekic acid (BA). These data suggest that a phenomenon that has been studied extensively, the so-called permeability transition (PT), which is inhibited by CsA and BA (30,31), might account for the $\Delta\Psi_m$ collapse observed during preapoptosis. PT involves the formation of proteaceous pores ("PT pores" or "megachannels"), probably by apposition of inner and outer mitochondrial membrane proteins, allowing for the diffusion of solutes <1500 kDa (and *ipso facto* dissipation of the $\Delta\Psi_m$). CsA is one of the best-studied inhibitors of PT. Its PT inhibitory effect is mediated via a conformational change in a mitochondrial CsA receptor, the matrix cyclophilin (32). In contrast, its immunosuppressive effect is mediated via an effect on calcineurin-dependent signaling. A CsA derivative that loses its immunosuppressive properties, *N*-methyl-4-CsA, still conserves its $\Delta\Psi_m$-stabilizing effect in apoptotic cells (29). This observation is again compatible with the implication of PT in apoptotic $\Delta\Psi_m$ disruption.

To demonstrate that PT might indeed be important for the apoptotic process, we have used three different approaches. First, we have used BA to evaluate the effect of PT inhibition in cells in long-term experiments (>120 min). (In this system CsA is not useful because it is a transient inhibitor of PT.) BA not only prevents the mitochondrial manifestations of apoptosis but also abolishes all later changes concerning the nucleus, the cytoplasm, and the plasma membrane (33). Thus, PT appears to constitute a central coordinating event of the apoptotic process. Second, we have constructed a cell-free system of apoptosis in which isolated mitochondria are cultured with purified nuclei in vitro. Using such an approach, we have shown that control mitochondria are not apoptogenic, whereas mitochondria induced to undergo PT acquire the capacity to induce nuclear apoptosis (chromatin condensation + DNA fragmentation) (34). Third, we have purified mitochondria from hepatocytes or lymphocytes undergoing apoptosis, showing that such mitochondria (but not those isolated from nonapoptotic controls) induce DNA fragmentation in healthy nuclei in vitro. This effect is partially antagonized by the PT inhibitor BA (34). Together, these data provide clear evidence that PT is involved in the apoptotic cascade.

THE MITOCHONDRION AS A SENSOR OF OXIDATIVE STRESS

In a number of different experimental systems of apoptosis induction, inhibition of ROS generation and/or action prevents apoptosis. This can be achieved by at least five different approaches: (i) culture in the absence of oxygen (12); (ii) utilization of cells lacking a functional respiratory chain owing to deletion of the mitochondrial DNA (35,36); (iii) addition of antioxidants (8,9,11,37); (iv) exogenous ROS-detoxifying enzymes such as catalase (13); or (v) transfection- or transgene-enforced overepression of ROS-detoxifying enzymes (38,39). However, the implication of reactive oxygen species is not a universal feature of apoptosis since hypoxia (which precludes the generation of ROS) suffices to trigger apoptosis (40, 41). Moreover, certain apoptosis inducers still function in the absence of oxygen: staurosporine, absence of growth factor (42), or cross-linking of Fas/Apo-1/CD95 (43). This suggests that ROS generation is critical in some private pathways of apoptosis, yet does not represent a universal requirement of apoptotis induction or execution.

ROS are among the most effective inducers of mitochondrial PT. According to current knowledge, the PT pore gating potential (i.e., the probability of PT pore opening at a given voltage) is regulated by cellular redox potentials. Thus, the PT pore functions as a thiol sensor. Oxidation (disulfide formation) of a critical mitochondrial dithiol increases the probability of pore opening (reduces the gating potential). The redox status of this dithiol is in equilibrium with that of glutathione (44). In addition, the PT pore functions as a sensor of the oxidation/reduction equilibrium of the pyridine nucleotide pool ($NADH_2$/NAD + $NADPH_2$/NADP). Oxidation of pyridine nucleotides reduces the gating potential (44). However, it would be an oversimplification to assume that ROS and/or cellular redox potentials would be the sole effectors regulating the PT pore. Indeed, a number of additional effects (concentrations of divalent cations, ATP/ADP concentrations, mitochondrial pH, peptides, and proteases) can regulate the PT pore gating potential (30,31) (Table 1). Pro-oxidants (H_2O_2, *t*-butylhydroperoxide, menadione) and the disulfide-cross-linking agent diamide (diazenedicarboxylic acid bis(5*N*,*N*-dimethylamide) cause PT both in isolated mitochondria and in cells. In cells, the oxidant or diamide-mediated disruption of the $\Delta\Psi_m$ results in apoptosis. In isolated mitochondria, oxidants and diamide cause, in addition to PT, the release of an apoptogenic protein capable of inducing nuclear apoptosis in a cell-free in

Table 1. Modulators of Mitochondrial Permeability Transition (PT) and Apoptosis[a]

Inducers of PT and apoptosis	Inhibitors of PT and apoptosis
Calcium	Calcium chelators
Oxidizing agents:	divalent cations:
Hydrogen peroxide	Zn^{2+}
t-butyl hydroperoxide	Butylhydroxytoluene
NO donors	*N*-acetylcysteine
Menadione	Thiol-derivatizing agents
$\Delta\Psi_m$-lowering agents:	Monochlorobimane
Protonophors	Thiol-reducing agents
Respiratory inhibitors	Dithiothreitol
Disulfide bridge formers:	ANT ligands:
Diamide	Bongkrekic acid
Thiol-derivatizing agents:	Polyamines:
Hg, Cd, Cu	spermine, carnitine
PBR ligands:	Calpain inhibitors
protoporphyrin IX	

[a] For references see main text.

vitro system (34). As is to be expected (45), monochlorobiman, which protects thiols against diamide-mediated cross-linking, prevents the diamide-induced PT in isolated mitochondria as well as in cells, thereby precluding apoptosis (Marchetti et al., unpublished; Table 1). These data underline the possibility that mitochondria can serve as sensors of ROS generation and/or thiol redox states and trigger apoptosis via PT, once oxidative stress or thiol oxidation has passed a critical threshold level.

The oncoprotein Bcl-2, which is present in the outer mitochondrial membrane, has previously been reported to inhibit ROS-induced apoptosis in cells (7). To study the functional impact of Bcl-2 on mitochondria, we compared the behavior of purified mitochondria from T cell hybridoma cells stably transfected with the human Bcl-2 gene or with a neomycin control vector. We observed that isolated mitochondria hyperexpressing

Table 2. Spectrum of Inhibitors of Mitochondrial Permeability Transition (PT)[a]

	PT Inducer				
PT inhbitor	Atractyloside	*t*-BHP	*m*-ClCCP	Ca^{2+}	Diamide
Bongkrekate	+[b]	+	+	−	−
Phosphotyrosine	+	+	+	−	−
Bcl-2 transfection	+	+	+	−	−
Zn^{2+}	+	−	−	−	+
Monochlorobimane	+	−	−	−	+
Cyclosporin A	+	−	−	−	−
Ruthenium red	−	−	−	+	−

[a] Tested on purified mouse liver mitochondria during an interval of 60 min. Data are compiled from Ref. 34.
[b] Positive signs indicate inhibition ≥50% of permeability transition as assessed by large-amplitude swelling.

Bcl-2 on their suface fail to undergo PT in response to a number of different PT inducers including the pro-oxidant *t*-butylhydroperoxide in conditions in which control mitochondria do readily undergo PT (34). Bcl-2, however, is not a universal inhibitor of PT. Thus, Bcl-2 does not confer protection against diamide, either on the level of the mitochondrion or on the level of the cell. In other words, Bcl-2-overexpressing mitochondria undergo PT and Bcl-2-transfected cells undergo apoptosis in response to diamide exactly as is the case for control organelles or cells, respectively (34). Thus, Bcl-2 has a PT inhibitory effect which explains its "antioxidant" apoptosis inhibitory effect (Table 2).

THE MITOCHONDRION AS AN EFFECTOR OF OXIDATIVE STRESS

As pointed out above, early apoptosis is associated with a collapse of the $\Delta\Psi_m$. In addition, apoptosis has been reported to be linked to an increased production of ROS (7). To investigate the relationship between enhanced ROS generation and $\Delta\Psi_m$ disruption, we developed a double staining technique using 3,3′-dihexyloxacarbocyanine iodide ($DiOC_6(3)$; fluorescence in green) for the detection of the $\Delta\Psi_m$ and hydroethidine (HE) for the detection of ROS (14). HE, which is membrane-permeant and not fluorescent, is oxidized by ROS to ethidium (Eth), which emits a red fluorescence and is trapped in the cell due to its hydrophilicity. Thus, the rate of HE→Eth conversion is a measure of ongoing ROS generation. Kinetic studies using the $DiOC_6(3)$/HE staining technique revealed that cells first disrupt their $\Delta\Psi_m$ and then hypergenerate ROS (14,20). Thus, within an asynchronous population undergoing apoptosis, only three types of cells can be detected: (i) $DiOC_6(3)^{high}(HE{\rightarrow}Eth)^{low}$ cells (normal phenotype); (ii) $DiOC_6(3)^{low}(HE{\rightarrow}Eth)^{low}$ cells (preapoptotic phenotype); and (iii) $DiOC_6(3)^{low}(HE{\rightarrow}Eth)^{high}$ cells (apoptotic phenotype). In contrast, no $DiOC_6(3)^{high}(HE{\rightarrow}Eth)^{high}$ cells are detectable. We have purified cells with the $DiOC_6(3)^{low}(HE{\rightarrow}Eth)^{low}$ (preapoptotic) phenotype and shown that they acquire a $DiOC_6(3)^{low}(HE{\rightarrow}Eth)^{high}$ phenotype after short-term in vitro culture (14,20). The transition between these two stages of the apoptotic process is inhibited by rotenone, an inhibitor of mitochondrial electron transport in complex I, as well as by the hexavalent cation ruthenium red, a specific inhibitor of the mitochondrial calcium uniport (14). In contrast, antimycin A, an agent that leads to increased superoxide anion generation at sites proximal to the complex III of electron transport, accelerates the $\Delta\Psi_m^{low}HE^-/\ \Delta\Psi_m^{low}HE^+$ transition. Control substances that affect nonmitochondrial pathways of ROS generation fail to affect the transition from the $DiOC_6(3)^{low}HE^-$ to the $DiOC_6(3)^{low}HE^+$ state. This applies to inhibitors of lipoxygenase, cyclooxygenase, monoamine oxidase, xanthine oxidase, nitric oxide synthase, pyrrolidinedithiocarbamate (which stops the Fenton reaction converting H_2O_2 into superoxide anions), and diphenylene iodonium (an inhibitor of NADPH oxidase and other flavin oxidases) (14). Altogether, these data suggest that ROS are overproduced by an uncoupled respiratory chain rather than by extramitochondrial sources. In this context, it appears plausible that inhibitors of PT such as Bcl-2 prevent the PT-dependent hypergeneration and/or action of ROS (6,7,14).

Although enhanced ROS production by mitochondria appears to be a constant feature of apoptosis, it is a sign rather than a mechanism of apoptosis. A number of observations argue in this sense. (i) As discussed above, apoptosis can occur in the absence of ROS (40–43). (ii) ρ° cells (which lack mitochondrial DNA and have a partially defective respiratory chain) can undergo apoptosis (35,46,47). (iii) Antioxidants do not neutralize the activity of the mitochondrial apoptogenic factor acting on isolated nuclei

(34). (iv) Some plasma membrane features of apoptosis, such as increased exposure of phosphatidylserine residues, become manifest after the $\Delta\Psi_m$ drop but before cells hyperproduce ROS (48). Thus, ROS hyperproduction is not necessary for cell death to occur and is not indispensable for the manifestation of common nuclear and plasma membrane features of apoptosis. It is, hower, a constant by-product of PT, which, in turn, appears to be an obligatory event of apoptosis.

CONCLUSIONS AND PERSPECTIVES

From the findings discussed in this paper it emerges that mitochondrial PT is likely to constitute a decisive coordinating event of the mitochondrial execution phase. If this premise is accepted, then ROS may play a dual role in the apoptotic cascade: (i) as a facultative inducer of PT; and (ii) as an obligate consequence of PT that does not, however, affect other sequelae of PT such as liberation of mitochondrial apoptogenic proteins. This concept would be compatible with the finding that ROS hypergeneration and drastic changes in redox homeostasis appear to be universal features of apoptosis yet participate in the regulation of apoptosis only in special cases. In those cases in which ROS participate in (private) signal transduction pathways of the apoptosis initiation phase, antioxidants can prevent the ROS-driven mitochondrial PT. However, if PT is induced by different pathways not involving ROS, interventions on cellular redox potentials will remain without any impact on apoptosis. In addition, the concept of a sensor/effector function of mitochondria with regard to oxidative stress would also explain hitherto apparently contradictory data on Bcl-2. Bcl-2 augments the resistance of mitochondria to the ROS-induced induction of PT. In addition, while preventing PT, it also abolishes mitochondrial ROS hyperproduction secondary to PT. In other words, Bcl-2 will affect both the sensor and–indirectly–the effector function of mitochondria with respect to oxidative stress.

In spite of the heuristic value of these speculations, a number of issues remain elusive. In particular, very little is known on the exact molecular composition and genetics of the mitochondrial PT pore, its putative association with oncoproteins from the Bcl-2 family, and its physiological function. Similarly, little is known about the relationship between apoptosis regulatory proteases and mitochondria. More extensive studies will be required to confirm or dismiss our current working hypothesis on apoptosis regulation via mitochondria.

ACKNOWLEDGMENTS

This work has been partially supported by ARC, ANRS, CNRS, FRM, INSERM, NATO, Leo Foundation, Ministère de la Recherche et de I'Industrie (France), and Sidaction (to G.K.)

REFERENCES

1. Cohen JJ. Apoptosis. Immunol Today 1993; 14: 126–130.
2. Martin SJ, Green DR, Cotter TG. Dicing with death: dissecting the components of the apoptosis machinery. Trends Biochem Sci 1994; 19: 26–30.

3. Kroemer G. The pharmacology of T cell apoptosis. Adv Immunol 1995; 58: 211–296.
4. Kroemer G. Martínez-A C. (eds.). Apoptosis of immune cells. Curr Top Microbiol Immunol 1995; 200: 175.
5. Kroemer G, Petit PX, Zamzami N, Vayssière, J-L, Mignotte B. The biochemistry of apoptosis. FASEB J 1995; 9: 1277–1287.
6. Kane DJ, Sarafian TA, Anton R, Hahn H, Gralla EB, Valentine JS, et al. Bcl-2 inhibition of neural death: decreased generation of reactive oxygen species. Science 1993; 262: 1274–1277.
7. Hockenbery DM, Oltvai ZN, Yin X-M, Milliman CL, Korsmeyer SJ. Bcl-2 functions in an antioxidant pathway to prevent apoptosis. Cell 1993; 75: 241–251.
8. Sandstrom PA, Mannie MD, Buttke TM. Inhibition of activation-induced death in a T cell hybridoma by thiol antioxidants–oxidative stress as a mediator of apoptosis. J Leukocyte Biol 1994; 55: 221–226.
9. Slater AF, Noberl CS, Maellaro E, Bustamante J, Kimland MS O. Nitrone spin traps and a nitroxide antioxidant inhibit a common pathway of thymocyte apoptosis. Biochem J 1995; 306: 771–778.
10. Busciglio J, Yankner BA. Apoptosis and increased generation of reactive oxygen species in Down's syndrome neurons in vitro. Nature 1995; 378: 776–779.
11. Fernandez A, Kiefer J, Fosdick L, McConkey DJ. Oxygen radical production and thiol depletion are required for Ca^{2+}-mediated endogenous endonuclease activation in apoptotic thymocytes. J Immunol 1995; 155: 5133–5139.
12. Stefanelli C, Stanic I, Bonavita F, Muscari C, Pignatti C, Rossoni C, et al. Oxygen tension influences DNA fragmentation and cell death in glucocorticoid-treated thymocytes. Biochem Biophys Res Commun 1995; 212: 300–306.
13. Torres-Roca JF, Lecoeur H, Amatore C, Gougeon ML. The early intracellular production of a reactive oxygen intermediate mediates apoptosis in dexamethasone-treated thymocytes. Cell Death Differ 1995; 2: 309–319.
14. Zamzami N, Marchetti P, Castedo M, Decaudin D, Macho A, Hirsch T, et al. Sequential reduction of mitochondrial transmembrane potential and generation of reactive oxygen species in early programmed cell death. J Exp Med 1995; 182: 367–377.
15. Deckwerth TL, Johnson EM. Temporal analysis of events associated with programmed cell death (apoptosis) of sympathetic neurons deprived of nerve growth factor. J Cell Biol 1993; 123: 1207–1222.
16. Vayssière J-L, Petit PX, Risler Y, Mignotte B. Commitment to apoptosis is associated with changes in mitochondrial biogenesis and activity in cell lines conditionally immortalized with simian virus 40. Proc Natl Acad Sci USA 1994; 91: 11752–11756.
17. Zamzami N, Marchetti P, Castedo M, Zanin C, Vayssière J-L, Petit PX, et al. Reduction in mitochondrial potential constitutes an early irreversible step of programmed lymphocyte death in vivo. J Exp Med 1995; 181: 1661–1672.
18. Petit PX, LeCoeur H, Zorn E, Dauguet C, Mignotte B, Gougeon ML. Alterations of mitochondrial structure and function are early events of dexamethasone-induced thymocyte apoptosis. J Cell Biol 1995; 130: 157–167.
19. Macho A, Castedo M, Marchetti P, Aguilar JJ, Decaudin D, Zamzami N, et al. Mitochondrial dysfunctions in circulating T lymphocytes from human immunodeficiency virus-1 carriers. Blood 1995; 86: 2481–2487.
20. Castedo M, Macho A, Zamzami N, Hirsch T, Marchetti P, Uriel J, et al. Mitochondrial perturbations define lymphocytes undergoing apoptotic depletion in vivo. Eur J Immunol 1995; 25: 3277–3284.
21. Cossarizza A, Franceschi C, Monti D, Salvioli S, Bellesia E, Rivabene R, et al. Protective effect of *N*-acetylcysteine in tumor necrosis factor-alpha-induced apoptosis in U937 cells: the role of mitochondria. Exp Cell Res 1995; 220: 232–240.
22. Wolvetang EJ, Johnson KL, Krauer K, Ralph SJ, Linnane AW. Mitochondrial respiratory chain inhibitors induce apoptosis. FEBS Lett 1994; 339: 40–44.

23. Smets LA, Van den Berg J, Acton D, Top B, van Rooij H, Verwijs-Janssen M. BCL-2 expression and mitochondrial activity in leukemic cells with different sensitivity to glucocorticoid-induced apoptosis. Blood 1994; 5: 1613–1619.
24. Newmeyer DD, Farschon DM, Reed JC. Cell-free apoptosis in *Xenopus* egg extracts: inhibition by Bcl-2 and requirement for an organelle fraction enriched in mitochondria. Cell 1994; 79: 353–364.
25. Martin SJ, Newmeyer DD, Mathisa S, Farschon DM, Wang HG, Reed JC, et al. Cell-free reconstitution of Fas-, UV radiation- and ceramide-induced apoptosis. EMBO J 1995; 14: 5191–5200.
26. Tanaka S, Saito K, Reed JC. Structure–function analysis of the Bcl-2 oncoprotein. Addition of a heterologous transmembrane domain to portions of the Bcl-2β protein restores function as a regulator of cell survival. J Biol Chem 1993; 268: 10920–10926.
27. Nguyen M, Branton PE, Walton PA, Oltvai ZN, Korsmeyer SJ, Shore GC. Role of membrane anchor domain of Bcl-2 in suppression of apoptosis caused by E1B-defective adenovirus. J Biol Chem 1994; 269: 16521–16524.
28. Greenhalf W, Stephan C, Chaudhuri B. Role of mitochondria and C-terminal membrane anchor of Bcl-2 in Bax induced growth arrest and mortality in *Sacharomyces cerevisiae*. FEBS Lett 1996; 380: 169–175.
29. Zamzami N, Marchetti P, Castedo M, Hirsch T, Susin SA, Masse B, et al. Inhibitors of permeability transition interfere with the disruption of the mitochondrial transmembrane potential during apoptosis. FEBS Lett 1996; 384: 53–57.
30. Zoratti M, Szabò I. The mitochondrial permeability transition. Biochim Biophys Acta–Rev Biomembranes 1995; 1241: 139–176.
31. Bernardi P, Petronilli V. The permeability transition pore as a mitochondrial calcium release channel; a critical appraisal. J Bioenenerg Biomembr 1996; 28: 129–136.
32. Nicolli A, Basso E, Petronilli V, Wenger RM, Bernardi P. Interactions of cyclophilin with mitochondrial inner membrane and regulation of the permeability transition pore, a cyclosporin A-sensitive channel. J Biol Chem 1996; 271: 2185–2192.
33. Marchetti P, Castedo M, Susin SA, Zamzami N, Hirsch T, Haeffner A, et al. Mitochondrial permeability transition is a central coordinating event of thymocyte apoptosis. J Exp Med 1996; 184: 1155–1160.
34. Zamzami N, Susin SA, Marchetti P, Hirsch T, Castedo M, Kroemer G. Mitochondrial control of nuclear apoptosis. J Exp Med 1996; 183: 1533–1544.
35. Schulze-Osthoff K, Beyaert R, Vandevoorde V, Haegeman G, Fiers W. Depletion of the mitochondrial electron transport abrogates the cytotoxic and gene-inductive effects of TNF. EMBO J 1993; 12: 3095–3104.
36. Richter C. Pro-oxidants and mitochondrial Ca^{2+}: their relationship to apoptosis and oncogenesis. FEBS Lett 1993; 325: 104–107.
37. Forrest VJ, Kang YH, McClain DE, Robinson DH, Ramakrishnan N. Oxidative stress-induced apoptosis prevented by Trolox. Free Radicals Biol Med 1994; 16: 675–684.
38. Przedborski S, Kostic V, Jackson-Lewis V, Naini AB, Simonetti S, Fahn S, et al. Transgenic mice with increased Cu/Zn-superoxide dismutase activity are resistant to *N*-methyl-4-phenyl-1,2,3,6-tetrahydropyridine-induced neurotoxicity. J Neurosci 1992; 12: 1658–1667.
39. Greenlund LJS, Deckwert TL, Johnson EM. Superoxide dismutase delays neuronal apoptosis: a role for reactive oxygen species in programmed cell death. Neuron 1995; 14: 303–315.
40. Shimizu S, Eguchi Y, Kosaka H, Kamlike W, Matsuda H, Tsujimoto Y. Prevention of hypoxia-induced cell death by Bcl-2 and Bcl-xL. Nature 1995; 374: 811–813.
41. Graeber TG, Osmanian C, Jacks T, Housman DE, Koch CJ, Lowe SW, et al. Hypoxia-mediated selection of cells with diminished apoptotic potential in solid tumors. Nature 1996; 379: 88–91.
42. Jacobson MD, Raff MC. Programmed cell death and Bcl-2 protection in very low oxygen. Nature 1995; 374: 814–816.

43. Hug H, Enari M, Nagata S. No requirement of reactive oxygen intermediates in Fas-mediated apoptosis. FEBS Lett 1994; 351: 311–313.
44. Costantini P, Chernyak BV, Petronilli V, Bernardi P. Modulation of the mitochondrial permeability transition pore by pyridine nucleotides and dithiol oxidation at two separate sites. J Biol Chem 1996; 271: 6746–6751.
45. Costantini P, Chernyak BV, Petronilli V, Bernardi P. Selective inhibition of the mitochondrial permeability transition pore at the oxidation–reduction sensitive dithiol by monobromobimane. FEBS Lett 1995; 362: 239–242.
46. Jacobson MD, Burne JF, King MP, Miyashita T, Reed JC, Raff MC. Bcl-2 blocks apoptosis in cells lacking mitochondrial DNA. Nature 1993; 361: 365–369.
47. Marchetti P, Santos S, Decaudin D, Gamen S, Castedo M, Hirsch T, et al. Apoptosis-associated derangement of mitochondrial function in cells lacking mitochondrial DNA. Cancer Res 1996; 56: 2033–2038.
48. Castedo M, Hirsch T, Susin SA, Zamzami N, Marchetti P, Macho A, et al. Sequential acquisition of mitochondrial and plasma membrane alterations during early lymphocyte apoptosis. J Immunol 1996; 157: 512–521.

22

Antioxidants and the Regulation of Reactive Iron

John M. C. Gutteridge
Royal Brompton Hospital
London, England

IRON

Iron is the fourth most abundant element in the earth's crust and the second most abundant metal. It is the metallic iron at the earth's center which accounts for its magnetic field as well as for its overall mass density. This metallic iron was exploited many centuries ago for navigation purposes with the pioneering development of the magnetic compass.

Aerobic life forms evolved to use iron as a catalyst for oxygen utilization and eventually came to dominate the planet, although it appears that earlier primitive anaerobic life forms may also have used iron–sulfur redox chemistry for energy capture. The move to land of life forms from the seas and tidal pools occurred when a protective screen of oxygen and ozone was able to filter out most of the damaging solar radiation. Part of the marine environment is still visible today, for we see the distribution of major ions in blood and body fluids reflecting the oceanic environment, with Na^+, K^+, Ca^{2+}, Mg^{2+}, and Cl^- ions predominating. These ions are present in sea water at concentrations some 10^5–10^6 times greater than those of trace metals such as Fe, Cu, Zn, Mn, Mo, Sn, and V. Inspite of aluminum and iron being the most abundant metals in the earth's crust, they do not appear as major ions in surface waters, reflecting their poor solubility at neutral pH values. Iron is present in sea water mainly as a colloidal particle of hydrous ferric oxide (finely dispersed rust) representing a true solution concentration of around 0.1 nmol/kg. Ever-increasing industrialization is causing acidification of surface waters, with a consequential rise in the solubility of both iron and aluminum, allowing both to enter biological eco-chains with unknown long-term consequences.

OXYGEN

The element oxygen (O) exists in air as a molecule (O_2), and was first isolated and characterized between 1772 and 1779 by Priestley, Lavoisier, and Scheele. Molecular oxygen appeared in significant amounts on the surface of the earth some 2.5×10^9 years ago, and geological evidence supports a biological origin through the photosynthetic activity of microorganisms, the blue-green algae. The percentage of molecular oxygen in dry air is now around 21%, the major component of air being nitrogen (78%). Oxygen in the air is a negligible proportion of the total present in water molecules and in mineral reservoirs of the earth's crust, where it is by far the most abundant element. The *stable* nature of molecular oxygen with its characteristically poor reactivity is a result of its unusual molecular structure. Molecular oxygen contains two unpaired electrons each in their own orbital but with the same spin quantum number, and only when this spin restriction is overcome can the true reactivity of oxygen be expressed. In order to bypass this spin restriction, molecular oxygen prefers to accept electrons one at a time to complete its 4-electron reduction to water. The problem with this strategy is that the addition of single electrons produces free radicals such as superoxide ($O_2^{\bullet -}$) and the hydroxyl radical ($^{\bullet}OH$). In order to achieve greater reactivity, molecular oxygen uses iron and, to a lesser extent copper as the major catalyst for oxidase, oxygenase, and antioxidant reactions, and for molecular oxygen and electron transport by proteins.

ANTIOXIDANTS

Antioxidants are widely discussed in biological and food sciences but are rarely defined. The chain-breaking properties of a variety of chemicals have tended to dominate concepts concerning antioxidant mechanisms. A definition which accommodates diverse molecules that act as enzymes, chelators, stoichiometric scavengers, chain breakers, and cytoprotective agents has recently been introduced (1), and defines an antioxidant as any substance that, when present at low concentrations compared with those of the oxidizable substrate, considerably delays or inhibits oxidation of the substrate.

Antioxidants can act at several different stages in an oxidative sequence, as illustrated by considering lipid oxidation occurring in cell membranes or lipid-rich products. Antioxidants can act by (a) removing molecular oxygen or decreasing local oxygen concentrations; (b) removing catalytic metal ions; (c) removing key reactive oxygen species (ROS) such as superoxide and hydrogen peroxide; (d) scavenging initiating free radicals such as hydroxyl, alkoxyl, and peroxyl species; (e) breaking the chain of an initiated sequence; or (f) quenching or scavenging singlet oxygen.

Inside cells we find unique enzymes evolved to deal with ROS as well as low-molecular-mass scavengers such as reduced glutathione. Within the hydrophobic lipid interior of membranes, lipophilic radicals are formed that are different from those seen in the intracellular aqueous milieu. Lipophilic radicals require different types of antioxidants for their removal. Vitamin E (α-tocopherol), a fat-soluble vitamin, is a poor antioxidant outside a membrane bilayer but is very effective when incorporated into the membrane, for example (2). In addition, an important part of membrane stability and protection is how the membrane is assembled from its lipid components. This structural organization requires that the correct ratios of phospholipids to cholesterol are present and that the correct types of phospholipids and their fatty acids are attached (2,3).

Body extracellular fluids do not contain the same enzymes as intracellular fluids, and it appears that plasma is more concerned with the control of reactive iron and copper complexes.

Plasma also contains a variety of redox-active low-molecular-mass molecules and many of these have been ascribed primary antioxidant roles. Thus, using iron-independent generation of peroxyl radicals from an azo initiator, others have found vitamin E, uric acid, bilirubin, ascorbic acid, and thiol groups to be important plasma antioxidants (4). Several of these antioxidants are vitamins and the subject of considerable nutritional debate. An important question that is frequently considered is "What is the most important biological antioxidant?" There is, of course, no simple answer to this simple question, since the answer will entirely depend on what pro-oxidants were used to drive radical formation.

REGULATION OF INTRACELLULAR IRON BY ANTIOXIDANTS

The metabolism of oxygen takes place inside cells, and it is here we expect to find antioxidants evolved to deal speedily and specifically with reduced intermediates of oxygen. Enzymes such as the superoxide dismutases rapidly promote the dismutation of superoxide into hydrogen peroxide and oxygen considerably faster than it occurs uncatalyzed (5). Hydrogen peroxide, a product of the dismutation reaction, can be destroyed by two different enzymes, namely, catalase and glutathione peroxidase (a selenium-containing enzyme). During normal aerobic metabolism these enzymes function in concert to eliminate toxic reduction intermediates of oxygen inside the cell, thereby allowing a small pool of low-molecular-mass iron compounds to exist safely for DNA synthesis and the manufacture of iron-containing proteins. Delivery of iron to the cell is normally achieved by iron-loaded transferrin binding to the transferrin receptor (TfR). Iron is stored in the cell within the ferritin molecule (which can store up to 4500 iron atoms). The synthesis rate of TfR and ferritin is regulated at the post transcriptional level by cellular iron and coordinated by the iron-dependent binding of a cytosolic protein called "the iron-responsive element binding protein" (IRE-BP), which binds to specific sequences on its target mRNAs (6). Later work has shown that IRE-BP is identical in sequence to the cytosolic enzyme aconitase (7). The protein functions as an active aconitase when it has an Fe–S cluster present or as an RNA-binding protein when iron is absent (8). Switching between these two forms depends on cellular iron status (9) such that when iron is replete the protein is an active aconitase, whereas when deprived of iron it has only RNA-binding activity. Intracellular low-molecular-mass iron levels may also be regulated by the oxidative stress protein heme oxygenase, which can increase intracellular levels of ferritin and thereby decrease intracellular free iron.

When human skin fibroblasts were treated with UVA radiation, a 4-fold increase in microsomal heme oxygenase activity was observed concomitantly with a 40% decrease in heme content (10). In parallel with these changes was a 2-fold increase in ferritin levels. The authors proposed that the induction of ferritin was a protective antioxidant response to decrease intracellular levels of low-molecular-mass iron compounds, and thereby restrict iron-stimulated reactions during periods of oxidative stress (10).

As previously mentioned, cells normally accumulate iron via the binding of transferrin to high-affinity surface receptors. There is, however, a transferrin-independent pathway of cellular iron uptake that involves a membrane-based transport system (11). When non-transferrin-bound iron appears in plasma it triggers the induction of membrane transporters which remove low-molecular-mass iron compounds from the extracellular environment. As already mentioned, the intracellular environment can better cope with low-molecular-mass iron than can the extracellular compartments.

REGULATION OF EXTRACELLULAR IRON BY ANTIOXIDANTS

Body extracellular fluids contain little or no catalase activity, and extremely low levels of superoxide dismutase. Glutathione peroxidases, in both selenium-containing and non-selenium-containing forms, are present in plasma, but there is little glutathione in plasma (<1 μM) to satisfy an enzyme with a k_m for GSH in the millimolar range. "Extracellular" superoxide dismutases (EC-SOD) have been identified (12) and shown to contain copper and attached carbohydrate groups. The limited ability of extracellular fluids to remove ROS and so allow the survival of $O_2^{\bullet -}$, H_2O_2, lipid peroxides (LOOH), and hypochlorous acid (HOCl) makes it possible for the body to utilize these molecules, and others such as nitric oxide (NO), as messenger, signal, or trigger molecules (13–15). A key feature of such a proposal is that $O_2^{\bullet -}$, H_2O_2, LOOH, and HOCl do not meet with reactive iron or copper, and that extracellular antioxidant protection has evolved to keep iron and copper in poorly reactive or nonreactive forms.

The iron transport protein transferrin is normally one-third loaded with iron and keeps the concentration of "free" iron in plasma at effectively nil. Iron bound to transferrin will not participate in radical reactions, and the available iron-binding capacity gives transferrin a powerful antioxidant property toward iron-stimulated free-radical reactions (16). Similar considerations apply to lactoferrin (16) which, like transferrin, can bind two moles of iron per mole of protein, but holds onto its iron down to pH values as low as 4.0. Hemoglobin, myoglobin, and heme compounds can accelerate lipid peroxidation by at least two different mechanisms. The heme ring can react with peroxides to form active iron-oxo species such as perferryl (iron oxidation state V) and ferryl (IV) (17), and a molar excess of peroxide can cause fragmentation of the cyclic tetrapyrrole rings, releasing chelatable iron (18). Plasma also contains proteins such as the haptoglobins and hemopexin specifically to bind and conserve hemoglobin and heme iron, respectively. Binding to these proteins greatly diminishes the ability of heme and hemoglobin to accelerate lipid peroxidation (19, 20).

The major copper-containing protein of human plasma is ceruloplasmin, unique for its intense blue coloration. Apart from its known acute-phase reactant properties, its biological functions remain an enigma. I and my colleagues, however, have pointed out that the protein's ferroxidase activity makes a major contribution to extracellular antioxidant protection when iron-driven lipid peroxidation and Fenton chemistry occur (21). Ceruloplasmin rapidly removes ferrous ions from solution and simultaneously reduces molecular oxygen to water, with the transfer of four electrons at the enzyme's active center (a protective strategy by not releasing reactive forms of oxygen into the aqueous milieu).

Iron functions as an important signal molecule both inside and outside cells. Inside cells it has more freedom because intracellular antioxidants catalytically remove reduced intermediates of molecular oxygen. Extracellularly, however, iron is removed or inactivated by proteins in order to allow the limited survival of $O_2^{\bullet -}$, H_2O_2, LOOH, HOCl, and NO, which can then act as humoral signal molecules.

ACKNOWLEDGMENTS

I thank the British Heart Foundation, The British Lung Foundation, the British Oxygen Group plc, and the Dunhill Medical Trust for their generous support.

REFERENCES

1. Halliwell B, Gutteridge JMC. Free Radicals in Biology and Medicine. 2d ed. Oxford UK: Oxford University Press; 1989.
2. Gutteridge JMC. The membrane effects of vitamin E, cholesterol and their acetates on peroxidative susceptibility. Res Commun Chem Pathol Pharmacol 1977; 77: 379–386.
3. Gutteridge JMC, Halliwell B. The antioxidant proteins of extracellular fluids. In Chow CK, ed. Cellular Antioxidant Defense Mechanisms. Vol. 2. Boca Raton. FL: CRC Press; 1988: 1–23.
4. Frei B, Stocker R, Ames BN. Antioxidant defenses and lipid peroxidation in human blood plasma. Proc Natl Acad Sci USA 1988; 88: 9748–9752.
5. Fridovich I. Superoxide dismutases. Adv Enzymol 1974; 41: 35–48.
6. Klausner RD, Rouault TA, Harford JB. Regulating the fate of mRNA: the control of cellular iron metabolism. Cell 1993; 72: 19–28.
7. Kennedy MC, Mende-Mueller L, Blondin GA, Beinert H. Purification and characterisation of cytosolic aconitase from beef liver and its relationship to the iron-responsive element binding protein (IRE-BP). Proc Natl Acad Sci USA 1992; 89: 11730–11734.
8. Haile DJ, Rouault TA, Harford JB, Kennedy MC, Blondin GA, Beinert H, Klausner RD. Cellular regulation of the iron-responsive element binding protein: disassembly of the cubane iron–sulfur cluster results in high affinity RNA binding. Proc Natl Acad Sci USA 1992; 89: 11735–11739.
9. Haile DJ, Rouault TA, Tang CK, Chin J, Harford JB, Klausner RD. Reciprocal control of RNA-binding and aconitase activity in the regulation of the iron-responsive element binding protein: role of the iron–sulfur cluster. Proc Natl Acad Sci USA 1992; 89: 7536–7540.
10. Vile GF, Tyrrel RM. Oxidative stress resulting from ultraviolet A irradiation of human skin fibroblasts leads to heme oxygenase–dependent increase in ferritin. J Biol Chem 1993; 268: 14678–14681.
11. Kaplan J, Jordan I, Sturrock A. Regulations of the transferrin-independent iron transport system in cultured cells. J Biol Chem 1991; 266: 2997–3004.
12. Marklund SL, Holme E, Hellner L. Superoxide dismutase in extracellular fluids. Clin Chim Acta 1982; 126: 41–51.
13. Halliwell B, Gutteridge JMC. Oxygen free radicals and iron in relation to biology and medicine: some problems and concepts. Arch Biochem Biophys 1986; 246: 501–514.
14. Saran M, Bors W. Oxygen radicals acting as chemical messengers: a hypothesis. Free Radical Res Commun 1989; 7: 213–220.
15. Saran M, Bors W. Signalling by O_2^- and NO: how far can either radical, or any specific reaction product, transmit a message under in vivo conditions? Chem Biol Interact 1994; 90: 35–45.
16. Gutteridge JMC, Paterson SK, Segal AW, Halliwell B. Inhibition of lipid peroxidation by the iron-binding protein lactoferrin. Biochem J 1981; 199: 259–261.
17. Rice–Edwards CA, Diplock AT, Symons MCR. Techniques in Free Radical Research. Vol. 22. Amsterdam: Elsevier; 1991: 101–124.
18. Gutteridge JMC. Iron promoters of the Fenton reaction and lipid peroxidation can be released from haemoglobin by peroxides. FEBS Lett 1986; 201: 291–295.
19. Gutteridge JMC. The antioxidant activity of haptoglobin towards haemoglobin stimulated lipid peroxidation. Biochim Biophys Acta 1987; 917: 219–223.
20. Gutteridge JMC, Smith A. Antioxidant protection by haemopexin of haem-stimulated lipid peroxidation. Biochem J 1988; 256: 861–865.
21. Gutteridge JMC, Stocks J. Caeruloplasmin: physiological and pathological perspectives. CRC Crit Rev Clin Lab Sci 1981; 14: 257–329.

23
Redox Regulation by the Thioredoxin and Glutaredoxin Systems

Arne Holmgren, Elias S.J. Arnér, Fredrik Åslund, Mikael Björnstedt, Zhong Liangwei, Johanna Ljung, Hajime Nakamura, and Dragana Nikitovic
Karolinska Institute
Stockholm, Sweden

THE THIOREDOXIN SYSTEM

The thioredoxin system, comprising NADPH, thioredoxin reductase, and thioredoxin (1–3) operates as an efficient protein disulfide reductase as outlined in Figure 1. Thioredoxin is ubiquitous in living cells and has a large number of functions, apart from its original isolation as a hydrogen donor for ribonucleotide reductase, which is essential for DNA synthesis. By its general protein disulfide reductase activity, thioredoxin can regulate enzymes and transcription factors by thiol redox control (1–3). This is based on the reversible formation of a disulfide involving the sulfur of a critical cysteine SH group

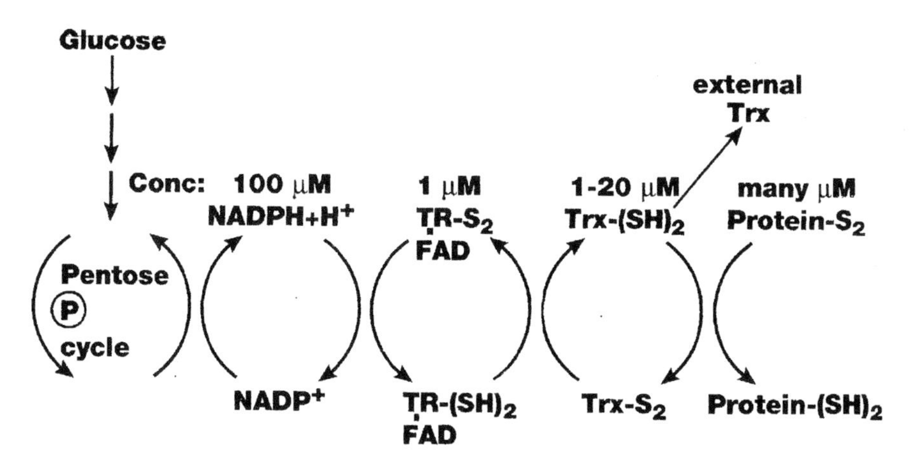

Figure 1. General disulfide reduction by the thioredoxin system. TR, thioredoxin reductase; $Trx\text{-}S_2$ oxidized thioredoxin; $Trx\text{-}(SH)_2$, reduced thioredoxin. "Conc" represents typical intracellular concentrations.

and another SH group either within the protein or to glutathione (GSH). Disulfide formation or oxidation generally leads to a loss of function for intracellular proteins which have many SH groups, whereas reduction of disulfides inactivates extracellular proteins such as coagulation factors, insulin, or PDGF (1–3). Thiol redox control thus operates as a covalent modification of SH groups in Cys residues analogous to phosphorylation–dephosphorylation of the OH groups of Ser, Thr, or Tyr residues in proteins.

In *Escherichia coli* or mammalian cells thioredoxin and thioredoxin reductase function as intracellular enzymes coupling the reducing power of NADPH to enzymatic substrate reductions catalyzed by ribonucleotide reductase, sulfate (PAPS) reductase, or methionine sulfoxide reductases (1–3). A secretory mechanism for thioredoxin has been revealed, initially from lymphocytes following their activation leading to generation of extracellular thioredoxin with cytokine or co-cytokine-like properties (4, 5). The secretion, which has been demonstrated in conditioned media from HTLV-1 virus-infected T cells with EBV-transformed B cells (5) or from activated B and T cells (6, 7) does not involve a classical ER–Golgi pathway. The details of the secretory mechanism of thioredoxin, also known from liver cells (8, 9), remain to be elucidated. Thioredoxin can act as a growth factor and is overexpressed and secreted in certain tumor cells (10–13), or act as a mediator of cell growth inhibition by interferon-γ (14).

STRUCTURE AND MECHANISM OF THIOREDOXIN

The structure of thioredoxin from *E. coli* and mammalian cells has been the subject of intense studies for over thirty years (1–3, 15). The three-dimensional structures of *E. coli* and human thioredoxin have the same general fold consisting of a core five-stranded β-sheet flanked by four α-helices (15). The conserved active site Trp-Cys-Gly-Pro-Cys, is located on a protrusion in the molecule and links the second β-strand to the second α-helix and forms the first turn in this helix. The thioredoxin fold is also present in glutaredoxin, protein disulfide-isomerase (PDI), *E. coli* DsbA, and glutathione *S*-transferase and glutathione peroxidase (16). A hydrophobic surface area surrounds the exposed N-terminal active-site cysteine residue (Cys-32) (15, 17). A local and minimal conformational change occurs upon reduction of oxidized thioredoxin to the dithiol form involving tilting the side-chain of Cys-32 toward the solution, as revealed from high-resolution multidimensional NMR studies of *E. coli* thioredoxin (15, 17). As a protein disulfide reductase reduced thioredoxin is between 10^4 and 10^6 times more efficient than the well-known strong reductant dithiothreithol (DTT) (1–3, 15). Understanding the mechanism of thioredoxin as an efficient disulfide reductase has involved studies of the structure of wild-type and mutant proteins, and the kinetics of the reaction with proteins like insulin (1–3, 15). A reaction mechansim for thioredoxin is illustrated in Figure 2. In reduced *E. coli* thioredoxin, a shared proton or a hydrogen bond between the sulfur of the exposed thiolate of Cys-32 and the buried SH group of Cys-35 stabilizes the reduced form (18). Thioredoxin-$(SH)_2$ initially binds noncovalently to form a complex with a protein with a target disulfide bond. In the hydrophobic surface area generated, a transition-state intermediate involves a mixed disulfide between thioredoxin and the target protein (15, 18). Rearrangements occur rapidly in the hydrophobic environment, leading to the disulfide form of thioredoxin and the reduced dithiol protein (Figure 2). Structures for oxidized and reduced human thioredoxin involving mutations at four sites, C62A, C69A, C73A, and M74T (15, 19) were suggested to show different behavior of the active site upon oxido-reduction compared to that of *E. coli*

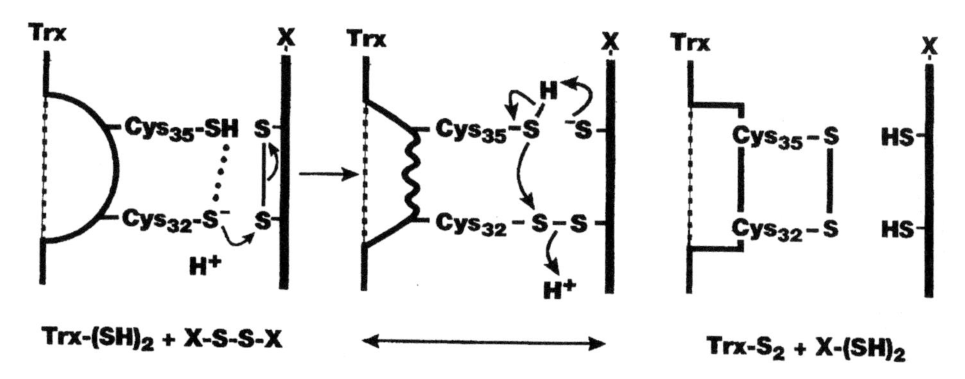

Figure 2. Mechanism of disulfide reduction by reduced thioredoxin. The dotted line indicates a shared proton or a hydrogen bond between the Cys-35 SH group and the thiolate of Cys-32 in *E. coli* thioredoxin.

thioredoxin. However, the NMR structures for human thioredoxin are not supported by recent high-resolution crystal structures of wild-type human thioredoxin (20). In these the active site is extremely similar to that of *E. coli* thioredoxin determined by NMR (17), including the presence of a hydrogen bond between the SH group of Cys-35 and the sulfur of Cys-32 in reduced thioredoxin (20).

Thioredoxin from *E. coli* with a redox potential (E'_0) of −270 mV (1–3) is a strong reductant, particularly when coupled to NADPH (E'_0-315 mV) through the thioredoxin reductase reaction (Figure 1). It should also be remembered that the redox potentials of individual dithiol–disulfide pairs in proteins span large differences determined by the folded conformation (21), meaning that thioredoxin in effect may catalyze both disulfide reductions or disulfide formation within a protein or between proteins by thiol redox control. In protein disulfide isomerase (PDI) and its isoenzymes the thioredoxin fold is also used, but the dithiol–disulfide couples in the thioredoxin domains are much more oxidizing and actually catalyze the reaction in the reverse direction to thioredoxin, forming disulfides in nascent proteins (22).

THIOREDOXIN AND OXIDATIVE STRESS

A number of redox regulatory roles for thioredoxin with transcription factors (23–25) have identified thioredoxin as a reductant of critical SH groups. We had previously characterized the cellular localization of thioredoxin in the calf and the rat (26, 27). Immunohistochemical studies of the distribution of thioredoxin in the rat brain demonstrate thioredoxin not only in the cytosol but also in the nucleus of the tetraploid big nerve cells from cortex (27). Earlier studies also showed that thioredoxin and thioredoxin reductase are present in nerve cell axons and move by axonal transport (28).

We have mapped the distribution of thioredoxin mRNA in the rat brain (29). Thioredoxin expression was widespread in different parts of the brain and its change during oxidative stress is illustrated in Figure 3. In this experiment the oxidative stress was generated by mechanical injury induced by partial unilateral hemitransection using a knife (29). As shown in Figure 3, this resulted in a strong upregulation at 24 h of

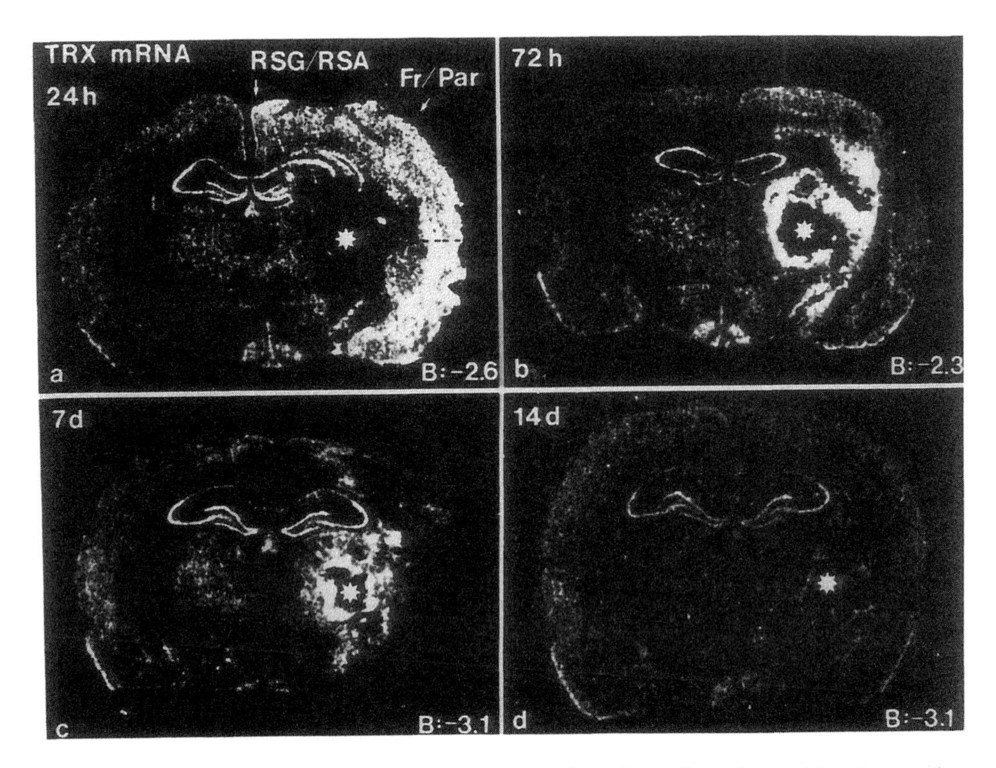

Figure 3. Upregulation of thioredoxin mRNA expression after unilateral partial hemitransection as revealed by in-situ hybridization. (a) 24 h after injury. RSG/RSA, retrosplinal granular and agranular cortices; Fr-Par,fronto-parietal cortex. (b–d): thioredoxin gene expression after 72 h (b), 7 days(c) and 14 days(d) after the lesion. Thioredoxin mRNA levels have been normalized, *denotes the lesion site. (From Ref. 29.)

thioredoxin mRNA around the lesion site and also in the surrounding cortical hemisphere and on the ipsilateral side. After 72 h increased mRNA signal was more restricted to the lesion site, and after 7 days the thioredoxin gene expression in the cortical areas was almost normal. However, cells in the area of the lesion were still showing an increased labeling. Fourteen days after surgery, the thioredoxin mRNA levels were normalized and back to the resting stage. These results clearly demonstrate a global induction of thioredoxin in a large area around the lesion. Probably, activated glial cells and activated astroglial and microglial cells contribute to the major upregulation of thioredoxin mRNA. The results point to important roles of thioredoxin in protection against oxidative stress. Such roles of thioredoxin may involve acting as an electron donor of methionine sulfoxide reductases (1–3) and the role of thioredoxin in reduction of mixed disulfides within proteins and artificial disulfides formed between proteins or with GSH (1–3). Also, upregulation of transcription factors including, for example, NFκB, EP-1, TF-3C, TZLF-1, and MYB should take place and require thioredoxin for activation. In this context, it should also be pointed out that thioredoxin has been identified as a growth factor from macrophages, which after release will stimulate the outgrowth of monamine neurons in vitro (30). Of course repair mechanisms involving DNA synthesis would also require thioredoxin as a hydrogen donor for ribonucleotide reductase (1–3).

THE GLUTAREDOXIN SYSTEM

Glutaredoxin was originally identified as a hydrogen donor for ribonucleotide reductase in an *E. coli* mutant lacking thioredoxin (31). As outlined in Figure 4, the key component in this system, glutaredoxin (Grx), may reduce protein disulfides directly via its active-site dithiol, which then is converted to a disulfide. Reduction of glutaredoxin is achieved by glutathione (GSH) (31,32), which usually is from 1 to 10 mM in cells (21). GSSG formed is reduced by the enzyme glutathione reductase via NADPH. Glutaredoxin (identical to thioltransferase in mammalian cells) has in our laboratory been purified and characterized from *E. coli* (33), calf thymus (34, 35), and human placenta (36). Glutaredoxin contains a conserved active-site sequence of Cys-Pro-Tyr-Cys (32) and is a member of the thioredoxin superfamily of proteins (16), with its active-site cysteine residues located between a β-strand and the beginning of an α-helix as shown by the structure of *E. coli* glutaredoxin 1 determined by NMR (37). The other important structural feature of glutaredoxin is a GSH binding site (Figure 5) as defined from the solution structure of *E. coli* glutaredoxin 1 with glutathione bound as a mixed disulfide (37). The binding site for GSH, which is antiparallel, is defined by several conserved regions of the protein sequence (32) making a site that shows similarities to the way glutathione is bound in other proteins with the thioredoxin fold, like glutathione *S*-transferases (16, 38).

The substrate specificity of glutaredoxin is either protein disulfides, as in ribonucleotide reductase, or glutathione mixed disulfides of proteins or low-molecular-mass compounds like cysteine (32, 33). Presumably, the GSH binding site in reduced glutaredoxin helps to position the substrate GSH mixed disulfide bond for cleavage. Assays of glutaredoxin use NADPH, glutathione reductase, GSH, and hydroxyethyl disulfide to record oxidation of NADPH (32). The gene for human glutaredoxin has been cloned from red blood cells (39) and placenta (36), and the human glutaredoxin gene has been located to chromosome 5q14 (40).

The viability of cells from *E. coli* lacking thioredoxin is explained by the presence of glutaredoxin, which itself is also nonessential (41). Recently, two additional glutaredoxins were identified in *E. coli* (42). These proteins, called glutaredoxins 2 and 3 (Grx 2 and Grx 3), may explain why a double mutant lacking both thioredoxin and glutaredoxin 1 is viable under certain conditions (41). The major function of glutaredoxin 1 in *E. coli* is to be a hydrogen donor for ribonucleotide reductase, a function that is shared

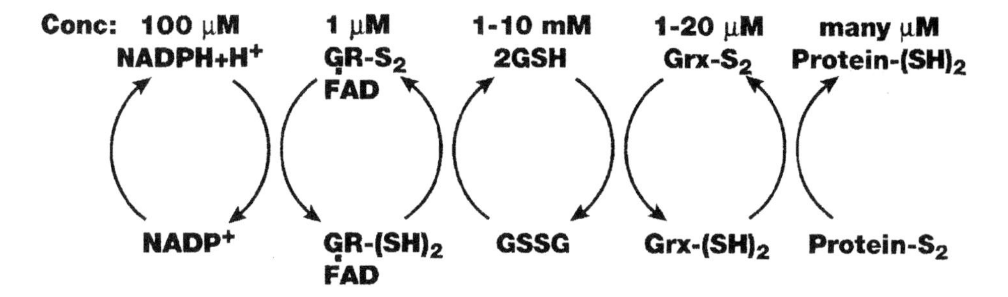

Figure 4. The glutaredoxin system as a general disulfide reductase. GR, glutathione reductase; Grx-S_2 and Grx-$(SH)_2$, oxidized and reduced glutaredoxin, respectively. "Conc" denotes representative intracellular concentrations.

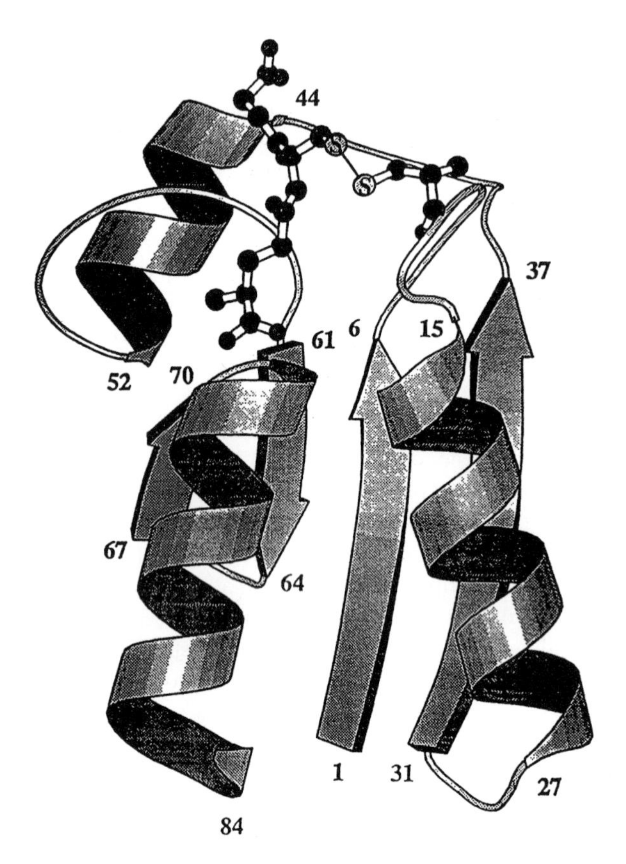

Figure 5. Structure of *E. coli* glutaredoxin C145 with GSH bound as a mixed disulfide. The Cys-11 and the three residues of glutathione are shown with all atoms. (From Ref. 38.)

with thioredoxin (32,41,42). The situation is not yet fully characterized in mammalian cells since the cellular localizations of ribonucleotide reductase and thioredoxin do not always coincide (1–3). Of particular interest with the glutaredoxin system is the fact that both the absolute concentration of glutathione and the GSH to GSSG ratio determine the redox state of SH groups on proteins such as transcription factors and receptors (20). Changes in the level of SH groups and disulfides in proteins relating to the glutathione redox buffer will be catalyzed by the glutaredoxins. We have determined the redox potentials of the active-site disulfide in *E. coli* glutaredoxin 1 and glutaredoxin 3, which are –0.23 and –0.20 V, respectively (Åslund, Berndt, Holmgren, unpublished). This indicates that these two *E. coli* glutaredoxins should operate as catalysts with different proteins and cover different redox states and environments in the cell in relation to the level of oxidative stress.

GLUTAREDOXIN AND OXIDATIVE STRESS

Reactions in both oxidative and reductive directions involving glutathione will be catalyzed by glutaredoxins. In particular, it should be noted that the formation of mixed disulfides eventually leading to the formation of proetin disulfides is related to the monothiol nature of glutathione in the glutaredoxin system. Thus, changes in both the absolute concentration of glutathione and its level of reduction will effect redox potential. With the system 2GSH/GSSG, changes in a reaction corresponding to an equilibrium constant of 10 will give rise to a change of E'_0 of 60 mV (20). This is in contrast to the situation with a dithiol system such as for thioredoxin or NADPH, where a change in the reaction equilibrium constant of 10 will correspond to an E'_0 change of 30 mV. Much research has concerned redox regulation by the glutathione system and intracellular enzymes (20). From the point of view of oxidative stress and redox regulation, the role of glutathione and thioredoxin in the activation of transcription factors like NFκB is particularly interesting (23, 43). The initial stage of activation of NFκB has been suggested to involve oxygen free radicals and a rise in GSSG (44), which presumably also involves functions of glutaredoxin. Experiments pertaining to these questions have still to be performed with pure proteins and in cells.

During severe oxidative stress or in certain physiological stimulation systems, when reactive oxygen species are produced, proteins will be modified by glutathionylation usually following the increase of GSSG in the cell (45, 46). Interestingly, recent results (45) have shown that a relatively limited number of proteins are modified and that reversal of this process involves both the thioredoxin and glutaredoxin systems. In particular, it should be noted that sometimes glutathionylation of a protein like carbonic anhydrase changes its properties. In the case of carbonic anhydrase III, this leads to a new enzyme specificity, which is as a tyrosine phosphatase (47). Clearly, much more work is required to understand the details of the use of covalent modification of a protein via glutathionylation and the role of glutaredoxin in regulation of signal transduction.

The synthesis of glutathione and the regulation of its levels in cells and the relation to potential deleterious effects on T cell function in HIV infection is the subject of intensive investigation (48, 49). Generation of inflammatory cytokines like TNF-α and IL-1 induces a general oxidative stress involving reactive oxygen species and loss of GSH from cells. Clearly, the glutaredoxins should play a role in transmitting changes in the redox buffer of glutathione to corresponding protein modifications with subsequent changes in function; this requires more detailed studies.

MAMMALIAN THIOREDOXIN REDUCTASE

As shown in Table 1, there are some major differences between the thioredoxin system of a prokaryote like *E. coli* and that of mammalian organisms. Thioredoxin reductase has changed remarkably in properties in mammalian cells (1–3). In contrast, as discussed above, thioredoxin is the same 12 kDa protein, and shows approximately 30% sequence identities with the active site and certain other residues absolutely conserved (50) and similar overall three-dimensional structures (17, 19, 20). Mammalian thioredoxins have two or three additional structural SH groups, which may form disulfides under oxidative conditions and act to regulate activity (1–3, 51–54).

Thioredoxin reductases from mammalian cells (calf liver or thymus (51,52) and rat (53) have a much higher molecular mass and a surprisingly wide substrate specificity. Our initial studies of thioredoxin reductase were complicated by the fact that some of the methods previously used for assays of the *E. coli* enzyme could not be applied to

Table 1. Properties of Thioredoxin Systems

	E. coli	Calf thymus
Thioredoxin	M_r = 12,000 108 aa 1 S–S bridge	M_r = 12,000 104 aa 1 S–S bridge + 2 SH groups inactivated by oxidation
Thioredoxin reductase	M_r = 70,000 2 subunits High specificity Stable	M_r = 116,000 2 subunits Broader specificity also PDI; inactivated by oxidation of SH groups

PDI = protein disulfide isomerase.

mammalian systems. Thus, DTNB (5, 5′-dithiobis(2-nitrobenzoic acid)), which is not a substrate for the *E. coli* thioredoxin reductase, is directly reduced by the mammalian enzymes and there is no need for reduced thioredoxin as an intermediate (51,52). In fact, addition of mammalian thioredoxin to an assay with low DTNB (<1 mM) leads to a rapid total inhibition of the enzyme (53, 54). New methods for measuring thioredoxin reductase were therefore developed involving the use of DTNB or reduction of insulin via thioredoxin (51–53). Thioredoxin reducase activity is readily measured at 412 nm using 5 mM DTNB reduction in the presence of NADPH (3).

The pure enzyme showed two subunits of 58 kDa (53) in sharp contrast to the well-characterized *E. coli* enzyme with subunits of 35 kDa and a known three-dimensional structure from X-ray crystallographic studies (55). During 1995–1996 major progress has been made in understanding the protein structure for the mammalian thioredoxin reductase with surprising and far-reaching implications.

Previous studies relating to oxidative stress utilized alloxan, a diabetogenic drug, which is a substrate for thioredoxin reductase and is also reduced by thioredoxin (56). Autoxidation of the product dialuric acid results in the formation of oxygen free radicals from NADPH. We suggested that the mechanism of alloxan-induced diabetes is by the production of oxygen free radicals via thioredoxin reductase. Immunohistochemical staining of the pancreas of rats showed that the β cells in the islets of Langerhans contained thioredoxin and thioredoxin reductase (57). The enzyme surrounded the granulae containing insulin and was also localized at the plasma membrane (57).

SELENIUM REDUCTION BY THE THIOREDOXIN SYSTEM

The fact that administration of selenium compounds like selenite (SeO_3^{2-}) caused inhibition of tumor cell proliferation in vivo (58) and the knowledge that thioredoxin reductase appeared to be more highly expressed in malignant cells prompted us to start a series of investigations on the reactions of selenium compounds with the mammalian thioredoxin systems (59). Contrary to expectation, we were able to demonstrate that selenite is a direct substrate for thioredoxin reductase as well as for thioredoxin (59–61). As shown in Figure 6, anaerobically selenite was reduced directly by calf thymus

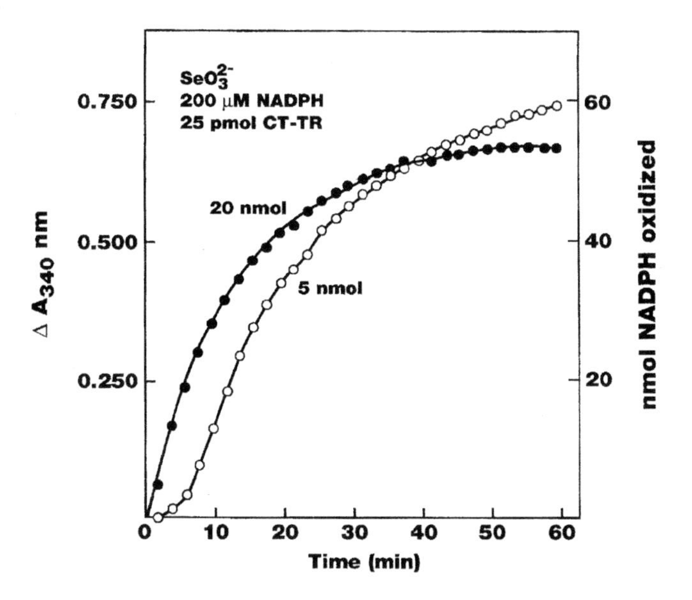

Figure 6. Reduction of selenite by calf thymus thioredoxin reductase (CT-TR). Oxidation of NADPH followed at 340 nm using two concentrations of selenite. (Taken from Ref. 60.)

thioredoxin reductase (TR) in a reaction giving rise to selenide with the consumption of three equivalents of NADPH. The total reaction is

$$SeO_3^{2-} + 3NADPH + 3H^+ \xrightarrow{TR} Se^{2-} + 3NADP^+ + 3H_2O \quad (1)$$

Under aerobic conditions, autoxidation of selenide causes a nonstoichiometiric reaction, leading to extensive consumption of NADPH (59–61). The reaction between thioredoxin reductase or thioredoxin is a mechanism for the production of selenide, which is required for the synthesis of selenocysteine (62). Another investigation of thioredoxin and thioredoxin reductase focused on its potential role as a hydrogen donor for the extracellular glutathione peroxidase present in blood plasma (63). This selenium-containing enzyme has been assumed to use glutathione as an electron donor. However, since there is less than 2 μM of free GSH in plasma and the enzyme has a K_m value for GSH in the range of 10 mM, this has strongly questioned how this plasma peroxidase, which is supposed to control peroxide tone in plasma and outside endothelial cells, can be functional. As shown in Figure 7, NADPH and thioredoxin reductase acted as direct hydrogen donors for the plasma peroxidase. The reaction was also stimulated by the presence of thioredoxin (63). GSH at 10 μM together with NADPH and glutathione reductase gave no direct reaction, but the addition of human glutaredoxin stimulated the reaction strongly (63). Since thioredoxin (26, 64) and glutaredoxin (65) are present in human plasma and thioredoxin and thioredoxin reductase also has been shown to be localized on the cell surface (66,67), there is a strong indication that these systems together with the plasma peroxidase may control the peroxide tone in the extracellular space.

Recently we became interested in the role of mammalian thioredoxin reductase in relation to lipid hydroperoxides. We showed that human thioredoxin reductase directly catalyzed reduction of lipid hydroperoxides like 15-(S)-HPETE (Figure 8) (68).

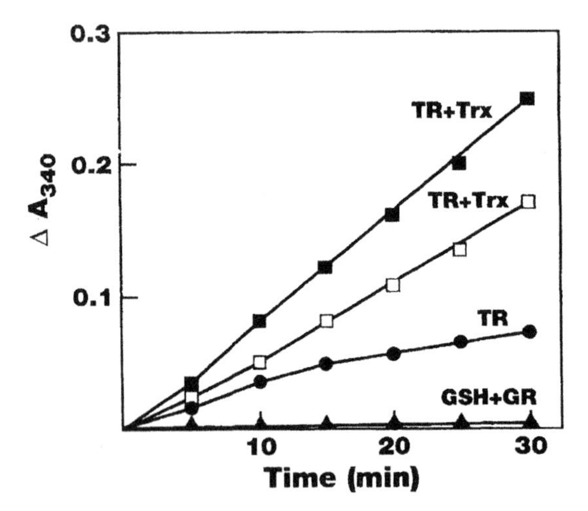

Figure 7. Activity of human 50 nM thioredoxin reductase (TR) alone (●) and with thioredoxin (Trx) 2.5 μM(□) or 5.0 μM (■) with human plasma glutathione peroxidase. The reaction was followed at 340 nm. Activity with only GSH (10 μM) and glutathione reductase (GR; ▲) is also shown. (From Ref. 61.)

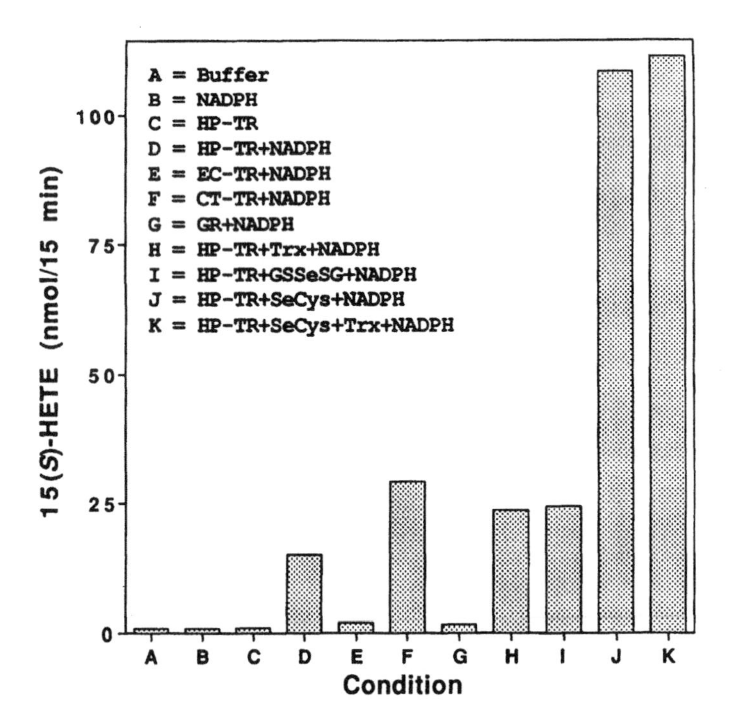

Figure 8. Human thioredoxin reductase ((HP-TR)-dependent reductions of 15-(*S*)-HPETE. The reactions were monitored by HPLC to determine the product 15-(S)-HETE. HP-TR 50 nM. (For further details see Ref. 68 from which this figure is taken.)

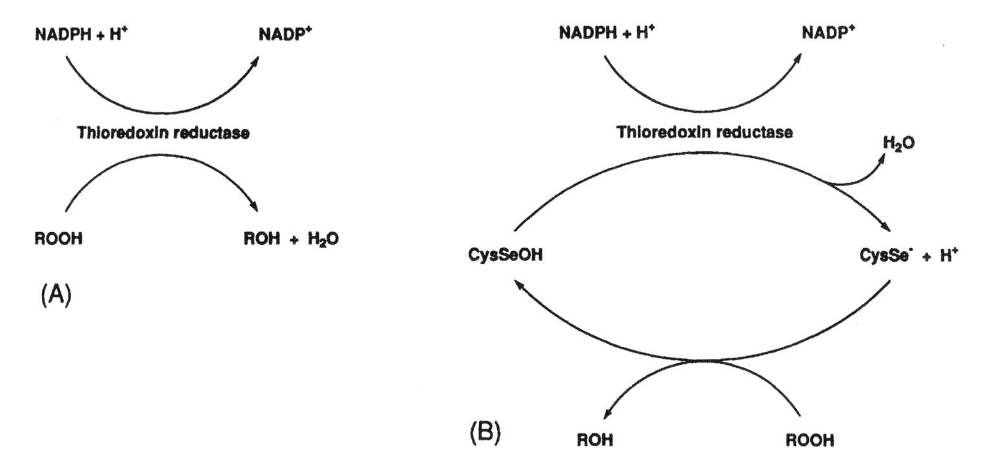

Figure 9. Two pathways for reduction of lipid hydroperoxides involving (A) thioredoxin reductase or (B) thioredoxin reductase plus selenocysteine. (From Ref.68.)

Interestingly, selenocystine, which was directly reduced by human thioredoxin reductase, also stimulated this reaction, leading to the postulation of two pathways for lipid hydroperoxide reduction by thioredoxin reductase as outlined in Figure 9. One is the direct reaction of a lipid hydroperoxide with thioredoxin reductase. The other is a reaction where by selenocysteine (either free or protein bound) stimulated the reaction, acting directly as a peroxidase. The fact that the human and calf thymus thioredoxin reductase displayed lipid hydroperoxide reductase activity and efficiently reduced selenocystine, clearly suggested that the enzyme had unusual properties. As will be explained below, this is of special significance in relation to the discovery that thioredoxin reductase is indeed itself a selenocysteine-containing enzyme (69).

THE THIOREDOXIN SYSTEM AND NITRIC OXIDE

Another reaction of importance in oxidative stress and redox regulation is catalyzed by the thioredoxin system. Upon stimulation of human neutrophils, induction of nitric oxide synthase generates a burst of nitric oxide (NO), which converts a large fraction of the intracellular GSH to *S*-nitrosoglutathione (GSNO) (70). GSNO has been shown to activate the hexosemonophosphate shunt, producing NADPH which is required for an efficient NADPH-oxidase activity (70), which in turn is used to produce superoxide for cytotoxic action of neutrophils. We found that GSNO is a direct substrate for human thioredoxin reductase, and also a substrate for reduced thioredoxin (71). Careful analysis of the reaction with thioredoxin demonstrated that one mole of GSNO is cleaved by one mole of NADPH to give rise to NO, GSH, and superoxide. The cleavage of GSNO to NO by thioredoxin-$(SH)_2$ (Figure 10) was demonstrated by spectral changes in oxymyoglobin, and spectroscopy with metmyoglobin did not indicate formation of nitroxylate anion (71). The direct reaction between GSNO and thioredoxin reductase has not yet been fully analyzed. Thus, the thioredoxin system cleaves GSNO and releases NO, prolonging the action of the short-lived free radical NO by its storage as a stable *S*-nitrosothiol.

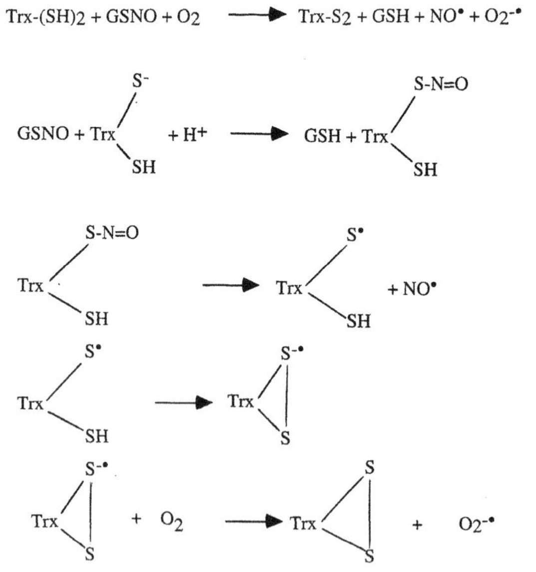

$$\text{Trx-(SH)}_2 + \text{GSNO} + \text{O}_2 \longrightarrow \text{Trx-S}_2 + \text{GSH} + \text{NO}^{\bullet} + \text{O}_2^{-\bullet}$$

$$\text{GSNO} + \text{Trx}\begin{matrix}\text{S}^-\\ \text{SH}\end{matrix} + \text{H}^+ \longrightarrow \text{GSH} + \text{Trx}\begin{matrix}\text{S-N=O}\\ \text{SH}\end{matrix}$$

$$\text{Trx}\begin{matrix}\text{S-N=O}\\ \text{SH}\end{matrix} \longrightarrow \text{Trx}\begin{matrix}\text{S}^{\bullet}\\ \text{SH}\end{matrix} + \text{NO}^{\bullet}$$

$$\text{Trx}\begin{matrix}\text{S}^{\bullet}\\ \text{SH}\end{matrix} \longrightarrow \text{Trx}\begin{matrix}\text{S}^{-\bullet}\\ |\\ \text{S}\end{matrix}$$

$$\text{Trx}\begin{matrix}\text{S}^{-\bullet}\\ |\\ \text{S}\end{matrix} + \text{O}_2 \longrightarrow \text{Trx}\begin{matrix}\text{S}\\ |\\ \text{S}\end{matrix} + \text{O}_2^{-\bullet}$$

Figure 10. Reduction of GSNO catalyzed by reduced thioredoxin. Potential intermediates in the reaction involving reoxidation of radical thioredoxin giving rise to superoxide are also shown. (From Ref. 71.)

Furthermore, when thioredoxin and thioredoxin reductase react with GSNO, there will be production of NO at their localizations, which may be of importance in signal transduction. In other experiments, we have found that GSNO strongly inhibits the disulfide reductase activity of the complete thioredoxin system, again suggesting novel aspects in redox regulation (71).

MAMMALIAN THIOREDOXIN REDUCTASE STRUCTURE AND SELENOCYSTEINE CONTENT

As summarized in Table 2, mammalian thioredoxin reductases show a broad substrate specificity radically different from that of the well-characterized *E. coli* thioredoxin reductase (1–3, 55). We have made major progress in understanding the structure of thioredoxin reductase (72). Our initial studies focused on the enzyme from calf thymus.

Table 2. NADPH-dependent Reactions of Mammalian Thioredoxin Reductase

Trx-S_2 reduction
SeO_3^{2-} and selenocystine reductase
GSNO reductase
Electron donor to plasma GSH-peroxidase
Lipid hydroperoxide reductase
NK-lysin disulfide reduction and inactivation of cytotoxicity

We have sequenced 28 internal peptides covering more than 60% of the protein sequence (72). The peptides displayed no or low homology to *E. coli* thioredoxin reductase but showed high homology to glutathione reductase, including one peptide with identical active-site sequence. The peptide sequences from the calf thymus enzyme were used to identify a rat cDNA clone coding for thioredoxin reductase (originally identified as a putative glutathione reductase). We used the gene from the clone and tried to express the protein in *E. coli*. We obtained a strong 20 kDa band, positive with polyclonal antibodies to rat liver thioredoxin reductase, and the protein obviously had no activity. We then sequenced the complete cDNA for the rat clone containing 2193 nucleotides and identified the coding sequence for rat liver thioredoxin reductase, which consists of 498 residues with a clear homology to human glutathione reductase. The N-terminal of the isolated enzyme was blocked. However, after treatment with 30% acetic acid, the N-terminal sequence corresponding to the -3 position in the enzyme was observed. While our work was in progress, the sequence of a putative clone for human thioredoxin reductase with homology to glutathione reductase, including an identical active site involving two half-cysteine residues separated by four amino acid residues, was published (73). However, no active protein was obtained when expression was attempted in *E. coli* (73).

The coding sequence of our clone showed high homology to the putative human clone (73), confirming that this codes for thioredoxin reductase. However, in the C-terminal end, as shown in Figure 11, there are two putative stop codons separated by a codon for glycine. Since TGA codes for selenocysteine (Sec) provided that a selenocysteine insertion sequence (SECIS) is present, we looked for this in the sequence of our clone and identified a conserved SECIS sequence both in the rat (nt1862–1903) and in the published (73) human clone (nt 2181–2222), both generating typical stem loop structures (74). We also identified a lysine endopeptidase peptide P38 with the sequence Ser-Gly-Gly-Asn-Ile-Leu-Gln-Thr-Gly-Cys-Sec-Gly-OH from calf thymus thioredoxin reductase. Analysis of rat and our calf thymus thioredoxin reductase by atomic absorption spectroscopy showed 0.6 Se per 60 kDa subunit, confirming a selenocysteine content. Thus, the sequences of human, rat, and calf thymus thioredoxin reductase all end with the C-terminal sequence Gly-Cys-Sec-Gly. The sequence published for human thioredoxin reductase (73) should include the two additional residues, Sec and Gly (Figure 12). The reason for the lack of activity in the human protein expressed in *E. coli* is probably absence of insertion of the selenocysteine residue and incomplete translation (73). We have obtained evidence for an essential role of the selenocysteine residue using carboxypeptidase digestion experiments (72).

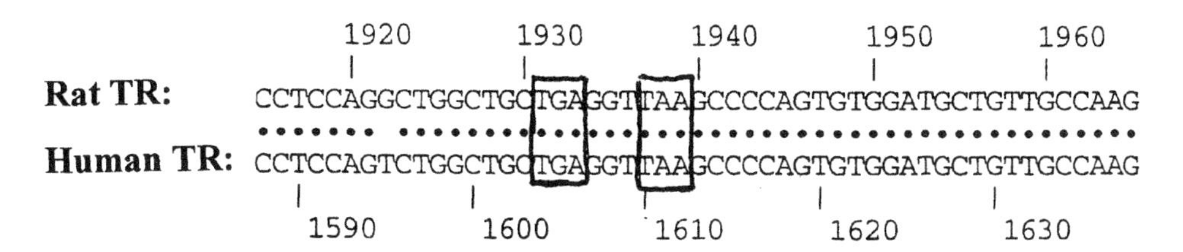

Figure 11. Part of the nucleotide sequences of the putative human thioredoxin reductase (TR) gene (taken from Ref. 73) and the sequence of the rat thioredoxin reductase gene. The two boxes denote two consecutive potential stop codons. TGA and TAA, respectively. Numbers refer to the nucleotide sequences of the cDNA clones.

Rat: -Ser-Gly-Gly-Asp-Ile-Leu-Gln-Ser-Gly-Cys-Sec-Gly-COOH

Calf: -Ser-Gly-Gly-Asn-Ile-Leu-Gln-Thr-Gly-Cys-Sec-Gly-COOH

Human: - Ser-Gly-Ala-Ser-Ile-Leu-Gln-Ala-Gly-Cys-Sec-Gly-COOH

Figure 12. The C-terminal sequence of rat, human and calf thymus thioredoxin reductases. The human sequence has been translated from Ref. 73. The boxed sequence with selenocysteine (Sec) is conserved in all species.

In conclusion, mammalian thioredoxin reductases contain a selenocysteine residue in an unusual C-terminal position and have evolved from glutathione reductase by elongation with some 20 residues. Selenite is an essential constituent of tissue culture media (75, 76), and, since thioredoxin reductase is involved in producing selenide and the enzyme requires selenide for its own synthesis, this may explain the known essential requirement for selenium in cell growth. Obviously the content of selenocysteine may also explain the broad substrate specificity of mammalian thioredoxin reductase, in particular its action as a lipid hydroperoxide reductase. Further studies of the enzyme should provide new insights into protection against oxidative stress as well as into signal transduction.

ACKNOWLEDGEMENTS

We thank Dr. Theresa Stadtman for communicating results regarding human thioredoxin reductase. This work was supported by grants from The Swedish Medical Research Council, The Swedish Cancer Society and the Knut and Alice Wallenberg Foundation.

REFERENCES

1. Holmgren A. Thioredoxin. Annu Rev Biochem 1985; 254: 237–271.
2. Holmgren A. Thioredoxin and glutaredoxin systems. J Biol Chem 1989; 264: 13963–13966.
3. Holmgren A, Björnstedt M. Thioredoxin and thioredoxin reductase. Methods Enzymol 1996; 252: 199–208.
4. Wakasugi N, Tagaya Y, Wakasugi A, Mitsui M, Maeda M, Yodoi J, Tursz T. Adult T-cell leukemia-derived factor/thioredoxin produced by both human T-lymphotropic virus type 1 and Epstein–Barr virus-transformed lymphocytes acts as an autocrine growth factor and synergizes with interleukin-1 and interleukin-2. Proc Natl Acad Sci USA 1990; 87: 8282–8286.
5. Yodoi J, Tursz T. ADF, a growth-promoting factor derived from adult T cell leukemia and homologous to thioredoxin: involvement in lymphocyte immortalization by HTLV-1 and EBV. Adv Cancer Res 1991; 57: 381–411.
6. Ericson ML, Hörling J, Wendel-Hansen V, Holmgren A, Rosén A. Secretion of thioredoxin after in vitro activation of human B cells. Lymphokine Cytokine Res 1992; 11: 201–207.

7. Rosén A, Lundman P, Carlsson M, Bhavani K, Srinivasa BR, Kjellström G, Nilsson K, Holmgren A. A secreted CD4+ T cell line-derived growth factor for cytokine activated normal and leukemic B-cells, identified as thioredoxin. Int Immunol 1995; 7: 625–633.
8. Rubartelli A, Bajetto A, Allavena G, Wollman E, Sitia R. Secretion of thioredoxin by normal and neoplastic cells through a leaderless secretory pathway. J Biol Chem 1992; 267: 24161–24164.
9. Rubartelli A, Bonifaci N, Sitia R. High rates of thioredoxin secretion correlate with growth arrest in hepatoma cells. Cancer Res 1995; 55: 675–680.
10. Nakamura H, Masutani H, Tagaya Y, Yanauchi A, Inamoto T, Nanbu Y, Fujii S, Ozawa K, Yodi J. Expression and growth promoting effect of adult T-cell leukemia derived factor: a human thioredoxin homologue in hepatocellular carcinoma cancer. 1992; 69: 2091–2097.
11. Gasdaska PY, Oblong JE, Cotgreave IA, Powis G. The predicted amino acid sequence of human thioredoxin is identical to that of the autocrine growth factor human adult T-cell derived factor (ADF): thioredoxin m-RNA is elevated in some human tumors. Biochim Biophys Acta 1994; 218: 292–296.
12. Gasdaska JR, Berggren M, Powis G. Cell growth stimulation by the redox protein thioredoxin coccurs by a novel helper mechanism. Cell Growth Differ 1995; 6: 1643–1650.
13. Biguet C, Wakazugi N, Mishal Z, Holmgren A, Chouaib S, Tursz T, Wakazugi H. Thioredoxin increases the proliferation of human B-cell lines through a protein kinase C dependent mechanism. J Biol Chem 1994; 269: 28865–28870.
14. Deiss LP, Kimchi A. A genetic tool used to identify thioredoxin as a mediator of a growth inhibitory signal. Science 1991; 252: 117–120.
15. Holmgren A. Thioredoxin structure and mechanism: conformational changes on oxidation of the active site sulfhydryls to a disulfide. Structure 1995; 3: 239–243.
16. Martin JF. Thioredoxin–a fold for all reasons. Structure 1995; 3: 245–250.
17. Jeng M-F, Campbell AP, Begley T, Holmgren A, Case DA, Wright PE, Dyson HJ. High-resolution solution structures of oxidized and reduced *E. coli* thioredoxin. Structure 1994; 2: 853–868.
18. Jeng M-F, Holmgren A, Dyson HJ. Proton sharing between cysteine thiols in *E. coli* thioredoxin: implications for the mechanism of protein disulfide reduction. Biochemistry 1995; 34: 10101–10105.
19. Qin J, Clore GM, Gronenborn AM. The high-resolution three-dimensional solution structures of the oxidized and reduced states of human thioredoxin structure. Structure 1994; 2: 503–522.
20. Weichsel A, Gaskaska JR, Powis G, Montfont WR. Crystal structures of reduced, oxidized and mutated human thioredoxins: evidence for a regulatory homo-dimer. Structure 1996; 4: 735–751.
21. Gilbert HF. Molecular and cellular aspects of thiol–disulfide exchange. Adv Enzymol 1990; 63: 69–172.
22. Lundström J, Holmgren A. Determination of the reduction-oxidation potential of the thioredoxin-like domains of protein disulfide-isomerase from the equilibrium with glutathione and thioredoxin. Biochemistry 1993; 32: 6649–6655.
23. Matthews JR, Wakazugi N, Virelizier J-L, Yodoi J, Hart RT. Thioredoxin regulates the DNA binding activity of NF-κB by reduction of a disulfide bond involving cysteine 62. Nucleic Acids Res 1992; 20: 3821–3830.
24. Grippo JF, Holmgren A, Pratt WB. Proof that the endogenous, heat-stable glucocorticoid receptor-activating factor is thioredoxin. J Biol Chem 1985; 260: 93–97.
25. Xanthoudakis S, Miao G, Wang F, Pan Y-C E, Curran T. Redox activation of Fos-Jun DNA binding activity is mediated by a DNA repair enzyme. EMBO J 1992; 11: 3323–3335.
26. Holmgren A, Luthman M. Tissue distribution and subcellular localization of bovine thioredoxin determined by radioimmunoassay. Biochemistry 1978; 17: 4071–4077.
27. Rozell B, Hansson H-A, Luthman M, Holmgren A. Immunohistochemical localization of thioredoxin and thioredoxin reductase in adult rats. Eur J Cell Biol 1985; 38: 79–86.

28. Stemme S, Hansson H-A, Holmgren A, Rozell B. Axoplasmic transport of thioredoxin and thioredoxin reducatase in rat sciatic nerve. Brain Res 1985; 359: 140–146.
29. Lippoldt A, Padilla C, Gerst H, Andbjer B, Richter E, Holmgren A, Fuxe K. Localization of thioredoxin in the rat brain and functional implications. J Neurosci 1995; 15: 6747–6756.
30. Endoh M, Kunishita T, Tabira T. Thioredoxin from activated macrophages as a trophic factor for central cholinergic neurons in vitro. Biochem Biophys Res Commun 1993; 192: 760–765.
31. Holmgren A. Hydrogen donor system for *E. coli* ribonucleoside-diphosphate reductase dependent upon glutathione. Proc Natl Acad Sci USA 1976; 73: 2275–2279.
32. Holmgren A, Åslund F. Glutaredoxin. Methods Enzymol 1995; 252: 283–292.
33. Holmgren A, Glutathione-dependent synthesis of deoxyribonucleotides. Purification and characterization of glutaredoxin from *E. coli.* J Biol Chem 1979; 254: 3664–3671.
34. Luthman M, Eriksson S, Holmgren A, Thelander L. Glutathione-dependent hydrogen donor system for calf thymus ribonucleoside diphosphate reductase. Proc Natl Acad Sci USA 1979; 76: 2158–2162.
35. Luthman M, Holmgren A. Glutaredoxin from calf thymus. I. Purification to homogeneity. J Biol Chem 1982; 257: 6686–6690.
36. Padilla CA, Martinez-Galisteo E, Bárcena T, Spyrou G, Holmgren A. Purification from placenta, amino acid sequence, structure comparisons and cDNA cloning of human glutaredoxin. Eur J Biochem 1995; 227: 27–34.
37. Xia T-H, Bushweller JH, Sodano P, Billeter M, Björnberg O, Holmgren A, Wüthrich K. N.M.R. structure of oxidized *E. coli* glutaredoxin: comparison with reduced *E. coli* glutaredoxin and functionally related proteins. Protein Sci 1992; 1: 310–321.
38. Bushweller JH, Billeter M, Holmgren A, Wüthrich K. The NMR solution structure of the mixed disulfide between *E. coli* glutaredoxin (C14S) and glutathione. J Mol Biol 1994; 235: 1585–1597.
39. Chrestensen CA, Eckman CB, Starke DW, Mieyal JJ. Cloning, expression and characterization of human thioltransferase (glutaredoxin) in *E. coli.* FEBS Lett 1995; 374: 25–28.
40. Padilla CA, Lagercrantz J, Bajalica S, Holmgren A. The gene for human glutaredoxin is localized to human chromosome 5q14. Genomics 1996; 32: 455–457.
41. Russel M, Holmgren A. Construction and characterization of glutaredoxin-negative mutants of *E. coli.* Proc Natl Acad Sci USA 1988; 85: 990–994.
42. Åslund F, Ehn B, Miranda-Vizuete A, Pueyo C, Holmgren A. Two additional glutaredoxins exist in *E. coli*: glutaredoxin-3 is a hydrogen donor for ribonucleotide reductase in a thioredoxin–glutaredoxin-1 double mutant. Proc Natl Acad Sci USA 1994; 91: 9813–9817.
43. Schenk H, Klein M, Erdbrügger W, Dröge W, Schulze-Osthoff K. Distinct effects of thioredoxin and other antioxidants on the activation of NF-κB and API. Proc Natl Acad Sci USA 1994; 91: 1672–1677.
44. Galter P, Mihm S, Dröge W. Distinct effects of glutathione disulfide on the nuclear transcription factors kappaB and the activator protein I. Eur J Biochem 1994; 221: 639–648.
45. Thomas JA, Zhao W, Hendrich S, Haddock P. Analysis of cells and tissues for *S*-thiolation of specific proteins. Methods Enzymol 1995; 251: 423–429.
46. Meister A. Glutathione metabolism. Methods Enzymol 1995; 251: 3–7.
47. Cabiscol E, Levine RL. The phosphatase activity of carbonic anhydrase III is reversibly regulated by glutathionylation. Proc Natl Acad Sci USA 1996; 93: 4170–4174.
48. Kinscherf R, Fischbach T, Mihm S, Roth S, Hohenhaus-Sievert E, Weiss C, Edler L, Bärtsch P, Dröge W. Effect of glutathione depletion and oral *N*-acetyl-cysteine treatment on $CD4^+$ cells. FASEB J 1994; 8: 448–451.
49. Staal FJ, Roederer M, Herzenberg LA. Intracellular thiols regulate activation of nuclear factor kappa-B and transcription of human immunodeficiency virus. Proc Natl Acad Sci USA 1990; 87: 9943–9947.

50. Eklund H, Gleason FK, Holmgren A. Structural and functional relations among thioredoxins of different species. Proteins 1991; 11: 13–28.
51. Engström NE, Holmgren A, Larsson A, Söderhäll S. Isolation and characterization of calf liver thioredoxin. J Biol Chem 1974; 249: 205–210.
52. Holmgren A. Bovine thioredoxin system. Purification of thioredoxin reducatase from calf liver and thymus and studies of its function in disulfide reduction. J Biol Chem 1977; 252: 4600–4606.
53. Luthman M, Holmgren A. Rat liver thioredoxin and thioredoxin reductase: Purification and characterization. Biochemistry 1982; 21: 6628–6633.
54. Ren X, Björnstedt M, Shen B, Ericson M, Holmgren A. Mutagenesis of structural half-cystine residues in human thioredoxin and effects on regulation of activity by selenodiglutathione. Biochemistry 1993; 218: 327–334.
55. Naksman G, Krishna TSR, Williams Jr CH, Kuriyan J. Crystal structure of *E. coli* thioredoxin reductase refined at 2 åA resolution. J Mol Biol 1994; 236: 800–816.
56. Holmgren A, Lyckeborg C. Enzymatic reduction of alloxan by thioredoxin and NADPH-thioredoxin reductase. Proc Natl Acad Sci USA 1980; 77: 5149–5152.
57. Hansson H-A, Holmgren A, Rozell B, Täljedal I-B. Immunohistochemical localization of thioredoxin and thioredoxin reductase in mouse exocrine and endocrine pancreas. Cell Tissues Res 1986; 245: 189–195.
58. Greeder GA, Milner JA. Factors influencing the inhibitory effect of selenium on mice inoculated with Ehrlich ascites tumor cells. Science 1980; 209: 825–827.
59. Holmgren A, Kumar S. Reactions of the thioredoxin system with selenium. In Wendel A et al., eds. Proceedings of the 4th International Symposium on Selenium. in Biology and Medicine. Berlin: Špringer-Verlag; 1982: 47–51.
60. Kumar S, Björnstedt M, Holmgren A. Selenite is a substrate for calf thymus thioredoxin reductase and thioredoxin and elicits a large non-stoichiometric oxidation of NADPH in the presence of oxygen. Eur J Biochem 1992; 207: 435–439.
61. Björnstedt M, Kumar S, Holmgren A. Selenodiglutathione is a highly efficient oxidant of reduced thioredoxin and a substrate for mammalian thioredoxin reductase. J Biol Chem 1992; 267: 8030–8034.
62. Stadtman TC. Selenium biochemistry Annu Rev Biochem 1990; 59: 111–127.
63. Björnstedt M, Xue J, Huang W, Åkesson B, Holmgren A. The thioredoxin and glutaredoxin systems are efficient electron donors to human plasma glutathione peroxidase. J Biol Chem 1994; 269: 29382–29384.
64. Nakamura H, Roederer M, Yodoi J, Holmgren A, Herzenberg LA, Herzenberg LA. Elevation of plasma thioredoxin levels in HIV-infected individuals. In Immunol 1996; 8: 603–611.
65. Nakamura H, Padilla CA, Waage J, Valen G, Björnstedt M, Holmgren A. Glutaredoxin remains constant and thioredoxin levels increase in human plasma during cardiac surgery with cardiopulmonary bypass; Submitted. J Clin Invest.
66. Schallrenter KU, Wood JM. The role of thioredoxin reductace in the reduction of free radicals at the surface of the epidermis. Biochem Biophys Res Commun 1986; 136: 630–637.
67. Martin H, Dean M. Identification of a thioredoxin-related protein associated with plasma membranes. Biochem Biophys Res Commun 1991; 175: 123–128.
68. Björnstedt M, Hamberg M, Kumar S, Xue J, Holmgren A. Human thioredoxin reductase directly reduces lipid hydroperoxides by NADPH and selenocystine strongly stimulates the reaction via catalytically generated selenols. J Biol Chem 1995; 270: 11761–11764.
69. Tamura T, Stadtman TC. A new selenoprotein from human lung adenocarcinoma cells: purification, properties, and thioredoxin reductase activity. Proc Natl Acad Sci USA 1996; 93: 1006–1011.
70. Clancy RM, Levartovsky D, Leszczynska-Piziak J, Yegudin J, Abramson SB. Nitric oxide reacts with intracellular glutathione and activates the hexosemonophosphate shunt in human neutrofiles: evidence for *S*-nitrosoglutathione as a bioactive intermediary. Proc Natl Acad Sci

USA 1994; 91: 3680–3684.

71. Nikitovic D, Holmgren A. *S*-Nitrosoglatathione is cleaved by the thioredoxin system with liberation of glutathione and redox regulating nitric oxide. J Biol Chem 1996; 271: 1918--19185.
72. Liangwei Z, Arnér ESJ, Ljung J, Åslund F, Holmgren A. Mammalian thioredoxin reductase contains an essential selenocysteine residue and is structurally similar to glutathione reductase. Manuscript in preparation.
73. Gasdaska PY, Gasdaska JR, Cochran S, Powis G. Cloning and sequencing of a human thioredoxin reductase. FEBS Lett 1995; 373: 5–9.
74. Berry MJ, Banu L, Harney JW, Larsen RP. Functional characterization of the eukaryotic SECIS elements which direct selenocysteine insertion at UGA codons. EMBO J 1993; 12: 3315–3322.
75. McKeehan WL, Hamilton WG, Ham RG. Selenium is an essential trace nutrient for growth of WI-38 diploid human fibroblasts. Proc Natl Acad Sci USA 1976; 73: 2023–2037.
76. Guilbert LJ, Iscove NN. Partial replacement of serum by selenite, transferrin, albumin and lecithin in haemopoietic cell cultures. Nature 1976; 263: 594–595.

24

Thioredoxin/Adult T Cell Leukemia-Derived Factor as the Key Redox Regulator of Signaling

Junji Yodoi, Yoshihisa Taniguchi, Tetsuro Sasada, and Kiichi Hirota
Kyoto University
Kyoto, Japan

INTRODUCTION

Growing evidence has indicated that cellular redox status regulates various aspects of cellular function. Oxidative stress can elicit positive responses, such as cellular proliferation or activation, as well as negative responses, such as growth inhibition or cell death. For example, the observation that hydrogen peroxide activates a transcription factor, NFκB, suggests that reactive oxygen intermediates act as one of the intracellular second-messengers (1). In addition, we and others have demonstrated that oxidative agents can modulate cellular responses via activation of protein tyrosine kinases or inactivation of protein tyrosine phosphatases (2). These findings strongly indicate the importance of cellular redox status as a regulator of various cellular functions.

Thioredoxin (TRX) is a small multifunctional and ubiquitous protein having a redox-active disulfide/dithiol within the conserved active-site sequence -Cys-Gly-Pro-Cys- (3). Since its original isolation from *Escherichia coli* as a hydrogen donor for ribonucleotide reductase, TRX has been purified and characterized from a wide variety of prokaryotic and eukaryotic species. Adult T cell leukemia-derived factor (ADF), which we originally defined as an IL-2 receptor α-chain/Tac inducer produced by human T cell lymphotropic virus-I (HTLV-I)-transformed T cells, has been identified as human TRX (4). TRX/ADF has been reported to have a strong reducing activity through its dithiol group and to possess multiple biological functions via the modulation of intracellular redox status (5). For example, we have demonstrated that TRX/ADF acts as an autocrine growth factor in HTLV-I- or Epstein–Barr virus (EBV)-transformed cells (6,7) and protects cells against various oxidative stresses, such as hydrogen peroxide- or TNF-α-induced cytotoxicity (8,9). In addition, TRX/ADF has been shown to be implicated in the regulation of some redox-sensitive molecules, including NFκB, AP-1, and the glucocorticoid receptor (10–12). Here we discuss the importance of TRX/ADF as one of the key redox regulators in the cellular responses against various stresses.

TRX/ADF AS A STRESS-INDUCIBLE PROTEIN

Originally, the expression of TRX/ADF was reported to be upregulated in HTLV-I- and EBV-transformed lymphoid cells (6,7). However, TRX/ADF has been shown to be a stress-inducible protein whose expression is enhanced not only by viral infections but also by a variety of cellular stresses (9, 13–15).

In particular, TRX/ADF can be strongly induced by oxidative stresses at the transcriptional level. We found that oxidative agents, such as hydrogen peroxide, diamide, and ultraviolet (UV) irradiation, enhanced the mRNA level of TRX/ADF in the Jurkat human T cell line. In addition, the analysis using a CAT expression vector under the control of the TRX/ADF promoter region revealed that the CAT activity was markedly increased after exposure to these oxidative agents. Furthermore, precise analyses of the TRX/ADF promoter region, including CAT assays with a series of deletion mutants of the TRX/ADF promoter and DNase I footprint analysis, showed a novel *cis*-regulatory element reactive with oxidative agents, in addition to the binding sites for other transcription factors such as AP-1, CREB, and NFκB. The sequence in this region had no homology with the consensus sequences of any known DNA-binding factors. Therefore, it is suggested that the expression of TRX/ADF is enhanced through a novel *cis*-regulatory element for oxidative stresses, and that new DNA-binding factors may be involved in this pathway. The precise mechanism of TRX/ADF induction is now under investigation.

More recently, we have found that various other cellular stimuli, such as a viral gene product (HTLV-I Tax), sex hormone (estrogen), cytokine (IL-2), prostaglandins, and anticancer agents (cisplatin and bleomycin) (14), induce the enhanced expression of TRX/ADF. These observations suggest that TRX/ADF has important roles in the cellular responses against various cellular stresses.

REDOX REGULATION OF INTRACELLULAR SIGNALING BY TRX/ADF

TRX/ADF has been found to translocate from the cytosol into the nucleus as a result of various stimuli. For example, we demonstrated that UV irradiation induced nuclear translocation of TRX/ADF in human keratinocytes. Although normal and transformed human keratinocytes (HSC-1) expressed TRX/ADF mainly in the cytosol, they showed diffuse nuclear localization of TRX/ADF after UV irradiation. In addition, nuclear translocation of TRX/ADF was observed in human cervical cells (HeLa) after treatment with phorbol myristate acetate. Although the precise mechanism of nuclear transport of TRX/ADF remains to be clarified, these observations suggest that TRX/ADF functions as one of the key regulators of signaling in the cellular responses against various stresses.

It is suggested that TRX/ADF is involved in various cellular processes as a potent endogenous thiol-related reducing agent by modulating protein–protein or protein–nucleic acid interactions through the reduction/oxidation of protein cysteine residues. We have recently demonstrated that TRX/ADF can regulate the biological activities of its target molecules, such as NFκB, p50, and redox factor-1 (Ref-1)/APEX/HAP-1, through direct interaction. Therefore, TRX/ADF, which translocates from the cytosol into the nucleus by cellular stresses, may be involved in the expression of various genes through the redox regulation of its target molecules, such as some transcription factors. Further studies to dissect the regulatory roles of TRX/ADF and its target molecules may clarify intracellular signaling pathways in the responses against various stresses.

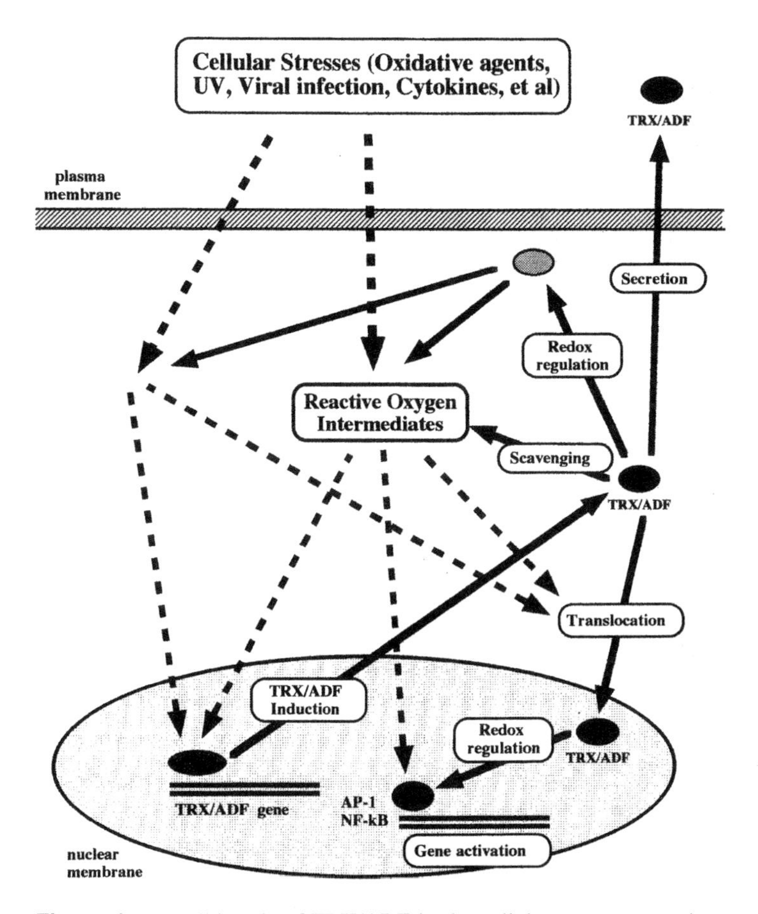

Figure 1. Possible role of TRX/ADF in the cellular responses against various stresses.

CONCLUSION

TRX/ADF is a stress-inducible protein with multiple biological functions. TRX/ADF translocates from the cytosol into the nucleus by a variety of cellular stresses to regulate the biological activities of its target molecules, suggesting that TRX/ADF may be involved in the cellular responses against various stresses (Figure 1). Further studies to clarify the regulatory roles of TRX/ADF and its target molecules may facilitate the elucidation of intracellular signaling pathways.

REFERENCES

1. Schreck R, Rieber P, Baeuerle PA. Reactive oxygen intermediates as apparently widely used messengers in the activation of the NF-κB transcription factor and HIV-1. EMBO J 1991; 10: 2247–2258.
2. Nakamura K, Hori T, Sato N, Sugie K, Kawakami T, Yodoi J. Redox regulation of a src family protein tyrosine kinase p56^{lck} in T cells. Oncogene 1993; 8: 3133–3139.
3. Holmgren A. Thioredoxin. Annu Rev Biochem 1985; 54: 237–271.

4. Tagaya Y, Maeda Y, Mitsui A, Kondo N, Matsui H, Hamuro J, Brown N, Arai K-i, Yokota T, Wakasugi H, Yodoi J. ATL-derived factor (ADF), an IL-2 receptor/Tac inducer homologous to thioredoxin: possible involvement of dithiol-reduction in the IL-2 receptor induction. EMBO J 1989; 8: 757–764.
5. Yodoi J, Uchiyama T. Diseases associated with HTLV-I: virus, IL-2 receptor dysregulation and redox regulation. Immunol Today 1992; 13: 405–411.
6. Yamauchi A, Masutani H, Tagaya Y, Wakasugi N, Mitsui A, Nakamura H, Inamoto T, Ozawa K, Yodoi J. Lymphocyte transformation and thiol compounds; the role of ADF/thioredoxin as an endogenous reducting agent. Mol Immunol 1992; 29: 263–270.
7. Wakasugi H, Rimsky R, Mahe Y, Kamel AM, Fradelizi D, Tursz T, Bertoglio J. Epstein–Barr virus-containing B-cell line produced an interleukin 1 that it uses as a growth factor. Proc Natl Acad Sci USA 1987; 84: 804–808.
8. Matsuda M, Mastutani H, Nakamura H, Miyajima S, Yamauchi A, Yonehara S, Uchida A, Irimajiri K, Horiuchi A, Yodoi J. Protective activity of ATL-derived factor (ADF) against tumor necrosis factor-dependent cytotoxicity on U937 cells. J Immunol 1992; 147: 3837–3841.
9. Nakamura H, Matsuda M, Furuke K, Kitaoka Y, Iwata S, Toda K, Inamoto T, Yamaoka Y, Ozawa K, Yodoi J. Adult T cell leukemia-derived factor/human thioredoxin protects endothelial F-2 cell injury caused by activated neutrophils or hydrogen peroxide. Immunol Lett 1994; 42: 75–80.
10. Okamoto T, Ogiwara H, Hayashi T, Mitsui A, Kawabe T, Yodoi J. Human thioredoxin/adult T cell leukemia-derived factor activates the enhancer binding protein of human immunodeficiency virus type 1 by thiol redox control mechanism. Int Immunol 1992; 4: 811–819.
11. Abate C, Patel L, Rauscher FJ III, Curran T. Redox regulation of Fos and Jun DNA-binding activity in vitro. Science 1990; 249: 1157–1161.
12. Grippo JF, Tienrungroj W, Dahmer MK, Housley PR, Pratt WB. Evidence that the endogenous heat-stable glucocorticoid receptor-activating factor is thioredoxin. J Biol Chem 1993; 258: 13658–13664.
13. Sachi Y, Hirota K, Masutani H, Toda K, Okamoto T, Takigawa M, Yodoi J. Induction of ADF/TRX by oxidative stress in keratinocytes and lymphoid cells. Immunol Lett 1995; 44: 189–193.
14. Sasada T, Iwata S, Sato N, Kitaoka Y, Hirota H, Nakamura K, Nishiyama A, Taniguchi Y, Takabayashi A, Yodoi J. Redox control of resistance to *cis*-diamminedichloroplatinum (II) (CDDP): protective effect of human thioredoxin against CDDP-induced cytotoxicity. J Clin Invest; 1996; 97: 2268–2276.
15. Nakamura H, De Rosa S, Roederer M, Anderson MT, Dubs JG, Todoi J, Holmgren A, Herzenberg LA, Herzenberg LA. Elevation of plasma thioredoxin levels in HIV-infected individuals. Int Immunol 1996; 8: 603–611.

25

Therapeutic Potential of the Antioxidant and Redox Properties of α-Lipoic Acid

Chandan K. Sen, Sashwati Roy, and Lester Packer
University of California–Berkeley
Berkeley, California

Thiols, ubiquitously distributed in aerobic cells, have been linked with a variety of key physiological functions (1,2). Disulfide linkages are critical determinants of protein structure. The cellular gutathione pool not only serves antioxidant functions but also serves as a reservoir of cysteine for protein synthesis. Fundamental physiological properties of cells such as growth and proliferation are also known to be regulated by cellular glutathione (1,2). The oxidation–reduction, or redox, states of protein-thiols are suggested to regulate a number of key signal transduction processes (3). Much of our current interest in thiols with respect to clinical disorders has stemmed from consistent observations that several pathophysiological conditions, such as certain forms of cancer and AIDS, are associated with lowered thiol status in some cells such as lymphocytes (4–7). Correction of such perturbations in cellular thiol homeostasis have resulted in beneficial clinical measures of outcome. Here the antioxidant and redox regulatory properties of the thiol-replenishing drug α-lipoic acid and discussed, highlighting the drug's remarkable therapeutic potential.

α-Lipoate, also known as thioctic acid, 1,2-dithiolane-3-pentanoic acid, 1,2-dithiolane-3-valeric acid, or 6,8-thioctic acid, has generated considerable clinical interest as a thiol-replenishing and redox-modulating agent. Oxidant–antioxidant reactions are essentially oxidation–reduction reactions in which reactive forms of oxygen are reduced and thus scavenged by electrons from antioxidants; in the process, the antioxidant is oxidized to its functionally inert form. Effective functioning of redox antioxidants requires the recycling of the oxidized form of antioxidant to its potent reduced form. The reducing power of cells is determined by the reducing equivalent status. The redox cycling antioxidant defense network involving the role of lipoate has been reviewed (8,9).

Cellular reducing equivalents such as NADH or NADPH, produced as a result of cellular metabolism, serve as cofactors of enzymes such as reductases or dehydrogenases in bioreduction processes. For physiological antioxidants like glutathione (GSH), glutathione reductase utilizes NADPH to reduce oxidized glutathione disulfide (GSSG). The problem, however, with GSH is that it is poorly bioavailable to mammalian cells and tissues and does not serve as an effective supplement per se (10). Among the several

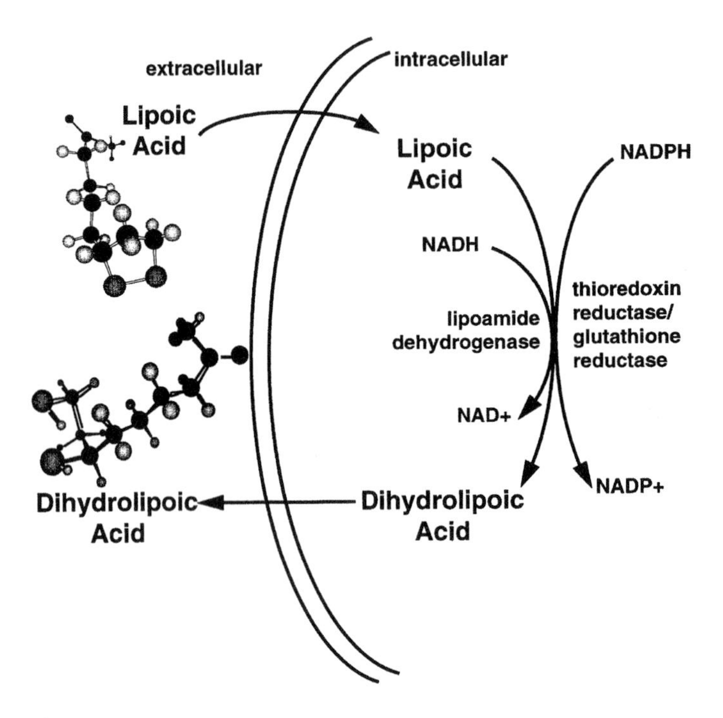

Figure 1. Cellular reduction of exogenous α-lipoic acid to dihydrolipoic acid. Lipoate rapidly enters the cell and is reduced by NADH- and/or NADPH-dependent enzymatic mechanisms to dihydrolipoate; this strong reductant is then released to the extracellular medium.

agents that have been tested for their efficacy as thiol-replenishing drugs, *N*-acetylcysteine (NAC) and lipoate have been proven to be safe and effective even when clinical measures of outcome are considered (5,9,11–17). A distinct property of lipoate is that it is a "metabolic antioxidant" in that enzymatic systems in human cells treat it as a substrate for bioreduction (8) (Figure 1). Thus, supplemented lipoate is reduced to dihydrolipoate at the expenses of cellular reducing equivalents, e.g., NADH and NADPH. As more of these reducing equivalents are utilized, the rate of cellular metabolism is increased to cater to the enhanced demand. Thus, a unique property of lipoate as a supplement is that it can harness the power of the cell's own metabolic processes for its recycling and potency.

LIPOYL RESIDUES: INTEGRAL ROLE IN OXIDATIVE METABOLISM

As early as the 1950s, α-lipoate had been identified as an essential cofactor in oxidative metabolism (18). Biologically, lipoate exists as lipoamide in at least five proteins, where it is covalently linked to a lysyl residue. Four of these proteins are found in α-keto-acid dehydrogenase complexes, the pyruvate dehydrogenase complex, the branched chain keto-acid dehydrogenase complex, and the α-ketoglutarate dehydrogenase complex. Three of the lipoamide-containing proteins are present in the E2 enzyme dihydrolipoyl acyltransferase, which is different in each of the complexes and specific for the substrate of the complex. One lipoyl residue is found in protein X, which is the same in each complex. The fifth lipoamide residue is present in the glycine cleavage system (19).

Lipoamide-containing enzymes of the α-keto-acid dehydrogenase complexes oxidatively decarboxylate their substrates, producing NADH. Human pyruvate dehydrogenase complex contains two lipoyl domains. Protein X is suggested to play a structural role in allowing a proper functioning of the pyruvate dehydrogenase complex (20). Each complex comprises about six copies of protein X that contain a lipoate-bound domain. The glycine cleavage system catalyzes the oxidation of glycine to CO_2 and NH_3, producing NADH and 5,10-methylenetetrahydrofolate as a result (8).

Dihydrolipoamide dehydrogenase, an integral component of α-keto-acid dehydrogenase complexes and the glycine cleavage system, oxidizes dihydrolipoamide to lipoamide. In the presence of lipoate, this reaction is also driven in the reverse direction at the expenses of NADH (9). It seems likely that dihydrolipoamide dehydrogenase may also act independently of the α-keto-acid dehydrogenase complex and the glycine cleavage system. Plasma membrane-associated dihydrolipoamide dehydrogenase has been found in *Trypanosoma brucei* and rat adipocytes (21,22). The exact function of this membrane-associated enzyme is as yet unclear; however, in eukaryotes the enzyme has been connected with insulin-stimulated hexose transport. Although the possible role of dihydrolipamide dehydrogenase in hexose transport is not yet clear, several studies have reported enhanced glucose uptake following exogenous lipoate treatment (23–27).

REDUCTION OF EXOGENOUS LIPOATE TO DIHYDROLIPOATE

In developing an understanding of the therapeutic potential of lipoate supplementation, much of the current interest is focused on the fate of exogenously supplemented non-protein-bound lipoate in cell culture, animal, and human experimental systems. Lipoate, in its native form, contains a disulfide bond. Reduction of this disulfide results in the conversion of lipoate to the corresponding thiol, dihydrolipoate (DHLA). Exogenous lipoate has been observed to be reduced to DHLA in several biological systems including mitochondria and bacteria, perfused rat liver, isolated hepatocytes, erythrocytes, keratinocytes, and lymphocytes (28). The mitochondrial E3 enzyme dihydrolipoyl dehydrogenase reduces lipoate to DHLA at the expense of NADH. The enzyme shows a marked preference for the naturally occuring (*R*)-enantiomer of lipoate (29). Lipoate is also a substrate for the NADPH-dependent enzyme glutathione reductase (30). Glutathione reductase shares a high degree of structural homology with lipoamide dehydrogenase. Both are homodimeric enzymes with 50 kDa subunits conserved between all species. In contrast to dihydrolipoyl dehydrogenase, however, glutathione reductase exhibits a preference for the (*S*)-enantiomer of lipoate. Although lipoate is recognized by glutathione reductase as a substrate for reduction, the rate of reduction to DHLA is much slower than that of the natural substrate glutathione disulfide. Whether lipoate is reduced in a NADH or NADPH dependent mechanism is largely tissue specific.

Thioredoxin reductase catalyzes the NADPH-dependent reduction of oxidized thioredoxin. Recently it has been found that thioredoxin reductase from calf thymus and liver, human placenta, and rat liver efficiently reduces both lipoate and lipoamide with Michaelis–Menten type kinetics in NADPH-dependent reactions (31). Under similar conditions at 20°C, pH 8.0, mammalian thioredoxin reductase reduced lipoic acid 15 times more efficiently than the corresponding NADH-dependent lipoamide dehydrogenase. The biological significance, i.e., the relative contribution, of the three different enzymes found to reduce lipoate in mammalian cells is tissue and cell specific depending on the presence or absence of mitochondrial activity and of oxidized thioredoxin and GSSG.

Studies with Jurkat T cells have shown that, when added to the culture medium, lipoate readily enters the cell where its is reduced to its dithiol form, DHLA. DHLA accumulated in the cell pellet, and when monitored over a 2 h interval the dithiol was released to the culture medium (Figure 1). As a result of lipoate treatment of the Jurkat T cells and human neonatal fibroblasts, accumulation of DHLA in the culture medium was observed (28).

ANTIOXIDANT PROPERTIES OF LIPOATE AND DIHYDROLIPOATE

As early as in 1959, Rosenburg and Culik (32) suggested an antioxidant function of lipoate. They observed that administration of lipoate prevented symptoms of scurvy in vitamin C-deficient guinea-pigs. They also observed that lipoate was able to prevent vitamin E deficiency symptoms in rats fed with a tocopherol-deficient diet. Lipoate scavenges several reactive species including hydroxyl radicals, hydrogen peroxide, hypochlorous acid, and singlet oxygen. In addition, lipoate has transition-metal chelation properties by virtue of which it may avert the transformation of relatively weak radical such as the superoxide anion to the deleterious hydroxyl radical (9).

The redox potential of the lipoate–DHLA couple is –320 mV (33). Thus, DHLA is a strong reductant capable of chemically reducing GSSG, the redox potential of the GSSG–GSH couple being –240 mV (33). Similar to lipoate, DHLA also scavenges hypochlorous acid, peroxyl, superoxide, hydroxyl, and nitric oxide radicals.

Hydroxyl radicals generated by transition-metal catalysis have been effectively scavenged by lipoate at a 1 mM concentration (34). The rate constant for the scavenging of hydroxyl radicals by lipoate has been estimated to be $4.7 \times 10^{10}\,\mathrm{M}^{-1}\,\mathrm{s}^{-1}$ (35). UVA irradiation of NPIII [*N,N′*-bis(2-hydroperoxy-2-methoxyethyl)-1,4,5,8-napthalenetetra-carboxylic diimide] generates hydroxyl radicals (36). This experimental system has been used to confirm that lipoate is indeed an effective hydroxyl radical scavenger, and that results observed in previous experimentals were not simply due to the metal-chelation properties of lipoate. In this metal-independent experimental system, the rate constant for the scavenging of hydroxyl radicals by lipoate was estimated to be $1.92 \times 10^{10}\,\mathrm{M}^{-1}\,\mathrm{s}^{-1}$. Similar to lipoate, DHLA proved to be an effective scavenger of hydroxyl radical generated by NPIII. At 0.5 mM, DHLA eliminated hydroxyl radical generation as detected by the DMPO-OH (DMPO = 5,5-dimethylpyrroline*N*-oxide) electron spin resonance signal (37).

The presence of free transition-metal ions in biological systems may catalyze Fenton and Haber–Weiss reactions that lead to the production of highly reactive hydroxyl radicals from relatively less reactive superoxides or hydrogen peroxide. Lipoate protected against Cd^{2+}-induced toxicity in isolated hepatocytes (38). Whether lipoate or DHLA acted as the metal chelator in this system was not clear however. Inconsistent results were obtained when the iron-chelating property of lipoate was tested. Lipoate protected against singlet oxygen-induced DNA strand breaks (39). This effect of lipoate was less pronounced in the presence of EDTA, indicating that at least partially the protective effect of lipoate was transition-metal ion chelation-dependent. In another independent study, lipoate protected against site-specific degradation of deoxyribose by a $FeCl_3/H_2O_2$/ ascorbate system (35). This effect of lipoate could be attributed to its ability to bind and remove iron ions attached to deoxyribose. In contrast to the above-mentioned observations, where it seemed probable that lipoate might have iron-chelating properties, lipoate failed to protect rat liver microsomal lipid peroxidation that was triggered by Fe_3SO_4 (40,41). More direct studies addressing the bivalent ion binding ability of lipoate revealed that it forms stable complexes with Mn^{2+}, Cu^{2+}, and Zn^{2+} (42). Because these complexes were formed almost entirely with the carboxylate group, it may be assumed that the metal-chelation property of lipoate will not be influenced by its redox state. Lipoate effectively prevented Cu^{2+}-induced ascorbic acid oxidation and liposomal peroxidation (43).

Hypochlorous acid has strong oxidant properties and at concentrations as low as 50 μM can completely inactivate α_1-antiproteinase enzyme. Independent studies have consistently shown that at a low concentration (50 μM) lipoate can protect α_1-antiproteinase from HOCI-induced inactivation (35). Singlet oxygen-scavenging properties of lipoate have been observed consistently in a number of different experimental systems including rubrene autoxidation (44), photosensitized methylene oxidation (45), and endoperoxide thermolysis (46). Under physiologically relevant conditions, singlet oxygen generated by thermolysis of endoperoxide resulted in DNA strand breaks. Lipoate clearly protected by scavenging singlet oxygen with a rate constant of $1.38 \times 10^8\,M^{-1}\,s^{-1}$ (46).

The interaction of lipoate with nitric oxide has been studied in RAW 264.7 macrophages that were activated with lipopolysaccharide in the presence or absence of lipoate. At 0.2 mM, lipoate clearly decreased lipopolysaccharide-induced nitrite production. Almost 50% inhibition of nitrite generation was observed in the presence of 0.5 mM lipoate. In this experimental system it is not possible to conclude whether lipoate influenced nitric oxide synthase enzyme expression or reacted with nitric oxide directly. However, in another experiment where nitric oxide was generated in a cell-free system by sodium nitroprusside, lipoate decreased nitrite generation as well suggesting a possible direct interaction of lipoate and NO. Similar to lipoate, DHLA appears to react with nitric oxide as well (H. Kobuchi, personal communication). Superoxides react with nitric oxide to form peroxynitrite that can initiate lipid peroxidation reactions. Lipoate has been shown to scavenge peroxynitrite as well (47).

Using the azo-esters AAPH [2,2′-azobis(2-amidinopropane) hydrochloride] or AMVN [2,2′-azobis(2,4-dimethylvaleronitrile)] as sources for peroxyl radical generation in aqueous or lipid phases, it has been shown that DHLA efficiently scavenges this radical (35,48). The rate constant for the scavenging of $CCl_3O_2^{\bullet}$ has been estimated to be $2.7 \times 10^7\,M^{-1}\,s^{-1}$ (35).

DHLA effectively scavenged superoxides generated by a xanthine–xanthine oxidase system. The rate constant for this reaction has been estimated to be $3.3 \times 10^5\,M^{-1}\,s^{-1}$ (34) This effect was confirmed in another independent experiment where superoxides where generated by a xanthine–xanthine oxidase system and the efficiency of DHLA to scavenge superoxides was tested by competition with superoxide-induced epinephrine oxidation (49). In this system the rate constant for the scavenging of superoxides by DHLA was estimated to be $7.3 \times 10^5\,M^{-1}\,s^{-1}$, in agreement with the previous estimation. In contrast to the observations described above, DHLA proved to be ineffective in scavenging superoxide generated by hypoxanthine–xanthine oxidase as measured by superoxide-dependent reduction of nitrobule tetrazolium (35). More recently, the superoxide scavenging activity of DHLA was evaluated using 2-methyl-6-[*p*-methoxyphenyl]-3,7-dihydroimidazo[1,2-a]pyrazine-3-one as a chemiluminescent superoxide probe (S. Matsugo et al., unpublished). The reaction rate of DHLA toward superoxide was estimated to be $1.43 \times 10^5\ \mathrm{M}^{-1}\ \mathrm{s}^{-1}$.

REPLENISHING OTHER ANTIOXIDANT POOLS

During the oxidation–reduction interaction between an antioxidant and reactive oxygen, electrons are donated by the antioxidant to reduce and scavenge the reactive species. As a result, the antioxidant itself becomes oxidized and is required to be recycled to its native reduced form for functional potency. DHLA is a strong reductant and is thus capable of recycling some such oxidized antioxidants. Experimental evidence suggests that DHLA can directly regenerate ascorbate and indirectly regenerate vitamin E from their respective oxidized radical forms (9).

Vitamin E is a strong lipid-phase antioxidant that protects biological membranes from lipid peroxidation reactions. In biological membranes, vitamin E is present in a low molar ratio compared to phospholipids, which are highly susceptible to oxidative damage. On average, for every 1000–2000 molecules of phospholipid only one molecule of vitamin E is present for antioxidant defense. Lipid peroxyl radicals–reactive oxygen species with sufficient energy to initiate lipid peroxidation chain reactions–are continuously produced in membranes at an estimate rate of 1–5 nmol of membrane protein per minute. So how does the membrane handle this much threat and still maintain its integrity especially when the availability of lipid-phase antioxidant is so limited? This apparent paradox may be explained by the fact that vitamin E is continuously recycled as it acts as an antioxidant. In this way, the vitamin E defense pool in the membrane is continuously rejuvenated. It is known that vitamin C, ubiquinols, and thiols are capable of recycling vitamin E (50–51).

Microsomal lipid peroxidation is a commonly used model to study oxidative lipid damage. DHLA protected against microsomal lipid peroxidation, but only when vitamin E was present (52). This suggests that DHLA itself could not directly break the lipid peroxidation chain reaction, but potentiated the ability of vitamin E to do so. Lipoate was not effective in this system. In a separate study it has been observed that DHLA decreased tocopheroxyl radical ESR signal in liposomes exposed to UV irradiation (unpublished data). Other studies have suggested that DHLA may also recycle vitamin E by reducing oxidized glutathione chemically. GSH thus regenerated recycles vitamin E. In support of this, a combination of DHLA and GSSG, but not DHLA alone, protected against Fe^{2+}–ascorbate-induced lipid peroxidation (41). Additionally, DHLA also seems to be able to recycle vitamin E by recycling ascorbate. Electron spin resonance spectroscopic studies with DOPC liposomes revealed that ascorbyl radical, generated as a result of ascorbate oxidation by chromanoxyl radical, was recycled by DHLA (48). In this way more ascorbate is made available for the direct recycling of vitamin E. It was also demonstrated that DHLA interacts with NADPH- or NADH-dependent electron transport chains to recycle vitamin E (48). Ascorbate-mediated regeneration of vitamin E has also been observed in human low-density lipoprotein and erythrocyte membranes (48,53).

Apart from GSH and ascorbate-mediated recycling of vitamin E, DHLA may also recycle vitamin E in a ubiqinol-dependent mechanism (9). Lipoate supplementation increased tissue ubiquinol content even in an oxidative stress situation. It is known that ubiquinol is effective in recycling vitamin E. Thus, it may be concluded that DHLA does recycle vitamin E and may therefore enhance lipid-phase antioxidant defenses. There are a number of parallel mechanisms of doing this, and the relative participation of each of these individual mechanisms in vivo is yet unclear.

That lipoate may protect against vitamin E or vitamin C deficiency was observed as early as 1959 (32). Similar results were observed in another recent study where lipoate supplementation protected tocopherol-deficient hairless mice (54). These observations may be explained by the ascorbate and tocopherol recycling ability of DHLA, and by the ability of lipoate to directly scavenge reactive oxygen and thus spare vitamins E and C consumption during oxidant challenge. Following supplementation, all major tissues showed the presence of lipoate and DHLA in detectable amounts.

Lipoate has proven to be one of the best feasible means to bolster cellular and tissue glutathione pools. Glutathione is the master physiological antioxidant that orchestrates the biological interaction of several other antioxidants. Pharmacological enhancement of tissue glutathione levels is a challenging task that can be satisfactorily achieved only by a very few agents. Among these agents that have been clinically tested are esterified GSH, *N*-acetylcysteine and lipoate. When evaluated from a concentration–effect standpoint, lipoate has a clear advantage.

When lipoate was infused into the liver, thiols were detected in the perfusate (55). The exact nature of these thiols was not identified, however. When lipoate was added to murine neuroblastoma and melanoma cell lines, a dose-dependent increase in cellular GSH content was observed in the range of 30–70% compared to untreated controls (56). Consistent results were obtained in vivo where mice were injected with doses of 4,8, or 16 mg/kg lipoate for 11 days. Glutathione levels in lung, liver, and kidney of these mice increased significantly (56). Lipoate is ability to increase cellular glutathione has also been studied extensively in Jurkat T cells as well as human peripheral blood lymphocytes. Lipoate treatment increased cellular glutathione content of Jurkat T cells to 1.5-fold of the control value. Recently we adapted a flow cytometric method to estimate cellular glutathione in lipoate-treated Jurkat T cells and peripheral blood lymphocytes. In Jurkat cells, the effect of lipoate on increasing cellular GSH was observed even at a concentration as low as 10 μM (our unpublished observations)

The rate-limiting step in the biosynthesis of glutathione is the availability of cysteine as a substrate within the cell. Cysteine, in its reduced form, is highly unstable and so cell culture media contain only the oxidized form of cysteine, i.e. cystine. Cystine is taken up by the cell by specific transport mechanisms and serves as a precursor of cysteine, which is finally utilized for GSH synthesis. Studies in our laboratory have shown that, following treatment of Jurkat T cells, lipoate is rapidly reduced to DHLA within the cell. After this, DHLA accumulated inside the cell is released to the extracellular cell culture medium where, because of its strong, reducing power, DHLA reduces cystine in the medium to cysteine. The cellular uptake mechanism for cysteine is approximately 10 times faster than that for cystine. Thus, DHLA markedly improves cysteine availability within the cell, resulting in acceleration of glutathione synthesis.

NFκB REGULATORY EFFECTS AND ITS THERAPEUTIC RELEVANCE

Redox changes in cells, as during oxidative stress, trigger molecular responses (3). NF κB is a well-characterized redox-sensitive transcription factor, the function of which has been related to a number of clinical disorders. The activity of NFκB is inducible in response to a wide range of stimuli including peroxide, cytokines, phosphatase inhibitors, and viral products (57). It is suggested that reactive oxygen species may serve as a common intracellular messenger for NFκB activation in response to a diverse range of stimuli (3,57).

A member of the Rel family of transcription factors, NFκB in its dormant form is localized in the cytosol as hetero- or homodimeric proteins associated with an inhibitory protein, IκB. Following appropriate stimulation of cells IκB is phosphorylated, dissociated from the dimer, and degraded. As a result, NFκB proteins are able to translocate to the nucleus under the guidance of a nuclear localization signal. In the nucleus, NFκB binds to a consensus κB site under reducing conditions. Depending on the nature of the NFκB protein bound to the DNA, transcription is switched on. A large number of genes, including those of cytokines, adhesion molecules, growth factors, immunoreceptors, NO synthase, some viral genes, and IκB, are regulated by NFκB activity (57).

Several antioxidants and reducing agents including ebselen, β-mercaptoethanol, pyrrolidinedithiocarbamate, deferroxamine, dithiolthione, *N*-acetylcysteine, vitamin E derivatives, and butylated hydroxyanisole have proven effective for their ability to inhibit NFκB activation (58–64). Treatment of Jurkat T cells with lipoate suppressed phorbol ester or tumor necrosis factor-α (TNF-α)-induced activation of NFκB in a dose-dependent manner (61,64). This NFκB inhibitory effect was also seen with DHLA. Direct addition of DHLA-to the cell culture medium suppressed TNF-α-induced NFκB activation (65). Both (*R*)- and (*S*)-enantiomers of lipoate were effective with respect to the NFκB inhibitory function (61,64). It has been observed that the ability of lipoate to inhibit NFκB activation is not dependent on its ability to increase cellular glutathione. Simultaneous treatment of cells with the glutathione synthesis inhibitor buthionine sulfoximine and lipoate for 18 h produced cells in which the glutathione pool was decreased by 95%. Lipoate treatment could not increase glutathione levels in these cells' however, lipoate was able to inhibit NFκB activation induced by phorbol ester, TNFκα or hydrogen peroxide (3). Thus, the molecular effects of lipoate are not simply mediated by enhanced cellular glutathione (Figure 2). Lipoate appears to act on certain specific molecular loci to produce these NFκB regulatory effects.

Although reactive oxygen species have been suggested to function as a common intracellular messenger in the NFκB activation cascade in response to a variety of stimuli, little is known about the precise mode of action. In our laboratory we have tested different cell systems as tools to address this issue. Jurkat T cells are not responsive to hydrogen peroxide with respect to NFκB activation; however, in a subclone of these cells developed by Dr. Patrick Baeuerle (Frieburg, Germany) and named Wurzburg cells, hydrogen peroxide treatment results is marked activation of NFκB (17,66). We used these two related cell lines with contrasting peroxide sensitivity to reveal the possible factors that are responsible for the oxidant sensitivity of Wurzburg cells.

Flow cytometric determination of intracellular Ca^{2+} concentration ($[Ca^{2+}]_i$) revealed that 0.25 mM hydrogen peroxide treatment results in a marked calcium flux within the cell (66). Using extracelluar calcium chelators we observed that this flux is mainly contributed by calcium released from intracellular stores. Although Wurzburg cells are derived from Jurkat T cells, a marked differences in the nature of this oxidant-induced calcium flux was noted in the two cell types. In Jurkat, the calcium flux was rapid and transient. Within 10–15 min after oxidant treatment, intracellular calcium concentration was restored to pretreatment levels. In contrast, the calcium response in Wurzburg cells was slower in kinetics and was sustained for a longer time (66).

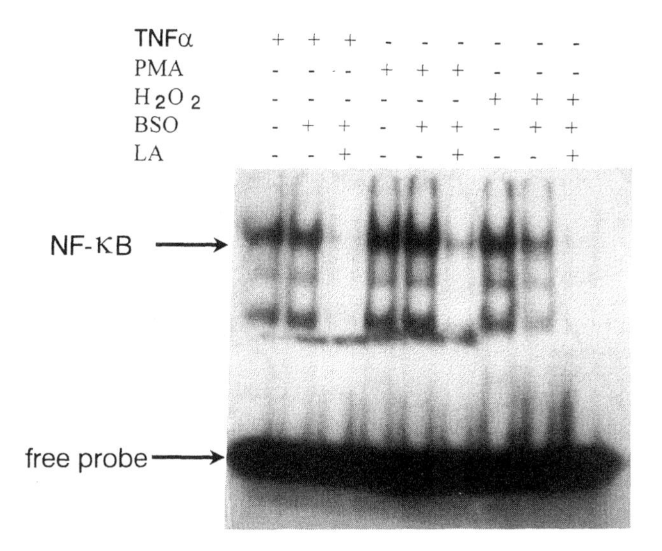

Figure 2. Inhibition of NFκB activation by lipoate is not dependent on its ability to enhance cellular GSH level. In the presence of BSO, lipoate (LA) fails to enhance cellular GSH but is still effective in inhibiting NFκB activation in response to TNF-α, PMA, or hydrogen peroxide. TNF, tumor necrosis factor; PMA, phorbol myristate acetate; BSO, buthionine sulfoximine; LA, lipoate. (Reprinted with permission from Ref.3.)

Two major steps in the activation of NFκB are the phosphorylation and degradation of IκB (Figure 3). Although at the time we did not have any specific knowledge about the phopsphorylation of IκB, as a general rule calcium is known to be a requirement for most protein phosphorylation reactions. IκB contains a PEST sequence of amino acids and is thus highly susceptible to proteolytic cleavage (67). It has been shown that degradation of such PEST-containing sequences may be catalyzed by proteases such as *m*-clapain (68), the activity of which is calcium dependent. Because intracellular calcium could potentially influence both phosphorylation and degradation of IκB, we tested the hypothesis that the peroxide-induced differential calcium response in Jurkat and Wurzburg cells is linked to their respective NFκB responses (66).

In Wurzburg cells that were loaded with the lipophilic esterified calcium chelator EGTA-AM, hydrogen peroxide failed to activate NFκB. This observation provided the first clue that intracellular calcium flux in response to hydrogen peroxide treatment may be involved in the NFκB activation process. We observed that a slow and sustained flux of calcium within the cell is a significant factor in oxidant-induced NFκB activation (66).

In order to substantiate our conclusion we tested whether hydrogen peroxide-would be able to trigger NFκB activation under conditions in which intracellular free calcium levels were maintained high on a sustained basis. Such manipulation of the intracellular calcium level was possible by treating the cells with 1 μM of the sarcoendoplasmic reticulum calcium pump inhibitor, thapsigargin. The sarcoendoplasmic reticulum serves as a major storehouse of intracellular calcium. Calcium is sequestered from the cytosol and retained in this organelle against a high concentration gradient by the active functioning of the sarcoendoplasmic reticular calcium pumps. Inhibition of these pumps resulted in a

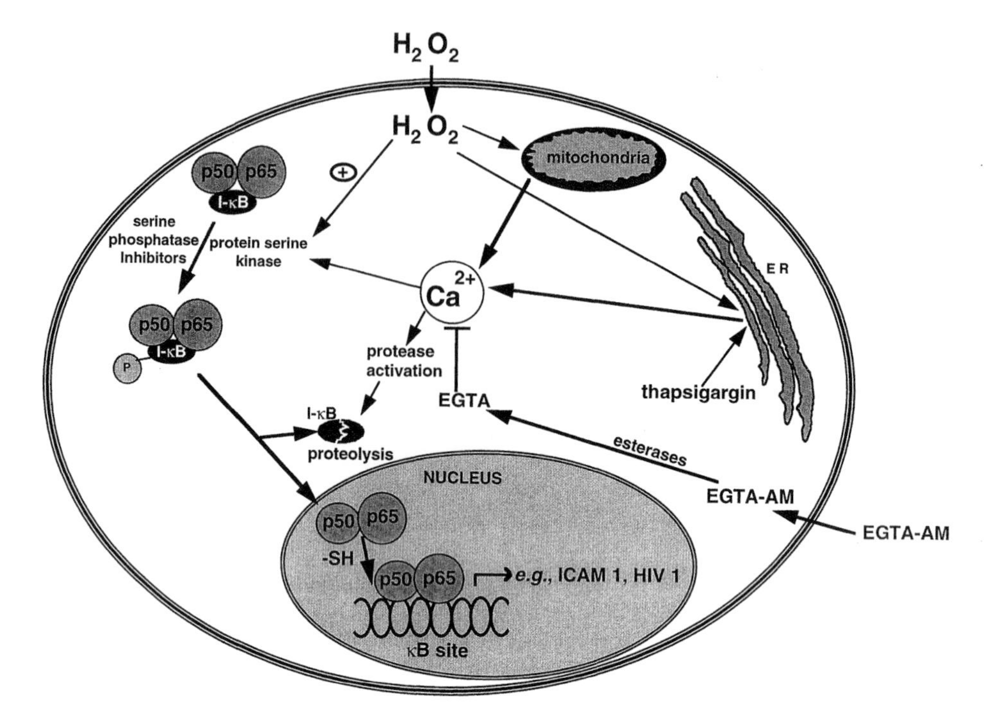

Figure 3. Involvement of intracellular calcium in oxidant-induced NFκB activation. A sustained increase in intracellular free calcium appears to be an important factor in oxidant-induced NFκB activation proceses. Hydrogen peroxide is membrane permeable. Elevated cell calcium may contribute to IκB phosphorylation or/and degradation. Thapsigargin blocks the sarcoendoplasmic reticulum calcium pump and thus results in release of calcium from the organelle to the cytosol. EGTA-AM chelates intracellular calcium and thus blocks calcium-dependent response.

release of stored calcium to the cytosol, resulting in a high level of intracellular free calcium for at least 1 h. Thapsigargin treatment resulted in a weak NFκB activation. This activation was markedly potentiated by hydrogen peroxide treatment of the Jurkat cells. Thus, NFκB activation in Jurkat T cells did respond to hydrogen peroxide under conditions of elevated intracellular free calcium levels. This activation could be completely inhibited by the intracellular calcium chelator EGTA-AM (Figure 4). Thus, for the first time we have observed the involvement of intracellular calcium in oxidant-induced NFκB activation (66).

Both lipoate and NAC are known to be able to suppress NFκB activation in response to a wide variety of activation stimuli. We tested whether pretreatment of cells with these antioxidants alters their oxidant-induced calcium response. Studies with indo-1-loaded cells revealed that such pretreatment certainly decreases oxidant-induced perturbation of intracellular calcium homeostasis (3,66). Thus, the calcium response regulatory effects of these antioxidants may in part be responsible for their NFκB suppressive effects.

In addition to suppression of cytosolic activation of NFκB, lipoate also influences the binding of activated NFκB proteins to the consensus κB site In vitro DNA binding

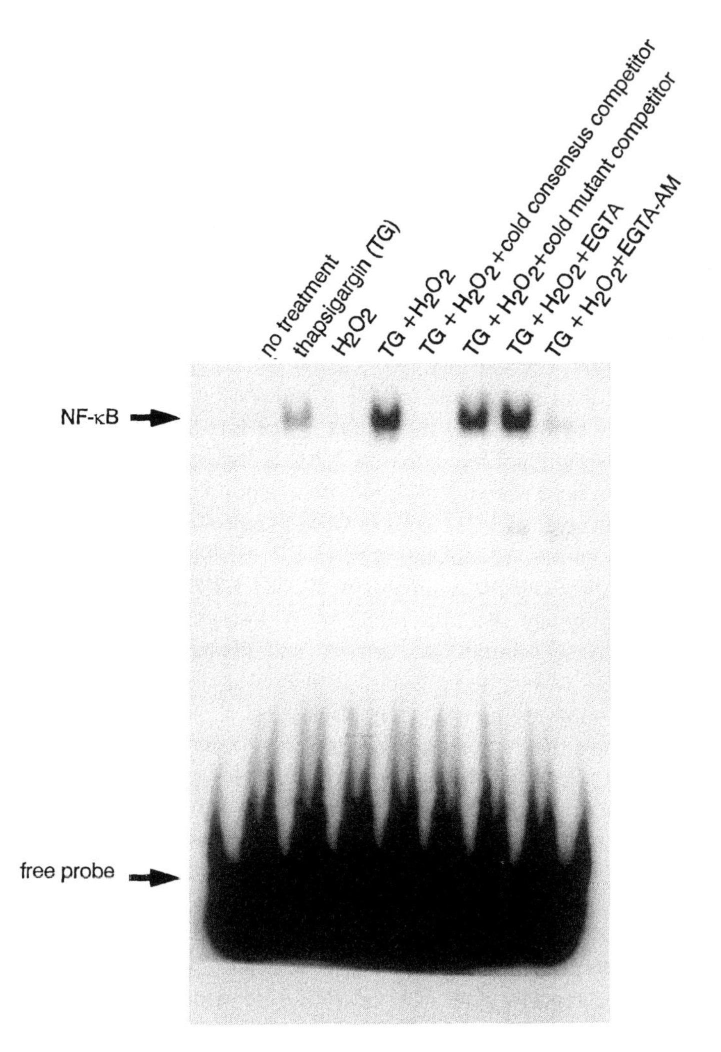

Figure 4. NFκB activation in Jurkat T cells by thapsigargin (1 μM) and hydrogen peroxide (0.25 mM). Treatment of cells with esterfied EGTA-AM 15 min before hydrogen peroxide treatment inhibited NFκB activation. This effect was not observed in the presence of 1 mM extracellular EGTA (lane 7, from left), suggesting that extracellular calcium did not influence the activation process. (Reprinted with permision from Ref. 66.)

studies showed that DHLA enhances such DNA binding (65). This effect of DHLA may be attributed to its strong reducing properties because reductants are well known to enhance NFκB DNA binding. It has previously been hypothesized that NFκB-dependent *trans*-activation requires an oxidizing environment in the cytosol for NFκB activation, and a reducing environment in the nucleus for DNA binding (69). During electrophoretic mobility shift assay of NFκB this reducing atmosphere of the DNA binding mixture is achieved by the use of millimolar concentrations of the potent reductant dithiothreitol.

ACQUIRED IMMUNODEFICIENCY SYNDROME (AIDS)

Human immunodeficiency virus (HIV) infection eventually leads to a substantial fall in the helper T cell ($CD4^+$) count in peripheral circulation. Thus, patients yield to opportunistic infections, widespread immune dysregulation, and certain neoplasms. The long terminal repeat (LTR) region of HIV-1 proviral DNA contains two NFκB binding sites (70,71). DNA binding of NFκB can activate HIV transcription. Hence strategies to suppress NFκB-dependent *trans*-activation may be of therapeutic importance to delay the onset and progression of AIDS.

Previously we have discussed that NAC is able to suppress NFκB activation in response to a wide variety of stimuli. Consistently, NAC also inhibited HIV LTR-directed expression of the β-galactosidase gene in response to TNF-α and phorbol ester (16). NAC facilitates cellular GSH synthesis by improving the supply of the rate-limiting substrate cysteine. However, whether transcriptional regulation of NAC is linked to the pro-GSH effect of NAC is yet unclear. It has been shown that certain transcription-dependent effects of NAC such as the enhanced survival of PC12 cells is GSH independent (72).

Lipoate has been found to inhibit the replication of HIV-1 in cultured lymphoid T cells (73). Jurkat, SupT1, and Molt-4 cells were infected with HTLV-IIB and HIV-1 Wal, and these cells were treated with lipoate 16 h after infection. A dose-dependent effect of lipoate on the inhibition of reverse transcriptase activity and plaque formation was observed. At 70 μg/ml, or 350 μM, reverse transcriptase activity was decreased by 90%, and at 35 μg/ml plaque-forming units were completely eliminated.

We have previously discussed that high intracellular calcium is involved in oxidant-induced NFκB activation (Figure 3). Using thapsigargin, cytosolic levels of free calcium can be elevated. Such a condition resulted in oxidant-induced NFκB activation in Jurkat T cells that are otherwise known to be insensitive to such activation (17). Independent studies suggest that thapsigargin treatment of T lymphocyte cells results in a marked activation of HIV production (74). Viral activation was manifest by increases in soluble viral core p24 production, increases in cellular immunofluorescent staining for viral antigens, and increased viral transcription as measured by HIV LTR-directed expression of the chloramphenicol acetyltransferase reporter gene. This calcium-dependent activation of the transcription of proviral HIV may be mediated by NFκB activation as observed in our study. Pretreatment of cultured T cells with lipoate or NAC diminished oxidant-induced perturbation of intracellular calcium homeostasis. This mechanism could partly explain some of the beneficial properties of both lipoate and NAC against HIV infection.

Characteristically, HIV-positive individuals have decreased levels of acid-soluble thiols–cysteine and GSH in particular–in their plasma and leukocytes (4). This observation, first made by the research group of Wolf Droge, generated substantial interest in AIDS and blood cell thiols. Because decreased thiol status is a typical marker of oxidative stress, this observation was the first to associate HIV infection with oxidative stress. One other critical observation made by the same group was that the plasma levels of glutamate of HIV-infected patients were significantly higher (75). Availability of cysteine is the rate-limiting step in cellular glutathione synthesis. Because of its marked instability in the reduced form, more than 90% of extracellular cysteine is present as cystine (76). Glutamate competitively inhibits cystine uptake by cells (77). In this way, substrate for glutathione synthesis within the cell is limited in the presence of high concentrations of extracellular glutamate. In Jurkat T cells we have observed that supplementation of 5 mM glutamate to the culture medium results in a 50% reduction in cellular GSH level (our unpublished observations).

Studies of Herzenberg and associates (78) using a flow cytometric approach to estimate thiol status of peripheral blood mononuclear cells have also shown that GSH levels in HIV-infected individuals are decreased. Interestingly, this decrease is mainly confined within T cells. B cells and monocytes did show some decrease, but the effects were not statistically significant. The decrease in cell GSH did not correlate with the absolute CD4 number and was only slightly correlated with the stage of the disease, indicating that the perturbation of cellular GSH homeostasis occurs very early after HIV infection. An interesting trend in these observations was that a group of unusually thiol-rich T cells are the most affected (78). This subpopulation of T cells was decreased or absent in infected patients. In a clinical trial with AIDS patients using 3.2–8 g of NAC supplementation per day it was observed that survival of patients was higher in supplemented individuals (see article by Herzenberg and associates in this volume).

Our flow cytometric studies revealed that lipoate treatment can increase cellular GSH levels in a dose-dependent manner from 10 to 100 μM. A very interesting observation was that lipoate can bypass glutamate inhibition of glutathione synthesis. On the basis of results obtained in our laboratory we have developed the following mechanistic explanation. When cells are treated with it, lipoate rapidly enters the cells and is reduced to DHLA. DHLA expelled from the cells to the culture medium reduces extracellular cystine to cysteine. Cysteine can bypass glutamate inhibition of cystine uptake, and can provide sufficient substrate for cellular GSH synthesis. Because these effects of lipoate are observed at concentrations of 100 μM or below, they should be considered to be clinically relevant.

Among the clinically relevant thiol-replenishing agents tested so far, undoubtedly NAC and lipoate hold most promise. Although in many respects the effect of NAC are quite similar to that of lipoate, much higher concentrations of NAC are required to produce comparable effects. Usually, under experimental conditions, 10–30 mM NAC is used to obtain NFκB or some other effects. In contrast, 1 mM lipoate is clearly effective in suppressing NFκB activation in response to a large number of stimuli. In Jurkat T cells, although 100 μM NAC failed to enhance cellular GSH level, lipoate could do so in a dose-dependent manner from 10 to 100 μM, as observed in our flow cytomeric assays (our unpublished observations). Much of these concentration differences may be explained by the mode of action of lipoate. As discussed at the outset, lipoate harnesses the metabolic power of the cell to continuously regenerate its potent reduced form. In this way, the lipoate pool can be continuously renewed at the expense of the cell's metabolic power. This is not possible for NAC, and thus higher concentrations should be necessary. A recent HIV-related study directly compared the efficacy of NAC and lipoate with respect to NFκB mediated gene expression (79). The study was performed using THP-1 cells that were stably transfected with a plasmid bearing a hygromycin B resistance gene under the control of HIV-1 LTR promoter. At 200 μM lipoate treatment resulted in a 40% decrease in HIV-1 p24 antigen expression in TNF-α-stimulated OM 10.1 cells latently transfected with HIV-1. In contrast, 10 mM NAC was required to produce comparable effects. Currently available evidence lends firm support to the contention that lipoate has a remarkable potential for AIDS therapy.

ACKNOWLEDGMENT

This research was supported by National Institutes of Health DK 50430.

REFERENCES

1. Iwata S, Hori T, Sato N, Ueda-Taniguchi Y, Yamabe T, Nakamua H, Masutani H, Yodoi J. Thiol mediated redox regulation of lymphocyte proliferation. Possible involvement of adult T cell leukemia-derived factor and glutathione in transferrin receptor expression. J Immunol 1994; 152: 5633–5642.
2. Sato N, Iwata S, Nakamura K, Hori T, Mori K, Yodoi J. Thiol mediated redox regulation of apoptosis. Possible role of cellular thiols other than glutathione in T cell apoptosis. J Immunol 1995; 154: 3194–3203.
3. Sen CK, Packer L. Antioxidant and redox regulation of gene transcription. FASEB J. 1996; 10: 709–720.
4. Eck HP, Gmunder H, Hartmann M, Petzoldt D, Daniel V, Droge W. Low concentrations of acid soluble thiol (cysteine) in blood plasma of HIV-1 infected patients. Biol Chem Hoppe-Seyler 1989; 370: 101–108.
5. Roederer M, Raju PA, Staal FJT, Herzenberg LA, Herzenberg LA. *N*-Acetylcysteine inhibits latent HIV expression in chronically infected cells. AIDS Res Hum Retroviruses 1991; 7: 563–567.
6. Staal FJT, Roederer M, Herzenberg LA, Herzenberg LA. Glutathione and immunophenotypes of T and B lymphocytes in HIV infected individuals. Ann NY Acad. Sci 1991; 651: 453–463.
7. Staal FJT, Roederer M, Israelski DM, Bubp J, Mole LA, McShane D, Deresinski SD, Ross W, Sussman H, Raju PA, Anderson MT, Moore W, Ela SW, Herzenberg LA, Herzenberg LA. Intracellular glutathione in T cell subsets decrease in HIV infected individuals. AIDS Res Hum Retroviruses 1992; 2: 305–311.
8. Packer L, Roy S, Sen CK. α-Lipoic acid: a metabolic antioxidant and potential redox modulator of transcription. Adv Pharmacol 1996; 38: 79–101.
9. Packer L, Witt EH, Tritschler HJ. Alpha-lipoic acid as a biological antioxidant. Free Radicals Biol Med 1995; 19: 227–250.
10. Sen CK, Atalay M, Hanninen O. Exercise-induced oxidative stress: glutathione supplementation and deficiency. J Appl Physiol 1994; 77: 2177–2187.
11. Ho W-Z, Douglas SD. Glutathione and *N*-acetylcysteine suppression of human immunodeficiency virus replication in human monocyte/macrophages in vitro. AIDS Res Hum Retroviruses 1992; 8: 1249–1253.
12. Kalebic T, Kinter A, Poli G, Anderson ME, Meister A, Fauci AS. Suppression of HIV expression in chronically infected monocytic cells by glutathione, glutathione ester, and *N*-acetylcysteine. Proc Natl Acad Sci USA 1991; 88: 986–990.
13. Levy EM, Wu J, Salibian M, Black PH. The effect of changes in thiol subcompartments on T-cell colony formation and cell cycle progression: relevance to AIDS. Cell Immunol 1992; 140: 370–380.
14. Mihm S, Ennen J, Pessara U, Kurth R, Droege W. Inhibition of HIV replication and NF-kB activity by cysteine and cysteine derivatives. AIDS 1991; 5: 497–503.
15. Raju PA, Herzenberg LA, Herzenberg LA, Roederer M. Glutathione precursor and antioxidant activities of *N*-acetylcysteine and oxothiazolidine carboxylate compared in in vitro studies of HIV replication. AIDS Res Hum Retroviruses 1994; 10: 961–967.
16. Roederer M, Staal FJT, Raju PA, Ela SW, Herzenberg LA, Herzenberg LA. Cytokine stimulated HIV replication is inhibited by *N*-acetylcysteine. Proc Natl Acad Sci USA 1990; 87: 4884–4888.
17. Staal FJT, Roederer M, Herzenberg LA, Herzenberg LA. Intracellular thiols regulate activation of nuclear factor kappa B and transcription of human immunodeficiency virus. Proc Natl Acad Sci USA 1990; 87: 9943–9947.
18. Reed LJ, DeBusk BG, Gunsalus IC, Hornberger J. Crystalline α-lipoic acid: a catalytic agent associated with pyruvate dehydrogenase. Science 1951; 114: 93–94.

19. Patel MS, Smith RL. Biochemistry of lipoic acid containing proteins: past and present. In Schmidt K, Diplock AT, Ulrich H, eds. The Evolution of Antioxidants in Modern Medicine. Stuttgart: Hippocrates Verlag; 1994: 65–77.
20. Lawson JE, Behal RH, Reed RJ. Disruption and mutagenesis of the *Saccharomyces cervisiae* PDX1 gene encoding the protein X component of the pyruvate dehydrogenase complex. Biochemistry 1991; 30: 2834–2839.
21. Danson MJ. Dihydrolipoamide dehydrogenase: a "new" function for an old enzyme? Biochem Soc Trans 1987; 16: 87–89.
22. Danson MJ, Eisenthal R, Hall S, Kessell SR, Williams DL. Dihydrolipoamide dehydrogenase from halophilic archaebacteria. Biochem J 1984; 218: 811–818.
23. Haugaard N, Haugaard ES. Stimulation of glucose utilization by thioctic acid in rat diaphragm incubated in vitro. Biochim Biophys Acta 1970; 222: 583–586.
24. Singh HPP, Bowman RH. Effect of D,L-alpha lipoic acid on the citrate concentration and phosphofructokinase activity of perfused hearts from normal and diabetic rats. Biochem Biophys Res Commun 1970; 41: 555–561.
25. Henriksen EJ, Jacob S, Tritschler H, Wessel K, Augustin HJ, Dietze GJ. Chronic thioctic acid treatment increases insulin-stimulated glucose transport activity in skeletal muscle of obese Zucker rats. Diabetes Suppl 1994; 1: 122A.
26. Bashan N, Burdett E, Klip A. Effect of thioctic acid on glucose transport. In Abstract book of the Third International Thioctic Acid Workshop, Frankfurt, Germany. Universimed Verlag; 1993; 3: 218–223.
27. Jacob S, Henriksen EJ, Schiemann AL, Simon I, Clancy DE, Tritschler HJ, Jung WI, Augustin HJ, Dietze GJ. Enhancement of glucose-disposal in patients with type 2 diabetes by alpha-lipoic acid. Arzneimittel-Forschung 1995; 45: 872–874.
28. Handelman GJ, Han D, Tritschler H, Packer L. Alpha-lipoic acid reduction by mammalian cells to the dithiol form, and release into the culture medium. Biochem Pharmacol 1994; 47: 1725–1730.
29. Haramaki N, Han D, Handelman GJ, Tritschler HJ, Packer L. Cytosolic and mitochondrial systems for NADH- and NADPH-dependent reduction of α-lipoic acid. Free Radical Biol Med; in press.
30. Pick U, Haramaki N, Constantinescu A, Handelman GJ, Tritschler HJ, Packer L. Glutathione reductase and lipoamide dehydrogenase have opposite stereospecificities for alpha-lipoic acid enantiomers. Biochem Biophys Res Commun 1995; 206: 724–730.
31. Arne ESJ, Nordberg J, Holmgren A. Efficient reduction of lipoamide and lipoic acid by mammalian thiredoxin reductase. Biochem Biophys Res Commun 1996; in press.
32. Rosenberg HR, Culik R. Effect of α-lipoic acid on vitamin C and vitamin E deficiencies. Arch Biochem Biophys 1959; 80: 86–93.
33. Jocelyn PC. The standard redox potential of cysteine–cystine from the thiol–disulphide exchange reaction with glutathione and lipoic acid. Eur J Biochem 1967; 2: 327–331.
34. Suzuki YJ, Tsuchiya M, Packer L. Thioctic acid and dihydrolipoic acid are novel antioxidants which interact with reactive oxygen species. Free Radical Res Commun 1991; 15: 255–263.
35. Scott BC, Aruoma OI, Evans PJ, O'Neill C, Van der Vliet A, Cross CE, Tritschler H, Halliwell B. Lipoic and dihydrolipoic acids as antioxidants. A critical evaluation. Free Radical Res 1994; 20: 119–133.
36. Matsugo S, Yan L-J, Han D, Tritschler HJ, Packer L. Elucidation of antioxidant activity of alpha-lipoic acid toward hydroxyl radical. Biochem Biophys Res Commun 1995; 208: 161–167.
37. Matsugo S, Yan LJ, Han D, Tritschler HJ, Packer L. Elucidation of antioxidant activity of dihydrolipoic acid toward hydroxyl radical using a novel hydroxyl radical generator NP-III. Biochem Mol Biol Int 1995; 37: 375–383.
38. Müller L, Menzel H. Studies on the efficacy of lipoate and dihydrolipoate in the alteration of $cadmium^{2+}$ toxicity in isolated hepatocytes. Biochim Biophys Acta 1990; 1052: 386–391.

39. Devasagayam TPA, Subramanian M, Pradhan DS, Sies H. Prevention of singlet oxygen-induced DNA damage by lipoate. Chem Biol Interact 1993; 86: 79–92.
40. Bast A, Haenen GRMM. Interplay between lipoic acid and glutathione in the protection against microsomal lipid peroxidation. Biochem Biophys Acta 1988; 963: 558–561.
41. Bast A, Haenen GRMM. Regulation of lipid peroxidation of glutathione and lipoic acid: involvement of liver microsomal vitamin E free radical reductase. In Emerit I, Packer L, Auclair M, eds. Antioxidants in Therapy and Preventive Medicine. New York: Plenum Press; 1990: 111–116.
42. Sigel H, Prijs B, McCormick DB, Shih JCH. Stability of binary and ternary complexes of α-lipoate and lipoate derivatives with Mn^{2+}, Cu^{2+}, and Zn^{2+} in solution. Arch Biochem Biophys 1978; 187: 208–214.
43. Ou P, Tritschler H, Wolff SP. Thioctic (lipoic) acid: a therapeutic metal-chelating antioxidant? Biochem Pharmacol 1995; 50: 123–126.
44. Stevens B, Perez SR, Small RD. The photoperoxidation of unsaturated organic molecules–IX. Lipoic acid inhibition of rubrene autoperoxidation. Photochem Photobiol 1974; 19: 315–316.
45. Stary FE, Jindal SJ, Murray RW. Oxidation of α-lipoic acid. J Org Chem 1975; 40: 58–62.
46. Kaiser S, Di Mascio P, Sies H. Lipoat und Singulettsauerstoff. In Borbe HO, Ulrich H, eds. Thioctsäure. Frankfurt: PMI Verlag GmbH; 1989: 69–76.
47. Whiteman M, Tritschler H, Halliwell B. Protection against peroxynitrite dependent tyrosine nitration and alpha-1-antiproteinase inactivation by oxidized and reduced lipoic acid. FEBS Lett 1996; 379: 74–76.
48. Kagan VE, Serbinova EA, Forte T, Scita G, Packer L. Recycling of vitamin E in human low density lipoproteins. J Lipid Res 1992; 33: 385–397.
49. Suzuki YJ, Packer L. Inhibition of NF-kB activation by vitamin E derivatives. Biochem Biophys Res Commun 1993; 193: 277–283.
50. Packer L. New horizons in vitamin E research–the vitamin E cycle, biochemistry and clinical applications. In Ong ASH, Packer L, eds. Lipid-Soluble Antioxidants: Biochemistry and Clinical Applications. Boston: Birkhauser Verlag; 1992: 1–16.
51. Sies H. Strategies of antioxidant defense. Eur J Biochem 1993; 215: 213–219.
52. Scholich H, Murphy ME, Sies H. Antioxidant activity of dihydrolipoate against microsomal lipid peroxidation and its dependence on α-tocopherol. Biochim Biophys Acta 1989; 1001: 256–261.
53. Constantinescu A, Han D, Packer L. Vitamin E recycling in human erythrocyte membranes. J Biol Chem 1993; 268: 10906–10913.
54. Podda M, Tritschler HJ, Ulrich H, Packer L. α-Lipoic acid supplementation prevents symptoms of vitamin E deficiency. Biochem Biophys Res Commun 1994; 204: 98–104.
55. Peinado J, Sies H, Akerboom TPM. Hepatic lipoate uptake. Arch Biochem Biophys 1989; 273: 389–395.
56. Busse E, Zimmer G, Schopohl B, Kornhuber B. Influence of alpha-lipoic acid on intracellular glutathione in vitro and in vivo. Arzneimittel-Forschung 1992; 42: 829–831.
57. Baeuerle PA, Henkel T. Function and activation of NF-kB in immune system. Annu Rev Immunol 1994; 12: 141–179.
58. Meyer R, Caselmann WH, Schluter V, Schreck R, Hofschneider PH, Baeuerle PA. Hepatitis B virus transactivator MHBst: activation of NF-kappa B, selective inhibition by antioxidants and integral membrane localization. EMBO J 1992; 11: 2992–3001.
59. Schreck R, Meier B, Maennel DN, Droge W, Baeuerle PA. Dithiocarbamates as potent inhibitors of nuclear factor κB activation in intact cells. J Exp Med 1992; 175: 1181–1194.
60. Sen CK, Traber K, Packer L. Inhibition of NF-kB activation in human T-cell lines by anetholdithiolthione. Biochem Biophys Res Commun 1996; 218: 148–153.
61. Suzuki YJ, Packer L. Alpha-lipoic acid is a potent inhibitor of NF-kappa B activation in human T cells: does the mechanism involve antioxidant activities. In Packer L, Cardenas E, eds. Biological Oxidants and Antioxidants. Stuttgart: Hippokrates Verlag; 1994.

62. Suzuki YJ, Packer L. Inhibition of NF-kB DNA binding activity by alpha-tocopheryl succinate. Biochem Mol Biol Int 1994; 31: 693–700.
63. Suzuki YJ, Packer L. Inhibition of NF-kB transcription factor by catechol derivatives. Biochem Mol Biol Int 1994; 32: 299–305.
64. Suzuki YJ, Agarwal BB, Packer L. Alpha-lipoic acid is a potent inhibitor of NF-kB activation in human T cells. Biochem Biophys Res Commun 1992; 189: 1709–1715.
65. Suzuki YJ, Mizuno M, Tritschler HJ, Packer L. Redox regulation of NF-kB DNA binding activity by dhydrolipoate. Biochem Mol Biol Int 1995; 36: 241–246.
66. Sen CK, Roy S, Packer L. Involvement of intracellular Ca^{2+} in oxidant-induced NF-kB activation. FEBS Lett 1996; 385: 58–62.
67. Thanos D, Maniatis T. NF-kappa B: a lesson in family values. Cell 1995; 80: 529–532.
68. Watt F, Molloy PL. Specific cleavage of transcription factors by thiol protease, *m*-calpain. Nucleic Acids Res 1993; 21: 5092–5100.
69. Droge W, Schulze-Osthoff K, Mihm S, Galter D, Schenck H, Eck H-P, Roth S, Gmunder H. Functions of glutathione and glutathione disulfide in immunology and immunopathology. FASEB J 1994; 8: 1131–1138.
70. Nabel G, Baltimore D. An inducible transcription factor activates expression of human immunodeficiency virus in T cells. Nature 1987; 326: 711–713.
71. Nabel G, Rick SA, Knipe DM, Baltimore D. Alternative mechanisms for activation of human immunodeficiency virus enhancer in T cells. Science 1988; 239: 1299–1302.
72. Yan CHI, Ferrari G, Greene LA. *N*-Acetylcysteine-promoted survival of PC12 cells is glutathione-independent but transcription-dependent. J Biol Chem 1995; 270: 26827–26832.
73. Baur A, Harrer T, Peukert M, Jahn G, Kalden JR, Fleckenstein B. Alpha-lipoic acid is an effective inhibitor of human immuno-deficiency virus (HIV-1) replication. Klin Wochenschr 1991; 69: 722–724.
74. Papp B, Bryn RA. Stimulation of HIV expression by intracellular calcium pump inhibition. J Biol Chem 1995; 270: 10275–10283.
75. Droege W, Eck H-P, Naher H, Pekar U, Daniel V. Abnormal amino-acid concentrations in blood of patients with acquired immunodeficiency syndrome (AIDS) may contribute to the immunological defect. Biol Chem Hoppe-Seyler 1988; 369: 143–148.
76. Droege W, Eck H-P, Mihm S. HIV-induced cysteine deficiency and T cell dysfunction–a rationale for treatment with *N*-acetylcysteine. Immunol Today 1992; 13: 211–214.
77. Watanabe H, Bannai S. Induction of cysteine transport activity in mouse peritoneal macrophages. J Exp Med 1987; 165: 628–640.
78. Roederer M, Staal FJT, Anderson ME, Rabin R, Raju PA, Herzenberg L.A, Herzenberg LA. Disregulation of leukocyte glutathione in AIDS. Ann NY Acad Sci 1993; 677: 113–125.
79. Merin JP, Matsuyama M, Kira T, Baba M, Okamoto T. α-Lipoic acid blocks HIV-1 LTR-dependent expression of hygromycin resistance in THP-1 stable transforms. FEBS Lett; in press.

26

Glutathione Compartmentation and Oxidative Stress

Melani K. Savage and Donald J. Reed
Oregon State University
Corvallis, Oregon

INTRODUCTION

The inner membrane of mitochondria isolated from a variety of tissues becomes permeable following activation or deregulation of a Ca^{2+}-dependent inner membrane pore (1–4). This process, referred to as permeability transition, usually requires the presence of a second agent such as inorganic phosphate (P_i), *t*-butylhydroperoxide, or sulfhydryl reagents (4). The mechanism(s) by which the various agents induce permeability transition is not completely understood; however, it is postulated that they may all act on a similar protein or pore structure, resulting in a loss of inner membrane integrity (1,4).

Opening or activation of this inner membrane pore with Ca^{2+} and P_i results in the release of small (<1500 Da) matrix solutes such as reduced glutathione (GSH) (307 Da) and Ca^{2+}, large-amplitude swelling, and a loss of coupled functions (4–7). Cyclosporin A (CsA), an immunosuppressive cyclic peptide, is a potent inhibitor of the permeability transition and its consequences (8,9). The mechanism of CsA's inhibition is thought to involve the binding of CsA to cyclophilins located in the mitochondrial matrix (10–12). Although the physiological relevance of such a pore remains uncertain, there is increasing evidence suggesting that the opening and closing of this inner membrane pore is a highly regulated process with several different control sites. This may explain why many structurally and functionally different inducing agents, including P_i, sulfhydryl reagents, oxidants, and heavy metals, induce the Ca^{2+}-dependent permeability transition (4).

Inducing agents such as *t*-butylhydroperoxide and hydrogen peroxide are thought to mediate permeability transition due to the oxidation of NADH and NADPH (4,13–14). The inducing agent P_i was not previously thought to oxidize pyridine nucleotides and therefore has often been studied in comparison to pyridine nucleotide-oxidizing agents, such as *t*-butylhydroperoxide (15–17).

We have since reported that during Ca^{2+}- and P_i-induced permeability transition both NADH and NADPH are extensively oxidized, but not glutathione, and ATP and ADP are depleted within several minutes (18,19). In conjunction with these findings we have also observed conditions in which permeability transition occurs in the absence of large-amplitude swelling (19), a parameter often used as a determinant of permeability transition.

We wished (1) to determine whether NAD(H) or NADP(H) (700–835 Da) were oxidized and then released from isolated mitochondria following Ca^{2+}- and P_i-induced permeability transition, as these molecules are normally impermeable to the inner membrane and relatively small (<1500 Da) and (2) to determine the redox status of glutathione and pyridine and adenine nucleotides following treatment with the respiratory inhibitors antimycin, oligomycin, or sulfide during Ca^{2+} and P_i-induced permeability transition.

METHODS

Mitochondrial Preparation

Isolated mitochondria were prepared by differential centrifugation according to the method of Schnaitman and Greenwalt (20). Briefly, livers of male Sprague–Dawley rats (325–375 g) (Simonsen Labs, Gilroy, CA) were excised and homogenized in isolation buffer containing 220 mM mannitol, 70 mM sucrose, 2 mM Hepes, 0.5 mM EGTA, and 0.5 mg/ml bovine serum albumin, pH 7.4. Mitochondria were washed twice in isolation buffer devoid of EGTA, pH 7.0.

After the final wash, the mitochondrial pellets were suspended (25–30 mg/ml) in incubation buffer containing 213 mM mannitol, 71 mM sucrose, 10 mM succinate, and 3 mM Hepes, pH 7.0. Mitochondrial protein was determined spectrophotometrically by the method of Bradford (21) with bovine serum albumin as the standard.

Mitochondrial Incubations

Mitochondria were added to 10 ml Erlenmeyer flasks (1 mg protein/ml) containing incubation buffer plus the various treatments described in the figure legends. After briefly swirling the flask, 1.0 ml samples were taken immediately following the addition of mitochondria. These samples constitute the initial time points. Mitochondrial suspensions were shaken gently and exposed to atmospheric air at room temperature throughout the remainder of the incubation period. Aliquots (1 ml) of mitochondrial suspensions, collected at the times indicated in the figure legends, were prepared for the analysis of pyridine and adenine nucleotides as described below.

Biochemical Analysis

GSH was determined by HPLC analysis according to the method of Reed et al. (22) with modifications as described previously (7). Intra- and extramitochondrial pyridine and adenine nucleotides were measured according to the HPLC method of Jones (23) with the following modifications to accommodate mitochondrial samples. Briefly, 1 ml samples containing 1 mg protein were removed from incubation flask at the appropriate time points, transferred to microcentrifuge tubes, and centrifuged at 13,000*g* for 30 s to separate the suspended mitochondria from the incubation buffer. A sample (0.500 ml) of the supernatant (extramitochondrial) was transferred to microcentrifuge tubes containing either 0.100 ml 10% perchloric acid (PCA) for the analysis of ATP, ADP, AMP, $NADP^+$, and NAD^+ or 0.100 ml 0.5 M KOH in 50% ethanol and 0.35% cesium chloride for the analysis of NADH and NADPH, and then frozen immediately at –80°C. The remaining supernatant was carefully discarded by aspiration. The mitochondrial pellets were suspended in either 0.500 ml 10% PCA for extraction of intramitochondrial ATP, ADP, AMP, $NADP^+$, and NAD^+ or 0.100 ml 0.5 M KOH in 50% ethanol and 0.35% cesium chloride for extraction of NADPH and NADH. The samples were immediately iced for 15 min, sonicated for 5 s, and then frozen at –80°C. Prior to analysis, samples were thawed and the acid-extracted samples were neutralized with 10 mM KOH and 1 M KH_2PO_4, centrifuged at 13,000*g* for 30 s to remove insoluble debris, and assayed for either NADH and NADPH (base extraction) or NAD^+, $NADP^+$, ATP, ADP, and AMP (acid extraction). All samples were assayed by HPLC within 24 h of collection.

The spectrophotometric determinations of pyridine nucleotides were monitored at 340 nm and 370 nm in an Aminco DW2000 spectrophotometer operated in dual beam mode.

RESULTS

Ca^{2+}- and P_i-induced permeability transition was accompanied by the rapid and extensive oxidation of NADH and NADPH (18). Since pyridine nucleotides are relatively small molecules (<850 Da) and normally not able to permeate the inner membrane, we hypothesized that release of pyridine nucleotides occurred through the putative inner membrane pore which allows passage of molecules <1500 Da. To test this hypothesis, both intra- and extramitochondrial NAD(H) and NADP(H) redox levels were measured during permeability transition induced with 70 μM Ca^{2+} and 3 mM P_i (CaP_i).

Mitochondrial NADH and NADPH levels decreased nearly 50% within 5 min following treatment with CaP_i (Figures 1A and 2A). The loss of intramitochondrial NADH and NADPH could not be accounted for extramitochondrially (Figures 1A and 2A), but was nearly all accounted for as NAD^+ and $NADP^+$, respectively, which were detected both intra- and extramitochondrially (Figures 1BC and 2BC). The addition of 0.5 μM CsA, an inhibitor of permeability transition, prevented the oxidation of NAD(P)H and subsequent release of NAD^+ and $NADP^+$ (Figures 1A,B,C and 2A,B,C).

To discount the possibility that NADH and NADPH were released from the matrix and then rapidly oxidized extramitochondrially, we monitored spectrophotometrically the redox state of endogenous and exogenous NADH and NADPH during CaP_i-induced permeability transition. As shown in Figure 3, the addition of CaP_i to mitochondrial suspensions resulted in a decrease in absorbance at 340 and 370 nm, indicating the oxidation of pyridine nucleotides. The subsequent addition of exogenous NADH and NADPH increased the absorbance at 340 and 370 nm, and was maintained throughout the incubation, indicating that NAD(P)H was not oxidized extramitochondrially (Figure 3).

We have previously reported (19) that combined treatment with CaP_i plus either oligomycin (3 μM), sulfide (0.5 mM), or antimycin (1 μM) of mitochondrial suspensions prevented large-amplitude swelling associated with permeability transition but did not prevent inner membrane permeability. This was demonstrated by the CsA-sensitive release of matrix GSH and Ca^{2+} in the absence of large-amplitude swelling (19), a

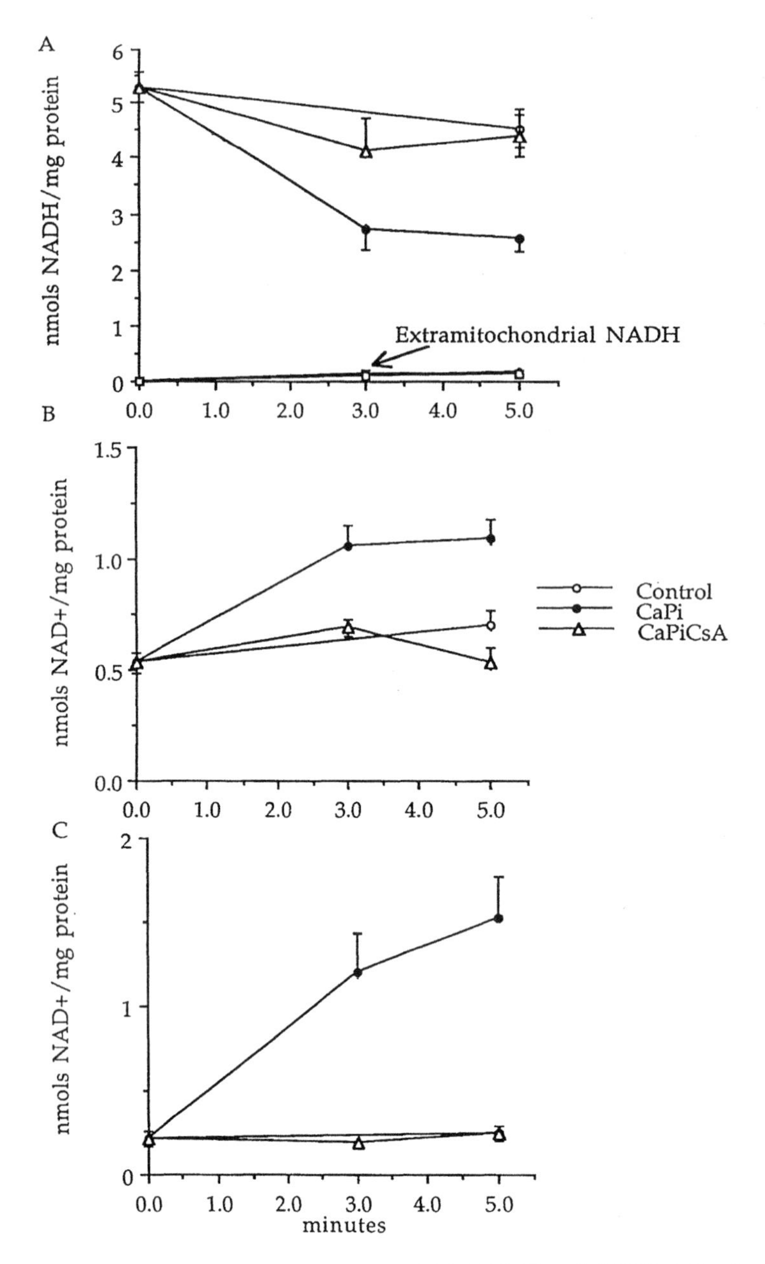

Figure 1. Intra- and extramitochondrial NADH and NAD^+. Mitochondria (1 mg/ml) were incubated in the absence (control) or presence of 70 μM Ca^{2+} and 3 mM P_i (CaP_i) or CaP_i plus 0.5 μM cyclosporin A (CaP_iCsA). Samples (1 ml) were taken at the times indicated and analyzed for (A) intra- and extramitochondrial NADH, (B) intramitochondrial NAD^+, or (C) extramitochondrial NAD^+ as described in Experimental Procedures; $n = 3–4 \pm$ S.E.

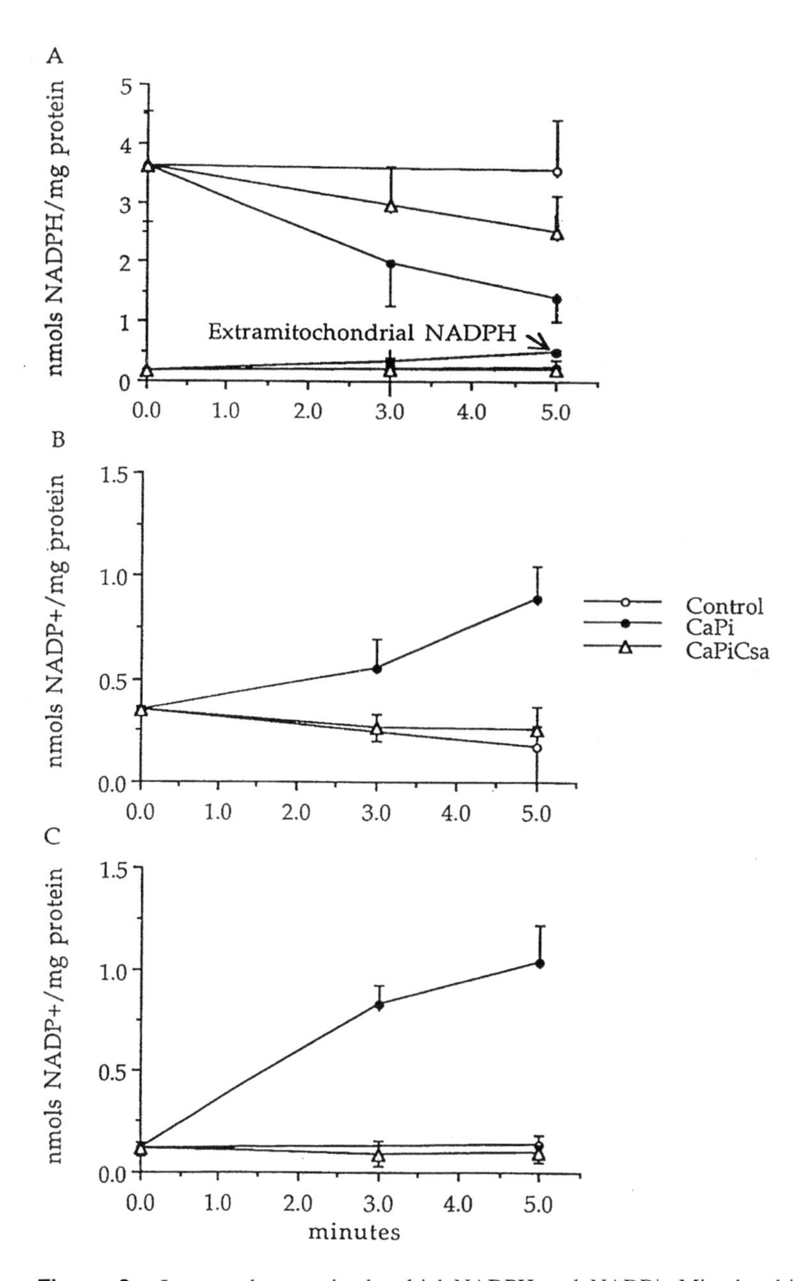

Figure 2. Intra- and extramitochondrial NADPH and $NADP^+$. Mitochondria (1 mg/ml) were incubated in the absence (control) or presence of 70 μM Ca^{2+} and 3 mM P_i (CaP_i) or CaP_i plus 0.5 μM cyclosporin A (CaP_iCsA). Samples (1 ml) were taken at the times indicated and analyzed for (A) intra- and extramitochondrial NADPH, (B) intramitochondrial $NADP^+$, or (C) extramitochondrial $NADP^+$ as described in Experimental Procedures; $n = 3–4 \pm$ S.E.

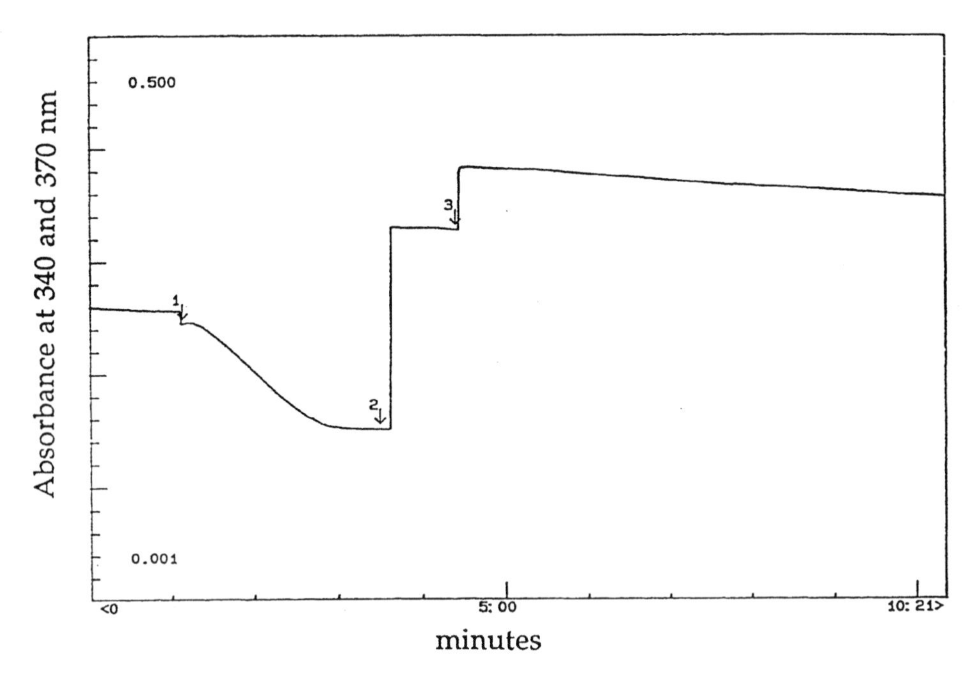

Figure 3. Spectrophotometric determination of the redox state of exogenous NADPH and NADH. Pyridine nucleotide oxidation occurred in mitochondrial suspensions (1 mg/ml) treated with (1) 70 μM Ca^{2+} and 3 mM P_i. Exogenous NADH (2) and NADPH (3) were added following the oxidation of endogenous pyridine nucleotides and monitored for redox changes.

response generally associated with permeability transition. Under these conditions, the membrane potential is lost almost immediately (19). We therefore investigated the status of NAD(P)H, ATP, and ADP under these conditions, since these molecules also relate to the energy status of mitochondria.

Treatment with the respiratory inhibitor oligomycin (3 μM) and sulfide (0.5 mM) preserved NADH at levels 40% and 20%, respectively, of CaP_i treatment alone, whereas the addition of antimycin potentiated the loss of NADH by 20% (Figure 4A). Similarly, the addition of oligomycin and sulfide maintained NADPH at levels 32% and 11%, respectively, of CaP_i treatment alone, whereas the addition of antimycin potentiated the loss of NADPH by 30% (Figure 5A).

The depletion of NADH and NADPH was not accounted for by their recovery extramitochondrially (Figures 4A and 5A). Instead, NADH and NADPH (Figures 4A and 5A) were recovered as NAD^+ and $NADP^+$, respectively, which were detected both intra- and extramitochondrially (Figures 4B,C and 5B,C). The levels of NAD^+ or $NADP^+$ released from the matrix (Figures 4C and 5C) did not necessarily correlate with the levels of NADH and NADPH oxidized (Figures 4B,C and 5B,C). For example, treatment of mitochondria with antimycin plus CaP_i resulted in an increased oxidation of both NADH and NADPH as compared to CaP_i treatment alone (Figures 4B,C and 5B,C); however, mitochondria treated with antimycin plus CaP_i released less NAD^+ and $NADP^+$ than mitochondria treated with CaP_i alone (Figures 4C and 5C).

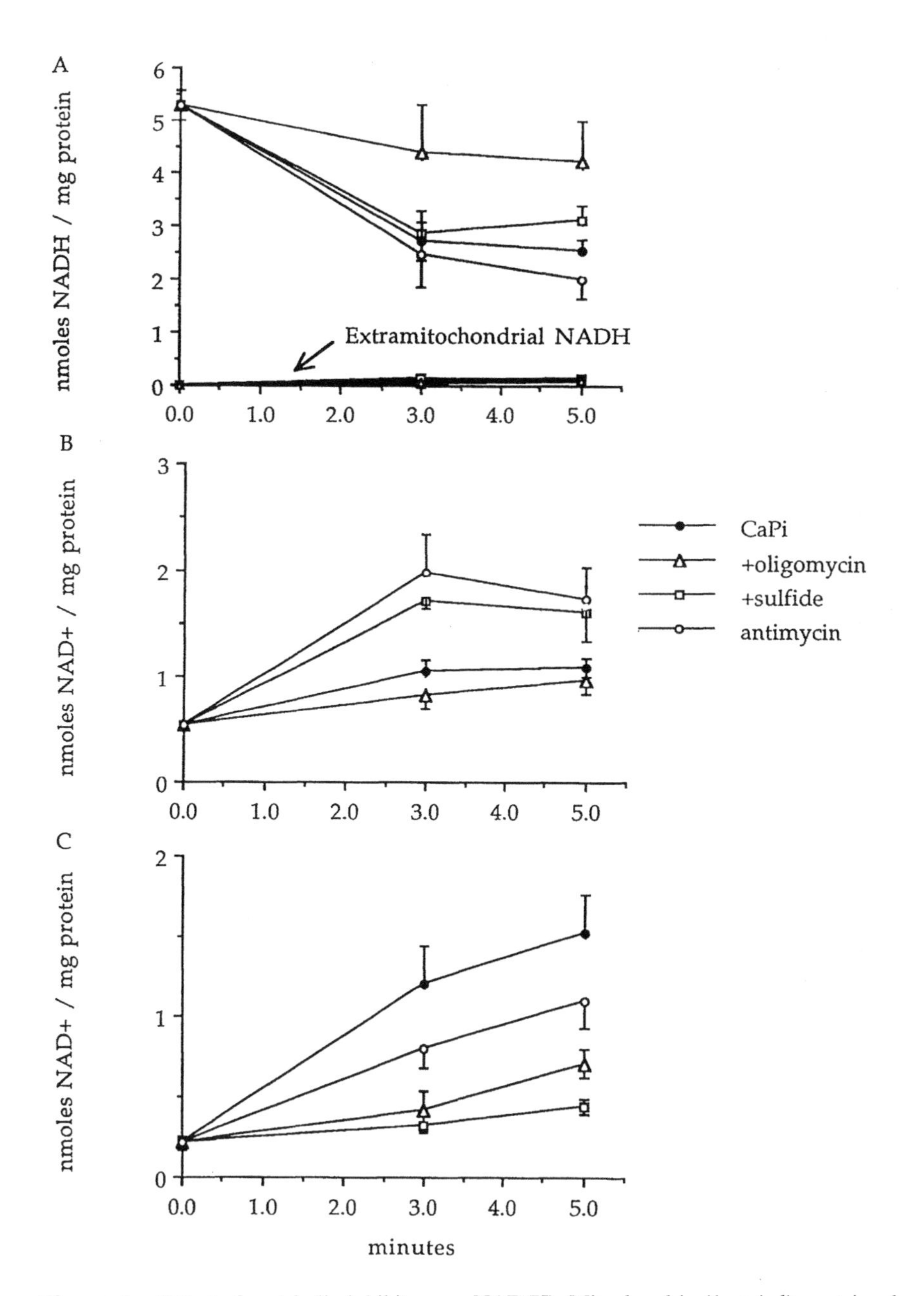

Figure 4. Effect of metabolic inhibitors on NAD(H). Mitochondria (1 mg/ml) were incubated in the presence of 70 μM Ca^{2+} and 3 mM P_i (CaP_i) alone or with the addition of either 3 μM oligomycin, 0.5 mM sulfide, or 1 μM antimycin. Samples (1 ml) were taken at the times indicated and analyzed for (A) intra- and extramitochondrial NADH, (B) intramitochondrial NAD^+, or (C) extramitochondrial NAD^+ as described in Experimental Procedures; n = 3–4 ± S.E.

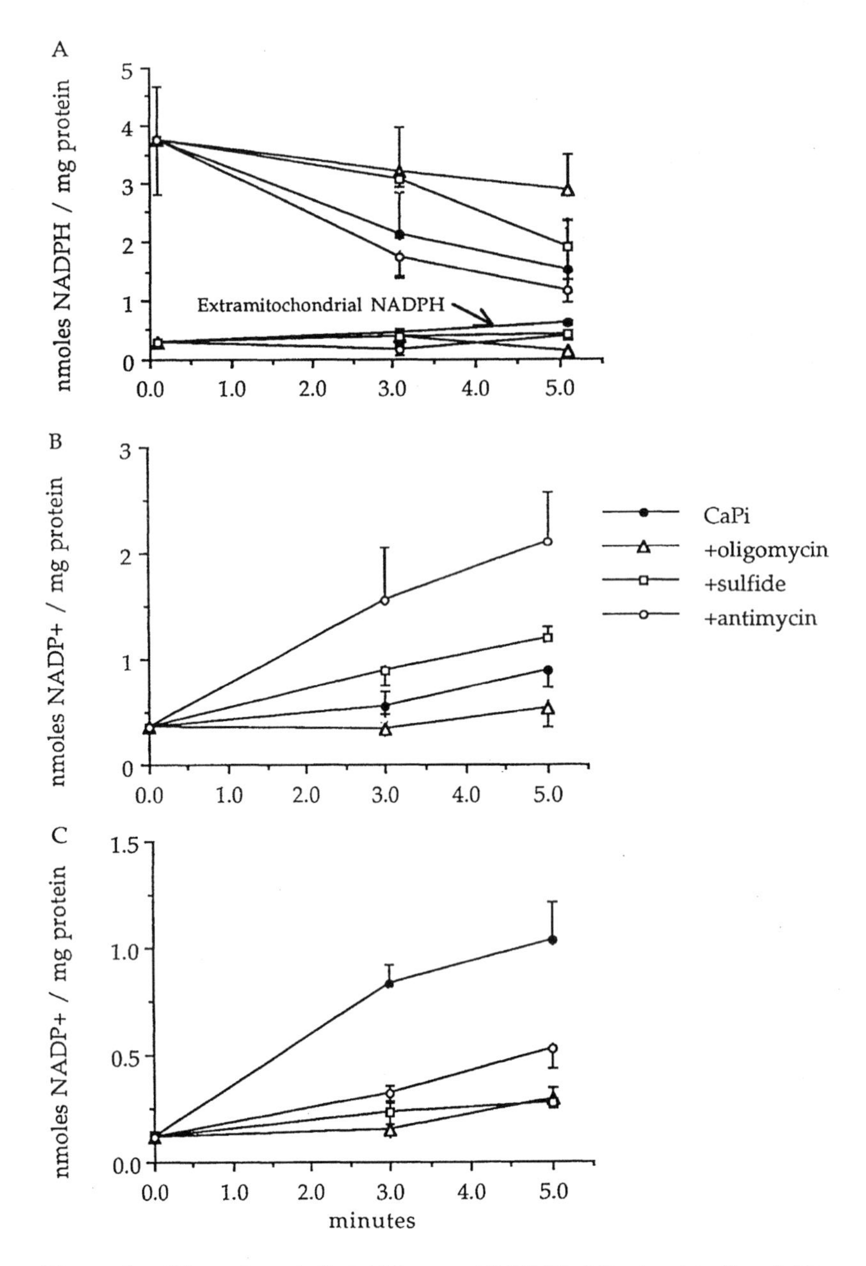

Figure 5. Effect of metabolic inhibitors on NADP(H). Mitochondria (1 mg/mL) were incubated in the presence of 70 μM Ca^{2+} and 3 mM P_i (CaP_i) alone or with the addition of either 3 μM oligomycine, 0.5 mM sulfide, or 1 μM antimycin. Samples (1 ml) were taken at the times indicated and analyzed for (A) intra- and extramitochondrial NADPH, (B) intramitochondrial $NADP^+$, or (C) extramitochondrial $NADP^+$ as described in Experimental Procedures; $n = 3–4 \pm$ S.E.

Table 1. Glutathione, NADH:NAD$^+$ and NADPH:NADP$^+$ Redox Ratios

	% GSH released at 10 min	NADH:NAD$^+$	NADPH:NADP$^+$
Control	0	4.97	13.50
CaP_i	95	1.00	0.96
+CsA	0	5.81	7.88
+Oligomycin	65	2.56	3.25
+Sulfide	27	1.58	1.43
+Antimycin	73	0.74	0.49

NADH:NAD$^+$ and NADPH:NADP$^+$ ratios were determined following a 5 min incubation in the absence (control) or presence of either 70 μM Ca^{2+} and 3 mM P_i (CaP_i) or CaP_i plus 0.5 μM CsA (+CsA), CaP_i plus 3 μM oligomycin (+Oligomycin), CaP_i plus 0.5 μM sulfide (+Sulfide), or CaP_i plus 1 μM antimycin (+Antimycin) (24).

Table 1 lists glutathione loss after 10 min and the pyridine nucleotide redox couples of NADH:NAD$^+$ and NADPH:NADP$^+$ following a 5 min incubation. Treatment with CaP_i resulted in greatly diminished pyridine nucleotide redox ratios as compared to both untreated control and CsA-treated mitochondrial samples. The addition of oligomycin or sulfide to CaP_i-treated mitochondria preserved the pyridine nucleotide redox couples to varying degrees as compared with CaP_i treatment alone. In contrast, the addition of antimycin to CaP_i-treated mitochondria resulted in the redox ratios of both NADH:NAD$^+$ and NADPH:NADP$^+$ being less than observed with CaP_i treatment alone. Measurements of ATP, ADP, and AMP with the various treatments are reported in Table 2. Within 5 min after treatment, CaP_i resulted in the depletion of ATP and ADP with 75% of the adenine pool recovered as AMP (Table 2). The addition of 0.5 μM CsA eliminated this effect, resulting in increased levels of ATP, slightly increased levels of ADP, and decreased AMP levels compared with untreated control samples (Table 2).

The addition of the metabolic inhibitors preserved ATP and/or ADP levels as compared with CaP_i treatment alone (Table 2). The presence of 3 μM oligomycin with CaP_i-treated mitochondria maintained ATP levels similar to untreated control values following a 5 min incubation and ADP loss was minimal. However, the addition of sulfide (0.5 mM) or antimycin (1 μM), in conjunction with CaP_i, did not sustain ATP levels but maintained ADP (Table 2).

The energy charge [(ATP + 0.5ADP)/(ATP + ADP + AMP)], a designation of energy status, suggested that the addition of the respiratory inhibitors preserved mitochondrial energy to varying degrees as compared with CaP_i treatment alone. Following treatment with CaP_i, the energy charge dropped to 0.158 as compared with untreated controls (0.351). Treatment of mitochondria with either oligomycin, sulfide, or antimycin plus CaP_i resulted in energy charge values of 0.409, 0.209, and 0.169, respectively, following a 5 min incubation.

Since treatment with the respiratory inhibitors maintained ADP levels at least 40% greater than observed with CaP_i treatment alone (Table 2) and prevented large-amplitude swelling (19), we tested whether exogenous ADP prevented large-amplitude swelling during CaP_i treatment. Mitochondria were incubated with increasing concentrations of

Table 2. Adenine Nucleotides

	ATP		ADP		AMP		
	I[a]	E[a]	I	E	I	E	TOTAL
Control							
0 min	2.2 (0.34)	1.7 (0.18)	5.5 (0.55)	0.42 (0.04)	7.6 (0.84)	0.39 (0.12)	17.8
5 min	1.8 (0.38)	1.5 (0.16)	5.0 (0.80)	0.32 (0.03)	8.2 (0.61)	0.17 (0.10)	17.0
CaP_i							
3 min	1.2 (0.12)	2.0 (0.59)	2.7 (0.28)	1.4 (0.20)	7.8 (1.4)	2.8 (0.72)	17.9
5 min	0.9 (0.09)	0.3 (0.04)	1.8 (0.18)	1.1 (0.11)	6.6 (1.1)	6.0 (0.33)	16.8
CaP_i + CsA							
3 min	2.9 (0.28)	1.9 (0.53)	7.6 (0.51)	0.1 (0.09)	4.4 (1.1)	0.03	17.0
5 min	2.7 (0.13)	2.9 (0.72)	7.1 (0.53)	0.4 (0.04)	6.3 (0.82)	0.1 (0.05)	19.5
CaP_i + oligomycin							
3 min	2.3 (0.20)	1.8 (0.14)	4.2 (0.18)	0.9 (0.56)	5.2 (0.85)	0.5 (0.42)	15.0
5 min	2.4 (0.32)	1.3 (0.31)	4.2 (0.80)	1.4 (0.29)	5.0 (0.80)	1.6 (1.1)	15.9
CaP_i + sulfide							
3 min	1.1 (0.24)	0.5 (0.17)	3.6 (0.44)	0.5 (0.02)	6.9 (0.77)	3.3 (0.48)	15.9
5 min	1.0 (0.22)	0.3 (0.01)	3.3 (0.43)	0.5 (0.08)	6.0 (1.3)	4.2 (0.44)	15.3
CaP_i + antimycin							
3 min	1.0 (0.22)	0.0	3.3 (0.62)	0.3 (0.11)	6.6 (1.3)	4.4 (0.47)	15.6
5 min	1.0 (0.19)	0.0	3.0 (0.60)	0.4 (0.03)	5.6 (0.63)	5.9 (0.66)	16.0

Intra- and extramitochondrial ATP, ADP, and AMP were measured in the absence (control) or presence of either 70 μM Ca^{2+} and 3 mM (CaP_i) or CaP_i plus 0.5 μM CsA CaP_i plus 3 μM oligomycin, CaP_i plus 0.5 mM sulfide or CaP_i plus 1 μM antimycin. Samples were taken at the times indicated and analyzed by HPLC as described in Experimental Procedures ($n = 3–5 \pm$ S.E.).

[a]I = Intramitochondrial; E = extramitochondrial.

Modified from Ref. 24.

ADP in the presence of CaP_i (Figure 6). The addition of ADP in the presence of CaP_i inhibited large-amplitude swelling in a dose-dependent manner (Figures 6B–D), with the highest concentration of ADP (150 μM) completely preventing large-amplitude swelling (Figure 6D).

DISCUSSION

Our findings demonstrate that CaP_i-induced permeability transition is accompanied by rapid intramitochondrial oxidation of the pyridine nucleotide pool, with the selective release of NAD^+ and $NADP^+$ and glutathione as GSH. Although the mechanism by which CaP_i treatment results in the oxidation of NADH and NADPH is not understood (18), it does not appear to involve the oxidation of GSH or lipid peroxidation, as no evidence of either has been detected (Table 1) (7,24). Under these conditions, GSH, a small peptide (307 Da), was rapidly and nearly completely released from the matrix by a CsA-sensitive mechanism following a 5 min or 10 min incubation with no increase in oxidized glutathione (GSSG) levels (Table 1) (7,24).

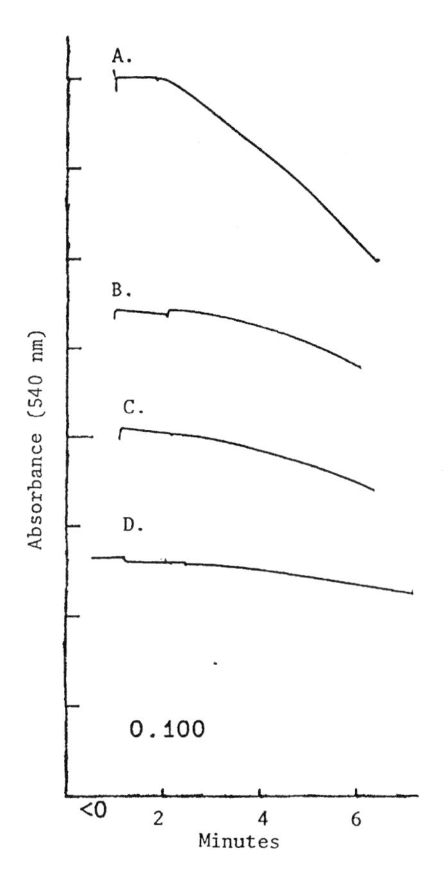

Figure 6. Dose–response inhibition of large-amplitude swelling by exogenous ADP. Mitochondria (1 mg/ml) were treated with (A) 70 μM Ca^{2+} and 3 mM P_i (CaP_i), (B) 60 μM ADP plus CaP_i, (C) 80 μM ADP plus CaP_i and (D) 150 μM ADP plus CaP_i and monitored for large-amplitude swelling.

Since NAD(H) and NADP(H) are also relatively small molecules (<1200 Da) and normally impermeable to the inner membrane, they could possibly be released through the Ca^{2+}-dependent, CsA-sensitive pore of the permeability transition (8). Minimal levels of NADH and NADPH were detected extramitochondrially during CaP_i-induced permeability transition; the levels detected were similar to those observed in untreated mitochondria, suggesting that either (1) NADH and NADPH are not released from the mitochondrial matrix during permeability transition, or (2) they are released and then rapidly oxidized extramitochondrially. To test the latter possibility, mitochondrial suspensions treated with CaP_i were monitored spectrophotometrically for redox changes in NADH and NADPH as indicated by a decrease in absorbance at 340 nm. The addition of CaP_i to mitochondrial suspensions resulted in a decrease in absorbance, as a result of the oxidation of endogenous NADH and NADPH. Subsequent additions of exogenous NADH and NADPH did not result in oxidation of the added pyridine nucleotide, indicating that oxidation of pyridine nucleotides occurred intramitochondrially as opposed to extramitochondrially.

It is unclear why $NAD(P)^+$, but not NAD(P)H, effluxed through the inner membrane pore during CaP_i-induced permeability transition. One possible explanation is that NADH and NADPH are protein bound and therefore unavailable to traverse the inner membrane pore; NAD^+ and $NADP^+$ are perhaps unbound and therefore efflux through the permeability transition pore. Another possible explanation centers on the issue of pore selectivity, with the pore selectively releasing NAD^+ and $NADP^+$, possibly owing to the positive charge associated with the oxidized pyridine nucleotide. Although the Ca^{2+}-dependent pore has generally been regarded as being nonselective to small (<1200 Da) matrix solutes (1), other studies have suggested that inner membrane permeability may be selective (25).

We also measured the status of pyridine and adenine nucleotides in mitochondria treated with either oligomycin, sulfide, or antimycin in the presence of CaP_i. We have previously observed that the addition of oligomycin, sulfide, or antimycin to CaP_i-treated mitochondria prevented mitochondrial large-amplitude swelling but not the CsA-sensitive release of GSH and Ca^{2+} associated with permeability transition (19). Treatment of mitochondria with oligomycin or sulfide in the presence of CaP_i preserved the pyridine nucleotide redox couple ratios as compared to CaP_i alone, but less so than in untreated control samples. On the other hand, antimycin, in the presence of CaP_i, potentiated the decrease in pyridine nucleotide redox couple ratios.

The increased oxidation of NADH and NADPH observed with antimycin treatment may be due to an increase in H_2O_2 production by the respiratory chain. Inhibition of electron transport with antimycin is thought to result in an increase in H_2O_2 formation due to the electron leakage at ubiquinone (26–27). However, since GSSG was not detected under these conditions, which consequently result in the rapid release of matrix GSH, it is possible that mitochondria have an extensive capacity to reduce H_2O_2, as shown for *t*-butylhydroperoxide (28).

Inducing agents such as *t*-butylhydroperoxide or hydrogen peroxide are thought to induce permeability transition due to the oxidation of pyridine nucleotides (4,13–14). However, based on the findings we report, the redox status of pyridine nucleotides alone does not seem to regulate permeability transition. This observation is based on the finding that treatment with antimycin resulted in an increased oxidation of both NADH and NADPH as compared to CaP_i treatment alone; but unlike CaP_i treatment, the addition of antimycin abolished large-amplitude swelling and diminished solute release (19). Although these findings do not challenge the significance of the pyridine nucleotide redox status in contributing to or regulating permeability transition, they do seem to suggest that pyridine nucleotide redox status alone does not determine the degree of permeability transition status, that is, permeability transition with or without large-amplitude swelling as well as the rate of solute release.

In addition to the inhibition of large-amplitude swelling (19), treatment with the respiratory inhibitors also resulted in the preservation of ADP. Oligomycin, sulfide, and antimycin added to CaP_i-treated mitochondria maintained ADP levels respectively 57%, 46%, and 40% of controls, greater than CaP_i treatment alone following a 5 min incubation. Previous studies have reported that ADP prevents permeability transition (10,29–30). In these studies, the end point for determining permeability transition was large-amplitude swelling rather than solute release. The release of solutes may be a more sensitive indicator of permeability transition rather than larger-amplitude swelling (19). In the current study, treatment with metabolic inhibitors coincided with preserved ADP levels and prevented large-amplitude swelling, although inner membrane permeability was not inhibited (19). These findings suggest that CsA-sensitive permeability transition occurs to varying degrees in mitochondria, both with and without swelling, and with varying levels of solute release.

To determine whether ADP prevented large-amplitude swelling of the mitochondria in the presence of CaP_i, we monitored large-amplitude swelling in the presence of increasing concentrations of ADP. ADP prevented large-amplitude swelling in a concentration-dependent manner in the presence of CaP_i (Figure 6). A recent report by Lapidus and Sokolove (31) concluded that P_i triggers the permeability transition by lowering the matrix concentration of ADP. Their findings also suggest that the presence of ADP inhibits only the swelling induced by P_i (31). Finally, Bernardi and coworkers (32) have described the modulation of the mitochondrial permeability transition pore by the oxidation of pyridine nucleotide(s) and/or dithiol(s) at two separate sites. Thus, the nature of the mitochondrial permeability transition has become more complex and appears to require the consideration of two distinctly different sites that are responsive to oxidation of protein thiols or the oxidation of NADH and/or NADPH. Our observations with both the oxidation and loss of the oxidized pyridine nucleotides through the pore to the outside of mitochondria undergoing permeability transition is of considerable interest concerning the pathological consequences to intact cells.

In conclusion, we report that during CaP_i-induced permeability transition, GSH, NADH, NADPH, ATP, and ADP are rapidly depleted. Under these conditions, NADH and NADPH undergo rapid and extensive oxidation with the subsequent release of NAD^+ and $NADP^+$ from the mitochondrial matrix within a 5 min incubation. However, GSH release occurred without oxidation as no increase in GSSG was observed. This release possibly occurs through the putative Ca^{2+}-dependent inner membrane pore as the oxidation and release of GSH and pyridine nucleotides were prevented by CsA. ATP and ADP are also rapidly depleted, with the concomitant formation of AMP during CaP_i treatment. The addition of either oligomycin, sulfide, or antimycin preserved GSH, NADH, NADPH, ATP, and ADP to varying degrees. Maintenance of ADP by the respiratory inhibitors may contribute to the partial protection observed against permeability transition.

ACKNOWLEDGMENTS

This research was supported by NIEHS grants ES-01978 and ES-00210.

REFERENCES

1. Halestrap AP, Griffiths EJ, Connern CP. Mitochondrial calcium handling and oxidative stress. Biochem Soc Trans 1993; 21: 353–358.
2. Crompton M, Costi A, Hayat L. Evidence for the presence of a reversible Ca^{2+}-dependent pore activated by oxidative stress in heart mitochondria. Biochem J 1987; 245: 915–918.
3. Crompton M, Costi, A. Kinetic evidence for a heart mitochondrial pore activated by Ca^{2+}, inorganic phosphate and oxidative stress. A potential mechanism for mitochondrial dysfunction during cellular Ca^{2+} overload. Eur J Biochem 1988; 178: 489–501.
4. Gunter TE, Pfeiffer DR. Mechanisms by which mitochondria transport calcium. Am J Physiol 1990; 258: C755–C786.
5. Hunter DR, Haworth RA. The Ca^{2+} induced membrane transition in mitochondria. III. Transitional Ca^{2+} release. Arch Biochem Biophys 1979; 195: 468–477.

6. Palmer JW, Pfeiffer DR. The control of Ca^{2+} release from heart mitochondria. J Biol Chem 1981; 256: 6742–6750.
7. Savage MK, Jones DP, Reed DJ. Calcium- and phosphate-dependent release and loading of glutathione by liver mitochondria. Arch Biochem Biophys 1991; 290: 51–56.
8. Crompton M, Ellinger H, Costi A. Inhibition by cyclosporin A of a Ca^{2+}-dependent pore in heart mitochondria activated by inorganic phosphate and oxidative stress. Biochem J 1988; 255: 357–360.
9. Broekemeier KM, Dempsey ME, Pfeiffer DR. Cyclosporin A is a potent inhibitor of the inner membrane permeability transition in liver mitochondria. J Biol Chem 1989; 264: 7826–7830.
10. Halestrap AP, Davidson AM. Inhibition of Ca2(+)-induced large-amplitude swelling of liver and heart mitochondria by cyclosporin is probably caused by the inhibitor binding to mitochondrial-matrix peptidyl-prolyl *cis-trans* isomerase and preventing it interacting with the adenine nucleotide translocase. Biochem J 1990; 268: 153–160.
11. Griffiths EJ, Halestrap AP. Further evidence that cyclosporin A protects mitochondria from calcium overload by inhibiting a matrix peptidyl-prolyl *cis-trans* isomerase. Implications for the immunosuppressive and toxic effects of cyclosporin. Biochem J 1991; 274: 611–614.
12. Connern CP, Halestrap AP. Purification and N-terminal sequencing of peptidyl-prolyl *cis-trans*-isomerase from rat liver mitochondrial matrix reveals the existence of a distinct mitochondrial cyclophilin. Biochem J 1992; 284: 381–385.
13. Rizzuto R, Pitton G, Azzone GF. Effect of Ca^{2+}, peroxides, SH reagents, phosphate and aging on the permeability of mitochondrial membranes. Eur J Biochem 1987; 162: 239–249.
14. Lê-Qûoc D, Lê-Qûoc K. Relationships between the $NAD(P)^+$ redox state, fatty acid oxidation, and inner membrane permeability in rat liver mitochondria. Arch Biochem Biophys 1989; 273: 466–478.
15. Schlegel J, Schweizer M, Richter C. "Pore" formation is not required for the hydroperoxide-induced Ca^{2+} release from rat liver mitochondria. Biochem J 1992; 285: 65–69.
16. Lehninger AL, Carafoli E, Rossi CS. Energy-linked ion movements in mitochondrial systems. Adv Enzymol 1967; 29: 259–320.
17. Carbonera D, Azzone GF. Permeability of inner mitochondrial membrane and oxidative stress. Biocheim Biophys Acta 1988; 943: 245–255.
18. Savage MK, Reed DJ. Oxidation of pyridine nucleotides and depletion of ATP and ADP during calcium- and inorganic phosphate-induced mitochondrial permeability transition. Biochem Biophys Res Commun 1994; 200: 1615–1620.
19. Savage MK, Reed DJ. Release of mitochondrial glutathione and calcium by a cyclosporin A-sensitive mechanism occurs without large amplitude swelling. Arch Biochem Biophys 1994; 315: 142–152.
20. Schnaitman C, Greenwalt JW. Enzymatic properties of the inner and outer membranes of rat liver mitochondria. J Cell Biol 1968; 38: 158–175.
21. Bradford M. A rapid and sensitive method for the quantitation of microgram quantities of protein utilizing the principle of protein–dye binding. Anal Biochem 1976; 72: 248–254.
22. Reed DJ, Babson JR, Beatty PW, Brodie AB, Ellis EE, Potter DW. High-performance liquid chromatography analysis of nanomole levels of glutathione, glutathione disulfide, and related thiols and disulfides. Anal Biochem 1980; 106: 55–62.
23. Jones DP. Determination of pyridine dinucleotides in cell extracts by high-performance liquid chromatography. J Chromatogn 1981; 225: 446–449.
24. Reed DJ, Savage MK. Influence of metabolic inhibitors on mitochondrial permeability transition and glutathione status. Biochim Biophys Acta 1995; 1271: 43–50.
25. Richter C, Schlegel J. Mitochondrial calcium release induced by prooxidants. Toxicol Lett 1993; 67: 119–127.
26. Loschen G, Azzi A, Flohé L. Mitochondrial H_2O_2 formation: relationship with energy conservation. FEBS Lett 1973; 33: 84–88.

27. Boveris A, Cadenas E, Stoppani AOM. Role of ubiquinone in the mitochondrial generation of hydrogen peroxide. Biochem J 1976; 156: 435–444.
28. Olafsdottir K, Reed DJ. Retention of oxidized glutathione by isolated rat liver mitochondria during hydroperoxide treatment. Biochim Biophys Acta 1988; 964: 377–382.
29. Novgorodov SA, Gudz TI, Milgrom YM, Brierley GP. The permeability transition in heart mitochondria is regulated synergistically by ADP and cyclosporin A. J Biol Chem 1992; 267: 16274–16282.
30. Haworth RA, Hunter DR. Allosteric inhibition of the Ca^{2+}-activated hydrophilic channel of the mitochondrial inner membrane by nucleotides. J Membrane Biol 1980; 54: 231–236.
31. Lapidus RG, Sokolove PM. The mitochondrial permeability transition. Interactions of spermine, ADP, and inorganic phosphate. J Biol Chem 1994; 269: 18931–18936.
32. Constantini P, Chernyak BV, Petronilli V, Bernardi P. Modulation of the mitochondrial permeability transition pore by pyridine nucleotides and dithiol oxidation at two separate sites. J Biol Chem 1996; 271: 6746–6751.

27
Flavonoid Functions In Vivo: Are They Predominantly Antioxidants?

Wolf Bors
Institut für Strahlenbiologie
GSF Forschunsgzentrum für Umwelt und Gesundheit, Neuherberg
Oberschleissheim, Germany

INTRODUCTION

Flavonoids are polyphenolic secondary plant metabolites which commonly exist as multiple *O*-and *C*-glycosidic derivatives (1,2), but also may be present as aglycones (3). They are an important part of the human diet (4–6) –consider the recent discussions on the connection between red wine consumption and reduced risk of heart diseases, the so-called "French paradox" (7–9) or the studies on the proanthocyanidin ingredients in green tea (10,11). They are also considered the active principle in a number of medicinal plants (12,13). Owing to these potential benefits to human health, polyphenolic plant metabolites are the major class of the recently popular *phytochemicals* (14,15). Early in the beginning of the research into the structures and functions of flavonoids, their antioxidative capacities, particularly with respect to stabilizing foodstuffs, was recognized (16,17). Up to now, this has remained the most important topic of investigation, despite the fact that various other functions have been attributed to them over the years. For instance, they are mutagenic yet they are also anticarcinogens; they exhibit biocidal effects and have antifertility properties, yet express beneficial effects in inflammatory and immunomodulatory systems and interact with signal transduction processes (Table 1).

In this presentation we shall review the available evidence of all functions that are *not* correlated with antioxidative and/or radical-scavenging properties. Reports on the biocidal effects are also omitted, as this property, along with mutagenicity, is probably based on the formation of reactive oxygen species (ROS). It may be worth mentioning, however, that there is evidence that ROS are not only formed by autoxidation of flavonoids and polyphenols, but that oxidative reactions may be elicited in cell systems by these compounds (18). An interesting feature among antiviral activities is the synergism between quercetin and cytokines (19) or other compounds (20,21). This suggests that the antiviral function is different from cytotoxic processes involving ROS formation and DNA cleavage.

Table 1. Flavonoid Functions

Function	In vitro	In plants	In animals
Antioxidant	+	+	(+)
Cytotoxic (biocidal)	+	–	–
Mutagen	+	–	–
UV-screen	+	+	–
Feeding repellant	+	+	–
Phytoalexin	–	+	–
Antiulcerogen	–	–	+
Anticarcinogen	–	–	+
Anti-inflammatory	(+)	–	+
Immunomodulatory	(+)	–	+
Signal-modifying (gene expression)	(+)	+	+
Radioprotective/anticlastogenic	–	–	+

There exists quite a controversy concerning the bioavailability of flavonoids in animals, but nevertheless we shall demonstrate that the in vivo functions are distinct from and at least equally important as those under in vitro conditions, comprising mainly the antioxidative properties. We base this statement on the references cited in two reviews covering this area quite extensively (22,23), with additional evidence from the literature up to spring of 1996.

BIOCHEMISTRY

Metabolism

Flavonoids, as ubiquitous plant metabolites, are exclusively catabolized in animal and human organs. While the biochemical and biological functions can, in most cases, be investigated in cell culture systems, any in vivo function of these compounds hinges on their metabolism in the digestive system (8). Three arguments highlight this field:

1. Hydrolysis of flavonoid glycosides by intestinal bacteria (24) causes activation, i.e., formation of potentially mutagenic flavonoid aglycones (25).
2. The intestinal microflora is capable of cleaving the heterocyclic C-ring of flavonoid aglycones, rendering the products completely inactive with respect to antioxidative or radical-scavenging potential (26–29).
3. Since the early statement of Clark and Mackay (30) has never been clearly refuted, we have to assume that flavonoids–aglycones to a higher extent than glycosides (31)–are barely transferred from the digestive system into the bloodstream (32,33).

It is so far unknown to what extent an oxidative metabolism, as modeled with peroxidase (34,35), occurs in vivo. A recently described enzymatic activity, causing sulfation of quercetin and catechin in rat liver (36), probably effected by sulfotransferases reactive with xenobiotics (37,38), awaits further elucidation. It is unclear whether it activates these compounds or facilitates excretion. Gastroprotective/antiulcerogenic effects of flavonoids have been observed after oral (39,40) or intraperitoneal (41) delivery. The mechanism is still unknown, with the damage to the gastric mucosa being a multicomponent phenomenon, on the one hand, and there being a lack of structure–activity relationship studies, on the other. Both involvement of the platelet-activating factor (41) and of nonprotein sulfhydryl groups (40) have been proposed.

Because of the rather insufficient information on the metabolic fate of flavonoids, it is not surprising that drugs with flavonoids as the major pharmacologically active ingredients remain elusive. Nevertheless, there are a few examples where the pharmacological functions have been studied extensively: the aglycone (+)-cyanidanol [(+)-catechin, (42)], the glycosides *O*-(β-hydroxyethyl)rutoside (hydroxyethylated derivatives of rutin; (43), EGb761 (*Ginkgo biloba* extracts; (44), and a mixture of both aglycone and glycoside, Daflon 500® (45). Flavonoid glycosides are considered more hydrophilic than the respective aglycones, but improved resorption of such glycosides was investigated in detail only for the rutin derivatives, hydroxyethyl rutosides (31). Furthermore, these studies may bear no relevance to the fate of genuine flavonoid glycosides, as the various hydroxyethyl groups may merely protect the rutinose moiety. Daflon 500®, containing the micronized glycosides diosmin (5,7,3′-trihydroxy-4′-methoxyflavon-7-*O*-rhamnoglucoside) and its 2,3-saturated analog hesperidin in a ratio of 9:1, has shown promising anti-inflammatory and venotropic activity (46). It is interesting to note that cyanidanol, subsequently taken off the market owing to serious side-effects (47,48), is the only aglycone among these potential drugs. Could it be that slow metabolic release of the aglycone from the other glycosides is more beneficial than an acute dose of the active principle?

Silybinin, a nonglycosylated flavonoid and derivative of dihydroquercetin isolated from *Silybium marianum* (Mary's thistle), and a topic of research since the late 1960s (49), has been shown to be an effective hepatoprotective agent (50). Cyanidanol (51), as well as other flavonoids (52–54), has similarly shown hepatoprotective properties. This is in obvious contradiction to the statement that flavonoid aglycones are not (or only poorly) resorbed from the digestive system into the bloodstream (30–33): the liver is supplied via the portal vein or the lymph system from the small intestine, whereas nonresorbed substances are metabolized and excreted via the large intestine. Another route of flavonoid drug delivery which would not be subject to metabolic degradation is topical application onto skin. This technique has frequently been applied in studies of anticarcinogenic effects of flavonoids involving skin tumors in rodents (55–58), but so far on an experimental basis.

The major pharmacological efforts are consequently directed toward stabilization of flavonoids against digestive degradation. This approach has recently been realized with IdB 1016, a 1:1 complex of silybinin and phosphatidylcholine (59) with high potential as an antioxidative and anti-inflammatory drug. In view of the intense efforts, major breakthroughs may be anticipated in the area of traditional medicine (60). For instance, studies on the active principles of Chinese herbal potions are considerably advanced (12,61,62). There is also strong interest in the anti-HIV activity of flavonoids (63–67), as both antioxidative (64), antiviral (65), and enzyme-inhibiting activities (63,67) may be involved.

Protein Interactions

The aforementioned difficulties for an unambiguous interpretation of in vivo functions of flavonoids is probably due to the fact that these compounds exhibit an extraordinarily wide range of activities toward proteins and enzymes. Polyphenols may react unspecifically with proteins (68,69)–e.g., the astringency of some flavonoids is explained by this mechanism (70)–but more specific effects, such as intracellular interaction with calmodulin (71), are also known. Special topics are the interactions with receptors, altering their functions, or with enzyme activities, which are discussed separately.

Effect on Receptor Functions

Receptors as membrane-bound or membrane-penetrating proteins have the important functions of transferring chemical signals between two different organelles. As such they are an important part of the signaling process, whose influence by flavonoids will be discussed in more detail later. Both inhibition (72,73) and stimulation (74,75) of receptor functions have been observed. The most detailed studies involved ligand affinity with the benzodiazepine (76) and the adenosine (77) receptors of the central nervous system. The structure–activity relationships (SAR) of both studies demonstrated that hydroxyl groups are not essential and methoxylation as opposed to acetylation enhances affinity: criteria quite in contrast to those for antioxidative or radical-scavenging functions. While the benzodiazepine receptor involves primarily γ-aminobutyric acid-driven (GABAergic) neurons (76), three subtypes of adenosine receptors with different functions are known, corresponding to distinct specificities of the flavonoids investigated (77).

Effect on Enzyme Activities

Most of these studies can be or were undertaken under in vitro conditions, for example, with pure enzyme peparations. Since it it can be expected that many of the in vivo effects observed for flavonoids are related to specific enzyme interactions, it is unfortunate that in this research area there are the most glaring deficiencies in the study of flavonoids: less than 20 percent of the studies pertain to SAR comparisons, and most of them are limited to one isoenzyme. Those few studies which take into consideration both aspects reveal dichotomous effects of flavonoids: (i) enzymes are either inhibited or activated depending on flavonoid structure (78–80) or (ii) isoenzymes are differently affected by individual flavonoids, as shown for some drug- and xenobiotic-metabolizing cytochrome P450 enzymes (81–83). Emphasizing the first point, an early comparison of the effects of polyphenols on various enzyme activities (84) demonstrated that both stimulatory and inhibitory effects can be observed but, more importantly, examples from all classes of enzymes are affected (85). A combination of effects on specific enzyme activation and inhibition may be the basis of the rarely reported *regulation* of enzyme activities (86,87). Evidence for inhibitory effects by far exceeds that for stimulatory activities and will be discussed first.

Inhibition of Enzyme Activities

Lacking a detailed description of the enzymes involved, mitochondrial uncoupling may describe a rather general inhibitory effect on the metabolic pathways in this organelle (88,89). In a review, Middleton and Kandaswami (22) have specified the crucial metabolic functions of a number of enzymes inhibited by flavonoids. A complete list of the affected enzymes appeared in 1992 (85) and was later expanded considerably (23). Studies with SAR or mechanistic implications were preferentially cited, yet, as noted before, most reports are confined to only one enzyme and/or one individual flavonoid. When only single flavonoids were examined, the favored compounds were quercetin, genistein and catechin.

Quercetin. Quercetin was examined as inhibitor of xanthine oxidase and xanthine dehydrogenase (90), of myeloperoxidase (91), of lipoxygenase and cyclooxygenase (92), of cytochrome P450 (93), of the constitutive nitric oxide synthase (94) [this activity has since been corroborated for other flavonoids (95)], of phenol sulfotransferase (96), of protein kinase C (97), of Ca^{2+} -ATPase (98,99), and of ornithine decarboxylase (100). The inhibition of phosphodiesterases, leading to the accumulation of cAMP, by quercetin further potentiated the inhibitory effect of prostacyclin on aggregation of human platelets (101).

Genistein. This isoflavone is routinely described as a "specific" inhibitor of tyrosine protein kinase (102). Inhibition has also been observed so far for histidine protein kinase (103), for Ras-MAP kinase (104), for topoisomerases (105), and for antioxidant enzymes (106). Less specific functions involve the blockage of thromboxane A_2-mediated platelet function (107), in vitro angiogenesis (108), a general anticarcinogenic effect (109,110), and induction of apoptosis in Jurkat T leukemic cells (111). It provides also one of the few examples where a distinction between in vitro and in vivo effects was found: c.f. the formation of micronuclei in mouse splenocytes (112).

Catechin. Catechin has been investigated both in its drug form, (+)-cyanidanol (113,114) and as a component of green tea (57,115).

Considering that enzymes of all classes are inhibited, for example, by quercetin, a common mechanism seems to be questionable. Most SAR studies describe similar structural requirements for optimal activity (e.g., a 2,3 double bond conjugated with the 4-oxo group, catechol group in the B-ring, and 3- or 5-hydroxy groups) and the same criteria enhance autoxidation as well as radical-scavenging efficiencies. It thus appears conceivable that an aroxyl radical intermediate might be the common denominator. However, this mechanism is unlikely to be applicable to the other enzyme classes, except perhaps for oxidoreductases. A mechanism common to the most active flavonoids may involve the coplanarity of all three rings–which evidently would not include the various effects reported for the flavanols, i.e., catechin and its derivatives–or they may have structures similar to the actual enzyme substrate or cosubstrates. The latter effect seems to be the case for purine-dependent proteases (116). A good example for structural elucidation of inhibiting substrate analogs has been given for the flavonoid-related aurones in the iodothyronine deiodinase reaction (117), which is also inhibited by flavonoids themselves (118).

Stimulation of Enzyme Activities

General stimulatory effects have been reported in two cases: reductive iron release from ferritin (119) and increase of cytoplasmic free calcium (120). The latter observation is in apparent contradiction to the various reports on the inhibition of Ca^{2+}-ATPases (121,122). While the evidence for stimulation of enzyme activities is considerably outpaced by the reports on inhibitory effects of flavonoids, it is worth noting that stimulation mainly involves detoxifying enzymes, and thus could contribute to an indirect antioxidative effect. The stimulation of superoxide dismutase (SOD) activity in patients with liver cirrhosis (123) is an early example of in vivo activities of flavonoids and may also partially explain their hepatoprotective effects. Other examples in which detoxifying enzymes are activated by flavonoids are the stimulation of cyclooxygenase (prostaglandin synthase) at low flavonoid concentrations (79,124), of NADPH diaphorase (125) [with an opposing claim of inhibition by quercetin (126)], and of the drug-metabolizing system, located primarily in hepatic tissue (81–83,127,128), which according to Obermeier and colleagues (129) may be confined to catechins. The reported induction of transferase activity is rather episodic (130) and is not supported by other studies (131–133). Anthrones show an inducing effect on the activity of ornithine decarboxylase (134) that is in contrast to the inhibitory effect of flavonoids (57,135).

BIOLOGY

The function of flavonoids in plants, the site of their biosynthesis, will not be discussed here. Limiting the discussion to their effects in animals and humans, where these compounds are subject to catabolic degradation, however, brings us to the uncertainty concerning the metabolic fate of these compounds when taken up in the diet (8). For example, the assumption of loss of antioxidative function of metabolites (26–29) is opposed by the retention of their activity in liver (50–54) and a profusion of data on in vivo functions of flavonoids, leaving aside the relevance of cell culture studies for interpretation of effects in whole animals. The few studies on the effects of flavonoids after oral intake, apart from the in vivo studies with Daflon 500®(45), *Ginkgo biloba* (44) or silymarin (49), concern anticlastogenic properties in radiation-exposed animals (136) and humans (137). While this is evidently correlated with the antioxidative properties (136), the antimetastatic effects in lungs of melanoma-bearing mice (138) still lack an explanation. Considering the human health aspect, the correlation of flavonoid intake with food may be of special importance (5,6,8,139,140), as exemplified by the effect of wine (7,9,141) or green tea consumption (10) on evaluation of long-term risk of coronary heart disease and cancer.

Two independent studies provided evidence, that flavonoids affected rat blood vessels and surrounding tissue. Potentiation of contractile responses (142) was found to be dependent on the presence of three adjacent hydroxy groups in either the A- or the B-ring and was considered a postsynaptic process of catecholamine response. The vasodilatory effects on aortic smooth muscle (143), in contrast, rated flavonols > flavones > flavans and proposed the inhibition of protein kinase C (and possibly also phosphodiesterase or Ca^{2+} uptake) as the basic mechanism. An in vitro SAR study of flavonoids and phenolic acids affecting sperm motility (144) helps to explain the antifertility effect of these compounds on the basis of ROS formation. The effect was more pronounced for hydroxycinnamic acid derivatives than the three flavonoids tested with decreasing efficiency in the order rutin > morin > quercetin. Hepatoprotective functions may also be considered a physiological role of flavonoids and these aspects have already been discussed earlier. At present, most of these studies do not allow a clear distinction between in vitro and in vivo studies.

Anticarcinogenicity

Unlike the antagonistic modes of antioxidative versus biocidal functions of flavonoids, the dichotomy between mutagenicity versus anticarcinogenicity has only recently been discussed (145,146). Despite the multiple reports on anticarcinogenic properties of flavonoids (5,58,147–151), it is surprising that these compounds are still considered potential carcinogens in view of their in vitro mutagenicity. A recent SAR study described differential inhibitory effects of polyhydroxylated (quercetin, dihydroquercetin) or polymethoxylated (5,6,7,8,4′-pentamethoxyflavone = tangeretin; 5,6,7,8,3′,4′-hexamethoxyflavone = nobiletin) flavonoids on human squamous tumor cells, solid gliosarcoma cells, and fibroblast cells (152). Fibroblasts were relatively insensitive toward all flavonoids tested. In contrast, proliferation of both types of tumor cells were effectively inhibited by the polymethoxylated compounds, whereas quercetin and dihydroquercetin were active against the squamous tumor cells only at higher concentrations. In a thorough investigation, the radical intermediates of the potent synthetic antitumor drug flavone-8-acetic acid were investigated by pulse radiolysis (153). Yet, that a radical mechanism is indeed the basis of the antitumor activity of this compound may be doubtful in view of reports claiming its interaction with cytokines (154,155), or with respect to the stimulation of nitric oxide formation by the drug (156).

Further insight into the anticarcinogenic mechanism of flavonoids might be gained by looking at the individual steps in the multistage hypothesis of cancer development, which are all affected by these compounds (157), namely, initiation, promotion, and progression/proliferation.

Initiation either by directly acting carcinogens (56,158–160) or after metabolic activation of procarcinogens (55,161–163) can be prevented both by flavonoids and by other polyphenols. Particularly efficient polyphenols are tannic (164) and ellagic acids (55,164). The inhibitory effect on the metabolic activation may be related to the inactivation of drug-metabolizing cytochrome P450 isoenzymes (81).

Anti-tumor-*promoting* activity of flavonoids is most often connected with the inhibition of the action of phorbol esters (100,155,165–167) and/or the inactivation of ornithine decarboxylase (57,168). The possible correlation with oxidative events, indicated by the proposed involvement of lipoxygenase inhibition (168), has recently been brought into perspective (169). Further support for a cytotoxic, i.e., DNA-fragmenting effect, of flavonoids on tumor cells comes from a SAR study (170) and another investigation involving the isoflavone genistein (171).

Inhibition of directly acting carcinogens may be rationalized if we take into account the various reports that propose *antiproliferative* effects of flavonoids (152,165,172), even though no obvious correlation is evident at present. The observations that the flavonol quercetin (157), the flavone apigenin (173), and the isoflavone genistein (174) all arrest cell cycle progression in late G_1 and G_2-M stages are interesting in this context.

These manifold interactions of various flavonoids with tumor cells suggest that different mechanisms may occur, only some of them involving ROS. Compounds such as polymethoxylated flavonoids effectively inhibit tumor cell proliferation (152). These substances have been shown to exhibit antioxidative capacity, albeit with different SAR than those for polyhydroxylated flavonoids (175), and probably do not react via radical scavenging.

Phagocytes and Inflammatory Systems

This second major topic on physiological effects of flavonoids has also been well reviewed (22,176). The majority of reports again involve individual flavonoids, frequently with quercetin as the favorite compound, and only few SAR studies have attempted to establish some structural perspective. It is also evident that these studies are confined to rather specific aspects of inflammatory systems, for example, effects on luminol-enhanced chemiluminescence of activated neutrophils (177,178); formation of ROS by stimulated neutrophils (179); and aggregation and secretion of platelets (107,180,181).

Mechanistic implications can be drawn from the inhibition by flavonoids of almost all enzymes involved in arachidonic metabolism (182): (i) Inhibition of arachidonic acid release (183) can probably most easily be traced to interference with phospholipase A_2 (184). (ii) Inhibition of eicosanoid/prostanoid production (181,185) is obviously related to the inhibitory effect of flavonoids on cyclooxygenase/prostaglandin synthetase (79,186,187). The latter activity is closely correlated with the inhibition of 5-lipoxygenase (186–188), converting arachidonic acid into leukotrienes.

It may seem surprising that the ischemia/reperfusion syndrome has only recently gained attention as an inflammatory system subject to interference by flavonoids (189–191), even though formation of radicals and prevention by other antioxidants is well established (192). Synthetic chalcones and flavones have now been prepared as specific inhibitors of the generation of inflammatory mediators, e.g., leukotriene B_4 preferentially to thromboxane B_2 (193). This can be extrapolated to the inhibition of phospholipase A_2 and 5-lipoxygenase (see above).

Immunomodulation

Immunomodulation is the major research area of Middleton and coworkers and thus only the most pertinent reviews (22,194) and SAR studies (195) published by this group are mentioned here. It is worth noting, however, that most of these studies are confined to histamine release (196) and T cell functions (195,197,198). A more specific effect, e.g., binding of flavonoids to erythrocytes and subsequent induction of antibody formation, has been reported for (+)-cyanidanol (199). Inhibition of the complement system, preferentially of the classical pathway involving IgG or IgM antibodies, is another example of the immunomodulatory function of flavonoids (200).

Gene Expression and Signal Transduction

During the previous discussion of the biological effects of flavonoids we repeatedly alluded to potential interactions of flavonoids with cytokines, and thus indirectly to signal transduction processes. Under the topic of protein interaction, we discussed involvement with receptor functions, another aspect of signaling events. Although there already exist a considerable number of reports implicating flavonoids in such reactions, this topic has never been reviewed and SAR studies have yet to be performed. The list of pertinent publications may be further expanded if we include the many reports of the inhibitory effects of flavonoids with protein kinases (201–205) or in tumor promotion/proliferation studies (152,165,172).

The involvement of flavonoids in gene expression is overwhelmingly represented by reports on the regulation of the heat shock response by quercetin (206–209)–with one early study involving other flavonoids (210)–and induction of apoptosis by quercetin (207) and other flavonoids (211). Only two other gene response have been studied thus far: the multidrug resistance gene 1 (212) and the cell adhesion protein (213).

Both inhibition (214) and stimulation (215) of intercellular (gap junction) communication by green tea catechins have been reported. Apigenin and tangeretin also show a stimulatory effect (216), thus making this the more likely response. The studies concerning signal transduction and the involvement of cytokines are still too unspecific to allow definite mechanistic conclusions (217–220).

CONCLUDING REMARKS

Taking together all the evidence of flavonoid functions in animals and humans together– i.e., radical scavenging versus antioxidative potential; mutagenicity and biocidal activities versus anticarcinogenicity; metabolic fate versus physiological effects on inflammatory, immune, and/or signal transduction system–a rather confusing picture may emerge. This is, hopefully, an expression of the extremely vigorous research activities involving flavonoids, particularly in the field of traditional medicine, that will probably require further efforts before the mechanisms of in vivo functions of flavonoids become clearer. Nevertheless, we consider, this topic one of the most exciting areas for future research, in contrast to the antioxidative and even anticarcinogenic effects of flavonoids which have been investigated extensively and where few further advances can be expected. Nevertheless, we can already state with certainty that flavonoids play a much wider role than acting merely as antioxidants, which function was the first to be demonstrated for this extremely versatile class of plant secondary metabolites that appear to be important for human health.

ACKNOWLEDGMENTS

The stimulating discussions with Werner Heller, Christa Michel, and Manfred Saran are greatly appreciated.

REFERENCES

1. Harborne JB (ed.). The Flavonoids. Advances in Research since 1980. London: Chapman & Hall; 1988.
2. Harborne JB (ed.). The Flavonoids. Advances in Research since 1986. London: Chapman & Hall; 1993.
3. Wollenweber E, Dietz VH. Occurrence and distribution of free flavonoid aglycones in plants. Phytochemistry 1981; 20: 896–932.
4. Kühnau J. The flavonoids. A class of semi-essential food components: their role in human nutrition. World Rev Nutr Diet 1976; 24: 117–191.
5. Stavric B, Matula TI. Flavonoids in foods: their significance for nutrition and health. In Ong ASH, Packer L, eds. Lipid-Soluble Antioxidants: Biochemistry and Clinical Applications. Basel: Birkhäuser; 1992: 274–294.
6. Hertog MGL, Hollman PCH, Katan MB. Content of potentially anticarcinogenic flavonoids of 28 vegetables and 9 fruits commonly consumed in the Netherlands. J Agric Food Chem 1992; 40: 2379–2383.
7. Waterhouse AL. Wine and Heart Disease. Chem Ind 1995: 338–341.
8. Formica JV, Regelson W. Review of the biology of quercetin and related bioflavonoids. Food Chem Toxical 1995; 33: 1061–1080.
9. Muldoon MF, Kritchevsky SB. Flavonoids and heart disease–evidence of benefit still fragmentary. Br Med J 1996; 312: 458–459.
10. Imai K, Nakachi K. Cross sectional study of effects of drinking green tea on cardiovascular and liver diseases. Br Med J 1995; 310: 693–696.
11. Yen GC, Chen HY. Relationship between antimutagenic activity and major components of various teas. Mutagenesis 1996; 11: 37–41.

12. Homma M, Oka K, Yamada T, Niitsuma T, Ihto H, Takahashi N. A strategy for discovering biologically active compounds with high probability in traditional Chinese herb remedies: an application of Saiboku-to in bronchial asthma. Anal Biochem 1992; 202: 179–187.
13. Perusquia M, Mendoza S, Bye R, Linares E, Mata R. Vasoactive effects of aqueous extracts from five Mexican medicinal plants on isolated rat aorta. J Ethnopharmacol 1995; 46: 63–69.
14. Huang MT, Osawa T, Ho CT, Rosen RT (eds.). Food Phytochemicals for Cancer Prevention. I. Fruits and Vegetables. Washington, D.C.: ACS Symposium Series, 1994;546.
15. Ho CT, Osawa T, Huang MT, Rosen RT (eds.). Food Phytochemicals for Cancer Prevention. II. Teas, Spices and Herbs. Washington, D.C.: ACS Symposium Series 547; 1994.
16. Richardson GA, El-Rafey MS, Long ML. Flavones and flavone derivatives as antioxidants. J Dairy Sci 1947; 30: 397–411.
17. Simpson TH, Uri N. Hydroxyflavones as inhibitors of the aerobic oxidation of unsaturated fatty acids. Chem Ind 1956: 956–957.
18. Schmalle HW, Jarchow OH, Hausen BM, Schulz KH. Aspects of the relationships between chemical structure and sensitizing potency of flavonoids and related compounds. In Cody V, Middleton E, Harborne JB, eds. Plant Flavonoids in Biology and Medicine. New York: Alan R Liss; 1986: 387–390.
19. Ohnishi E, Bannai H. Quercetin potentiates TNF-induced antiviral activity. Antiviral Res 1993; 22: 327–331.
20. Mucsi I. Combined antiviral effects of flavonoids and 5-ethyl-2′-deoxyuridine on the multiplication of herpes virus. Acta Virol 1984; 28: 395–400.
21. Vrijsen R, Everaert L, Boeyé A. Antiviral activity of flavones and potentiation by ascorbate. J Gen Virol 1988; 69: 1749–1751.
22. Middleton E, Kandaswami C. The impact of plant flavonoids on mammalian biology: implications for immunity, inflammation and cancer. In Harborne JB, ed. The Flavonoids: Advances in Research since 1986. London: Chapman & Hall; 1993: 619–652.
23. Bors W, Heller W, Michel C, Stettmaier K. Flavonoids and polyphenols: chemistry and biology. In Cadenas E, Packer I, eds. Handbook of Antioxidants. New York: Marcel Dekker; 1996: 409–466.
24. Bokkenheuser VD, Winter J. Hydrolysis of flavonoids by human intestinal bacteria. In Cody V, Middleton E, Harborne JB, Beretz A, eds. Plant Flavonoids in Biology and Medicine II. New York: Alan R Liss. [Progr Clin Biol Res 1988; 280: 143–145.]
25. Brown JP, Dietrich PS. Mutagenicity of plant flavonols in the Salmonella/mammalian microsome test. Activation of flavonol glycosides by mixed glycosidases from rat fecal bacteria and other sources. Mutat Res 1979; 66: 223–240.
26. Barz W. Microbial degradation of flavanoids, isoflavones and isoflavonoid phytoalexins. Bull Liaison, Groupe Polyphenols 1978; 8: 63–90.
27. Hackett AM. The metabolism of flavonoid compounds in mammals. In Cody V, Middleton E, Harborne JB, eds. Plant Flavonoids in Biology and Medicine. New York: Alan R Liss; 1986: 177–194.
28. Ibrahim ARS, Abul-Hajj YJ. Microbiological transformation of flavone and isoflavone. Xenobiotica 1990; 20: 363–373.
29. Manach C, Morand C, Texier O, Favier ML, Agullo G, Demigné C, Régérat F, Rémésy C. Quercetin metabolites in plasma of rats fed diets containing rutin or quercetin. J Nutr 1995; 125: 1911–1922.
30. Clark WG, Mackay EM. The absorption and excretion of rutin and related flavonoid substances. J Am Med Assoc 1950; 143: 1411–1455.
31. Balant LP, Wermeille M, Griffith LA. Metabolism and pharmacokinetics of hydroxyethylated rutosides in animals and man. Drug Metab Drug Interact 1984; 5: 1–24.
32. Hollman PCH, de Vries JHM, van Leeuwen SD, Mengelers MJB, Katan MB. Absorption of dietary quercetin glycosides and quercetin in healthy ileostomy volunteers. Am J Clin Nutr 1995; 62: 1276–1282.

33. Okushio K, Matsumoto N, Kohri T, Suzuki M, Nanjo F, Hara Y. Absorption of tea catechins into rat portal vein. Biol Pharm Bull 1996; 19: 326–329.
34. Schreier P, Miller E. Studies on flavonol degradation by peroxidase (donor: H_2O_2 oxidoreductase, EC 1.11.1.7.): Part 2–Quercetin. Food Chem 1985; 18: 301–317.
35. Oszmianski J, Lee CY. Enzymatic oxidative reaction of catechin and chlorogenic acid in a model system. J Agric Food Chem 1990; 38: 1202–1204.
36. Shali NA, Curtis CG, Powell GM, Broy AB. Sulphation of the flavonoids quercetin and catechin by rat liver. Xenobiotica 1991; 21: 881–893.
37. Mulder GJ, Jacoby WB. Sulphation. In Mulder GJ, ed. Conjugation Reactions in Drug Metabolism. London: Taylor & Francis, 1990: 108–161.
38. Falany CN. Molecular enzymology of human liver cytosolic sulfotransferases. Trends Pharmacol Sci 1991; 12: 255–259.
39. Attaguile G, Caruso A, Pennisi G, Savoca F. Gastroprotective effect of aqueous extract of *Cistus incamus* L in rats. Pharmacol Res 1995; 31: 29–32.
40. La Casa C, Martin Calero MJ, Alarcon de la Lastra C, Motilva V, Ayuso MJ, Martin Cordero C, Lopez Vergara A. Role of mucus secretion and sulfhydryl groups in gastroprotection mediated by a flavonoid fraction of *Bidens aurea*. Z Naturforsch 1995; 50c: 854–861.
41. Izzo AA, di Carlo G, Mascolo N, Capasso F, Autore G. Antiulcer effect of flavonoids. Role of endogenous PAF. Phytother Res 1994; 8: 179–181.
42. Conn HO. (ed.). (+)-Cyanidanol-3 in diseases of the liver. Proceedings of an International Workshop. San Francisco: Grune & Stratton; 1981.
43. Voelter W, Jung G. (eds.).*O*-(ß-Hydroxyethyl)-rutosid–Experimentelle und klinische Ergebnisse. Berlin: Springer; 1978.
44. Ferradini C, Droy-Lefaix MT, Christen Y (eds.). *Ginkgo biloba* extract (Egb761) as a free-radical scavenger. Adv Ginkgo biloba Extract Res 1993; 2.
45. Balas P (ed.). The pharmacological, pharmacodynamic and clinical properties of a new vasoactive agent: Daflon 500. Int Angiol 1988; 7 (Suppl. 2).
46. Labrid C, Mallet C, Freyria JL. Interférences du S5682 avec divers médiateurs de la réaction inflammatoire péri-veineuse. In Davy A, Stemmer R, eds. Phlébologie 89. Paris: Libbey Eurotext; 1989: 683–685.
47. Daniel PT, Holzschuh J, Berg PA. The pathogenesis of cyanidanol-induced fever. Eur J Clin Pharmacol 1988; 34: 241–247.
48. Jaeger A, Wälti M, Neftel K. Side effects of flavonoids in medical practice. In Cody V, Middleton E, Harborne JB, Beretz A, eds. Plant Flavonoids in Biology and Medicine II: Biochemical, Cellular, and Medicinal Properties. New York: Alan R Liss; 1988: 379–394.
49. Hahn G, Lehmann HD, Kürten M, Übel H, Vogel G. Zur Pharmakologie und Toxikologie von Silymarin, des antihepatotoxischen Wirkprinzipes aus *Silybum marianum* (L.) Gaertn. Arzneim Forsch 1968; 18: 698–703.
50. Ferenci P, Dragosics B, Dittrich H, Frank H, Benda L, Lochs H, Meryn S, Base W, Schneider B. Randomized controlled trial of silymarin treatment in patients with cirrhosis of the liver. J Hepatol 1989; 9: 105–113.
51. Altorjay I, Dalmi L, Sari B. In vitro and in vivo demonstration of the cytoprotective effect of (+)-cyanidanol-3. Acta Physiol Hung 1984; 64: 471–474.
52. Wagner H. Antihepatotoxic flavonoids. In Cody V, Middleton E, Harborne JB, eds. Plant Flavonoids in Biology and Medicine. New York: Alan R Liss; 1986: 545–558.
53. Davila JC, Lenherr A, Acosta D. Protective effect of flavonoids on drug-induced hepatotoxicity in vitro. Toxicology 1989; 57: 267–286.
54. Anon MT, Ubeda A, Alcaraz MJ. Protective effects of phenolic compounds on CCl_4-induced toxicity in isolated rat hepatocytes. Z Naturforsch 1992; 47c: 275–279.
55. Chang RL, Huang MT, Wood AW, Wong CQ, Newmark HL, Yagi H, Sayer JM, Jerina DM, Conney AH. Effect of ellagic acid and hydroxylated flavonoids on the tumorigenicity of

benzo[a]pyrene and (±)-7ß, 8a-dihydroxy-9a,10a-epoxy-7,8,9,10-tetrahydrobenzo [a]pyrene on mouse skin and in the newborn mouse. Carcinogenesis 1985; 6: 1127–1133.
56. Vijayaraghavan R, Sugendran K, Pant SC, Husain K, Malhotra RC. Dermal intoxication of mice with bis(2-chloroethyl)sulphide and the protective effect of flavonoids. Toxicology 1991; 69: 35–42.
57. Agarwal R, Katiyar SK, Zaidi SIA, Mukthar H. Inhibition of skin tumor promoter-caused induction of epidermal ornithine decarboxylase in SENCAR mice by polyphenolic fraction isolated from green tea and its individual epicatechin derivatives. Cancer Res 1992; 52: 3582–3588.
58. Perchellet JP, Gali HU, Perchellet EM, Lakas PE, Bottari V, Hemingway RW, Scalbert A. Antitumor-promoting effects of gallotannins, ellagitannins and flavonoids in mouse skin in vivo. In Huang MT, Osawa T, Ho CT, Rosen RT, eds. Food Phytochemicals for Cancer Prevention. I. Washington, D.C.: ACS Symposium Series; 1994; 546: 303–329.
59. Carini R, Comoglio A, Albano E, Poli G. Lipid peroxidation and irreversible damage in the rat hepatocyte model. Protection by the silybin–phospholipid complex IdB 1016. Biochem Pharmacol 1992; 43: 2111–2115.
60. Anton R. Flavonoïds and traditional medicine. In Cody V, Middleton E, Harborne JB, Beretz A, eds. Plant Flavonoids in Biology and Medicine II: Biochemical, Cellular, and Medicinal Properties. New York: Alan R Liss; 1988: 423–439.
61. Liu XF, Liu ML, Iyanagi T, Legesse K, Lee TD, Chen S. Inhibition of rat liver NAD(P)H: quinone acceptor oxidoreductase (DT-diaphorase) by flavonoids isolated from the Chinese herb *Scutellariae radix* (Huang Qin). Mol Pharmacol 1990; 37: 911–915.
62. Chang CW, Lin MT, Lee SS, Liu KCSC, Hsu FL, Lin JY. Differential inhibition of reverse transcriptase and cellular DNA polymerase-alpha activities by lignans isolated from Chinese herbs, *Phyllanthus myrtifolius* Moon, and tannins from *Lonicera japonica* Thunb and *Castanopsis hystrix*. Antiviral Res 1995; 27: 367–374.
63. Brinkworth RI, Stoermer MJ, Fairlie DP. Flavones are inhibitors of HIV-1 proteinase. Biochem Biophys Res Commun 1992; 188: 631–637.
64. Müller F. Reactive oxygen intermediates and human immunodeficiency virus (HIV) infection. Free Radical Biol Med 1992; 13: 651–657.
65. Hu CQ, Chen K, Shi Q, Kilkuskie RE, Cheng YC, Lee KH. Anti-AIDS agents. 10. Acacetin-7-*O*-ß-D-galactopyranoside, an anti-HIV principle from *Chrysanthemum morifolium* and a structure–activity correlation with some related flavonoids. J Nat Prod 1994; 57: 42–51.
66. Hayashi K, Kamiya M, Hayashi T. Virucidal effects of the steam distillate from *Houttuynia cordata* and its components on HSV-1, influenza virus, and HIV. Planta Medica 1995; 61: 237–241.
67. Pengsuparp T, Cai L, Constant H, Fong HHS, Lin LZ, Kinghorn AD, Pezzuto JM, Cordell GA, Ingolfsdottir K, Wagner H, Hughes SH. Mechanistic evaluation of new plant-derived compounds that inhibit HIV-1 reverse transcriptase. J Nat Prod 1995; 58: 1024–1031.
68. McManus JP, Davis KG, Lilley TH, Haslam E. The association of proteins with polyphenols. J Chem Soc, Chem Commun 1981: 309–311.
69. Hagerman AE, Butler LG. The specificity of proanthocyanidin–protein interactions. J Biol Chem 1981; 256: 4494–4497.
70. Butler LG. Polyphenols and herbivore diet selection and nutrition. In Scalbert A, ed. Polyphenolic Phenomena. Paris: INRA Editions; 1993: 149–154.
71. Nishino H, Naito E, Iwashima A, Tanaka K, Matsuura T, Fujiki H, Sugimura T. Interaction between quercetin and Ca^{2+}-calmodulin complex: possible mechanism for antitumor-promoting action of the flavonoid. Gann 1984; 74: 311–316.
72. Shisheva A, Shechter Y. Quercetin selectively inhibits insulin receptor function in vitro and the bioresponses of insulin and insulinomimetic agents in rat adipocytes. Biochemistry 1992; 31: 8059–8063.

73. Ramassamy C, Naudin B, Christen Y, Clostre F, Costenin J. Prevention by *Gingko biloba* extract (EGb 761) and trolox C of the decrease in synaptosomal dopamine or serotonin uptake following incubation. Biochem Pharmacol 1992; 44: 2395–2401.
74. Morita K, Hamano S, Oka M, Teraoka K. Stimulatory actions of bioflavonoids on tyrosine uptake into cultured bovine adrenal chromaffin cells. Biochem Biophys Res Commun 1990; 171: 1199–1204.
75. Kuppusamy UR, Das NP. Potentiation of ß-adrenoceptor agonist-mediated lipolysis by quercetin and fisetin in isolated rat adipocytes. Biochem Pharmacol 1994; 47: 521–529.
76. Häberlein H, Tschiersch KP, Schäfer HL. Flavonoids from *Leptospermum scoparium* with affinity to the benzodiazepine receptor characterized by structure activity relationships and in vivo studies of a plant extract. Pharmazie 1994; 49: 912–922.
77. Ji XD, Melman N, Jacobson KA. Interactions of flavonoids and other phytochemicals with adenosine receptors. J Med Chem 1996; 39: 781–788.
78. Kyriakidis SM, Sotiroudis TG, Evangelopoulos AE. Interaction of flavonoids with rabbit muscle phosphorylase kinase. Biochim Biophys Acta 1986; 871: 121–129.
79. Kalkbrenner F, Wurm G, von Bruchhausen F. In vitro inhibition and stimulation of purified prostaglandin endoperoxide synthase by flavonoids: structure–activity relationship. Pharmacology 1992; 44: 1–12.
80. Kuppusamy UR, Das NP. Effects of flavonoids on cyclic AMP phospho-diesterase and lipid mobilization in rat adipocytes. Biochem Pharmacol 1992; 44: 1307–1315.
81. Huang MT, Johnson EF, Muller-Eberhard U, Koop DR, Coon MJ, Conney AH. Specificity in the activation and inhibition by flavonoids of benzo[a]pyrene hydroxylation by cytochrome P-450 isozymes from rabbit liver microsomes. J Biol Chem 1981; 256: 10897–10901.
82. Yamamoto K, Kato S. Steric and electronic requirements for chloroflavone congeners as hepatic microsomal monooxygenase inducers. Biol Pharm Bull 1994; 17: 1404–1408.
83. Canivenc-Lavier MC, Bentejac M, Miller ML, Leclerc J, Siess MH, Latruffe N, Suschetet M. Differential effects of nonhydroxylated flavonoids as inducers of cytochrome P450 1A and 2B isozymes in rat liver. Toxicol Appl Pharmacol 1996; 136: 348–353.
84. Bors W. Bedeutung und Wirkungsweise von Antioxidanzien. In Elstner EF, Bors W, Wilmanns W, eds. Reaktive Sauerstoffspezies in der Medizin. Heidelberg: Springer; 1986: 161–183.
85. Bors W, Heller W, Michel C, Saran M. Structural principles of flavonoid antioxidants. In Csomos G, Feher J, eds. Free Radicals and the Liver. Berlin: Springer; 1992: 77–95.
86. Graziani Y, Chayoth R, Karny N, Feldman B, Levy J. Regulation of protein kinases activity by quercetin in Ehrlich ascites tumor cells. Biochim Biophys Acta 1981; 714: 415–421.
87. Ferrer MA, Pedreno MA, Munoz R, Ros Barceló A. Constitutive isoflavones as modulators of indole-3-acetic acid oxidase activity of acidic cell wall isoperoxidases from lupine hypocotyls. Phytochemistry 1992; 31: 3681–3684.
88. Ravanel P, Tissut M, Douce R. Uncoupling activities of chalcones and dihydrochalcones on isolated mitochondria from potato tubers and mung bean hypocotyls. Phytochemistry 1982; 21: 2845–2850.
89. Hodnick WF, Milosavljevic EB, Nelson JH, Pardini RS. Electrochemistry of flavonoids. Relationships between redox potentials, inhibition of mitochondrial respiration, and production of oxygen radicals by flavonoids. Biochem Pharmacol 1988; 37: 2607–2611.
90. Bindoli A, Valente M, Cavallini L. Inhibitory action of quercetin on xanthine oxidase and xanthine dehydrogenase activity. Pharm Res Commun 1994; 17: 831–839.
91. Pincemail J, Deby C, Thirion A, de Bruyn-Dister M, Goutier R. Human myeloperoxidase activity is inhibited in vitro by quercetin. Comparison with three related compounds. Experientia 1988; 44: 450–453.
92. Kingston WP. The actions of quercetin and rutin on cyclo oxygenase and 15-lipoxygenase. Br J Pharmacol 1983; 80: 515P.

93. Siess MH, Brouard C, Vernevaut MF, Suschetet M. Effects of feeding quercetin and flavone on hepatic drug metabolizing enzymes of rat. In Cody V, Middleton E, Harborne JB, Beretz A, eds. Plant Flavonoids in Biology and Medicine II. New York: Alan R Liss [Progr Clin Biol Res 1988; 280: 147–150].
94. Chiesi M, Schwaller R. Inhibition of constitutive endothelial NO-synthase activity by tannin and quercetin. Biochem Pharmacol 1995; 49: 495–501.
95. Krol W, Czuba ZP, Threadgill MD, Cunningham BDM, Pietsz G. Inhibition of nitric oxide production in murine macrophages by flavones. Biochem Pharmacol 1995; 50: 1031–1035.
96. Walle T, Eaton EA, Walle UK. Quercetin, a potent and specific inhibitor of the human P-form phenolsulfotransferase. Biochem Pharmacol 1995; 50: 731–734.
97. Gschwendt M, Horn F, Kittstein W, Marks F. Inhibition of the Ca- and phospholipid-dependent protein kinase activity from mouse brain cytosol by quercetin. Biochem Biophys Res Commun 1983; 117: 444–447.
98. Wuethrich A, Schatzmann HJ. Inhibition of the red cell Ca pump by quercetin. Cell Calcium 1980; 1: 21–35.
99. Shoshan V, MacLennan DH. Quercetin interaction with the (Ca^{2+} + Mg^{2+})-ATPase of sacroplasmic reticulum. J Biol Chem 1981; 256: 887–892.
100. Kato R, Nakadate T, Yamamoto S, Sugimura T. Inhibition of 12-*O*-tetradecanoylphorbol-13-acetate-induced tumor promotion and ornithine decarboxylase activity by quercetin: possible involvement of lipoxygenase inhibition. Carcinogenesis 1983; 4: 1301–1305.
101. Beretz A, Stierle A, Anton R, Cazenave JP. Role of cyclic AMP in the inhibition of human platelet aggregation by quercetin, a flavonoid that potentiates the effect of prostacyclin. Biochem Pharmacol 1982; 31: 3597–3600.
102. Akiyama T, Ishida J, Nakagawa S, Ogawara H, Watanabe S, Itoh N, Shibuya M, Fukami Y. Genistein, a specific inhibitor of tyrosine specific protein kinases. J Biol Chem 1987; 262: 5592–5595.
103. Huang J, Nasr M, Kim Y, Matthews HR. Genistein inhibits protein histidine kinase. J Biol Chem 1992; 267: 15511–15515.
104. Thorburn J, Thorburn A. The tyrosine kinase inhibitor, genistein, prevents alpha-adrenergic-induced cardiac muscle cell hypertrophy by inhibiting activation of the Ras-MAP kinase signaling pathway. Biochem Biophys Res Commun 1994; 202: 1586–1591.
105. Okura A, Arakawa H, Oka H, Yoshinari T, Monden Y. Effect of genistein on topoisomerase activity and on the growth of [val 12]Ha-ras-transformed NIH 3T3 cells. Biochem Biophys Res Commun 1988; 157: 183–189.
106. Cai QY, Wei HC. Effect of dietary genistein on antioxidant enzyme activities in SENCAR mice. Nutr Cancer 1996; 25: 1–7.
107. Nakashima S, Koike T, Nozawa Y. Genistein, a protein tyrosine kinase inhibitor, inhibits thromboxane-A_2-mediated human platelet responses. Mol Pharmacol 1991; 39: 475–480.
108. Fotsis T, Pepper M, Adlercreutz H, Fleischmann G, Hase T, Montesano R, Schweigerer L. Genistein, a dietary-derived inhibitor of in vitro angiogenesis. Proc Natl Acad Sci USA 1993; 90: 2690–2694.
109. Giri AK, Lu LJW. Genetic damage and the inhibition of 7,12-dimethylbenz[alpha] anthracene-induced genetic damage by the phytoestrogens, genistein and daidzein, in female ICR mice. Cancer Lett 1995; 95: 125–133.
110. Lamartiniere CA, Moore JB, Brown NM, Thompson R, Hardin MJ, Barnes S. Genistein suppresses mammary cancer in rats. Carcinogenesis 1995; 16: 2833–2840.
111. Spinozzi F, Pagliacci MC, Migliorati G, Moraca R, Grigna C, Nicoletti I. The natural tyrosine kinase inhibitor genistein produces cell cycle arrest and apoptosis in Jurkat T- leukemia cells. Leukemia Res 1994; 18: 431–439.
112. Record IR, Jannes M, Dreosti IE, King RA. Induction of micronucleus formation in mouse splenocytes by the soy isoflavone genistein in vitro but not in vivo. Food Chem Toxical 1995; 33: 919–922.

113. Baumann J, von Bruchhausen F. Cyanidanol-3 as inhibitor of prostaglandin synthetase. Arch Pharmacol 1979; 306: 85–87.
114. Blazovics A, Vereckei A, Cornides A, Feher J. The effect of (+)cyanidanol-3 on the Na^+, K^+-ATPase and Mg^{2+}-ATPase activities of the rat brain in the presence and absence of ascorbic acid. Acta Physiol Hung 1989; 73: 9–14.
115. Wang ZY, Das M, Bickers DR, Mukhtar H. Interaction of epicatechins derived from green tea with rat hepatic cytochrome P-450. Drug Metab Dispos 1988; 16: 98–103.
116. Ferrell JE, Chang Sing PDG, Loew G, King R, Mansour JM, Mansour TE. Structure/activity studies of flavonoids as inhibitors of cAMP phosphodiesterase and relationship to quantum chemical indices. Mol Pharmacol 1979; 16: 556–568.
117. Aufmkolk M, Köhrle J, Hesch RD, Cody V. Inhibition of rat liver iodothyronine deiodinase. Interaction of aurones with the iodothyronine ligand-binding site. J Biol Chem 1986; 261: 11623–11630.
118. Spanka M, Köhrle J, Irmscher K, Hesch RD. Flavonoids specifically inhibit iodothyronine-deiodinase in rat hepatocytes. In Cody V, Middleton E, Harborne JB, Beretz A, eds. Plant Flavonoids in Biology and Medicine II. New York: Alan R Liss; [Progr Clin Biol Res 1988; 280: 341–344.]
119. Boyer RF, McArthur JS, Cary TM. Plant phenolics as reductants for ferritin iron release. Phytochemistry 1990; 29: 3717–3719.
120. Tomonaga T, Mine T, Kojima I, Taira M, Hayashi H, Isono K. Isoflavonoids, genistein, PSI-tectorigenin, and orobol, increase cytoplasmic free calcium in isolated rat hepatocytes. Biochem Biophys Res Commun 1992; 182: 894–899.
121. Bohmont C, Aaronson LM, Mann K, Pardini RS. Inhibition of mitochondrial NADH oxidase, succinoxidase, and ATPase by naturally occurring flavonoids. J Nat Prod 1987; 50: 427–433.
122. Thiyagarajah P, Kuttan SC, Lim SC, Teo TS, Das NP. Effect of myricetin and other flavonoids on the liver plasma membrane Ca^{2+} pump. Kinetics and structure–function relationships. Biochem Pharmacol 1991; 41: 669–675.
123. Feher J, Lang I, Nekam K, Müzes G, Deak G. Effect of free radical scavengers on superoxide dismutase (SOD) enzyme in patients with alcoholic cirrhosis. Acta Med Hung 1988; 45: 265–276.
124. Robak J, Shridi F, Wolbis M, Krolikowska M. Screening of the influence of flavonoids on lipoxygenase and cyclooxygenase activity, as well as on nonenzymic lipid oxidation. Pol J Pharmacol Pharm 1989; 40: 451–458.
125. de Long MJ, Prochaska HJ, Talalay P. Induction of NAD(P)H: quinone reductase in murine hepatoma cells by phenolic antioxidants, azo dyes and other chemoprotectors: a model system for the study of anticarcinogens. Proc Natl Acad Sci USA 1986; 83: 787–791.
126. Tamura M, Kagawa S, Tsuruo Y, Ishimura K, Morita K. Effects of flavonoid compounds on the activity of NADPH diaphorase prepared from the mouse brain. Jpn J Pharmacol 1994; 65: 371–373.
127. Siess MH, LeBon AM, Suschetet M. Dietary modification of drug-metabolizing enzyme activities: dose–response effect of flavonoids. J Toxicol Environ Health 1992; 35: 141–152.
128. Firozi PF, Bhattacharya RK. Effects of natural polyphenols on aflatoxin B-1 activation in a reconstituted microsomal monooxygenase system. J Biochem Toxicol 1995; 10: 25–31.
129. Obermeier MT, White RE, Yang CS. Effects of bioflavonoids on hepatic P450 activities. Xenobiotica 1995; 25: 575–584.
130. Siess MH, Guillermic M, LeBon AM, Suschetet M. Induction of monooxygenases and transferase activities in rat by dietary administration of flavonoids. Xenobiotica 1989; 19: 1379–1386.
131. Schwabe KP, Flohé L. Catechol-*O*-methyltransferase, III. Beziehung zwischen der Struktur von Flavonoiden und deren Eignung als Inhibitoren der Catechol-*O*-Methyltransferase. Hoppe-Seyler's Z Physiol Chem 1972; 353: 476–482.

132. Zhang K, Das NP. Inhibitory effects of plant polyphenols on rat liver glutathione *S*-transferases. Biochem Pharmacol 1994; 47: 2063–2068.
133. Eaton EA, Walle UK, Lewis AJ, Hudson T, Wilson AA, Walle T. Flavonoids, potent inhibitors of the human P-form phenolsulfotransferase–potential role in drug metabolism and chemoprevention. Drug Metab Dispos 1996; 24: 232–237.
134. DiGiovanni J, Kruszewski FH, Coombs MM, Bhatt TS, Pezeshk A. Structure–activity relationships for epidermal ornithine decarboxylase induction and skin tumor promotion by anthrones. Carcinogenesis 1988; 9: 1437–1443.
135. Gali HU, Perchellet EM, Perchellet JP. Inhibition of tumor promoter-induced ornithine decarboxylase activity by tannic acid and other polyphenols in mouse epidermis in vivo. Cancer Res 1991; 51: 2820–2825.
136. Shimoi K, Masuda S, Furugori M, Esaki S, Kinae N. Radioprotective effect of antioxidative flavonoids in gamma-ray irradiated mice. Carcinogenesis 1994; 15: 2669–2672.
137. Emerit I, Oganesian N, Sarkisian T, Arutyunyan R, Pogosian A, Asrian K, Levy A, Cernjavski L. Clastogenic factors in the plasma of Chernobyl accident recovery workers: Anticlastogenic effect of *Ginkgo biloba* extract. Radiat Res 1995; 144: 198–205.
138. Menon LG, Kuttan R, Kuttan G. Inhibition of lung metastasis in mice induced by B16F10 melanoma cells by polyphenolic compounds. Cancer Lett 1995; 95: 221–225.
139. Hertog MGL, Kromhout D, Aravanis C, Blackburn H, Buzina R, Fidanza F, Giampaoli S, Jansen A, Menotti A, Nedeljkovic S, Pekkarinen M, Simic BS, Toshima H, Feskens EJM, Hollman PCH, Katan MB. Flavonoid intake and long-term risk of coronary heart disease and cancer in the Seven Countries Study. Arch Intern Med 1995; 155: 381–386.
140. Knekt P, Järvinen R, Reunanen A, Maatela J. Flavonoid intake and coronary mortality in Finland: a cohort study. Br Med J 1996; 312: 478–481.
141. Fuhrman B, Lavy A, Aviram M. Consumption of red wine with meals reduces the susceptibility of human plasma and low-density lipoprotein to lipid peroxidation. Am J Clin Nutr 1995; 61: 549–554.
142. Berger ME, Golub MS, Chang CT, Al-Kharouf JA, Nyby MD, Hori M, Brickman AS, Tuck ML. Flavonoid potentiation of contractile responses in rat blood vessels. J Pharmacol Exp Ther 1992; 263: 78–83.
143. Duarte J, Vizcaino FP, Utrilla P, Jimenez J, Tamargo J, Zarzuelo A. Vasodilatory effects of flavonoids in rat aortic smooth muscle structure-activity relationships. Gen Pharmacol 1993; 24: 857–862.
144. Pardeep KG, Laloraya M, Laloraya MM. The effect of some of the polyphenolic compounds on sperm motility in vitro: a structure–activity relationship. Contraception 1989; 39: 531–539.
145. Sahu SC. Dual role of flavonoids in mutagenesis and carcinogenesis. Environ Carcin Ecotox Rev 1994; 12: 1–21.
146. Das A, Wang JH, Lien EJ. Carcinogenicity, mutagenicity and cancer preventing activities of flavonoids: a structure–system–activity relationship (SSAR) analysis. Progr Drug Res 1994; 42: 133–166.
147. Edwards JM, Raffauf RF, Le Quesne PW. Antineoplastic activity and cytotoxicity of flavones, isoflavones and flavanones. J Nat Prod 1979; 42: 85–91.
148. Heo MY, Yu KS, Kim KH, Kim HP, Au WW. Anticlastogenic effect of flavonoids against mutagen-induced micronuclei in mice. Mutat Res 1992; 284: 243–249.
149. Okuda T. Natural polyphenols as antioxidants and their potential use in cancer prevention. In Scalbert A, ed. Polyphenolic Phenomena. Paris: INRA Editions; 1993: 221–235.
150. de Vincenzo R, Scambia G, Benedetti Panici P, Ranelletti FO, Bonanno G, Ercoli A, Delle Monache F, Ferrari F, Piantelli M, Mancuso S. Effect of synthetic and naturally occurring chalcones on ovarian cancer cell growth: structure–activity relationships. Anticancer Drug Design 1995; 10: 481–490.
151. Adlercreutz H. Phytoestrogens: epidemiology and a possible role in cancer protection. Environ Health Perspect 1995; 103(Suppl. 7): 103–112.

152. Kandaswami C, Perkins E, Drzewiecki G, Soloniuk DS, Middleton E. Differential inhibition of proliferation of human squamous cell carcinoma, gliosarcoma and embryonic fibroblast-like lung cells in culture by plant flavonoids. Anti-Cancer Drugs 1992; 3: 525–530.
153. Candeias LP, Everett SA, Wardman P. Free radical intermediates in the oxidation of flavone-8-acetic acid–possible involvement in its antitumour activity. Free Radical Biol Med 1993; 15: 385–394.
154. Mace KF, Hornung RL, Wiltrout RH, Young HA. Correlation between in vivo induction of cytokine gene expression by flavone acetic acid and strict dose dependency and therapeutic efficacy against murine renal cancer. Cancer Res 1990; 50: 1742–1747.
155. Mahadevan V, Malik STA, Meager A, Fiers W, Lewis GP, Hart IR. Role of tumor necrosis factor in flavone acetic acid-induced tumor vasculature shutdown. Cancer Res 1990; 50: 5537–5542.
156. Thomsen LL, Ching LM, Zhuang L, Gavin JB, Baguley BC. Tumor dependent increased plasma nitrite concentrations as an indication of the antitumor effect of flavone-8-acetic acid and analogues in mice. Cancer Res 1991; 51: 77–81.
157. Rotstein JB, Slaga TJ. Anticarcinogenesis mechanisms, as evaluated in the multistage mouse skin model. Mutat Res 1988; 202: 421–427.
158. Imaida K, Hirose M, Yamaguchi S, Takahashi S, Ito N. Effects of naturally occurring antioxidants on combined 1.2-dimethylhydrazine- and 1-methyl-1-nitrosourea-initiated carcinogenesis in F344 male rats. Cancer Lett 1990; 55: 53–59.
159. Xu Y, Ho CT, Amin SG, Han C, Chung FL. Inhibition of tobacco-specific nitrosamine-induced lung tumorogenesis in A/J mice by green tea and its major polyphenol antioxidants. Cancer Res 1992; 52: 3875–3879.
160. Edenharder R, von Petersdorff I, Rauscher R. Antimutagenic effects of flavonoids, chalcones and structurally related compounds on the activity of 2-amino-3-methylimidazo [4,5-f] quinoline (IQ) and other heterocyclic amine mutagens from cooked food. Mutat Res 1993; 287: 261–274.
161. Shah GM, Bhattacharya RK. Modulation by plant flavonoids and related phenolics of microsome catalyzed adduct formation between benzo(a)pyrene and DNA. Chem-Biol Interact 1986; 59: 1–15.
162. Bartsch H, Ohshima H, Pignatelli B. Inhibitors of endogenous nitrosation. Mechanisms and implications in human cancer prevention. Mutat Res 1988; 202: 307–324.
163. LeBon AM, Siess MH, Suschetet M. Inhibition of microsome-mediated binding of benzo(a)pyrene to DNA by flavonoids either in vitro or after dietary administration to rats. Chem-Biol Interact 1992; 83: 65–71.
164. Kuo ML, Lee KC, Lin JK. Genotoxicities of nitropyrenes and their modulation by apigenin, tannic acid, ellagic acid and indole-3-carbinol in the Salmonella and CHO systems. Mutat Res 1992; 270: 87–95.
165. Ramanathan R, Das NP, Tan CH. Inhibition of tumour promotion and cell proliferation by plant polyphenols. Phytother Res 1994; 8: 293–296.
166. Yasukawa K, Takido M, Takeuchi M, Sato Y, Nitta K, Nakagawa S. Inhibitory effects of flavonoid glycosides on 12-*O*-tetradecanoylphorbol-13-acetate-induced tumor promotion. Chem Pharm Bull 1990; 38: 774–776.
167. Katiyar SK, Agarwal R, Wood GS, Mukhtar H. Inhibition of 12-*O*-tetradecanoyl-phorbol-13-acetate-caused tumor promotion in 7,12-dimethylbenz[a]-änthracene-initiated Sencar mouse skin by a polyphenolic fraction isolated from green tea. Cancer Res 1992; 52: 6890–6897.
168. Nakadate T, Yamamoto S, Aizu E, Kato R. Effects of flavonoids and antioxidants on 12-*O*-tetradecanoylphorbol-13-acetate-caused epidermal ornithine decarboxylase induction and tumor promotion in relation to lipoxygenase inhibition by these compounds. Gann 1984; 75: 214–222.

169. Wei H, Frenkel K. Relationship of oxidative events and DNA oxidation in Sencar mice to in vivo promoting activity of phorbol ester-type tumor promoters. Carcinogenesis 1993; 14: 1195–1201.
170. Ramanathan R, Tan CH, Das NP. Cytotoxic effect of plant polyphenols and fat-soluble vitamins on malignant human cultured cells. Cancer Lett 1992; 62: 217–224.
171. Yamashita Y, Kawada SZ, Nakano H. Induction of mammalian topoisomerase II dependent DNA cleavage by nonintercalative flavonoids, genistein and orobol. Biochem Pharmacol 1990; 39: 737–744.
172. Post JFM, Varma RS. Growth inhibitory effects of bioflavonoids and related compounds on human leukemic CEM-Cl and CEM-C7 cells. Cancer Lett 1992; 67: 207–213.
173. Sato F, Matsukawa Y, Matsumoto K, Nishino H, Sakai T. Apigenin induces morphological differentiation and G2-M arrest in rat neuronal cells. Biochem Biophys Res Commun 1994; 204: 578–584.
174. Matsukawa Y, Marui N, Sakai T, Satomi Y, Yoshida M, Matsumoto K, Nishino H, Aoike A. Genistein arrests cell cycle progression at G2-M. Cancer Res 1993; 53: 1328–1331.
175. Mora A, Paya M, Rios JL, Alcaraz MJ. Structure–activity relationships of polymethoxyflavones and other flavonoids as inhibitors of non-enzymic lipid peroxidation. Biochem Pharmacol 1990; 40: 793–797.
176. Middleton E, Kandaswami C. Effects of flavonoids on immune and inflammatory cell functions. Biochem Pharmacol 1992; 43: 1167–1179.
177. ‘t Hart BA, Ip Vai Ching TRAM, van Dijk H, Labadie RP. How flavonoids inhibit the generation of luminol-dependent chemiluminescence by activated human neutrophils. Chem-Biol Interact 1990; 73: 323–335.
178. Krol W, Shani J, Czuba Z, Scheller S. Modulating luminol-dependent chemiluminescence of neutrophils by flavones. Z Naturforsch 1992; 47c: 889–892.
179. Limasset B, Le Doucen C, Dore JC, Ojasoo T, Damon M, Crastes de Paulet A. Effects of flavonoids on the release of reactive oxygen species by stimulated human neutrophils. Multivariate analysis of structure-activity relationships (SAR). Biochem Pharmacol 1993; 46: 1257–1271.
180. Beretz A, Cazenave JP, Anton R. Inhibition of aggregation and secretion of human platelets by quercetin and other flavonoids: structure–activity relationships. Agents Actions 1982; 12: 382–387.
181. Petroni A, Blasevich M, Salami M, Papini N, Montedoro GF, Galli C. Inhibition of platelet aggregation and eicosanoid production by phenolic components of olive oil. Thromb Res 1995; 78: 151–160.
182. Welton AF, Hurley J, Will P. Flavonoids and arachidonic acid metabolism. In Cody V, Middleton E, Harborne JB, Beretz A, eds. Plant Flavonoids in Biology and Medicine II. New York: Alan R Liss. [Progr Clin Biol Res 1988; 280: 301–312.]
183. Tordera M, Ferrandiz ML, Alcaraz MJ. Influence of anti-inflammatory flavonoids on degranulation and arachidonic acid release in rat neutrophils. Z Naturforsch 1994; 49c: 235–240.
184. Alcaraz MJ, Hoult JRS. Effects of hypolaetin-8-glucoside and related flavonoids on soybean lipoxygenase and snake venom phospholipase A_2. Arch Int Pharmacodyn 1985; 278: 4–12.
185. Beil W, Birkholz C, Sewing KF. Effects of flavonoids on parietal cell acid secretion, gastric mucosal prostaglandin production and *Helicobacter pylori* growth. Arzneimittel-Forschung 1995; 45-1: 697–700.
186. Hoult JRS, Moroney MA, Paya M. Actions of flavonoids and coumarins on lipoxygenase and cyclooxygenase. Methods Enzymol 1994; 234: 443–454.
187. Abad MJ, Bermejo P, Villar A. The activity of flavonoids extracted from *Tanacetum microphyllum* DC (Compositae) on soybean lipoxygenase and prostaglandin synthetase. Gen Pharmacol 1995; 26: 815–819.

188. Voß C, Sepulveda-Boza S, Zilliken FW. New isoflavonoids as inhibitors of porcine 5-lipoxygenase. Biochem Pharmacol 1992; 44: 157–162.
189. Rump AFE, Schüssler M, Acar D, Cordes A, Theisohn M, Rösen R, Klaus W, Fricke U. Functional and antiischemic effects of luteolin-7-glucoside in isolated rabbit hearts. Gen Pharmacol 1994; 25: 1137–1142.
190. Friesenecker B, Tsai AG, Intaglietta M. Cellular basis of inflammation, edema and the activity of Daflon 500 mg. Int J Microcirc: Clin Exp 1995; 15: 17–21.
191. van Jaarsveld H, Kuyl JM, Schulenburg DH, Wild NM. Effect of flavonoids on the outcome of myocardial mitochondrial ischemia reperfusion injury. Res Commun Mol Pathol Pharmacol 1996; 91: 65–75.
192. Hall ED. Cerebral ischaemia, free radicals and antioxidant protection. Biochem Soc Trans 1993; 21: 334–339.
193. Ballesteros JF, Sanz MJ, Ubeda A, Miranda MA, Iborra S, Paya M, Alcaraz MJ. Synthesis and pharmacological evaluation of 2′-hydroxy-chalcones and flavones as inhibitors of inflammatory mediators generation. J Med Chem 1995; 38: 2794–2797.
194. Gambhir SS, Oandey BL, Devi KS, Banerjee RS, DasGupta G. Autocoid-immunopharmacology of flavonoids. In Ong ASH, Packer L, eds. Lipid-Soluble Antioxidants: Biochemistry and Clinical Applications. Basel: Birkhäuser; 1992: 307–319.
195. Schwartz A, Middleton E. Comparison of the effects of quercetin with those of other flavonoids on the generation and effector function of cytotoxic T lymphocytes. Immunopharmacology 1984; 7: 115–126.
196. Middleton E, Fujiki H, Savliwala M, Drzewiecki G. Tumor promoter induced basophil histamine release: effect of selective flavonoids. Biochem Pharmacol 1987; 36: 2048–2052.
197. Baum CG, Szabo P, Siskind GW, Becker CG, Firpo A, Clarick CJ, Francus, T. Cellular control of IgE induction by a polyphenol-rich compounds. Preferential activation of Th2 cells. J Immunol 1990; 145: 779–784.
198. Lee SJ, Choi JH, Son KH, Chang HW, Kang SS, Kim HP. Suppression of mouse lymphocyte proliferation in vitro by naturally-occurring biflavonoids. Life Sci 1995; 57: 551–558.
199. Salama A, Mueller-Eckhardt C. Cianidanol and its metabolites bind tightly to red cells and are responsible for the production of auto- and/or drug-dependent antibodies against these cells. Br J Haematol 1987; 66: 263–266.
200. Shahat AA, Hammouda F, Ismail SI, Azzam SA, de Bruyne T, Lasure A, van Poel B, Pieters L, Vlietinck AJ. Anti-complementary activity of *Crataegus sinaica*. Planta Medica 1996; 62: 10–13.
201. Ferriola PC, Cody V, Middleton E. Protein kinase C inhibition by plant flavonoids. Kinetic mechanisms and structure activity relationships. Biochem Pharmacol 1989; 38: 1617–1624.
202. Hagiwara M, Inoue S, Tanaka T, Nunoki K, Ito M, Hidaka H. Differential effects of flavonoids as inhibitors of tyrosine protein kinases and serine/threonine protein kinases. Biochem Pharmacol 1988; 37: 2987–2992.
203. Cushman M, Nagarathnam D, Burg DL, Geahlen RL. Synthesis and protein-tyrosine kinase inhibitory activities of flavonoid analogues. J Med Chem 1991; 34: 798–806.
204. Polya GM, Wang BH, Foo LY. Inhibition of signal-regulated protein kinases by plant-derived hydrolysable tannis. Phytochemistry 1995; 38: 307–314.
205. Wang BH, Polya GM. Selective inhibition of cyclic AMP-dependent protein kinase by amphiphilic triterpenoids and related compounds. Phytochemistry 1996; 41: 55–63.
206. Elia G, Santoro MG. Regulation of heat shock protein synthesis by quercetin in human erythroleukaemia cells. Biochem J 1994; 300: 201–209.
207. Wei YQ, Zhao X, Kariya Y, Fukata H, Teshigawara K, Uchida A. Induction of apoptosis by quercetin: involvement of heat shock protein. Cancer Res 1994; 54: 4952–4957.
208. Nagai N, Nakai A, Nagata K. Quercetin suppresses heat shock response by down regulation of HSF1. Biochem Biophys Res Commun 1995; 208: 1099–1105.

209. Elia G, Amici C, Rossi A, Santoro MG. Modulation of prostaglandin A(1) induced thermotolerance by quercetin in human leukemic cells: role of heat shock protein 70. Cancer Res 1996; 56: 210–217.
210. Hosokawa N, Hirayoshi K, Nakai A, Hosokawa Y, Marui N, Yoshida M, Sakai T, Nishino H, Aoike A, Kawai K, Nagata K. Flavonoids inhibit the expression of heat shock proteins. Cell Struct Funct 1990; 15: 393–401.
211. Hirano T, Oka K, Mimaki Y, Kuroda M, Sashida Y. Potent growth inhibitory activity of a novel *Ornithogalum* cholestane glycoside on human cells: induction of apoptosis in promyelocytic leukemia HL- 60 cells. Life Sci 1996; 58: 789–798.
212. Kioka N, Hosokawa N, Komano T, Hirayoshi K, Nagata K, Ueda K. Quercetin, a bioflavonoid, inhibits the increase of human multidrug resistance gene (MDR1) expression caused by arsenite. FEBS Lett 1992; 301: 307–309.
213. Gerritsen ME, Carley WW, Ranges GE, Shen CP, Phan SA, Ligon GF, Perry CA. Flavonoids inhibit cytokine-induced endothelial cell adhesion protein gene expression. Am J Pathol 1995; 147: 278–292.
214. Ruch RJ, Cheng S, Klaunig JE. Prevention of cytotoxicity and inhibition of intercellular communication by antioxidant catechins isolated from Chinese green tea. Carcinogenesis 1989; 10: 1003–1008.
215. Sigler K, Ruch RJ. Enhancement of gap junctional intercellular communication in tumor promoter-treated cells by components of green tea. Cancer Lett 1993; 69: 15–19.
216. Chaumontet C, Bex V, Gaillard-Sanchez I, Seillan-Heberden C, Suschetet M, Martel P. Apigenin and tangeretin enhance gap junctional intercellular communication in rat liver epithelial cells. Carcinogenesis 1994; 15: 2325–2330.
217. DasGupta G, Gambhir SS. Bioflavonoids and vasoactive mediator release from mast cells. Indian J Physiol Pharmacol 1988; 32: 29–36.
218. Namgoong SY, Son KH, Chang HW, Kang SS, Kim HP. Effects of naturally occurring flavonoids on mitogen-induced lymphocyte proliferation and mixed lymphocyte culture. Life Sci 1994; 54: 313–320.
219. Kunizane H, Ueda H, Yamazaki M. Screening of phagocyte activators in plants; enhancement of TNF production by flavonoids. Yakugaku Zasshi–J Pharm Sci 1995; 115: 749–755.
220. Gescher A. Modulators of signal transduction as cancer chemotherapeutic agents–novel mechanisms and toxicities. Toxicol Lett 1995; 82–83: 159–165.

28
Immunoregulatory Properties of Superoxide Dismutase

E. Postaire
CRE
Paris, France

B. Dugas
Institut de Recherches Fractales
Bois-Colombes, France

P. Debré
URA CNRS 625
Paris, France

C. Gudin
Héliosynthèse
Aix-en-Provence, France

Superoxide dismutase (SOD), the first enzyme with a free radical as its substrate to be identified (1) was defined by McCord and Fridovich in 1969 (2) as a natural system of defense that limits the toxic effects of oxygen free radicals.

It is now well established that oxygen-derived free radicals play a fundamental role in immunity and in inflammatory mechanisms involved in most human diseases (3): in inflammatory phenomena, the oxidizing catabolism of eicosanoids, particularly prostaglandins and thromboxane via the cyclooxygenase pathway and leukotrienes via the lipoxygenase pathway. Thus, conventional therapy of inflammation was generally limited to the inhibition of prostaglandin biosynthesis. Anti-inflammatory treatment using SOD seems to be a promising alternative to these conventional anti-inflammatory therapies.

Visner and coworkers (4) reported that the induction of Mn-SOD by interleukin-1 (IL-1), tumor necrosis factor (TNF), and lipopolysaccharide occurred in pulmonary epithelial cells. Mn-SOD in cultured monocytes was also induced by the addition of TNF; the pathway that transduces a signal from the corresponding receptors to the Mn-SOD gene is not clearly understood. However, it has been showed that phorbol 12-myristate 13-acetate (TPA) also induces Mn-SOD in various cell lines which are all resistant to TNF. This give a clue to the investigation of the intracellular signal transduction pathway, since TPA enhanced Mn-SOD mRNA expression in TNF-resistant cell lines in which it is conceivable that protein kinase C is involved in the gene expression by TNF through phosphorylation of certain substrates. One likely candidate for this substrate is nuclear factor κB (NFκB). Indeed NFκB can be activated by releasing inhibitory protein IκB after protein kinase C-dependent phosphorylation. This activated NFκB may enhance Mn-SOD gene expression like other TNF-responsive genes. Another possibility is that AP-1 (Fos/Jun heterodimer) is responsible for this gene expression. Ho and colleagues (5) found the consensus sequence for AP-1 enhancer binding protein in the 5′-flanking region of the rat Mn-SOD gene. TPA acts both to activate AP-1 protein and to enhance proto-oncogene Jun/AP-1 expression. The activation of NFκB and AP-1 by this pathway may account for the induction of inducible NO synthase (iNOS) transcription, as responsive elements to both transcription factors have been reported in the promoter region of murine and human iNOS. The production of TNF following CD23 (low-affinity IgE receptor) ligation may further potentiate iNOS expression through a specific TNF-responsive element present in the promoter of the gene encoding human hepatic iNOS (6) or, indirectly, through its ability to activate NFκB. Together, the data suggest that TNF and NO potentiate each other during the immune response and enhance Mn-SOD mRNA expression.

We can so summarize the immunoregulatory and anti-inflammatory properties of SOD as follows.

- SOD inhibits the production of free radicals and thus protects against molecular and cellular toxicity.
- SOD regulates the expression of iNOS and thus limits the toxicity of NO.
- SOD regulates TNF production and thus limits the development of inflammatory and immune processes.
- SOD regulates the production of growth factors.

SUPEROXIDE DISMUTASE AND HIV-1 INFECTION

Oxidative Stress and Human Immunodeficiency Virus Type 1 (HIV) Infection

HIV infection is associated with oxidative stress. Glutathione levels are decreased in both infected individuals and cell culture. Exogenous reducing agents such as *N*-acetylcysteine and glutathione suppress HIV expression in chronically infected monocytes. Conversely, ultraviolet irradiation or cytokine treatment, which are known to increase oxidative stress, activate HIV expression. Cells acutely infected with HIV have been reported to express less Mn-SOD and to lose their ability to induce this antioxidant enzyme in response to TNF. Some viruses appear to have evolved mechanisms to control cellular oxidant status. A recent study suggest that HIV takes control of the cell's redox status (7). Exposure of cells to oxidants results in activation of transcriptional factors and/or increased proliferation–conditions necessary for successful viral replication. One of these cellular transcriptional factor is NFκB, whose activity is modulated by the cellular redox status. The recognition that the *trans*-acting transcriptional activator (Tat) may be instrumental in viral regulation of host oxidant status underscores the potential importance of this HIV protein. A fascinating property of Tat is that it appears to be secreted by infected cells and can be taken up by noninfected cells. That is, Tat can effect transcellular *trans*-activation. The Tat protein has been reported to interact with the *trans*-activation response (TAR) region contained at the 5′ end of all HIV transcripts, but the mechanisms by which this interaction regulates HIV gene expression are controversial. The fact that Tat has been shown to bind only to RNA makes it somewhat difficult to envision a mechanism for transcriptional regulation, yet the data published in (7) are consistent with this view. Clearly, the steady-state levels of Mn-SOD mRNAs are lower in Tat-producing cells, probably reflecting a net decrease in transcription. On the other hand, results support an interaction between Tat and Mn-SOD mRNA. It seems highly unlikely that this interaction is coincidental and independent of the downregulation of SOD.

Antiretroviral Activity of SOD

The antiretroviral activity of Cu,Zn-SOD has been tested in Molt-4 cells infected with the HIV-1 and compared to the anti-HIV-1 activity of reverse transcriptase inhibitors (AZT, ddC, ddU), the HIV protease inhibitor (retronavir), and as the CD4-masking compound aurintricarboxylic acid. SOD at 300 nM reduced the release of the viral antigen gp120 of HIV-1_{NDK}-infected Molt-4 cells by 50% (EC_{50}). The EC_{50} of SOD reaches 10% of AZT's anti-HIV-1_{NDK} activity and exceeds that of all tested antiretrovirals 40- to 3000-fold. SOD also inhibits dose-dependently the oxidative stress-induced depletion of sulfhydryls, which are crucially involved in the NKκB-controlled HIV transcription (8).

Unpublished data demonstrate that SOD, added exogenously, penetrates the cellular membrane, increases total SOD activity, inhibits cell lipoperoxidation, prevents cellular glutathione consumption, and inhibits HIV replication. The more important inhibitory effects of SOD were observed in HIV acute infection and in HIV transmission from HIV-infected to uninfected PBMCs, strongly suggesting the important involvement of superoxide anion on HIV expression (9).

MN-SOD AND INTERFERON-γ and-α

Interferons (IFNs) are pleiotropic cytokines having antiviral, antiproliferative, and immunomodulatory effects (10). Using Edman microsequencing of proteins separated by two-dimensional gel electrophoresis, it has been demonstrated that Mn-SOD is one of the proteins induced by INF-γ in different cell lines (11). To determine whether Mn-SOD plays a role in the antiviral action of IFN-γ, Raineri and coworkers employed an antisense strategy to inhibit the expression of Mn-SOD in the human melanoma cell line A375 (12). Three antisense-containing clones that exhibited expression of Mn-SOD have been investigated with respect to their response to the antiviral protective effects of IFN-γ and IFN-α. They have observed a striking decrease in the ability of IFN-γ to protect antisense clones from vesicular stomatitis virus infection (VSV). The IFN-α-induced antiviral state is also impaired, but to a lesser degree than observed with IFN-γ. They have excluded possibility that these effects were caused by a higher sensitivity of the antisense cells to VSV itself and have found that the antisense clones were less sensitive to VSV. Therefore, they conclude that Mn-SOD is involved in the establishment of the IFN-γ-induced antiviral state and to a lesser degree in the antiviral activity of IFN-α. It is possible, of course, that the reduced responsiveness of the antisense clones to the IFNs might be unrelated to a redox state-related mechanism. IFNs exert their antiviral activities by conferring protection against virus in the IFN-treated cells (13).

SUPEROXIDE DISMUTASE AND BRAIN IMMUNITY

Distribution of Mn-SOD in Brain

A heterogeneous distribution of Mn-SOD is observed in the human brain. In the forebrain, numerous immunostained neurons are detected in the striatum, thalamus, pallidal complex, and the nucleus basalis of Meynert. In the cerebellum, only granular and Purkinje cells are immunostained. Various nuclei from the brainstem displayed SOD reactivity. The heterogeneous but not ubiquitous distribution of cells expressing Mn-SOD suggests that not all cells in human brain are protected to the same extent against the deleterious-effects of superoxide (14,15). Recent results indicate that the striatal cholinergic and parvalbumin interneurons are enriched in SOD, whereas striatal projection neurons and neuropeptide Y/somastatin interneurons express lower levels of SOD than those in the striosome compartment. Since projection neurons have been reported to be more vulnerable than interneurons, and striosome neurons more vulnerable than matrix neurons to neurodegenerative processes, the results are consistent with the notion that superoxide free radicals are at least partly involved in producing the differential neuron loss observed in the striatum following global brain ischemia or in Huntington's disease (16).

Regulation of Mn-SOD and iNOS Gene in Neuronal and Glial Cells

Bidirectional communication occurs between neuroendocrine and immune systems through the action of various cytokines. Response to various inflammatory mediators includes increases in intracellular reactive oxygen species (ROS), notably superoxide anion and nitric oxide (NO). Neurotoxicity mediated by NO may result from the reaction of NO with superoxide anion, leading to formation of peroxynitrite. ROS are highly toxic, potentially contributing to extensive neuronal damage. Kifle and coworkers evaluated the effects of a variety of inflammatory mediators on the regulation of mRNA levels for Mn-SOD and iNOS in primary cultures of rat neuronal and glial cells (17). To determine age-dependent variation of mRNA expression, they used glial cells derived from newborn, 3-, 21-, and 95-day-old rat brains. Interleukin-1β, IFN-γ, bacterial lipopolyscaccharide (LPS), and TNF showed significant induction of Mn-SOD in both glial and neuronal cells. However, only LPS and IFN-γ increased iNOS mRNA. These data demonstrate that these two genes are similarly regulated in two kinds of cells in the nervous system, suggesting that the oxidative state of a cell may dictate a neurotoxic or neuroprotective outcome.

Oxidative Stress and Neurodegenerative Diseases

Oxidative stress is associated with most of the neurodegenerative diseases: radioinduced neurodegeneration (18), Alzheimer's disease (19), Parkinson's disease (20), AIDS dementia (21), spongiform encephalopathies (22), sporadic amyotrophic lateral sclerosis (23), and Downs syndrome (24). This give a rationale for the use of SOD as pharmacological agent.

SUPEROXIDE DISMUTASE AND NITRIC OXIDE

Superoxide, nitric oxide, and peroxynitrite have all been proposed to mediate cellular damage under conditions of shock, inflammation, and oxidative stress. Multiple interactions of NO and superoxide have been described. Initially, it was proposed that superoxide acts as an inactivator of NO, since SOD prolongs the biological half-life of NO. Along the lines of this concept, it has been demonstrated that NO can limit the cytotoxicity of superoxide. On the other hand, the reaction of NO and superoxide has been shown to yield peroxynitrite, a reactive oxidant species, and an important mediator of cell damage under conditions of inflammation and oxidant stress. Current data suggest that NO can act as an inactivator of the biological activity of peroxynitrite, and that the biological activity and decomposition of peroxynitrite is very much dependent on cellular or chemical environment (presence of proteins, thiols, glucose, carbon dioxide, and other factors).

Another level of regulation of NO production by superoxide exists at the level of the regulation of the expression of iNOS. Oxidant stress upregulates iNOS mRNA and is involved in the activation of NFκB expression. The direct investigation of the role of endogenously produced superoxide in the regulation of the expression of iNOS and in

Table 1. Effect of SOD on TSST-1-induced TNF Production by Human PBL

Challenge	SOD (30 U/ml)	TNF (pg/ml)
Medium	–	<10
Medium	+	<10
TSST-1	–	955 ± 15
TSST-1	+	108 ± 12

Human PBL were stimulated or not by 200 μg/ml of TSST-1 in the presence or in the basence of 30 U/ml of SOD. Cell-free supernatants were then collected after 24 h and their TNF contents were evaluated by specific ELISA. Data are the mean ± S.D. of one representative experiment out of 3

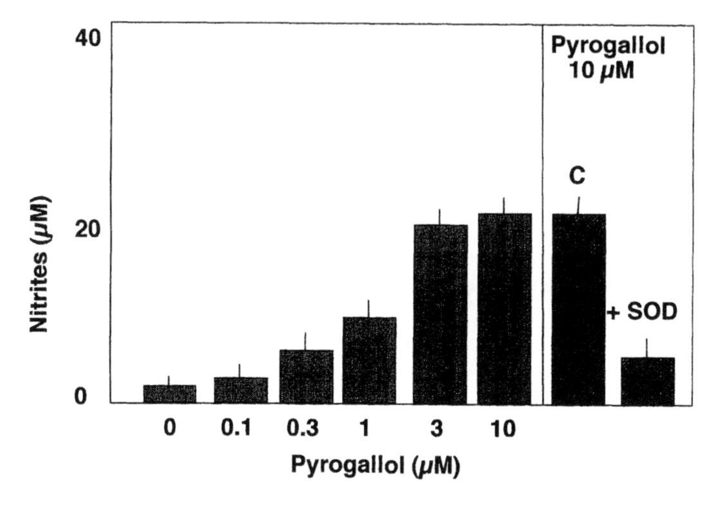

Figure 1. NO production induced by oxidative stress and suppressed in the presence of SOD. Human PBL were stimulated for 3 to 4 days in the presence of a chemical donor of superoxide anion (pyrogallol from 0.1 to 10 μM), and nitrogen derivatives were measured by the Griess reaction. As control, SOD was added to block this induction effect. Data are mean ± SD of one representative experiment out of three.

biological activity of NO action has been hampered by the lack of potent, biologically useful cell-permeable superoxide dismutase mimetics.

We have characterized the actions of Mn-SOD on superantigen-induced TNF-α production by human PBL (Table 1) and on NO production by human PBL in presence of a chemical donor of superoxide anion (Figure 1), and showed that Mn-SOD is a potent inhibitor of a peroxynitrite-induced oxidative reaction.

SOD AS THERAPEUTIC AGENT AFTER ORAL ADMINISTRATION

The kinetic behavior of bovine erythrocyte Cu,Zn-SOD has been investigated in rats after subcutaneous and oral administrations of doses ranging from 0.5 to 20 mg/kg. Studies have been carried out with SOD and with SOD encapsulated into liposomes containing or not containing ceramides. The maximum concentration in blood cell pellets ranges from 8.65 to 11.03 U/mg Hb after subcutaneous injection, and from 4.48 to 8.23 U/mg Hb after oral administration. The maximum concentrations are reached in 5 h for the two routes. Comparison between the areas under the curves obtained after subcutaneous and oral administration allows the calculation of relative bioavailability. The maximum bioavailability after oral administration is 14% for free SOD, 22% for SOD encapsulated ito liposomes, and 57% when ceramides are added to liposomes (25). Anti-inflammatory properties have also been investigated by the oral route. Evaluation consisted of measurement of paw edema volume with determination of prostaglandin E_2, thromboxane B_2, and 6-keto prostaglandin $F_1\alpha$ levels. Polymorphonuclear oxidative metabolism has been evaluated by measurement of superoxide anion production; levels of SOD have been determined in cells and pleural exudates. Greater anti-inflammatory effects are obtained after eight administrations of encapsulated forms (0.5 mg/kg), whereas free SOD has shown no effects. Ceramides enhanced, in the anti-inflammatory effects (26), in accordance with the pharmacokinetic data.

REFERENCES

1. Fridovich I. The toxicology of oxygen radicals. Annu Rev Pharmacol Toxicol 1983; 23: 239–257.
2. McCord JM, Fridovich I. Superoxide dismutase: an enzymatic function for eythrocuprein (hemocuprein). J Biol Chem 1969; 244: 6049–6055.
3. McCord JM. A superoxide activated chemotactic factor and its role in inflammatory process. Agents Action 1980; 10: 522–527.
4. Visner GA, Dougall WC, Wilson JM, Burr IA, Nick HS. Regulation of manganese superoxide dismutase by lipopolysaccharide, interleukin-1 and tumor necrosis factor. J Biol Chem 1990; 265: 2856–2864.
5. Ho YS, Howard AJ, Crapo JD. Molecular structure of a functional rat gene for manganese-containing superoxide dismutase. Am J Respir Cell Mol Biol 1991; 4: 278–286.
6. Dugas B, Mossalayi MD, Damais C, Kolb JP. Nitric oxide production by human monocytes: evidence for a role of CD23. Immunol Today 1995; 16: 574–580.
7. Flores SC, Marecki JC, Harper KP, Bosc SK, Nelson SK, McCord JM. Tat protein of human immunodeficiency virus type 1 repress expression of manganese superoxide dismutase in HeLa cells. Proc Natl Acad Sci USA 1993; 90: 7632–7636.
8. Miesel N, Nahmood N, Weser U. Activity of Cu_2Zn_2 superoxide dismutase against the human immunodeficiency virus type 1. Redox Report 1995; 1: 99–103.

9. Edeas M. Oxidative stress and HIV infection. Ph.D. dissertation, Paris XI University, Chatenay-Malabry, France; 1996.
10. Pestka S, Langer JA, Zoon KC, Samuel CE. Interferons and their actions. Annu Rev Biochem 1987; 56: 727–777.
11. Harris CA, Derbin KS, Hunte MB, Krauss MR, Chen KT, Smith DM, Epsten LB. Manganese superoxide dismutase is induced by interferon-gamma in multiple cell types. Synergistic induction by interferon-gamma and tumor necrosis factor or interleukin-1. J Immunol 1991; 147: 149–154.
12. Raineri I, Huang TT, Epstein CJ, Epstein LB. Antisense manganese superoxide dismutase mRNA inhibits the antiviral action of interferon-γ and interferon-α. J Interferon Cytokine Res 1996; 16: 61–68.
13. Demayer E, Demayer-Guignard I. The antiviral activity of interferons. In Interferons and Other Regulatory Cytokines. New York: Wiley; 1988: 114–133.
14. Zhang P, Damier P, Hirsch EC, Agid Y, Ceballos-Picot I, Sinet PM, Nicole A, Laurent M, Javoy-Agid F. Preferential expression of superoxide dismutase messenger RNA in melanized neurons in human mesencephalon. Neuroscience 1993; 55: 167–175.
15. Zhang P, Anglade P, Hirsch EC, Javoy-Agid F, Agid Y. Distribution of manganese-dependent superoxide dismutase in the human brain. Neuroscience 1994; 61: 317–330.
16. Medina L, Figueredo-Cardenas G, Reiner A. Differential abundance of superoxide dismutase in interneurons versus projection neurons and in matrix versus striosome neurons in monkey striatum. Brain Res 1996; 708: 59–70.
17. Kifle Y, Monnier J, Chesrown SE, Raizada MK, Nick HS. Regulation of the manganese superoxide dismutase and inducible nitric oxide synthase gene in rat neuronal and glial cells. J Neurochem 1996; 66: 2128–2135.
18. Lefaix JL. Pathological features of cerebral radiation necrosis. Bull Cancer Radiother 1992; 79: 125–135[Part I]; 1992; 79: 251–270 [Part II].
19. Harris ME, Hensley K, Butterfield DA, Leedle RA, Carney JM. Direct evidence of oxidative injury produced by the Alzheimer's β-amyloid peptide (1–40) in cultured hippocampal neurons. Exp Neurol 1995; 131: 193–202.
20. Ebadi M, Srinivasan SK, Baxi MD. Oxidative stress and antioxidant therapy in Parkinson's disease. Prog Neurobiol 1996; 48: 1–19.
21. Lipton SA, Gendelman HE. Dementia associated with the acquired immunodeficiency syndrome. N Engl J Med 1995 (April 6); 934–940.
22. Brown DR, Schmidt B, Kretzschmer HS. Role of microglia and host prion protein in neurotoxicity of a prion protein fragment. Nature 1996; 380 (28 March): 345–347.
23. Ihara Y, Mori A, Hayabara T, Kawai M, Namba R, Nobukuni K, Sato K, Kibata M. Superoxide dismutase and free radicals in sporadic amyotrophic lateral sclerosis; relationship to clinical data. J Neurol Sci 1995; 134: 51–56.
24. Busciglio J, Yankner BA. Apoptosis and increased generation of reactive oxygen species in Down's syndrome neurons in vitro. Nature 1995; 378: 776–779.
25. Regnault C, Soursac M, Roch-Arveiller M, Postaire E, Hazebroucq G. Pharmacokinetics of superoxide dismutase in rats after oral administration. Biopharm Drug Dispos 1996; 17: 165–174.
26. Regnault C, Roch-Arveiller M, Tissot M, Sarfati G, Giroud JP, Postaire E, Hazebroucq G. Effect of encapsulation on the anti-inflammatory properties of superoxide dismutase after oral administration. Clin Chim Acta 1995; 240: 117–127.

29

Pharmacological Inhibition of Endothelial Cytoskeleton Alterations Induced by Hydrogen Peroxide and TNF-α

P. d'Alessio and C. Marsac
INSERM U75
Paris, France

M. Moutet and J. Chaudière*
OXIS International SA
Bonneuil sur Marne, France

INTRODUCTION

Endothelial cells line the vascular wall. Because of their topological situation between blood and tissue, they are critically involved in the tissue recruitment of circulating leukocytes. The normal transit time of leukocytes through the endothelium is short, unlike that of activated leukocytes whose adhesion to endothelial cells is much stronger. Such activated leukocytes release substantial amounts of cytokines and reactive oxygen species, two families of molecules that are critically involved in the endothelial expression of cell adhesion molecules.

There are also indications that structural modifications of the cytoskeleton may play an important role in the topological distribution of adhesion molecules at the cell surface (1,2).

The inflammatory response can result in the overproduction of cytotoxic peroxides owing to the joint contribution of polymorphonuclear leukocytes, macrophages, and the endothelial cells themselves. Thus, it is likely that reactive oxygen species play an important role in the structural and functional alterations of the vascular wall that are often observed in acute episodes of inflammation, for example, during infection, allergic responses, or ischemia–reperfusion, allograft rejection, or following vascular trauma.

Among the cytokines that are released by leukocytes, tumor necrosis factor-α (TNF-α) induces the expression of adhesion molecules which amplify the vascular recruitment of the parent leukocytes. TNF-α also stimulates the endothelial production of peroxides (3) and that of hydrogen peroxide (H_2O_2) in particular. Hydrogen peroxide is the most abundant hydroperoxide produced by our cells. It acts as a redox regulator of cell signaling (4), but it also induces oxidative damage in endothelial cells when its concentration exceeds a certain level. Therefore, our initial hypothesis was that some of the pathogenic and/or cytotoxic

**Current affiliation*: Université de Marne la Vallée, Noisy-le-Grand, France.

effects of TNF-α might be prevented by a pharmacological effector that would protect endothelial cells from the toxicity of hydrogen peroxide.

Glutathione peroxidases (GPx) are selenoenzymes that degrade hydrogen peroxide and organic hydroperoxides in the cytosolic and mitochondrial compartments of the cell. In a previous study, we found that a marked protection of human endothelial cells could be obtained with new synthetic GPx mimics (5). More specifically, we found that among a group of GPx mimics, which included ebselen (5,6) and new selenium-containing molecules, the new selenazine BXT-51072 provided major and much stronger protection of endothelial cells from the toxicity of either hydroperoxides, TNF-α-activated neutrophils or TNF-α alone than that observed with some other compounds.

The aim of the present study was to look for ultrastructural correlates of the protection that could be obtained with compound BXT-51072, using either hydrogen peroxide or TNF-α alone as mediators of endothelial cell alteration.

This study shows that this GPx mimic is indeed able to inhibit the structural modifications of the endothelial cytoskeleton which are induced by hydrogen peroxide or by TNF-α. Moreover, this compound partially inhibits the TNF-α-induced expression of the adhesion molecule ICAM-1 by endothelial cells.

MATERIALS AND METHODS

The experiments were performed with human umbilical vein endothelial cells (HUVEC) cultured according to Jaffe (7), using a complete medium supplemented with 20% fetal calf serum (FCS) and 2% growth factor (ECGF, Sigma), between the second and fifth passages.

Incubation in the Presence of H_2O_2 and the Effect of Compound BXT-51072. Endothelial cells were pretreated or not (0.1% ethanol) with 4 μM BXT-51072 (FCS 2%) for 1 h, washed with PBS, then incubated for 1 h with 140 μM H_2O_2. After further washing with PBS, cells were incubated for 22 h in the culture medium.

Stimulation with TNF-α and Effect of Compound BXT-51072. Endothelial cells were incubated with 1 ng/ml TNF-α for 2 and 6 h. In parallel, cells were preincubated for 1 h with 10 μM BXT-51072 (2% FCS) and then coincubated with BXT 51072 (10 μM) and TNF-α for 2 and 6 h.

Visualization of Endothelial Cytoskeleton. Polymerized actin visualization was achieved by means of phalloidine–rhodamine staining of endothelial monolayers. The latter were first washed with PBS, then fixed with 3% paraformaldehyde in PBS for 20 min. The fixed monolayers were then saturated with 50 mM ammonium chloride in PBS for 10 min and treated with 0.2% Triton × 100 for 5 min. After further washing in PBS, monolayers were stained with rhodamine-labeled phalloidine (TRITC-phalloidine, Sigma) for 20 min, then washed with PBS and fixed with glycerol (50% in PBS) for observation by means of optical or confocal microscopy.

The expression of ICAM-1 was measured by means of ELISA, using a monoclonal antibody from R&D Systems (UK) for the detection of ICAM-1. For confocal microscopy studies the same antibody was labeled with fluorescein.

RESULTS

When they are close to confluence, normal endothelial cells exhibit multidirectional filaments of polymerized actin which form a cortical network with some geodesic structures. There is no stress fiber (Figure 1A).

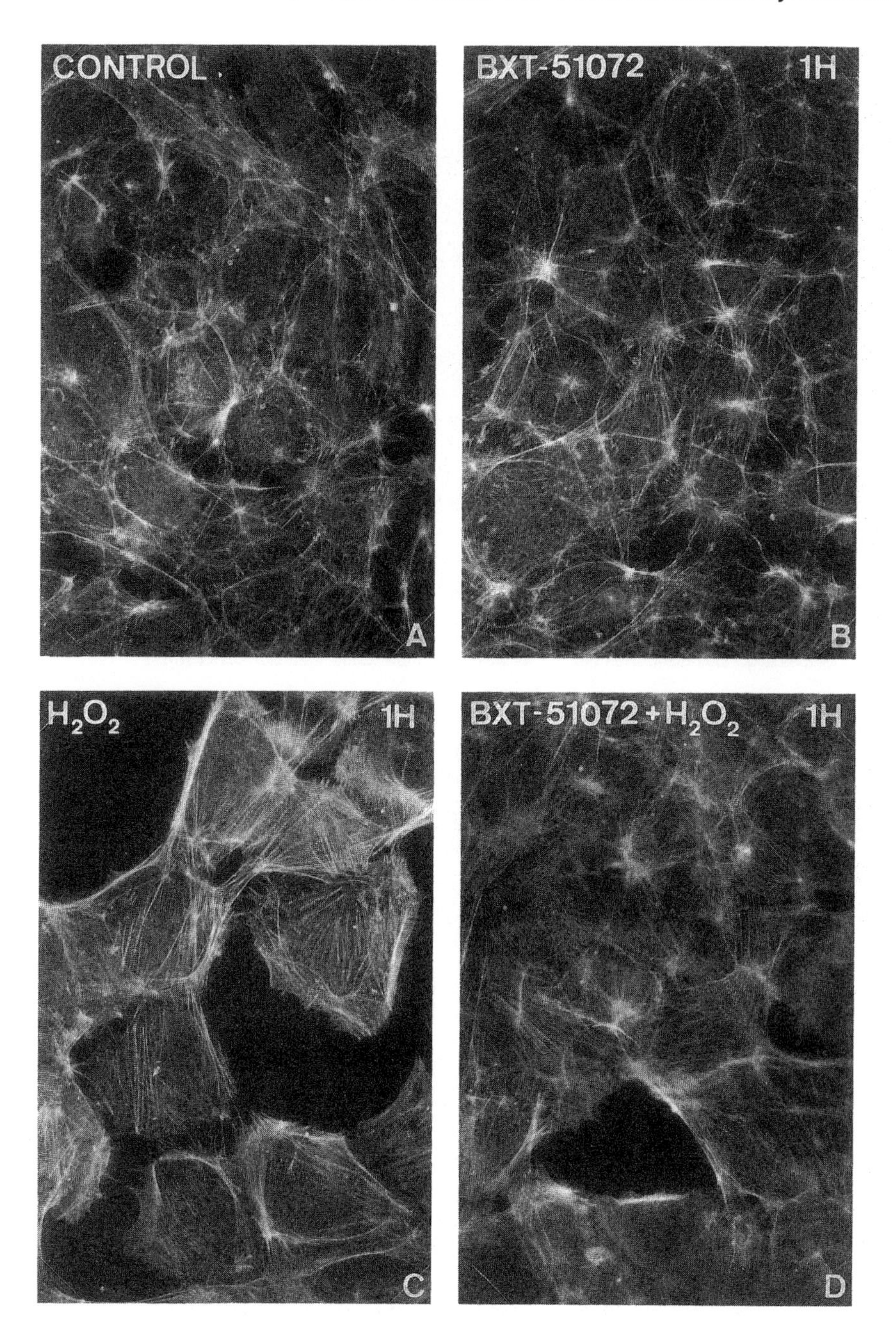

Figure 1. (A–D) Structure of the endothelial cytoskeleton observed by optical microscopy ×440. Fluorescent visualization of phalloidine/rhodamine complexed with actin filaments. (A) Control cells, close to confluence. (B) Incubation of cells with 4 μM compound BXT-51072 for 1 h, washing with PBS, and addition of complete culture medium for 22 h. (C) Incubation of cells with 140 μM hydrogen peroxide for 1 h, washing with PBS, and addition of complete culture medium for 22 h. (D) Preincubation of cells with 4 μM compound BXT-51072 for 1 h, washing with PBS, and incubation of cells with 140 μM H_2O_2 for 1 h, followed by further washing with PBS and addition of complete culture medium for 22 h.

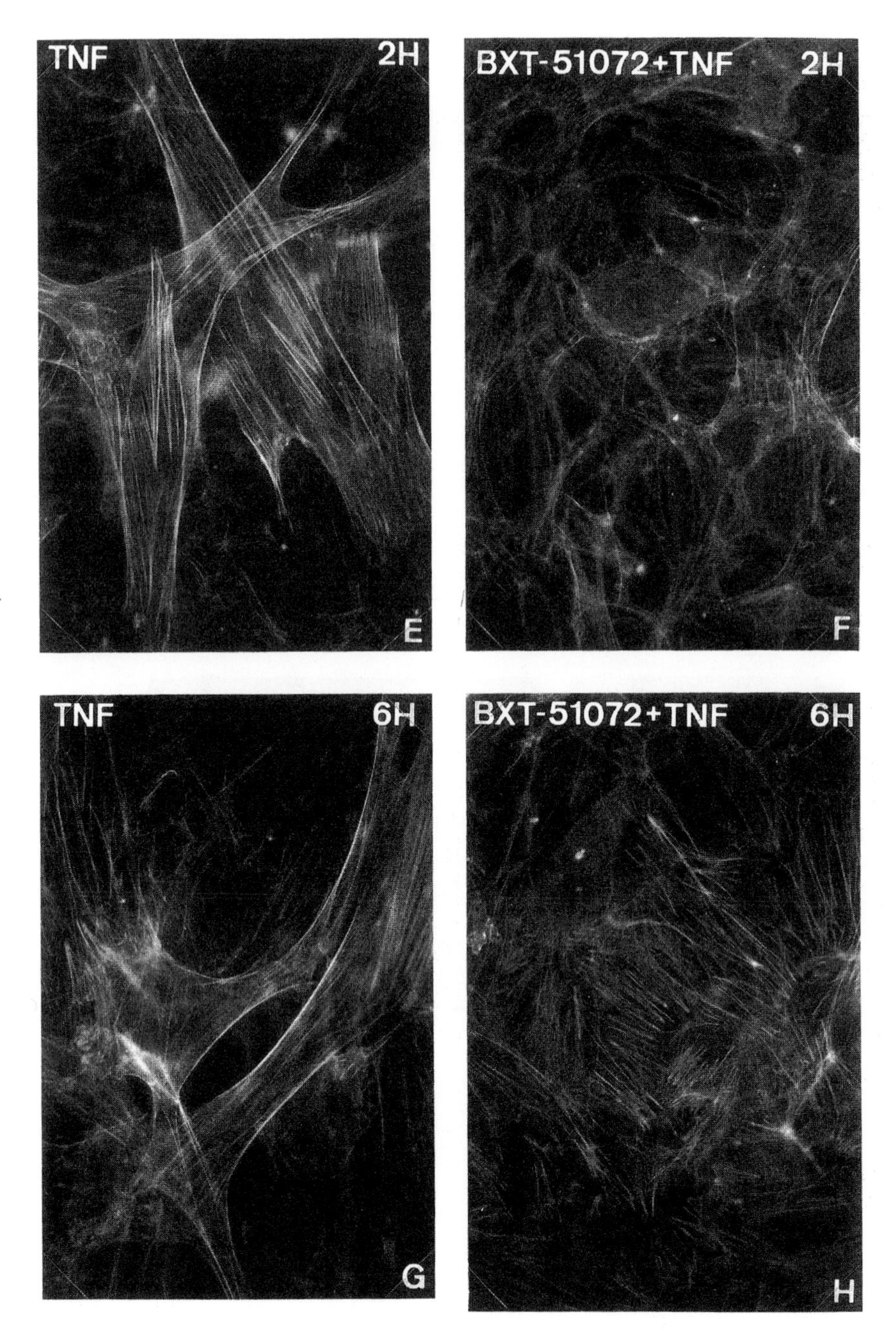

Figure 1. (E–H) (E) Incubation of cells with 1 ng/ml TNF-α for 2 h and washing with PBS. (F) Preincubation of cells with 10 μM compound BXT-51072 for 1 h, washing with PBS, followed by further incubation with 1 ng/ml TNF-α for 2 h in the presence of 10 μM compound BXT-51072. (G) Incubation of cells with 1 ng/ml TNF-α for 6 h and washing with PBS. (H) Preincubation of cells with 10 μM compound BXT-51072 for 1 h, washing with PBS, followed by further incubation with 1 ng/ml TNF-α plus 10 μM compound BXT-51072 for 6 h.

Hydrogen Peroxide-induced Alterations of the Endothelial Cytoskeleton. The incubation of the endothelial monolayer with 140 μM hydrogen peroxide for 1 h (Figure 1C) induced a decrease of cell density and an extensive reorganization of actin filaments. The cortical network, the geodesic structures and the intercellular connections tended to disappear. Individual cells were flattened and appeared as compact masses of stress fibers whose density was increased near the plasma membrane.

Effect of the GPx Mimic BXT-51072 on the Endothelial Cytoskeleton. When HUVEC were incubated with 4 μM GPx mimic for 1 h (Figure 1B), the organization of the actin network did not differ significantly from that of control cells, although a slightly greater number of geodesic structures could not be excluded.

Hydrogen Peroxide-induced Alterations of the Endothelial Cytoskeleton are Prevented by the GPx Mimic. The preincubation of endothelial cells with 4 μM GPx mimic preserved the integrity of cellular monolayer and inhibited morphological alterations induced by subsequent treatment with hydrogen peroxide (Figure 1D). The distribution of actin filaments between the cortical network and the stress fibers appeared intermediate between that of control cells and hydrogen peroxide-treated cells. The intercellular connections were preserved.

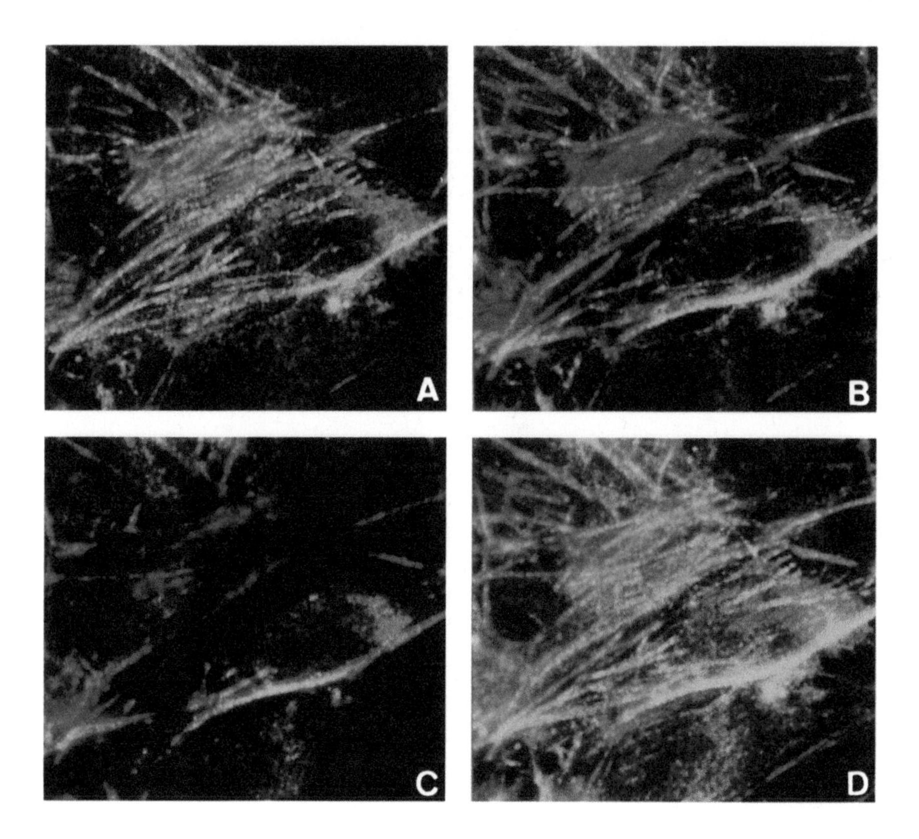

Figure 2. Fluorescent covisualization of actin filaments and ICAM-1 in endothelial cells treated with TNF-α by confocal microscopy ×280. (A) cell surface; (B) cell center; (C) cell base; (D) computerized integration of the three scanned levels.

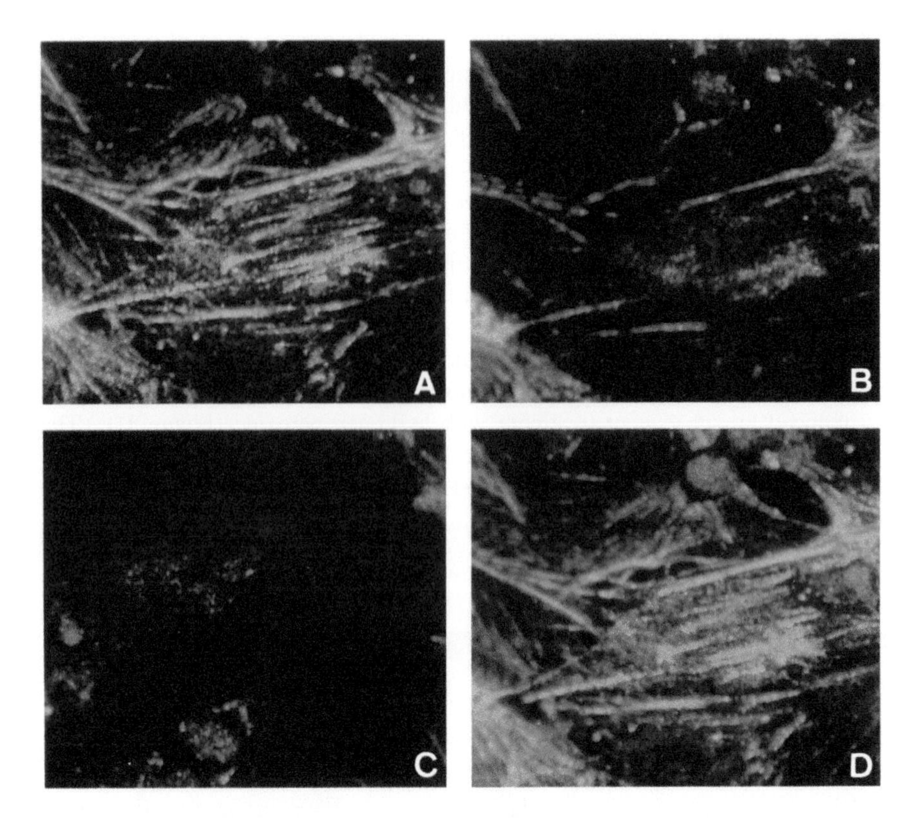

Figure 3. Fluorescent covisualization of actin filaments and ICAM-1 in endothelial cells treated with TNF-α and 10 μM compound BXT-51072 by confocal microscopy ×280. (A) cell surface; (B) cell center; (C) cell base; (D) computerized integration of the three scanned levels.

TNF-α-induced Alterations of the Endothelial Cytoskeleton. The incubation of endothelial monolayer with TNF-α for 2 h induced an extensive reorganization of actin filaments with the appearance of huge bundles of roughly unidirectional fibers (Figure 1E). These stress fibers were very long and therefore markedly distinct from those observed upon treatment of the cells with hydrogen peroxide. Individual cells were spread and spindle-shaped. When endothelial cells were incubated with TNF-α for 6 h, the resulting morphological modifications and the structural reorganization of actin filaments was essentially the same (Figure 1G).

TNF-α-induced Alterations of the Endothelial Cytoskeleton are Prevented by the GPx Mimic. When endothelial cells were preincubated with 10 μM GPx mimic and then coincubated with TNF-α and 10 μM GPx mimic, most of the morphological alterations described above were prevented. As shown in Figure 1F, following 2 h of incubation with TNF-α in the presence of the GPx mimic, the organization of actin filaments was very similar to that of control cells, and little cell spreading had occurred. Only very few stress fibers could be seen. Under the optical microscope, the cell monolayer maintained its typical cobblestone morphology (data not shown).

Following 6 h of incubation with TNF-α in the presence of the GPx mimic (Figure 1H), some reorganization of the actin network could be observed and short stress fibers were scattered throughout the monolayer.

As shown in Figure 2 (A to D), incubation of endothelial cells with TNF-α for 6 h induced the expression of ICAM-1 as well as the polymerization of actin in the form of stress fibers. The double-labeling technique demonstrates that the expression of ICAM-1 is maximal in regions of maximal density of polymerized actin. The densities of polymerized actin and ICAM-1 decrease from the base to the apex of the cell (from A to C).

In contrast, following 6 h of incubation with TNF-α in the presence of 10 μM GPx mimic, the expression of ICAM-1 was almost entirely abolished (Figure 3). It is noteworthy that the GPx mimic almost entirely inhibited the detectable expression of ICAM-1 on the apical side of the cell, which corresponds to the luminal side of the endothelium.

These results have been confirmed by ELISA: the percentage inhibition of ICAM-1 expression is 53.0% ± 15.3% (n=9) relative to control cells (no GPx mimic).

In summary, the incubation of endothelial cells with either hydrogen peroxide or TNF-α induces a drastic reorganization of the endothelial actin network and destroys the intercellular tightness, whereas the GPx mimic that was used in this study can prevent these alterations to a large extent.

DISCUSSION

The vascular endothelium is situated in a strategic position to regulate the traffic of leukocytes. The access of leukocytes to tissues depends on the expression of adhesion molecules by the activated endothelium. Once proinflammatory cytokines such as TNF-α or interleukin-1 (Il-1) have been secreted by circulating leukocytes, their transient or prolonged adhesion to the endothelium will depend on the pattern of production of such adhesion factors (8).

Appropriate regulation of the peroxide tone of endothelial cells plays a key role in the production of cell adhesion molecules and other signals that regulate the amplitude and the nature of cellular interactions supporting the inflammatory response. An overproduction of endogenous peroxides–i.e., of mitochondrial or cytosolic origin– could markedly affect this endothelial control of inflammation.

Endothelial antioxidant enzymes might be critically involved in the preservation of the functional and morphological integrity of the vascular endothelium. This integrity would be affected below certain thresholds of enzyme activities, thereby compromising the controlled migration of leukocytes. Among such antioxidant enzymes, glutathione peroxidases (Se-GPx) play a prominent role and endothelial cells are known to efficiently recycle the reduced form of glutathione (9,10).

In this study, exogenous hydrogen peroxide was used as a tool that would enable comparisons, but we certainly do not claim that this model is physiologically realistic. Exogenous hydrogen peroxide induces specific alterations of the endothelial cell morphology. At the concentrations that were used in our experiments, ultrastructural alterations were indeed extensive, and it should be emphasized that their similarity with those induced by TNF-α might be restricted to the production of stress fibers and the loss of intercellular connections.

We focused our investigations on TNF-α because the endothelial impact of this proinflammatory cytokine is well documented (11). TNF-α can have direct and indirect effects on endothelial cells in vitro or in vivo. Among other effects, TNF-α induces the endothelial expression and release of Il-1 and IL-8, the immediate release of von Willebrand factor, the early expression of P- and E-selectin, and subsequently, the expression of cell adhesion molecules of the immunoglobulin superfamily such as ICAM-1 and VCAM-1. Moreover, it is known that TNF-α induces an increase in the production of intracellular peroxides (3) as well as marked alterations of the cytoskeleton of its target cells (12).

Interestingly, we had found that the GPx mimic BXT-51072 was able to protect endothelial cells from the toxicity of hydrogen peroxide and to downregulate the TNF-α mediated expression of Il-8 and cell adhesion molecules (6,13). Such results suggested that an increase in hydrogen peroxide production could mediate or regulate some of the effects of TNF-α. One hypothesis was that the extent of structural reorganization of the cytoskeleton of the target cell might depend on the level of hydrogen peroxide.

In the present study, we could not monitor the intracellular fluxes of hydrogen peroxide, but our results are compatible with the above hypothesis.

It is indeed of great interest that a GPx mimic that is able to protect endothelial cells from the toxic effects of hydrogen peroxide (6) and from its deleterious effects on the cytoskeletal structure also counteracts the effects of TNF-α on the cytoskeletal structure as well as on the expression of ICAM-1.

In this study, we observed fast and drastic effects of TNF-α in cell populations whose viabilities were always higher than 90% TNF-α alone does induce cell death, this is seen only many hours later (results not shown). An additional effect of TNF-α is due to stimultaneous activation of leukocytes such as polymorphonuclear neutrophils and their subsequent interactions with endothelial cells. This could not have occurred in our model since leukocytes were not present. However, we showed previously that compound BXT-51072 prevented the adhesion of neutrophils to endothelial cells in the presence of TNF-α (6,13) and we showed that at least part of this protection could be due to the strong inhibition of selectins. Here we have confirmed that inhibition of the expression of ICAM-1 is significant.

It has been shown that H_2O_2 and TNF-α activate the expression of ICAM-1 through distinct regulatory effectors of the ICAM-1 promoter involving the transcription factors NF-κB and AP-1 (14). NFκB mediates the effect of H_2O_2, as well as some of the effects of TNF-α. Interestingly, NFκB and AP-1 are both redox-regulated *trans*-activation factors. Finally, it has been shown that ICAM-1 specifically associated with the α-actinin component of the cytoskeleton (15). Ongoing experiments should help to assess the possibility that the expression of ICAM-1 is actually linked to actin reorganization.

Compound BXT-51072 was selected for its high GPx activity and its cytoprotective effects against inflammatory agonists. However, we do not have information on its intracellular distribution and one cannot exclude the possibility that other properties might be involved. Ongoing in vivo studies lend hope that the GPx mimic BXT-51072 could be of therapeutic utility in pathologies in which the vascular recruitment of activated leukocytes plays a critical role.

ACKNOWLEDGMENTS

We thank Dr. Evelyne Coudrier (Laboratoire de Morphogénèse et Signalisation Cellulaire), Institut Curie, for sharing her experience with us and for providing suggestions in the analysis of cytoskeleton pictures; Dr. Raymond Hellio (Station de Microscopie Confocale), Institute Pasteur for his technical collaboration; and Mrs. Sophie Darquenne (OXIS International SA) for her excellent contribution to cell culture experiments.

REFERENCES

1. Helander TS, Carpen O, Turunen O, Kovanen PE, Vaheri A, Timonen T. ICAM-2 redistributed by ezrin as a target for killer cells. Am Soc Cell Biol 1995; abstr H37.
2. Sellak H, Franzini E, Hakim J, Pasquier C. Reactive oxygen species rapidly increase endothelial ICAM-1 ability to bind neutrophils without detectable upregulation. Blood 1994; 83: 2669–2677.
3. Schulze-Osthoff K, Bakker AC, Vanhaesebroek B, Beyaert R, Jacob WA, Fiers W. Cytotoxic activity of tumor necrosis factor is mediated by early damage of mitochondrial function. J Biol Chem 1992; 267: 5317–5323.
4. Schreck R, Bauerle PA. Assessing oxygen radicals as mediators in activation of inducible eukaryotic transcription factor NF-κB. Methods Enzymol 1994; 234: 151–163.
5. Chaudière J, Yadan JC, Erdelmeier I, Tailhan-Lomont C, Moutet M. Design of new selenium-containing mimics of glutathione peroxidase. In Paoletti R, ed. Oxidative Processes and Antioxidants. New York: Raven Press; 1994: 165–184.
6. Chaudière J, Moutet M, d'Alessio P. Protection antioxydante et anti-inflammatoire des cellules endothéliales vasculaires par de nouveaux mimes synthétiques de la glutathion peroxydase. CR Soc Biol 1996; 189: 861–882.
7. Jaffe EA. Culture and identification of large vessel endothelial cells. In Jaffe EA, ed. Biology of Endothelial Cells. Boston: Martinus Nijhoff; 1984: 1–13.
8. Kishimoto TK, Rothlein R. Integrins, ICAMs and selectins: role and regulation of adhesion molecules in neutrophil recruitment to inflammatory sites. Adv Pharmacol 1994; 25: 117–169.
9. Jongkind G, Verkerk A, Baggen RG. Glutathione metabolism of human vascular endothelial cells under peroxidative stress. Free Rad Biol Med 1989; 7: 507–512.
10. Schuppe-Koistinen I, Gerdes R, Moldeus P, Cotgreave IA. Studies on the reversibility of protein-*S*-thiolation in human endothelial cells. Arch Biochem Biophys 1994; 315: 226–234.
11. Strieter R, Kunkel SL, Bone RC. Role of tumor necrosis factor-α in disease states and inflammation. Crit Care Med 1993; 21: S447–S463.
12. Goldblum SE, Ding X, Campbell-Washington J. TNF-α induces endothelial cell F-actin depolymerization, new actin synthesis, and barrier dysfunction. Am J Physiol 1993; 264: C894–905.
13. Moutet M, d'Alessio P, Malette P, Devaux V, Chaudière J. New GPx mimics protect endothelial cells from the toxic effects of TNF-α and activated neutrophils (manuscript in preparation).
14. Robuck KA, Rahman A Lakshminarayanan V, Janakidevi K, Malik AB. H_2O_2 and TNF-α activate intercellular adhesion molecule-1 (ICAM-1) gene transcription through distinct *cis*-regulatory elements within the ICAM-1 promoter. J Biol Chem 1995; 270: 18966–18974.
15. Carpén O, Pallai P, Stanton DE, Springer PA. Association of intercellular adhesion molecule-1 (ICAM-1) with actin-containing cytoskeleton and α-actinin. J Cell Biol 1992; 118: 1223–1234.

30

Perinatal Development of Superoxide Dismutase and Other Antioxidant Enzyme Activities in Rat-brain Mitochondria

Ingrid Wiswedel, Sigrid Hoffmann, and Wolfgang Augustin
Otto-von-Guericke-Universität Magdeburg
Institut für Klinische Chemie, Bereich Pathologische Biochemie
Magdeburg, Germany

Heiko Noack
Otto-von-Guericke-Universität Magdeburg
Institut für Medizinische Neurobiologie
Magdeburg, Germany

INTRODUCTION

The fetal–neonatal transition during birth is a period of important biochemical changes and vulnerability for the fetus (1), especially concerning the energy metabolism (2). Perinatal maturation of mitochondrial activities, such as complexes of the respiratory chain, the adenine nucleotide translocator (ANT), and active respiration, is well documented in brain and other organs (3–5). A close relationship between the perinatal development of ADP-stimulated respiration and the ANT protein has been described for liver and brain (5,6) that plays an important role in the onset of oxidative phosphorylation after birth.

The question arose whether the mitochondrial complement of antioxidant enzymes as major cellular defenses against acute oxygen toxicity also exhibits peculiar developmental profiles.

Accordingly, enzyme activities of superoxide dismutase (SOD), glutathione peroxidase (GPX), glutathione reductase (GR), catalase, and glutathione *S*-transferase (GT) were followed in brain during perinatal development and compared with those of mitochondria from heart and liver, as reported earlier (7). The time course of maturation of SOD was followed in brain by activity assay and also by electrophoretic techniques including activity staining and western blots, which can clearly differentiate between Cu/ Zn- and Mn-SOD enzymes.

METHODS

Preparation of Mitochondria

Wistar rats were kept under constant environmental conditions (temperature 22 ± 1°C, circadian rhythm 12 h light and 12 h dark) and fed a normal laboratory diet as pellets (Haltungsdiät Altrumin) with water ad libitum. They were sacrified by decapitation and used for preparation of functional intact brain mitochondria (8) according to standard differential centrifugation procedures. Fetuses were obtained from term-pregnant anesthetized rats by cesarean section.

Homogenates from rat brain (10% or 20% w/v in 50 mM phosphate buffer, pH 7.4) were obtained by homogenizing the brain, treating the homogenate ultrasonically in an ice bath (total time 6 min), and centrifugation at 10,000*g*.

The age of the prenatal rats was estimated from the time of conception and by weighing the fetuses (9) starting at day 17 after conception, i.e., 4 days before birth. The freshly prepared mitochondria were frozen immediately and stored in liquid nitrogen. The protein content was determined by a modified biuret method (10) as well as by the method of Lowry (11). Before enzymatic assay, mitochondria were thawed and frozen three times to ensure complete mitochondrial lysis.

Assay of Antioxidant Enzymes

Total superoxide dismutase activity was measured as reduction rate of cytochrome *c* at 550 nm with the xanthine/xanthine oxidase system (12). For determination of catalase activity, the method of Aebi (13) was used, which records the initial rate of hydrogen peroxide decomposition at 240 nm. Glutathione peroxidase activity was measured in a coupled optical test, in which the GSSG (oxidized glutathione) produced drives the oxidation of NADPH at 340 nm with *t*-butylhydroperoxide and cumene hydroperoxide as substrates (14). For glutathione reductase activity, the NADPH oxidation at 340 nm was also used (15) and the activity of the glutathione transferase was measured as nucleophilic attack of glutathione on the electrophilic center of the substrate 1-chloro-2,4-dinitrobenezene (16).

Enzyme activities were expressed as mean ± SD of 4 to 6 different mitochondrial preparations. In some cases (as indicated in the figures) activities in single mitochondrial preparations are shown.

Electrophoretic Techniques

Equal amounts of the solubilized and reduced proteins (~ 60 μg) were separated by Tricine–SDS electrophoresis according to Schaegger and von Jagow (17). For separation of cytochrome *c* oxidase subunits, the gels contained 6 M urea. After semidry transfer of the proteins to nitrocellulose membranes (18), these were washed and blocked for 30 min in Tris-buffered saline (20 mM Tris, pH 7.6, 137 mM NaCl) containing 0.1% Tween 20 (TBS-Tween) and 5% nonfat dry milk. After washing (three times with TBS-Tween), the membranes were incubated overnight at 4°C with the primary antibodies diluted in TBS-Tween (anti-rat-Mn-SOD 1:40,000; anti-rat-Cu/Zn-SOD 1:40,000; anti-cytochrome *c* oxidase 1:1000).

Membranes were washed and incubated with peroxidase-coupled anti-rabbit secondary antibody (anti-rabbit IgG POD FAB-fragment, Boehringer) diluted in TBS-Tween containing 0.5% nonfat dry milk for 2 h at room temperature. Proteins were visualized by DAB/H_2O_2.

For semiquantitative determination of protein, the developed blots were scanned with a laser densitometer (Pharmacia) and the resulting peak areas were determined by the GSXL-densitometer software.

Anti-rat-Mn-SOD (rabbit) and anti-rat-Cu/Zn-SOD (rabbit) were a kind gift of Dr. K. Asayama, Yamanashi-University, Japan, and anti-rat-cytochrome *c* oxidase (rabbit) was a kind gift of Dr. H. Schaegger, University of Frankfurt/Main.

RESULTS AND DISCUSSION

Initial activities of total superoxide dismutase in brain mitochondria (Figure 1) remain nearly unchanged during the prenatal period until day 10 after birth and increase strongly thereafter to about 5-fold higher final levels at the age of 6 weeks. This is in contrast to activities in liver and heart mitochondria, as previously reported (7). In these organs mitochondrial SOD activities are generally smaller than in brain and increase steadily to plateau values immediately after birth–i.e., much earlier than in brain mitochondria. Catalase activities in brain mitochondria (Figure 2) are very low and show

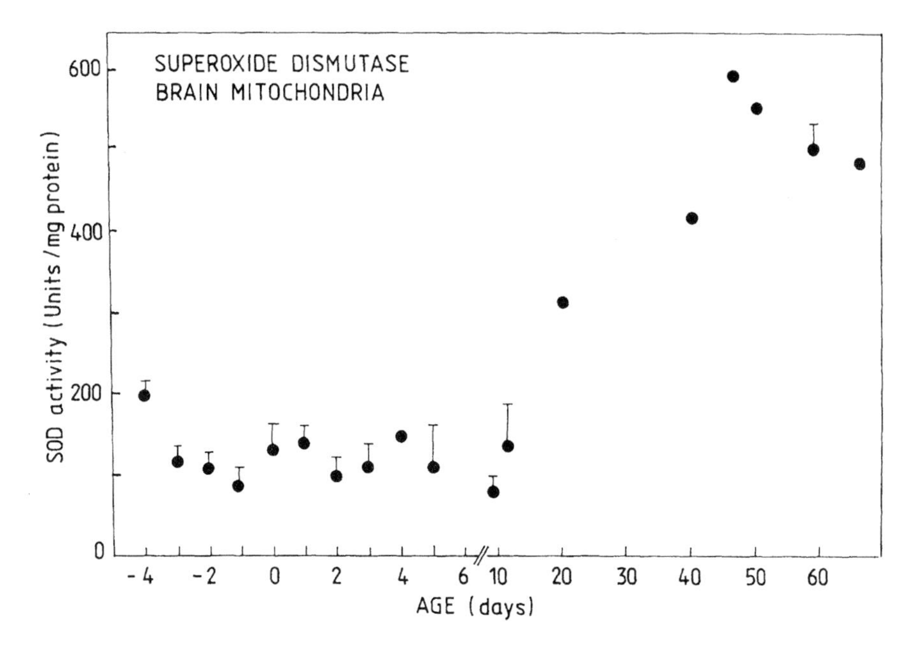

Figure 1. Total superoxide dismutase activities in brain mitochondria as a function of age. SOD activities were expressed as units/mg protein, where one unit of SOD is defined as that amount of enzyme which inhibits the rate of cytochrome *c* reduction by 50%. A calibration curve was constructed using standard SOD.

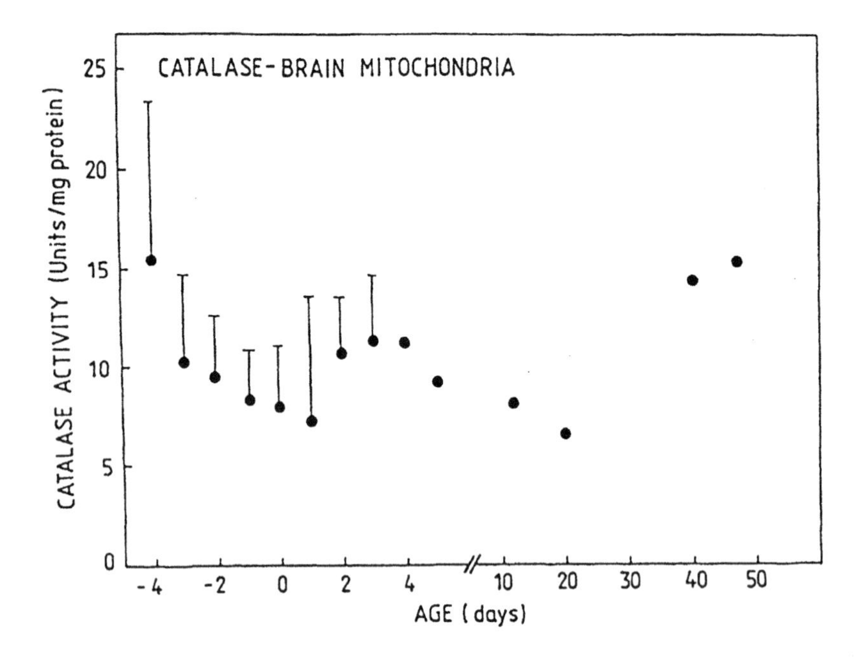

Figure 2. Catalase activities in brain mitochondria. Units of catalase are defined in terms of initial velocity of hydrogen peroxide decomposition per minute at 240 nm.

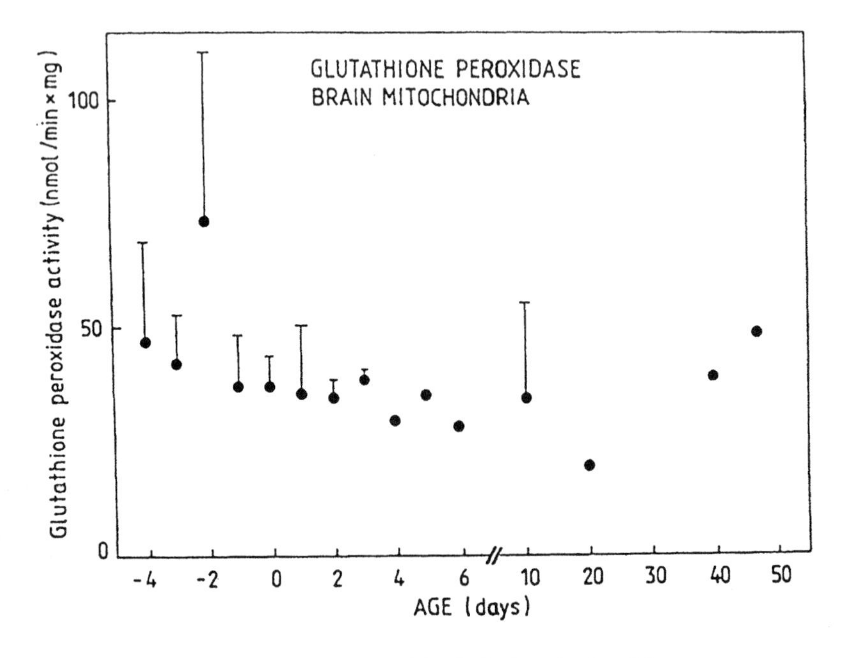

Figure 3. Glutathione peroxidase activity in brain mitochondria.

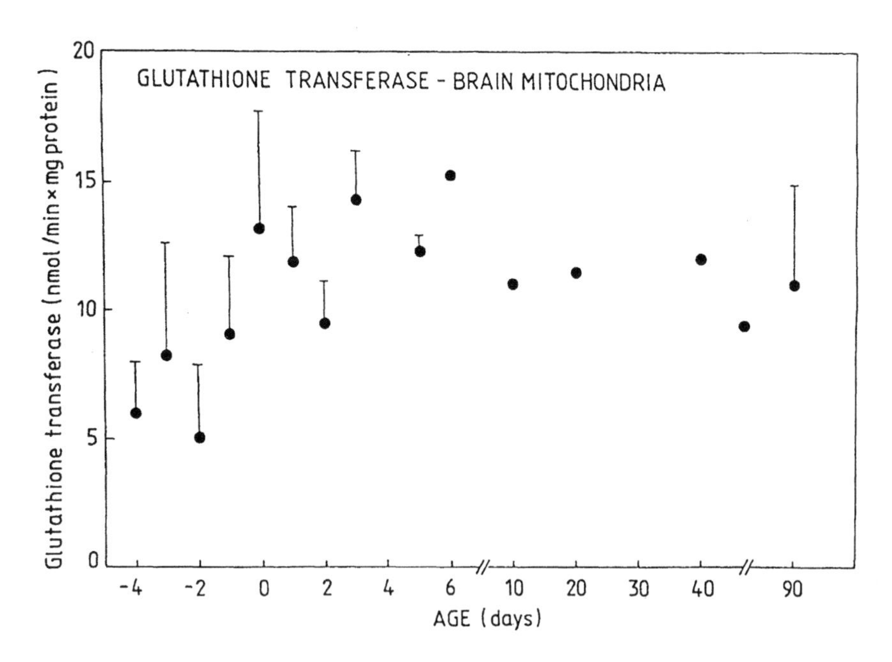

Figure 4. Glutathione *S*-transferase activity in brain mitochondria.

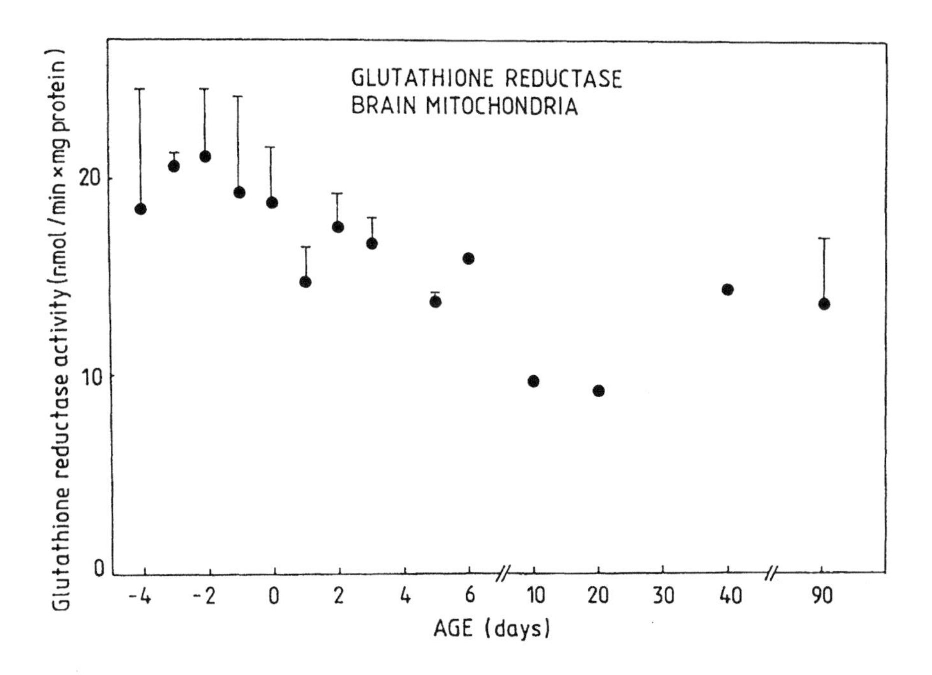

Figure 5. Glutathione reductase activity in brain mitochondria.

nonsignificant developmental changes. Although the question of the functional role and localization of catalase in mitochondria in general remains open (19), one may assume that in brain catalase may be of minor importance. Most H_2O_2 detoxification very likely occurs via glutathione peroxidase in view of its low K_m for hydrogen peroxide (20). GPX, however, exhibits low activities in brain mitochondria (Figure 3) that are practically unchanged during the life span considered. This is in contrast to the developmental profile of glutathione peroxidase in liver mitochondria (7), where GPX exhibits pronounced alterations within the perinatal period. GT activities in brain mitochondria (Figure 4) increase only in the fetal period and attain 2-to 3-fold higher plateau values at birth or immediately thereafter; they are as low as in liver mitochondria and could not be detected in heart mitochondria (7). Activities for GR in brain are as low as in liver mitochondria (Figure 5) and decline to a minimum 3 weeks after birth, followed by a small increase.

A "redox index" was calculated, defined as glutathione reductase activity devided by peroxidase activity multiplied by 100 (21). The redox index in brain mitochondria (Figure 6) remained constant until 6 days after birth and declined slowly, whereas in liver mitochondria (7) a constant decline was observed with a minimum 20 days after birth. The different behavior in brain and liver mitochondria may reflect the different antioxidant capacities in these tissues during the perinatal development. Thus a better antioxidative defense of brain mitochondria by the glutathione system in the very early phase of life appears likely.

Obviously most important for brain mitochondria is the late expression of SOD, which is in accordance with the development of the ANT protein and the full capacity of the respiratory chain in brain (6).

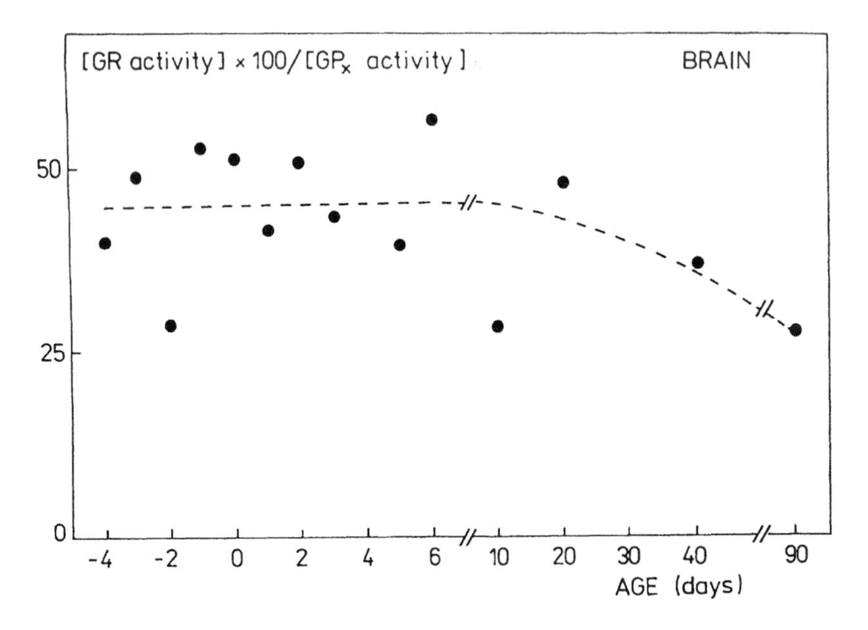

Figure 6. Glutathione "redox index" calculated from glutathione peroxidase and reductase activities (21).

The late increase of mitochondrial SOD activities during brain development led us to conduct a more detailed study of the two SOD enzymes in parallel with cytochrome *c* oxidase in the rat brain during perinatal development. Cu/Zn-SOD, Mn-SOD, and cytochrome *c* oxidase were also studied in total brain homogenates (Figure 7) and isolated

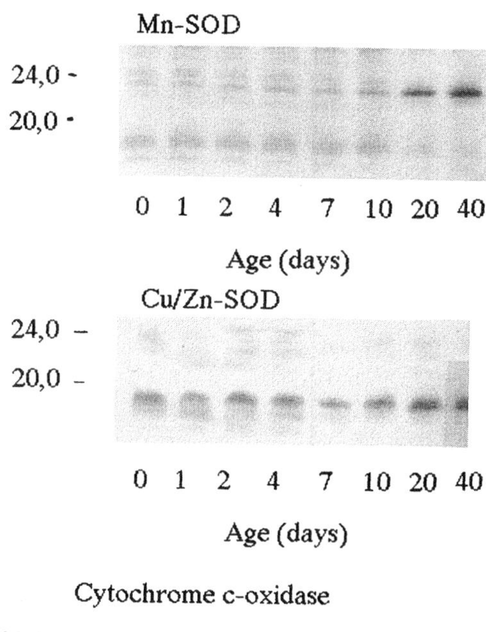

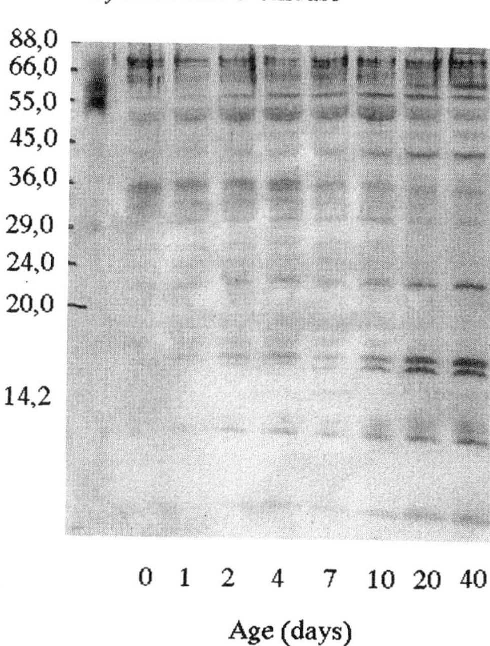

Figure 7(A). Enzyme proteins (Mn-SOD, Cu/Zn-SOD, and cytochrome oxidase) in rat brain during postnatal development: (A) western blots

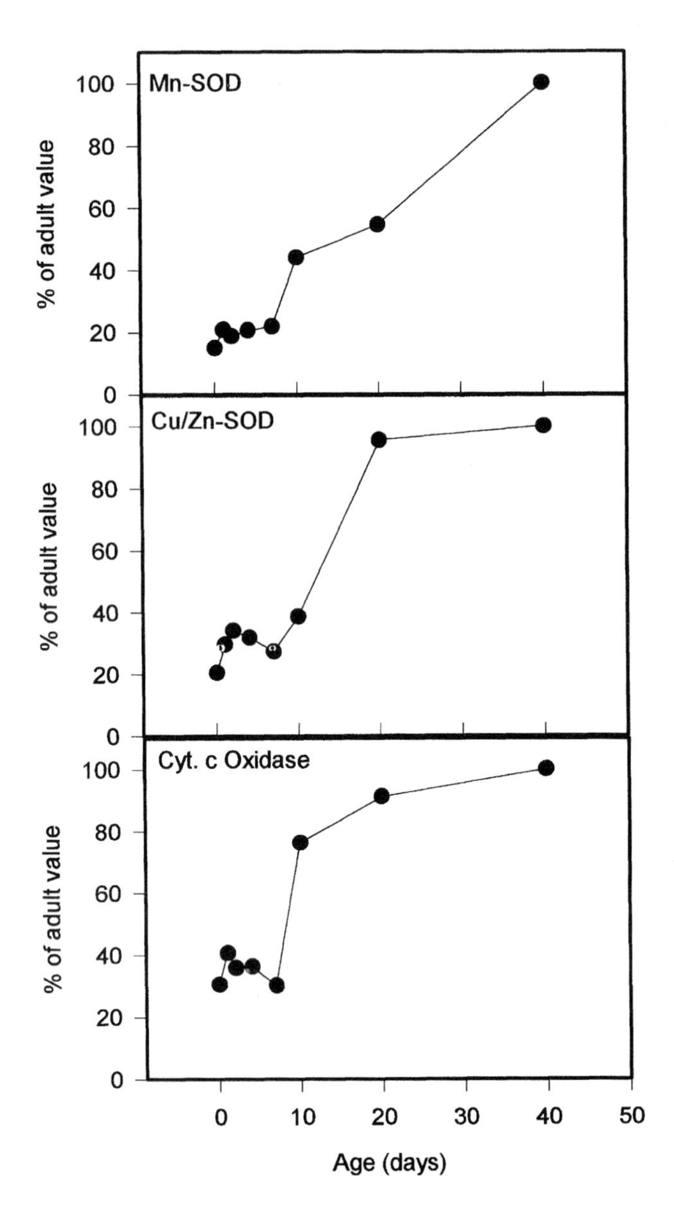

Figure 7(B). Enzyme proteins (Mn-SOD, Cu/Zn-SOD, and cytochrome oxidase) in rat brain during postnatal development: (B) densitometric analysis.

mitochondria (not shown). The results of densitometric scanning (Figure 7b) of the blots show that the increase of both SOD proteins fully corresponds to the SOD activities (Figure 1) and the expression of oxidative capacities of mitochondria as deduced from cytochrome oxidase data. In summary, it can be concluded that in the developing rat brain there is a close parallelism between the expression of antioxidative defense systems (in terms of SOD) and the expression of mitochondrial oxidative capacities.

REFERENCES

1. Clark JB, Bates TE, Cullingford T, Land JM. Development of enzymes of energy metabolism in the neonatal mammalian brain. Dev Neurosci 1993; 15: 174–180.
2. Clark JB, Bates TE, Almeida A, Cullingford T, Warwick J. Energy metabolism in the developing mammalian brain. Biochem Soc Trans 1994; 22: 980–983.
3. Bates TE, Almeida A, Heales SJR, Clark JB. Postnatal development of the complexes of the electron transport chain in isolated rat brain mitochondria. Dev Neurosci 1994; 16: 321–327.
4. Aprille JR. Perinatal development of mitochondria in rat liver. In Fiskum G, ed. Mitochondrial Physiology and Pathology. New York: Van Nostranol Reinhold; 1986: 66–99.
5. Schönfeld P, Fritz S, Halangk W, Bohnensack R. Increase in the adenine nucleotide translocase protein contributes to the perinatal maturation of respiration in rat liver mitochondria. Biochim Biophys Acta 1993; 1144: 353–358.
6. Schönfeld P, Bohnensack R. Developmental changes of the adenine nucleotide translocation in rat brain. Biochim Biophys Acta 1995; 1232: 75–80.
7. Wiswedel I, Hoffmann S, Augustin W. Perinatal development of antioxidant enzyme activities in rat brain, heart and liver mitochondria. Proceedings of the International Society for Natural Antioxidants. Beijing: AOCS Press; 1995.
8. Wustmann C, Petzold D, Fischer HD, Kunz W. ATP metabolizing enzymes in suspensions of isolated coupled rat brain mitochondria. Biomed Biochim Acta 1987; 46: 331–340.
9. Von Goltzsch W, Bittner R, Didt L, Sparmann G, Böhme HJ, Hofmann E. Bestimmung des Gestationsalters der Ratte. Z Versuchstierkd 1980; 22: 1–7.
10. Steinbrecht I, Augustin W. An improved modification of the biuret method for the determination of protein in turbid materials with high lipid and hemoglobin content. Biomed Biochim Acta 1983; 42: 335–342.
11. Lowry OH, Rosenbrough NJ, Farr AL, Randall RJ. Protein measurement with the Folin reagent. J Biol Chem 1951; 193: 265–275.
12. Flohé L, Ötting F. Superoxide dismutase assays. Methods Enzymol 1984; 105: 93–104.
13. Aebi H. Catalase in vitro. Methods Enzymol 1984; 105: 121–126.
14. Paglia DE, Valentine WN. Studies on the quantitative and qualitative characterization of erythrocyte glutathione peroxidase. J Lab Clin Med 1967; 70: 158–169.
15. Bergmeyer HU. Methoden der enzymatischen Analyse. Berlin: Akademie-Verlag, 1970; 2 Auflage, Bd 1: 424–425.
16. Habig WH, Pabst MJ, Jakoby WB. Glutathione *S*-transferases. J Biol Chem 1974; 249: 7130–7139.
17. Schaegger H, von Jagow G. Tricine–sodium dodecyl sulfate polyacrylamide gel electrophoresis for the separation of proteins in the range from 1 to 100 kDa. Anal Biochem 1987; 166: 368–379.
18. Kyhse-Anderson J. Electroblotting of multiple gels, a simple apparatus without buffer tank for rapid transfer of proteins from polyacrylamide to nitrocellulose. J Biochem Biophys Methods 1984; 10: 203–209.
19. Radi R, Turrens JF, Chang LY, Bush KM, Crapo JD, Freeman BA. Detection of catalase in rat heart mitochondria. J Biol Chem 1991; 266: 22028–22034.
20. Jones DP, Thor H, Andersson B, Orrenius S. Detoxification reactions in isolated hepatocytes. J Biol Chem 1978; 253: 6031–6037.
21. Benzi G, Pastoris O, Marzatico F, Villa RF. Cerebral enzyme antioxidant system. Influence of aging and phosphatidylcholine. J Cereb Blood Flow Metab 1989; 9: 373–380.

31

Coexistence of a "Reactive Oxygen Cycle" with "Q-Cycle" and "H^+-Cycle" in the Respiratory Chain: A Hypothesis for the Generation, Partitioning, Targeting, and Functioning of Superoxide in Mitochondria

Shu-sen Liu
Chinese Academy of Science
Beijing, China

INTRODUCTION

It is now commonly recognized that oxygen radicals, and free radicals in general, are involved in a variety of physiological and pathological processes in organisms, including cellular signaling, cell proliferation and differentiation, apoptosis, cancer, and AIDS, as well as other diseases such as ischemia–reperfusion injury, inflammation, and degenerative diseases, and in senescence (1–3,33,34). One of the consistent sources of oxygen radicals among tissues is the mitochondrial respiratory chain, especially at the flavoprotein and ubiquinone-cytochrome *b* segments. When electron transfer from substrate to dioxygen proceeds along the respiratory chain, not all the oxygen is tetravalently reduced to form water via cytochrome oxidase. Instead, a small portion of oxygen molecules can accept single-electron transfer to form superoxide radicals ($O_2^{\bullet-}$) by the so-called "electron univalent leak," or "electron leak," pathway. In general, only 2–4% of oxygen consumed is partially reduced to $O_2^{\bullet-}$ (and H_2O_2) (1,10–13), and the daily yield of $O_2^{\bullet-}$ might reach 10^7 molecules per mitochondrion (2). Under normal physiological conditions, the superoxide radicals in the mitochondria can be destroyed by Mn^{2+}-superoxide dismutase (Mn-SOD) and other scavenging enzymes. However, the activities of the enzymes decrease with the development of disease and with aging, and the $O_2^{\bullet-}$ are accumulated in mitochondria, resulting in dysfunction of mitochondria and mtDNA alteration (1–3). It is also known that more than one hundred human mitochondrial diseases are related to mitochondrial dysfunction and mutations of mtDNA, in which reactive oxygen species might be involved (4–6). It is therefore of importance to study the mechanism of $O_2^{\bullet-}$ generation and targeting and its pathophysiological functions in mitochondria.

We have proposed a model of coexistence of a "reactive oxygen cycle" with the Q-cycle and H^+-cycle in the mitochondrial respiratory chain to combine the processes of electron leak and proton leak (7–9). This model emphasizes that $O_2^{\bullet -}$ generated by the electron leak pathway in the mitochondrial respiratory chain, and the cycling of the radical ions across the inner mitochondrial membrane in connection with the Q-cycle and H^+-cycle, may serve as an endogenous protonophore inducing proton leak. The functional role of these cycles may be involved in the regulation and partition of energy transduction and heat production, as well as in targeting of macromolecules and mtDNA, resulting in the pathogenesis of mitochondrial diseases. In this chapter we present a more detailed description of this model and some supporting experimental evidence.

HYPOTHETICAL MODEL

From the point of view of free-radical biology and medicine, the electron leak pathway for generation of superoxide in the mitochondrial respiratory chain is of particular importance, because $O_2^{\bullet -}$ generation by this pathway constitutes a consistent source of oxygen free radicals in the living body, from which other reactive oxygen species, such as $HO_2^{\bullet}$, H_2O_2, $^1O_2$, and $OH^{\bullet}$ can be formed (Figure 1). Therefore, $O_2^{\bullet -}$ generated in mitochondria may be considered as an intrinsic source of the harmful free radicals that induce various human diseases (4,14,16).

In terms of cell bioenergetics and cell biology, the electron leak pathway for generation of $O_2^{\bullet -}$ may have different physiological significance. First, it has been shown that $O_2^{\bullet -}$ may be involved in cyanide-resistant respiration of plant and animal mitochondria, which constitutes a branch of the respiratory chain not associated with ATP formation at the level of the Q pool before the substrate side (23,24,35). Second, under state 4 respiration (e.g., under high protonmotive force) the ubisemiquinone or reduced cyt b_{566} formed in the Q-cycle may directly donate a single electron to the oxygen molecule to make $O_2^{\bullet -}$; this constitutes an additional branch of the electron transfer pathway in connection with the Q-cycle. Superoxide radical anions can form H_2O_2 via SOD or by self-dismutation reaction, and the H_2O_2 formed can be further subjected by the action of glutathione peroxidase to regenerate molecular oxygen and H_2O. The Q-cycle is known to occupy the central part of the electron transfer chain and present a very unique and complicated mechanism for the coupling of electron transfer with H^+ translocation in mitochondria. Both electron transfer branches of the respiratory chain are believed to be involved in regulating and partitioning the energy transduction of the protonmotive force and in heat production (14,24,31). However, the mechanisms are far from clear and have not been studied extensively. Third, the proton leak process in mitochondria occurs only during state 4 respiration, and it is not only involved in regulating and partitioning the energy transduction of the protonmotive force and heat production but could also be involved in the development of some other physiopathological states (14,16). In addition, reactive oxygen species have been shown to be involved in a variety of cellular signaling processes (33,34).

In considering these points, it is logical to assume that both electron leak and proton leak processes occurring at the same location (the inner mitochondrial membrane) in association with the respiratory chain and under the same conditions (state 4 respiration) should have some intrinsic connection with each other physically and/or functionally. Also, both processes are involved in the same functional process of heat production and

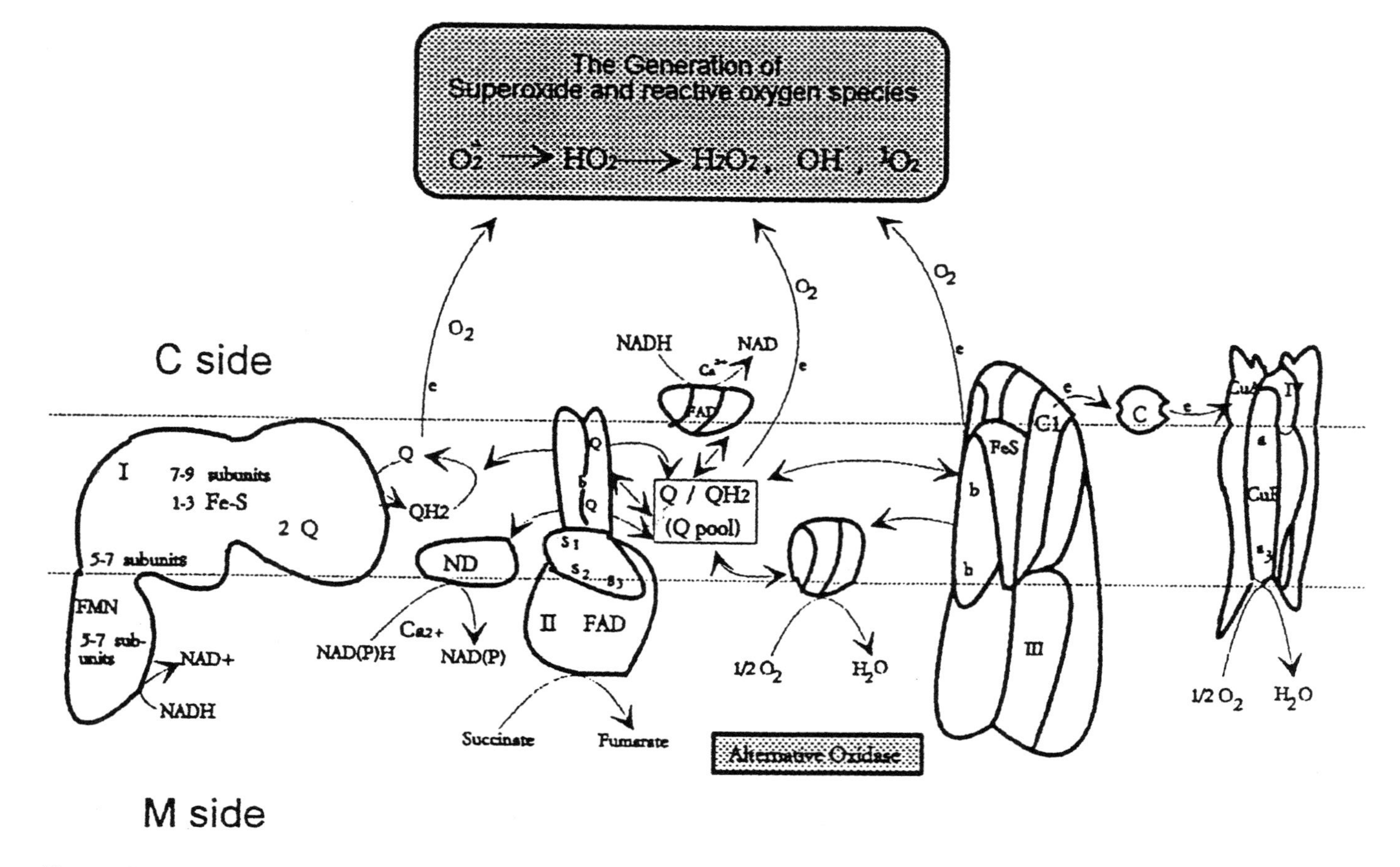

Figure 1. Mitochondrial respiratory chain. (From Liu SS, The electron transfer chain in mitochondria. In Tsou CL, ed. Frontiers in Biosciences, 1996, in press.)

make contributions to the basic metabolic rate and to the regulation of physiopathological processes in the living body. This idea constitutes the foundation of our present hypothesis.

Figure 2 summarizes the main points of our hypothetical model. It emphasizes that during state 4 respiration an interaction between electron leak (a branch of electron transfer directly from the respiratory chain to form $O_2^{\bullet -}$ but not H_2O) and proton leak (a branch pathway in the utilization of the protonmotive force to produce heat, but not ATP) may take place in connection with the Q-cycle and proton cycle in mitochondria through the consumption of H^+ by $O_2^{\bullet -}$ anions to form protonated perhydroxyl radicals, $HO_2^{\bullet}$, which are permeable across the inner mitochondrial membrane directly and induce the leakage of high-energy protons.

The first postulate of the model is that a higher value of $\Delta\Psi$ (about 200 mV) across the inner mitochondrial membrane under state 4 respiration may constitute a physical condition making it easier for ubisemiquinone or reduced cyt b_{566} to leak a single electron directly onto an oxygen molecule to produce $O_2^{\bullet -}$ at the Q_o site of the bc_1 complex facing the C side of the inner mitochondrial membrane, but not allowing donation of an electron to the cyt b_{562} at the Q_i site near the M side of the membrane. The distance between the Q_o site and the surface of C side of the membrane is very small (about 15 Å), and the

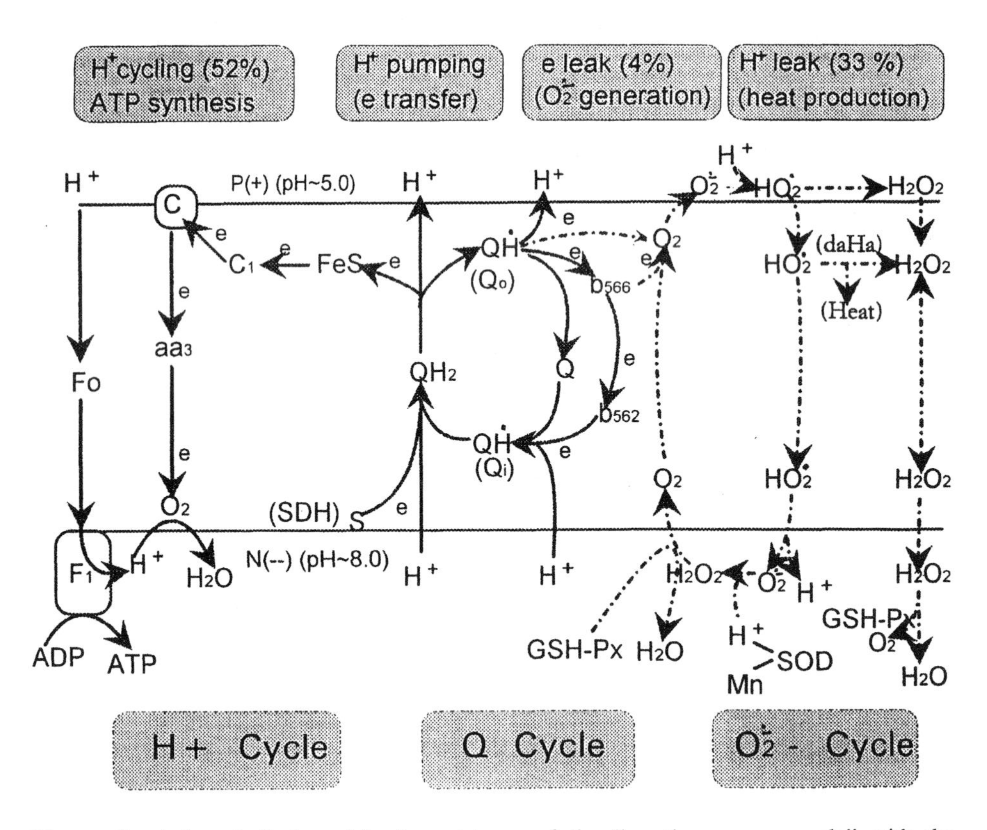

Figure 2. A hypothetical model of coexistence of the "reactive oxygen cycle" with the "Q-cycle" and the "H^+-cycle" in connection with the mechanism of proton leak induced by electron leak in the mitochondrial respiratory chain. (From Ref. 9.)

C side of inner mitochondrial membrane is more electropositive relative to the M side. The latter is more electronegative, especially under the conditions of state 4 respiration. The evidence for this model is presented in the first set of experiments discussed in this chapter.

The second postulate of the model is that under state 4 respiration the ΔpH component of protonmotive force may provide a suitable chemical environment, and the energized H^+ from the redox proton pumps is primarily localized on the surface of the C side of the inner mitochondrial membrane, where $O_2^{\bullet -}$ anions may consume the energized protons and conduct them across the membrane to the matrix side through a p*K*-dependent formation of a protonated species of $HO_2^{\bullet}$, perhydroxyl radicals (18–22,26). The reaction is

$$H^+ + O_2^{\bullet -} \leftrightarrow HO_2^{\bullet} \qquad (1)$$

with p*K* = 4.8 (1,22,26). Under acidic conditions–for example, on the C side of the inner mitochondrial membrane under state 4 respiration, as demonstrated in our previous papers (17–20)–the direction of equilibrium of the reaction will be shifted toward the right side of the equation and more $HO_2^{\bullet}$ will be formed. The permeability of the neutralized $HO_2^{\bullet}$ across the membrane ought to be much higher than that of $O_2^{\bullet -}$ anions. The results presented in the second set of experiments provide evidence to confirm that the second postulate may be the case. In addition, it has been reported that, using EPR assay, the signal of $HO_2^{\bullet}$ has has been detected in preparations of rat cardiac mitochondria under state 4 respiration (13,22).

The third postulate is that under state 4 respiration the pH in the matrix of mitochondria becomes more alkaline (about 8.0) (21) and the $HO_2^{\bullet}$ formed dissociates again into H^+ and $O_2^{\bullet -}$ anions as soon as $HO_2^{\bullet}$ radicals are transported to the matrix. Usually either the $O_2^{\bullet -}$ are subject to dismutation by SOD and glutathione peroxidase (GSH-Px) in the matrix of mitochondria to regenerate oxygen molecule, which could re-enter the "reactive oxygen cycle" as described in our model, or the $O_2^{\bullet -}$ in the matrix may damage the functions of mitochondria or the mtDNA structure when the activity of the scavenging enzyme system in mitochondria decreases (1,3,6). The H^+ regenerated in a dismutation reaction of $O_2^{\bullet -}$ in the matrix can also re-enter the energy-dependent H^+-cycle of the redox proton pumps of mitochondria. Alternatively, the formed $HO_2^{\bullet}$ in the membrane lipid phase may directly attack the double allylic H atoms (daHa) in polyunsaturated fatty acids (PUFA) of membrane phospholipid of mitochondria to release heat about 80 kcal/mol with formation of H_2O_2, from which molecular oxygen could be regenerated (25). Therefore, $O_2^{\bullet -}$ generation in the mitochondrial respiratory chain and its cycling across the inner mitochondrial membrane in connection with the Q-cycle and H^+-cycle may serve as an endogenous protonophore in regulating and partitioning energy transduction and heat production as well as pathogenesis of mitochondrial diseases. The data in the following two sections are presented as supporting evidence.

EXPERIMENTAL EVIDENCE

The Generation of Superoxide Anions in Mitochondria is Protonmotive Force ($\Delta\mu H^+$)-dependent

The accumulated evidence indicates that the univalent reduction of dioxygen through the electron leak pathway may occur at different sites of the respiratory chain. All the sites are located before the substrate side of cyt b_{562}; however, the highest rates of superoxide generation were observed when reducing equivalents were permitted to equilibrate with ubisemiquinone ($Q^{\bullet}$) and *b*-type cytochromes in the presence of antimycin A or under state 4 respiration (12,13,16). These results show that $O_2^{\bullet-}$ generation is mainly connected with the operation of the Q-cycle of the respiratory chain and may also be controlled by protonmotive force.

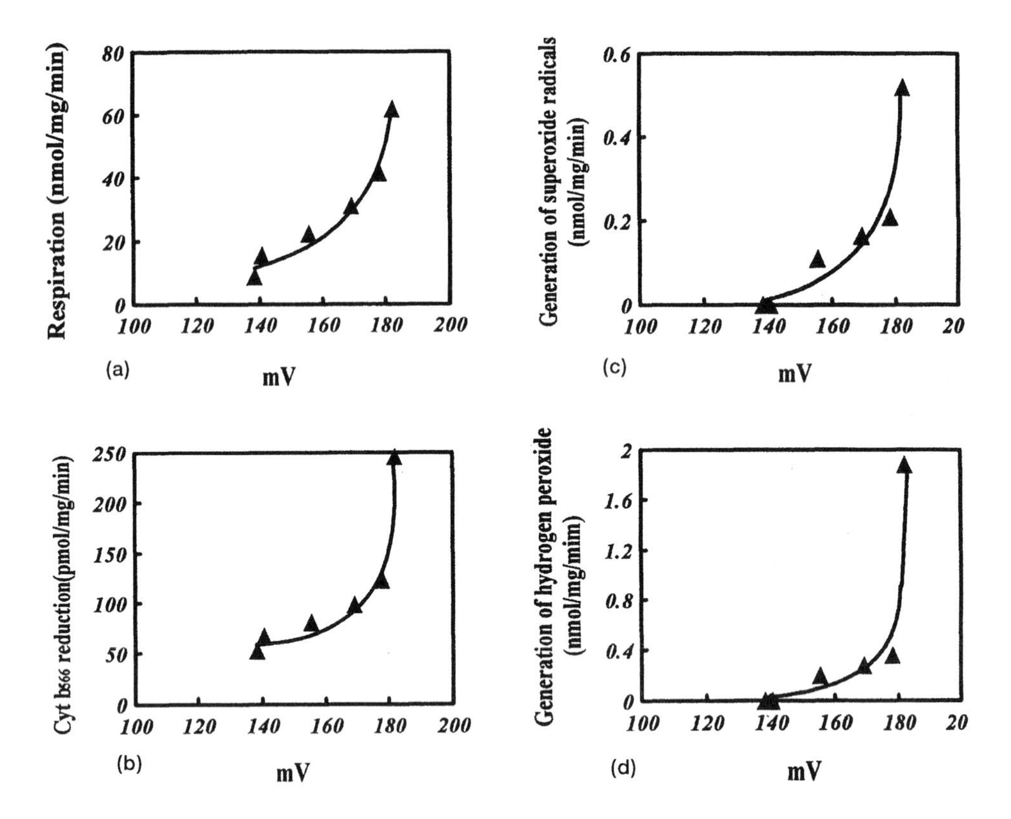

Figure 3. Correlation between the rates of respiration, superoxide, hydrogen peroxide generation, and cytochrome b_{566} reduction and the transmembrane potential in rat heart mitochondria. (a) The respiration rate of mitochondria was measured as in Ref. 9 using a Clark oxygen electrode on a YSI model 53 oxygen monitor (United States). The 2 ml of reaction medium contained sucrose 225 mmol/L, KCl 20 mmol/L, Tris-HCI 15 mmol/L, pH 7.2, KH_2PO_4 17 mmol/L, $MgCl_2$ 7 mmol/L, BSA 1 mg/ml, mitochondria protein 1 mg/ml; temperature 25°C. Succinate (2.5 mmol/L) was added to initiate the reaction; molanate was 0–5 nmol/L when added. (b) Cytochrome b_{566} reduction was measured on a Hitachi 557 double-wave length spectrophotometer following the absorbance changes at 566–575 nm according to Ref. 9. (c) Superoxide radicals were assayed on a Luminometer using MCLA as a probe according to Nakano (9). The reaction medium (1 ml) contained sucrose 300 mmol/L; Tris-HCl 10 mmol/L, pH 7.5, potassium phosphate 10 mmol/L, KCI 10 mmol/L, $MgCl_2$ 0.05 mmol/L, rotenone 3 μmol/L, MCLA, 4 μmol/L, mitochondria 0.1 mg/ml. Succinate (5 mmol/L) was added to initiate the reaction; temperature 25°C. (d) Hydrogen peroxide was determined according to Ref.9. Reaction medium (2 ml) contained sucrose 75 mmol/L, mannitol 225 mmol/L, Tris-HCl 20 mmol/L, pH 7.3, mitochondria 1 mg/ml, horseradish peroxidase (HPR) 1.7 μmol/L. Succinate (5 mmol/L) was added to initiate the reaction; temperature 25°C. The activity of the enzyme was monitored by following the absorbance changes at 417–407 nm ($\Delta\epsilon$ = 50 (mmol/L/cm).

We compared the steady-state rates of $O_2^{\bullet -}$ and H_2O_2 generation as well as cyt b_{566} reduction in rat heart mitochondria with different $\Delta\Psi$ titrated by succinate with molanate in order to determine the mechanism of superoxide generation in the respiratory chain, and to distinguish the possible different roles between the components of the protonmotive force, $\Delta\Psi$ and ΔpH, in the process of generation of superoxide and its relevant active oxygen species during state 4 respiration. The experimental results obtained are depicted in Figure 3.

It can be seen that there is a clear nonlinear relationship between respiration rate and $\Delta\Psi$ (Figure 3a), as first discovered by Nicholls (15) and subsequently confirmed by different groups (27,28). Interestingly, the relationship between the changes in values of $\Delta\Psi$ aross mitochondrial membrane and the rates of $O_2^{\bullet -}$ and H_2O_2 generation as well as the reduction of cyt b_{566} are also nonlinear and show a pattern almost identical to that of $\Delta\Psi$ versus respiration (Figure 3c–d). Both CCCP, and nigericin, typical uncouplers that *m*-chlorocarbonyl cyanide phenylhydrazone respectively collapse $\Delta\Psi$ and ΔpH across the inner mitochondrial membrane, were found to inhibit the generation of $O_2^{\bullet -}$ and H_2O_2 (Figure 4).

Antimycin A is known to inhibit the electron transfer from cyt b_{562} to Q and prevent Q reduction at the site of center Q_i of the bc_1 complex, and was found to increase cyt b_{566} reduction at the Q_o center and also to stimulate $O_2^{\bullet -}$ production (Figure 4 and 5).

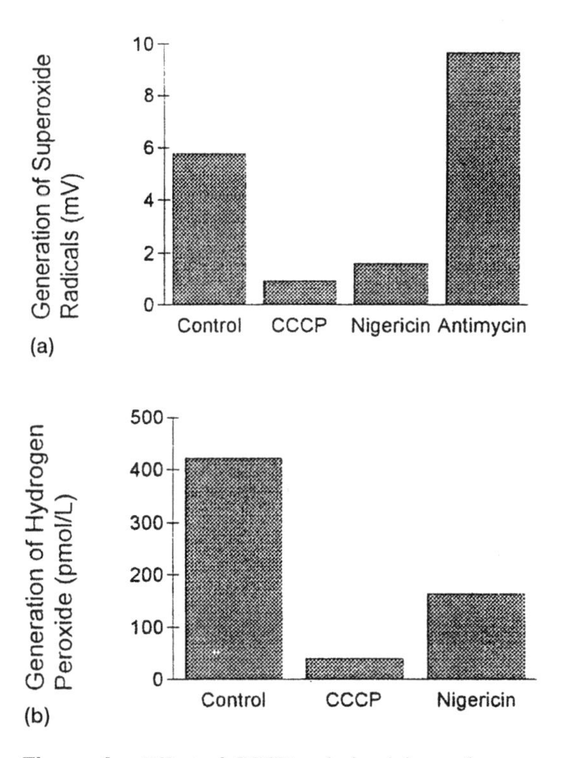

Figure 4. Effect of CCCP and nigericin on the superoxide and hydrogen peroxide formation in (a) rat liver and (b) rat heart mitochondria. The conditions for measurement of superoxide and hydrogen peroxide were as in Figure 3. The concentrations of CCCP and nigericin were 4 μmol/L and 1.2 μmol/L, respectively.

Myxothiazol is known to be the inhibitor of QH_2 oxidation at the Q_o center of the bc_1 complex to prevent the formation of ubisemiquinone and the reduction of cyt b_{566}; it was also found to inhibit $O_2^{\bullet-}$ generation (Figure 5).

Superoxide dismutase (SOD) was shown to inhibit $O_2^{\bullet-}$ production completely in our system (Figure 5). Since reduced cyt b_{566} is favorable for the formation of ubisemiquinone at the Q_o site, both reduced cyt b_{566} and ubisemiquinone in the Q-cycle may be essential sites for the formation of $O_2^{\bullet-}$ at higher values of protonmotive force.

We do not know the precise molecular mechanism of $O_2^{\bullet-}$ generation before the side of cyt b_{562} of the respiratory chain, especially in connection with the operation of the Q-cycle. The experimental results reported from this laboratory could constitute a basis on which to propose that a higher value of $\Delta\Psi$ across the inner mitochondrial membrane may create a suitable physical condition for generation of $O_2^{\bullet-}$ (7–9,17–20).

The Q_o center is very close to the surface of the cytosolic (C) side of the inner mitochondrial membrane–approximately 15 Å (31)–and the C side of the mitochondrial membrane is electropositive relative to the matrix (M) side, especially under state 4 respiration. A high value of $\Delta\Psi$ about 200 mV across the inner mitochondrial membrane during state 4 respiration may have a strong effect on ubisemiquinone or reduced cyt b_{566}, which may donate electrons more easily to the oxygen molecule near the C side of membrane than to cyt b_{562} at the Q_i center near the electronegative M side of the inner mitochondrial membrane. In addition, there was evidence that a higher value of

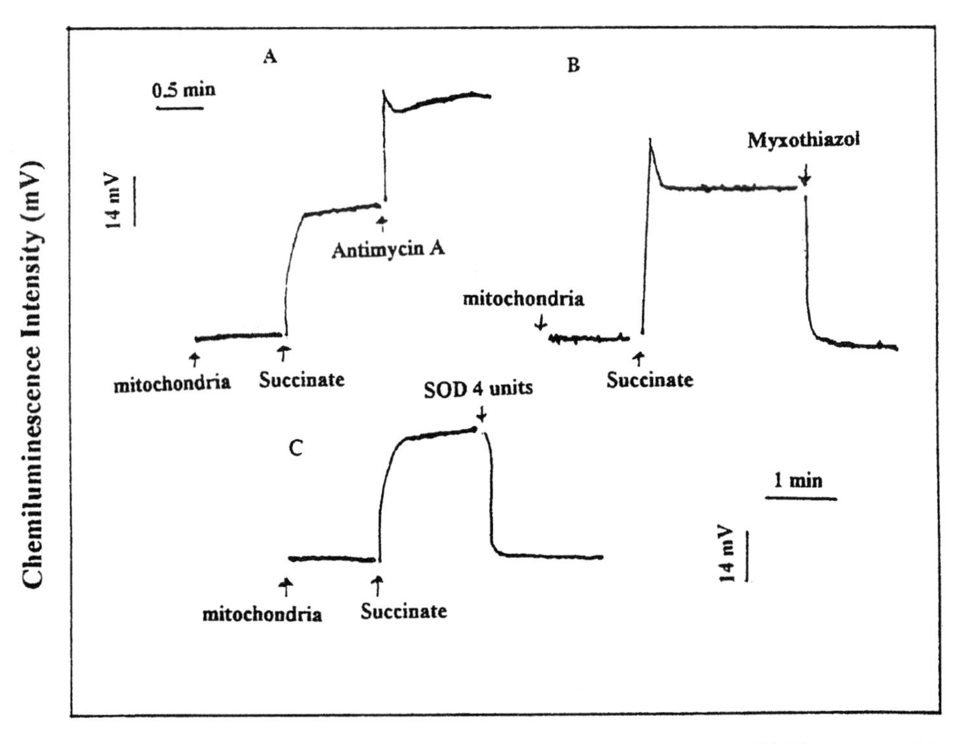

Figure 5. Influence of myxothiazol, antimycin A, and superoxide dismutase (SOD) on superoxide generation in rat heart mitchondria. Experimental conditions were the same as in Figure 3. The concentrations of myxothiasol and and antimycin A were 5 μg/ml, and 4 units of SOD were added to 1.0 ml of reaction medium.

protonmotive force, particularly of transmembrane potential in mitochondria caused cyt b_{566} to become reduced relative to cyt b_{562}. The apparent midpoint potential (Eh-7) of cyt b_{566} changed from –40 mV to +70 mV. This means that a membrane potential causes the displacement of electrons from cyt b_{562} on the matrix side to cyt b_{566} on the cytosolic side (15).

As mentioned previously, Nicholls first found that the proton leak rates in mammalian mitochondria increased disproportionately with an increase of mitochondrial $\Delta\Psi$ (15), and the increase in proton conductance across mitochondrial membrane was also non-ohmic with $\Delta\Psi$ increase (27,28). This indicates that the disproportionality of the increase in proton conductance with the increasing value of $\Delta\Psi$ may relate to the disproportionality of proton leak rate with the increase of production of $O_2^{\bullet-}$.

The data in Figure 3 clearly show the nonlinearity between the values of $\Delta\Psi$ and the rates of $O_2^{\bullet-}$ and H_2O_2 generation, and the yield of these two reactive oxygen species is also shown to be non-ohmic with $\Delta\Psi$, suggesting that the increased proton leak at a high value of $\Delta\Psi$ across inner mitochondrial membrane is due to the higher yield of $O_2^{\bullet-}$ generation at a higher value of $\Delta\Psi$ during state 4 respiration of mitochondria. The evidence supporting this idea also comes from our two reports, one of which showed that the singlet oxygen, $^1O_2$, generated in hematoporphyrin photosensization led to an increase of proton leak across mitochondrial membrane, resulting in dissipation of $\Delta\mu H^+$ and uncoupling oxidative phosphorylation (29). It is known that $^1O_2$ species are excited molecules with strong electrophilicity and can probably be formed by superimposing the reaction of $O_2^{\bullet-}$ and H_2O_2 in a biological system (16). Another paper of ours (17) has presented evidence that the nonlinearity between $\Delta\Psi$ and proton leak rates is also observed in rat liver mitochondria.

Superoxide Radicals as Endogenous Protonophores Induce Mitochondrial Proton Leakage

We have shown that the membrane surface pH of mitochondria or mitoplasts can be decreased by 0.6 or 1.0 pH unit during state 4 respiration with succinate (18,19). By inserting a pH fluorescence probe [fluorescein-phosphatidyl ethanolamine (FPE) or β-methylumbelliferyl undeanate (4-MU-C_{11}] into the mitoplast membrane surface (for FPE) or the interface between headgroup and fatty acid chain of phospholipids (for 4-MU-C_{11}), we observed a correlation between the extent of protonation (acidification) on the membrane surface (or interface) and the rates of state 4 respiration with succinate (9,19,20,30). We also reported that the energized H^+ on the mitoplast surface (interface) due to redox H^+ pumps of the respiratory chain can constitute a part of the coupling proton current of H^+, and serve as an energy source for ATP synthesis (19) or for a proton pumping-dependent membrane fusion processes (20). By labeling the outer surface (interface) of the mitoplast membrane with a fluorescent probe, we were able to show that under state 4 respiration succinate oxidation induces fluorescence quenching (a pH decrease); under this condition, addition of an $O_2^{\bullet-}$ generating system (X/XO) could reverse the reaction and increase the fluorescence intensity of the probe on the mitoplast surface (increase pH) rapidly, with no simultaneous change in pH of the medium. The extent of pH increase on the mitoplast membrane interface is proportional to the amount of X/XO added (9). These results are considered as evidence supporting the contention that $O_2^{\bullet-}$ acts as an H^+ carrier to induce proton leakage back into the mitochondrial matrix.

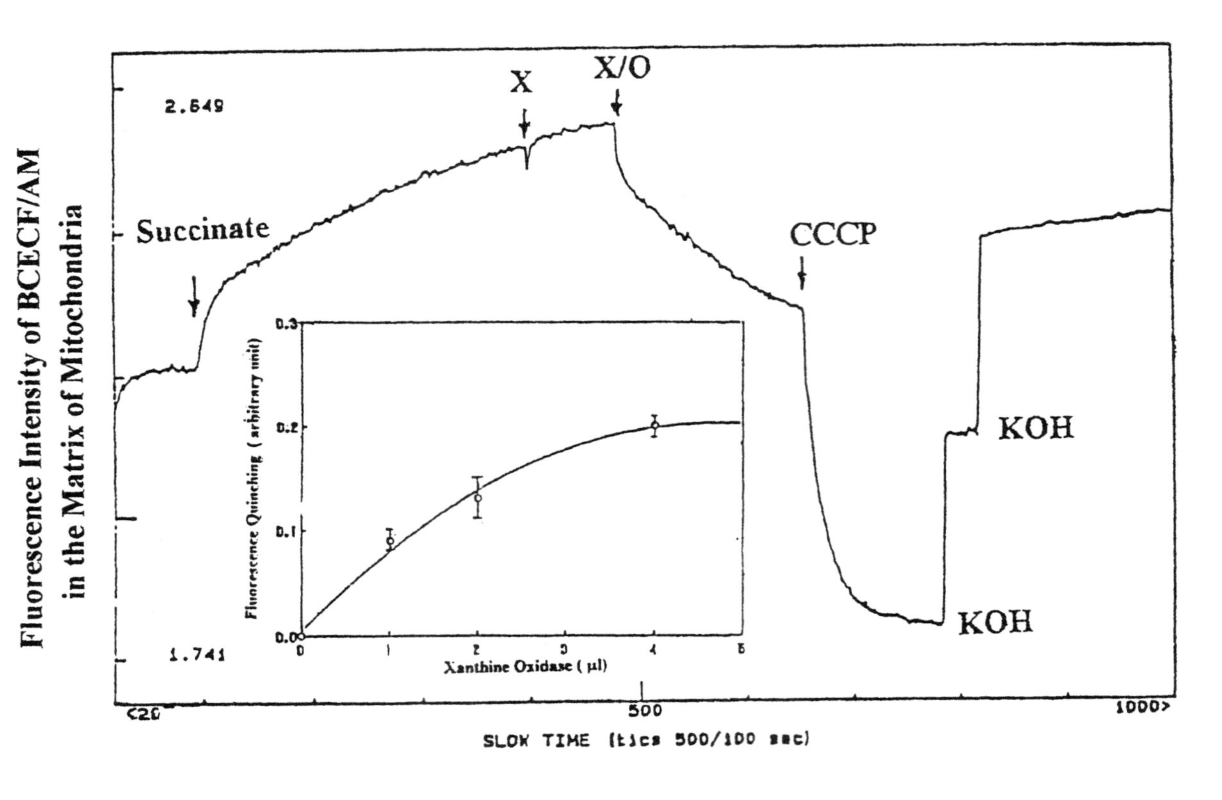

Figure 6. Succinate oxidation increases the pHm (pH of matrix) and xanthine/xanthine oxidase (X/XO) decreases the pHm in the matrix of mitochondria. The reaction medium contained 100 mmol/L KCl, 80 mmol/L K^+-Mops, pH 7.0, 100 nmol/L rotenone, mitochondria (1 mg protein/ml) preloaded with BCECF/AM, 5 mmol/L succinate, 240 μmol/L xanthine, 0.028 U xanthine oxidase, 1.0 μmol/L CCCP. The inset shows that the extent of fluorescence quenching of BCECF/AM in the matrix of mitochondria was proportional to the amount of X/XO added to the medium outside of the mitochondria.

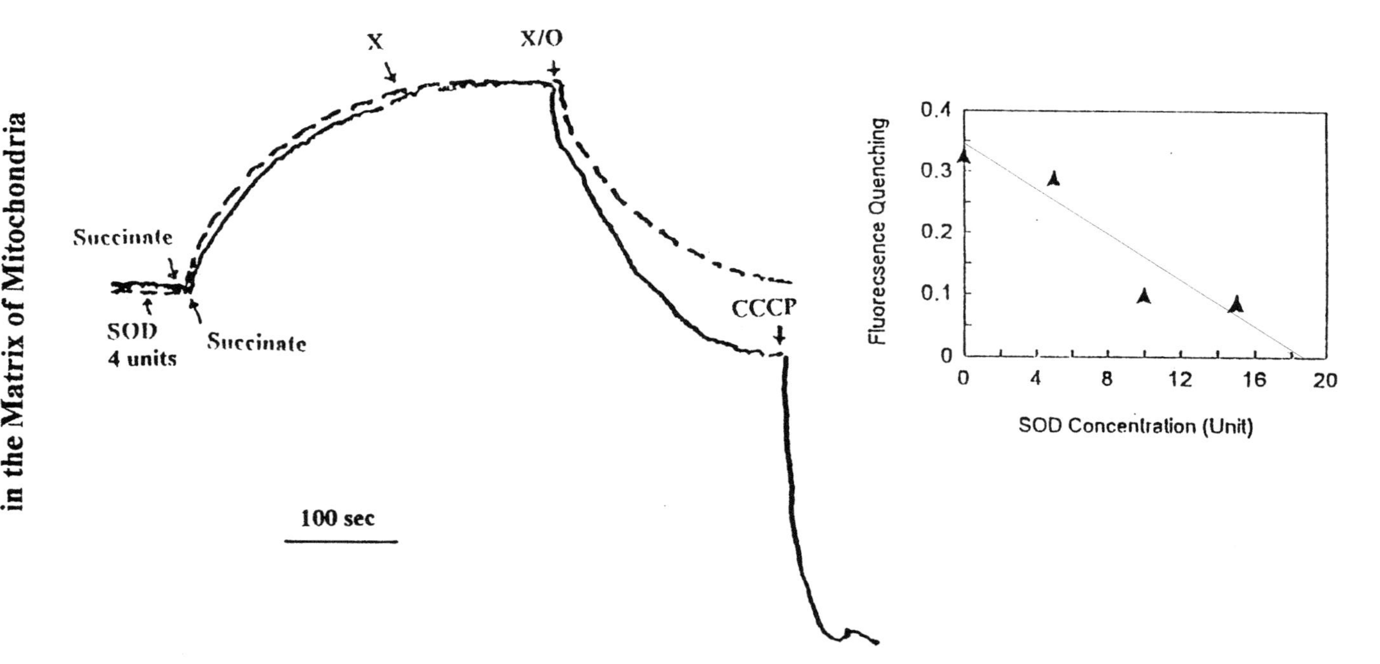

Figure 7. The preventive effect of superoxide dismutase (SOD) on the changes in pHm of the matrix of mitochondria lowered by xanthine/xanthine oxidase (X/XO). Experimental conditions were the same as in Figure 6. The inset shows that the extent of prevention by SOD of fluorescence quenching of BCECF/AM preloaded in the matrix of mitochondria was proportional to the amount of SOD added to the reaction medium.

Another set of experimental results that supports this idea is shown in Figures 6 and 7 and described in Ref. 9. These data indicate that $O_2^{\bullet -}$ generated outside of mitochondria by the X/XO system could induce proton translocation from outer surface of the mitochondrial inner membrane (C side) into the matrix (M side). The pH change in the matrix of mitochondria was detected by changes in fluorescence intensity of a soluble pH-sensitive probe, BCECF/AM preloaded in the mitochondrial matrix.

The pH in matrix was found to increase rapidly by proton pumping activity associated with succinate oxidation or ATP hydrolysis. Also, inhibitors of the respiratory chain–malonate, KCN, or CAT (the inhibitor of ADP/ATP translocase)–, were found to reverse the pH increase (fluorescence decrease) in mitochondrial matrix induced by ATP or succinate energization (9). Therefore, the pH-sensitive changes in fluorescence of BCECF/AM could reflect the pH changes in matrix associated with proton translocation across the inner mitochondrial membrane

Interestingly, the increase of pH in matrix of mitochondria during succinate oxidation could be reduced by an $O_2^{\bullet -}$ generating system, X/XO, external to mitochondria, and the extent of pH decrease in matrix was proportional to the amount of X/XO added (Figure 7). This indicates that the superoxide generated outside of mitochondria could cause H^+ to leak back to the mitochondrial matrix, resulting in a H^+ concentration increase in the matrix.

Incubating mitochondria with an ATP generation system to drive the H^+ pumping of ATP hydrolysis first acidifies the surface of the inner mitochondrial membrane. Adding succinate and antimycin A, instead of adding the X/XO system, stimulates $O_2^{\bullet -}$ production,

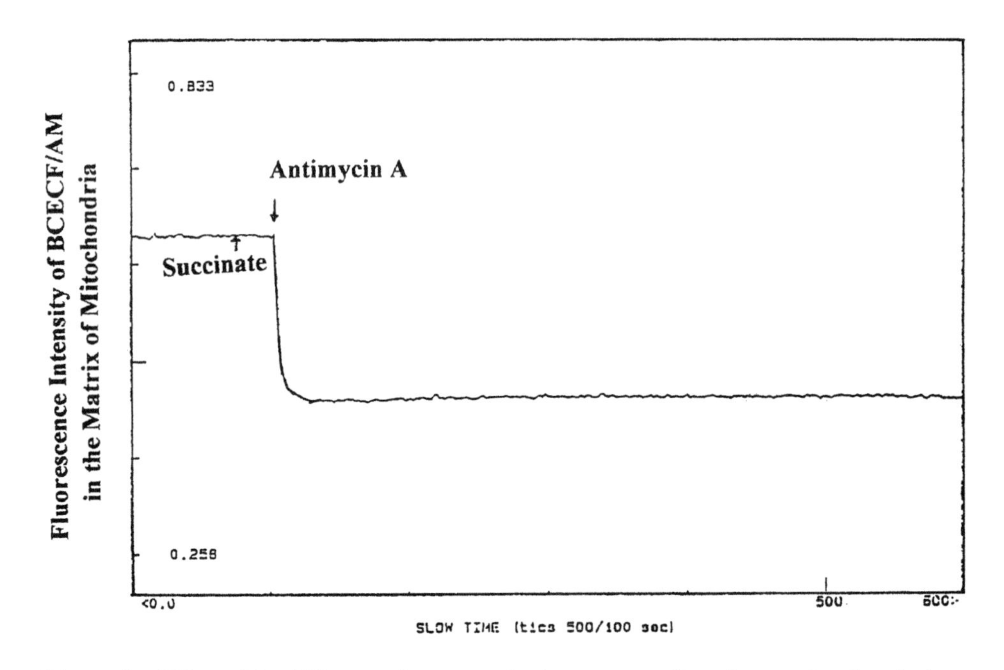

Figure 8. Effect of the ATP generation system in the presence of succinate and antimycin A on the pH changes in the matrix of mitochondria. The reaction medium contained 50 mmol/L Tris-HCl buffer, pH 7.0, 2 mmol/L NADH, 2 mol/L ATP, 2.5 mmol/L $MgCl_2$, 2 mmol/L phosphoenol pyruvate (PEP), 50 U pyruvate kinase, 50 U LDH, 1 mg/ml mitochondria, 2 μmol/L rotenone. Succinate (5 mmol/L) and antimycin A (4 μg/ml) was added to initiate the reaction.

probably by reduced cyt b_{566} or ubisemiquinone at the Q_o center in mitochondria. Under such conditions, addition of succinate with antimycin A was found to induce a significant and rapid pH decrease in the matrix of mitochondria (Figure 8), the reverse of the conditions in the absence of the ATP hydrolysis system, where succinate and antimycin A were found to increase pH in matrix of mitochondria (Figure 7). This set of experiments strongly confirms that $O_2^{\bullet-}$ generated in mitochondria (in vivo superoxide generation) exhibit the same effect of decreasing pH in the matrix of mitochondria as for the exogenously generated $O_2^{\bullet-}$ system, X/XO (in vitro superoxide generation). Adding SOD to the reaction medium prevented $O_2^{\bullet-}$ induced H^+ translocation from the outer surface of inner mitochondrial membrane into the matrix. The extent of the prevention effect of SOD is proportional to the amount of SOD added (Figure 7). All the results from in vivo and in vitro experiments confirm that $O_2^{\bullet-}$ anions generated during mitochondrial state 4 respiration may serve as an endogenous protonophore that induces proton leakage across the inner mitochondrial membrane.

Superoxide Radicals Induce Decreases of the $H^+/2e^-$ Ratio and Transmembrane Potential Values across Mitochondrial Membranes

According to our hypothesis concerning the coexistence of a "reactive oxygen cycle" with the Q-cycle and H^+-cycle in connection with a mitochondrial proton leakage mechanism, and from the results mentioned in previous sections, one may ask whether the $O_2^{\bullet-}$ generated during state 4 respiration could dissipate some portion of the free energy stored in the protonmotive force. This reduction of the free energy would lead to a decrease of proton pumping activity and of the $H^+/2e^-$ ratio associated with mitochondrial respiration.

A series of experiment was designed and performed in our laboratory to answer these questions (7–9). The results indicated that the $O_2^{\bullet-}$ generating system, X/XO, decreases the "energized" transmembrane potential, $\Delta\Psi_E$, of mitochondria by succinate oxidation. However, the "rest" transmembrane potential, $\Delta\Psi_R$, stemming from the fixed charge differences between the surfaces of the inner mitochondrial membrane in the absence of added oxidative substrate, was not affected (8). The superoxide radicals generated in cardiac mitochondria during ischemia–reperfusion were also shown to decrease the "energized" transmembrane potential, $\Delta\Psi_E$, but not $\Delta\Psi_R$ (8). Superoxide scavengers (3,4-dihydroxyphenyllactate) added to the reperfusion solution were found to restore the $\Delta\Psi_E$ values of mitochondria isolated from ischemia–reperfusion injured rat heart to levels close to that of control mitochondria.

These results clearly confirm the idea that superoxide radicals generated both in vitro and in vivo have a specific inhibitory effect on $\Delta\Psi_E$ across the mitochondrial membrane. Consistent with this mode of action of $O_2^{\bullet-}$ on the mitochondria, the $H^+/2e^-$ ratio and proton pumping activity of mitochondria oxidizing succinate were also decreased by adding the X/XO system (8).

Based on all this, we can conclude that the physiopathological action of $O_2^{\bullet-}$ on the mitochondria is due mainly to its promoting effect on proton leaks, by a mechanism described as a "proton leak induced by $O_2^{\bullet-}$ generated from the univalent reduction of oxygen through the electron leak pathway in the respiratory chain," as I proposed in 1991 (7). In other words, the protonmotive force, $\Delta\mu H^+$, may also possess a novel functional role in scavenging oxygen free radicals generated by respiration, and provide an antioxidative defense mechanism against $O_2^{\bullet-}$ at the mitochondrial level (8,9).

Superoxide Radicals Enhance Proton Leakage and Heat Production in Hyperthyroid Rat Liver Mitochondria

On the basis of the experimental results shown in previous sections, one may ask whether superoxide radicals induce proton leakage in mitochondria, leading to an increase of heat production. It is well known that mitochondria from hyperthyroid animals possess higher levels of respiration and of proton leakage, and so can be used as a suitable experimental model for studying this problem (32). Using Calvet MS-80 microcalorimetry, a comparative assay on heat production in mitochondria with succinate as oxidation substrate between hyperthyroid and euthyroid rat liver mitochondria was carried out. Also, the effect of SOD on thermogenesis by mitochondrial respiration was also examined.

As shown in Figure 9, the total amount of heat production in hyperthyroid mitochondria supported by 5 mM succinate was determined to be 13.03×10^{-2} J/mg

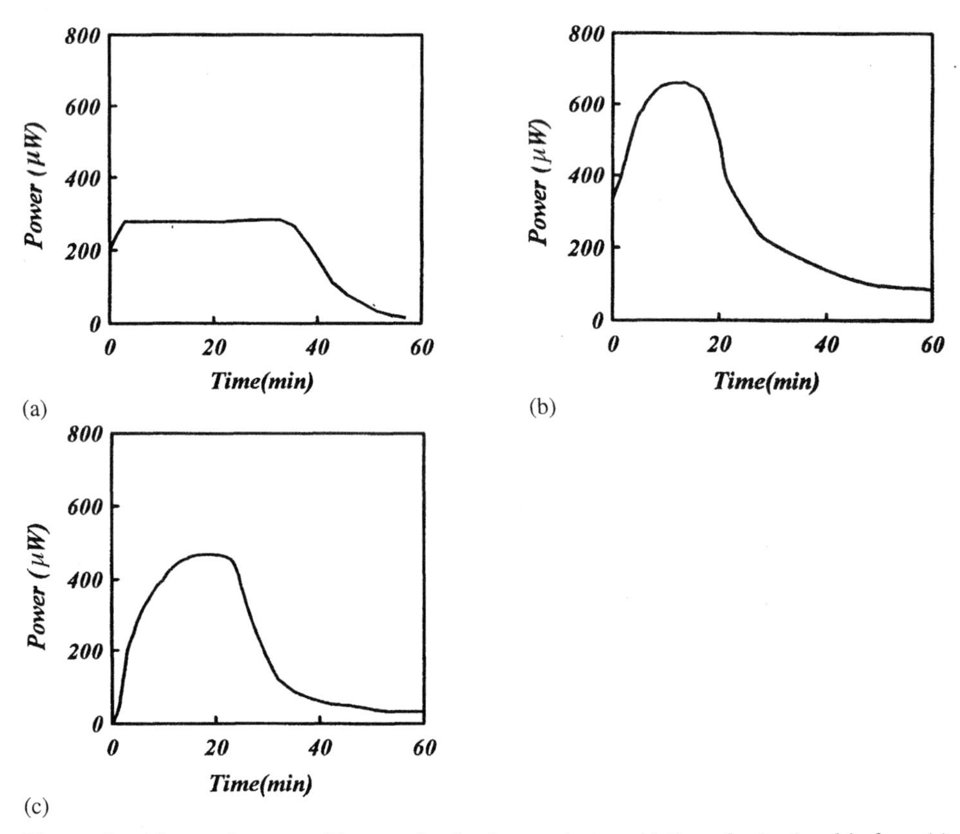

Figure 9. Measured curves of heat production by succinate oxidation of mitochondria from (a) euthyroid rat liver and (b) hyperthyroid rat liver, and (c) the effect of SOD. The Calvet MS-80 microcalorimeter was used. The reaction medium (5 ml) contained 250 mmol/L sucrose, 2 mmol/L Hepes, pH 7.4), 0.1 mmol/L EDTA, 0.004 mmol/L rotenone, 2 mg/ml oligimycin, 1 mg/ml mitochondrial protein, 5 mmol/L succinate; temperature 25°C. The euthyroid curve reflects the total amount of heat production by oxidation of 5 mmol/L succinate in euthyroid mitochondria (7.67×10^{-2} J/mg protein). The hyperthyroid curve reflects the total amount of heat production of hyperthyroid mitochondria (13.03×10^{-2} J/mg protein). The SOD curve reflects the total amount of heat production of hyperthyroid mitochondria with 25 U SOD/ml (7.55×10^{-2} J/mg protein).

protein of mitochondria, and is about 70% higher than that of euthyroid mitochondria (7.67×10^{-2} J/mg protein of mitochondria). These data are consistent with the recent observation in this laboratory that the state 4 respiration of hyperthyroid mitochondria was also 71% higher than that of euthyroid mitochondria (32). This means that the portion of the elevated state 4 respiration in hyperthyroid mitochondria is totally associated with nonphosphorylating respiration and almost completely contributes to heat production. Interestingly, addition of SOD to the reaction medium completely repressed the "extra" heat production and reduced theromogenesis of hyperthyroid mitochondria to a level equal to that of euthyroid mitochondria (7.55×10^{-2} J/mg protein of mitochondria). A parallel experiment showed that SOD per se has no effect on heat production (not shown). This set of experiments provides strong evidence confirming that superoxide radicals generated in mitochondrial state 4 respiration are the main cause of the enhanced rate of proton leak across the mitochondrial membrane, and consequently of the elevated heat production in hyperthyroid mitochondria. They are probably also responsible for the elevation of basic metabolic rate in the living body under pathological states of hyperthyroidism.

All of the results presented here provide direct experimental evidence supporting our hypothetical model, and show that the generation, partitioning, targeting, and functioning of superoxide in mitochondria may have significance not only for the biology of free radicals and biomedicine but also for the basic aspects of cell bioenergetics as well as the biosciences in general. Moreover, they also represent the first report on the role of oxygen free radicals in the pathological mechanism of hyperthyroidism, which may provide new ideas for further research and the clinical treatment of hyperthyroidism.

ACKNOWLEDGMENTS

This work was supported by the China National Natural Science Foundation and the foundation for basic research of Chinese Academy of Science. It was carried out in collaboration with my colleagues J.P. Huang, Y.H. Hu, X.J.Li, T. Xie, and X.H. Su.

REFERENCES

1. Hillward B, Gutteridge JMC. Free Radicals in Biology and Medicine. 2d ed. Oxford: Clarendon Press, 1989: 367–415.
2. Hillward B. Free radicals and antioxidants: a personal view. Nutr Rev 1994; 52(8): 253–265.
3. Yu BP. Cellular defenses against damage from reactive oxygen species. Physiol Rev 1994; 74(1): 139–162.
4. Luft R. The development of mitochondrial medicine. Proc Natl Acad Sci USA 1994; 91: 8731–8738.
5. Shigenaga MK, Hagen TM, Bruce NA. Oxidative damage and mitochondrial decay in aging. Proc Natl Acad Sci USA 1994; 91: 10771–10778.
6. Lawen A, Martiinus RD, McMullen GL, Nagley P, Vaillant F, Wolvetang EJ, Linnane AW. The universality of bioenergetic disease: the role of mitochondrial mutation and the putative interrelationship between mitochondria and plasma membrane NADH oxidoreductase. Mol Aspects Med 1994; 15(suppl): s13–27.

7. Liu SS, Huang JP. The relationship between reactive oxygen species and oxidative phosphorylation in mitochondria: an attempt to define the interaction of electron leakage and proton leakage. Report at the National Conference on Free Radical Biology. Beijing: The Commission for Free Radical Biology and Medicine, Chinese Biophysical Society; 1991: 10.
8. Liu SS, Wang XM, Jiao XM, Zhang L. Interaction of electron leak and proton leak in respiratory chain: proton leak induced by superoxide from an electron leak pathway of univalent reduction of oxygen [in Chinese]. Scientia Sinica 1995; 25(6): 596–603. [Republished in English in Science in China, Life Sciences, series C, 1996; 39(2): 168–178.]
9. Liu SS, Huang JP. Coexistence of a "reactive oxygen cycle" with "Q cycle" in the respiratory chain: a hypothesis for generating, partitioning and functioning of superoxide in mitochondria. Chapter 53. Proceedings of the International Symposium on Natural Antioxidants: Molecular Mechanisms and Health Effects. Champaign, IL: AOCS Press; 1996: 513–529.
10. Chance B, Sies H, Boveris A. Hydroperoxide metabolism in mammalian organs. Physiol Rev 1979; 59: 527–605.
11. Boveris A, Cadens E. Production of superoxide radicals and hydrogen peroxide in mitochondria. In Superoxide Dismutase. 1982; 11: 15–30.
12. Cross AR, Jones OTG. Enzymic mechanism of superoxide production. Biochim Biophys Acta 1991; 1057: 281–298.
13. Nohl H. A novel superoxide radical renerator in heart mitochondria. FEBS Lett 1987; 214(2): 269–273.
14. McCord JM, Turrens JF. Mitochondrial injury by ischemia and reperfusion. Curr Topics Bioenerg 1994; 17: 173–195.
15. Nicholls D. Bioenergetics: An Introduction to Chemiosmotic Theory. London: Academic Press; 1982: 119–121.
16. Halliward B, Gutteridge JMC. Role of free radicals and catalytic metal ions in human disease: an overview. Methods Enzymol 1990; 186: 1–85.
17. Wei YY, Liu SS. Coupling between electron transfer and proton pumping in respiratory chain complex II + III of rat liver mitochondria. Chinese Biochem J 1995; 11(5): 593–599.
18. Lian JP, Liu SS. Fluorescence measurement of the membrane surface charge and surface pH of mitochondria upon energization. Biochim Biophys Acta Sinica 1989; 21(5): 394–400.
19. Liu SS, Xiong JW, Jiao XM. Evidence for ATP synthesis coupled with energizing protons on membrane surface of mitoplast upon succinate oxidation. 7th European Bioenergetics Conference. EBEG Short Report 1992; 7: 104.
20. Liu SS. A new function of proton pumps in membrane fusion. In Zhou CL, ed. Current Biochemical Research in China. New York: Academic Press, 1989: 15–26.
21. Lippman RD. Site-specific chemiluminescence probe used in the analysis of pH, peroxidation and free radicals intensity in metabolically active organelles. In DeLuca MA, McElroy WD, eds. Bioluminescence and Chemiluminescence: Basic Chemistry and Analytical Applications. New York: Acadamic Press, 1981: 373–381.
22. Green MJ, Hill HAO. Chemistry of dioxygen. Methods Enzymol 1984; 105: 3–35.
23. Rustin P. The nature of the terminal oxidation step of the alternative electron transport pathway. In Moore AL, Beechy RB, eds. Plant Mitochondria: Structural, Functional, and Physiological Aspects. New York: Plenum Press, 1987: 37–46.
24. Moore AL, Seidow JN. The regulation and nature of the cyanide-resistant alternative oxidase of plant mitochondria. Biochim Biophys Acta 1991; 1059: 121–140.
25. Wagner BA, Buettner GR, Burns CP. Free radical-mediated lipid peroxidation in cell: oxidizability is a function of cell lipid bis-allylic hydrogen content. Biochemistry 1994; 33: 4449–4453.
26. Benon HJ, Ross AB. Reactivity of HO_2/O_2: radicals in aqueous solution. J Phys Chem Ref Data 1985; 14(4): 1041–1051.

27. Murphy MP. Slip and leak in mitochondrial oxidative phosphorylation. Biochim Biophys Acta 1989; 977: 123–141.
28. Brand MD, Chien LF, Ainscow EK, Rolfe DFS, Porter RK. The causes and functions of mitochondrial proton leak. Biochim Biophys Acta 1994; 1187: 132–139.
29. Zhou TQ, Jiao XM, Wei YY, Liu SS. Proton leakage is the mechanism for hematoporphorin photosensitization-promoted inactivation of rat liver mitochondria. Progress in Natural Science: Communication from State Key Laboratories of China 1994; 4(3): 313–318.
30. Liu SS, Xian T, Qu LC. Synthesis and characterization of β-methyl umbelliferyl undecanate (4-MU-C11) and its application in labeling membrane surface as a pH sensitive fluorescent probe [abstr]. Fifth National Symposium on Biological Membrane Research in China, Haikao, China, April 21–24, 1993: 50.
31. Jakow GV, Link TA, Ohnish T. Organization and function of cytochrome *b* and ubiquinone in cristae membrane of beef heart mitochondria. J Bioenerg Biomembr 1986; 18(3): 157–179.
32. Liu SS, Hu YH. Superoxide induced proton leak in hyperthyroid rat liver mitochondria: the pathological mechanism of hyperthyroidism. International Symposium on Natural Antioxidants: Molecular Mechanism and Health Effects, June 20–24, 1995, Beijing, China, 1996: 274.
33. Rusting R. Les causes du viellssement. Pour la Science 1993 (Feb.); 184: 54–62.
34. Kroemer G, Pett P, Zamzami N, Vayssiere J-L, Mignotte B. The biochemistry of programmed cell death, Reviews. FASEB 1995; 9: 1277–1287.
35. Xiao Y, Liu SS. Experimental evidence for the existence of cyanide-resistant respiration in animal mitochondria. 1996; submitted.
36. Nakano H. Determination of superoxide radicals and singlet oxygen based on chemiluminescence of luciferin analogs. Methods Enzymol 1990; 186: 585–591.
37. Emaus RK, Graunward R, Lemaster JJ. Rhodamine 123 as a probe of transmembrane potential in rat liver mitochondria: spectral and metabolic properties. Biochim Biophys Acta 1986; 850: 436–448.

32

Endogenous Oxidative DNA Damage and Spontaneous Mutagenesis: Role of 8-Oxoguanine

Serge Boiteux
UMR217 Centre National de la Recherche Scientifique
Commissariat à l'Energie Atomique
Fontenay aux Roses, France

INTRODUCTION

Considerable interest has arisen in recent years in the formation and consequences of oxidative damage to DNA. Reactive oxygen species (ROS) and free radicals have been suggested to play a role in biological processes such as carcinogenesis and aging (1,2). This interest is derived in part from the realization that ROS form inside cells as a by-product of normal cell metabolism (3). ROS such as hydroxyl radical ($OH^{\bullet}$) or singlet oxygen ($^1O_2$) are highly reactive and produce a complex pattern of DNA modifications (4). For example $OH^{\bullet}$ radical causes a variety of DNA damage including base modifications, abasic sites, DNA strand breaks, and cross-links (5). Several lines of evidence suggest that an oxidatively damaged form of guanine, 7,8-dihydro-8-oxoguanine (8-OxoG) threatens the integrity of genetic information and may be used as a model lesion to study the biological impact of endogenous oxidative stress.

The aim of this review is to summarize data which suggest that endogenous oxidative stress is a major cause of spontaneous mutations in prokaryotes and eukaryotes and to discuss possible correlation with degenerative human diseases.

RESULTS AND DISCUSSION

Formation of 8-OxoG in DNA

Among DNA base damage, 7,8-dihydro-8-oxoguanine is probably the most frequent base modification after reaction of DNA with ROS. Identification and quantification of 8-OxoG in DNA has been facilitated by the development of a sensitive method using HPLC coupled with electrochemical detection (HPLC-ECD) (6,7). Steady-state levels of 8-OxoG in bacterial and mammalian cell DNA have been determined by this method. For example, values determined in rat liver DNA range between 50 and 100 residues per 10^7 base pairs (Table 1 and Ref.9). In contrast, 8-OxoG values determined in *Escherichia coli*

Table 1. Steady-state Levels of 8-OxoG in DNA by Means of HPLC and Electrochemical Detection and Fpg-sensitive Sites

Origin of DNA	Detection	8-OxoG/10^7 bp	Reference
Calf thymus[a]	HPLC-ECD	180	(7)
E. coli[b]	HPLC-ECD	12	(10)
E. coli[c]	HPLC-ECD	24	(10)
Rat liver	HPLC-ECD	60	(34)
Human lymphoblasts	HPLC-ECD	10	(35)
Mouse cells[d]	Fpg^s-sites	4	(8)
Human fibroblasts	Fpg^s-sites	2	(36)

[a] Calf thymus DNA was of commercial origin.
[b] *E. coli* wild-type strain.
[c] *E. coli* strain (*fpg/mut*M) is defective in the repair of 8-OxoG.
[d] L1210 leukemia cells.

or human cells DNA are 10-fold lower (Table 1). An enzymatic method using Fpg protein and the alkaline elution technique was also developed to quantify 8-OxoG in DNA (8). Values obtained by the enzymatic assay are lower than values by HPLC-ECD (Table 1). To date neither an underestimation of Fpg-sensitive sites nor artifactual 8-OxoG formation during work-up procedures for HPLC-ECD can be excluded (9). Quantification of 8-OxoG in an *E. coli* mutant that has no 8-OxoG DNA glycosylase activity (*fpg/mut*M) shows a steady-state level of 8-OxoG which is 2- to 3-fold higher in the mutant compared with the wild-type strain (Table 1 and Ref.10). These results show that 8-OxoG forms in DNA as a consequence of endogenous stress in *E. coli*. The relatively modest accumulation of 8-OxoG in the *fpg* mutant suggests either that the steady-state level in the wild-type strain is overestimated or that *E. coli* possesses alternative repair pathways that permit elimination of this lesion. One such alternative pathway may be that of nucleotide excision repair (11).

Mutagenesis by 8-OxoG

The coding properties of 8-OxoG were investigated using oligodeoxyribonucleotide templates containing a single 8-OxoG at a defined position and purified DNA polymerases. Shibutani and Coworkers (12) reported that both prokaryotic and eukaryotic DNA polymerases incorporate dAMP or dCMP opposite 8-OxoG in the template strand. In contrast, they did not observe incorporation of dTMP and dGMP opposite 8-OxoG (12). DNA polymerases associated with replication such as pol α, pol ϵ and pol III predominantly incorporate dAMP opposite 8-OxoG (13). On the other hand, DNA polymerases associated with repair such as pol β and pol I preferentially incorporate dCMP opposite 8-OxoG (13). Molecular modeling and biophysical studies indicate that the 8-OxoG:A base pair resembles a normal Watson–Crick base pair. Furthermore, the 8-OxoG:A pair is not recognized as a mismatch by the 3′→5′ exonucleolytic proofreading activity of *E. coli* DNA polymerase I (12,13). The facts that DNA polymerases frequently incorporate dAMP opposite 8-OxoG and that they do not recognize the 8-OxoG:A pair as a mismatch imply that the presence of 8-OxoG in DNA should be mutagenic, yielding GC→TA transversions.

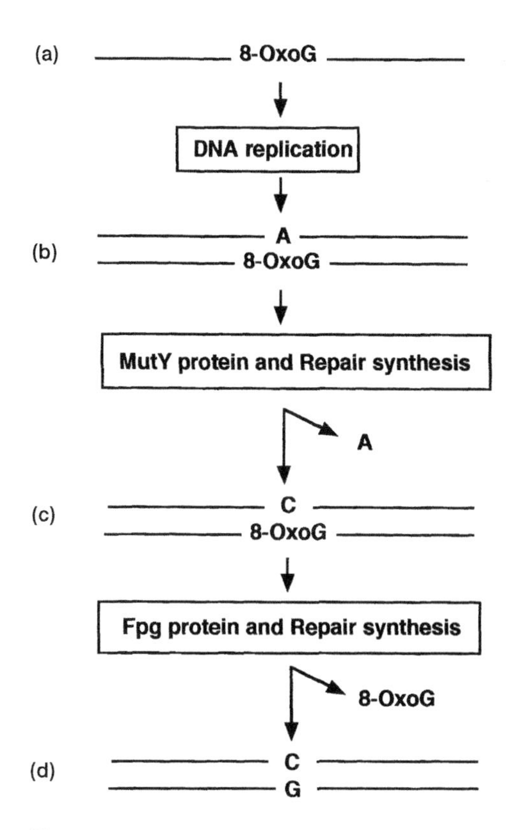

Figure 1. Base excision repair of 8-OxoG in single-stranded DNA. (a) Oxidative stress generates 8-OxoG in single-stranded region of the chromosomal DNA. (b) Replication results in the incorporation of adenine opposite 8-OxoG, yielding the 8-OxoG:A base pair. (c) MutY protein excises the adenine residue and DNA repair synthesis preferentially incorporates cytosine opposite 8-OxoG, yielding the 8-OxoG:C base pair. (d) Fpg protein excises 8-OxoG and DNA repair synthesis regenerates a normal G:C base pair. This scheme explains why 8-OxoG in single-stranded DNA vectors apears to be poorly mutagenic when transfected into a wild-type strain (14,32).

The mutagenic potential of 8-OxoG was investigated in vivo using M13 single-stranded phage DNA containing a single lesion and transfected into bacterial hosts of different repair capabilities (14). The data show that less than 5% of the phage progeny are mutated after transfection into a wild-type strain. In contrast, when M13 phage DNA containing 8-OxoG is transfected into the *fpg mut*Y double mutant of *E. coli*, 65% of the phage progeny are mutated (14). Sequence analysis indicates that the vast majority of these mutations are targeted at the position of 8-OxoG and more than 95% are GC→TA transversions (14). These results lead to the conclusions that (i) 8-OxoG is highly mutagenic in vivo and (ii) 8-OxoG is repaired in single-stranded DNA by base excision repair (Figure 1).

Repair of 8-OxoG in Prokaryotes

Two DNA glycosylases are implicated in the repair of 8-OxoG in *E. coli*: the Fpg protein which excises 8-OxoG when paired with a cytosine and the MutY protein which excises adenine when paired with 8-OxoG (13,15,16).

The studies of the properties and biological function of the Fpg protein were facilitated by the cloning of the *fpg* gene of *E. coli* (17), overexpression, and purification of the protein (18). The Fpg protein is a metalloprotein of 269 amino acids with a single zinc atom per molecule probably associated with a zinc finger motif at the C-terminal end of the protein (19). The Fpg protein is a multifunctional repair enzyme endowed with DNA glycosylase, AP-lyase and dRPase activities (16). The DNA glycosylase activity of the Fpg protein excises damaged purine residues including imidazole ring-opened purines and 8-oxopurines (20). The Fpg protein efficiently removes 8-OxoG paired with cytosine. In contrast, Fpg protein removes 8-OxoG at a 200-fold reduced rate when paired with adenine (21,22). The isolation of *fpg*$^-$ mutant of *E. coli* has also contributed to the elucidation of the biological role of the Fpg protein. The *fpg* mutant of *E. coli* is not unusually sensitive to various oxidative stress as compared with the wild-type strain (23). However, an *fpg* (*mut*M) mutant displays a moderate spontaneous GC→TA mutator phenotype (24,25). The substrate specificity of the Fpg protein and the specificity of the mutator effect of *fpg*$^-$ mutant strongly suggest that spontaneous mutagenesis in this mutant is mostly due to a defect in the repair of 8-OxoG in DNA.

A second level of defense mediated by the MutY protein is used to prevent mutagenesis by 8-OxoG that have been replicated by DNA polymerase. The MutY protein of *E. coli* is a monomer of 350 amino acids with a molecular mass of 39 kDa (5,14,15,26). The Muty protein possesses a DNA glycosylase activity that catalyses the removal of adenine residues incorporated opposite 8-OxoG. This is a unique example of a repair process where the repair enzyme removes not the lesion (8-OxoG) but a normal base (adenine) placed opposite the lesion. The *mut*Y$^-$ mutant exhibits a strong spontaneous GC→TA mutator phenotype (27). The substrate specificity of the MutY protein and the GC→TA mutator phenotype displayed by *mut*Y$^-$ mutant again argue in favor of the direct implication of 8-OxoG as a cause of spontaneous mutations.

The *fpg mut*Y double mutant of *E. coli* shows an extremely strong mutator phenotype which corresponds to 10- to 20-fold the sum of the mutation rates of single mutants (25,28). The synergistic increase in mutation rate in the double mutant is explained by the substrate specificity of the Fpg and MutY protein, which cooperate to prevent mutagenesis by 8-OxoG (13,15,25,28). These results in prokaryotes strongly suggest that enzyme activities (Fpg and MutY) involved in the repair of 8-OxoG play an important role in the maintenance of genetic integrity.

Repair of 8-OxoG in Eukaryotes

In our laboratory, we chose to investigate the repair of 8-OxoG in a lower eukaryote, the yeast *Saccharomyces cerevisiae*. Recently we have cloned the *OGG1* gene which encodes a DNA glycosylase activity that excises 8-OxoG from DNA (29). The expression of the *OGG1* gene in the (*fpg mut*Y) mutant of *E. coli* partially suppresses the spontaneous mutator phenotype of the bacterial mutant. The *OGG1* gene of *S. cerevisiae* codes for a protein of 376 amino acids, the Oggl protein, which has been purified to apparent homogeneity (29). The Oggl protein readily removes 8-OxoG in the 8-OxoG:C base pair but not in the 8-OxoG:A base pair as described for the Fpg protein of *E. coli* (29). Table 2 shows a comparison of the properties of the Fpg protein of *E. coli* and of the Oggl protein of *S. cerevisiae*. The *OGG1* gene has been disrupted and the viability of the haploid *oggl*::TRP1 mutant apears normal. The *oggl* mutant exhibits a spontaneous mutator phenotype and specifically accumulates GC→TA transversion events (30). The

Table 2. Comparison of the Properties of the Fpg Protein of *E. coli* and Ogg1 Protein of *S. cerevisiae*

	Fpg protein	Ogg1 protein
Gene level		
Structural gene	*fpg*	*OGG1*
Localization	82 min	Chromosome XIII
Protein level		
Size (amino acids)	269	376
Molecular mass (kDa)	30	43
Metal ion	Zn	ND[a]
Zinc-finger motif	Yes	No
Cross-reactivity of antibodies	No	No
Enzyme activity level		
DNA glycosylase	Yes	Yes
Excision of 8-OxoG	+++	+++
Excision of Fapy	+++	+
AP-lyase activity	Yes	Yes
β-elimination	Yes	Yes
δ-elimination	Yes	No
Mutant level	*fpg-1*	*ogg1::TRP1*
Viability	Yes	Yes
Mutator	Yes	Yes
Cross-complementation	Yes	ND[a]

[a] ND, not determined.

similarity of enzymatic activities and of mutant properties suggests that Ogg1 protein is the functional homologue of the Fpg protein (Table 2). However, comparison of the amino acid sequence of the Ogg1 protein and of the Fpg protein does not reveal an obvious conserved region. Both the highly conserved [PELPEVE] sequence at the amino-terminal and the zinc finger motif at the carboxy-terminal ends of bacterial Fpg proteins (25) are absent in the Ogg1 protein. In mammalian cells, enzyme activities that repair 8-OxoG have been partially characterized. The results show that human cells possess two enzymes that repair 8-OxoG (31). Furthermore, transfection of single-stranded vectors containing 8-OxoG in mammalian cells does not result in a large increase in mutagenesis among progeny (32). By analogy with the bacterial system (Figure 1), this observation implies that mammalian cells possess both Fpg/Ogg1 and MutY homologues. An enzyme activity that cleaves DNA at adenine residues when paired with 8-OxoG has been identified and partially purified in human cells (33).

CONCLUSIONS

The results reported in this review, using 8-OxoG as a model of oxidative damage to DNA, lead us to conclude that endogeneous oxidative stress is a major cause of mutations in bacteria that exhibit repair deficiencies. However, the chemical nature of the reactive species which attack DNA and result in the formation of 8-OxoG remains unidentified. In bacteria, the elimination of 8-OxoG from DNA is based on two repair proteins that act in synergism: the Fpg and the MutY proteins. A homologue, *OGG1*, of the bacterial *fpg* gene has been cloned in *S. cerevisiae*. The *ogg1* mutant of *S. cerevisiae* displays a spontaneous mutator phenotype. The homologue of the *mut*Y gene has not yet been identified in *S. cerevisiae*. These results indicate that endogenous oxidative stress is a cause of genetic alterations in eukaryotes. Recent data strongly suggest that homoloues of Fpg/Ogg1 and MutY proteins may exist in human cells. The elucidation of the biological role of these repair activities in mammalian cells will await the cloning of the corresponding genes from murine and human origin. The construction of mutant cell lines and of transgenic animals defective in the repair of oxidative damage to DNA will permit us to address the role of endogenous oxidative stress in degenerative diseases such as cancer and aging.

REFERENCES

1. Ames BN. Dietary carcinogens and anticarcinogens. Oxygen radicals and degenerative diseases. Science 1983; 221: 1256–1264.
2. Breimer LH. Molecular mechanisms of oxygen radical carcinogenesis and mutagenesis, the role of base damage. Mol Carcinogen 1990; 3: 188–197.
3. Halliwell B, Gutteridge JM. Free Radicals in Biology and Medicine. 2nd ed. Oxford: Oxford University Press; 1989.
4. Dizdaroglu M. Chemical determination of free-radical induced damage to DNA. Free Radical Biol Med 1991; 10: 225–242.
5. Demple B, Harrisson L. Repair of oxidative damage to DNA: enzymology and Biology. Annu Rev Biochem 1994; 63: 915–948.
6. Floyd RA, Watson JJ, Wong PK, Altmiller DH, Rickard RC. Hydroxyl free radical adduct of deoxyguanosine: sensitive detection and mechanisms of formation. Free Radical Res Commun 1986; 1: 163–172.
7. Ravanat JL, Turesky RJ, Gremaud E, Trudel LJ, Stadler RH. Determination of 8-oxoguanine in DNA by gas chromatography–mass spectrometry and HPLC electrochemical detection. Chem Res Toxicol 1995; 8: 1039–1045.
8. Pflaum M, Boiteux S, Epe B. Visible light generates oxidative DNA base modification in high excess of strand breaks in mammalian cells. Carcinogenesis 1994; 15: 297–300.
9. Epe B. DNA damage profiles induced by oxidizing agents. Rev Physiol Biochem Pharmacol 1995; 127: 223–249.
10. Tajiri T, Maki H, Sekiguchi M. Functional cooperation of MutT, MutM and MutY proteins in preventing mutations caused by spontaneous oxidation of guanine nucleotide in *Escherichia coli*. Mutat Res 1995; 336: 257–267.
11. Czeczot H, Tudek B, Lambert B, Laval J, Boiteux S. *Escherichia coli* Fpg protein and UvrABC endonuclease repair DNA damages induced by methylene blue plus visible light in vivo and in vitro. J Bacteriol 1991; 173: 3419–3424.
12. Shibutani S, Takeshita M, Grollman AP. Insertion of specific bases during DNA synthesis past the oxidation-damaged base 8-OxodG. Nature 1991; 349: 431–434.
13. Grollman AP, Moriya M. Mutagenesis by 8-oxoguanine: an enemy within. Trends Genet 1993; 9: 246–249.
14. Moriya M, Grollman AP. Mutation in the *mut*Y gene of *Escherichia coli* enhance the frequency of targeted GC→TA transversions induced by a single 8-oxoguanine residue in single-stranded DNA. Mol Gen Genet 1993; 239: 72–76.

15. Michaels ML, Miller JH. The GO system protects organisms from the mutagenic effect of the spontaneous lesion 8-hydroxyguanine (7,8-dihydro-8-oxoguanine). J Bacteriol 1992; 174: 6321–6325.
16. Boiteux S. Properties and biological functions of the Nth and Fpg proteins of *Escherichia coli*: two DNA glycosylases that repair oxidative damage in DNA. Photochem Photobiol B 1993; 19: 87–96.
17. Boiteux S, O'Connor TR, Laval J. Formamidopyrimidine DNA glycosylase of *Escherichia coli*: cloning and sequencing of the structural *fpg* gene and overproduction of the protein. EMBO J 1987; 6: 3177–3183.
18. Boiteux S, O'Connor TR, Lederer F, Gouyette A, Laval J. Homogeneous Fpg protein, a DNA glycosylase which excises imidazole ring-opened purines and nicks DNA at abasic sites. J Biol Chem 1990; 265: 3916–3922.
19. O'Connor TR, Graves R, De Murcia G, Castaing B, Laval J. Fpg protein of *Escherichia coli* is a zinc finger protein whose cysteine residues have a structural and/or functional role. J Biol Chem 1993; 268: 9063–9070.
20. Boiteux S, Gajewski E, Laval J, Dizdaroglu M. Substrate specificity of *Escherichia coli* Fpg protein: excision of purine lesions in DNA produced by ionizing radiation and photosensitization. Biochemistry 1992; 31: 106–110.
21. Tchou J, Kasai H, Shibutani S, Chung MH, Laval J, Grollman AP, Nishimura S. 8-Oxoguanine glycosylase and its substrate specificity. Proc Natl Acad Sci USA 1991; 88: 4690–4694.
22. Castaing B, Geiger A, Seliger H, Nehls P, Laval J, Boiteux S. Cleavage and binding of a DNA fragment containing a single 8-oxoguanine by wild type and mutant Fpg proteins. Nucleic Acids Res 1993; 21: 2889–2905.
23. Boiteux S, Huisman O. Isolation of a formamidopyrimidine DNA glycosylase (*fpg*) mutant of *Escherichia coli* K12. Mol Gen Genet 1989; 215: 300–305.
24. Cabrera M, Nghiem Y, Miller JH. MutM a second mutator locus in *E. coli* that generates GC→TA transversions. J Bacteriol 1988; 170: 5405–5407.
25. Duwat P, De Oliveira R, Ehrlich DS, Boiteux S. Repair of oxidative DNA damage in Gram-positive bacteria: the *Lactococcus lactis* Fpg protein. Microbiology 1995; 141: 411–417.
26. Tsai-Wu JJ, Liu HF, Lu AL. *Escherichia coli* MutY has both *N*-glycosylase and apurinic/apyrimidinic endonuclease activities on A/G and A/C mispairs. Proc Natl Acad Sci USA 1992; 89: 8779–8783.
27. Nghiem Y, Cabrera M, Cuppless CG, Miller JH. The *mut*Y gene: a mutator locus in *Escherichia coli* that generates GC→TA transversions. Proc Natl Acad Sci USA 1988; 85: 2709–2713.
28. Michaels ML, Cruz C, Grollman AP, Miller JH. Evidence that MutY and MutM combine to prevent mutations by an oxidatively damaged form of guanine. Proc Natl Acad Sci USA 1992; 89: 7022–7025.
29. Auffret Van der Kemp P, Thomas D, Barbey R, De Oliveira R, Boiteux S. Cloning and expression in *E. coli* of the *OGG1* gene of *S. cerevisiae* which codes for a DNA glycosylase that excises 7,8-dihydro-8-oxoguanine and 2,6-diamino-4-hydroxy-5-*N*-methylformamidopyrimidine. Proc Natl Acad Sci USA 1996; 93: 5197–5202.
30. Thomas D, Scott AD, Barbey R, Padula M, Boiteux S. Inactivation of *OGG1* increases the incidence of GC→TA transversions in *Saccharomyces cerevisiac*: evidence for endogenous oxidative damage to DNA in eukaryotic cells. Mol Gen Genet 1997; in press.
31. Bessho T, Tano K, Kasai H, Ohtsuka E, Nishimura S. Evidence for two DNA repair enzymes for 8-hydroxyguanine in human cells. J Biol Chem 1993; 268: 19416–19421.
32. Moriya M. Single stranded shuttle phagemid for mutagenesis studies in mammalian cells: 8-oxoguanine in DNA induces GC→TA transversions in simian kidney cells. Proc Natl Acad Sci USA 1993; 90: 1122–1126.
33. McGoldrick JP, Yeng-Chen Yeh, Solomon M, Essigman J, Lu AL. Characterization of a human homolog of the *E. coli* MutY repair protein. Mol Cell Biol 1995; 15: 989–996.

34. Wagner JR, Hu CC, Ames BN. Endogeneous oxidative damage of deoxycitidine in DNA. Proc Natl Acad Sci USA 1992; 89: 3380–3384.
35. Takeuchi T, Morimoto K. Increased formation of 8-hydroxyguanosine, an oxidative DNA damage, in lymphoblasts of Fanconi's anemia patient due to possible catalase deficiency. Carcinogenesis 1993; 14: 1115–1120.
36. Czene S, Harms-Ringdahl M. Detection of single-strand breaks and formamidopyrimidine-sensitive sites in DNA of cultured human fibroblasts. Mutat Res 1995; 336: 235–242.

33

The Elevated Serum Level of Thioredoxin in Patients with Malignant Disease and Chronic Inflammatory Diseases

Hiro Wakasugi and Masaaki Terada
National Cancer Research Institute
Tokyo, Japan

Kunihisa Miyazaki and Michio Miyata
Jichi Medical School
Omiya Medical Center
Omiya, Japan

INTRODUCTION

Thioredoxin (TRX) is a ubiquitous protein with two redox-active cystine residues in the active center, having the amino acid sequence (-Cys-Gly-Pro-Cys-) (1). TRX was first isolated in 1964 as the hydrogen donor for the enzymatic synthesis of deoxyribonucleotides by ribonucleotide reductase in *Eschevichia coli* (2). In addition, TRX is an essential subunit of phage T7 DNA polymerase (3). The human TRX is known as the autocrine growth factor derived from the human Epstein–Barr virus (EBV)-containing B lymphoblastoid cell lines 3B6, and was termed "3B6-interleukin-1" (3B6-IL-1) (4). An inducer of the interleukin-2 receptor (IL-2R) on T cell lines transformed by human T cell lymphotrophic virus-I (HTLV-I) was named "adult T-cell leukemia (ATL)-derived factor (ADF)" (5,6). It was found that 3B6-IL-1 and ADF were identical (7). Moreover, it has been reported that TRX is identical to the human eosinophil cytotoxicity-enhancing factor (ECEF) which enhances human eosinophil cytotoxic function in vitro (8), and the early pregnancy factor (EPF) which has a vital role in the development and immune protection of the embryo (9). TRX has various biological activities as a hydrogen donor, including reduction of insulin (10), activation of glucocorticoid receptor (11), upregulation of nuclear factor κB (NFκB) binding (12), protein disulfide isomerase (PDI) activity (13), activation of protein kinase C (14), and radical scavenging activity (15, 16). In addition, TRX plays an important role in protection against various stresses including TNF-α (17), lethal doses of irradiation (18,19), and ultraviolet (UV) irradiation (20). Some studies have been reported wherein TRX expressed strongly in malignant cells associated with viral infections including human papilloma virus-associated cervical intraepithelial neoplasia (21–24), hepatitis virus (type B or C)-infected hepatocellular carcinomas (25), and EBV-associated nasopharyngeal carcinoma (Wakasugi, H. et al., unpublished). There however, has been, no report showing changes of the serum level of TRX in patients with these virus-associated cancers.

Helicobacter pylori infection has been reported as a new risk factor for atrophic gastritis and also for gastric cancer (26,27) *H. pylori* directly or indirectly damages gastric mucosae, leading to atrophic gastritis and possibly gastric cancer (28).

To measure TRX in serum and tissues, we established an enzyme-linked immunosorbent assay (ELISA). We here report that the serum level of TRX in patients with hepatocellular carcinoma (HCC) was significantly higher when compared not only with the level in normal volunteers but also with that in patients with liver cirrhosis (LC) or chronic hepatitis (CH) without HCC. It is well established that hepatocellular carcinoma is frequently associated with the hepatitis B virus or hepatitis C virus. On the assumption that the serum level of TRX might be increased in microbial-associated diseases, we also investigated the serum level of TRX in patients with chronic gastritis with or without *H. pylori* infection.

METHODS

Serum Samples and Liver Tissues

Serum samples were taken before breakfast from 65 healthy volunteers (46 males, 19 females; age range 22–68 years, mean ± S.D. 44 ± 9.1 years), 38 LC/CH patients without HCC (21 males, 17 females; age 60 ± 9.1 years), 38 patients with HCC (31 males, 7 females; age 65 ± 8.4 years), and 69 patients with chronic gastritis (43 males, 26 females; age 56.6 ± 9.17 years). Of 38 non-HCC patients, 29 were hepatitis C virus (HCV) positive, 6 were hepatitis B virus (HBV) positive, and 3 were hepatitis virus negative. Thirty-three of 38 patients with HCC were HCV positive, while two were HBV positive and 3 were hepatitis virus negative. The patients with HCC were further divided into two groups. In one group, the maximum tumor size was 3 cm or under and the number of tumors was 2 or less. This was termed "early stage." In the other group, which was termed "advanced stage," the maximum tumor size was over 3 cm or the number of tumors was greater than 2. We also selected 24 (14 males, 10 females; age 51.9 ± 8.9 years) gastritis patients with *H. pylori* infection, and 45 (27 males, 18 females; age 59.1 ± 9.1 years) gastritis patients without *H. pylori* infection to study the serum TRX levels. The presence or absence of *H. pylori* infection was determined by the two methods: (i) ELISA for the anti-*H. pylori* IgG antibody in the serum of peripheral blood; (ii) culture of gastric mucosa biopsy sample obtained by the use of sterilized endoscopes. To evaluate the correlation between the serum level of TRX and the degree of mucosal atrophy, the serum concentrations of pepsinogen I (PG-I) and pepsinogen II (PG-II) were measured. Twenty-four patients with *H. pylori* infection and 45 patients without *H. pylori* infection were divided into two groups: 33 mild (24 males, 9 females) and 36 severe (19 males, 17 females) atrophic gastritis patients, according to the PG-I concentration and/or PG-I/ PG-II ratio as follows: mild atrophic gastritis group, PG-I ≥ 40 ng/ml and PG-I/II ratio ≥3.0; severe atrophic gastritis group

PG-I < 40 ng/ml or PG-I/II ratio <3.0 (27). Sixty-five healthy volunteers in the study had not been screened for *H. pylori* infection. All patients had been diagnosed as having chronic gastritis without gastric cancer. Each serum sample was stored frozen at –80°C after separation from whole blood by centrifugation at 3000 rpm for 10 min. During the course of the study, we found that erythrocytes contain TRX. It was found that serum contaminated with more than 4 mg/dl hemoglobin showed an increased concentration of serum TRX due to the release of TRX from the erythrocytes (Miyazaki K., et al., unpublished). Serum samples with high levels of hemoglobin concentrations (>4 mg/dl) were excluded. Normal liver tissue was obtained at surgery from a patient with gastric cancer, and a surgical specimen of HCC was obtained from a patient with HCC. Each sample was measured in duplicate at 3 different dilutions (×2, ×4, ×8) by our ELISA system. In this assay system an anti-TRX monoclonal antibody was used as the immoblized antibody, and polyclonal antibodies were used as enzyme-labeled antibodies (K. Miyazaki et al., unpublished data).

Immunoblotting Analysis

Immunoblotting analysis was performed according to established methods (29).

RESULTS

Standard Curve for the Sandwich ELISA of TRX

We investigated which combination of 6 anti-TRX monoclonal antibodies (moAbs) and 2 polyclonal antibodies (poAbs) showed the highest sensitivity in the sandwich ELISA system. The highest sensitivity was obtained with the combination of 5t1 moAb and polyclonal antibodies against N-terminal peptides.

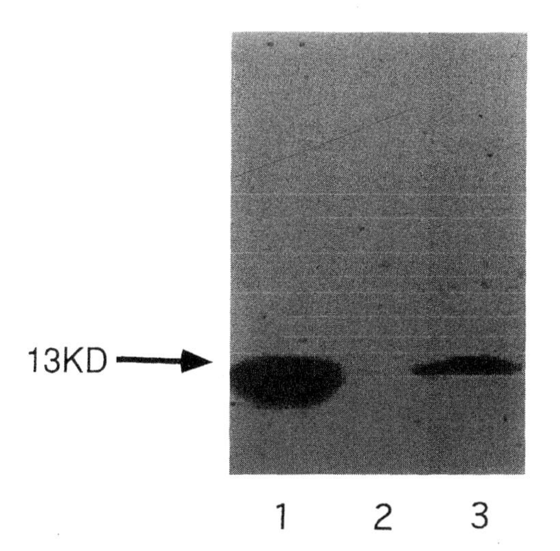

Figure 1. Immunoblotting analysis for detecting TRX in human serum. Lane 1. rTRX (1 μg) was applied. Lane 2: 5 ml of the effluent before eluting, which was condensed into 20 μl, was applied as a negative control. Lane 3: 20 μl of 50 μl of the eluate from 5t1 anti-TRX monoclonal antibody column in which 5 ml of human serum passed through twice was applied.

Detection of TRX in Serum by Immunoblotting Analysis

To confirm whether the molecule which reacted with the 5t1 anti-TRX moAb used in the ELISA system is truly TRX, the molecule purified from serum by the 5t1 antibody affinity column was examined by immunoblotting analysis. As shown in Figure 1, only one band corresponding to molecular mass of 13 kDa, corresponding to that of TRX, was detected (lane 3), whereas no band was observed in the effluent (lane 2). It should be noted that the molecular mass of the sample (lane 3) was slightly greater than that of rTRX (lane 1).

Measurement of the Serum Level of TRX in Healthy Volunteers

The serum level of TRX in 65 healthy volunteers as a normal control, 88.9 ± 35.8 ng/ml (mean ± S.D.) (Table 1(A)). Interestingly, the mean value for males (94.9 ± 37.9 ng/ml) was higher than that for females (74.9 ± 26.0 ng/ml) (p=0.026). There was no significant difference among any age (data not shown).

Measurement of the Serum Level of TRX in LC/CH Patients

Next, we compared the serum level of TRX in LC/CH patients with or without HCC. Contrary to expectation, the mean value in LC/CH patients without HCC was 90.6 ± 45.6 ng/ml (Table 1(B)), which showed no significant difference from that in normal volunteers. Furthermore, there was no difference in the serum level of TRX between the LC group and CH group (data not shown).

Table 1. Serum Level of Thioredoxin in Healthy Volunteers and in Patients with Chronic Liver Diseases

		TRX (ng/ml, mean ± S.D.)	*p*-value	
(A)	Healthy volunteers			
	Total (*n*=65)	88.9 ± 35.8		
	Male (*n*=46)	94.9 ± 37.9	0.0026	
	Female (*n*=19)	74.9 ± 26.0		
(B)	LC/CH patients			
	Healthy volumteers (*n*=65)	88.9 ± 35.8		
	Without HCC (*n*=38)	90.6 ± 45.6	–	
	With HCC (*n*=38)	159.0 ± 103.3	0.0004	<0.0001
(C)	HCC patients			
	Healthy volunteers (*n*=65)	88.9 ± 35.8		
	LC/CH patients (*n*=38)	90.6 ± 45.6		
	HCC patients			
	Early stage (*n*=19)	120.2 ± 50.2	0.003	0.029
	Advanced stage (*n*=19)	197.9 ± 127.4	<0.0001	<0.0001

Each sample was measured in duplicate at 3 different dilutions (×2, ×4, ×8) and values in the range of 5–150 ng/ml were considered as the serum level of TRX. Early stage means maximum tumor size 3 cm or below tumor number 2 or less. Advanced stage means tumor size bigger than 3 cm or tumor number more than 2.

Measurement of the Serum Level of TRX in Patients with HCC

The mean value of serum TRX in patients with HCC was 159.0 ± 103.3 ng/ml, which was significantly higher than that in the normal volunteers ($p<0.0001$) and that in the patients without HCC ($p=0.0004$) (Table 1(B)). In order to investigate the serum level of TRX in patients with HCC at different stages, we divided 38 patients with HCC into "early stage" and "advanced stage". In 19 early-stage patients, the mean value was 120.2 ± 50.2 ng/ml, which was significantly higher than that in LC/CH patients ($p=0.029$) and that in normal volunteers ($p=0.003$) (Table 1(C)). In 19 patients in the advanced stage, the mean value was 197.9 ± 127.4 ng/ml, which was significantly higher than in early stage patients ($p=0.03$), in LC/CH patients ($p<0.0001$), and in normal volunteers ($p<0.0001$) (Table 1(C)). We tentatively set the upper limit of the serum level of TRX at 150 ng/ml since most normal samples (62/65) were under 150 ng/ml. Sixteen of 38 patients with HCC (11 of 19 in advanced stage; 5 of 19 in early stage) had levels exceeding 150 ng/ml.

Measurement of the Serum Level of TRX in Patients with HCC after Surgical Removal of Tumors

In 4 patients with HCC where serum TRX had exceeded 150 ng/ml before treatments, the serum level of TRX was measured 4–6 months after surgical removal of tumors. All four values decreased to a normal level below 150 ng/ml (Table 2).

Measurement of the Serum Level of TRX in Patients with Chronic Gastritis

Both the mean (±S.D.) serum concentration of TRX of the 24 gastritis patients infected by *H. pylori* (193.3 ± 188.0 ng/ml) and that of the 45 gastritis patients without *H. pylori* infection (176.5 ± 106.6 ng/ml) were significantly higher than that of the 65 normal volunteers (88.9 ± 35.8 ng/ml) ($p<0.0001$ and $p<0.0001$, respectively) (Table 3(A)). The serum levels of TRX of the gastritis patients with *H. pylori* infection (193.3 ± 188.0 ng/ml) were higher than those of the gastritis patients without *H. pylori* infection (176.5 ± 106.6 ng/ml), but the increase was not statistically significant (Table 3).

Table 2. Serum Level of Thioredoxin in Patients with HCC Before and After Surgical Removal of Tumors

	TRX (ng/ml)	
Patients	Before[a]	After[a]
1	256	114
2	156	104
3	275	60
4	178	57

[a] Before and after surgical removal of tumors. In four patients with HCC who had shown the high level of serum TRX before treatment, the serum level of TRX was measured 4–6 months after removal of cancers surgically. All values of serum TRX had decreased below 150 ng/ml.

Table 3. Serum Level of Thioredoxin in Patients with Chronic Gastritis

	TRX (ng/ml, mean ± S.D.)	*p*-value		
Patients				
Healthy volunteers (*n* = 65)	88.9 ± 35.8			
Chronic gastritis				
H. pylori infection (–) (*n* = 45)	176.5 ± 106.6	<0.0001	–	
H. pylori infection (+) (*n* = 24)	193.3 ± 188.0	<0.0001		
Chronic gastritis patients				
Healthy volunteers (*n* = 65)	88.9 ± 35.8			
H. pylori infection (–) (*n* = 45)	90.6 ± 45.6			
Mild (*n* = 21)	146.7 ± 75.1	<0.0001	–	–
Severe (*n* = 24)	202.6 ± 123.7	<0.0001		–
Healthy volunteers (*n* = 65)	88.9 ± 35.8			
H. pylori infection (+) (*n* = 24)	193.3 ± 188.0			
Mild (*n* = 12)	140.9 ± 81.8	0.0006	–	
Severe 9 (*n* = 12)	241.2 ± 260.6	<0.0001		

Twenty-four patients with *H. pylori* infection and 45 patients without *H. pylori* infection were divided into two groups: mild and severe atrophic gastritis, according to the PG-1 concentration and/or PG-1/PG-II ratio as indicated in the following: Mild atrophic gastritis group means PG-1≥40 ng/ml and PG-I/II ratio≥3.0. Severe atrophic gastritis group means PG -l <40 ng/ml or PG-I/II ratio<3.0. Each sample was measured in duplicate at 3 different dilutions (×2, ×4, ×8) and values in the range 5–150 ng/ml were considered as the serum concentration of TRX.

The serum concentration of TRX in 12 mild atrophic gastritis patients with *H. pylori* infection (140.9 ± 81.8 ng/ml) or that of 21 patients without *H. pylori* infection (146.7 ± 75.1 ng/ml) was much higher than that of the normal volunteers (88.9 ± 35.8 ng/ml) (p=0.0006 and p<0.0001, respectively) (Table 3(B)). The serum concentration of TRX in 12 severe atrophic gastritis patients with *H. pylori* infection (241.2 ± 260.6 ng/ml) or that of 24 patients without *H. pylori* infection (202.62 ± 123.7 ng/ml) was also much higher than that of the normal volunteers (88.9 ± 35.8 ng/ml) (p<0.0001 and p<0.0001, respectively)(Table 3(B)). Regardless of the presence or absence of *H. pylori* infection, the serum level of TRX in the severe atrophic gastritis patients was higher than that of the mild atrophic gastritis patients both with and without *H. pylori* infection. However, the increases were not statistically significant (Table 3(B)). Eleven of 24 *H. pylori*-infected patients with chronic gastritis were treated with antimicrobial drugs to remove *H. pylori*. Following the successful treatments, the elevated serum concentration of TRX in 4 patients declined to a normal level of less than 150 ng/ml (Table 4). The 7 patients with normal serum levels of TRX did not show any significant changes in the serum TRX concentrations even after the treatment for *H. pylori*.

DISCUSSION

A great deal of attention has focused on the role that TRX may play in malignant diseases associated with viral infections, such as hepatocellular carcinoma (25), nasopharyngeal

Table 4. Serum Level of Thioredoxin in Patients Infected with *H. pylori* Before and After Treatment with Antimicrobial Drugs

	TRX (ng/ml)	
Patients	Before[a]	After[a]
1	241	94
2	566	92
3	194	106
4	157	46
5	97	55
6	87	98
7	32	132
8	42	86
9	82	75
10	16	112
11	56	52

[a] Before and after treatment with antimicrobial drugs.

Eleven out of 24 *H. pylori*-infected patients were treated by antimicrobial drugs. Negative results for the presence of *H.pylori* by the culture of gastric mucosa confirmed the successful treatment of antibiotics in all 11 patients. Four out of the 11 patients treated by antimicrobial drugs, in whom the serum level of TRX exceeded 150 ng/ml before treatment, showed the normal level of TRX (less than 150 ng/ml).

carcinoma (Wakasugi H. et al., unpublished), cervical intraepithelial neoplasia (24), and HTLV-I(6). It is possible that serum TRX levels in patients with chronic microbial infections or microbial-associated cancers might be increased. Thus, the serum level of TRX could be a useful marker for the diagnosis and clinical follow-up of patients with chronic infection, patients with high risk status for the development of malignancy, and patients with malignant diseases. Accordingly, we established a sandwich ELISA for determining the amounts of TRX in serum.

We concluded that antibodies used for the ELISA assay in the present report specifically recognized serum TRX on the basis of the following two criteria. First, immunoblotting analysis demonstrated only one band corresponding to TRX. It was observed that the molecular mass of TRX from serum was slightly greater than that of rTRX, suggesting that the small difference was due to the glycosylation of TRX in serum. Second, an insulin-reducing assay demonstrated that the sample recognized by the monoclonal antibody exerted TRX enzymatic activity. The percentage recovery of added TRX was approximately 98%, showing that there were no interfering factors for measurement of serum TRX by this method and that serum could be directly applied to the ELISA system.

There was a significant difference between the mean value of serum TRX in male normal volunteers and that in females. The male value (94.9 ng/ml) was a little higher than the female (74.9 ng/ml) (p=0.026). The reason for this difference remains to be clarified. It develops frequently in patients with advanced stages of chronic liver diseases (LC/CH).

Most patients with LC/CH are infected with hepatitis virus B or C (30). TRX has been reported as a growth factor in malignant cells associated with EBV and HTLV-I (4,6). Nakamura and colleagues reported that TRX enhanced the proliferation of transformed hepatocytes in vitro. They also reported that TRX was expressed strongly in HCC tissues compared with that in the normal liver (25), and we confirmed their results by an immunohistochemical study (Figure 4). Patients with HCC but not those with LC/CH having no carcinoma showed significantly higher levels of serum TRX. Furthermore, we divided patients with HCC into "early stage" and "advanced stage" according to tumor size and tumor number, and were able to demonstrate that the level of serum TRX in the advanced stage of HCC was significantly higher than in the early stage, suggesting that the serum level of TRX rose with the increased stage of HCC. Immunohistochemical staining of TRX in HCC indicated that the high level of serum TRX in patients with HCC might be reflected the high production of TRX in HCC cells. It was demonstrated that the high concentrations of serum TRX were reduced remarkably after surgical removal of HCC, suggesting further that the high level of serum TRX was due to the high production of TRX in HCC.

Gastric cancer is one of the most common cancers in the world (31). Recently, at least two microorganisms have been reported to be involved directly or indirectly in the development of gastric diseases, including gastric cancer. One is *H. pylori* infection (26, 27). *H. pylori* damages gastric mucosae, leading to atrophic gastritis and possibly gastric cancer (28). Second, a great deal of attention is nowadays paid to the fact that EBV plays an important role in the development of gastric cancer (32). EBV has been found in the cancer cells of 6–8% of gastric cancer patients in Japan and in around 15% of gastric cancer patients in Western countries, including the United States and Europe (32, 33).

It was suspected that the serum level of TRX might be increased in microbial-associated gastric diseases. We thus investigated serum levels of TRX in patients with chronic gastritis, which is a high risk status for development of gastric cancer.

In this study we demonstrate clearly that the serum level of TRX is significantly elevated in patients with chronic gastritis. This abnormal elevation was already observed even in patients with mild atrophic gastritis, and the level of the serum TRX increased in patients with severe atrophic gastritis. Contrary to our expectation, the concentration of serum TRX apparently was not influenced by the presence of *H. pylori* infection. However, a marked reduction of the serum concentration of TRX was observed when *H. pylori* was removed by antimicrobial-drugs, suggesting that the improvement of gastritis induced by the removal of *H. pylori* led to the reduction of the serum level of TRX.

Our results provide new insight into the possible role of an elevated level of serum TRX as a useful biological marker for both nonmalignant diseases and malignant diseases.

ACKNOWLEDGMENTS

This work was supported by a 2nd-Term Grant-in-Aid for a Comprehensive 10-Year Strategy for Cancer Control from the Ministry of Health and Welfare of Japan; by a Grant-in-Aid from the Ministry of Education, Science, Sports and Culture of Japan; and by the Bristol-Myers Squibb Foundation.

We thank Dr. Noriyuki Ihara and Dr. Tsuneo Yamanaka for assistance in obtaining samples, and Dr. Setsuo Hirohashi and Dr. Yoshihiro Matsuno for their advice in the pathological analysis.

Part of this study was presented at the 54th Annual Meeting of the Japanese Cancer Association, October 1995, Kyoto, Japan.

REFERENCES

1. Luthman M, Holmgren A. Rat liver thioredoxin and thioredoxin reductase: purification and characterization. Biochemistry 1982; 21: 6628–6633.
2. Laurent TC, Moore EC, Reichard P. Enzymic synthesis of deoxyribonucleotides. J Biol Chem 1964; 239: 3436–3444.
3. Mark DF, Richardson CC. *Eschevichia coli* thioredoxin: a subunit of bacteriophage T7 DNA polymerase. Proc Natl Acad Sci USA 1976; 73: 780–784.
4. Wakasugi H, Rimsky L, Mahe Y, Kamel AM, Fradelizi D, Tursz T, et al. Epstein–Barr virus-containing B-cell line produces an interleukin 1 that it uses as a growth factor. Proc Natl Acad Sci USA 1987; 84: 804–808.
5. Yodoi J, Takatsuki K, Masuda T. Letter: Two cases of T-cell chronic lymphocytic leukemia in Japan. N Engl J Med 1974; 290: 572–573.
6. Teshigawara K, Maeda M, Nishino K, Nikaido T, Uchiyama T, Tsudo M, et al. Adult T leukemia cells produce a lymphokine that augments interleukin 2 receptor expression. J Mol Cell Immunol 1985; 2: 17–26.
7. Wakasugi N, Tagaya Y, Wakasugi H, Mitsui A, Maeda M, Yodoi J, et al. Adult T-cell leukemia-derived factor/thioredoxin, produced by both human T-lymphotropic virus type I- and Epstein–Barr virus-transformed lymphocytes, acts as an autocrine growth factor and synergizes with interleukin 1 and interleukin 2. Proc Natl Acad Sci USA 1990; 87: 8282–8286.
8. Balcewicz-Sablinska MK, Wollman EE, Gorti R, Silberstein DS. Human eosinophil cytotoxicity-enhancing factor. II. Multiple forms synthesized by U937 cells and their relationship to thioredoxin/adult T cell leukemia-derived factor. J Immunol 1991; 147: 2170–2174.
9. Tonissen KF, Wells JR. Isolation and characterization of human thioredoxin-encoding genes. Gene 1991; 102: 221–228.
10. Holmgren A. Thioredoxin catalyzes the reduction of insulin disulfides by dithiothreitol and dihydrolipoamide. J Biol Chem 1979; 254: 9627–9632.
11. Grippo JF, Holmgren A, Pratt WB. Proof that the endogenous, heat-stable glucocorticoid receptor-activating factor is thioredoxin. J Biol Chem 1985; 260: 93–97.
12. Matthews JR, Wakasugi N, Virelizier JL, Yodoi J, Hay RT. Thioredoxin regulates the DNA binding activity of NF-kappa B by reduction of a disulphide bond involving cysteine 62. Nucleic Acids Res 1992; 20: 3821–3830.
13. Pigiet VP, Schuster BJ. Thioredoxin-catalyzed refolding of disulfide-containing proteins. Proc Natl Acad Sci USA 1986; 83: 7643–7647.
14. Biguet C, Wakasugi N, Mishal Z, Holmgren A, Chouaib S, Tursz T, et al. Thioredoxin increases the proliferation of human B-cell lines through a protein kinase C-dependent mechanism. J Biol Chem 1994; 269: 28865–28870.
15. Schallreuter KU, Wood JM. The role of thioredoxin reductase in the reduction of free radicals at the surface of the epidermis. Biochem Biophys Res Commun 1986; 136: 630–637.
16. Spector A, Yan GZ, Huang RR, McDermott MJ, Gascoyne PR, Pigiet V. The effect of H_2O_2 upon thioredoxin-enriched lens epithelial cells. J Biol Chem 1988; 263: 4984–4990.
17. Matsuda M, Masutani H, Nakamura H, Miyajima S, Yamauchi A, Yonehara S, et al. Protective activity of adult T cell leukemia-derived factor (ADF) against tumor necrosis factor-dependent cytotoxicity on U937 cells. J Immunol 1991; 147: 3837–3841.

18. Hill HZ, Cathcart KN, Bargellini J, Trizna Z, Hill GJ, Schallreuter KU, et al. Does melanin affect the low LET radiation response of Cloudman S91 mouse melanoma cell lines? Pigment Cell Res 1991; 4: 80–86.
19. Lunn CA, Pigiet VP. The effect of thioredoxin on the radiosensitivity of bacteria. Int J Radiat Biol Relat Stud Phys Chem Med 1987; 51: 29–38.
20. Sachi Y, Hirota K, Masutani H, Toda K, Okamoto T, Takigawa M, et al. Induction of ADF/TRX by oxidative stress in keratinocytes and lymphoid cells. Immunol Lett 1995; 44: 189–193.
21. Durst M, Gissmann L, Ikenberg H, zur Hausen H. A papillomavirus DNA from a cervical carcinoma and its prevalence in cancer biopsy samples from different geographic regions. Proc Natl Acad Sci USA 1983; 80: 3812–3815.
22. Schwarz E, Freese UK, Gissmann L, Mayer W, Roggenbuck B, Stremlau A, et al. Structure and transcription of human papillomavirus sequences in cervical carcinoma cells. Nature 1985; 314: 111–114.
23. Tsunokawa Y, Takebe N, Kasamatsu T, Terada M, Sugimura T. Transforming activity of human papillomavirus type 16 DNA sequence in a cervical cancer. Proc Natl Acad Sci USA 1986; 83: 2200–2203.
24. Fujii S, Nanbu Y, Nonogaki H, Konishi I, Mori T, Masutani H, et al. Coexpression of adult T-cell leukemia-derived factor, a human thioredoxin homologue, and human papillomavirus DNA in neoplastic cervical squamous epithelium. Cancer 1991; 68: 1583–1591.
25. Nakamura H, Masutani H, Tagaya Y, Yamauchi A, Inamoto T, Nanbu Y, et al. Expression and growth-promoting effect of adult T-cell leukemia-derived factor. A human thioredoxin homologue in hepatocellular carcinoma. Cancer 1992; 69: 2091–2097.
26. Warren JR, Marshall BJ. Unidentified curved bacilli on gastric epithelium in active chronic gastritis. Lancet 1983; i: 1273–1275.
27. Fukuda H, Saito D, Hayashi S, et al. *Helicobacter pylori* infection, serum pepsinogen level and gastric cancer: a case-control study in Japan. Jpn J Cancer Res 1995; 86: 64–71.
28. Correa P. *Helicobacter pylori* and gastric carcinogenesis. [Review]. Am J Surg Pathol 1995; 19: S37–43.
29. Immunoblotting and immunodetectyion. In Coligan E, et al., eds. Current Protocols in Immunology New York: Wiley; National Institutes of Health; 1995: 8.10.1–8.10.17.
30. Tanaka K, Hirohata T, Koga S, Sugimachi K, Kanematsu T, Ohryohji F, et al. Hepatitis C and hepatitis B in the etiology of hepatocellular carcinoma in the Japanese population. Cancer Res 1991; 51: 2842–2847.
31. Correa P. Human gastric carcinogenesis: a multistep and multifactorial process. First American Cancer Society Award Lecture on Cancer Epidemiology and Prevention. Cancer Res 1992; 52: 6735–6740.
32. Imai S, Koizumi S, Sugiura M, et al. Gastric carcinoma: monoclonal epithelial malignant cells expressing Epstein–Barr virus latent infection protein. Proc Natl Acad Sci USA 1994; 91: 9131–9135.
33. Shibata D, Weiss LM. Epstein–Barr virus-associated gastric adenocarcinoma. Am J Pathol 1992; 140: 769–774.

34

The Role of Free Radicals in Antitumor Effects of Cancer Treatments

Toshikazu Yoshikawa, Satoshi Kokura, and Motoharu Kondo
Kyoto Prefectural University of Medicine
Kyoto, Japan

INTRODUCTION

Oxygen free radicals and similar species have been implicated in various diseases (1–4), as well as carcinogenesis (5–9) and aging (10–12). Oxygen radicals are highly cytotoxic, and we have demonstrated the importance of radical-induced lipid peroxidation as a mechanism of that toxicity in various experimental animal models (13–16). Oxygen radicals give rise to lipid peroxidation, producing a variety of pathologies. As a result, free radical research has focused on disease treatment with radical scavengers. However, by utilizing these potent actions effectively, that is, by producing large quantities of active oxygen species at the site of a cancer, an antitumor effect might be expected. Today, radiation therapy, chemotherapy, and photochemical reactions are well known as cancer treatments that make use of the cytotoxicity of reactive oxygen species and free radicals (17, 18). We have focused our experimental studies on rabbit VX2 carcinoma and rat AH 109A carcinoma treated with free-radical reactions.

MATERIALS AND METHODS

VX2 carcinoma cells were transplanted to the right hind leg of rabbits and AH109A carcinoma cells to the same portion of rats, and adriamycin or a transient embolic agent (Degradable Starch Microspheres; Pharmacia, Sweden) was given through a transfemoral arterial catheter. Hyperthermia was induced by capacitive 8 MHz radio frequencies using a Thermotron RF I.V. (Yamamoto Vinyter, Japan) for 20 min for rabbits and 15 min for rats. The size of tumors was measured by ultrasonic tomography.

Superoxide dismutase (CuZn-SOD), catalase, and DMSO were used as scavengers of reactive oxygen species, and the concentration of thiobarbituric acid-reactive substances (TBARS), an index of lipid peroxidation, was measured in the tumor tissue.

The tumor tissues were removed, and homogenized with 1.5 ml of 10 mM potassium phosphate buffer (pH 7.8) containing 30 mM KCl in a Teflon Potter–Elvehjem homogenizer. The level of TBARS in the tumor tissue homogenates was expressed as nmoles of malondialdehyde per mg protein using 1,1,3,3-tetramethoxypropane as the standard.

RESULTS

Adriamycin Generated Reactive Oxygen Species In Vivo

Adriamycin (ADM) was given to rabbits with VX2 carcinoma via transarterial catheter, and marked reduction of the tumor was observed. To investigate whether the antitumor effect of ADM might be related the generation of reactive oxygen species, rabbits were pretreated with DMSO or SOD and catalase. Regression of tumors by ADM was prevented by DMSO, but not by SOD and catalase. It was strongly suggested that hydroxyl radicals played a part in the antitumor effect of ADM in vivo.

Ischemia–Reperfusion Injury Induced by a Transient Embolization Using DSM

Degradalde starch microspheres (DSM), obtained from hydrolyzed potato starch, are easily digested by amylase in plasma. Arterial infusion of DSM alone causes transient ischemia and subsequent reperfusion of tumor tissue. Transient occlusion of a blood vessel followed by recanalization is known to induce so-called ischemia–reperfusion injury to the tissue. We therefore conducted a study to clarify the effect on tumor tissue of transient ischemia and subsequent reperfusion induced by the transient embolic agent DSM by examining the production of reactive oxygen species. Rabbits with VX2 carcinoma received regional infusion of DSM under transcatheter angiography, and it was confirmed that DSM occluded tumor vessels completely. Blood flow in the tumor decreased rapidly immediately after DSM, and returned to the original level within 1 h. The size of tumors did not change with a single infusion of DSM, while DSM repeated five times showed a significant reduction in the size, suggesting that reperfusion injury might occur in the tumor and destroy tumors to some extent. This reduction of the tumor was prevented by the pretreatment of rabbits with SOD and catalase, indicating the generation of reactive oxygen species in the tumor during reperfusion by DSM. TBARS also increased in the tumor with repeated infusion of DSM, and the increase was inhibited by the pretreatment with SOD and catalase.

Hyperthermia

VX2 carcinoma were transplanted into rabbit hind leg, and the effect of hyperthermia on tumor growth was measured at 7 and 14 days after heating. As an index of lipid peroxidation, TBARS in the tumor were measured prior to hyperthermia and 3, 6, 12, and 24 h after hyperthermia. Tumor growth in rabbits treated with hyperthermia was significantly reduced, and TBARS in the tumor treated with hyperthermia were significantly increased until 6 h after hyperthermia. The antitumor effect of hyperthermia and the increase of TBARS in the tumor treated with hyperthermia were significantly inhibited by the administration of DMSO or SOD and catalase.

DISCUSSION

Among various anticancer agents, ADM is one of the agents which generate hydroxyl radical, an important oxygen radical exerting tissue damage, during its autoxidation process. Since the antitumor effect of ADM on VX2 carcinoma was protected by DMSO, a potent scavenger of hydroxyl radical, and not by SOD and catalase, hydroxyl radical might be a factor responsible for the destruction of tumors.

Ischemia–reperfusion injury, which is induced by the generation of reactive oxygen species during the ischemic period followed by an influx of oxygen during recanalization, has been documented in various organs (19–21). If ischemia–reperfusion induced by DSM damages tumor tissues, it can be expected to have antitumor effects mediated through reactive oxygen species and lipid peroxidation. We demonstrated that repeated daily injections of DSM induced regression of VX2 carcinomas in a rabbit model. Furthermore, the participation of reactive oxygen species was demonstrated by the protective effect of the radical scavengers SOD and catalase. In addition, the tumor tissue TBARS, which are an index of lipid peroxidation, increased significantly after the infusion of DSM, and that increase was significantly inhibited by the simultaneous injection of SOD and catalase. Consequently, active oxygen species produced by DSM-induced ischemia–reperfusion apparently act as initiators of lipid peroxidation in tumor tissues, which suggests that lipid peroxidation is one of the important mechanisms of the antitumor effect of DSM.

Hyperthermia, used for the treatment of cancer, was investigated with regard to its therapeutic efficacy from the standpoint of reactive oxygen species. The present study strongly supports the idea that excessive generation of superoxide anion radical in the tumor mediation of lipid peroxidation by active oxygen may play an important part in the antitumor effect of hyperthermia.

In summary, these treatments are closely related to the generation of free radical-mediated tumor cell killing, and our basic data suggest that free radicals may play an important role in the mechanism of the antitumor effects of these treatments.

REFERENCES

1. Smith LL. The response of the lung to foreign compounds that produce free radicals. Annu Rev Physiol, 1986; 48: 681–692.
2. Asayama K, Dohashi K, Hayashibe H. Lipid peroxidation and free radical scavengers in thyroid dysfunction in the rat. Endocrinology 1987; 121: 2112–2118.
3. Said SI, Foda HD. Pharmacologic modulation of lung injury. Am Rev Respir Dis 1989; 139: 1553–1564.
4. Blasig IE, Shuter S, Garlick P, Slater T. Relative time-profiles for free radical trapping, coronary flow, enzyme leakage, arrhythmias, and function during myocardial reperfusion. Free Radical Biol Med 1994; 16: 35–41.
5. Ames BN. Dietary carcinogens and anti carcinogens. Oxygen radicals and degenerative diseases. Science 1983; 221: 1256–1264.
6. Troll W, Weisner R. The role of oxygen radicals as a possible mechanism of tumor promotion. Annu Rev Pharmacol Toxicol 1985; 25: 509–528.
7. Floyd RA. The role of 8-hydroxyguanosine in carcinogenesis. Carcinogenesis (London) 1990; 11: 1447–1450.
8. Emerit I. Reactive oxygen species, chromosome mutation, and cancer; possible role of clastogenic factors in carcinogenesis. Free Radical Biol Med 1994; 16: 99–109.

9. Haedele AD, Briggs SP, Thompson HJ. Antioxidant status and dietary lipid unsaturation modulate oxidative DNA damage. Free Radical Biol Med 1994; 16: 111–115.
10. Harman D. Aging; a theory based on free radical and radiation chemistry. J Gerontol 1956; 11: 293–300.
11. Yoshikawa M, Hirai S. Lipid peroxide formation in the brain of aging rats. J Gerontol 1967; 22: 162–165.
12. Uchiyama M, Mihara M. Determination of malondialdehyde precursor in tissues by thiobarbituric acid. Analy Biochem 1978; 86: 271–278.
13. Yoshikawa T, Naito Y, Ueda S, Ichikawa H, Takahashi S, Yasuda M, Kondo M. Ischemia–reperfusion injury and free radical involvement in gastric mucosal disorders. Adv Exp Med Biol 1992; 316: 231–238.
14. Yoshikawa T, Takahashi S, Naito Y, Ueda S, Tanigawa T, Yoshida N, Kondo M. Effects of a platelet-activating factor antagonist, CV-5209, on gastric mucosal lesions induced by ischemia-reperfusion. Lipids 1992; 27: 1058–1060.
15. Yoshikawa T, Takano H, Takahashi S, Ichikawa H, Kondo M. Changes in tissue antioxidant enzyme activities and lipid peroxides in endotoxin-induced multiple organ failure. Circ Shock 1994; 42: 53–58.
16. Yoshikawa T, Naito Y, Kishi A, Tomii T, Kaneko T, linuma S, Ichikawa H, Yasuda M, Takahashi S, Kondo M. Role of active oxygen, lipid peroxidation, and antioxidants in the pathogenesis of gastric mucosal injury induced by indomethacin in rats. Gut 1993; 34: 732–737.
17. Kokura S, Yoshikawa T, Kishi A, Tomii T, Tujigiwa M, Yasuda M, Ichikawa H, Takano H, Takahashi S, Naito Y, Ueda S, Oyamada H, Tainaka K, Kondo M. Role of oxygen derived free radicals for antitumor effects of intra-arterial injection with adriamycin. Jpn J Cancer Chemother 1990; 17: 1711–1714.
18. Weishaupt KD, Gomer CJ, Dougherty TJ. Identification of singlet oxygen as the cytotoxic agent in photoinactivation of a murine tumor. Cancer Res 1976; 36: 2326–2329.
19. Granger DN, Rutili G, McCord JM. Superoxide radicals in feline intestinal ischemia. Gastroenterology 1981; 81: 22–29.
20. Perry MA, Wadhwa S, Parks DA, Pickard W, Granger DN. Role of oxygen radicals in ischemia-induced lesions in the cat stomach. Gastroenterology 1986; 90: 362–368.
21. Sanfey H, Bulkley GB, Cameron JL. The role of oxygen-derived free radicals in the pathogenesis of acute pancreatitis. Ann Surg 1984; 200: 405–413.

35

The Role of Cysteine and Glutathione in the Pathogenesis of HIV Infection: Effects of Treatment with N-Acetylcysteine

Wulf Dröge, Hans-Peter Eck, Heike Schenk, Klaus Schulze-Osthoff, Dagmar Galter, George Shubinsky, Sabine Mihm, Volker Hack, and Ralf Kinscherf
Deutsches Krebsforschungszentrum
Heidelberg, Germany

HIV-infected persons were found to have, on average, abnormally low intracellular glutathione levels (1,2), low plasma cystine levels (3,4), and low plasma glutathione levels (5). At face value, the low glutathione and cysteine levels suggested the possibility that HIV-infected patients might suffer from various manifestations of oxidative damage due to an insufficient antioxidative defense. At the time, it was a quite common belief that oxidative processes in biological systems are generally disadvantageous, and antioxidants accordingly beneficial. In the meantime, it has become widely appreciated (i) that oxidative processes play, among other roles, an important positive role in the regulation of signal transduction and gene expression in the immune system, and (ii) that the cysteine supply and intracellular glutathione level determine not only the capacity of one of the most powerful antioxidant defense systems but also the level of glutathione disulfide (GSSG), an important biological oxidant with regulatory function. It is, therefore, not appropriate simply to equate a glutathione and cysteine deficiency with an insufficient antioxidant defense system.

The regulation of the DNA binding activity of the nuclear transcription factor κB (NFκB) is a classic example of this principle. The DNA binding activity of this transcription factor was found to be inhibited in cell-free systems by physiologically relevant concentrations of GSSG and in intact cells by several procedures that increase the intracellular level of GSSG (6,7). As the DNA binding activity is reactivated quite effectively by thioredoxin but not by reduced glutathione (GSH), an increased supply of cysteine and increased intracellular GSH level ultimately exert a negative effect on the DNA binding by increasing the intracellular GSSG level (6–8). Thus, in this case, the intracellular GSH level has a regulatory function that is not related to its biological activity as an antioxidant.

Another even more striking example of this principle is the regulation of the insulin receptor β-chain (9). The β-chain of the insulin receptor is a tyrosine kinase which shows strong autophosphorylation after binding of insulin to the insulin receptor α-chain. Insulin-like effects were also seen after exposure of cells to oxidants such as hydrogen peroxide, and this has been interpreted previously as an oxidative attack on the extracellular domains of the β- and α-chains in view of the accumulation of many cysteine residues in these domains. However, a strong increase of insulin receptor β-chain autophosphorylation is also seen if the intracellular GSSG level is increased by incubation of the cells (CHOT) with the glutathione reductase inhibitor 1,3-bis-(2-chloroethyl)-1-nitrosourea (BCNU), indicating that the tyrosine kinase activity of the insulin receptor β-chain is under regulatory control of the intracellular redox status. The phenolic antioxidant BHA inhibited the stimulating effect of BCNU in the absence and presence of insulin. To determine whether GSSG enhances the phosphorylation of the insulin receptor β-chain by direct interaction with the insulin receptor complex, the insulin receptor was immunoprecipitated and then treated with GSSG. As expected, the tyrosine kinase activity of the insulin receptor β-chain was substantially increased by physiologically relevant micromolar concentrations of GSSG. Surprisingly, however, a substantial increase was also obtained with GSH, although slightly higher concentrations were needed in this case. The phosphorylation of myelin basic protein (MBP) was increased in a similar way. The structural basis of this regulatory effect remains to be investigated. It is clear, however, from these results that GSH and GSSG function not merely as antioxidant and oxidant, respectively. The simplest interpretation is that these compounds may derivatize the β-chain by a disulfide exchange reaction with a pre-existing disulfide in the insulin receptor β-chain. The crystal structure of the protein tyrosine kinase domain of the (nonactivated) human insulin receptor has recently been published (10). There is substantial structural similarity and sequence homology with the tyrosine kinases of the Src family including a combination of two cysteines spaced 11 amino acids apart (Cys-1245 and Cys-1234). Mutation of Cys-1245 and the equivalent residue Cys-475 in $P56^{lck}$ led in either case to a substantial loss of enzymatic activity (11,12). As the tyrosine kinase activities of several members of the Src family, notably $P59^{fyn}$, $P56^{lck}$, and $P60^{src}$, are similarly regulated in cell-free systems by GSSG and GSH and in intact cells by BCNU and BHA, it is reasonable to assume that these homologous cysteine residues may be involved in these regulatory processes. Oxidizing agents such as hydrogen peroxide, which were shown to activate the insulin receptor β-chain and the Src kinases in intact cells, are believed to provide the oxidizing equivalents which are needed to facilitate their interaction with glutathione.

Most HIV-infected individuals remain disease free for many years and maintain for relatively long periods of time stable numbers of $CD4^+$ T cells, strong cytotoxic T cell responses, and low numbers of HIV-infected cells in the blood, indicating that the virus is essentially under immune control. In "long-term nonprogressors," the immunological control of the virus is maintained over very long periods of time. In most of the infected individuals, however, the immune system eventually fails, and the disease starts to progress toward manifest AIDS. Exactly why the immune system eventually fails to control HIV infection remains controversial. It is conceivable that it may take only a minor insult to the immune system to shift the balance in favor of the virus. There are several lines of argument to suggest that the sudden decrease of the plasma cystine and glutamine levels may be the decisive insult to the immune system that shifts this balance. First, the cysteine and glutamine supply were both found to be particularly important in lymphocyte cultures. Second, whereas a decrease of the mean intracellular glutathione level has been found already in the earliest stages of HIV infection and also in rhesus macaques two weeks after infection with SIV, the decrease of plasma cystine and glutamine levels was found to coincide in both cases with a decrease of the $CD4^+$ T cell numbers in the late asymptomatic stage. Only a relatively small proportion of healthy human subjects were found to have plasma cystine levels <40 μM and glutamine levels <550 μM. In the early asymptomatic stage with $CD4^+$ counts >400/mm^3, the plasma cystine levels are, on average, higher than normal but the plasma glutamine level is generally lower than in healthy human subjects. In contrast, more than 40% of the asymptomatic persons with $CD4^+$ T cell counts <400/mm^3 have plasma cystine levels <40 μM and again abnormally low plasma glutamine levels. A similar decrease of plasma cystine and glutamine levels has been found in patients with chronic fatigue syndrome, i.e., in the absence of the virus (13). This syndrome is associated with several immunological dysfunctions including a striking loss of natural killer (NK) cell activity. Studies of several laboratories have shown that the decrease of $CD4^+$ T cell numbers in HIV infection is indeed preceded by a functional deficiency of the immune system which includes amongst other effects a decrease of NK cell activity. As NK cells play an important role in the defense against virus infections and support $CD8^+$ and $CD4^+$ T cell functions, it is reasonable to assume that the striking decrease of the plasma cystine level at the time when $CD4^+$ T cell counts start to decrease in the late asymptomatic stage of HIV infection may be the decisive pathogenetic factor that drives disease progression. As a considerable proportion of HIV^+ persons with $CD4^+$ T cell counts >400/mm^3 already had low cystine levels in combination with very low plasma glutamine levels, it appears that these biochemical changes may precede the loss of $CD4^+$ T cells. As cystine is a disulfide and capable of converting GSH into GSSG via disulfide exchange reaction, there is a strong possibility that the decrease of the plasma cystine levels indirectly affects the intracellular level of GSSG. Studies on healthy human subjects with different plasma cystine levels indicate that this is indeed the case (V. Hack and W. Dröge, unpublished observation).

Last but not least, the plasma cystine level plays a regulatory role in the conversion of protein into other forms of chemical energy and the concomitant production of urea. As such, it is critically involved in the regulation of the nitrogen balance, which is seriously dysregulated in the process of cachexia. It is well established that the threshold for the conversion of amino acids into glucose is essentially determined by the rate at which ammonium ions in the liver are converted either into urea or into glutamine and, in the latter case, retained in the amino acid pool. The rate-limiting step for urea biosynthesis is the production of carbamoyl phosphate. This, in turn, is limited by the availability of hydrogencarbonate anions (HCO_3^-). Studies on SIV-infected rhesus macaques and tumor-bearing mice, together with studies on the effects of *N*-acetylcysteine on the body cell mass and body fat ratios in healthy human volunteers, suggest strongly that the hepatic cysteine catabolism into sulfate and protons downregulates the availability of HCO_3^- and thereby favors glutamine biosynthesis (14,15).

The skeletal muscle tissue of SIV-infected rhesus macaques showed significantly increased sulfate levels, indicating that an increased intramuscular cysteine catabolism into sulfate may be at least partly responsible for the decrease of the systemic cystine levels. The hepatic sulfate levels are accordingly decreased together with the glutamine/urea ratios and the glutamine/glutamate ratios. A similar pattern was found in tumor-bearing mice, and it was shown in this case that the exogenous administration of cysteine not only increased the hepatic sulfate levels but also reversed the changes of the glutamine/urea and glutamine/glutamate levels in these mice (14).

In view of the decreased plasma cystine and intracellular glutathione levels, we proposed to treat these patients with a cysteine derivative. We favored *N*-acetylcysteine (NAC) because it was already a well established drug for the treatment of chronic bronchitis in several European countries with well-documented toxicology and pharmacology. Longitudinal observations on 4 patients for up to 4 years revealed that NAC causes not only a substantial increase of the plasma cystine level but also, as expected, a concomitant significant increase of the plasma glutamine and arginine levels. In view of the dysregulation of these amino acids in catabolic conditions, we have reason to believe that NAC treatment may prevent the protein catabolism in these patients as well as the immunological dysfunctions that are seen in persons with abnormally low plasma cystine and glutamine levels.

ACKNOWLEDGMENTS

The assistance of Mrs. I. Fryson in the preparation of this manuscript is gratefully acknowledged.

REFERENCES

1. Eck HP, Gmünder H, Hartmann M, Petzoldt D, Dröge W. Low concentrations of acid-soluble thiol (cysteine) in the blood plasma of HIV-1 infected patients. Biol Chem Hoppe-Seyler 1989; 370: 101–108.
2. Roederer M, Staal FJT, Osada H, Herzenberg LA, Herzenberg LA. CD4 and CD8 T cells with high intracellular glutathione levels are selectively lost as the HIV infection progresses. Int Immunol 1991; 3: 933–937.
3. Dröge W, Eck HP, Näher H, Pekar U, Daniel V. Abnormal amino acid concentrations in the blood of patients with acquired immune deficiency syndrome (AIDS) may contribute to the immunological defect. Biol Chem Hoppe-Seyler 1988; 369: 143–148.
4. Hortin GL, Landt M, Powderly WG. Changes in plasma amino acid concentrations in response to HIV-1 infection. Clin Chem 1994; 40: 785–789.
5. Buhl R, Jaffe HA, Holroyd KJ, Wells FB, Mastrangeli A, Altini C, Cantin AM, Crystal RG. Systemic glutathione deficiency in symptom-free HIV-seropositive individuals. Lancet 1989; 2: 1294–1298.
6. Galter D, Mihm S, Dröge W. Distinct effects of glutathione disulfide on the transcription factors NF-κB and AP-1. Eur J Biochem 1994; 221: 639–648.
7. Mihm S, Galter D, Dröge W. Modulation of transcription factor NFκB activity by intracellular glutathione levels and by variations of the extracellular cysteine supply. FASEB J 1995; 9: 246–252.
8. Dröge W, Schulze-Osthoff K, Mihm S, Galter D, Schenk H, Eck H-P, Roth S, Gmünder H. Functions of glutathione and glutathione disulfide in immunology and immunopathology. FASEB J 1994; 8: 1131–1138.
9. Schenk H, Galter D, Kögl M, Klein G, Schulze-Osthoff K, Dröge W. Redox regulation of insulin receptor and Src family protein tyrosine kinases by glutathione and glutathione disulfide; manuscript in preparation.
10. Hubbard SR, Wie L, Ellis L, Hendrickson WA. Crystal structure of the tyrosine kinase domain of the human insulin receptor. Nature (London) 1994; 372: 746–754.
11. Macaulay SL, Polites M, Frenkel MJ, Hewish DR, Ward CW. Mutagenic structure/function analysis of the cytoplasmic cysteines of the insulin receptor. Biochem J 1995; 306: 811–820.

12. Veillette A, Dumont S, Fournel M. Conserved cysteine residues are critical for the enzymatic function of the lymphocyte-specific tyrosine protein kinase $p56^{lck}$. J Biol Chem 1993; 268: 17547–17553.
13. Aoki T, Miyakoshi H, Usuda Y, Herberman RB. Low NK syndrome and its relationship to chronic fatigue syndrome. Clin Immunol Immunopathol 1993; 69: 253–265.
14. Hack V, Groβ A, Kinscherf R, Bockstette M, Fiers W, Berke G, Dröge W. Abnormal glutathione and sulfate levels after interleukin-6 treatment and in tumor-induced cachexia. FASEB J 1996; 10: 1219–1226.
15. Kinscherf R, Hack V, Fischbach T, Friedmann B, Weiss C, Edler L, Bärtsch, Dröge W. Low plasma glutamine in combination with high glutamate levels indicate risk for loss of body cell mass (BCM) in healthy individuals: the effect of *N*-acetyl-cysteine on BCM. J Mol Med 1996; 74: 393–400.

36

Low Glutathione Levels in CD4 T Cells Predict Poor Survival in AIDS; *N*-Acetylcysteine May Improve Survival

Leonard A. Herzenberg, Stephen C. De Rosa, and Leonore A. Herzenberg
Stanford University Medical School, Stanford, California

INTRODUCTION

Glutathione (GSH) depletion impairs T cell function (1) and promotes cytokine-stimulated HIV expression (2,3). *N*-Acetylcysteine (NAC), which provides the cysteine necessary to replenish GSH (4), improves T cell function and blocks HIV expression (2,3,5–7). Since GSH levels are lower in HIV-infected individuals, particularly at later stages of HIV/AIDS (8–13), we and others (notably Droge) have suggested that replenishing GSH by administration of NAC or other nontoxic GSH prodrugs could slow the progress of HIV disease (5,14–17). However, despite the strong in vitro data supporting the importance of GSH replenishment in HIV disease, there has been no evidence to date directly linking the low GSH levels in HIV-infected individuals with the pathogenesis of HIV disease in these individuals.

Data summarized here, derived from a GSH monitoring study and an associated clinical trial testing oral NAC for GSH replenishment in subjects with AIDS, provide the first clear demonstration of the importance of GSH depletion in AIDS pathogenesis. In essence, findings for subjects with AIDS, defined by having CD4 T cell counts below 200/μl in this study, show that (i) the probability of surviving for 2 years is dramatically lower in subjects with low GSH levels, particularly in CD4 T cells; (ii) oral administration of NAC replenishes GSH, particularly in individuals with low GSH levels; and (iii) taking NAC for 8–32 weeks is associated with substantially improved survival for individuals with AIDS. Full reports of these studies will be published shortly (18; Dubs et al., unpublished findings).

This work was conducted in collaboration with Drs. J. Gregson Dubs, Mario Roederer, Stanley Deresinski, Michael Anderson, Stephen W. Ela, and Malcolm Zaretsky.

METHODS

GSH Measurements

We used HPLC analyses to measure GSH levels in rapidly processed whole blood samples (19; M.T. Anderson, unpublished findings) and multiparameter fluorescence activated cell sorter (FACS) analyses to measure intracellular GSH levels in T cell subsets in peripheral blood mononuclear cells (PBMCs) reacted with monochlorobimane to form the fluorescent glutathione-*S*-bimane (GSB) conjugate (20). We determined the median GSB level for each PBMC subset for each subject and used these median values in subsequent analyses, for example, to compute means for groups of subjects, to display distributions, to group subjects, and so on. GSB levels are expressed relative to the lymphocyte GSB level in a frozen PBMC standard measured in parallel with the PBMC samples.

Bivariate analyses demonstrate a significant correlation between the FACS measurements of GSH levels (i.e., GSB levels) and HPLC-measured whole blood GSH levels, which mainly reflect GSH levels in erythrocytes. For example, comparison of CD4 T cell GSB levels against HPLC-measured whole blood GSH for 47 subjects in a bivariate analysis generates an r-value of 0.53 and p-value of 0.0001. In a least-squares model, CD4 GSB levels significantly predict whole blood GSH; however, the model is greatly improved by including hematocrit level, which corrects for variation due to the volume of erythrocytes in the blood sample (adjusted R2 = 0.4; $p = 0.004$ for CD4 GSB and 0.002 for hematocrit).

Survival Analyses

We collected baseline data, including GSH levels, T cell subset counts, and clinical laboratory measurements for over 200 HIV-infected subjects, 83 of whom were enrolled into a double-blind, placebo-controlled trial testing the functional bioavailability of NAC. One to two years later, we surveyed the survival status of all subjects and evaluated the relationships between survival and GSH levels. Since all but 2 of the deaths we recorded occurred in subjects with a diagnosis of AIDS (defined as having CD4 T cell counts below 200/μl), we restrict analysis here to the 96 subjects in this group.

GSH Replenishment Following Oral Administration of NAC

Subjects were enrolled in a randomized, double-blind, placebo-controlled trial and given either NAC (6.9 ± 1.1 g/day on average) or placebo for 8 weeks. All subjects who qualified for enrollment had low GSB levels, were free of active opportunistic infections, and were otherwise relatively healthy, as judged by Karnofsky score and professional assessment. Subjects were also required to have maintained a stable reverse-transcriptase inhibitor regimen for the previous 4 months and were limited with respect to the taking of drugs that deplete GSH (e.g., acetaminophen) or diminish oxidative stress (e.g., high doses of vitamins C or E).

Statistical Analyses

We used the JMP Macintosh statistical package produced by the SAS Institute (Carey, NC) for all statistical analyses.

RESULTS

Survival Is Dramatically Lower in Subjects with Low CD4 T Cell GSH (GSB) Levels

CD4 T cell GSH levels, referred to hereafter simply as GSB levels, and measured at time 0 or baseline, tend to be lower in subjects with AIDS (defined for this study as subjects with CD4 T cell counts below 200/μl) (see Table 1). The mean GSB level for these subjects is below the GSB levels observed in over 80% of uninfected control subjects. In addition, because trial enrollment criteria so specified, GSB levels for subjects with AIDS who qualified for the NAC trial were substantially lower than the group with AIDS as a whole. The full report of this study (18) presents data for additional groups of subjects and detailed data for the AIDS subjects shown here.

The low GSB levels in subjects with AIDS were associated with poor survival. Roughly 40% (37/96) of these subjects died within 2 years of baseline data collection. The great majority of these deaths occurred among subjects with GSB levels below the mean for the group as a whole; very few deaths occurred among subjects with the highest GSB levels. Furthermore, the frequency of deaths was substantially higher in subjects with the lowest GSB levels.

Logistic regression analysis in Figure 1 shows the sharp increase in survival as a function of increasing baseline GSB levels in subjects with AIDS. Since, as we show below, subjects with AIDS who took NAC for more than 8 weeks survived longer than comparable subjects who did not take NAC, logistic regression analysis for the subgroup of monitored subjects who were not enrolled in the NAC trial results in an even greater survival differential ($p < 0.0001$, curve not shown). In essence, only 27% of subjects in the lowest quartile of this GSB distribution survived the 2-year

Table 1. CD4 GSB Levels are Lower in Subjects with AIDS

Subjects[a]		n[b]	CD4 GSB[c] (Mean±SD)
Uninfected		47	1.14±0.28
All HIV$^+$		203	0.97±0.28
CD4>200	All	107	1.05±0.25
CD4≤200 (AIDS)	All	96	0.88±0.29
	Trial subjects[d]	37	0.72±0.16

[a]Groups of subjects studied. CD4 ≤200 = subjects with CD4 T cell counts less than or equal to 200/μl, used in this study as synonymous with subjects with AIDS. Percentage of survivors at the end of the 2-year observation period: CD4 >200 = 97%; CD4 ≤200 = 65%.

[b]Number of subjects for whom CD4 GSB, absolute CD4 T cell counts, and survival status were recorded and on which computations are based. Overall study group composition: total, 203; male, 194; Caucasian, 151; mean age, 40.4±7.8 years, range 23–68 years.

[c]CD4 GSB values were normally distributed for all groups and differed significantly for all combinations of distinct groups (Anova t-test): $p \leq 0.0001$ for uninfected vs. all HIV$^+$, CD4>200 vs. CD4≤200 and No-TS vs Trial subjects; $p=0.002$ for No-TS vs. CD4>200. Standard error of the mean (SEM) for CD4 GSB means for the groups shown = 0.02–0.04. SD=standard deviation.

[d]Trial subject group includes all NAC and placebo arm subjects with CD4 T cell counts below 200/μl (37/55 in the trial as a whole). Low CD4 GSB levels for these subjects reflect the trial enrollment requirement for low GSB levels.

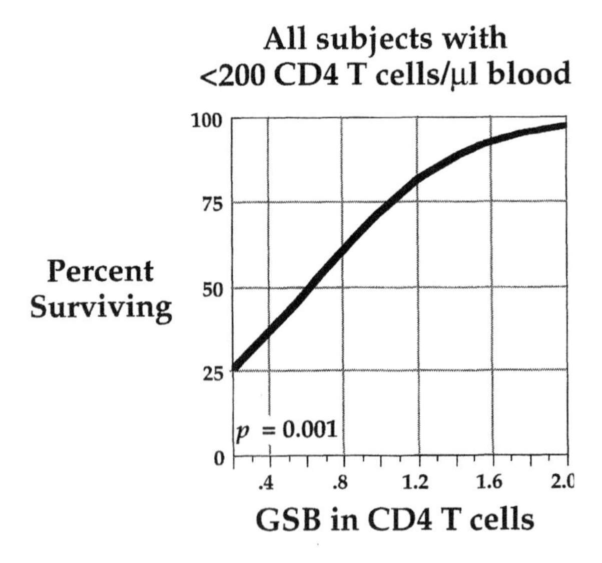

Figure 1. Probability of surviving for 2 years increases with increasing CD4 GSB levels. Logistic regression analysis for all subjects with CD4 counts below 200/μl (AIDS), including those in the NAC trial (most of whom took NAC). Survival status of subjects was determined 2 years after baseline data collection.

observation period of this study, whereas 87% of those in the highest quartile survived.

Logistic regression analyses demonstrate the importance of baseline GSB levels for survival by reporting the effect of baseline GSB levels on the survival status of subjects 2 years after baseline data collection. Kaplan–Meier analyses, which report survival as a

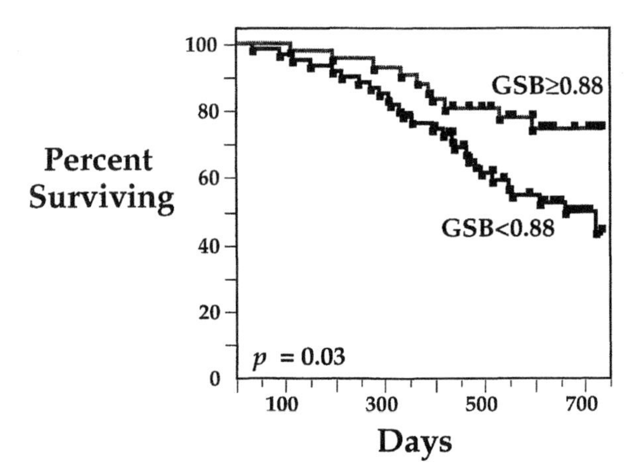

Figure 2. Low CD4 GSB levels are associated with poor survival in AIDS. Kaplan–Meier analysis for all subjects with CD4 counts below 200/μl (AIDS), including those in the NAC trial (most of whom took NAC). Subjects are divided at the mean baseline CD4 GSB level (0.88) for the this group of subjects. Survival times are computed from the date of the baseline visit.

function of time for subjects grouped according to GSB level (above or below the mean for subjects with AIDS), similarly show that higher GSB levels are associated with longer survival (Figure 2). Finally, proportional hazard analyses, which take both survival time and baseline GSB levels into account, show that GSB levels predict survival for the subjects with AIDS cohort (p = 0.0002) and report a 2-fold increase in survival for a difference of 0.3 GSB units (the standard deviation of the GSB levels in the cohort).

GSB Levels Versus CD4 T Cell Counts

CD4 T cell counts show a significant ability to predict survival in subjects with AIDS (p = 0.001); however, the significance of this prediction in proportional hazard analyses is greatly diminished when GSB levels are added to the model (p = 0.01 for CD4 T cell count; p = 0.003 for GSB). Since CD4 T cell counts are loosely correlated with GSB levels (Pearson's r = 0.33), this loss of significance suggests that for relatively healthy subjects with AIDS (i.e., with CD4 T cell counts below 200 μl), GSB levels are as good or a better predictor of survival than CD4 T cell counts. In addition, it also suggests that low GSB levels are an important contributor to the poor survival of HIV-infected individuals with low CD4 T cell counts.

Although the AIDS CD4 T cell count threshold (200/μl blood) provides a strong predictor of survival in overall HIV-infected populations, data presented above indicate that the prognostic value of counts under this threshold is more questionable. Combining GSB levels and CD4 T cell counts improves the accuracy of this survival prediction and thus is potentially of greater value for the group of HIV-infected subjects with low CD4 T cell counts. The combined measurement may also provide a better method for subject selection or stratification in AIDS clinical trials, since low GSB levels identify a high-risk

Table 2. NAC-dependent Increase in Whole Blood GSH

Combination of variables tested	Contribution to prediction of whole blood GSH levels at the end of an 8-week trial[a]		
	p-value	R^{2b}	RMSE[c]
0 week GSH	<0.0001	0.45	0.14
NAC vs. placebo	0.0008		
0 week GSH	0.0005		
NAC vs. placebo	0.0003	0.53	0.13
0 week CD4 GSB[d]	0.004		

[a]Whole-blood GSH principally reports GSH levels in erythrocytes in blood. Blood samples from 47 subjects were tested at the beginning and end of an 8-week randomized double-blind placebo- controlled trial testing the ability of orally administered NAC to raise GSH levels. One outlier was excluded. Data show the standard least-squares model fit. Subjects took 3200–8000 mg of NAC per day (median 5400 mg) for up to 8 months, supplied as 800 mg effervescent tablets.

[b]R^2 adjusted for number of variables added.

[c]Root mean square error.

[d]The significant contribution of initial CD4 GSB values, which are loosely correlated with initial whole-blood GSH levels, reflects the tendency for NAC ingestion to result in a greater increase in GSH levels in subjects with low initial GSB values (Dubs et al., unpublished findings).

subgroup of subjects, only a small percentage of whom are likely to survive longer than 2 years in the absence of intervention.

Oral Administration of NAC Replenishes GSH, Particularly in Individuals with Low GSB Levels

Data from the randomized, double-blind, placebo-controlled trial that we conducted show an average dose of that administration of NAC at 6900 mg/day for 8 weeks significantly elevates whole blood GSH (p = 0.0008) (Dubs et al., unpublished findings). Covariate analyses demonstrate that NAC increases whole-blood GSH levels more effectively in the subjects who had the lowest GSB levels at the start of the trial (Table 2). Thus, as might be expected, NAC is most effective in raising GSH levels when those levels are substantially depleted.

NAC Ingestion is Associated with Improved Survival in Subjects with AIDS

The NAC trial discussed above was not designed to test NAC efficacy in prolonging survival. However, after the initial double-blind placebo-controlled phase of the trial (8 weeks duration), all subjects were offered open-label NAC for up to 6 months during the continuation phase of the trial. As part of our overall monitoring study, we compared the fate of these subjects over the next 2 years with the fate of otherwise similar subjects who were not enrolled in the trial and did not have an opportunity to take NAC.* To our surprise, given the relatively short time (8–32 weeks) that NAC was administered, we found that NAC ingestion was associated with substantially longer survival (Table 3 and Ref. 18).

Since the enrollment criteria for the trial specifically excluded subjects with higher GSB levels, GSB levels in subjects in the NAC and No-NAC groups were amongst the

Table 3. Taking NAC is Associated With Better Survival

NAC history of subjects[a]	n[b]	Survival at 2.5-years[c] %	CD4 GSB (Mean at baseline ± SD)
NAC (8–32 weeks)	25	76	0.73±0.14
No-NAC (matched to NAC group)[d]	19	42	0.72±0.17

[a]All subjects had CD4 T cell counts below 200/μl and GSB levels low enough to qualify for entry into the NAC trial (18). All subjects who took NAC were enrolled in the NAC trial. NAC subjects took NAC for 8–32 weeks. Some were randomized to the NAC arm of the trial; others to open-label NAC during the continuation phase.

[b]Number of subjects in each group.

[c]Proportional hazard calculation for NAC: No-NAC survival yields a survival risk ratio of 1.8 (1.1–3.0, 95% confidence interval), p=0.018. Survival time in the model is computed from the time each subject began taking NAC. Table shows percentage surviving at the end of the 2.5 year observation period. Kaplan–Meier analyses comparing survival in the NAC and No-NAC groups is shown in Figure 3.

[d]There were no significant differences (p>0.1) in baseline measurements between the NAC and the No-NAC group for all parameters tested, including the following: absolute CD4 and CD8 counts; naive and memory T cell subset counts; hematocrit and other clinical laboratory tests; Karnofsky score; age; weight; GSB levels in B cells, monocytes, NK cells and all T cell subsets

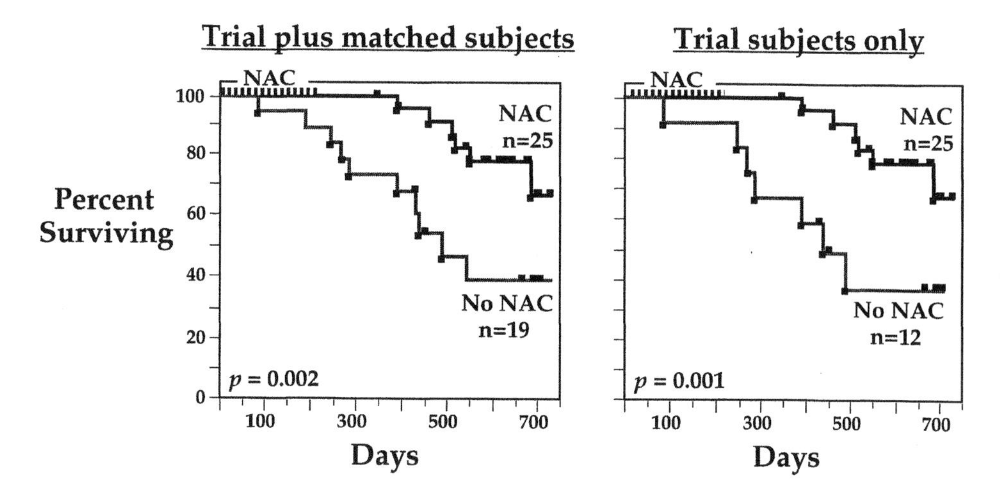

Figure 3. Taking NAC is associated with increased survival. Kaplan–Meier survival analyses compare survival of subjects who took NAC (NAC group) with survival of similar subjects who did not take NAC (No-NAC groups). The NAC group (both panels) includes 25 subjects who took NAC for 8–32 weeks (median 24 weeks; interquartile range 12–27 weeks), initiated either during the randomized, double-blind phase of the NAC replenishment trial (13 subjects) or during the open-label trial phase (12 subjects). The No-NAC group shown in the left panel includes all screened subjects (19) who met the basic criteria for entry into the NAC trial. The No-NAC group shown in the right panel includes only those subjects (12) who were actually enrolled in trial. Survival times for subjects who took NAC are computed from initiation of NAC ingestion (0 week for NAC arm; 8 weeks after the trial began for placebo arm). Survival times for subjects who did not take NAC are computed from the trial entry or screening date.

lowest in the study. Therefore, based on data presented above (Figures 1 and 2), the probability that these subjects would survive the 2-year observation period in our study was very low. Consistent with this, 11 of the 19 subjects in the No-NAC group died before the end of the observation period. In contrast, however, only 6 of the 25 subjects in the NAC group succumbed. Furthermore, half of the deaths in the No-NAC group occurred before the first death in the NAC group (Figure 3).

Proportional hazard analyses show the significant improvement of survival associated with oral administration of NAC. In addition, these analyses demonstrate that recruitment and trial management factors that contributed to determining whether and for how long NAC was taken did not significantly influence survival outcomes. In essence, there was no significant effect attributable to the reasons subjects took NAC (randomized to the NAC arm, elected NAC during open-label; $p > 0.2$) or the reasons subjects did not take NAC (not enrolled in the trial, left the trial, refused open-label; $p > 0.2$). However,

*In essence, although we lacked a proper placebo control group for survival comparison, we had a total of 19 subjects who did not take NAC but whose history indicated that they were very similar to those who did. We confirmed the comparability of these two groups (NAC and "No-Nac" groups) in analyses that failed to reveal any significant differences between them for a wide variety of clinical and FACS measurements, including GSB levels and CD4 T cell counts (18).

NAC ingestion significantly improved survival ($p = 0.019$) and indicated a roughly 2-fold survival advantage for subjects in the NAC group, i.e., NAC:No-NAC risk ratio = 1.8, 95% confidence interval = 1.1–3.0.

The association of prolonged survival with oral administration of NAC in this study is very dramatic. However, it is suspect because NAC was not administered in the context of a prospective trial in which survival was an endpoint. Since subjects in both the NAC and placebo arms of the trial in which NAC was administered were offered open-label NAC, the increased survival associated with taking NAC could be explained by factors associated with whether subjects took NAC rather than with the ingestion of the NAC itself. Although we found no indication of such bias, it cannot be excluded. Therefore, our findings basically argue for the initiation of a prospective placebo-controlled trial designed specifically to determine the therapeutic value of NAC in AIDS (14,16) and/or that of other GSH-replenishing drugs (21–25). Since NAC is nontoxic and could be used where medical services are limited, our findings indicate that such a trial should be initiated as rapidly as possible. Other pharmaceuticals that replenish GSH should also be tried for the same purpose. In any event, the poor survival that we have demonstrated in GSH-depleted subjects with AIDS underscores the importance of finding ways to replenish GSH in these individuals and ways to prevent this GSH depletion earlier in the disease.

DISCUSSION

We have shown that GSH depletion is associated with impaired survival: the greater the depletion, the worse the prospects for survival (Ref. 18, summarized here). These findings complement earlier preclinical data indicating that GSH depletion may play important roles in AIDS pathogenesis, for example, impairment of T cell function, facilitation of NFκB activation and HIV replication (1–3). By replenishing GSH, NAC or other agents may be able to modulate such adverse effects of GSH depletion. However, HIV-infected individuals would be better served if we could identify the mechanisms that underlie the GSH depletion and intervene, if possible, to prevent its occurrence. If ways could be found to do this on a long-term basis, HIV disease progression might be controlled in a way that would prevent the worst aspects of the disease.

Is such intervention possible? We have been struck by recent data indicating that HIV-TAT induces oxidative stress (Chapter 1; 26–29). We wonder whether the production and release of TAT could play a major role in the progressive depletion of GSH in HIV disease. If so, then targeting TAT for intervention (e.g., production of an anti-TAT vaccine) could be an important new strategy for controlling HIV disease.

At a more immediate level, certain rather simple precautions might help to slow the progress of HIV disease. Since studies presented here associate GSH depletion and oxidative stress with poor survival in AIDS, we believe that HIV-infected individuals should avoid excessive exposure to sun and UV irradiation and excessive use of drugs such as acetaminophen (Tylenol) that are known to deplete GSH. Physicians treating HIV-infected individuals should similarly consider exercising caution in prescribing formulations or recommending over-the-counter preparations containing such GSH-depleting drugs. These conservative measures could eliminate some of the more accessible causes of GSH depletion and thus could potentially prevent decrease of GSH to the level that predicts death within the following 2 years.

ACKNOWLEDGMENTS

We are first and foremost appreciative of the ready cooperation granted by the large number of HIV-infected individuals in the San Francisco area who generously contributed blood samples, information and, sadly, survival data demonstrating how vulnerable we are to HIV.

We thank Dr. Byron Brown, Stanford University Medical School, and Dr. John Sall, Senior Vice-President of the SAS Institute, for the time-consuming statistical advice they generously gave us. We also thank Ms. S. O'Leary, Dr. N. Watanabe, Dr. S. Mitra, Dr. H. Nakamura, Dr. F. Staal, Mr. I. Tijoe, Ms. N. Martin, Mr. E. Wunderlich, and the members of the Herzenberg laboratory, D. Hahn (Pharmaquest Corporation, San Rafael, CA), the Stanford Center for AIDS Research (directed by Dr. T. Merigan), and the Stanford Shared FACS Facility (directed by Dr. D. Parks) for helping to make this study a success. Dr. E. Winger and the Immunodiagnostic Laboratory (IDL), San Francisco, and Drs. C-M. Huang and Susan Wormsley (PharMingen, La Jolla, CA) deserve special mention; and we also thank Dr. C. Myers (University of Virginia) and Dr. S. Broder (IVAC Corporation, Miami, FL) for help and advice in the early stages of this work.

The Elan Pharmaceutical Corporation (Gainesville, GA), made this study possible by generously providing the palatable NAC and placebo for the trial and open-label follow-up. This work was largely supported by a grant from the National Cancer Institute (CA-42509), National Institutes of Health (Bethesda, MD). Dr. S. De Rosa was supported by NIH Immunology training grant 5T32 AI-07290. The Unicorn Foundation (Chicago) and Project Inform (San Francisco) also provided key financial support.

REFERENCES

1. Staal FJT, Anderson MT, Staal GEJ, Herzenberg LA, Gitler C, Herzenberg LA. Redox regulation of signal transduction: tyrosine phosphorylation and calcium influx. Proc Natl Acad Sci USA 1994; 91:3619–3622.
2. Staal FJT, Roederer M, Herzenberg LA, Herzenberg LA. Intracellular thiols regulate activation of nuclear factor kappaB and transcription of human immunodeficiency virus. Proc Natl Acad Sci USA 1990; 87:9943–9947.
3. Mihm S, Ennen J, Pessara U, Kurth R, Droge W. Inhibition of HIV-1 replication and NF-kappaB activity by cysteine and cysteine derivatives. AIDS 1991; 5:497–503.
4. Thomas SH. Paracetamol (acetaminophen) poisoning. Pharmacol Ther 1993; 60:91–120.
5. Roederer M, Staal FJT, Raju PA, Ela SW, Herzenberg LA, Herzenberg LA. Cytokine-stimulated human immunodeficiency virus replication is inhibited by *N*-acetyl-L-cysteine. Proc Acad Natl Sci USA 1990; 87:4884–4888.
6. Eylar EH, Baez I, Vazquez A, Yamamura Y. *N*-Acetylcysteine (NAC) enhances interleukin-2 but suppresses interleukin-4 secretion from normal and HIV$^+$ CD4$^+$ T-cells. Cell Mol Biol 1995; 41(supplement I):S35–S40.
7. Jeannin P, Delneste Y, Lecoanet-Henchoz S, et al. Thiols decrease human interleukin (IL) 4 production and IL-4-induced immunoglobulin synthesis. J Exp Med 1995; 182:1785–1792.
8. Eck H-P, Gmunder H, Hartmann M, Petzoldt D, Daniel V, Droge W. Low concentrations of acid-soluble thiol (cysteine) in the blood plasma of HIV-1-infected patients. Biol Chem Hoppe-Seyler 1989; 370:101–108.
9. Buhl R, Holroyd KJ, Mastrangeli A, et al. Systemic glutathione deficiency in symptom-free HIV-seropositive individuals. Lancet 1989; ii:1294–1298.

10. Roederer M, Staal FJT, Osada H, Herzenberg LA, Herzenberg LA. CD4 and CD8 T cells with high intracellular glutathione levels are selectively lost as the HIV infection progresses. Int Immunol 1991; 3:933–937.
11. de Quay B, Malinverni R, Lauterburg BH. Glutathione depletion in HIV-infected patients: role of cysteine deficiency and effect of oral *N*-acetylcysteine. AIDS 1992; 6:815–819.
12. Staal FJT, Roederer M, Israelski DM, et al. Intracellular glutathione levels in T cell subsets decrease in HIV infected individuals. AIDS Res Hum Retrovin 1992; 8:305–314.
13. Helbling B, Von Overbeck J, Lauterburg BH. Decreased release of glutathione into the systemic circulation of patients with HIV infection. Eur J Clin Invest 1996; 26:38–44.
14. Droge W, Eck H-P, Gmunder H, Mihm S. Modulation of lymphocyte functions and immune responses by cysteine and cysteine derivatives. Am J Med 1991; 91(supplement 3C):140S–144S.
15. Ruffmann R, Wendel A. GSH rescue by *N*-acetylcysteine. Klin Wochenschr 1991; 69:857–862.
16. Staal FJT, Ela SW, Roederer R, Anderson MT, Herzenberg LA, Herzenberg LA. Glutathione deficiency and human immunodeficiency virus ïnfection. Lancet 1992; 339:909–912.
17. Roederer M, Ela SW, Staal FJT, Herzenberg LA, Herzenberg LA. *N*-Acetylcysteine: a new approach to anti-HIV therapy. AIDS Res Hum Retrovir 1992; 8:209–217.
18. Herzenberg LA, De Rosa SC, Dubs JG, et al. Glutathione deficiency is associated with impaired survival in HIV disease. Proc Natl Acad Sci USA, 1997; 94:1967–1972.
19. Anderson ME. Determination of glutathione and glutathione disulfide in biological samples. In Meister A, ed. Methods in Enzymology, vol. 113. Orlando: Academic Press; 1985:548–555.
20. Roederer M, Staal FJT, Anderson M, et al. Disregulation of leukocyte glutathione in AIDS. In Landay AL, Ault KA, Bauer KD, Rabinowitch PS, eds. Clinical Flow Cytometry, vol. 677. New York: NY Acad Sci; 1993:113–125.
21. Giorgi G, Micheli L, Fiaschi AI, et al. L-2-Oxothiazolidine-4-carboxylic acid and glutathione in human immunodeficiency virus. Curr Ther Res 1992; 52:461–467.
22. Williamson JM, Boettcher B, Meister A. Intracellular cysteine delivery system that protects against toxicity by promoting glutathione synthesis. Proc Natl Acad Sci USA 1982; 79:6246–6249.
23. Meister A. Metabolism and function of glutathione. In Dolphin D, Avramovic O, Poulson R, eds. Coenzymes and Cofactors, vol. 3A. Glutathione: Chemical, Biochemical, and Medical Aspects. New York: Academic Press; 1989:367–474.
24. Kalebic T, Kinter A, Poli G, Anderson ME, Meister A, Fauci AS. Suppression of human immunodeficiency virus expression in chronically infected monocytic cells by glutathione, glutathione ester, and *N*-acetylcysteine. Proc Natl Acad Sci USA 1991; 88:986–990.
25. Levy EJ, Anderson ME, Meister A. Transport of glutathione diethyl ester into human cells. Proc Natl Acad Sci USA 1993; 90:9171–9175.
26. Flores SC, Marecki JC, Harper KP, Bose SK, Nelson SK, McCord JM. Tat protein of human immunodeficiency virus type 1 represses expression of manganese superoxide dismutase in HeLa cells. Proc Natl Acad Sci USA 1993; 90:7632–7636.
27. McCord J, Flores S. The human immunodeficiency virus and oxidative balance. In: Paoletti R, ed. Oxidative Processes and Antioxidants. New York: Raven Press; 1994: 13–23.
28. Westendorp MO, Shatrov VA, Schulze-Osthoff K, et al. HIV-1 Tat potentiates TNF-induced NF-kappaB activation and cytotoxicity by altering the cellular redox state. EMBO J 1995; 14:546–554.
29. Li CJ, Wang CL, Friedman DJ, Pardee AB. Reciprocal modulations between p53 and Tat of human immunodeficiency virus type 1. Proc Natl Acad Sci USA 1995; 92:5461–5464.

37

Modulation of HIV-1 Long Terminal Repeat by Arachidonic Acid

Simonetta Camandola, Tiziana Musso, Gabriella Leonarduzzi, Rita Carini, and Giuseppe Poli
University of Torino, Torino, Italy

Luigi Varesio
Istituto G. Gaslini, Genoa, Italy

Patrick A. Baeuerle
Tularik Inc., South San Francisco, California

INTRODUCTION

Arachidonic acid (AA), 20 carbon atoms, four double bonds, is one of the major polyunsaturated fatty acids present in the mammalian cell membrane. This omega-6 (n–6) fatty acid and its metabolites, released by cells responding to a wide range of stimuli, can play an important role in many physiopathological processes such as chemotaxis, inflammation, gene transcription, and signal transduction (1–3).

Arachidonic acid, introduced with diet, is never present in a free form in the cells, being immediately esterified to arachidonyl phospholipids, from which it is either consumed directly or released in the cytosol mainly through the action of phospholipase A_2. Unesterified arachidonate is degradated along the lipid peroxidation pathway and/or enzymatically oxygenated with production of eicosanoids, i.e., prostaglandins (PGs_2) and leukotrienes (LTs_4). Formation of eicosanoids in cells and tissues is a ubiquitous process, which may indicate that these mediators are of major importance in regulation of different physiological functions (2).

Eicosapentaenoic acid (EPA), 20 carbon atoms, 5 double bonds, is an omega-3 (n–3) fatty acid found in fish oil, clinically important because is metabolically incorporated into mammalian tissues and gives rise to a different series of prostaglandins (PGs_3) and leukotrienes (LTs_5) (2), that have, among others (4), anti-inflammatory properties (5). Numerous studies have shown that the composition of membrane phospholipids can be altered by modification of fatty acid dietary uptake (6–8).

The suggested role of AA in signal transduction (1) prompted us to investigate whether the relative concentration of membrane unsaturated lipids and their oxidative modification or metabolism may influence the human immunodeficiency virus (HIV-1) genome transcription.

We then analyzed the activity of HIV-1 long terminal repeat (LTR) inserted into U937 human promonocytic cells (renamed U938) undergoing incubation with micromolar amounts of AA or EPA in the culture medium so as to induce only a moderate increase in their membrane content, as checked by gas-chromatographic analysis, and not to interfere with cell viability. A cell line of the macrophage lineage was chosen because monocytes/macrophages represent an important reservoir of latent infection in which virus may lie dormant for years, with eventual activation by host or environmental factors. It is essential, therefore, to understand the events that can be responsible for inducing or enhancing HIV-1 replication in these cells. The effect of both AA and EPA supplementation on NFκB nuclear translocation in the monocytic cell line was also investigated because of the well-known implication of the transcription factor in this viral infection.

METHODS

Cell Line and Culture

Human promonocyte cell line U937 stably infected with a recombinant retrovirus containing the LTR region of the HIV-1 genome linked to the chloramphenicol acyltransferase (CAT) gene, designated as U938 cells (9), were used. The cells were cultured in RPMI 1640 (Gibco Laboratories, Grand Island, NY) containing 10% FCS (Gibco Laboratories), 100 U/ml penicillin (Sigma Chemical Co., St. Louis, MO), 100 μg/ml streptomycin (Sigma.), and 2 mM L-glutamine (Sigma) under an atmosphere of 5% CO_2 and 95% air.

LTR-CAT Assay

Cells were pelleted and dispensed at 10^6/ml. In each experiment, cultures of 10^7 cells were untreated or treated for 24 h with 45 μM AA (Sigma) or EPA (Sigma) in order to enrich the cell membrane with the fatty acids. Some samples were treated with ARA in presence of either 0.8 μg/ml indomethacin (Sigma) or 37 μM nordihydroguaiaretic acid (NDGA) (Sigma), or in the absence of both. Other samples were treated with arachidonic acid metabolites, i.e., prostaglandin E_2 (PGE_2) (1 μg/ml) (Sigma) and 5-hydroxyeicosatetrenoic acid (5-HETE) (1 μg/ml) (Sigma). Stimulation of LTR-CAT by tumor necrosis factor (TNF-α) (80–100 U/ml) (Sigma) was reproduced. After 24 h, the culture medium was removed and replaced with serum-free RPMI 1640; the cells were then subjected to the treatment described above, again for 24 h. CAT activity was measured by diffusion of butyrylated chloramphenicol into scintillation fluid. Briefly, cells were pelleted, washed with phosphate-buffered saline (PBS, calcium/magnesium free), lysed by sonication, resuspended in 150 μl of 1 M Tris buffer, heated to 65°C for 5 min to inactivate any potential inhibitors of enzyme expression, and incubated overnight at 37°C with 32 mM chloramphenicol substrate (Sigma), and 0.5 μCi [^{14}C]butyryl-coenzyme A (Dupont NEN Products, Boston, MA). Radiolabeled chloramphenicol was captured in nonaqueous Econofluor (Du Pont De Nemours, Bruxelles, Belgium) and ^{14}C was measured by β-scintillation (Beckman LS 1801, Beckman Analytical, USA). The activity of CAT was expressed as a percentage of [^{14}C]butyryl-CoA conversion compared to the control, adjusted per mg of protein.

Nuclear Extracts

Cells were plated in serum-free medium and treated as indicated above for 1 h. Nuclear extracts were prepared according to a modification of the method described by Parker and Topol (19). Cells were washed twice with cold PBS without calcium and magnesium and resuspended in 1 ml of buffer A (15 mM KCl, 10 mM Hepes pH 7.6, 2 mM $MgCl_2$, 0.1 mM EDTA, 1 mM dithiothreitol (DTT), 0.5% Nonidet P-40), incubated for 10 min on ice, mixed briefly, and centrifuged for 10 min at 4°C. The nuclear pellet was lysed by incubation for 15 min at 4°C in 20 μl of buffer B (1 M KCl, 25 mM Hepes pH 7.6, 0.1 mM EDTA, 1 mM DTT), and centrifuged at 14,000 rpm for 20 min. The supernatant was diluted with 75 μl of buffer C (20% glycerol, 25 mM Hepes pH 7.6, 0.1 mM EDTA, 1 mM DTT) and stored at –70°C.

Oligonucleotides

Oligonucleotides were synthesized by the phosphoramidite method on a Cyclone automated DNA synthesizer (MilliGen-Biosearch, Burlington, MA). The synthetic oligonucleotides were purified through Pure Pack cartridge columns (Perkin-Elmer, Norwalk, CT). Complementary strands were denatured at 85°C for 10 min and annealed at room temperature overnight. The HIV-NFκB nucleotide is 5′-TCGACAAGG-GACTTTCCGCTGGGGACTTTCCAGGGC-3′. As an unrelated oligonucleotide, we used the *Sp*1 binding site 5′-GATCGGGAGGCGTGGCCTGGGACTGGGG AGTGGCGA-3′. The radiolabeled double-stranded probes were labeled with [α-^{32}P]dCTP using the Klenow fragment of DNA polymerase I.

Electrophoretic Mobility Shift Assay (EMSA)

Binding reactions were carried out in 25 mM Hepes pH 7.5, 50 mM NaCl, 10% glycerol, 0.05% Nonidet P-40, 1 mM DTT, and 2 μg poly (dI-dC) for 30 min at room temperature. A typical binding reaction (25 μl) contained 10,000 cpm (0.2–0.5 ng) of end-labeled DNA and 5 μg of nuclear protein extracts. The mixture was then electrophoresed at 11 V/cm in 1× Tris-borate pH 8 through a pre-electrophoresed 5% polyacrylamide gel. After 90 min, the gel was dried and radioactivity was detected by exposure to Kodak XAR-5 films (Eastman Kodak Company, Rochester, NY). For competition experiments the conditions were as above except that the specific and competing DNAs were included in the mixture as indicated in the Results section. The mixture was further incubated with ^{32}P-labeled probe.

Statistical Analysis

To test the null hypothesis concerning normality of distribution, the Shapiro–Wilk test was used. The obtained *w*-test did not reject the null hypothesis at the level of $p = 0.05$. Consequently, statistical analysis for multiple comparison was performed by one-way ANOVA test with Bonferroni's corrections for multiple comparisons. A p-value < 0.05 was considered significant. Statistical calculations were performed with the Statistical Analysis System.

Table 1. LTR-CAT Activity in U938 Cells Treated with TNF-α, AA or EPA

Treatment	CAT activity (% increase over control)
Control	100 ± 61
TNF-α	341 ± 51*
EPA	133 ± 64
AA	335 ± 70*

Data are means of 6 experiments ± SD.
U938 cells at a concentration of 1×10^6/ml were maintained for 24 h in RPMI 1640 containing 10% serum in presence of 80–100 U/ml TNF-α, 45 μM AA or 45 μM EPA or in absence of any stimulus. At the end of this treatment, the culture medium was removed and replaced with a serum-free RPMI 1640 and the cells were subjected to the same treatment for further 24 h. None of these treatments affected cell viability as measured by Trypan Blue exclusion.
*Significant versus control and EPA groups ($p<0.05$).

RESULTS

Table 1 shows that supplementation with AA consistently increases the constitutive LTR-CAT activity in U938, while membrane enrichment with EPA does not lead to any significant modulation of the gene construct. As in the case of EPA, membrane enrichment with either *n*–6 linoleic or *n*–3 docosaesenoic acid did not affect the HIV-1 promoter activity (data not shown). The well-known enhancing effect of TNF-α on the HIV-1 promoter was checked to validate the experimental system.

The marked upregulation reported for AA can be partly prevented by cell incubation with NDGA, a specific inhibitor of lipoxygenase-dependent metabolism of AA, or else by cell incubation with indomethacin, an active inhibitor of prostaglandin synthase-dependent metabolism of AA (Table 2).

Table 2. LTR-CAT Activity in U938 Cells: Inhibition by NDGA or Indomethacin of AA-induced Activation and Effect of Treatment with PGE_2 and 5-HETE

Treatment	CAT activity (% increase over control)
Control	100 ± 61
AA	335 ± 70*
AA+NDGA	176 ± 57
AA+indomethacin	166 ± 52
PGE_2	189 ± 35
5-HETE	182 ± 18

Data are means of 6 different experiments ± SD.
U938 cells were treated with 45 μM AA in presence or in absence of 37 μM NDGA or 0.8 μg/ml indomethacin or with the arachidonic acid metabolites PGE_2 (1 μg/ml) and 5-HETE (1 μg/ml). None of these treatments affected cell viability as measured by Trypan Blue exclusion.
*Difference statistically significant vs control, AA+NDGA and AA+indomethacin ($p<0.05$).

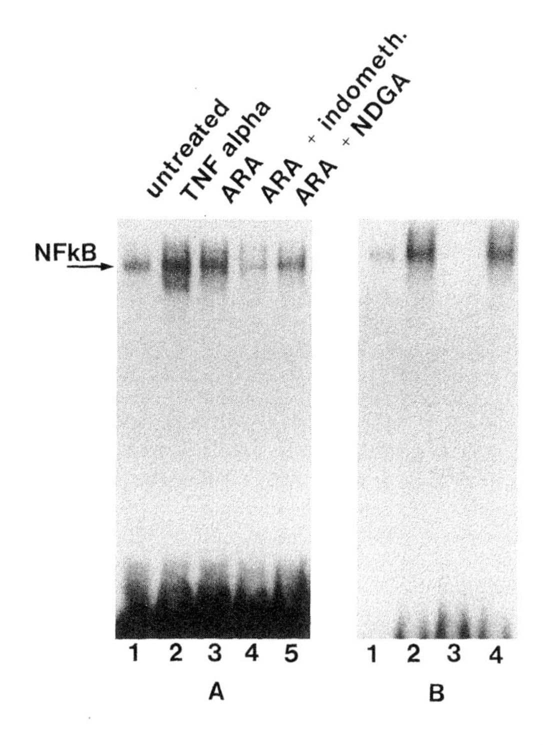

Figure 1. Effect of the treatment with ARA in presence or in absence of indomethacin or NDGA on DNA binding activity of NFκB. U938 cells were suspended in serum-free RPMI 1640 and treated for 1 h with TNF-α (80–100 U/ml), 45 μM AA, 45 μM AA plus 37 μM NDGA, or 45 μM AA plus 0.8 μg/ml indomethacin. In a different experiment, U938 cells undergoing the same incubation conditions were treated with TNF-α (80–100 U/ml), 45 μM AA or 45 μM EPA. Nuclear extracts were then subjected to EMSA. (A) Lane 1, untreated cells; Lane 2, TNF-α treated cells; Lane 3, AA-treated cells; lane 4, AA+indomethacin-treated cells; lane 5, AA + NDGA-treated cells. (B) Nuclear extracts from untreated U938 (lane 1) or ARA-treated U938 (lanes 2 to 4) were incubated with HIV-NFκB in the absence of competitor (lanes 1 and 2) or in the presence of 100-fold-excess of unlabeled HIV-NFκB (lane 3) or unlabeled unrelated fragment (lane 4).

Two intermediate compounds of the oxidative metabolism of AA, namely PGE_2 and 5-HETE were also tested for their effect on the HIV promoter. As reported in Table 2, both metabolites afforded significant stimulation of LTR-CAT activity.

Using the electrophoretic mobility shift assay (EMSA), we then analyzed the effect of AA supplementation on NFκB binding to the HIV LTR and the results of two representative experiments are shown in Figures 1A and 2. In control U938, we detected one shifted band, whose intensity was markedly increased by treatment with AA. The addition of 100-fold molar excess of unlabeled HIV-NFκB, but not of an unrelated fragment, blocked the formation of the complex (Figure 1B), thus confirming the specificity of the DNA–protein interaction. Simultaneous cell treatment with AA and NDGA or indomethacin afforded a significant down-modulation of the AA-induced level of NFκB activity (Figure 1A). As in the case of HIV LTR-CAT activity, cell membrane enrichment with EPA did not lead to any significant variation of NFκB nuclear translocation (Figure 2).

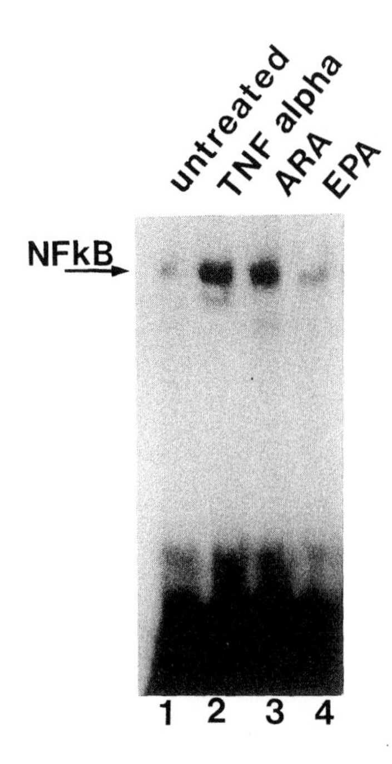

Figure 2. Lack of effect of the treatment with EPA on DNA binding activity of NFκB. U938 cells were suspended in serum-free RPMI 1640 and incubated for 1 h without treatment (lane 1) or in the presence of TNF-α (80–100 U/ml) (lane 2), 45 μM AA (lane 3), or 45 μM EPA (lane 4). Nuclear extracts were then subjected to EMSA.

DISCUSSION

The *n*-6 fatty acid AA appears able to upregulate HIV-1 expression by favoring its promoter activity. In contrast, supplementation of promonocytic cells with the *n*–3 fatty acid EPA does not lead to any significant modulation of HIV-1 LTR. Even if a direct gene effect of AA cannot be excluded, the prevention exerted by indomethacin and NDGA point to AA metabolites in the activation of the virus promoter. The fact that both PGE_2 and 5-HETE stimulate LTR-CAT activity indicates that prostaglandin synthase and 5-lipoxygenase are involved in the process. In principle, other arachidonate products can also exert modulation of HIV-1 promoter. Arachidonic acid metabolites have been shown to regulate gene expression in different types of cells. From the use of the same metabolic inhibitors, but on smooth-muscle cells and 3T3 fibroblasts, respectively, a role of AA metabolites in the regulation of c-*jun* (10) and c-*fos* gene transcription (11,12) has been suggested. Further, PGE_2 was shown to stimulate c-*fos* mRNA in 3T3 cells (12) and 12(*R*)-hydroxyeicosatetrenoic acid was shown to be involved in the expression of c-*fos*, c-*jun*, and c-*myc* in microvessel endothelial cells (13). However, the molecular mechanisms involved in gene regulation by AA metabolites are still largely speculative: the reported effect of PGE_2 might be mediated by protein kinase C activation (12). In the case of 12(*R*)-hydroxyeicosatetrenoic acid, an increased nuclear translocation of NFκB has been demonstrated (13).

The HIV-1 LTR contains two binding sites for NFκB that are required for HIV gene expression in activated T cells and mature monocytes (14–16).

The present research shows that AA, but not EPA, supplementation increases the binding to the HIV-NFκB site, suggesting that NFκB factor can be important in the modulation of HIV expression by AA.

The observed effect of AA on HIV LTR upregulation can be achieved either directly or through the involvement of TNF-α. It is known that TNF-α expression depends on the activation of NFκB (15) and we have shown that this transcription factor can be activated by AA. Hence, the possibility exists that the fatty acid can induce the autocrine production of TNF-α, which can in turn stimulate the HIV LTR. On the other hand, AA could induce HIV LTR activation by directly inducing NFκB nuclear translocation.

In any case, arachidonate appears to be the initial molecule of a signal transduction pathway, which probably plays a primary role in HIV-1 expression through the activity of defined eicosanoids and NFκB upregulation.

This conclusion is supported by the observed inhibition of NFκB nuclear translocation by inhibitors of the eicosanoid cascade, namely indomethacin, NDGA (this paper), and aspirin (17). In addition, one of these nonsteroidal anti-inflammatory drugs, i.e., indomethacin, was proved to exert strong inhibiting activity on HIV replication in MT-4 lymphocytes (18), and the anti-HIV drug Avarol confirms that inhibitors of cyclooxygenase and lipoxygenase could be leading compounds in the therapy of the disease (19).

But the most suggestive evidence for a critical role of AA in the maintenance and activation of HIV infection is the demonstration that gp120 protein of the viral envelope can induce the production of arachidonic acid metabolites in human monocytes (20) but for some reason not in T cells (21). Permissiveness to HIV replication in T lymphocytes and monocytic cells, though, differs considerably. We believe that AA-induced activation of NFκB and the pre-eminent role of macrophages in inflammation explain, at least in part, the self-perpetuation that the virus is able to activate. The membrane ARA utilization by the virus probably results in both an autocrine and a paracrine loop. The activation of AA metabolism by a noxious stimulus could stimulate HIV-1 replication from latency, and the virus itself, by inducing a further production of ARA metabolites, could perpetuate its replication. A paracrine loop could also be hypothesized since AA cleaved from the membrane is not all metabolized but can permeate and affect other cells in the proximity (2).

Studies employing the whole virus HIV-1 system are under way to confirm the suggested relationship between AA-induced activation of HIV LTR and actual HIV-1 replication.

In conclusion, the possible regulation of gene expression through the modulation of the *n*–6/*n*–3 unsaturated fatty acid ratio in cell membranes is important for its general implication. The present report strongly suggests a careful definition of the dietary supply in terms of *n*–3 and *n*–6 polyunsaturated fatty acids, especially to HIV-infected nonsymptomatic individuals. Also, inhibitors of the prostaglandin cascade could contribute in delaying the progression of HIV infection. Overall, the regulation of NFκB nuclear translocation by polyunsaturated fatty acids can markedly influence the expression and synthesis of a variety of growth factors, in this way interfering with their role in inflammation, tissue remodeling, and proliferation.

ACKNOWLEDGMENTS

This work was supported by grants from the Italian Ministry for University, Scientific and Technological Research, from the Italian Ministry of Health VI AIDS Project, from the National Research Council Targeted Project "ACRO," and from the Italian Association for Cancer Research.

REFERENCES

1. Sumida C, Graber R, Nunez E. Role of fatty acids in signal transduction: modulators and messengers. Prostaglandins Leukotrienes Essential Fatty Acids 1993; 48:117–122.
2. Marcus AJ, Hajjar DA. Vascular transcellular signaling. J Lipid Res 1993; 34:2017–2031.
3. Rao GN, Bass AS, Glasgow WC, Eling TE, Runge MS, Alexander RW. Activation of mitogen-activated protein kinases by arachidonic acid and its metabolites in vascular smooth muscle cells. J Biol Chem 1994; 269:32586–32591.
4. Johnston PV, Marshall LA. Dietary fat prostaglandins and immune response. Prog Food Nutr Sci 1984; 8:3–25.
5. Bjorneboe A, Soyland E, Bjorneboe GEA, Rajka G, Drevon CA. Effect of dietary supplementation with eicosapentaenoic acid in the treatment of atopic dermatitis. Br J Dermatol 1987; 117:463–470.
6. Meydani SN, Lichtenstein AH, Cornwall S, et al. Immunologic effects of national cholesterol education panel step-2 diets with and without fish-derived *n*-3 fatty acid enrichment. J Clin Invest 1993; 92:105–113.
7. National Cholesterol Education Program Expert Panel. Detection, evaluation and treatment of high blood cholesterol in adults. Arch Intern Med 1988; 148:36–39.
8. Meydani SN, Lichstein A, White PJ, et al. Food uses and health effect of soybean and sunflower oil. J Am Coll Nutr 1991; 10:406–428.
9. Latham PS, Lewis AN, Varesio L, et al. Expression of immunodeficiency virus long terminal repeat in the human promonocyte cell line U937: effect of endotoxin and cytokines. Cell Immunol 1990; 129:513–518.
10. Rao GN, Lassegue B, Griendling KK, Alexander RW. Hydrogen peroxide stimulates transcription of c-*jun* in vascular smooth muscle cells: role of arachidonic acid. Oncogene 1993; 8:2759–2764.
11. Rao GN, Lassegue B, Griendling KK, Alexander RW, Berk BC. Hydrogen peroxide-induced c-*fos* expression is mediated by arachidonic acid release: role of protein kinase C. Nucleic Acids Res 1993; 21:1259–1263.
12. Danesch U, Weber PC, Sellmayer A. Arachidonic acid increases c-*fos* and *Egr*-1 mRNA in 3T3 fibroblasts by formation of prostaglandin E_2 and activation of protein kinase C. J Biol Chem 1994; 269:27258–27263.
13. Laniado-Schwartzman M, Lavrovsky Y, Stoltz R, et al. Activation of nuclear factor kB and oncogene expression by 12(*R*)-hydroxyeicosatetrenoic acid, an angiogenic factor in microvessel endothelial cells. J Biol Chem 1994; 269:24321–24327.
14. Duh EJ, Maury WJ, Folks TM, Fauci AS, Rabson AB. Tumor necrosis factor alpha activates human immunodeficiency virus type 1 through induction of nuclear factor binding to the NF-κB sites in the long terminal repeat. Proc Natl Acad Sci USA 1989; 86:5974–5978.
15. Grilli M, Chiu JJS, Lenardo MJ. NF-kB and Rel: participants in a multiform transcriptional regulatory system. Int Rev Cytol 1993; 143:1–62.
16. Meltzer MS, Skillman DR, Hoover DL, et al. Macrophages and the human immunodeficiency virus. Immunol Today 1990; 11:217–223.

17. Kopp E, Ghosh S. Inhibition of NF-kB by sodium salicylate and aspirin. Science 1994; 265:956–959.
18. Bourinbaiar AS, Lee-Huang S. The non-steroidal anti-inflammatory drug, indomethacin, as an inhibitor of HIV replication. FEBS Lett 1995; 360:85–88.
19. Schoroeder HC, Begin ME, Klocking R, et al. Avarol restores the altered prostaglandin and leukotriene metabolism in monocytes infected with human immunodeficiency virus type 1. Virus Res 1991; 21:213–223.
20. Wahal LM, Corcoran ML, Pyle SW, Arthur LO, Haren-Bellen A, Farren WL. Human immunodeficiency virus glycoprotein (gp120) induction of monocyte arachidonic metabolites and interleukin 1. Proc Natl Acad Sci USA 1989; 86:621–625.
21. Kaufmann R, Laroche D, Buchner K, et al. The HIV-1 surface protein gp120 has no effect on transmembrane signal transduction in T cells. J Acquir Immune Defic Syndr 1992; 5:7

38

Protein Degradation in Lymphocytes as an Indicator of Oxidative Stress in HIV Infection

Giuseppe Piedimonte
Università di Messina, Messina, Italy

Mauro Magnani and Dario Corsi
Università di Urbino, Urbino, Italy

Javier F. Torres Roca, Denise Guetard, and Luc Montagnier
Insitut Pasteur, Paris, France

INTRODUCTION

A role for oxidative stress in the pathogenesis of AIDS is suggested by a large body of data from in vitro and biochemical–clinical studies (1–10).

Recent reports have implicated the intracellular excess of reactive oxygen species (ROS) in the induction of HIV expression (1–4) and in the initiation of apoptotic cell death (11–13). Proof of a metabolic alteration leading to decreased ability to counteract oxidative stress came from studies which showed that glutathione is decreased in peripheral blood mononuclear cells from symptom-free individuals (10). Other studies showed alterations of biochemical indicators of systemic oxidative damage (7–9), thus suggesting an increase in tissue degeneration with the progression of AIDS. These findings together give a rational basis for antioxidant treatments to reduce viral multiplication and apoptotic cell death (14,15). However, the origin of these cellular and systemic oxidative alterations and whether, and to what extent, they influence immune functions are still far from clear.

Mechanisms of oxidative molecular damage of DNA and lipids in human diseases are known (16). Another constitutive element of oxidative tissue injury is the perturbation of cell protein metabolism: oxidatively modified proteins are selectively degraded after ubiquitination by multicatalytic proteinase complexes called proteasomes (17,18). Targets of such structural modifications can include various enzyme molecules (19–21) whose oxidative inactivation may be of physiological significance as part of a regulatory system. Alterations of the homeostatic equilibrium between synthesis and oxidative inactivation of cell proteins can be involved in the pathogenesis of a variety of diseases (21).

On the basis of these premises, we thought it interesting to study whether such alterations, attributable to an altered cell redox status, can occur in protein synthetic balance during HIV infection. We studied peripheral blood lymphocytes, either resting or committed to enter the cell cycle, purified from healthy and infected asymptomatic donors. We investigated (i) cell ability to maintain a normal oxidant/antioxidant balance in different proliferative conditions and (ii) reciprocal relationships between cell redox status, protein synthetic balance, and cell growth.

LYMPHOCYTE ACTIVATION; HYDROETHYDINE OXIDATION; MnSOD EXPRESSION AND ACTIVITY

The recruitment in G1 phase of the cell cycle has been evaluated by cytometric measures of cell DNA content (22). This method is useful for rapidly screening a wide number of lymphocyte cultures for cell cycle position. However, since cells in G0 and in G1 phase have the same DNA content, this method is inadequate for recognizing activated from

Table 1

	Hours after Con A addition											
	0		5		10		16		36		48	
	Control	HIV^+	Control	HIV^+	Control	HIV^+	Control	HIV^+	Control	HIV^+	Control	HIV^+
CD3	87	78	–		–		482	401	411	415	458	421
CD25	4	4	–		–		311	317	301	298		
Proline uptake							0.21	0.34	1.1	1.3	1.2	1.2
Leucine incorporation							0.22	0.32	1.23	0.99	0.41	0.51
IL-2 production			0.3	0.2	0.5	0.7	1.9	2.1	0.2	0.15	nd	
Cell cycle												
G0 G1 (%)	98		98		98		95		90		81	
S (%)	–		–		–		–		8		12	
Thymidine incorporation	–		–		–		5.5	4.2	6.5	6.8	12.5	8.9

Lymphocytes from controls and patients are purified, cultured, and stimulated as follows: PBMCs isolated by Ficoll–Hypaque (Pharmacia LKB Biotechnology Inc., Uppsala, Sweden) were cultured in complete RPMI 1640 medium (supplemented with 10% FCS) at an initial density of 1×10^6 cells/ml. Concanavalin A was added at a final concentration of 5 μg/ml. Percentage distribution of cells in the different phases of cycle was determined daily by flow cytometry and growth rate quotient was measured. CD3 and CD25 fluorescence intensity was expressed in arbitrary units (MFI) on an X299 channel scale with 4 decades. Proline uptake (amino acid transport system A) was determined as described in Ref.34 and values are expressed as pmol/min/10^6 cells. IL-2 production was measured by ELISA and values are expressed as ng IL-2 produced by 10^6 cells. Cell cycle measurements were carried out by cytometric methods. Thymidine incorporation is expressed as cpm/10^3 cells.

resting lymphocytes. Thus, we integrated cytometry with measures of IL-2 production, surface molecule expression, and uptake of small metabolites, all sequential events typically associated to the G0/G1 transition. Results are summarized in Table 1.

Cell cultures shifted from G0 to G1 early after Con A stimulation as indicated by the appearance of a small, but well defined, production of IL-2 5 h after stimulation. IL-2 production reached a peak 16 h after stimulation and rapidly decreased at 36 h. It should be noted that, although IL-2 production was quantitatively comparable to that of controls, IL-2 biological activity associated with lymphocytes from infected individuals was considerably reduced (data not shown).

The induction of CD25 antigen expression, the increase in CD3 expression, and the stimulation of amino acid transport system A (proline), are all functional modifications of cell membrane found in the lapse of time between 16 and 36 h. The initiation of DNA synthesis (S phase of cell cycle) was observed between 36 and 48 h.

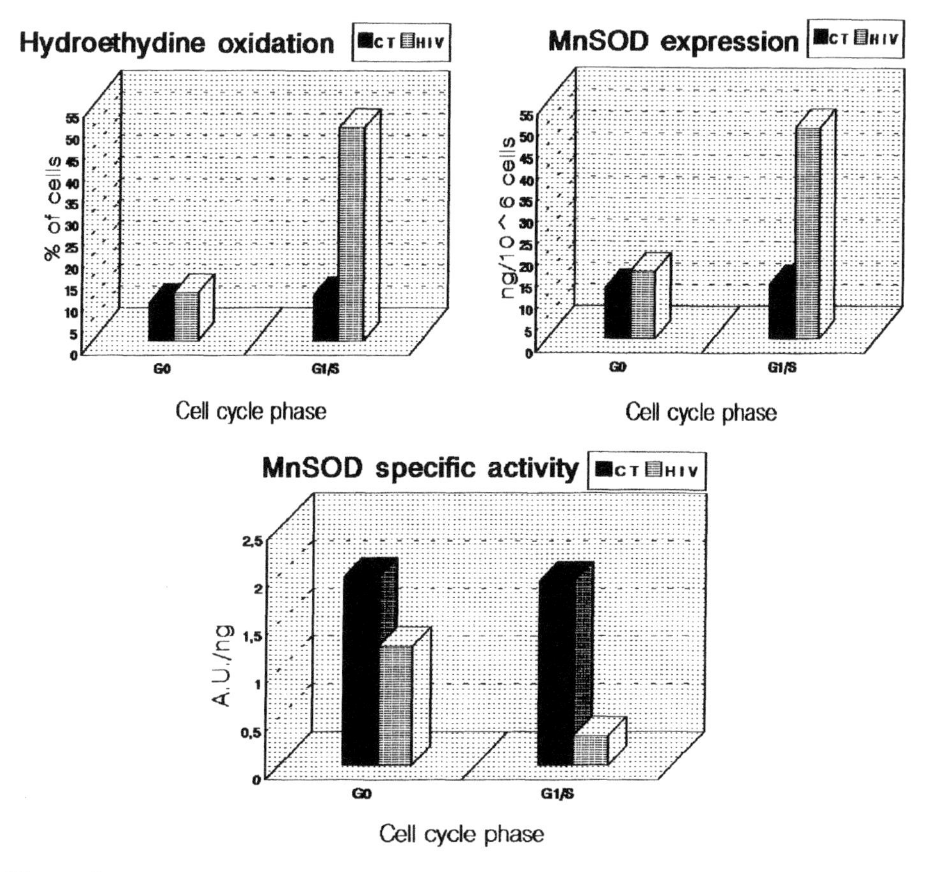

Figure 1. Intracellular concentration of oxygen metabolites (mostly O_2^-) was evaluated by measuring the level of hydroethydine oxidation in resting and activated lymphocytes by flow cytometric analysis as described in (23) and values are expressed as percentage of positive cells. MnSOD expression and activity was evaluated by methods described in Refs 24–26. and is expressed as ng/10^6 cells. MnSOD specific activity is expressed as arbitrary units (A. U.)/ng of enzyme. Experiments were carried out in cultured cells from 12 patients and 10 controls.

In lymphocytes from infected individuals cell cycle-related functional modifications are comparable to those of controls and appear at the same time during the synchronous recruitment in G1.

Cell death, as assessed by Trypan Blue exclusion, was comparable in cultures from HIV^+ up to 48 h of Con A stimulation.

In such proliferative conditions as those obtained after 36 h of Con A stimulation, we measured hydroethydine (HE) oxidation and manganese superoxide dismutase expression and activity (Figure 1). Methods for quantitation of oxygen free radical production by cells are frequently based on measurements of extracellular release of ROS. The flow cytometric method used by us is based on evaluation of intracellular hydroethydine oxidation and allows us to establish directly the amount of ROS, particularly of O_2^-, in the cytoplasm and/or in other subcellular structures (23). After activation, the percentage of cells able to oxidize HE to the red fluorescent product ethidium bromide (EB) increased in HIV^+ subjects and remained relatively constant in controls.

Manganese superoxide dismutase is an inducible enzyme whose cellular availability rises adaptively in response to the increased intracellular concentration of oxygen free radicals. Similar adaptive events are found in cells exposed to molecules such as dioxygen (24), paraquat (25), or TNF (26), whose toxicity is due to alterations of cell redox balance. In HIV^+ subjects, cell content of MnSOD increases markedly when lymphocytes enter the G1 phase of the cell cycle (15.7 ng/10^6 cells in G0 vs. 48.9 ng in G1/S phase). This adaptive response, found only in lymphocytes from HIV^+ individuals, is accompanied by a strong decrease in specific activity (from 1.24 AU/ng in G0 to 0.31 in G1/S: see Figure 1).

The contemporaneous raising of the number of cells able to oxidize hydroethydine and of the MnSOD content indicates that the intracellular oxidant/antioxidant balance shifts together with the mitogenic activation of lymphocytes from HIV-infected individuals. Furthermore, lymphocyte activation in these subjects involves parallel occurrence of induction and inactivation of this scavenger enzyme.

CELL PROLIFERATION AND INITIAL RATE OF PROTEIN SYNTHESIS, PROTEIN OXIDATION, AND DEGRADATION

It is known that an unbalance between ROS production and cellular antioxidant defenses can induce oxidative modifications of proteins with consequent increase in protein degradation (27). Protein synthesis and degradation were measured in three different conditions of cell culture:

1. Lymphocytes freshly isolated from peripheral blood were cultured in the presence of 10% fetal calf serum without any addition of mitogens. As expected, in these conditions the majority of lymphocytes are synchronized in G0 phase. The growth in these culture conditions lasts 24 h.
2. At this time concanavalin A was added to cell cultures. This mitogenic stimulus brought the majority of cells to the G1/S phase 36/48 h after addition.
3. Finally, to syncronously commit lymphocytes in G2/M phase, IL-2 was added to the cultures. After 18 h of stimulation by IL-2, the majority of cells have been recruited in G2/M and some of these (about the 30% of the total population)

have completed the mitosis as evaluated by calculating the growth rate quotient. At the end point of IL-2 stimulation, the majority of cells (about 80%) had completed mitosis as indicated by the increase (similar in normal and HIV$^+$ cells) in growth rate quotient.

Values of protein synthesis and protein half-life in these proliferative conditions are reported in Table 2.

Protein synthesis reached a peak during the G1 phase of cell cycle with 1.23 nmol leucine incorporated by 10^6 cells for 30 min in control PBL. In PBL from infected individuals this value is 0.99 nmol of leucine incorporated. As expected, an important decrease of protein synthetic activity, with absolute values of synthesis similar in HIV$^+$ and control subjects, was found during G2/M phases.

Half life of newly synthesized proteins was 59 h in controls and markedly decreased (16 h) in HIV$^+$ individuals during G1/S phase of the cell cycle. The evidence of a decrease

Table 2. Cell Cycle and Protein Synthesis, Oxidation, and Degradation in Peripheral Blood Lymphocytes from Control and HIV-infected Individuals

	Cell cycle phase					
	G0		G1/S		G2/M	
	Control	HIV$^+$	Control	HIV$^+$	Control	HIV$^+$
Protein new synthesis (nmol leu/10^6 cells in 30 min)	ND	ND	1.23 ±0.139	0.99 ±0.087	0.231 ±0.035	0.287 ±0.025
Half-life ($t_{1/2}$) of newly synthesized proteins (h)	ND	ND	59±2.1	16±3.5	61±5.8	18±2.1
Protein carbonyl content (nmol/mg protein)	0.24	0.31	0.27	2.91	0.31	3.41

Protein synthesis was measured in initial velocity by [^{3}H]leucine incorporation (2 μCi/ml of RPMI 1640, 10% FCS) in TCA-precipitable fractions of cultured PBMCs. Briefly, at the peak of protein synthesis, aliquots of 0.5×10^6 cells washed three times with complete RPMI are incubated for 30 min in fresh medium (1×10^6 cells/ml) suplemented with labeled leucine. In these cultures, at least 85% of cells from both normal and HIV$^+$ donors, are in G1 phase of the cell cycle as documented by cytometric analysis. At the end of incubation, the pellets obtained by centrifuging at 800*g* for 10 min and washed three times with cold buffer (PBS), were treated with 5% TCA (20 min; 4°C; 0.5×10^6 cells/ml solution). Leucine incorporation in cell proteins was calculated by measuring the radioactivity associated with the TCA-precipitable fraction obtained by centrifuging (12,000*g*, 10 min) TCA-treated cells. Values are expressed as nmol Leu incorporated in the TCA-precipitable fraction of 10^6 cells in 30 min. The half-life of newly synthesized proteins is determined as follows: (1) the cells are labeled with [^{3}H]leucine as described above. (2) Labeled cells are washed (three times) and incubated at a density of 1×10^6/ml in fresh medium supplemented with Con A in the absence of labeled Leu. (3) Aliquots of 1×10^6 cells are collected every 10 h for 4 days and radioactivity still linked to the TCA-precipitable fraction is determined (4) Values, expressed as nmol leucine still associated to TCA-precipitable fraction, are plotted against the time spent by the cells in the fresh medium. The equation which best describes the trend of the decrease of radioactivity associated with the TCA-precipitable fraction is a linear regression. Based on the parameters which define this equation, values of the half-lines of newly synthesized proteins are established for each HIV$^+$ and normal individual under study. Carbonyl content in TCA-precipitable fractions was determined by the method described in Ref.21. ND = not determined in protein half-life with a parallel increase in ROS production suggested the rationale for a series of measures of protein carbonyl content, whose increase normally reveals oxidative alterations to proteins (27), in resting and activated lymphocytes from normal and infected individuals.

Starting from similar values found in resting cells (0.24 and 0.31 nmol/mg protein for controls and infected cells, respectively), protein carbonyl content reached values 12–14 times higher in HIV^+ lymphocytes, while remaining constant in controls, during the G1/S and G2/M phases of cell cycle.

In lymphocytes from HIV^+ individuals, the increase in protein carbonyl content appears abruptly just after entry into the early G1 phase and seems to be a part of the early pleiotropic events associated with the anomalous commitment of these cells in the cycle. If a causal relationship exists between the increased carbonyl content and the increased protein turnover, these measures can be considered as reliable intracellular indicators of oxidative cell damage.

PROTEIN UBIQUITINATION: INCREASED DEGRADATION OF PROTEIN CONJUGATES IN LYMPHOCYTES FROM HIV+ INDIVIDUALS

Most proteins in eukaryotic cells are degraded by an ATP- and ubiquitin-dependent proteolytic system (18). Degradation of a protein via the ubiquitin system involves two distinct steps: the covalent conjugation of multiple ubiquitin molecules to the protein substrate, and the degradation of the targeted protein by an ATP-dependent high-

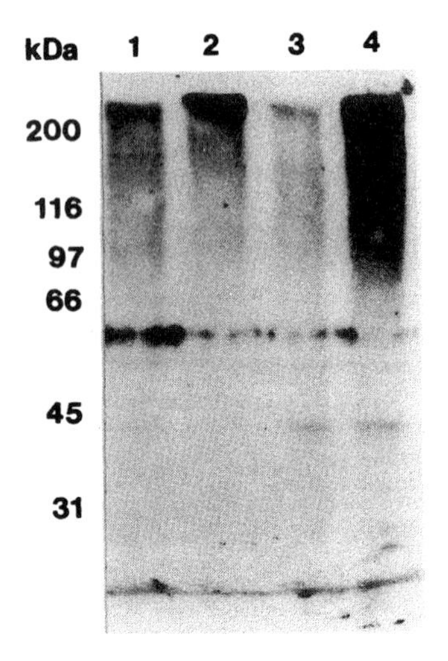

Figure 2. Total proteins from blood lymphocytes were separated by SDS-PAGE on 10% acrylamide gels and transferred by western blotting onto a nitrocellulose sheet that was probed with a rabbit anti-ubiquitin antibody. Detection was carried out with a goat anti-rabbit IgG horseradish peroxidase conjugate and ECL. Lane 1, resting cells from HIV-infected individuals; lane 2, resting cells from controls; lanes 3 and 4, cycling cells (G1 phase) from HIV-infected patients and controls, respectively. Each lane received a protein content of 1.7×10^5 cells. The experiment shown is representative of four.

molecular-mass protease (the proteasome). Having shown that proteins in activated lymphocytes from HIV-infected patients are degraded more rapidly than in controls, we next evaluated protein ubiquitination (Figure 2):

Freshly isolated lymphocytes from patients and controls show the same protein profile in SDS-PAGE (not shown) and a similar ubiquitination as detected by western blotting with a specific antiubiquitin antibody (see lanes 1 and 2 in Figure 2). When committed to enter G1 phase of cell cycle, control cells show an increase in protein ubiquitination (lane 4). The content of ubiquitinated proteins is strongly reduced in cells from HIV-infected individuals (lane 3). This is likely due to a faster proteolysis of the protein conjugates (see data on protein turnover) although other possibilities cannot be excluded.

DECREASE IN PROTEIN CELL MASS AS A RESULT OF INCREASED PROTEIN DEGRADATION

Cell protein content results from a balance between protein synthesis and degradation. Cells reach a certain size, critical for G2/M transition, by increasing their protein, DNA, and RNA content and by consequent arrangements of the intracellular water. The attainment of this critical mass occurs mostly during the G1 phase. Such a wide perturbation of the turnover of newly synthesized proteins early after G0/G1 transition can determine damage to these control mechanisms. In Figure 3 we show the sequence of modifications in lymphocyte protein mass with progression in cycle: starting from similar values, control cells constantly increased in protein content during the G1 phase, and their mass was twice that in G0, while the increase in mass was only 29% in lymphocytes frm HIV^+ individuals. During the G2 phase, protein mass increased slightly in control lymphocytes, in comparison to that of G1/S phase, and remained relatively constant in cells from HIV-infected individuals, with absolute values again lower than controls. At the endpoint of IL-2 stimulation, protein content per cell returned to the initial values and was comparable in controls and HIV^+ individuals.

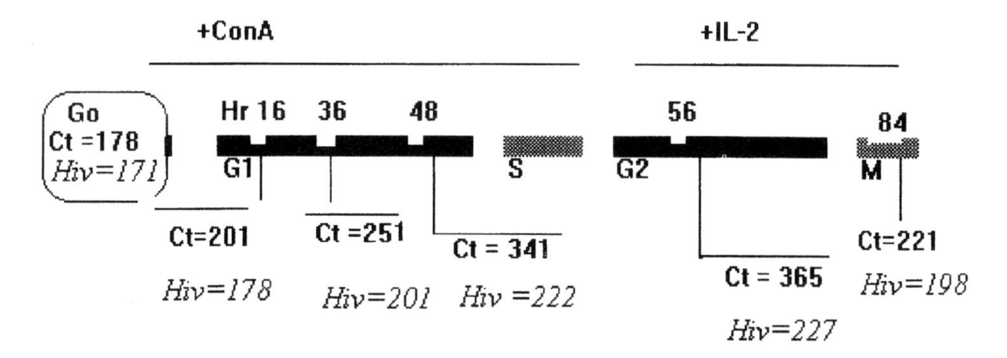

Figure 3. Cultured lymphocytes were synchronously committed in G1/S and G2/M phases of cell cycle by addition of Con A or IL-2 at the indicated times (hours). Protein cell mass (i.e., protein content per cell) was evaluated, as in Ref. 22, in 18 patients and 21 controls. Protein content per cell is expressed as $\mu g/10^6$ cells; Ct = controls; Hiv = HIV^+ individuals; Hr = hours after mitogen addition.

CONCLUSIONS

Clearly, in lymphocytes from HIV-infected patients the discrepancy between proliferative activity (similar to that of controls) and cell protein mass (decreased) seems to be the final outcome of oxidative damage of proteins leading to an inappropriate activation of proteolytic systems. Normally, the progression in the cycle, particularly the G1/S and G2/M transitions, requires a programmed degradation of specific regulatory proteins. For example, cyclins are degraded at critical checkpoints by an ubiquitin–proteasome system (28). Since both cycling and oxidatively damaged proteins are specific targets of the ubiquitin–proteasome system, it could be speculated that the activation of this enzyme pathway can lead to a nonspecific degradation of cyclins at inappropriate times during the cell cycle.

Cells enter a definite phase of cell cycle only after completion of the metabolic processes corresponding to the previous stage (29). Alterations of the temporal order of these sequential events can determine an increasing incidence of lethal mitosis and/or activation of cellular programs of apoptotic death (30,31). For example, the dependence of M on S phase is strictly controlled by the balance between synthesis and ubiquitin-dependent degradation of $p34^{cdc2}$ kinase–B-cyclin complex (32), whose premature activation during the cell cycle seems to be a prerequisite for apoptosis (33).

In mammalian cells the attainment of a critical mass is required for transition from G1 to S phase and is the way in which modulation of protein synthetic activity and formation of cell machinery for cell division and DNA synthesis are coordinated. One can reasonably suppose that loss of ability to reach a critical protein content at the end of G1 phase results in a loss of coordination between protein synthetic activity in G1 and initiation of DNA synthesis in S phase. Thus, in HIV infection, oxidatively dependent degradation of newly synthesized proteins, caused by T cell activation, leads to cells that are smaller than normal at inappropriate, and perhaps lethal, mitosis.

REFERENCES

1. Schreck R, Rieber P, Baeuerle, PA. Reactive oxygen intermediates as apparently widely used messengers in the activation on NF-kB transcription factor and HIV-1. EMBO J 1991; 10:2247–2258.
2. Legrand-Poels S, Vaira D, Pincemail J, Van de Vorst A, Piette J. Activation of human immunodeficiency virus type I by oxidative stress. AIDS Res Hum Retrovir 1990; 6:1389–1397.
3. Kalebic T, Kinter A, Poli G, Anderson, ME, Meister, A, Fauci AS. Suppression of human immunodeficiency virus expression in chronically infected monocytic cells by glutathione, glutathione ester and *N*-acetylcysteine. Proc Natl Acad Sci USA. 1991; 88:986–990.
4. Roederer M, Staal FJT, Raju PA, Ela SW, Herzenberg LA. Cytokine-stimulated human immunodeficiency virus replication is inhibited by *N*-acetyl-L-cysteine. Proc Natl Acad Sci USA 1990; 87:4884–4888.
5. Baruchel S, Gao Q, Wainberg MA. Desferrioxamine and HIV. Lancet 1991; 337:1356.
6. Flores SC, Marecki JC, Haroer KP, Bose SK, Nelson SK, McCord JM. Tat protein of human immunodeficiency virus type I represses expression of manganese superoxide dismutase in HeLa cells. Proc Natl Acad Sci USA 1993; 90:7632–7636
7. Cirelli A, Ciardi M, De Simone C, et al. Serum selenium concentration and disease progress in patients with HIV infection. Clin Biochem 1991; 24:211–214.

8. Leff JA, Oppegard MA, Curiel TJ, Brown KS, Schooley RT, Repine JE. Progressive increase in serum catalase activity in advancing human immunodeficiency virus infection. *Free Radical Biol Med* 1992; 13:143–149.
9. Revillard PJ, Vincent CMA, Favier AE, Richard MJ, Zittoun M, Kazatchkine MDJ. Lipid peroxidation in human immunodeficiency virus infection. J Acquir Immune Defic Syndr 1992; 5:637–638.
10. Staal FJT, Ela SW, Roederer M, Anderson MT, Herzenberg LA, Herzenberg LA. Glutathione deficiency and human immunodeficiency virus infection. Lancet 1992; 339:909–912.
11. Levine A, Tenhaken R, Dixon R, Lamb C. H_2O_2 from the oxidative burst orchestrates the plant hypersensitive disease resistance response. Cell 1994; 79:583–593.
12. Hockenbery DM, Oltvai N, Yin X M, Milliman, C L, Korsmeyer S J. 1993. Bcl-2 functions in an antioxidant pathway to prevent apoptosis. Cell 75:241–251.
13. Veis DJ, Sorenson CM, Shutter JR, Korsmeyer SJ. Bcl-2 deficient mice demonstrate fulminant lymphoid apoptosis, polycystic kidneys, and hypopigmented hair. Cell 1993; 75:229–240
14. Buttke TM, Sandstrom PA. Oxidative stress as a mediator of apoptosis. Immunol Today 1994; 15:7–10.
15. Droge W, Eck HP, Mihm S. HIV-induced cysteine deficiency and T-cell dysfunction: a rationale for treatment with *N*-acetylcysteine. Immunol Today 1992; 214:211–213
16. Halliwell B, Gutteridge JMC. Role of free radicals and catalytic metal ions in human disease: an overview. Methods Enzymol 1990; 186:1–85.
17. Jentsch S, Schlenker S. Selective protein degradation: a journey's end within the proteasome. Cell 1995; 82:881–884
18. Goldberg AL. Functions of the proteasome: the lysis at the end of the tunnel. Science 268; 1995:522–523
19. Salo DC, Pacifici RE, Lin SW, Giulivi C, Davies KJA. Superoxide dismutase undergoes proteolysis and fragmentation following oxidative modification and inactivation. J Biol Chem 1990; 265:11919–11927.
20. Dean RT, Thomas SM, Garner A. Free-radical-mediated fragmentation of monoamine oxidase in the mitochondrial membrane. Biochem J. 1986; 240:489–494.
21. Oliver CN, Starke-Reed PE, Stadtman ER, Liu GJ, Carney JM, Floyd RA. Oxidative damage to brain proteins, loss of glutamine synthetase activity, and production of free radicals during ischemia/reperfusion-induced injury to gerbil brain. Proc Natl Acad Sci USA 1990; 87:5144–5147.
22. Di Rienzo AM, Petronini PG, Guetard D, et al. Modulation of cell growth and host protein synthesis during HIV infection. J Acquir Immune Defic Syndr 1992; 5:921–929.
23. Rothe G, Valet G. Flow cytometric analysis of respiratory burst activity in phagocytes with hydroethidine and 2′,7′-dichlorofluoresein. J Leukocyte Biol 1990; 47:440–448.
24. Gregory EM, Fridovich I. The induction of superoxide dismutase by molecular oxygen. J Bacteriol 1973; 114:543–548.
25. Bagley AC, Krall J, Lynch RE. Superoxide mediates the toxicity of paraquat for chinese hamster ovary cells. Proc Natl Acad Sci USA 1986; 83:3189–3193.
26. Wong GHW, Goeddel DV. Induction of manganous superoxide dismutase by tumor necrosis factor: possible protective mechanism. Science 1988 242:941–944.
27. Davies KJA. Protein damage and degradation by oxygen radicals. J Biol Chem 1987; 262:9895–9901.
28. Murray A. Cyclin ubiquitination: the destructive end of mitosis. Cell 1995; 81:149–152.
29. Nurse P. Ordering S phase and M phase in the cell cycle. Cell 1994; 79:547–550.
30. Li JJ, Deshaies RJ. Exercising self-restraint: discouraging illicit acts of S and M in eukaryotes. Cell 1993; 74:223–226.
31. Heichman KA, Roberts JM. Rules to replicate by. Cell 1994; 79:557–562
32. Hayles J, Fisher H, Woollard A, Nurse P. Temporal order of S phase and mitosis in fission yeast is determined by the state of the $p34^{cdc2}$ – mitotic B cyclin complex. Cell 1994; 78:813–822.

33. Shi L, Nishioka WK, Th'ng J, Bradbury EM, Litchfield DW, Greenberg AH. Premature $P34^{cdc2}$ activation required for apoptosis. Science 1994; 263:1143–1145.
34. Dall'Asta V, Gazzola GC, Franchi-Gazzola R, Bussolati O, Longo N, Guidotti G. Pathways of L-glutamic acid transport in cultured human fibroblasts. J Biol Chem 1983; 258:6371–6

39

Oxidative Stress and AIDS: One-year Supplementation of HIV-positive Patients with Selenium or β-Carotene

C. Sergeant, C. Hamon, and M. Simonoff
CNRS-URA 451, Gradignan, France

J. Constans and C. Conri
Hôpital Saint-André, Bordeaux, France

E. Peuchant, M. C. Delmas, and M. Clerc
Université de Bordeaux II, Bordeaux, France

J. L. Pellegrin and B. Leng
Hôpital Haut-Lévêque, Pessac, France

I. Pellegrin and H. Fleury
Université de Bordeaux II, Bordeaux, France

INTRODUCTION

Human immunodefiency virus (HIV) infection is characterized by a progressive decrease of lymphocytes T CD4 level. Several hypotheses have been proposed to explain this decrease:

- Autoimmune mechanisms (1,2).
- More active viral replication in lymphoid nodes than in blood circulation (1), favored by early dysregulation of the secretion of cytokines by monocellular cells (3).
- Programmed cell death (apoptosis) favored by cofactors such as mycoplasma or oxidative stress (4).
- Dysregulation of the oxidant–antioxidant balance evoked as a promoter of athergenesis (5–7).

Oxidative stress may be evaluated in vivo in several ways but particularly from the endogenous antioxidant systems: glutathione, enzymes (superoxide dismutase, catalase, selenodependent glutathione peroxidase), vitamins A, E, and C, or as the result of lipid peroxidation such as production of malonic dialdehyde.

In our attempt to reach a better understanding of why premature vascular atherosclerosis occurs in HIV-seropositive persons (8,9), we examined the factors relating oxidant–antioxidant balance and plasma lipids.

DETERMINATION OF LIPIDIC PARAMETERS AND ANTIOXIDANT STATUS

In a first study on about 100 HIV patients we examined several biological parameters in plasma with respect to the level of CD4 T lymphocytes. The results obtained by our group (10–15) are presented in Table 1, and can be summarized as follows. A decrease in the CD4 count (>400 to less than 50, dividing our patients in four groups I, II, III, IV) is correlated with (i) an increase in TNF-α, triglycerides, lipoprotein (a) (Lp(a)), and plasma malonic dialdehyde; (ii) decrease in high-density lipoprotein cholesterol, retinol, and

Table 1. Lipids in HIV-positive Patients, Expressed as Mean (Standard Deviation)

Group	I	II	III	IV	Healthy
CD4/mm^3	<50	50–200	200–400	>400	
Patients (n)	32	25	22	15	20
TG (mmol/L)	1.97 (1.27)	1.32 (0.71)	1.56 (0.84)	1.42 (1.0)	1.0 (0.35)
p^a	0.002	NS	0.003	NS	
Ch (mmol/L)	4.0 (0.93)	4.6 (0.9)	4.9 (1.0)	5.2 (1.3)	5.2 (0.66)
p^a	0.002	0.01	NS	NS	
HDL-C (mmol/L)	0.75 (0.24)	1.12 (0.37)	1.01 (0.3)	1.09 (0.46)	1.38 (0.31)
p^a	0.0001	0.016	0.0001	0.05	
LDL-C (mmol/L)	2.79 (1.01)	2.89 (0.76)	3.15	3.52 (1.1)	3.73 (0.66)
p^a	0.0001	0.0001	(0.91)0.02	NS	
apoA1 (mmol/L)	1.05 (0.23)	1.38 (0.30)	1.36 (0.35)	1.39 (0.31)	1.85 (0.34)
p^a	0.0001	0.0001	0.0001	0.0001	
apoB (mmol/L)	1.13 (0.27)	1.0 (0.27)	1.14 (0.34)	1.17 (0.31)	1.05 (0.14)
p^a	NS	NS	NS	NS	
Lp(a)[b] (g/L)	0.26 (2.55)	0.32 (0.76)	0.33 (1.96)	0.27 (1.16)	0.20 (0.39)
p^a	0.04	0.02	0.03	NS	
p-PUFA (%)	9 (2)	11 (3)	10 (3)	11 (3)	11 (3)
p^a	0.05	NS	NS	NS	
TNF-α (pg/L)	45 (33)	36 (22)	36 (25)	19 (16)	5 (5)
p^a	0.0001	0.0001	0.0001	0.0001	
p-Retinol (mg/L)	0.26 (0.15)	0.28 (0.16)	0.40 (0.25)	0.29 (0.17)	0.51 (0.16)
p^a	0.0001	0.0001	NS	0.0001	
p-α-Toco (mg/L)	7.5 (3.1)	7.5 (4.4)	9.7 (6.4)	8.8 (5.2)	8.75 (3.0)
p^a	NS	NS	NS	NS	
Se (μg/L)	55 (14)	66 (17)	70 (16)	72 (9)	83 (17)
p^a	0.001	0.001	0.001	NS	

NS = nonsignificant difference; TG = triglycerides; Ch = cholesterol; HDL-C and LDL-C = HDL and LDL-cholesterol; p-PUFA = plasma polyunsaturated fatty acids; α-Toco = α-tocopherol. [a]Statistically significant compared to healthy seronegative (Student's or Mann–Whitney test). [b]Median value.

selenium. It appears, then, that an atherogenic situation for plasma lipids is observed in the presence of oxidative stress, which is accentuated as CD4 counts decrease.

Plasma Lipids

Triglycerides

Hypertriglyceridemia has been observed by others in HIV infection (16). Nevertheless, it is neither the only, nor the most frequent of the lipidic anomalies observed in these patients. These anomalies are able to influence the evolution of the disease by interacting with the immune system: lipids of the lymphocytic membranes can be modified and cytokine production stimulated.

These lipidic anomalies could also favor the replication and the cytopathogenic effect of the virus. Moreover, these patients with lipidic profile of atherogenic type could develop an accelerated atherosclerosis. A hypertriglyceridemia is present for 30% of the HIV-seropositive patients with a CD4 T lymphocyte level above 50/mm^3 and for 50–60% of symptomatic patients or patients with a very low CD4 T lymphocyte count (14,17). Opportunistic infections seem to be correlated with the triglyceride (TG) increase. Hypertriglyceridemia is connected with an excess of the peripheral fat degradation and with an increased VLDL hepatic synthesis (17). Increase of VLDL hepatic synthesis and inhibition lipoprotein lipase by TNF-α does not play any role in HIV-seropositive hypertriglyceridemia. On the other hand, interferon-α (IFN-α), closely correlated with triglycerides (14,16,17), might be decisive in their increase in hepatic synthesis and peripheral degradation (17,18). IFN-α is detectable in seropositive HIV patients' plasma only in a late phase of the infection when hypertriglyceridemia is frequent (14).

Plasma triglycerides have been used in a study as markers for therapeutic effectiveness in seropositive HIV patients (19).

Decrease in HDL Cholesterol and in Apolipoprotein A1

A decrease in HDL cholesterol and in apolipoprotein A1 occurs in the majority of HIV-seropositive patients, even in those whose CD4 T lymphocyte counts remain higher than 400/mm^3 (20,21). On the other hand, the LDL level is only slightly lower at stages prior to advanced infection, and plasma cholesterol does not decrease until patients reveal fewer than 50 CD4 T lymphocytes/mm^3. Reduction in cholesterolemia remains moderate, whereas the HDL-C level is reduced by nearly 50% in these patients (12). The reasons for anomalies in cholesterol metabolism in this case are not yet clear. A decrease in HDL occurs in certain inflammatory illnesses (22,23), although an inflammatory syndrome is infrequent at an early stage of HIV infection. A negative correlation between cholesterolemia and TNF-α has been observed, but there is a positive correlation between cholesterolemia and certain plasma antioxidants (e.g. vitamin A and selenium) (14,15). TNF-α promotes oxidative stress, thus contributing to anomalies in cholesterol metabolism. This oxidative stress can stimulate oxidative degradation of cholesterol by the macrophages, which explains the early and preferential decease in HDL-C that is responsable for uptake of cholesterol by the hepatic receptors.

Recent work has shown that an HDL-C value less than 0.6 mmol/L, together with an apolipoprotein A1 concentration less than 1 g/L, are better predictive factors for the occurrence of opportunistic infections or death than CD4 T lymphocyte counts lower than 200/mm^3 (24). HDL-cholesterol and apolipoprotein A1 may thus represent useful progress markers.

Increase in Lipoprotein(a)

Lipoprotein(a) is synthesized by the liver, and an increase in its concentration is considered to be an independent risk factor for atherosclerosis (25). Plasma levels of Lp(a) are genetically determined and are stable with time, but Lp(a) increase has also been observed in the course of certain acute inflammatory states (26).

The reasons for Lp(a) increases as found in HIV-positive patients (11) at a relatively early stage of infection have not, as yet, been elucidated. It appears possible that profound and early anomalies in the secretion of cytokines by the cells of the immune system in the course of HIV infection may be involved (3).

Peroxidation of Fatty Acids

HIV infection is accompanied by modifications in the plasma fatty acids. Our studies of these fatty acids in 95 HIV-positive patients show that there is a decrease in the amount of polyunsaturated fatty acids and a relative increase in the amounts of saturated and monounsaturated fatty acids, as the CD4 cell count decrease. These results are confirmed by others workers (27–29) and also by evidence in these patients for an increase in malonic dialdehyde plasma levels (14,29,30). On the whole, the values obtained suggest that a lipid peroxidation process exists in the course of HIV infection.

Consequences of Lipidic Anomalies

Lipid anomalies involve effects at various levels, particularly with respect to the virus itself, the lymphocytes, and the state of atherosclerosis.

The Virus. Lipid perturbations may produce a highly deleterious effect on the course of HIV infection. Apolipoprotein A1 inhibits in vitro the formation of syncitia by the HIV virus (31). A decrease in apolipoprotein A1 may favor the change of a nonsyncitializing phenotype to a phenotype that syncitializes the virus. This phenotype transformation is associated with very poor prognosis (32).

Lymphocytes. Modifications in membrane fatty acids have an inevitable repercussion on the cells, particularly in the case of the CD4 T lymphocytes. The latter have a very low capacity to incorporate cystein, which is a precursor of intracellular glutathione (33). Moreover, they have low concentrations of glutathione, which is an intracellular antioxidant that is synthesized from plasmatic cysteine. Accordingly, the CD4 cells are more sensitive than the other cells with respect to modifications connected with an oxidative stress that is likely to be influential in producing apoptosis of the CD4 T lymphocytes (34,35).

Atherosclerosis. Evidence for the development of atherosclerotic lesions in young HIV-positive patients who were not exposed to known risk factors has been presented (8,9,36). Among the possible mechanisms involved, it seems possible that disturbances in lipid metabolism may play an important part. It is known that HIV infection has a mean period of development (incubation period) of the order of 10 years before the appearance of symptoms. Throughout this period, the patients are exposed to metabolic anomalies that are potentially atherogenic, namely, a decrease in HDL cholesterol with normal cholesterolemia up until a late stage of the infection, an increase in Lp(a) and in triglycerides, saturation of the fatty acids, a decrease in certain natural antioxidants such as cysteine, selenium, glutathione, and vitamin A, and an increase in TNF-α (10, 37). Various workers have shown that HIV-positive patients reveal the presence of activation markers or indications of endothelial agression (thrombomodulin, Von Willebrand factor)

(38, 39). These factors produce in the endothelium a coagulant phenotype that may favor thrombosis.

Oxidative Stress

Our results provide evidence for a decrease in plasma levels of vitamin A and selenium, accompanied by an increase in MDA that indicates the presence of oxidative stress associated with the HIV infection.

Vitamin A

Levels of antioxidant plasma vitamin A, which inhibits the production of singlet oxygen ($^1O_2$), are decreased in HIV-positive patients (15,40). It has been shown by various authors that there exists a relation between vitamin A deficiency and an increased mortality rate. We have observed that the vitamin A level in 95 HIV patients was low even when the CD4 T lymphocyte count was higher than $400/mm^3$. This suggests that a decrease in vitamin A is associated with aggravation of the state of immune deficiency.

Vitamins E and C

It has not been established whether there is a clear diminution of these vitamins in the serum of HIV-positive patients (15,40). Tang (41) showed that administration of high doses of vitamin C seem to be associated with a decrease (with a slight degree of significance) of the risk of progression of the illness.

Selenium

Selenium is an indispensable cofactor in the action of cellular glutathione peroxidase. Plasma selenium levels have been found to be low in the course of HIV infection (14,15,42,43). We have established that there is a significant decrease of this element in subjects whose CD4 T lymphocyte count is less than $400/mm^3$ and that there is a correlation with the CD4 level. Moreover, assessment of the selenium level is as useful as that of CD4 as a means of prognosis for the appearance of opportunistic infections or life expectancy (44) at one year in patients with advanced HIV infection.

Lowering of Antioxidant Defense Mechanisms

There exists a quantitative diminution of natural antioxidant systems in the course of HIV infection. This may be due to increased consumption as a result of excessive production of free radicals (45). However, it is possible that nutritional factors are also responsible, as in the case of selenium and vitamin A.

ORAL SUPPLEMENTATION: SELENIUM OR β-CAROTENE

On the basis of these findings and in view of evidence for a high degree of oxidative stress in plasma of HIV seropositive persons, we next studied the effects on erythrocytes of supplementation with antioxidant in the form of selenium or β-carotene. Selenium was administered in the form of selenomethionine, 100 μg/day, and β-carotene, in natural vegetal form, was given at a dose of 60 mg/day. This supplementation was continued for one year with HIV patients at a highly advanced stage of illness, whose CD4 mean count was equal to or less than 200/mm^3. Groups that received Selenium and β-carotene supplements were compared with a control group that did not.

Patients

The Bordeaux Ethics Comitee approved this study and informed consent was obtained from the patients. Criteria for the selection of patients for this group study were highly restrictive and were based on CD4 values, age, sex, body weight, and similarity in retroviral therapy.

Fifty-two HIV-infected patients with a CD4 cell count <400/mm^3 were selected for this study and divided into three groups with the same CD4 cell count means at time M0. Most patients were under AZT antiviral treatment, while others were under DDI, D4T, or associated treatments. A control group, without any supplementation, included 22 patients. The selenium group (Se) included 15 patients who took an oral supplement with 250 μg L-selenomethionine (100 μg selenium) per day over one year. The β-carotene group (BC) comprised 15 patients supplemented with 60 mg β-carotene (100,000 IU provitamin A) per day over 1 year.

At the end of the follow-up (M12), 4 control, 4 Se, and 2 BC patients had died. Opportunistic infections had occurred in 3 control, 4 Se, and 7 BC patients.

The results of the patients staying alive at M12 and with complete follow-up are presented: 35 patients comprise 14 CTRL, 11 Se, and 10 BC. Their characteristics are presented in Table 2.

Experimental Techniques

Measurements were carried out on various biological parameters that are generally affected by a disturbance of the oxidant–antioxidant equilibrium. These parameters concerned vitamins, trace elements, fatty acids, enzymes, and lipid peroxidation products and were examined at the beginning of supplementation (month zero:M0) and at months six (M6) and twelve (M12) with respect to plasma, hemolysate, and erythrocyte membranes.

Clinical and immunological status were evaluated from CD4 and CD8 lymphocyte counts by flow cytometry on a FacScan (Becton Dickinson, San Jose, CA), β_2-microglobulin by nephelemetry, and p_{24} antigenemia by an ELISA assay (Abbott). Apoproteins (apo) A1 and B and Lp(a) were determined by nephelometry using a BNA instrument (Behring, France). Nutritional parameters were determined by weight; albuminemia and proteins were measured on a multiparameter automated system (Baxter, Paris).

Malonic dialdehyde (MDA) and liposoluble vitamins (A and E) were determined on erythrocyte membranes by high-performance liquid chromatography (HPLC) after

Table 2. Characteristics of the Patients (age and CD4 values are expressed as mean ± standard deviation)

	Control (*n*=14)	Se (*n*=11)	BC (*n*=10)
Age (years)	35 ± 6	35 ± 6	34 ± 7
Sex M/F	9/5	10/1	8/2
CD4 (/mm^3) at M0	136 ± 105	180 ± 83	136 ± 107

chemical separations (46–48). Selenium, copper, and zinc in plasma were measured by proton-induced X-ray emission (PIXE) after chemical pretreatments (49–50). Zinc in erythrocyte membranes was also determined.

The Bradford method (51) was used for erythrocyte membrane protein determination. Erythrocyte membrane fatty acids were extracted, separated, and quantified by gas chromatography (52). The antioxidant enzyme glutathione peroxidase (GPx) was measured by Randox kit method (Randox, Grumlin, U.K.) in red blood cells. Statistical analyses were performed using nonparametric Mann–Whitney and Wilcoxon tests.

Results and Discussion

Evolution of Plasma Trace Element Levels and Hemolysate Glutathione Peroxidase

Selenium. Plasma selenium at a concentration of 50–60 μg/L in the three groups at M0 was significantly lower than in the healthy subjects ($p < 0.01$); it did not change with time in the control and β-carotene-supplemented groups. On the other hand, the Se-supplemented group showed a significant rise in selenemia, especially at M6 corresponding to satisfactory intestinal absorption and biodisposal. This continued up to M12 (Figure 1).

Erythrocyte selenodependent GPx was higher at M0 in the control and Se-groups than in healthy reference persons. The question arises whether there exists a mobilization of selenium toward this enzyme at the expense of plasma selenium. Selenium supplementation does not cause a significant increase in the activity of selenodependent glutathione peroxidase, even though we observed a tendency to increase with time in the two supplemented groups and a tendency to decrease in the control group (Figure 2).

Biological functions of selenium are mediated by selenoproteins, most containing the selenocysteine. So far, about 30 proteins have been identified but only 8 have been characterized (53–55). Among them, the biological functions are partially known for

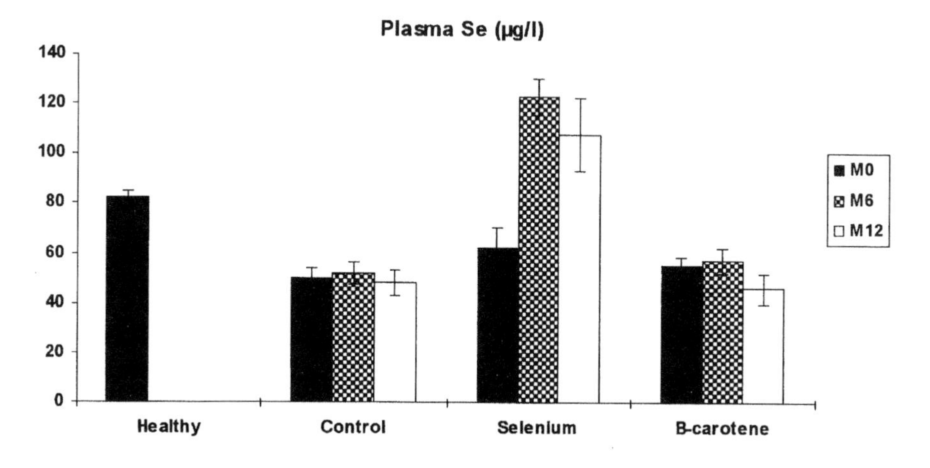

Figure 1. Plasma selenium concentrations in healthy seronegative subjects and in the three groups of patients at M0, M6, and M12, expressed as mean ± standard error of the mean.

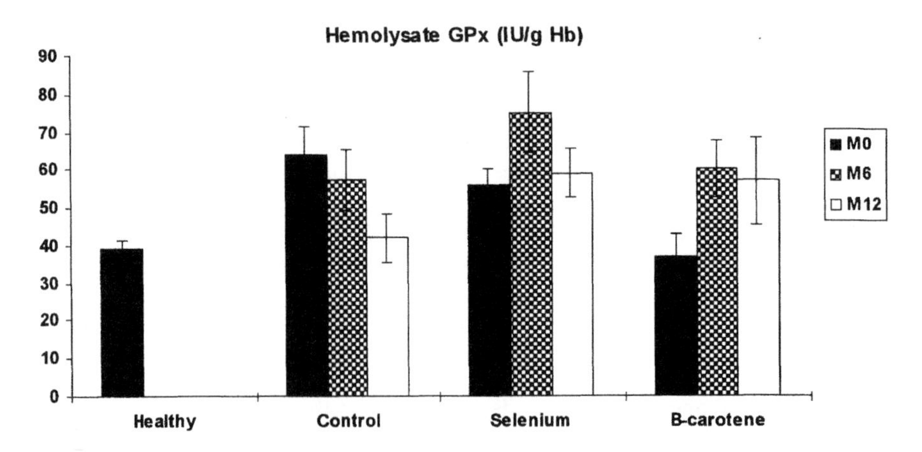

Figure 2. Hemolysate glutathione peroxidase concentrations in healthy seronegative subjects and in the three groups of patients at M0, M6, and M12, expressed as mean ± standard error of the mean.

selenoprotein P, glutathione peroxidase GPx, phospholipid hydroperoxide glutathione peroxidase PHGPx, and iodothyronine 5′-deiodinase.

Our results clearly show that selenium deficiency is present in HIV-1-infected patients and that this is correlated with a decrease in the CD4 lymphocyte count.

Dietary selenium has protective effects against a number of viral pathogens: murine mammary tumor virus (56), Rausher leukemia virus (57), coxsackie B virus (58), and hepatitis B virus (59).

Selenium has an important role for the maintenance of the immune functions and is a cofactor of enzymes that protect against toxic oxygen species. Selenium inhibits reverse trancriptase activity in RNA virus-infected animals (60). In these conditions, it appeared that selenium could have a beneficial effect in the treatment of HIV-infected subjects, possibly delaying the development of AIDS in newly HIV-infected subjects.

The mean daily selenium consumption is about 45 μg in France (61), corresponding to a plasma selenium level of about 80 μg/L. Our patients had a normal oral food intake but suffered from diarrhea and/or suspected malabsorption. The recommended intake is 55 μg and 70 μg/day, respectively, for women and men (62). Based on these considerations, we decided on supplementation with selenomethionine (Se 100 μg/day) for one year.

Such supplementation was found to increase plasma selenium from ~60 μg/L on average to ~120 μg/L between M0 and M6 (significantly higher than the controls). However, between M6 and M12 no further increase occurred and the plateau did not reach toxic level, which has been estimated as ~474 μg/L (6 μmol/L) by Rannem (63). The increase of erythrocyte selenium is gradual and the time required is about 4 months, corresponding to that for production and maturation of erythroid cells in the bone marrow.

From data on experimental animals it has been shown that selenium deficiency results in an impairment of immunoresponse, whereas selenium supplementation with appropriate nontoxic doses has an immunostimulatory effect (64). Few studies concern the human immune system and these are inconclusive and a high variability is observed in cell response (65–67). The chemical form of selenium is implicated in the differing bioavailability and distribution of selenium within the cell. Borella (68) had shown that low doses of selenium (0.5–2 μM), either as sodium selenite or as selenomethionine, did not alter the secretion of antibodies by cultured human lymphocytes in vitro. At higher levels (5 μM) a progressive increase in immunoglobulin production was observed with selenomethionine and an inhibitory effect with selenite. In long-term selenium supplementation in vivo, selenomethionine was more effective than inorganic forms even if the latter raised platelet glutathione peroxidase activity to a greater extend than did selenomethionine (69). Moreover, the correlation between selenium and GPx was poorer for subjects supplemented with selenomethionine, presumably because a significantly greater amount of selenium is associated with nonspecific protein such as hemoglobin in erythrocytes and albumin in plasma (70, 71). The use of cultured human lymphocytes and selenium supplementation appears to be a promising model for investigation of the effects of selenium on the immune system (68).

Results of trials on the effect of selenium supplementation in AIDS patients appear to be lacking. A study with 10 AIDS subjects (with nonobstructive cardiomyopathy) supplemented with sodium selenite over 3 weeks revealed that 8 of the patients showed improvement of their cardiac function during therapy (72). To establish whether intestinal absorption of dietary selenium is impaired in AIDS or ARC, a supplementary trial was performed over 70 days with 19 symptomatic HIV patients. The whole-blood selenium level of the treated subjects was found to increase compared to that of the healthy subjects, and the selenium supplements were well tolerated (73).

Schrauzer (60) reported a subjective improvement (appetite and intestinal functions), and a constant body weight with 100–300 μg sodium selenite during an observation period ranging from 3 to 8 months; nevertheless, CD4 still tended to decline during the 8 months of observation.

Zinc. The mean plasma zinc concentration in our patients was within the normal range for the supplemented groups compared to healthy seronegative subjects and significantly decreased ($p<0.01$) for controls. Individual values were widely scattered and some patients showed a pronounced hypozincemia (Figure 3).

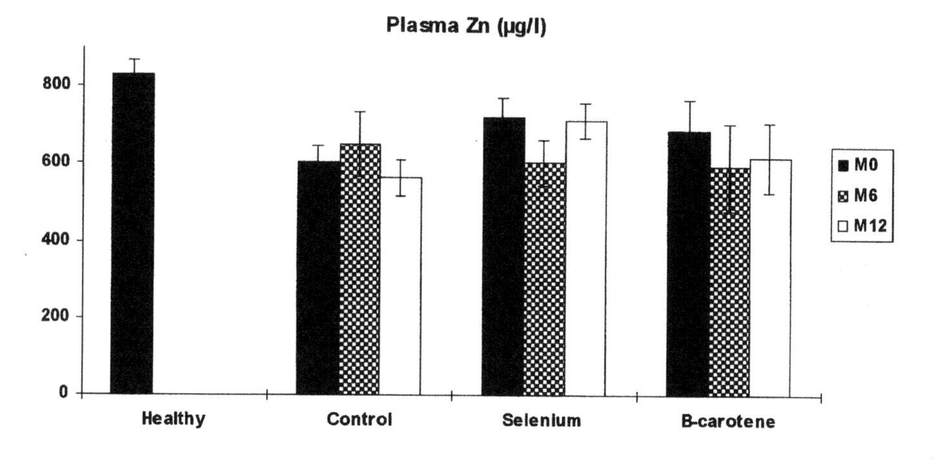

Figure 3. Plasma zinc concentrations in healthy seronegative subjects and in the three groups of patients at M0, M6, and M12, expressed as mean ± standard error of the mean.

These results agree with those of many other authors. Schuhmacher (74), Heise (75), Walter (76), Beck (77), Graham (78), Beach (79), and Dousset (80) observed no significant differences in the serum zinc between control subjects and patients with AIDS or asymptomatic HIV-1 individuals, whereas decreased levels of serum zinc in HIV patients have been reported by others (81–83). Several explanations might be found: there were very few patients in some studies; the CDC classification was not used by all authors; the results were presented as a function of the number of CD4 T lymphocytes instead of CDC stage; blood collections for control groups and patients were not made exactly at the same time and under identical conditions (sampling, vacutainers, methods of measurements, age-matching, etc.). In a study with drug addicts, Ruiz (84) demonstrated that plasma zinc levels were decreased in both HIV-1-positive and -negative subjects. Most circulating zinc is loosely bound to serum albumin, making interpretation of measured levels in severely hypoalbuminemic patients difficult. Individual serum zinc concentrations in AIDS (75) and albumin were correlated in some studies not in others (85). If hypozincemia is the consequence of reduction of serum albumin, a normalization of serum albumin would enhance zincemia. Reported hypozincemia might be due to the deficient nutritional status of most HIV-1-infected patients, resulting from excessive zinc loss from diarrhea. It is well known that an acute infection can rapidly lower circulating zinc levels (86). Thus the hypozincemia reported by some authors might be related not to the HIV-1 infection per se but to other infectious agents or to drug addition. Nevertheless, supplementation of AIDS patients with 125 mg zinc gluconate twice daily orally for 3 weeks had no significant effect on the numbers of total and CD4 lymphocytes (87), whereas in a study by Chandra (88) an excessive intake of zinc impaired immune responses, altering the functions of lymphocytes and neutrophils.

The assessment of zinc nutritional status and the detection of marginal deficiencies is difficult. The element is involved in more than one hundred metalloenzymes, and the use of plasma zinc to assess nutritional status has limitation because it is influenced by factors other than dietary intake: serum albumin; inflammatory response associated with infections; dietary ratios of minerals, particularly copper and iron. In this study, in spite

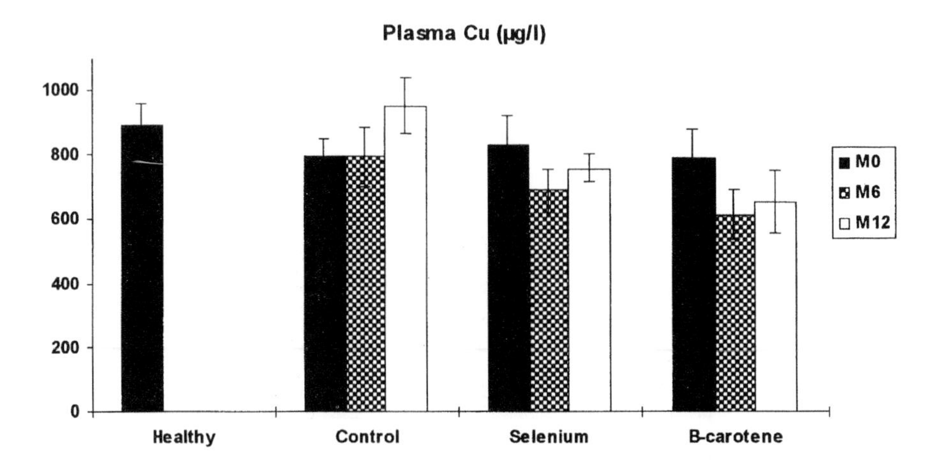

Figure 4. Plasma copper concentrations in healthy seronegative subjects and in the three groups of patients at M0, M6, and M12, expressed as mean ± standard error of the mean.

of the low CD4 lymphocyte counts, intensive therapy, and the opportunistic infections during the one-year follow-up, plasma zinc levels indicate acceptable nutritional status for this element without patent marginal zinc deficiency.

Copper. Concentrations of copper appeared to be unaffected by time in the three groups and not to differ from those of the healthy subjects at any time during the follow-up (Figure 4).

This normal level was also found in a one-year follow-up of 80 HIV-1-seropositive patients at stage IV of infection (89) and in a group of HIV-1-infected children at AIDS stage, even though copper seem higher in non-AIDS groups than in healthy persons (90) and higher in HIV-1-seropositive progressor patients than in the nonprogressors (78). The authors hypothesized that opportunistic infective processes that occur during HIV infection may play a role in the increase of serum copper levels.

Erythrocyte Membranes

The red blood cells were hemolyzed and the "ghosts" were washed repeatedly and used for determination of membrane zinc, membrane proteins, liposoluble vitamins A and E, malonic dialdehyde, and fatty acids.

Membrane Zinc and Proteins. Analysis of membrane zinc revealed that the zinc-containing proteins were stable with time within the three groups and that the levels were identical with those of healthy persons (Figure 5).

Erythrocytes are a convenient model for the study of membrane transport (91). It is not known whether the capability for zinc transport over the cell membrane of mature erythrocytes is of physiological significance. During the maturation of the erythrocytes, a relatively large quantity of zinc is required for the synthesis of the enzymes carbonic anhydrase (EC 4.2.1) and superoxide dismutase (EC 1.15.1.1). Ohno (92) found that 92% of a total amount of 170 μmol Zn/per liter of erythrocytes was accounted for by these two enzymes. We measured zinc in membrane ghosts washed several times with 5P8 (K_2HPO_4, 5 mM, pH 8); this must presumably be bound to proteins. Zinc level measured in erythrocyte membranes, as expressed per gram of protein, was remarkably constant with time in the

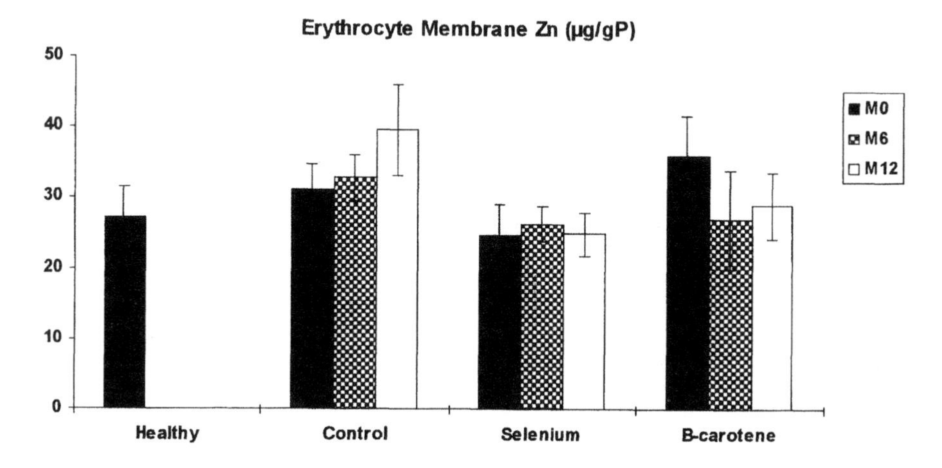

Figure 5. Erythrocyte membrane zinc concentrations in healthy seronegative subjects and in the three groups of patients at M0, M6, and M12, expressed as mean ± standard error of the mean.

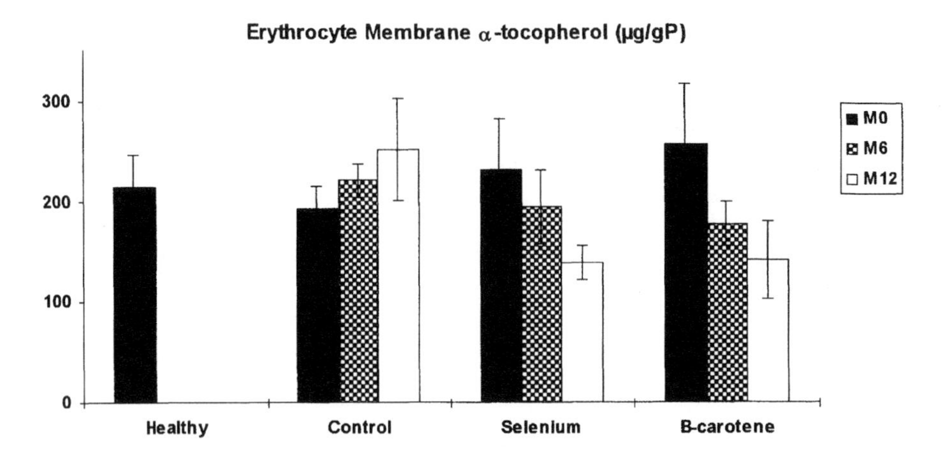

Figure 6. Erythrocyte membrane α-tocopherol concentrations in healthy seronegative subjects and in the three groups of patients at M0, M6, and M12, expressed as mean ± standard error of the mean.

three groups. In contrast, the range of the iron content was very large (data not shown), reflecting the binding of hemoglobin molecules to ghosts in spite of washing.

Copper levels in erythrocytic membranes were below our detection limit.

Vitamins. The antioxidant membrane vitamins, α-tocopherol and retinol, also showed remarkably similar values in all groups and in comparison to healthy subjects, together with stability in the course of time (Figures 6 and 7).

This normal level is important because structural stabilization of biological membranes by vitamin E is directly related to the amount of α-tocopherol present in the lipid bilayer of each membrane (93, 94).

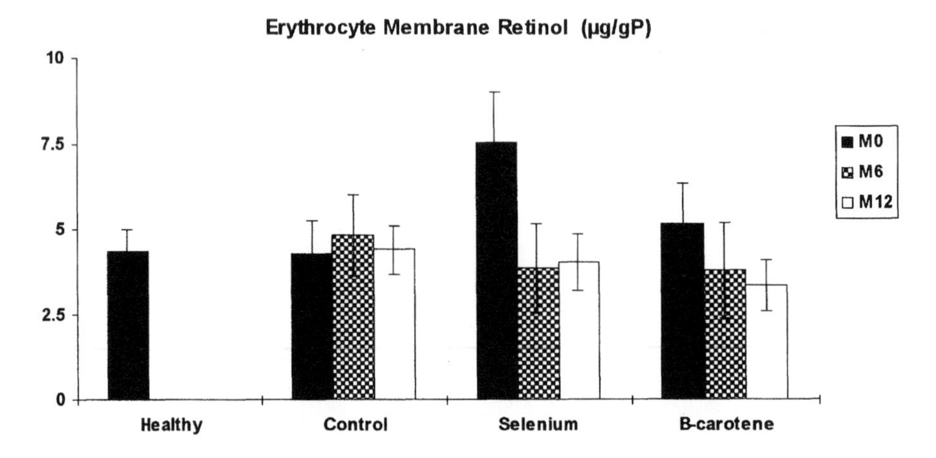

Figure 7. Erythrocyte membrane retinol concentrations in healthy seronegative subjects and in the three groups of patients at M0, M6, and M12, expressed as mean ± standard error of the mean.

Erythrocyte membrane liposoluble vitamins have not often been studied. Recently, with rats after acute iron overload, Galleano (95) showed that red blood cells from rats exposed to pro-oxidant conditions presented a significant increase in peroxidation products and that α-tocopherol pretreatment prevented this increase. This study also showed that neither the activity of antioxidant enzymes nor the content of α-tocopherol in red blood cells were affected by iron overload.

In erythrocyte membranes of children with cystic fibrosis, the vitamin E concentration was lower than in healthy children with a higher level of glutathione peroxidase activity (96). Sickle cell patients had significantly higher levels of vitamin E in erythrocytes relative to controls (97), whereas plasma levels of α-tocopherol were lower than in controls.

Our results seemed to show that the patients of this study, with their antiviral treatment, still had good antioxidant protection by liposoluble vitamins in the erythrocyte fraction. We may also note that some patients had very low values, attesting a vitamin carency.

Membrane Malonic Dialdehyde (MDA). Malonic dialdehyde was formed in vivo by lipid peroxidation of polyunsaturated fatty acids. We measured free MDA in the erythrocyte membranes by ion-pairing high-performance liquid chromatography; the values are expressed as μmol/g protein.

In our groups of patients, the MDA levels at the beginning of the supplementation protocol were not different from the healthy people. After 6 months, only the selenium-supplemented patients presented MDA levels lower than before the treatment ($p < 0.05$). At the end of the year, the three groups showed lower values than at MO ($p < 0.05$) and than the heathy controls ($p < 0.01$). We detect less free MDA in the erythrocytic membrane fraction in the three groups at the end of the follow-up than at the beginning (Figure 8).

Membrane Fatty Acids. We determined the concentrations of 22 fatty acids (FA) by lipid extraction and gas chromatography and grouped them by degree of unsaturation:

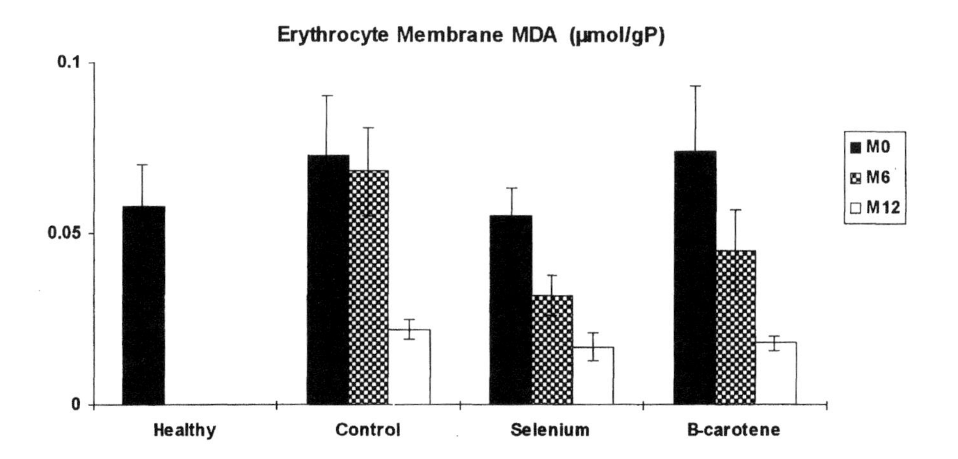

Figure 8. Erythrocyte membrane malonic dialdehyde concentrations in healthy seronegative subjects and in the three groups of patients at M0, M6, and M12, expressed as mean ± standard error of the mean.

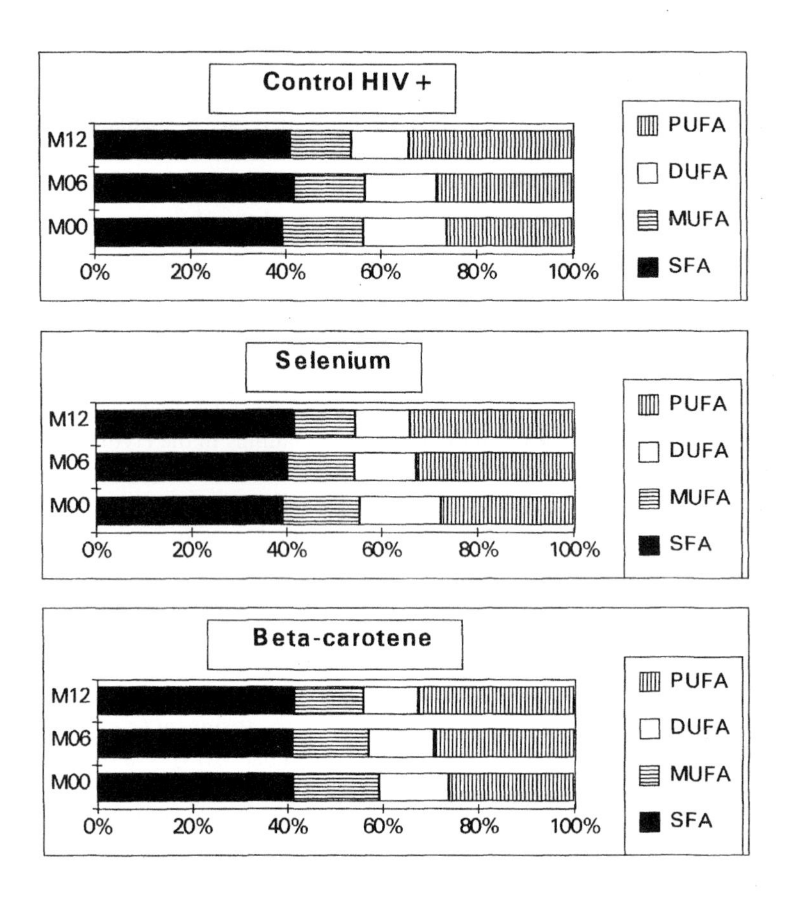

Figure 9. Erythrocyte membrane fatty acid concentrations in the three groups of patients at M0, M6, and M12, expressed as percentage of the total fatty acids.

- Saturated fatty acids (SFA): $C_{14:0}$, $C_{16:0}$, $C_{18:0}$, $C_{20:0}$, $C_{22:0}$, $C_{24:0}$
- Monounsaturated fatty acids (MUFA): $C_{16:1}$ (*n*–7), $C_{18:1}$ (*n*–9), $C_{20:1}$ (*n*–9), C_{22} (*n*–9), $C_{24:1}$ (*n*–9)
- Diunsaturated fatty acids (DUFA): $C_{18:2}$ (*n*–6), $C_{20:2}$ (*n*–6), $C_{22:2}$ (*n*–6)
- Polyunsaturated fatty acids (PUFA): $C_{18:3}$ (*n*–3), $C_{18:3}$ γ (*n*–6), $C_{20:3}$ γ (*n*–6), $C_{20:4}$ (*n*–6), $C_{20:5}$ (*n*–3), $C_{22:3}$ (*n*–3), $C_{22:4}$ (*n*–6), $C_{22:6}$ (*n*–3).

Expressing the results as percentages of total fatty acids in erythrocyte membranes, we noted a similar progression between M0 and M12 in the three groups, with an increase of PUFA ($p < 0.01$) and a decrease of MUFA and DUFA ($p < 0.01$). Further, a significant increase in SFA with respect to the total FA concentration is observed in the control and selenium-supplemented groups between M0 and M12 ($p < 0.01$) (Figure 9).

CONCLUSION

No modifications are observed between groups supplemented with selenium or β-carotene and nonsupplemented patients with respect to plasma trace elements (Copper and Zinc), intraerythrocyte enzyme selenium-dependent glutathione peroxidase, zinc proteins or liposoluble antioxidant vitamin A and E contents of the erythrocyte membranes, fatty acid contents of the erythrocyte membranes, or malonic dialdehyde (MDA) produced by lipid peroxidation of the polyunsaturated fatty acids. On the other hand, a similar development in the course of time is observed for the three groups comprising controls, selenium supplementation, and β-carotene supplementation, thus providing evidence for a significant decrease in MDA extracted from the erythrocyte membranes between months 0 and 12, and for a proportional increase in PUFA with respect to the total fatty acid concentrations for the three groups between months 0 and 12.

The erythrocyte membranes seem to be relatively unaffected by HIV infection in terms of liposoluble vitamins and zinc protein levels.

For HIV-positive patients at a higher advanced stage (<200 CD4 T lymphocytes/ mm^3 on average), the increases in selenemia and β-carotene plasma concentration provide evidence for a satisfactory degree of intestinal absorption and biodisposition of the orally administered supplements. However, supplementation carried out in the course of one year does not indicate a significantly beneficial effect in terms of survival or resistance to opportunistic infections.

For patients in an earlier phase of HIV infection, antioxidant supplementation over several years might be of greater efficacy in terms of peroxidation protection than for patients at an advanced stage of HIV infection.

REFERENCES

1. Fauci AS, Pantaleo G, Stanley S, Weissman D. Immunopathogenic mechanisms of human immunodeficiency virus (HIV) infection. Ann Intern Med 1996; 124:654–663.
2. Constans J, Conri C, Pellegrin JL, et al. Stress oxydatif et infection à VIH: un concept à préciser et une voie thérapeutique à explorer. Ann Med Int 1995; 146(7):514–520.
3. Clerici M, Shearer GM. Th1-Th2 switch is a critical step in the etiology of HIV infection. Immunol Today 1993; 14:107–111.
4. Gougeon ML, Montagnier L. Apoptosis in AIDS. Science 1993; 260:1269–1270.
5. Shaish A, Daugherty A, O'Sullivan F, Schonfeld G, Heinecke JW. Beta-carotene inhibits atherosclerosis in hypercholesterolemic rabbits. J Clin Invest 1995; 96:2075–2082.
6. Retsky KL, Frei B. Vitamin C prevents metal ion-dependent initiation and propagation of lipid peroxidation in human low-density lipoprotein. Biochim Biophys Acta 1995; 1257:279–287.
7. Walzem RL, Watkins S, Frankel EN, Hansen RJ, German JB. Older plasma lipoproteins are more susceptible to oxidation: a linking mechanism for the lipid and oxidation theories of atherosclerotic cardiovascular disease. Proc Natl Acad Sci USA 1995; 92:7460–7464.
8. Tabib A, Greenland T, Mercier I, Loire R, Mornex JF. Coronary lesions in young HIV-positive patients at necropsy. Lancet 1992; 340:730.
9. Capron L, Kim Y-U, Laurin C, Bruneval P, Fiessinger JN. Atheroembolism in HIV-positive individuals. Lancet 1992; 340:1039–1040.
10. Constans J, Pellegrin JL, Peuchant E, et al. Tumor necrosis factor alpha in HIV infected patients: correlations with opportunistic infections and immunological status. J Infect Dis 1993; 168:1327–1328.
11. Constans J, Pellegrin JL, Peuchant E, et al. High plasma lipoprotein(a) in HIV-positive patients. Lancet 1993; 341:1099–1100.
12. Constans J, Pellegrin JL, Peuchant E, et al. Plasma lipids in HIV-infected patients: a prospective study in 95 patients. Eur J Clin Invest 1994; 24:416–420.

13. Constans J, Pellegrin JL, Peuchant E, et al. Hypocholesterolemia and HIV-1 infection. Am J Med 1995; 98:518–519.
14. Constans J, Peuchant E, Pellegrin JL, et al. Fatty acids and plasma anti-oxidants in HIV-positive patients: correlation with nutritional and immunological status. Clin Biochem 1995; 28:421–426.
15. Sergeant C, Simonoff M, Hamon C, et al. Plasma antioxidant status (selenium, retinol and α-tocopherol) in HIV infection. In Pasquier C, et al., eds. Oxidative Stress, Cell Activation and Viral Infection. Basel: Birkauser Verlag; 1994:341–351.
16. Grunfeld C, Kotler DP, Shigenaga JK. Circulating interferon levels and hypertriglyceridemia in the acquired immunodeficiency syndrome. Am J Med 1991; 154–162.
17. Grunfeld C, Pang M, Doerrler W, Shigenaga JK, Jensen P, Feingold KR. Lipids, lipoproteins, triglyceride elearance, and cytokines in human immunodeficiency virus infection and the acquired imunodeficiency syndrome. J Clin Endocrinol Metab 1992; 74:1045–1052.
18. Grunfeld C, Soued M, Adi S, Moser AH, Dinarello CA, Feingold KR. Evidence for two classes of cytokines that simulate hepatic lipogenesis: relationships among tumor necrosis factor, interleukin-1 and interferon-alpha. Endocrinology 1990; 127:46–51.
19. Milvan D, Machado SG, Wilets I, Grossberg SE. Endogenous interferon and triglyceride concentrations to assess response to zidovudine in AIDS and advanced AIDS-related complex. Lancet 1992; 339:453–456.
20. Shor-Posner G, Basit A, Lu Y. Hypotholesterolemia is associated with immune dysfunction in early human immunodeficiency virus-1 infection. Am J Med 1993; 94:515–519.
21. Constans J, Pellegrin JL, Peuchant E, et al. Hypocholesterolemia and HIV-1 infection. Am J Med 1995; 98:518–519.
22. Adnaoui M, Dellinger A, Vital-Durand D, Bienvenu J, Sibille M, Levrat R. Changes in plasma apolipoproteins A1 and B during the inflammatory response. Eur J Intern Med 1991; 2:101–105.
23. Constans J, Rispal P, Lasseur C, et al. Plasma lipid abnormalities in acute bacterial infections: a possible error in assessing lipidic vascular risk factors. Int Angiol 1994; 13(supplement 1):26.
24. Constans J, Pellegrin JL, Brossard G, et al. Practical relevance of lipidic abnormalities in HIV infection: HDL-cholesterol and apolipoprotein A1 levels predict outcome at one year. Eur J Intern Med 1995; 6:183–186.
25. Scanu AM, Lawn RM, Berg K. Lipoprotein(a) and atherosclerosis. Ann Intern Med 1991; 115:209–218.
26. Maeda S, Abe A, Seishima M, Makino K, Noma A, Kawade M. Transient changes of serum lipoprotein(a) as an acute phase protein. Atherosclerosis 1989; 78:145–150.
27. Passi S, Picardo M, Morrone A, De Luco C, Ippolito F. Study on plasma polyunsaturated phospholipids and vitamin E, and on erythrocyte glutathione peroxidase in high risk HIV infection categories and AIDS patients. Clin Chem Enzyme Commun 1993; 5:169–177.
28. Begin ME, Manku MS, Horrobin DF. Plasma fatty acid levels in patients with acquired immune deficiency syndrome and in controls. Prostaglandins Leukotriens Essential Fatty Acids 1989; 37:135–137.
29. Revillard JP, Vincent CMA, Favier AE, Richard MJ, Zittoun M, Kazatchkine MD. Lipid peroxidation in human immunodeficiency virus infection. J Acquir Immune Defic Syndr. 1992;5:637–638.
30. Sonneborg A, Carlin G, Akerlund B, Jarstrand C. Increased production of malondialdehyde in patients with HIV infection. Scand J Infect Dis 1988; 20:287–290.
31. Owens RJ, Anabtgaramaiah GM, Kahlon JB, Srinivas RV, Compans RW, Segrest JP. Apolipoprotein A-I and its amphipathic helix peptide analogues inhibit human immunodeficiency virus-induced syncitium formation. J Clin Invest 1990; 86:1142–1150.
32. Tersmette M, De Goede R. Differential syncitium-inducing capacity of human immunodeficiency virus isolates: frequent detection of syncitium-inducing isolates in patients with

acquired immunodeficiency virus (AIDS) and AIDS-related complex. J Virol 1988; 62:2026–2032.
33. Droge W, Eck HP, Mihm S. HIV-induced cysteine deficiency and T-cell dysfunction–a rationale for treatment with *N*-acetylcysteine. Immunol Today 1992; 13:211–214.
34. Greenspan HC, Aruoma OI. Oxidative stress and apoptosis in HIV infection: a role for plant-derived metabolites with synergistic antioxidant activity. Immunol Today 1994; 15:209–213.
35. Gougeon ML, Boudet F, Lecoeur H, Heeney J. Apoptosis in HIV infection: influence of Bc12 and Fas molecules on CD4 and CD8 T cell deletions (abstract). Oxidative Stress and Redox Regulation: Cellular Signaling, Aids, Cancer and Other Diseases, Institut Pasteur, Paris; May 1996:84.
36. Constans J, Marchand JM, Conri C, et al. Asymptomatic atherosclerosis in HIV-positive patients: a case-controlled ultrasound study. Ann Med 1995; 27:683–685.
37. Odeh M. The role of tumor necrosis factor α in acquired immunodeficiency syndrome: review article. J Intern Med 1990; 228:549–556.
38. Lafeuillade A, Alessi MC, Poizot-Martin I, et al. Endothelial cell dysfunction in HIV infection. J AIDS 1992; 5:127–131.
39. Conri C, Seigneur M, Blann AD, Pellegrin JL, Boisseau MR, Constans J. Lésion de l'endothélium vasculaire au cours de l'infection à VIH. Revue de Méd Int 1995; 16(3):327.
40. Lack P, Livrozet JM, Bourgeay-Causse M, Fayol V, Saint-Marc T, Touraine JL. Vitamin status at the first blood test analysis in 120 HIV seropositive patients. IXth International Conference on AIDS, Berlin; 1993; B36:2357.
41. Tang A, Graham NMH, Kirby J. Dietary micronutrient intake and risk of progression to AIDS in HIV-1 infected homosexual men. Am J Epidemiol 1993; 138:937–951.
42. Dworkin BM. Selenium deficiency in HIV infection and the acquired immunodeficiency syndrome (AIDS). Chem-Biol Interact 1994; 91(2–3):181–186.
43. Cirelli A, Ciardi M, De Simone C, et al. Serum selenium concentration and disease progress in patients with HIV infection. Clin Biochem 1991; 24(2):211–214.
44. Constans J, Pellegrin JL, Sergeant C, et al. Serum selenium predicts outcome in HIV infection. J Acquir Immune Defic Syndr Hum Retrovir. 1995; 10(3):392.
45. Flores SC, Marecki JC, Harper KP, Bosc SK, Nelson SK, MaCord JM. Tat protein of human immunodeficiency virus represses expression of managanese superoxide dismutase in HeLa cells. Proc Natl Acad Sci 1993; 90:7632.
46. Garnier N. Etude de l'action antiradicalaire d'une spécialité pharmaceutique à base de sélénium et vitamine E chez le rat soumis à une atmosphère enrichie en oxygène. Thèse de l'Université de Bordeaux I; 1992.
47. Largillière C, Melancon SB. Free malondialdehyde determination in human plasma by high performance liquid chromatography. Anal Biochem 1988; 170:123–126.
48. Catignani GL, Bieri JG. Simultaneous determination of retinol and α-tocopherol in serum or plasma by liquid chromatography. Clin Chem 1993; 29(4):708–712.
49. Simonoff M, Hamon C, Moretto P, Llabador Y. Simonoff G. High sensitivity Pixe determination of selenium in food and biological samples using a preconcentration technique. Nucl Instrum Methods Phys Res 1988; B31:442–448.
50. Simonoff M, Sergeant C, Razafindrabe L, et al. Plasma trace element levels in 89 HIV-infected patients: correlation with nutritional and immunological status. In Anke M, Meissner D, Mills CF, eds. Trace Elements in Man and Animals 8. Gersdorf: Verlag Media; 1993:768–771.
51. Bradford MM. A rapid and sensitive method for the quantitation of microgram quantities of protein utilizing the principle of protein-dye binding. Anal Biochem 1976; 72:248–254.
52. Lepage G, Roy CC. Direct transesterification of all classes of lipids in a one-step reaction. J. Lipid Res 1987; 27:114–120.
53. Sunde RA. Intracellular glutathione peroxidases–structure regulation and function. In Burk RF, ed. Selenium in Biology and Human Health. New York: Springer-Verlag; 1994:45–77.

54. Arthur JR, Beckett GJ. New metabolic roles for selenium. Proc Nutr Soc 1994; 53:615–624.
55. Bermano G, Nicol F, Dyer JA, et al. Selenoprotein gene expression during selenium-repletion of selenium-deficient rats. Biol Trace Elem Res 1996; 51:211–224.
56. Medina D, Shepherd F. Selenium-mediated inhibition of mouse mammary tumorigenesis. Cancer Lett. 1980; 8:241–245.
57. Balansky RM, Argirova RM. Sodium selenite inhibition of some oncogenic RNA-viruses. Experientia 1981; 37:1194–1195.
58. Beck K, Kolbeck PC, Rohr LH. et al. Benign human enterovirus becomes virulent in selenium-deficient mice. J Med Virol 1994; 43:166–170.
59. Shu-Yu Y, Li WG, Zhu YJ, Yu WP, Hou C. Chemoprevention trial of human hepatitis with selenium supplementation in China. Biol Trace Elem Res 1989; 20:15–22.
60. Schrauzer GN, Sacher J. Selenium in the maintenance and therapy of HIV-infected patients. Chem-Biol Interact 1994; 91:199–205.
61. Simonoff M, Simonoff G. Le Sélénium et la Vie. Paris: Editions Masson; 1991.
62. NCR (National Research Council). Recommended daily allowances, 10th ed. Washington DC: National Academy Press; 1989.
63. Rannem T, Ladefoged K, Hylander E, Heghhoj J, Jarnum S. Selenium depletion in patients on home parenteral nutrition. Biol Trace Elem Res 1993; 39:81–90.
64. Turner RJ, Finch JM. Selenium and the immune response. Proc Nutr Soc 1991; 50:275–285.
65. Arvilommi H, Poikonen K, Jokinen I, et al. Selenium and immune functions in humans. Infect Immun 1983; 41:185–189.
66. Watson RR, Moriguchi S, McRae B, Tobin L, Mayberry JC, Lucas D. Effects of selenium in vitro on human T-lymphocytes functions and K-562 tumour cell growth. J Leukocyte Biol 1986; 39:447–457.
67. Harvima RJ, Jagerroos H, Kajander EO, et al. Screening of effects of selenomethionine-enriched yeast supplementation on various immunological and chemical parameters of skin and blood in sporiatic patients. Acta Dermatol Venereol 1993; 73:88–91.
68. Borella P, Bargellini A, Medici CI. Chemical form of selenium greatly affects metal uptake and responses by cultured human lymphocytes. Biol Trace Elem Res 1996; 51:43–54.
69. Thomson CD, Robinson MF, Butler JA, Whanger PD. Long-term supplementation with selenate and selenomethionine: selenium and glutathione peroxidase (EC 1.11.1.9) in blood components of New Zealand women. Br J Nutr 1993; 69:577–588.
70. Butler JA, Thomson CD, Whanger PD, Robinson MF. Selenium distribution in blood fractions of New Zealand women taking organic or inorganic selenium. Am J Clin Nutr 1991; 53:748–754.
71. Deagen JT, Butler JA, Zachara A, Whanger PD. Determination of the distribution of selenium between glutathione peroxidase, selenoprotein P, and albumin in plasma. Anal Biochem 1993; 208:176–181.
72. Zazzo JF, Chalas J, Lafont A, Camus F, Chappuis P. Is nonobstructive cardiomyopathy in AIDS a selenium deficiency-related disease? J Parenteral Enteral Nutr 1988; 12(5):537–538.
73. Olmsted L, Schrauzer GN, Flores-Arce M, Dowd J. Selenium supplementation of symptomatic human immunodeficiency virus infected patients. Biol Trace Elem Res 1989; 20(1–2):59–65.
74. Schuhmacher M, Peraire J, Domingo JL, Vidai F, Richart C, Corbella J. Trace elements in patients with HIV-1 infection. Trace Elem Electrolytes 1994; 11(3):130–134.
75. Heise W, Nehm K, L'Age M, Averdunk R, Günther T. Concentrations of magnesium, zinc and copper in serum of patients with acquired immuno-deficiency syndrome. J Clin Chem Clin Biochem 1989; 27:515–517.
76. Walter RM Jr, Oster MH, Lee TJ, Flynn N, Keen CL. Zinc status in human immunodeficiency

virus infection. Life Sci 1990; 46:1597–1600.
77. Beck K, Scramel P, Hedl A, Jaeger H, Kaboth W. Serum trace element levels in HIV-infected subjects. Biol Trace Elem Res 1990; 25:89–96.
78. Graham N, Sorensen D, Odaka N, et al. Relationship of serum copper and zinc levels to HIV-1 seropositivity and progression to AIDS. J AIDS 1991; 4:976–980.
79. Beach R, Mantero-Antienza E, Shor-Posner G. Specific nutrient abnormalities in asymptomatic HIV-1 infection. AIDS 1992; 6:701–708.
80. Dousset B, Hussenet F, May T, Dubois F, Canton P, Belleville F. A basis for new approaches to the chemotherapy of AIDS: novel genes in HIV-1 potentially encode selenoproteins expressed by ribosomal frameshifting and termination suppression. Ann Med de Nancy et de l'Est 1995; 34(2):81–83.
81. Allavena C, Dousset B, May T, Dubois F, Canton P, Nabet-Belleville F. Zinc and selenium blood values: other prognostic biological parameters in the progression of human immunodeficiency virus infection. In Galteau MM, Siest G, Henry J, eds. Biol. Perspective. Paris: John Libbey Eurotext; 1993:127–130.
82. Fabris N, Mocchegiani E, Galli M, Irato L, Lazzarin A, Moroni M. AIDS, zinc deficiency, and thymic hormone failure. J Am Med Assoc 1988; 259:839–840.
83. Falutz J, Tsoukas C, Gold P. Zinc as a cofactor in human immunodeficiency virus-induced immunosuppression. J Am Med Assoc 1988; 259:2850–2851.
84. Ruiz M, Gil B, Maldonado A, Cantero J, Moreno V. Trace elements in drug addicts. Klin Wochenschr 1990; 68:507–511.
85. Constans J, Sergeant C, Pellegrin JL, et al. Plasma selenium, zinc and copper in HIV-positive patients (abstract). 1st International Conference: Nutrition and HIV Infection, Cannes; April 1995.
86. Cousins RJ. Absorption, transport, and hepatic metabolism of copper and zinc: special reference to metalothionein and ceruloplasmin. Physiol Rev 1985; 65:238–309.
87. Math G, Misset JL, Gil-Delgado M, et al. From experimental to clinical attempts in immurestoration with bestatin and Zn. Biomed Pharmacother 1986; 40:383–385.
88. Chandra RK. Excessive intake of zinc impairs immune responses. J Am Med Assoc 1984; 252:1443–1446.
89. Allavena C, Dousset B, May T, Dubois F, Canton P, Belleville F. Relationship of trace element, immunological markers and HIV-1 infection progression. Biol Trace Elem Res 1995; 47(1–3):133–138.
90. Periquet BA, Jammes NM, Lambert WE, et al. Micronutrient levels in HIV-1 infected children. AIDS 1995; 9:887–893.
91. De Kok J, Van Der Schoot C, Veldhuizen M, Wolterbeek HTh. The uptake of zinc by erythrocytes under near-physiological conditions. Biol Trace Elem Res 1993; 38:13–26.
92. Ohno H, Doi R, Yamamura K, Yamashita K, Lizuka S, Taniguchi N. A study of zinc distribution in erythrocytes of normal humans. Blut 1985; 50:113.
93. Patel JM, Seckharam M, Block ER. Vitamin E distribution and modulation of the physical state and function of pulmonary endothelial cell membranes. Exp Lung Res 1991; 17:707–723.
94. Gomez-Fernandez JC, Villalain J, Aranda FJ, et al. Localization of α-tocopherol in membranes. Ann NY Acad Sci 1989; 570:109–120.
95. Galleano M, Puntarulo S. Role of antioxidants on the erythrocyte resistance to lipid peroxidation after acute iron overload in rats. Biochim Biophys Acta 1995; 1271:321–326.
96. Therond P, Couturier M, Navarro J, Demelier JF, Lemonnier F. Production of phospholipid hydroperoxides in erythrocyte membranes of children with cystic fibrosis (abstract). SFFR, Paris; 1993.
97. Natta CL, Tatum VL, Chow CK. Antioxidant status and free radical-induced oxidative damage of sickle erythrocytes. Ann NY Acad Sci 1992; 669:365–367.

40
Glutathione Oxidation and Mitochondrial DNA Damage in AIDS: Effect of Zidovudine

José Viña, José García-de-la-Asunción, María L. Del Olmo, Arantxa Millán, Juan Sastre, José A. Martín, and Federico V. Pallardó
Universidad de Valencia, Valencia, Spain

INTRODUCTION

AIDS patients who receive zidovudine (AZT) frequently suffer from myopathy (1). The typical features of this myopathy are ragged red fibers and paracrystalline inclusions in mitochondria. This has been attributed to damage to mitochondria, and specifically to mitochondrial DNA (2). In 1991 it was proposed that AZT causes oxidation of guanosine to 8-hydroxy-2′-deoxyguanosine in experimental animals (3). The role of mitochondrial DNA in cell physiology and medicine has been emphasized (4). We have developed a new method to determine accurately glutathione redox status in blood (5,6). Using this method, we have recently observed that there is a correlation between glutathione oxidation and damage to mitochondrial DNA both in rats and in mice and that damage to mitochondrial DNA can be prevented by administration of antioxidants (7). The aims of the present work were to test whether AIDS causes oxidation of blood glutathione and to determine whether the mitochondrial damage caused by AZT, both in humans and in experimental animals, is due to damage to mitochondrial DNA.

MATERIALS AND METHODS

Patients

We analyzed blood and urine from male adults aged between 25 and 40 years. Blood samples were taken from healthy controls, HIV-positive patients, and AIDS patients. Urine samples were collected from each individual's 24-h void and stored at –30°C until analysis. Four groups were studied: (a) healthy controls, (b) HIV-positive patients (who had not received AZT), (d) HIV-positive patients treated with AZT (Retrovir, 250 mg/12 h) for 12 to 24 months, and (d) HIV-positive patients treated with AZT and antioxidants (vitamin C 1 g/day, and vitamin E 0.6 g/day, for 1 month).

Animals

Male OF1 mice (from IFFA-Credo, Barcelona, Spain) were maintained on a 12 h/12 h light/dark cycle at 22°C. Mice were fed on a standard laboratory diet (containing 590 g carbohydrates, 30 g lipids, and 160 g protein per kilogram of diet) and tap water ad libitum. Animals treated with antioxidants were fed on the same diet but supplemented with vitamin C (10 g/kg per day) and vitamin E (0.6 g/kg per day) for 2 months before sacrifice, and their food intake was not significantly different from controls. Mice 4–5 months of age were anesthetized with sodium pentobarbital (60 mg/kg body wt) by intraperitoneal injection and killed by decapitation at 09:00–11:00 to minimize circadian variations of the parameters studied.

Animals were divided into three groups: (a) controls, (b) treated with AZT (3′-azido-3-deoxythymidine, or Zidovudine, from Sigma Chemical Co.), and (c) treated with AZT and with dietary antioxidants. AZT was administered in drinking water (10 mg/kg body wt per day) for 35 days.

Isolation of Mitochondria

After the animals were killed, their skeletal muscle (from a pool of hindquarters), hearts, livers, and brains were quickly removed. Isolation of mitochondria was performed using a standard differential centrifugation procedure as described by Rickwood et al. (8).

Measurement of Oxo8dG

This metabolite, 8-oxo-7,8-dihycho. 2′-deoxyguanosine, was measured as described by Ames et al. (9) based on reversed-phase HPLC combined with electrochemical detection. Oxo8dG present in DNA hydrolyzates was separated isocratically on a 25 × 0.46 cm ODS-2 Spherisorb column, with a mobile phase of 1% methanol in 50 mM potassium phosphate buffer pH 5.5, pumped at a flow rate of 1 ml/min. Electrochemical detection of oxo8dG was performed on an ESA Coulochem II (Bedford, MA) Model 5200 equipped with a 5011 analytical cell and a 5021 guard cell. The potentials set for the dual coulometric detector were +0.15 for detector 1 and +0.45 for detector 2.

Measurement of Glutathione

Reduced glutathione (GSH) was measured spectrophotometrically using glutathione *S*-transferase. Oxidized glutathione (GSSG) was assayed by a high-performance liquid chromatography (HPLC) method with UV-V detection which we recently developed to measure GSSG in the presence of a large excess of GSH (5,6).

Measurement of Malondialdehyde

Malondialdehyde (MDA) was measured in mitochondria from muscle, liver, heart, and brain. We have adapted an HPLC method described by Wong et al. (10) to measure MDA formed from mitochondrial lipoperoxides.

Statistics

Results are expressed as mean ± SD. Statistical analyses were performed by the least-significant difference test, which consists of two steps. First, an analysis of variance was performed. The null hypothesis was accepted for all numbers of those sets in which F was nonsignificant at the level of $p \leq 0.05$. Second, the sets of data in which F was significant were examined by the modified t-test using $p \leq 0.05$ as the critical limit.

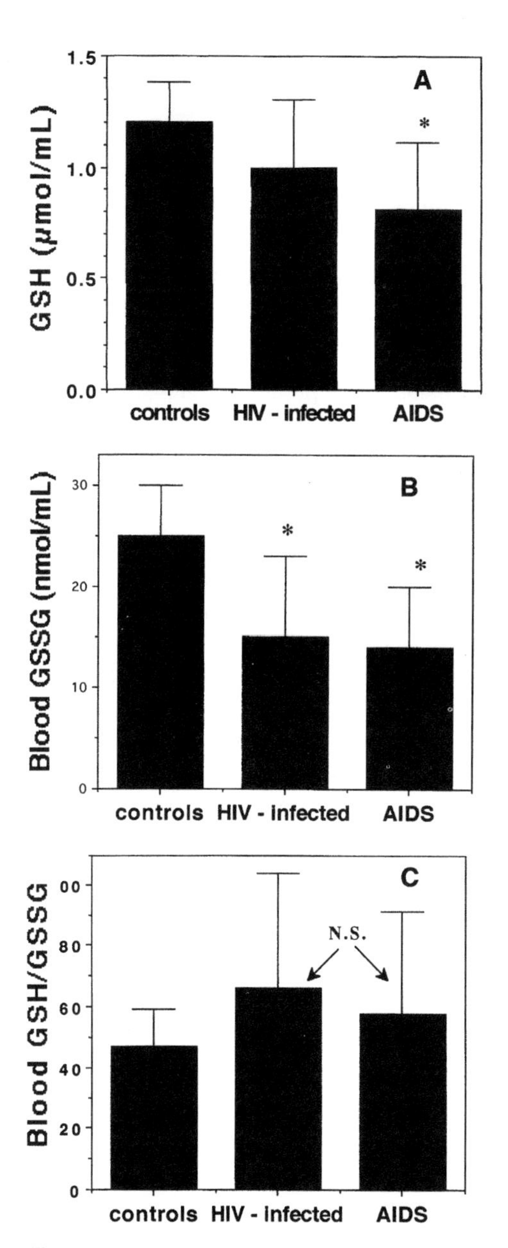

Figure 1. Glutathione levels in the blood of controls and HIV-infected and AIDS patients: (A) GSH; (B) GSSG; (C) GSH/GSSG. *$p < 0.05$; N.S. = not statistically significant.

RESULTS AND DISCUSSION

Glutathione Levels in Whole Blood of AIDS Patients

Figure 1 shows the levels of GSH and GSSG and the GSSG/GSH ratio in blood of healthy controls, of asymptomatic HIV-infected subjects and of AIDS patients. We found that blood GSH levels were decreased in AIDS patients. In asymptomatic HIV patients a small, nonsignificant decrease was also observed. GSSG levels were also decreased both in asymptomatic HIV and AIDS patients. The fall in GSSG was statistically significant in both cases. The GSH/GSSG ratio in whole blood was not affected. These results confirm those previously reviewed by Staal et al. (11), which showed that there is a deficiency of glutathione in AIDS patients. However, here we show that the low glutathione levels are not accompanied by oxidation of glutathione and may be due to impaired synthesis from precursors. Indeed, we have previously found that this is the case in other pathological situations such as cataract formation (12), surgical stress (13), or prematurity (14).

Urinary Excretion of Oxo8dg in HIV-Infected Patients Who Receive AZT

An important side-effect of AZT is that it causes mitochondrial myopathy. The fact that AZT might lead to an increased formation of oxo8dG was reported in one paper where it was studied in experimental animals (3). Here we report that asymptomatic HIV patients who receive AZT have a marked increase in oxo8dG excretion when compared with controls. This can be prevented by oral administration of antioxidant vitamins (Figure 2). Indeed, when asymptomatic HIV patients received oral antioxidants their excretion of oxo8dG was significantly lower than that of patients treated with AZT only. This may provide a biochemical basis for an indication of administration of antioxidants to patients who receive AZT.

Effect of AZT on Muscle Mitochondrial Glutathione Oxidation

We have recently observed that there is a direct relationship between mitochondrial glutathione oxidation and mitochondrial DNA damage as measured by formation of

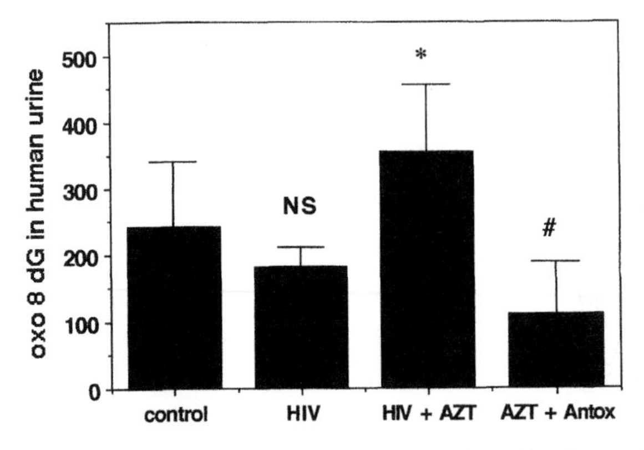

Figure 2. Elimination of oxo8dG in urine of patients. $*p < 0.05$ vs. controls; # $p < 0.05$ vs. patients not treated with antioxidants. NS = not statistically significant.

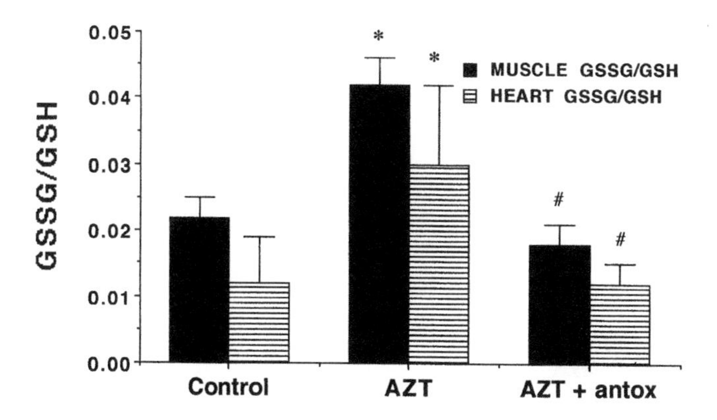

Figure 3. Effect of AZT on GSSG/GSH ratio in muscle and heart mitochondria from mice. $^*p < 0.05$ vs. controls; # $p < 0.05$ vs. mice not treated with antioxidants.

oxo8dG. The fact that we found increased excretion of oxo8dG prompted us to study the possible effect of AZT on glutathione levels in mitochondria from various tissues of mice treated with AZT. Mitochondria from heart or skeletal muscle have an increased level of GSSG and a decreased level of GSH, and thus an oxidized GSSG/GSH ratio (see Figure 3).

Effect of AZT on Muscle Mitochondrial Malondialdehyde Levels

We decided to test the effect of AZT on MDA levels in mitochondria and found that those from heart or skeletal muscle have an increased MDA level (see Figure 4). However, those from other organs, such as liver or brain, do not show an increased level of MDA (results not given). The fact that those organs that have an oxidized glutathione ratio also have increased lipid peroxidation indicates that the glutathione redox ratio is a good indicator of oxidative stress.

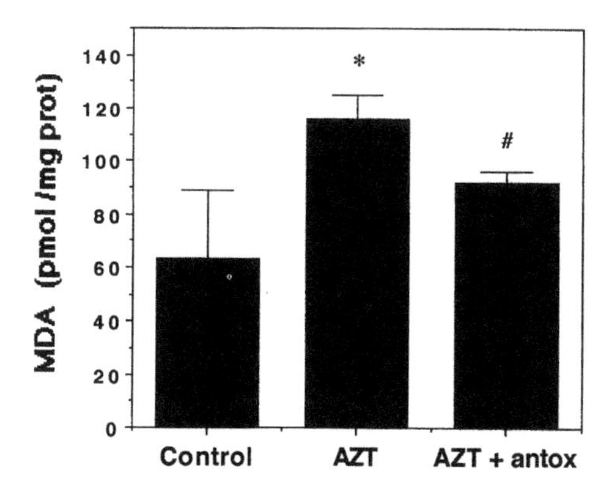

Figure 4. Effect of AZT on MDA levels in muscle mitochondria. $^*p < 0.05$ vs. controls; # $p < 0.05$ vs. mice not treated with antioxidants.

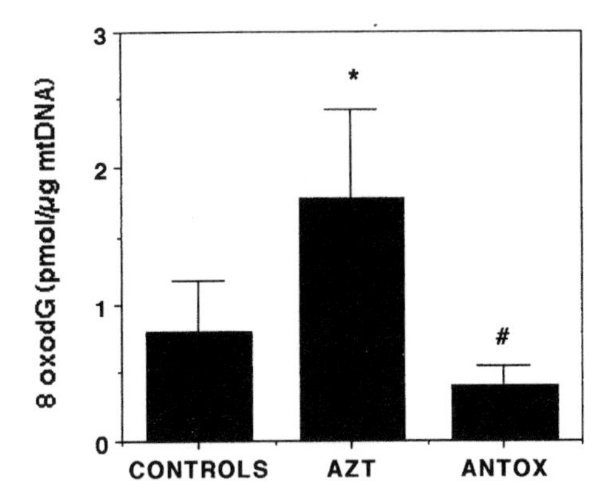

Figure 5. Effect of AZT on oxo8dG in muscle mitochondria. *$p < 0.05$ vs. controls; # $p < 0.05$ vs. mice not treated with antioxidants.

Effect of AZT on Oxidative Damage to Mitochondrial DNA

The fact that urinary excretion of 8oxodG is increased in patients who receive AZT prompted us to study the possible effect of this drug on oxidative damage to mitochondrial DNA in mice. We tested muscle mitochondria because we had found that muscle mitochondria, and not those from other organs, suffer from oxidative stress, as evidenced by changes in glutathione redox ratio or MDA levels (see above). Previous work from Ames and colleagues (9) showed that age-associated damage to mitochondrial DNA is extensive. We have recently reported that there is a direct relationship between mitochondrial DNA damage and oxidation of mitochondrial glutathione. Figure 5 shows that AZT caused an increase in the levels of oxo8dG of more than 100%. This change is prevented completely by oral administration of antioxidants.

CONCLUDING REMARKS

The relevant findings reported here are as follows.

1. There is a decrease in GSH and GSSG in both asymptomatic HIV-infected and AIDS patients, but there is no change in blood GSSG/GSH ratio.
2. Administration of AZT to HIV patients increases their excretion of oxo8dG in urine.
3. Administration of AZT to mice causes an increase in mitochondrial MDA levels, oxidation of glutathione, and oxidative damage to mitochondrial DNA in mitochondria from skeletal muscle and heart. These changes are not found in liver or brain.
4. Oral administration of antioxidants partially prevents the oxidative damage caused by AZT.

REFERENCES

1. Dalakas M, Illa I, Pezeshkpour GH, Laukatis JP, Cohen B, Griffin JL. Mitochondrial myopathy caused by long-term zidovudine therapy. N Engl J Med 1991; 323:1089–1105.

2. Arnaudo E, Dalakas M, Shanske S, Moraes CT, Dimauro S, Schon EA. Depletion of muscle mitochondrial DNA in AIDS patients with zidovudine-induced myopathy Lancet 1991; 337:508–510.
3. Hayakawa M, Ogawa T, Sugiyama S, Tanaka M, Ozawa T. Massive conversion of guanosine to 8-hydroxyguanosine in mouse liver mitochondrial DNA by administration of azidothymidine. Biochem Biophys Res Commun 1991; 176:87–93.
4. Luft R. The development of mitochondrial medicine. Proc Natl Acad Sci USA 1994; 91:8731–8738.
5. Asensi M, Sastre J, Pallardó FV, García de la Asunción J, Estrela J, Viña J. A high-performance liquid chromatography method for measurement of oxidized glutathione in biological samples. Anal Biochem 1994; 217:323–328.
6. Viña J, Sastre J, Miguel Asensi M, Packer L. Assay of blood glutathione oxidation during physical exercise. Methods Enzymol 1995; 251:237–243.
7. García De La Asunción J, Millán A, Plá R, et al. Mitochondrial glutathione oxidation correlates with age-associated oxidative damage to mitochondrial DNA. FASEB J 1996; 10:333–338.
8. Rickwood DW, Wilson MT, Darley-Usmar VM. Isolation and characteristics of intact mitochondria. In (darley-Ulmar VM, Wilson MT, Rickwood D, eds Mitochondria Practical Approach Oxford: IRL Press; 1987:4–6.
9. Richter C, Park JW, Ames B. Normal oxidative damage to mitochondrial and nuclear DNA is extensive. Proc Natl Acad Sci USA 1988; 85:6465–6467.
10. Wong SHY, Knight JA, Hopfer SM, Zaharia O, Leach CN, Sunderman FW. Lipoperoxides in plasma as measured by liquid-chromatographic separation of malondialdehyde–thiobarbituric acid adduct. Clin Chem 1987; 33:214–220.
11. Staal FJ, Ela SW, Roederer M, Anderson MT, Herzenberg LA, Herzenberg LA. Glutathione deficiency and human immunodeficiency virus infection. Lancet 1992; 339:909–912.
12. Ferrer JV, Gascó E, Sastre J, Pallardó FV, Asensi M, Viña J. Age-related changes in glutathione synthesis in the eye lens. Biochem J 1990; 269:531–534.
13. Viña J, Gimenez A, Puertes IR, Gascó E, Viña JR. Impairment of cysteine synthesis from methionine in rats exposed to surgical stress. Br J Nutrition 1992; 68:421–429.
14. Viña J, Vento M, García-Sala F, et al. L-Cysteine and glutathione metabolism in premature infants due to cystathionase deficiency. Am J Clin Nutr 1995; 61:1

41

Effect of L-Carnitine Treatment In Vivo on Apoptosis and Ceramide Generation by Peripheral Blood Lymphocytes from AIDS Patients

M.G. Cifone, E. Alesse, L. Di Marzio, P. Roncaioli, F. Zazzeroni, S. Moretti, G. Famularo, S. Marcellini, G. Santini, V. Trinchieri, E. Nucera and C. De Simone
University of L'Aquila, L'Aquila, Italy

INTRODUCTION

Viral replication and load are causally related to prognosis in subjects infected with the human immunodeficiency virus (HIV) (1–3), thus providing a rationale for antiviral therapy. However, despite the introduction of antiretroviral drugs, the morbidity and mortality associated with the infection remain high, suggesting that indirect mechanisms triggered by the virus are crucial to the pathogenesis of the disease.

There is a growing body of evidence that the progressive loss of T lymphocytes in HIV-infected subjects is associated with elevated apoptotic cell death of both infected and uninfected cells, ultimately resulting in the overt acquired immunodeficiency syndrome (AIDS) (4,5).

Recent data have demonstrated that the Fas–Fas ligand (FasL) system may contribute substantially to the functional defects and depletion of T lymphocytes in HIV-infected patients (6–8). The apoptotic signal through Fas involves the activation of an acidic sphingomyelinase (SMase), sphingomyelin breakdown, and ceramide generation (9). In addition, increased ceramide generation is associated with HIV infection (10) and ceramide itself enhances the replication of the virus (11,12).

Taken together, these studies suggest that treatment strategies directed at down-modulating the production of ceramide may slow the progression of the infection toward AIDS through reducing the rates of both apoptotic lymphocyte death and HIV replication. In this respect, preliminary work from our laboratory has pointed to L-carnitine as a candidate drug.

Fas AND APOPTOSIS

The stimulation through Fas may costimulate cellular activation and proliferation, but in turn it may actually induce a "death signal" (13). It is now clear that Fas can mediate opposite effects depending on the state of activation of the T cells (13). Signaling through Fas in the early stages of an immune response augments the generation of an effective immune response, whereas signaling through Fas in the later stages leads to downsizing of an immune response at the clonal level via apoptosis (13). Therefore, Fas and FasL have an important role in regulating normal immune responses and maintaining self-tolerance.

Lymphocytes normally begin to express Fas a few days after activation and so either become targets for killing by other FasL-expressing lymphocytes or commit suicide because they themselves express FasL. In contrast, lymphocytes from either lpr/lpr or gld/gld mice, which have defects in the genes encoding Fas and FasL, respectively, do not undergo activation-induced apoptotic cell death and the clonal downsizing of immune responses is impaired, thus explaining the progressive accumulation of lymphocytes in the periphery and the autoimmune features seen in these mice (14).

The human equivalent of the mouse mutants may be the families with childhood-inherited lymphadenopathy and autoimmunity who have been described recently, in whom the primary defect seems to be abnormal persistence of activated lymphocytes (15). These children have an autoimmune lymphoproliferative disease identical to that of lpr/lpr and gld/gld mice and exhibit mutations in the Fas gene (15). Furthermore, there is some evidence that dysregulated Fas–FasL interactions may have a role even in the pathogenesis of systemic lupus erythematosus, fulminant hepatitis, graft versus host disease, and some rare malignancies, such as large granular lymphocytic leukemia and natural killer cell lymphoma (15).

APOPTOTIC SIGNALING THROUGH Fas

Several apoptosis-inducing extracellular agents and agonists, such as tumor necrosis factor (TNF)-α, cause activation of sphingomyelinases in various cell lines and result in the generation of ceramide (16). It is now recognized that even apoptotic signaling through Fas involves sphingomyelin breakdown and the generation of cell-associated ceramide (9). Furthermore, this pathway has a central role even in the induction of other aspects of growth suppression, such as differentiation and cell cycle arrest (17).

Ceramide is generated through the hydrolysis of sphingomyelin (17). This occurs via the action of a sphingomyelin-specific form of phospholipase C, a SMase, which might initiate signaling leading to programmed cell death in response to ionizing radiation or activation of Fas or TNF receptor (17). SMase exists in two forms: a membrane-bound or a cytosolic variant of the neutral SMase with a neutral pH optimum, and a lysosomal acidic form (18–20). Nevertheless, the acidic SMase, rather than the neutral SMase, has been implicated in generation of ceramide relevant to mediating Fas-induced apoptosis, even if both these enzymes are activated after Fas cross-linking (9,21). It is of note that the absence of acidic SMase is central to the pathogenesis of Niemann–Pick disease and causes the widespread deposition of sphingomyelin seen in these patients (22).

Initial studies of apoptosis with cell-permeable ceramides showed that the cytotoxicity of these molecules closely mimicks the toxic effects of TNF-α (17).

Furthermore, ceramide-induced cytotoxicity displays significant structural specificity, in that molecules closely related to ceramide, such as dihydroceramide, lack any significant toxic effects (17). Apoptosis is induced in U937 leukemia cells and other cell lines by C2-ceramide, as shown by the appearance of specific DNA ladders on agarose gel electrophoresis as well as by other additional morphological and histochemical criteria (17). Similar results have been obtained using C8-ceramide in other cell systems but, in contrast, C2-dihydroceramide and other dihydroceramides proved ineffective in inducing apoptosis, thus confirming the specificity of the effects of ceramide (17).

The intracellular targets of ceramide action are as yet unknown. However, the search for potential immediate targets has led to the identification of several candidate ceramide-regulated enzymes, including a ceramide-activated protein kinase (CAPK), a ceramide-activated protein phosphatase (CAPP), and the isozyme zeta of PKC (17).

Fas AND HIV INFECTION

A growing body of data suggests that the Fas–FasL system may be involved in functional defects and depletion of T lymphocytes and, therefore, in the pathogenesis of HIV disease. The expression of Fas has been shown to be increased in both CD4 and CD8 cells from HIV-infected subjects and to correlate with spontaneous apoptosis in cultures in vitro (7,8). It has also been observed that expression of Fas by CD4 T cells from HIV-infected subjects correlates negatively with their CD4 counts and increases concomitantly with the progression of HIV infection to more advanced stages (7,8). Furthermore, it has been reported that Fas stimulation by cross-linking with monoclonal antibodies induced peripheral blood T cell apoptosis in HIV-infected individuals (6). Remarkably, the effect of HIV Tat protein on T cell apoptosis appears to be mediated by increased FasL expression at concentrations of Tat similar to those observed in the sera from HIV-infected patients (23).

Studies focusing on the expression of Fas and FasL by T helper (Th) 1 and Th2 lymphocyte populations have shown that, whereas both subsets express Fas, only Th1 cells express readily detectable amounts of FasL after activation (24). Interestingly, the differential expression of FasL by Th1 and Th2 correlates with their susceptibility to apoptosis (24) and this has been proposed to explain the shift from a Th1-type to a Th2-type response occurring in subjects progressing to AIDS (25,26).

CERAMIDE AND HIV INFECTION

Recent studies have demonstrated that the generation of ceramide is involved in the pathogenesis of HIV infection. Ceramide has indeed been reported to increase following the experimental infection of the CEM cell line (10) and to potently induce retroviral replication in chronically HIV-infected HL60, U-1IIB, and OM-10.1 cells (11,12). Moreover, the treatment of HIV-LTR transfected cells with ceramide strongly inceases the reporter CAT expression, and Rb hypophosphorylation induced by ceramide might remove the E2F-imposed transcriptional inhibition of the viral genome (27,28).

Patients with AIDS have significantly higher lymphocyte-associated ceramide levels in comparison to long-term nonprogressors, who remain clinically healthy and immunocompetent over an extended time despite chronic HIV infection (29,30). Both

AIDS patients and long-term nonprogressors, though, had elevated levels compared to healthy individuals (29,30). Remarkably, the frequency of lymphocytes undergoing apoptosis differed between AIDS patients and long-term nonprogressors, since a higher frequency of CD4 and CD8 lymphocytes undergoing apoptosis was measured in AIDS patients than in long-term nonprogressors (30). Furthermore, the difference in the levels of ceramide appeared to be associated with differences in lymphocyte counts and HIV viremia. In fact, CD4 and CD8 cell counts were higher in long-term nonprogressors than in AIDS patients, whereas the viral load was more elevated in the AIDS group (30). These data add weight to the in vitro observations reported above suggesting a critical role for ceramide in HIV replication and apoptotic loss of T lymphocytes throughout the course of the infection. It is conceivable that fluctuations in lymphocyte-associated ceramide content may accompany or precede an accelerated rate of lymphocyte apoptosis and viral replication. In turn, down-modulation of ceramide may have a strong impact on the clinical progression of the infection.

EFFECT OF L-CARNITINE TREATMENT IN VIVO ON APOPTOSIS AND CERAMIDE GENERATION IN PATIENTS WITH AIDS

In a preliminary study, we investigated the effects of short-term L-carnitine treatment on the apoptosis of CD4 and CD8 cells as well as on ceramide generation. Briefly, 10 male patients with overt AIDS received L-carnitine (Sigma Tau) (6 g) intravenously in normal saline over a 2 h period each day for 5 days. Blood samples for measuring CD4 and CD8 cell counts, the frequency of cells undergoing apoptosis, and the levels of peripheral blood mononuclear cell (PBMC)-associated ceramide were taken at baseline (T0), at day 3 (T1), and at day 6 (T2).

L-Carnitine treatment had a strong impact on the counts of CD4 and CD8 lymphocytes undergoing apoptosis. Apoptotic CD4 cells decreased significantly at both

Table 1. PBMC-associated Ceramide Levels in 10 Patients with AIDS at T0 T1 (day 3) and T2 (day 6) of L-Carnitine Treatment

	Ceramide (pmol/10^6 cells)		
Patient	T0	T1	T2
1	112	27	27
2	178	20	7.7
3	42	25	21
4	72	9	12
5	56	39	19
6	87	29	28
7	36	22	3
8	98	21	42
9	116	72	64
10	48	52	30
Mean	84.5	31.6*	25.4**
SD	41.4	17.4	16.9

$*p = 0.0039$, $**\ p = 0.002$ compared to T0.

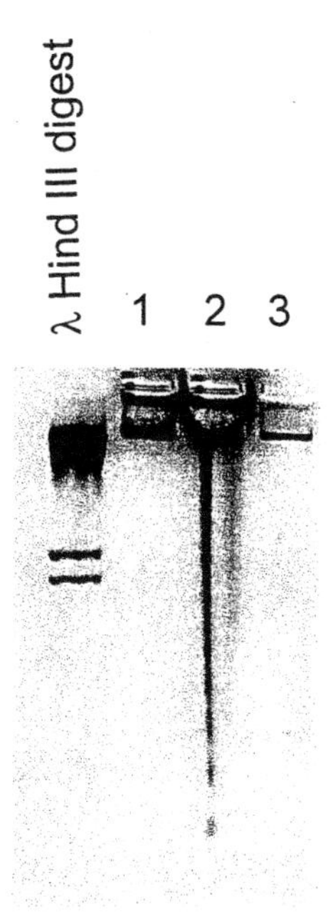

Figure 1. Effect in vitro of L-carnitine on Fas-induced apoptosis in Hut 78 cells. DNA fragmentation is shown. Hut 78 cells were treated with anti-Fas mAb (200 ng/ml) for 8 h in the presence or absence of L-carnitine (100 μg/ml). DNA was then isolated and electrophoresed according to standard methods. *Hin*dIII digest = markers. Lane 1, DNA from untreated cells; lane 2, DNA from cells treated with anti-Fas mAb for 8 h; lane 3, DNA from cells treated with anti-Fas mAb for 8 h in the presence of L-carnitine, which was added to cell culture 1 h before the addition of stimulus.

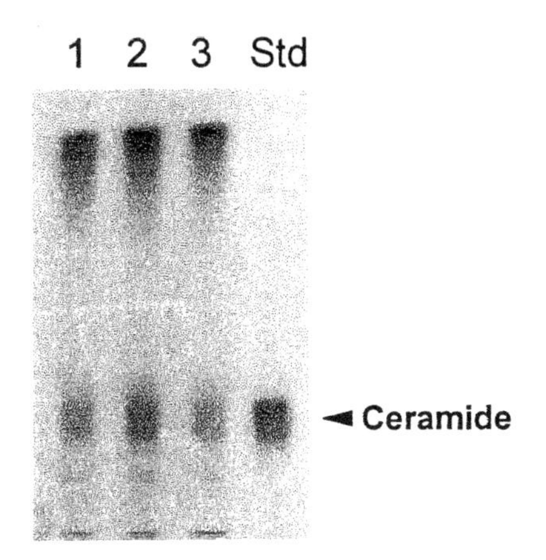

Figure 2. Effect of L-carnitine treatment in vitro on ceramide generation induced by anti-Fas mAb. Hut 78 cells were stimulated with anti-Fas mAb (200 ng/ml) for 10 min in the presence or absence of L-carnitine (100 μg/ml). Lipids were extracted, subjected to DAG kinase assay, and separated by TLC. Radioactive spots were visualized by autoradiography. Lane 1, untreated cells; lane 2, cells treated with anti-Fas mAb; lane 3, cells treated with anti-Fas mAb in the presence of L-carnitine, which was added to cell culture 1 h before the addition of stimulus; Std = ceramide authentic standard. Two different experiments gave similar results.

T1 (271±88 cells/μl) ($p = 0.014$) and T2 (243.7 ± 127 cells/μl) ($p = 0.02$) compared to T0 (473 ± 216 cells/μl). A decrease in the counts of apoptotic CD8 cells was observed at T1 compared to T0 (191 ± 144 and 201.4 ± 129 cells/μl, respectively), but this approached statistical significance only at T2 (132 ± 96 cells/μl) ($p = 0.06$). Remarkably, when the treatment was interrupted the counts of apoptotic CD4 and CD8 cells returned toward the baseline level within a week (not shown). Moreover, we observed throughout the study period a trend indicating an increase in the counts of total lymphocytes as well as CD4 and CD8 cells (not shown).

The L-carnitine administration resulted in decreased levels of PBMC-associated ceramide with respect to baseline and the decrease was statistically significant at both T1 ($p = 0.0039$) and T2 ($p = 0.002$) (Table 1). The decrease in ceramide levels from baseline to T2 was moderately correlated ($r = 0.467$; $p = 0.166$) with the changes in the frequency of CD4 cells undergoing apoptosis and well correlated ($r = 0.709$; $p = 0.024$) with the frequency of CD8 cells undergoing apoptosis.

IN VITRO EFFECTS OF L-CARNITINE ON Fas-MEDIATED APOPTOSIS AND CERAMIDE PRODUCTION

We investigated the effects of L-carnitine in vitro on apoptosis and ceramide production by Fas-sensitive cell lines (Hut 78 and U937) after Fas cross-linking. Furthermore, we assayed the effects of L-carnitine on the activity of acidic and neutral SMases. Briefly, Hut78 and U937 (5×10^6 cells/ml) were stimulated with anti-Fas monoclonal antibody (UBI, Lake Placid, NY) (200 ng/ml) for 10 min. When indicated, L-carnitine (Sigma Tau, Pomezia, Italy) was added to the cell cultures 1 h before stimulation. The measurement of cell-associated ceramide and the assay of SMases were performed by standard methods (9,21). Quantitative results for ceramide production are expressed as pmol ceramide 1-phosphate/mg protein. The activation of SMases was expressed as pmol sphingomyelin hydrolysed/mg protein.

These experiments demonstrate that L-carnitine inhibits the Fas-dependent apoptosis of Hut 78 cells (Figure 1) and U937 cells (not shown) by preventing in a dose-dependent way the generation of ceramide (Figure 2).

Investigation of the effects of L-carnitine in vitro on purified acidic and neutral SMases has demonstrated that L-carnitine down-modulates ceramide generation through the inhibition of the acidic SMase whereas Fas-activated neutral SMase is not significantly affected by L-carnitine treatment (Figure 3). The results demonstrate that L-carnitine directly inhibits the activity of the acidic enzyme in a dose-dependent manner. The inhibitory effects of L-carnitine on acidic SMase were maximally 30–40%, even at doses of the drug that were able to totally inhibit the enzyme activity in the cells. Transformation inside the cells of L-carnitine into acylcarnitines could account for this apparent discrepancy, according to experiments demonstrating that acylcarnitines are more effective than L-carnitine in inhibiting acidic SMase both in the cells and in the purified enzyme system (Cifone et al., unpublished observations).

The structural basis for the SMase inhibition by L-carnitine is still unknown. However, one can speculate that similarities between carnitine and sphingomyelin allow L-carnitine to be recognized as a substrate analog by acidic SMase. The inability of L-carnitine to affect neutral SMase either in the cellular extracts or in the purified enzyme system could be accounted for by the different ionization state of the drug at pH 7.0 in comparison to that at pH 5.0.

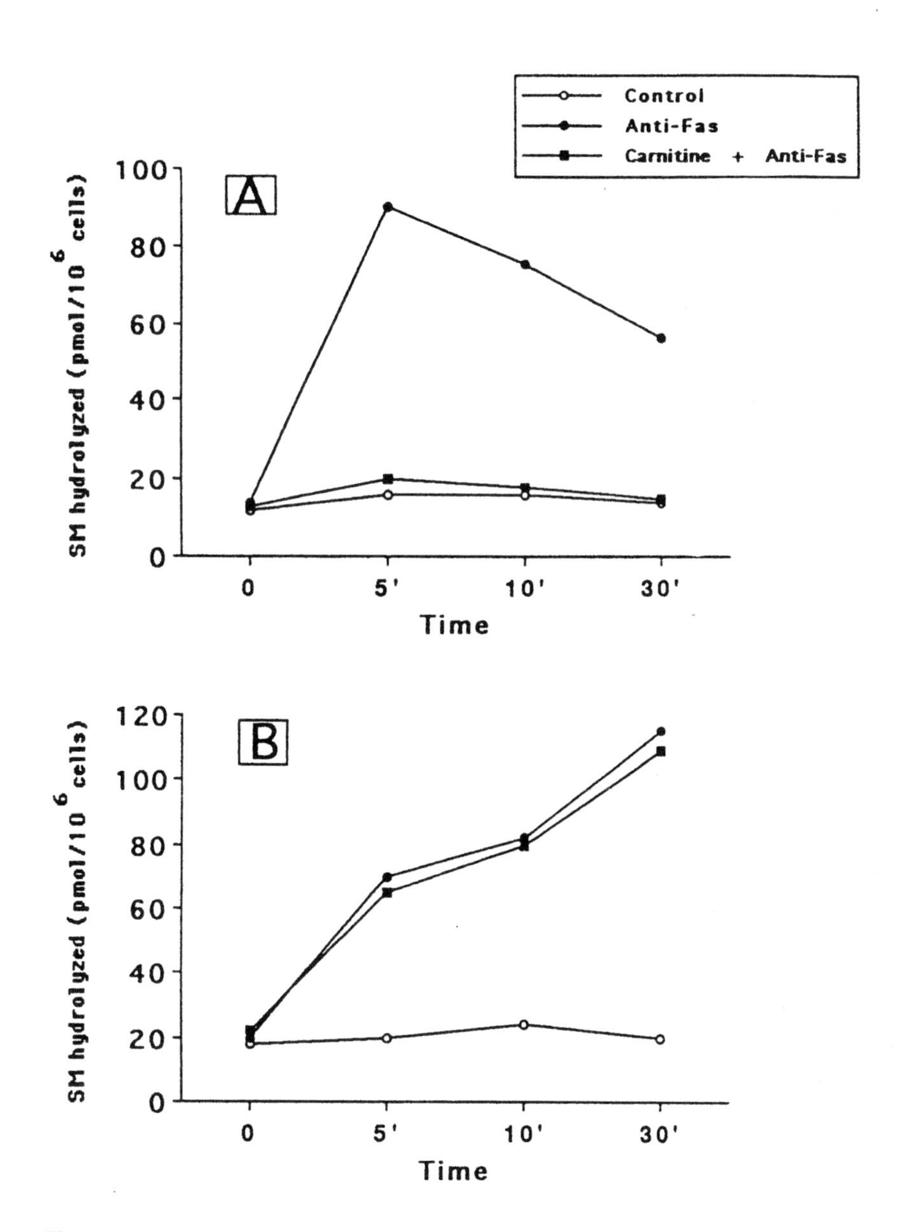

Figure 3. Effect of L-carnitine treatment in vitro on acidic and neutral SMases induced by anti-Fas mAb. Hut 78 cells were stimulated with anti-Fas mAb for different times and cell lysates were then reacted at pH 5.5 or 7.4 with labeled sphingomyelin (SM) vesicles to assay acidic SMase (A) or neutral SMase (B), respectively. After reaction of cell lysates with labeled SM vesicles in the proper buffer, phospholipids were extracted, separated by TLC, and visualized by autoradiography. Radioactive spots (SM) were scraped from the plate and counted by liquid scintillation. Data are expressed as pmol sphingomyelin hydrolyzed/10^6 cells. Two different experiments gave similar results.

PERSPECTIVES

Our data demonstrate that L-carnitine treatment in vivo may significantly reduce the apoptosis of CD4 and CD8 cells and the generation of ceramide in HIV-infected patients. These results are reinforced by experiments in vitro showing the ability of L-carnitine to prevent the Fas-dependent apoptosis through inhibiting the activity of acidic SMase and the consequent generation of ceramide. Thus, L-carnitine appears to bypass the initiation step of apoptosis and block the engagement of the apoptotic machinery through down-modulating ceramide.

These preliminary results suggest that L-carnitine treatment has the potential to slow the progression of HIV infection through reducing the levels of lymphocyte apoptosis. There is evidence that in vivo removal of cells by apoptosis is an extremely rapid process and, once this is considered, it becomes apparent that even a very small reduction in apoptosis may result in a substantial increase in lymphocyte counts (31). Our results are consistent with this hypothesis since we found that the reduced frequency of cells undergoing apoptosis in L-carnitine-treated patients was paralleled by an increase in CD4 and CD8 cell counts, despite the short-term course of L-carnitine therapy.

These encouraging results require confirmation in further studies. However, in this regard it should be noted that early clinical trials of high-dose L-carnitine supplementation aimed at correcting the carnitine deficiency commonly observed in AIDS patients have demonstrated that L-carnitine therapy may improve several immunologic and metabolic parameters of the disease (32–35).

REFERENCES

1. Verhofstede C, Reniers S, Van Wanzeele F, Plum J. Evaluation of proviral copy number and plasma RNA level as early indicators of progression in HIV-1 infection: correlation with virological and immunological markers of disease. AIDS 1994; 8:1421–1427.
2. Loverday C, Hill A. Prediction of progression to AIDS with serum HIV-1 RNA and CD4 count. Lancet 1995; 345:790–791.
3. Mellors JW, Kingsley LA, Rinaldo CR, et al. Quantitation of HIV-1 RNA in plasma predicts outcome after seroconversion. Ann Intern Med 1995; 122:537–579.
4. Gougeon ML, Montagnier L. Apoptosis in AIDS. Science 1993; 260:1269–1270.
5. Ameisen JC, Estaquier J, Idziorek T, De Bels F. The relevance of apoptosis to AIDS pathogenesis. Trends Cell Biol 1995; 5:27–40.
6. Katsikis PD, Wunderlich ES, Smith CA, Herzenberg LA. Fas antigen stimulation induces marked apoptosis of T lymphocytes in human immunodeficiency virus-infected individuals. J Exp Med 1995; 181:2029–2036.
7. Debatin KM, Fahrig-Faissner A, Enenkel-Stoodt S, Kreuz W, Benner A, Krammer PH. High expression of APO-1 (CD95) on T lymphocytes from human immunodeficiency virus-1-infected children. Blood 1994; 83:3101–3103.
8. Andrieu JM, Lu W. Viro-immunopathogenesis of HIV disease: implications for therapy. Immunol Today 1995; 16:5–7.
9. Cifone MG, De Maria R, Roncaioli P, et al. Apoptotic signaling through CD95 (Fas/Apo-1) activates an acidic sphingomyelinase. J Exp Med 1994; 177:1547–1552.
10. Van Veldhoven PP, Matthews TJ, Bolognesi DP, Bell RM. Changes in bioactive lipids, alkylacylglycerol and ceramide occur in HIV-infected cells. Biochem Biophys Res Commun 1992; 187:209–216.

11. Rivas CI, Golde DW, Vera JC, Kolesnick RN. Involvement of the sphingomyelin pathway in autocrine tumor necrosis factor signaling for human immunodeficiency production in chronically infected HL-60 cells. Blood 1994; 83:2191–2197.
12. Papp B, Zhang D, Groopman JE, Byrn RA. Stimulation of human immunodeficiency virus type 1 expression by ceramide. AIDS Res Hum Retrovir 1994; 10:775–780.
13. Lynch DH, Ramsdell F, Alderson MR. Fas and FasL in the homeostatic regulation of immune responses. Immunol Today 1995; 16:569–574.
14. Nagata S, Suda T. Fas and Fas ligand: lpr and gld mutations. Immunol Today 1995; 16:39–43.
15. Rowe PM. Glimmers of clinical relevance for Fas. Lancet 1996; 347:1398.
16. Pushkareva M, Obeid LM, Hannun YA. Ceramide: an endogenous regulator of apoptosis and growth suppression. Immunol Today 1995; 16:294–297.
17. Kolesnick RN, Fuks Z. Ceramide: a signal for apoptosis or mitogenesis. J Exp Med 1995; 181:1949–1952.
18. Kolesnick RN. Sphingomyelin and derivatives as cellular signals. Prog Lipid Res 1991; 30:1–38.
19. Schuchman EH, Suchi M, Takahahi T, Sandhoff K, Desnick RJ. Human acid sphingomyelinase. J Biol Chem 1991; 266:8531–8539.
20. Okazaki T, Bell RM, Hannun Y. Sphingomyelin turnover induced by vitamin D_3 in HL60 cells: role in cell differentiation. J Biol Chem 1989; 264:19076–19080.
21. Cifone MG, Roncaioli P, De Maria R, et al. Multiple pathways originate at the Fas/APO-1 (CD95) receptor: sequential involvement of phosphatidylcholine-specific phospholipase C and acidic sphingomyelinase in the propagation of the apoptotic signal. EMBO J 1995; 14:5859–5868.
22. Kolodny EH. Niemann–Pick disease. In Wyngaarden JB, Smith LH, Bennett JC, eds. Cecil Textbook of Medicine. Philadelphia: W.B. Saunders; 1992:1093–1094.
23. Westendorp MO, Frank R, Ochsenbauer C, et al. Sensitization of T cells to CD95-mediated apoptosis by HIV-1 Tat and gp120. Nature 1995; 375:497–500.
24. Ramsdell F, Seaman MS, Miller RE, Picha KS, Kennedy MK, Lynch DH. Differential ability of T(h)1 and T(h)2 cells to express Fas ligand and to undergo activation-induced cell death. Int Immunol 1994; 6:1545–1553.
25. Clerici M, Shearer GM. A Th1-Th2 switch is a critical step in the etiology of HIV infection. Immunol Today 1993; 14:107–111.
26. Clerici M, Shearer GM. The Th1-Th2 hypothesis of HIV infection: new insights. Immunol Today 1994; 15:575–581.
27. Venable ME, Lee JY, Smyth MJ, Bielawska A, Obeid LM. Role of ceramide in cellular senescence. J Biol Chem 1995; 270:30701–30708.
28. Kundu M, Srinivasan A, Pomerantz RJ, Khalili K. Evidence that a cell cycle regulator, E2F1, down-regulates transcriptional activity of the human immunodeficiency virus type I promoter. J Virol 1995; 69:6940–6946.
29. De Simone C, Cifone G, Roncaioli P, et al. Ceramide, AIDS, and long-term survivors. Immunol Today 1996; 17:48.
30. De Simone C, Cifone G, Alesse E, et al. Cell-associated ceramide in HIV-1-infected subjects. AIDS 1996, in press.
31. Clarke AR, Sphyris N, Harrison DJ. Apoptosis in vivo and in vitro: conflict or complementarity? Mol Med Today 1996; 2:189–191.
32. De Simone C, Tzantzoglou S, Famularo G, et al. High-dose L-carnitine improves immunologic and metabolic parameters in AIDS patients. Immunopharmacol Immunotoxicol 1993; 15:1–12.
33. De Simone C, Famularo G, Tzantzoglou S, et al. Carnitine depletion in peripheral blood mononuclear cells from patients with AIDS: effect of oral L-carnitine. AIDS 1994; 8:855–860.

34. Famularo G, De Simone C. A new era for carnitine? Immunol Today 1995; 16:211–213.
35. De Simone C, Famularo G, Cifone G, Mitsuya H. HIV-1 infection and cellular metabolism. Immunol Today 1996; 17:2

42

Nutriceutical Modulation of Glutathione with a Humanized Native Milk Serum Protein Isolate, IMMUNOCAL™: Application in AIDS and Cancer

Sylvain Baruchel and Ginette Viau
McGill University–Montreal Children's Hospital Research Institute, Montreal, Quebec, Canada

René Olivier
Pasteur Institute Paris, France

Gustavo Bounous
Montreal General Hospital, Montreal, Quebec, Canada

Mark A. Wainberg
Jewish General Hospital, Lady Davis Institute, Montreal, Quebec, Canada

NUTRITIONAL IMMUNOMODULATION AND ITS RELATION TO GLUTATHIONE SYNTHESIS

Fresh, raw milk includes the group of proteins that remain soluble in "milk serum." These proteins can be preserved in their native form if extracted carefully from their natural source.

In 1981 it was discovered that normal mice fed a milk serum protein concentrate (specially prepared under mild nondenaturing conditions) exhibited a marked increase in the humoral immune response to a T helper cell-dependent antigen (1). In the following years, numerous experiments confirmed the consistency of this phenomenon (2–10). Over a period of 12 years and based on these findings a humanized native milk serum protein isolate (HNMPI) named IMMUNOCAL™ was developed (Immunotec Research Corporation Ltd., Montreal, Quebec, Canada).

This property was found to be related, at least in part to a greater production of splenic glutathione (L-α-glutamylcysteinylglycine) (GSH) during the oxygen-requiring antigen-driven clonal expansion of the lymphocyte pool in animals fed with this bioactive HNMPI (9). Adequate levels of GSH are necessary for lymphocyte proliferation in the development of the immune response (11,12). Moderate but sustained elevation of cellular GSH was also found in the liver and the heart of healthy, old mice fed with this HNMPI for a prolonged period. In addition, HNMPI markedly increased their life expectancy in comparison to control animals fed nutritionally equivalent diets (13).

Glutathione is of major significance in cellular antioxidant activity in what Meister called the "GSH antioxidant system" because it participates directly in the destruction of reactive oxygen compounds and also because it maintains in reduced form ascorbate (vitamin C) and α-tocopherol (vitamin E), which also exerts an antioxidant effect (14).

FUNCTION OF HNMPI AS A CYSTEINE DELIVERY SYSTEM

What ingredient in IMMUNOCAL™ makes it an effective "cysteine delivery system"?

Systemic availability of oral GSH is negligible in man (15) and there is no evidence for transport of GSH into cells (16). Thus, it has to be synthesized intracellularly. This occurs in two steps: (a) glutamylcysteine synthesis; (b) glutathione synthesis. Even though the inflow of cysteine, glutamate, and glycine might prove somewhat limiting under selected circumstances, numerous observations have shown that it is the transport of cysteine (or cystine, which usually is promptly reduced to cysteine on cell entry) which tends to be the rate-limiting event in GSH synthesis. whereas free cysteine does not represent an ideal delivery system (17) because it is toxic and is spontaneously oxidized. Cysteine present as the disulfide cystine released during digestion in the gastrointestinal tract is more stable than free amino acid. GSH synthesis is submitted to negative feedback inhibition by the end-product GSH. The disulfide bond is pepsin- and trypsin-resistant, but may be split by heat and mechanical stress (9). Cystine accounts for about 90% of the low-molecular-mass cysteine in the blood plasma, while reduced cysteine is present only at extremely low concentration (18).

In a comparative study, we found that commercial milk serum concentrates exhibiting far less bioactivity, including less GSH promoting activity, contain about half the amount of serum albumin (9) and 4 times less lactoferrin than HNMPI, expressed as percentage of total milk serum protein. IMMUNOCAL™ is produced in a proprietary lenient process which results in the preservation of the most thermolabile proteins in their native conformation.

In the serum albumin, there are 17 cystine residues per 66 kDa molecule and 6 Glu-Cys dipeptides (19); in lactoferrin there are 17 per 77 kDa molecule and 4 Glu-Cys dipeptides (20); and in the α-lactalbumin there are 4 cystines in a 14,000 kDa molecule

Table 1.

	Molecular Mass (kDa)	Residues	Cysteine residues per molecule	Cysteine $(Cys)_2$ (disulfide)	Glu-$(Cys)_2$
β-Lactoglobulin	18,400	162	5	2	0
α-Lactabumin	14,200	125	8	4	0
Serum albumin	66,000	582	35	17	6
Lactoferrin	77,000	708	40	17	4

Source: Refs. 19, 20.

(19). On the other hand, β lactoglobulin has only 2 cystines in a 18,400 kDa molecule (19), and IgG1, the predominant immunoglobulin in cow whey, has only 4 disulfide bridges in a 166,000 kDa molecule (Table 1). In addition, Meister and colleagues (16) have demonstrated that the γ-glutamylcysteine (Glu-Cys) precursors of GSH can easily enter the cell and there be synthesized into GSH. It thus become noteworthy that the most labile milk proteins–, serum albumin and lactoferrin–are those which contain these putative GSH-promoting peptide components.

Finally, the bioavailibility of the presumed active component (cystine and Glu-Cys group) may be influenced by the coexistence of the other proteins throughout the digestive–absorptive process.

This newly discovered property of HNMPI was found to be independent of its nutritional value, as other proteins of similar nutritional efficiency do not exhibit this unique property (1–10). The concept that a specific biological activity can exist in addition to and independent of the systemic effect of IMMUNOCAL™ as a good protein source is further substantiated by recent in vitro assays (21).

The dietary provision of cystine is particularly relevant to the immune system. The coordinated response of macrophages and lymphocytes in the T cell-mediated immune response is regulated, in part, by macrophage cystine uptake and subsequent release of reduced cysteine into the local environment for uptake by lymphocytes. When the antigen-presenting macrophages come into close contact with antigen-specific T cells, they supply these cells with additional amounts of cysteine and thereby raise their intracellular GSH level (18).

The validity of this assumption is confirmed by the demonstration that the immunoenhancing and GSH-promoting (data not shown) effect of IMMUNOCAL™ is abolished by buthionine sulfoximine, which inhibits γ-glutamylcysteine synthetase, the initial step in GSH synthesis (17).

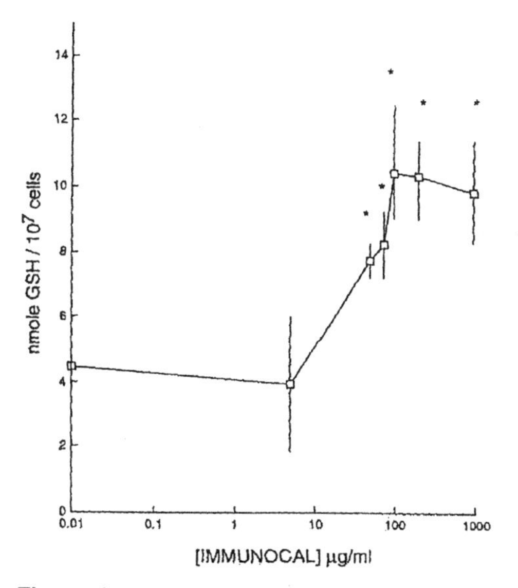

Figure 1. Incubation of PBMC for 72 h in the presence of various amounts of IMMUNOCAL™. Each point represents the mean ±SD of 3 measurements of intracellular glutathione. $*p < 0.05$.

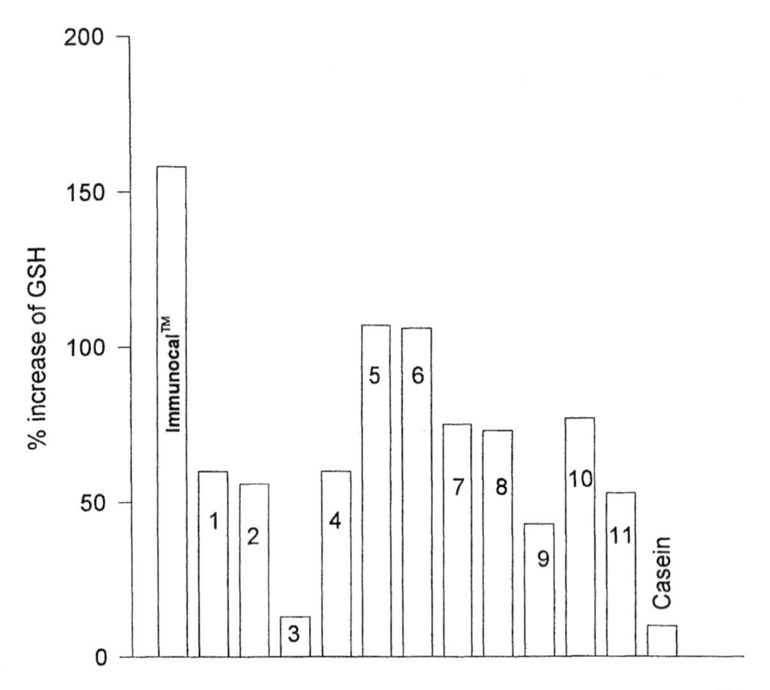

Figure 2 Incubation of PBMCs for 72 h in the presence of IMMUNOCAL™ and other serum milk products: Percentage increase in glutathione.

IN VITRO MODULATION OF INTRACELLULAR GLUTATHIONE BY IMMUNOCAL™

We demonstrated that normal human lymphocytes cultured for 3 days with HNMPI 100 μg/ml show an increase in intracellular GSH content from 4.5 ± 0.4 to 10.5 ± 3.4 nmol/10^6 cells, $p < 0.01$ (Figure 1). This increase in GSH correlates with an increase in cellular proliferation measured by thymidine incorporation (data not shown). The

Table 2 Presence of Cytopathic Effects in MT-4 Cells

	$TCID_{50}$/well[a]			
IMMUNOCAL™ (μg/ml)	2000	200	20	2
0	+++	++	+	–
1	+++	++	+	–
10	++	+	+	–
100	–	–	–	–
500	–	–	–	–
1000	–	–	–	–

[a]+ Presence of cytopathic effects; – absence of cytopathic effects.

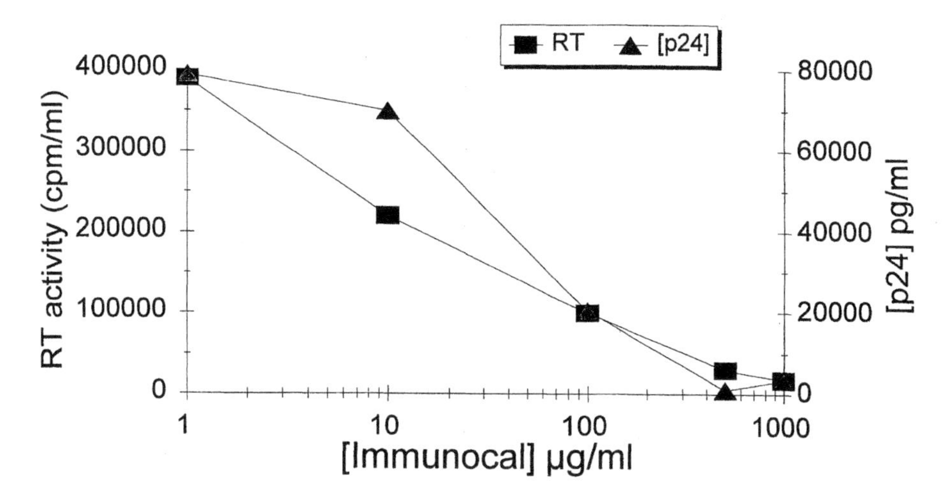

Figure 3. IMMUNOCAL™ has been shown to inhibit HIV replication.

increase in GSH is dose-dependent and has not been found for casein or for any commercially available milk serum protein concentrate (Figure 2).

IN VITRO ANTI-HIV and ANTIAPOPTOTIC ACTIVITY OF HNMPI

Clinically, there is direct evidence that HIV infection is associated with a GSH deficiency in the peripheral blood mononuclear cells (PBMC) (18). The depletion of intracellular GSH suggests an association between oxidative stress and HIV infection. Oxidative stress may be one of the mechanisms that contribute to disease progression and the wasting syndrome through mediators of inflammation such as TNF-α and IL-6. During this period of progression, glutathione is consumed owing to an increase in oxidative stress. GSH

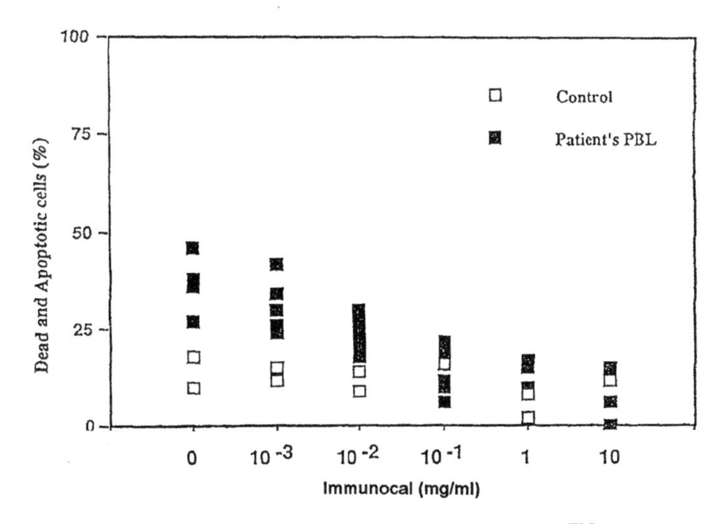

Figure 4. Inhibition of early cell death by IMMUNOCAL™.

depletion, a consequence of chronic oxidative stress, is part of the spectrum of HIV infection. GSH has, in addition, a crucial role in lymphocyte function and cell survival.

IMMUNOCAL™ functioning as a cysteine delivery system can enhance GSH synthesis in vitro (Figure 1) and inhibits HIV replication on a cord mononuclear cell system infected by HTL V-IIIB (Figure 3). IMMUNOCAL™ also inhibits the formation of syncitium between infected and noninfected cells. The inhibition of syncitium formation occurred at the same concentration as inhibition of HIV replication (Table 2). This viral inhibition was not associated with any cytotoxicity. IMMUNOCAL™, via its GSH-promoting activity, reduces apoptosis in HIV-infected cells. Apoptosis was evaluated by flow cytometry on PBMC from HIV-infected individuals (Dr. R. Olivier, AIDS and Retrovirus Department, Pasteur Institute). HIV-infected PBMC cultured at concentrations of IMMUNOCAL™ of 100 μg/ml or higher were less prone to die of apoptosis than untreated cells: 15% ± 2.6% vs. 37% ± 2.4, $p<0.001$ (Figure 4).

HNMPI SUPPLEMENTATION IN AIDS AND WASTING SYNDROME

Based on these preclinical data, we conducted a Canadian clinical trial (Canadian HIV Trials Network) with IMMUNOCAL™ in children with AIDS and wasting syndrome. The major objective was to evaluate the effect of oral supplementation with IMMUNOCAL™ on nutritional parameters and intracellular blood lymphocyte GSH concentration in children with AIDS and wasting syndrome. This was an open single-arm pilot study of 6 months duration. Wasting syndrome and severe weight loss within the 6 months preceding entry into the study was an absolute criterion for entry.

IMMUNOCAL™ was administered twice a day as a powder diluted in water. In some patients, IMMUNOCAL™ was administered via nasogastric tube when necessary. The administered starting dose was based on 20% of the total daily protein requirement and was increased by 5% each month over 4 months to reach 35% of the total protein intake at the end of the study. The total duration of the study was 6 months.

Weight, height, triceps skinfold and mid-arm muscle circumferences, CD4/CD8 counts, and peripheral lymphocyte GSH concentrations (measured by spectrophotometric assay) were measured monthly. Energy intake was assessed by the use of two independent 2-day food records with a 2–3 week period between the food records. Each food record included a weekday and a weekend, and the average of these records was calculated to reflect the daily nutritional intake. Out of 14 patients enrolled, 10 were evaluable. The ages of the patient were from 8 months to 15 years. The 10 patients studied were enrolled in four different centers across Canada: Montreal Children's Hospital (Dr. S. Baruchel), The Hospital For Sick Children Toronto (Dr. S. King), Children's Hospital for Eastern Ontario (Dr. U. Allen), and Centre Hospitalier Laval Quebec (Dr. F. Boucher). Of the 4 remaining patients, 2 lacked compliance after 2 months while the other 2 died of AIDS progressive disease within the first 2 months of entry into the study. None of the deaths was related to the tested product.

None of the patients experienced any major toxicity such as diarrhea or vomiting or manifestation of milk intolerance. One patient had to stop IMMUNOCAL™ transiently for minor digestive intolerance such as nausea and vomiting (< twice/day) at month 3 and was subsequently able to restart the treatment without any problem.

At the end of the study, all patients experienced a weight gain in the range of 3.2% to 22% from their starting weight. The mean weight gain for the group was 8.4% ± 5.7%. On analysis of the mean percentage of requirement nutrient intake (RNI) per month for all

Table 3 Changes from Baseline (expressed as percentage) at Weeks 24 and 36 in Weight, Anthropometric Measurements and GSH in Patients Treated with IMMUNOCAL™

	Weight change (%)		Mid-arm muscle circumference change		Triceps skinfold change (%)		PBMC GSH change (%)	
Patient no.	wk 24	wk 32	wk 24	wk 32	wk 24	wk 32	wk 24	wk 32
1	22.1	29.8	9.5	14.3	50.0	25.0	12.2	–9.0
2	14.0	17.3	18.7	25.3	20.0	–20.0	84.0	56.0
3	5.1	9.2	–3.0	–2.0	–17.0	–3.0	37.0	55.0
4	3.8	3.4	4.2	NA	–42.0	NA	305.0	550.0
5	7.1	4.5	13.1	11.4	–24.0	–16.0	–18.0	14.3
6	3.7	5.6	–2.0	–2.0	16.0	16.0	7.1	174.0
7	2.5	NA	5.0	NA	–13.0	NA	54.2	NA
8	14.2	18.2	–3.1	2.0	41.0	43.0	17.3	62.4
9	8.9	7.9	–4.0	–8.0	–30.0	–39.0	–6.6	50.9
10	7.0	NA	1.0	NA	41.0	NA	–1.6	NA

NA

the patients, no correlation was found between the weight gain and any significant increase in the mean percentage of RNI, suggesting reduced catabolism rather than an anabolic effect of IMMUNOCAL™. Six of ten patients have demonstrated an improvement in their anthropometric parameters such as triceps skinfold or mid-arm muscle circumference independently of an increase in energy intake (Table 3).

Two groups of patients were identified in terms of GSH modulation: responders and nonresponders. The responders were those who started the study with a low GSH level.

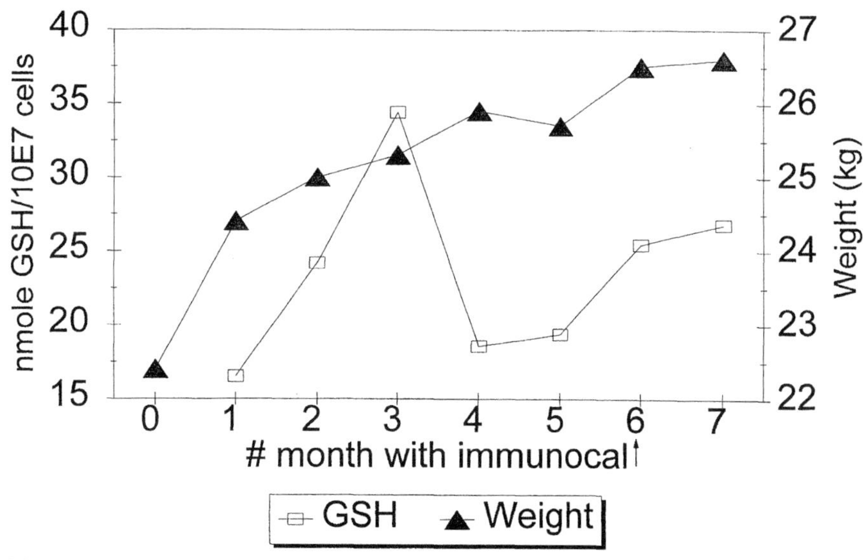

Figure 5. Intracellular glutathione in. HSC 4. Each point represents the mean ±SD of 3 measurements. ↑ indicates end of study.

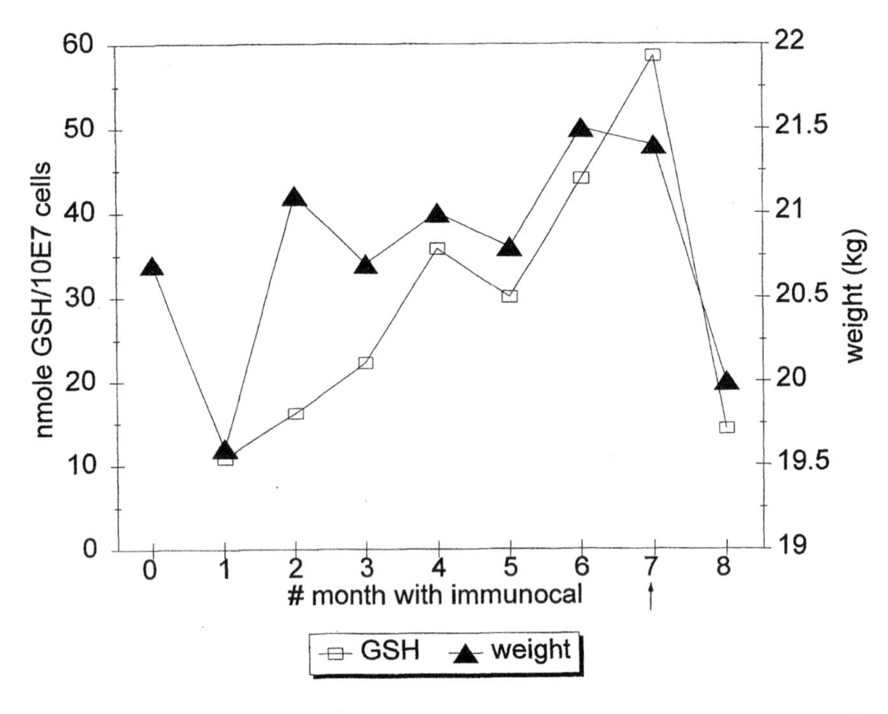

Figure 6. Intracellular glutathione in CHUL 1. Each point represents the mean ±SD of three measurements. ↑ indicates end of study.

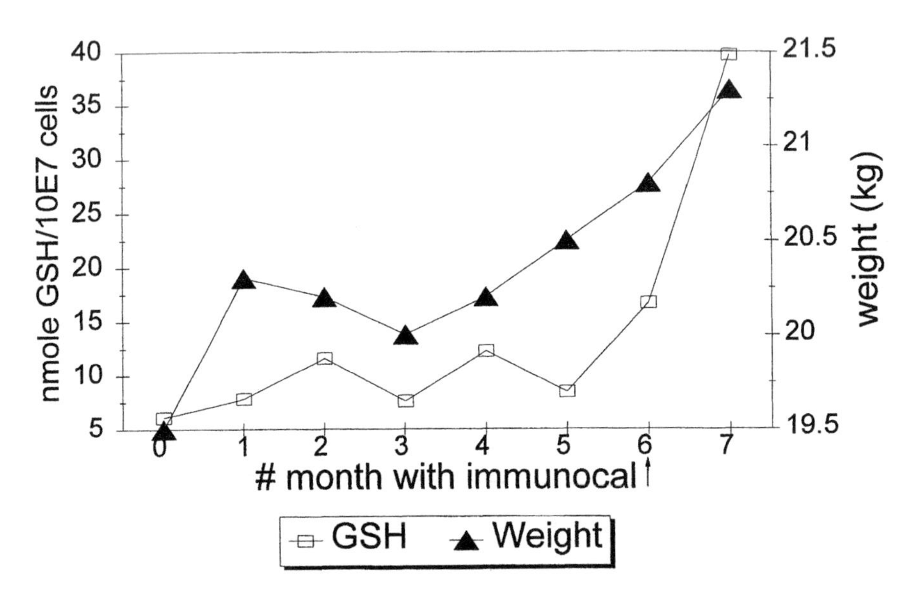

Figure 7. Intracellular glutathione in MCH 3. Each point represents the mean ±SD of three measurements. ↑ indicates end of study.

The nonresponders were those who started with a normal GSH level. A positive correlation was found between increase in weight and increase in GSH (Figures 5,6,7). No changes were found in terms of blood lymphocyte CD4 cell count, but 2 patients exhibited an increase in the percentage of their CD8 cells and 4 patients showed a trend toward an increase in the number of NK cells.

In conclusion, this pilot study demonstrates that IMMUNOCAL™ is very well tolerated in children with AIDS and wasting syndrome and is associated with an amelioration of the nutritional status of the patient as reflected by weight and antrhopometric parameters. Moreover, the GSH-promoting activity of IMMUNOCAL™ in vivo seem to be validated in 6 out of 10 patients. An international multicenter double-blind randomized study is currently under way in France and Canada in adults patients with AIDS and wasting syndrome.

SELECTIVE GLUTATHIONE MODULATION OF BREAST CANCER CELLS AND IMPACT ON CANCER CELL GROWTH

The specific involvement of GSH in the carcinogenic process is supported by the major role played by this compound in the detoxification of carcinogens by conjugation (26). We demonstrated that feeding GSH-promoting HNMPI to mice chronically treated with dimethylhydrazine (DMH) significantly reduces the number and size of colon carcinomas induced by DMH (27,28). These colon tumors appear to be similar to those found in the human insofar as the type of lesions and the chemotherapeutic response characteristics are concerned (26). HNMPI feeding appears to exert an inhibitory effect not only on the initiation (27) of cancer, but also on the progression of tumors (28).

Recently, a direct inhibitory effect of HNMPI in human cancer cell replication was confirmed (21,29,30). In other human cancer cell studies, the inhibitory effect was found to be related to the serum albumin component of milk serum (31) and most recently to α-lactalbumin (32). Feeding lactoferrin to mice inhibited the growth of solid tumors and in addition reduced lung colonization by melanomas (33). Unlike other proteins, serum albumin was found to exhibit a strong antimutagenic effect in an in vitro assay using hamster cells (34). It is therefore noteworthy that in this HNMPI we have succeeded in concentrating serum albumin, α-lactalbumin, and lactoferrin, all containing a significant number of GSH precursors. A possible explanation for these newly discovered properties of dietary milk serum protein may be found in recent findings on the role of GSH in tumor biology (35).

The search for ways to inhibit cancer cells without injuring normal cells has been based over the years on a vain effort to identify the metabolic parameters in which cancer cells are at variance with normal cells. One such function could well be the all-important synthesis of cellular GSH.

Recent experimental evidence has revealed an intriguing response of tumor versus normal cells to GSH synthesis-promoting compounds. Cellular GSH levels have been found to be several times higher in human cancer cells than in adjacent normal cells (35). This finding is presumably related to their proliferative activity. In fact, cancer is the only condition in which elevation of such a tightly regulated system as GSH has been reported. However, when a cysteine- and GSH-promoting compound such as 2-L-oxothiazolidine-4-carboxylate (OTZ) was added to cultured human lung cancer cells exhibiting very high levels of GSH at the outset, no intracellular increase was noted, whereas GSH increased substantially in normal cells (35). This differential response is even more pronounced in vivo. We demonstrated that in tumor-bearing rats, OTZ treatment was actually found to deplete GSH in the tumors (36).

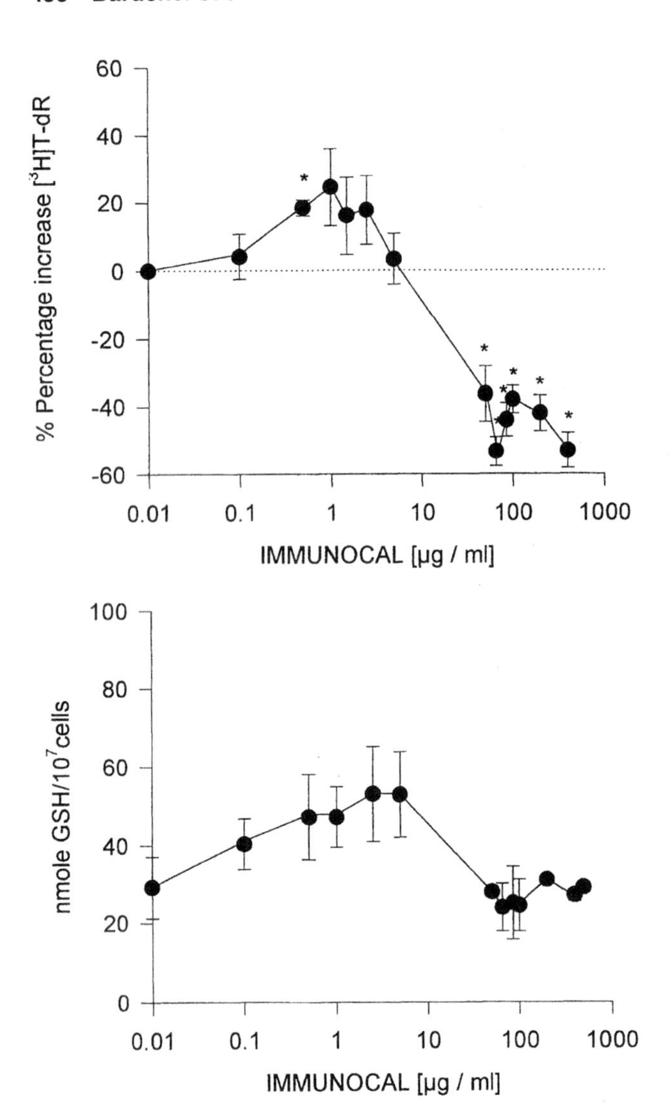

Figure 8. Intracellular glutathione in MATB WT. Each point represents the mean ±SE of three measurements. $*p < 0.05$.

More specifically, an in vitro assay showed that, at concentrations that induce GSH synthesis and proliferation in normal human cells (Figure 1), IMMUNOCAL™ caused GSH depletion and inhibition of proliferation of cells in a rat mammary carcinoma (Figure 8) and Jurkat T cells (Figure 9) (21).

The selectivity demonstrated in these experiments may be explained by the fact that GSH synthesis is negatively inhibited by its own synthesis and since, as mentioned, baseline intracellular GSH in tumor cells is much higher than in normal cells, it is easier to reach the level at which negative feedback inhibition occurs in this cellular system than in a nontumor cellular system.

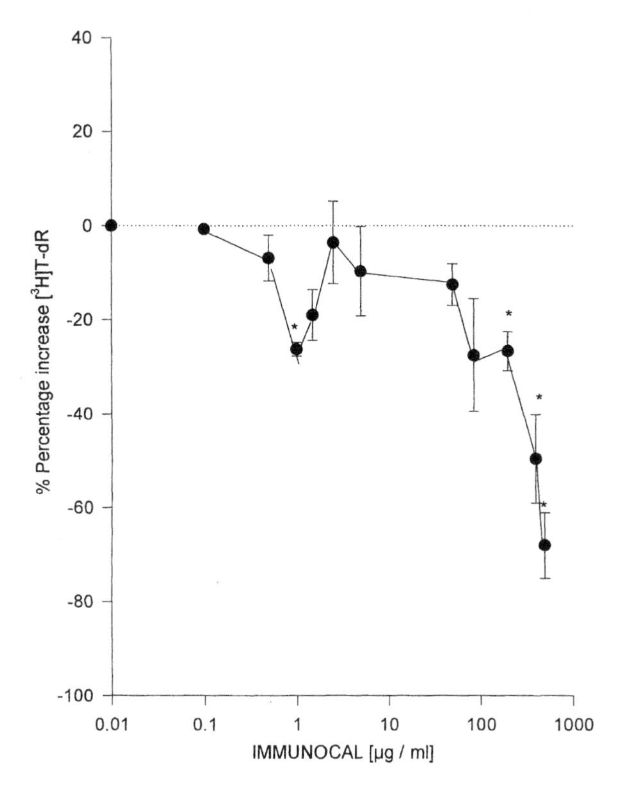

Figure 9. Intracellular glutathione in JURKAT. Each point represents the mean ±SE of three measurements.

HNMPI IN CANCER CLINICAL TRIALS

On the basis of these experiments, 5 patients with metastatic carcinoma of the breast, 1 of the pancreas, and 1 of the liver were fed 30 g of IMMUNOCAL™ daily for 6 months. In 6 patients, the blood lymphocyte GSH levels were substantially above normal at the outset, probably reflecting high tumor GSH levels. At completion of the 6 months of daily supplementation, 2 patients exhibited signs of tumor regression, normalization of hemoglobin and peripheral lymphocytes counts, and a sustained drop of lymphocyte GSH levels toward normal. Two patients showed stabilization of the tumor and increases in hemoglobin levels. In 3 patients, the disease progressed with a trend toward higher lymphocytes GSH levels (37).

A major problem in the use of chemotherapeutic agents in cancer therapy is the protection offered by the defense mechanisms of cancer cells. An important element of protection is represented by GSH, which is an effective detoxification agent that is relatively abundant in tumor cells. Indeed, when GSH synthesis is inhibited by buthionine sulfoximine (BSO), the activity of several chemotherapeutic agents such as alkylating agents is increased and drug resistance can be reversed (36–38). However, the concomitant depletion of GSH in normal cells greatly limits the practical usefulness of this modality of treatment.

We recently demonstrated that a selective GSH prodrug such as OTZ protects some normal tissue (36) but also potentiates the activity of some alkylating agents (38). The apparently selective depletion of tumor GSH levels by provision of a natural precursor of GSH as contained in IMMUNOCAL™ seems to be associated with inhibition of proliferation of cancer cells in vitro. This natural precursor of GSH favorably influences the GSH synthesis in normal cells. These in vitro and preliminary clinical results indicate that this newly discovered property of HNMPI may be a promising adjunct to the nutritional management of cancer patients undergoing chemotherapy. We are currently developing a phase II study in breast carcinoma, attempting to confirm that this selective depletion of GSH may, in fact, render tumor cells more vulnerable to chemotherapy and eventually protect normal tissue against the deleterious effect of chemotherapy.

ANALOGY BETWEEN HNMPI IMMUNOCAL™ AND HUMAN MILK

Human milk contain about 80% of whey protein and 20% of casein. The opposite is true for cow milk. An analysis of the mass ratio of casein to whey protein in milk from various mammals clearly indicates that human milk has the lowest ratio in any mammalian species (39). On the basis of our laboratory studies showing the immunoprotective and anticancer effects of cow whey protein concentrate, it is tempting to speculate that this predominance of whey proteins in human milk is advantageous and thus represent an evolutionary adaptation.

Scientific data based on the similarity between the bioactive components of this native milk protein isolate (HNMPI) of cow milk, IMMUNOCAL, and human whey protein appear to substantiate this theory, as will now be discussed in more detail.

It is well known that breast feeding is superior to the use of cow milk-based formulas of similar nutritional efficiency for the health of human babies. Breast feeding protect against otitis media, and pneumonia (40,41). Mothers milk also has a protective effect on the incidence of several types of childhood cancer including leukemia, lymphomas, bone tumors, and brain tumors (42). Children who are artificially fed or are breast fed for only a short period of time are more at risk for developing several types of cancer before the age of 15 years as compared to long-term breast feeders (43). Thus, the concept of a biological activity in addition to but independent of the nutritional efficiency, formulated to describe the immunoenhancing and GSH-promoting activity of the HNMPI IMMUNOCAL™, may indeed apply to the breast feeding of neonates and infants. Glutathione synthesis appears to be the crucial factor in the health benefit of HNMPI.

It may then be appropriate to identify the features common to HNMPI and human whey proteins that are capable of influencing GSH synthesis in the host. Cysteine, a crucial limiting factor in the synthesis of GSH, is about as abundant in cow's whey protein as it is in whole human milk proteins and several times more abundant than in cow's whole milk (39), since most caseins contain either no cysteine or one or two cysteine residues (19). As mentioned earlier, our studies showed that the most thermolabile milk proteins, namely, serum albumin, α-lactalbumin, and lactoferrin, are crucial to expression of the bioactivity of HNMPI. As shown in Table 1, these proteins are rich in cystine and glutamylcystine residues, natural precursors of GSH. The presence of these dipeptides in the product IMMUNOCAL™ is a characteristic shared with human milk (Table 4).

Traditionally, it has been advocated that "humanized" cow milk should contain more α-lactalbumin because this protein is twice as abundant in human milk. On the basis of our experimental findings, we propose instead that the principal health factor in human milk,

Table 4. Protein Composition of Cow and Human Milk Composition (g/litre)

Component	Cow milk	Human milk	
Casein (g/L)	26	3.2	(0 or 2 cysteine/molecule no disulfide bond)
β-Lactoglobulin (g/L)	3.2	Negligible	
α-Lactalbumin (g/L)	1.2	2.8	
Serum albumin (g/L)	0.4	0.6	
Lactoferrin (g/L)	0.14	2.0	
Total cystine (mol/L)	8.19×10^{-4}	13.87×10^{-4}	
Total cystine (mg/g protein)	6.4	38.7	

Source: Ref. 19; Jennes R. Inter-species comparison of milk proteins. In Fox, ed. Developments in dairy chemistry–1. New York: ASP; 1982:8

not denaturated by heat pasteurization, is due to the predominance of the thermolabile proteins rich in cystine and containing the Glu-Cys dipeptide which are characteristic of the bioactive HNMPI, namely, serum albumin, α-lactalbumin, and lactoferrin. This HNMPI differs from other commercially available milk serum protein concentrates in having a relatively high content of serum albumin (about 10%), lactoferrin (about 0.65%), and α-lactalbumin (about 28%). The variety of diseases against which breast feeding appears to be effective suggest a broader protective mechanism involving cellular GSH and its effect on free radicals, lymphocyte proliferation, and detoxification of carcinogens and other xenobiotics.

CONCLUSION

The biological activity of the proteins isolated from cow's milk in IMMUNOCAL™ depends on the preservation of those labile proteins which share with the predominant human milk proteins the same extremely rare GSH-promoting components. Cellular GSH depletion has been implicated in the pathogenesis of a number of degenerative conditions and disease states including Parkinson's, Alzheimer's, arteriosclerosis, cataracts, cystic fibrosis, malnutrition, aging, AIDS, and cancer (9).

This newly discovered nutriceutical modulation of GSH by the use of humanized native milk serum protein isolate of bovine origin in AIDS and cancer may well find other applications in disease where oxidative stress and pathology of GSH metabolism are largely implicated. Extensive pharmacoepidemiological study of GSH metabolism and standardized methods of measurement of intracellular GSH applicable in clinical trials are needed in order to better define the clinical application of this new type of therapy.

REFERENCES

1. Bounous G, Stevenson MM, Kongshavn PAL. Influence of dietary lactalbumin hydrolysate on the immune system of mice and resistance to Salmonellosis J Infect Dis 1981; 144:281.
2. Bounous G, Kongshavn PAL. Influence of dietary proteins on the immune system of mice. J Nutr 1982; 112:1747–1555.

3. Bounous G, Letourneau L, Kongshavn PAL. Influence of dietary protein type on the immune system of mice. J Nutr 1983; 113:1415–1421.
4. Bounous G, Kongshavn PAL. Differential effect of dietary protein type on the B-cell and T-cell immune response in mice. J Nutr 1985; 115:1403–1408.
5. Bounous G, Shenouda N, Kongshavn PAL, Osmond DG. Mechanism of altered B-cell response induced by changes in dietary protein type in mice. J Nutr 1985; 115:1409–1417.
6. Bounous G, Kongshavn PAL, Gold P. The immunoenhancing property of dietary whey protein concentrate. Clin Invest Med 1988; 11:271–278.
7. Bounous G, Kongshavn PAL. Influence of protein type in nutritionally adequate diets on the development of immunity. In Friedman M, ed. Absorption and utilization of amino acids. Boca Raton, Florida: CRC Press; 1989; 2:219–223.
8. Parker N, Goodrum KJ. A comparison of casein, lactalbumin, and soy protein effect on the immune response to a T-dependent antigen. Nutr Res 1990; 10:781–792.
9. Bounous G, Gold P. The biological activity of undenatured whey proteins: role of glutathione. Clin Invest Med 1991; 14:296–309.
10. Hirai R, Nakai S, Kikuishi H, Kawai K. Evaluation of the immunological enhancement activities of Immunocal. Otsuka Pharmaceutical Co. Cellular Technology Institute; Dec. 13, 1990.
11. Noelle RJ, Lawrence DA. Determination of glutathione in lymphocyte and possible association of redox state and proliferative capacity of lymphocytes. Biochem J 1981; 198:571–579.
12. Fidelus RK, Tsan MF. Glutathione and lymphocyte activation: a function of aging and auto-immune disease. Immunology 1987; 61:503–508.
13. Bounous G, Gervais F, Amer V, Batist G, Gold P. The influence of dietary whey protein on tissue glutathione and the diseases of aging. Clin Invest Med 1989; 12:343–349.
14. Meister A. The antioxidant effects of glutathione and ascorbic acid. In Pasquier et al., eds. Oxidative Stress. Cell Activation and Viral Infection. Basel: Birkauser Verlag; 1994: 101–110.
15. Williamson JM, Boettcher B, Meister A. Intracellular cysteine delivery system that protects against toxicity by promoting glutathione synthesis. Proc Natl Acad Sci USA 1982; 79:6246–6249.
16. Anderson ME, Meister A. Transport and direct utilisation of gamma-glutamylcyst(e)ine for glutathione synthesis. Proc Natl Acad Sci USA 1983; 80:707–711.
17. Bounous G, Batist G, Gold P. Immunoenhancing property of dietary whey protein in mice: role of glutathione. Clin Invest Med 1989; 12:154–161.
18. Droege W, Eck HP, Mimm S, Galter D. Abnormal Redox regulation in HIV infection and other immunodeficiency diseases. In Pasquier C et al., eds. Oxidative Stress, Cell Activation and Viral Infection. Basel: Birkauser Verlag; 1994: 285–301.
19. Eigel WM, Butler JE, Ernstrom CA, et al. Nomenclature of proteins of cow's milk, fifth revision; J Dairy Sci 1984; 67:1599–1631.
20. Goodman RE, Schanbacher FL. Bovine lactoferrin mRNA: sequence, analysis and expression in the mammary gland. Biochem Biophys Res Commun 1991; 180:75–84.
21. Baruchel S, Viau G. In vitro selective modulation of cellular glutathione by a humanized native milk protein isolate in mammal cells and rat mammary carcinoma model. Anticancer Res April, 1996; 15: 1095–1100.
22. Reynolds P, Jellinger K, Youdim MBH. Transition metals, ferritin, glutathione and ascorbic acid in Parkinsonian brains. J Neurochem 1989; 52:515–520.
23. Belleville F, Penin F, Cuny G. Lipid peroxidation and free radical scavengers in Alzheimer's disease. Gerontology 1989; 35:275–282.
24. Kuzuya M, Naito M, Funaki C, Hayashi T, Asai K, Kuzuya F. Protective role of intracellular glutathione against oxidized low density lipoprotein in cultured endothelial cells. Biochem Biophys Res Commun 1989; 163:1466–1472.

25. Calvin HI, Medvedovsky C, Worgul BV. Near total glutathione depletion and age-specific cataracts induced by buthionine sulfoximine in mice. Science 1986; 28:553–555.
26. Orrenius S, Thor H, Bellomo G, Moldeus P. Glutathione and tissue toxicity In Paton W, Mitchell I, eds. 9th International Congress of Pharmacology, London, England. London: MacMillan; 1984:57–68.
27. Bounous G, Papenburg R, Kongshavn PAL, Gold P, Fleiser D. Dietary whey protein inhibits the development of dimethylhydrazine induced malignancy. Clin Invest Med 1988; 11:213–217.
28. Papenburg R, Bounous G, Fleiszer D, Gold P. Dietary milk proteins inhibit the development of dimethylhydrazine-induced malignancy. Tumor Biol 1990; 11:129–136.
29. Bourtourault M, Buleon R, Samperes S, Jouans Effects des proteins du lactoserum bovin sur la multiplication de cellules cancereuses humaines. CR Soc Biol 1991; 185:319–323.
30. Barta O'Barta VD, Crisman LM, Akers RM. Inhibition of lymphocyte blastogenesis by whey. Am J Vet Dis 1991; 512:247–253.
31. Laursen L, Briand P, Lykkesfldt AE. Serum albumin as a modulator of growth of the human breast cancer cell line MCF-7. Anticancer Res 1990; 10:343–352.
32. Hakansson A, Zhivotovsky B, Orrenius S, Sabharwal H, Svangorg C. Apoptosis induced by a human milk protein. Proc Natl Acad Sci USA 1995; 92:8064–8068.
33. Bezault J, Bhimani R, Wiprovnich J, Furmanski P. Human lactoferrin inhibits the growth of solid tumours and development of experimental metastases in mice. Cancer Res 1994; 54:2310–2312.
34. Bosselaers IE, Caessens PW, Banboeket MA. Differential effects of milk proteins, BSA and soy on 4NOO-or MNNG-induced SCE's ub V79 cells. Food Chem Toxicol 1994; 32:905–909.
35. Russo A, Degraff W, Friedman N, Mitchell FB. Selective modulation of glutathione levels in human normal versus tumour cells and subsequent differential response to chemotherapy drugs. Cancer Res. 1986; 26:2845–2848.
36. Baruchel S, Wang T, Farah R, Batist G. In vivo selective modulation of tissue glutathione in a rat mammary carcinoma model. Biochem Pharmacol 1995; 50:1505–1508.
37. Kennedy RS, Konok GP, Bounous G, Baruchel S, Lec T. The use of a whey protein concentrate in the treatment of patients with metastatic carcinoma: phase 1–11 clinical study. Anticancer Res 1995; 15:2643–2650.
38. Jamali M, Wang T, Baruchel S, Lee T. Modulation of glutathione by a cysteine prodrug enhances in vivo tumor responses. J Pharm Exp Ther 1996; 276:1169–1173.
39. Bounous G, Kongshavn PAL, Taveroff A, Gold P. Evolutionary traits in human milk proteins. Medical Hypothesis 1988; 27:133–140.
40. Duncan B, Ey J, Holberg CJ, Wright AL, Martinez F, Taussig LM. Exclusive breast-feeding for at least 4 months protects against otitis media. Paediatrics 1993; 91:867–872.
41. Aniasson G, Alm B, Andersson B, Hakansson A. Prospective cohort study on breast feeding and otitis media in Swedish infants. Paediatrics 1982; 70:239–245.
42. Mather G, Gupta N, Mathur S, Gupta U, Pradan S. Breast feeding and childhood cancer. Indian Paediatr 1993; 30:652–657.
43. Davis MK, Savitz DA, Graubard BI. Infant feeding and childhood cancer. Lancet 1988; 1:3

43

Successful Antioxidant Therapy Including Superoxide Dismutase Associated with Antiretroviral Therapy in an HIV-Infected Patient with Hepatitis B-related Cirrhosis

J. Emerit, E. Postaire, and D. Bonnefont-Rousselot
Hôpital de la Salpêtrière, Paris, France

O. Lopez, and F. Bricaire
Oenobiol, Paris, France

INTRODUCTION

Liver cirrhosis occurs in response to chronic liver injury from many causes including alcohol, iron-overload and viruses, for example, hepatitis B (HB) virus, which is often associated with HIV. We describe an HIV-1-seropositive patient with a chronic hepatitis B-related cirrhosis which healed both clinically and histologically under interferon-α therapy associated with antiretroviral treatment and antioxidant therapy including superoxide dismutase (SOD) and deferoxamine. Indeed, antioxidants seem to have the property of inhibiting viral replication through their action on a cellular transcription factor, the NFκB (1), and SOD was proved efficient in an animal model of radio-induced fibrosis (2).

CASE REPORT

A 44-year-old patient consulted at the hospital on May 2, 1990, because of seropositivity for HIV and HBs antigen, which had just been discovered. The patient had never undergone transfusion and he had had unproctected homosexual intercourse in 1984.

The history of his disease began in May 1989 with a vesiculopustular eruption of the scalp and of the median part of the thorax. This eruption was cured by antibiotics and topical antiseptics, but relapsed 9 months later and became resistant to the same therapy. A 2-fold increase of alanine aminotransferase (ALT) was discovered at laboratory testing: HIV and HBS serologies were positive.

In May 1990 the patient was admitted at the hospital for the evaluation of these two infections. At entry, the clinical examination found a patient in good general condition

Table 1. Evolution of Knodell Score under Therapy

	Before IFN-α therapy	After IFN-α therapy	After 2 months SOD therapy	After 3 years antioxidant therapy including SOD
Liver biopsy date	07/13/90	02/11/91	05/13/91	10/04/93
Piecemeal necrosis	1	0	0	0
Intralobular hepatocyte necrosis	1	1	1	1
Portal inflammation	2	2	3	0
Intralobular mesenchymal inflammation	1	1	0	0
Fibrosis	4	4	3	1
Knodell score	9	8	7	2

IFN = interferon-α
SOD = superoxide dismutase

(weight 76 kg, height 1.76 m). There was no hepatosplenomegaly. Blood tests for the HIV infection showed a negative P24 antigenemia, a T4 count of 280/mm^3, T8 of 390/mm^3 (ratio 0.72). There was a leukopenia (2900/mm^3) and a thrombopenia at 87000/mm^3. IgG immunoglobulins were increased to 28 g/L. The patterns of hepatitis B markers clearly indicated a chronic stage (negative IgM HBc) with viral replication (positive HBe antigen, presence of viral DNA).

The liver biopsy (Figure 1) showed a moderately evolutive macro- and micronodular cirrhosis with a Knodell score estimated at 9. The decision was made to administer interferon treatment (3 million units, 3 times a week) in spite of the thrombopenia. This 6-month treatment caused the disappearance of viral DNA and HBe antigen, and the ALT returned to normal level, but the Knodell score was not improved (see Table 1) (Figure 2).

After the 6 months of interferon therapy, an antiretroviral therapy of AZT 400 mg/day was given. During the 6 years of follow-up, the antiretroviral treatment was a

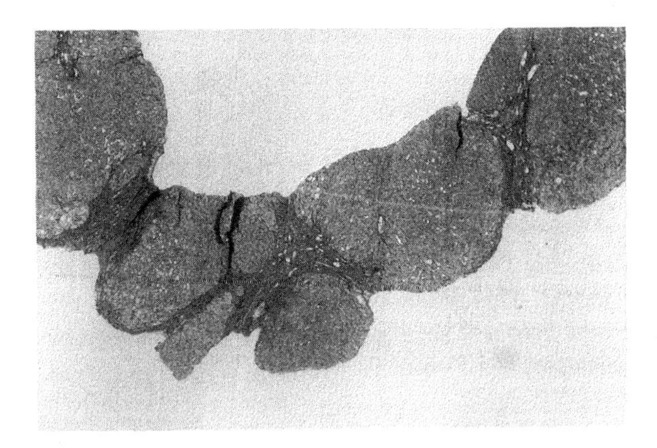

Figure 1. First liver biopsy (07/13/90), before interferon therapy.

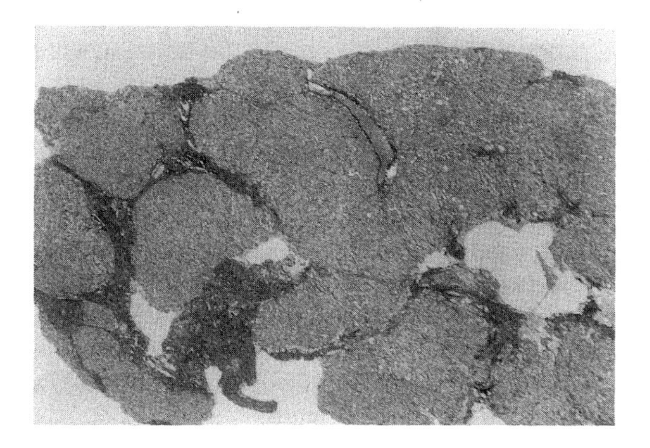

Figure 2. Second liver biopsy (02/11/91), after interferon-α therapy.

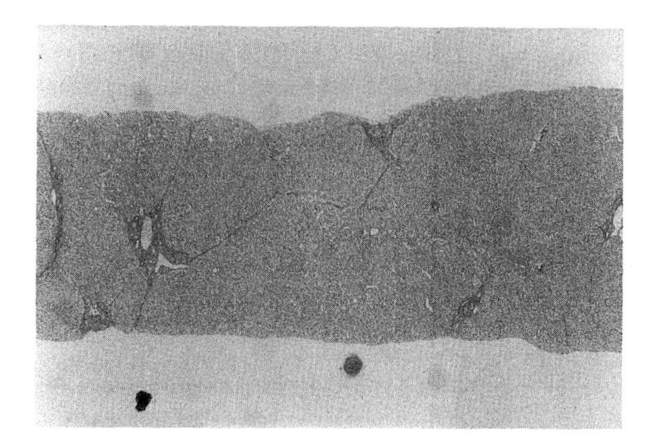

Figure 3. Third liver biopsy (05/13/91), after 2 months of SOD therapy.

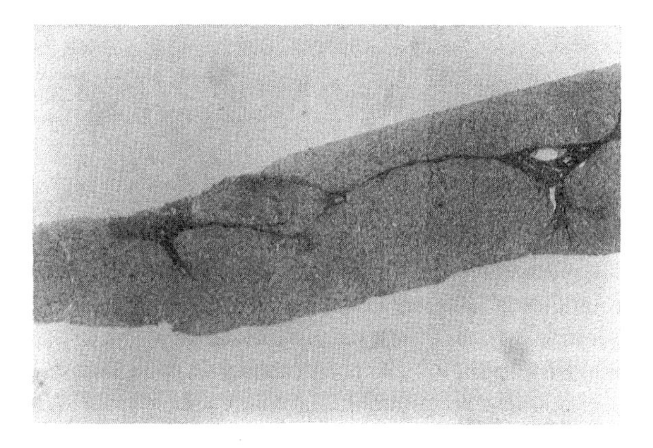

Figure 4. Fourth liver biopsy (10/04/93), after 3 years of antioxidant therapy including SOD.

monotherapy changing each year–AZT, DDI, 3TC, and bitherapy for the next 6 montths (AZT 3TC).

The antioxidant treatment comprised two parts. Part 1: Vitamin E 1000 mg/day, vitamin A 12,500 IU/day, vitamin C 500 mg/day, selenium 100 mg/day, β-carotene 30 mg/day, *N*-acetylcysteine 1200 mg/day. This regimen was continuous but was monitored by the evaluation of oxidative stress status assessed by plasma levels of thiobarbituric acid-reactive substances (TBARS), vitamin E, selenium, glutathione peroxidase (GPx), and erythrocyte levels of GPx and SOD (3). Part 2: SOD and deferoxamine (DFXO). The treatment schedule was, for SOD, one intramuscular injection of 6 mg twice a day, and, for deferoxamine, a subcutaneous injection of 500 mg once a week. This treatment was resumed for 2-month periods every 6 months. The Cu, Zn SOD of bovine erythrocyte origin. was supplied by Pharmacie Centrale des Hôpitaux de Paris.

The 6-year follow-up disclosed the healing of hepatitis B-related cirrhosis, both clinically (ALT in normal range, disappearance of viral DNA and HBe antigen) and histologically (Figure 4) (Knodell score of 2 in 1993). For the HIV infection: (a) The T4 cell count remained stable (250/ml) with a ratio T4/T8 = 0.6. (b) The P24 antigenemia remained negative (>20 pg/ml) except for a 4-month period in 1994 during which SOD–DFXO antiretroviral therapy was interrupted. The P24 antigenemia level was 45 pg/ml and returned to negative with SODD-DFXO treatment alone. (c) When available in 1995, the viremia for HIV was negative in lymphocyte culture: cellular viremia <0.2 NIU million cells, plasmatic viremia <0.4 NIU/ml. (d) Antibodies against P24 increasing: 836 in February 1994, 1097 in March 1995. (e) The oxidative stress status, under this combination of antioxidants, returned to normal values, especially the TBARS, the level of which is almost always high even in asymptomatic HIV patients. To obtain normal level of plasma and erythrocyte GPx, selenemia was maintained above normal by selenium supplementation. On withdrawal of selenium supplementation, selenemia decreased quickly.

The patient's quality of life was high in terms of social aspects (he returned to work full-time for five years) and family and psychological aspects. (Informed consent was obtained for this complex therapy.) The platelet count remained stable for the first 5 years between 90,000 and 130,000/ml, but averaged 50,000/ml in the last 6 months. The role of antibodies against platelets or the side-effects of an antiretroviral drug are under study, but there has been no clinical sign of thrompenia.

CONCLUSION

These observations raise a number of questions:

1. Does antioxidant treatment help to control HIV infection? What is the role of SOD in this combination of antioxidants? (4). This enzyme has been useful as an experimental model in inhibiting TNF action (5).
2. Is SOD active on hepatic fibrosis? (6). SOD has been proved to be active in one animal model of radio-induced fibrosis (2). It is interesting to note that this enzyme inhibited TGF-β expression in cultured fibroblasts (the role of this cytokine in HIV infection seems important).

Controlled studies are needed to answer these questions.

REFERENCES

1. Hayashi T, Ueno Y, Okamoto T. Oxireductive regulation of nuclear factor κB. J Biol Chem 1993; 268:11380–11388.
2. Delanian S, Lefaix JL, Huart J, Martin M, Daburon F. Successful treatment of radio-induced fibrosis using Cu-Zn SOD: experimental porcine study (part II). Bull Cancer 1993; 80:799–807.
3. Coutellier A, Bonnefont-Rousselot D, Delattre J, et al. Stress oxydatif chez 29 sujets séropositifs: résultats à 2 ans d'une étude en double aveugle diéthyldithiocarbamate versus placebo. Press Méd 1992; 21:1809–1812.
4. Wang Y, Watson RR. Potential therapeutics of vitamin E in AIDS and HIV. Drugs 1994; 48:3211–3238.
5. Wendel A, Niehorster M, Tiegs G. Interactions between reactive oxygen and mediators of sepsis and shock. In Sies H, ed. Oxidative Stress. London: Academic Press; 1991:585–591.
6. Friedman SL. The cellular basis of hepatic fibrosis. N Engl J Med 1993; 328:1

44

Parkinson's Disease, Apoptosis, and Oxidative Stress

Merle Ruberg, Valentine France-Lanord, Bernard Brugg, Stéphane Hunot, Philippe Anglade, Philippe Damier, Baptiste Faucheux, and Yves Agid
INSERM U.289, Hôpital de la Salpêtrière, Paris, France

INTRODUCTION

In the course of Parkinson's disease, massive degeneration of mesencephalic dopaminergic neurons occurs, in adults, over a period of several decades. Other neuronal populations, for example, the cholinergic neurons of the nucleus basalis of Meynert or the noradrenergic neurons of the locus coeruleus, may be affected as well, although to a lesser extent. It is not known why these neurons die, nor whether all the affected populations degenerate for the same reason. The following discussion will ignore the question of nondopaminergic neuronal degeneration, with the hope that once the mechanism underlying the death of the dopaminergic neurons is elucidated, the death of the others will also be explicable.

A number of hypotheses concerning the cause of parkinsonian neurodegeneration have been proposed, and for the most part excluded (reviews in Refs. 1–4). No indication of a viral infection has been found in idiopathic Parkinson's disease, although the virus of Von Economo was responsible for the epidemic of postencephalitic parkinsonism after World War I. It is unlikely that an exogenous toxin is responsible for Parkinson's disease, although manganese or MPTP (1-methyl-4-phenyl-1,2,3,6-tetrahydropyridine) can kill dopaminergic neurons; the disease is far too ubiquitous. Studies on monozygotic twins with Parkinson's disease do not strongly support the existence of a genetic anomaly predisposing to the disease, but hereditary forms are known (review in Ref. 5). Although the disease manifests itself in elderly persons, loss of the dopaminergic neurons in Parkinson's disease is not age-related; the dopaminergic neurons that die during aging predominate in the dorsolateral substantia nigra are not the same as those that die in parkinsonian patients, which are primarily in the ventrolateral part of the structure (6). A disease specific-degenerative process seems, therefore, to be involved.

There are some clues to the nature of this process, although we have not yet been able to decipher the message. The Lewy body, for example, is a highly structured histopathological inclusion that forms in dopaminergic neurons in patients with Parkinson's disease, but not in other diseases such as postencephalitic parkinsonism or progressive supranuclear palsy, where nigral dopaminergic neurons also die. The mechanism underlying its formation has not been elucidated.

A clue to the molecular mechanism might also be provided by the cellular specificity of neurodegeneration in Parkinson's disease (7), where only a subset of the already small population of mesencephalic dopaminergic neurons die following a

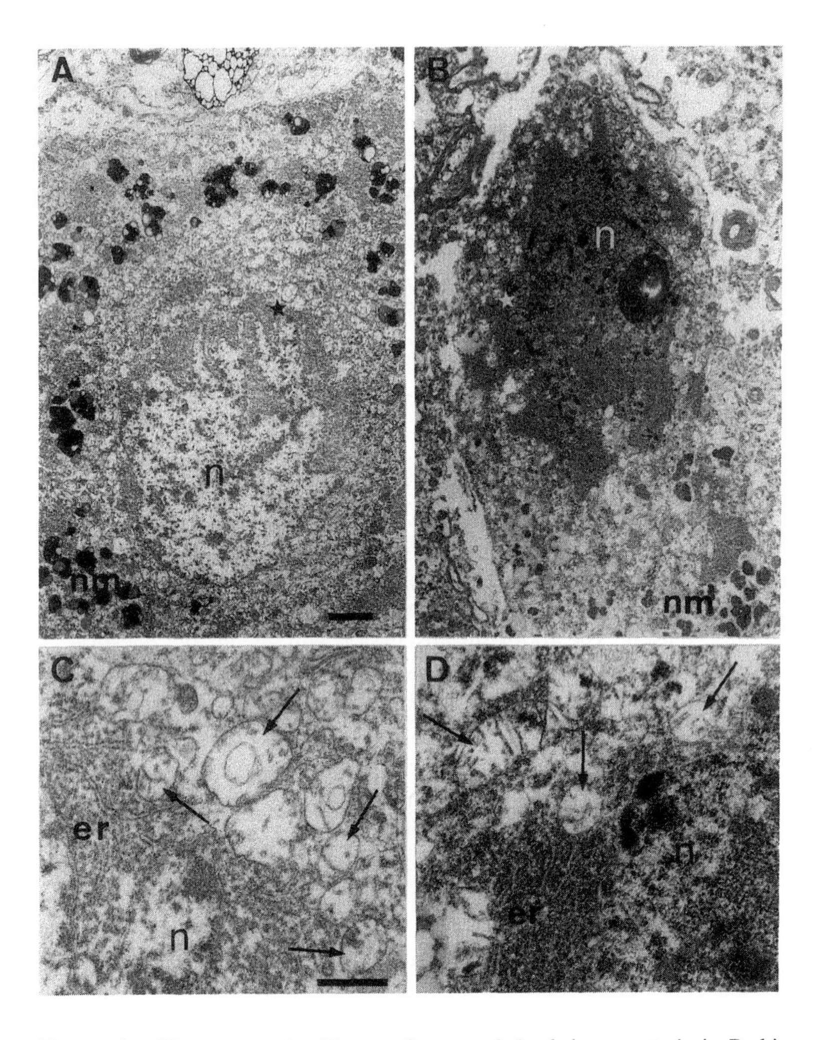

Figure 1. Ultrastructural evidence of neuronal death by apoptosis in Parkinson's disease. (A) Neuromelanin (nm)-containing dopaminergic neuron with normally distributed nuclear (n) chromatin dispersed in a fine network. Bar = 2 μm. (B) Dopaminergic neuron with apoptotic characteristics: shrunken cell body, condensed chromatin in a convoluted nucleus (n). Same magnification as (A). (C,D) Cytoplasmic organelles, such as endoplasmic reticulum (er) and mitochondria (arrows), in enlargements of starred areas in A and B, do not present morphological abnormalities in the apoptotic neuron (D) compared to the control (C), although they are more closely packed due to cell body shrinkage. Bar = 0.5 μm.

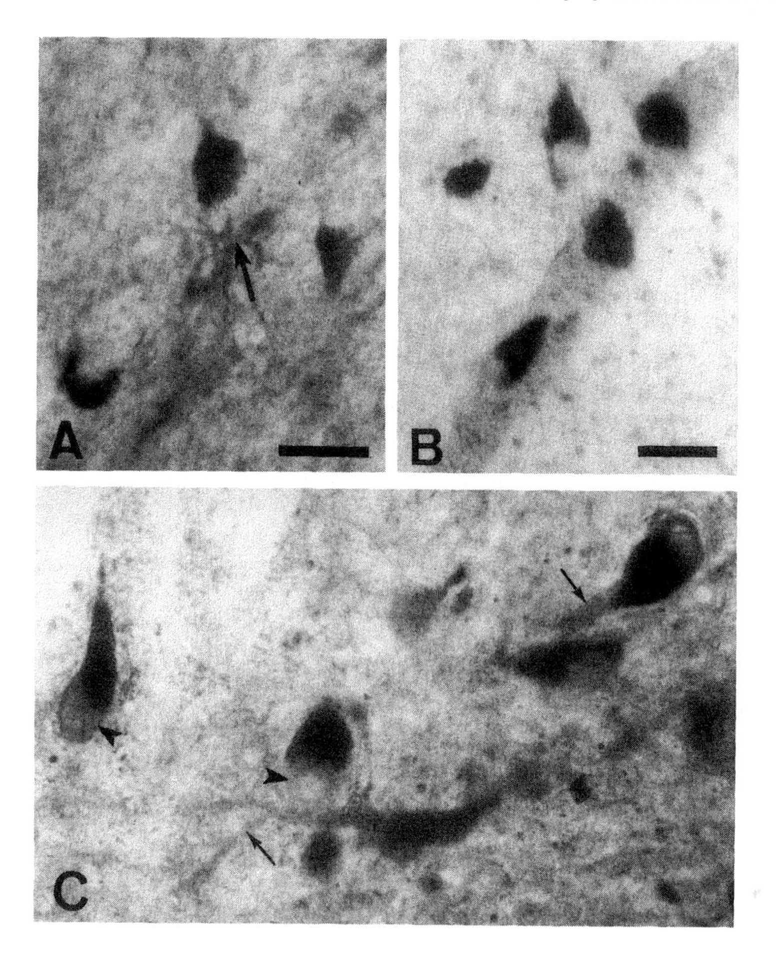

Figure 2. Immunocytochemical localization of tumor necrosis factor-α (TNF-α) and its receptor (TNFR p55) in the substantia nigra of a control subject and a patient with Parkinson's disease (PD). A TNF-positive microglial cell (arrow) surrounded by three immunonegative melanized dopaminergic neurons in the substantia nigra of a PD patient. (B) Absence of TNF-immunoreactivity in neurons or glial cells in normal substantia nigra. (C) TNFR-immunoreactive cell bodies (arrowheads) and processes (arrows) of melanized neurons in the substantia nigra of a PD patient. Also observed in normal subjects (not shown). Bars = 20 μm.

rostrocaudal, dorsoventral, mediolateral gradient of increasing severity within a heterogeneous tissue. In contrast to traumatic or ischemic brain damage, the brain tissue is not affected; only individual isolated neurons die, and this asynchronously over many years. This is evocative of cell autonomous neuronal apoptosis as described in the nematode *Caenorhabditis elegans* during development (8). Indeed, rare images of dopaminergic neurons undergoing apoptosis in the substantia nigra of patients with Parkinson's disease have been obtained (Figure 1). A putative cause of neurodegeneration in this disease must then be capable of activating a cell death programm.

A third possible index of the cause of parkinsonian neurodegeneration, if it is not a consequence, is the presence of activated microglial cells in the parkinsonian substantia nigra (9). Are they a response to the presence of dying neurons, or do they play a role in the induction of apoptosis? Some of these glial cells synthesize the cytokine TNF-α (10) (Figure 2), a known apoptogenic substance in the immune system. Does it play the same role in the central nervous system? Dopaminergic neurons express receptors for this cytokine (10), which could, therefore, quite plausibly activate a cell death program. This raises the question: What is the signal that activates glial cells to play this role?

Finally, dopaminergic neurons in the parkinsonian substantia nigra are hypothesized to suffer from oxidative stress (reviews in Refs. 11–16), a theory that relates a putative excess of free radicals in the dopaminergic neurons to an ensuing energy crisis attributed to mitochondrial dysfunction, although this theory is contested (17). The experimental data will be reviewed below. We will also present data, obtained in an in vitro model of neuronal apoptosis, that suggest that free radical production may not be simply a source of deleterious oxidative damage to cell components but may play an active role in a cell death program. Preliminary data from post-mortem parkinsonian substantia nigra indicate that the model may be relevant to the disease.

OXIDATIVE STRESS IN PARKINSON'S DISEASE

There are a number of reasons why dopaminergic neurons in patients with Parkinson's disease are thought to be vulnerable to oxidative stress. First of all, these neurons contain a large number of mitochondria predisposing to a high level of oxygen consumption, coupled with an as yet inexplicable accumulation of iron in the neurons in parkinsonian patients (18–24), which could result in the production of deleterious free radicals. This accumulation of iron is rather specific to Parkinson's disease since it is not observed in patients with progressive supranuclear palsy, where dopaminergic neurons also die (23). It may, however, be a consequence rather than a cause of the degenerative process, since it is also observed after MPTP-induced degeneration of dopaminergic in monkeys (25,26). Excess iron is indeed toxic to dopaminergic neurons in vitro (27) and in vivo (28). Not all dopaminergic neurons die in patients with Parkinson's disease, however, suggesting that neither oxygen consumption nor the presence of iron is sufficient to explain the specificity of the degenerative process.

Second, there is a correlation between the number of neuromelanin-containing neurons in the various regions containing dopaminergic neurons and the degree of neuronal loss observed in patients (7). The most vulnerable dopaminergic neurons, those of the substantia nigra where neuronal loss reaches 80% or more, contain neuromelanin, whereas the nonmelanized dopaminergic neurons of the central gray substance are barely affected. Quantitative assays of neuromelanin in subpopulations of dopaminergic neurons in control and parkinsonian subjects confirms that the more melanized neurons are those that die (29). Neuromelanin is the product of the potentially toxic, nonenzymatic degradation of dopamine, which produces free radicals and quinones (30,31). In cell cultures, the autoxidation of dopamine can indeed induce the death of such neurons (32), but by a mechanism that is independent of that caused by iron (27), and neuromelanin seems to contribute to the sensitivity of dopaminergic neurons to MPTP in monkeys (33). Is enough dopamine autoxidized in the dopaminergic neurons that die in Parkinson's disease to be lethal?

The subset of dopaminergic neurons that die in Parkinson's disease is also inversely correlated with the density of glial cells containing glutathione peroxidase (34), which perhaps provide a higher level of protection against oxidative stress in the neurons that are spared. The neurons that do not degenerate are also characterized by the selective presence of a calcium-binding protein, calbindin D28K (35–37), that might protect against calcium-mediated consequences of oxidative stress, as seen in some experimental models in culture (38) and in animals (37,39).

There is thus an accumulation of presumptive evidence that oxidative stress might be involved in parkinsonian neurodegeneration, but it is difficult to interpret. The evidence that oxidative stress does in fact occur in Parkinson's disease will be presented below. This data is also problematic, for several reasons (1) There are difficulties inherent in the use of post-mortem brain tissue from which most of the data derives. Post-mortem brains are for the most part from patients at highly evolved states of the disease. Peri-mortem factors can introduce artifactual modifications of the markers studied. Causes and consequences of the disease cannot be distinguished. (2) The population of neurons affected, is very small. The neurons that suffer represent perhaps only 5% of the cells in the substantia nigra of a normal subject (24), of which perhaps 80% will have been lost in post-mortem brain; among the neurons which remain are dopaminergic neurons that are variously at different stages of the degenerative process, not evidently affected, or completely resistant to the disease. Evidence from homogenates of whole tissue must therefore be interpreted cautiously, and quantifications at the level of individual cells should be sought where possible. (3) Finally, what does oxidative stress entail? Is it a chronic state of oxidant production, that is, a generally high basal level of reactive oxygen species that causes progressive accumulation of nonmetabolizable oxidation products that with time prevent the cell from functioning? Is it an energy crisis, due to mitochondrial damage? Is it a punctual event, as short lived as a free radical in an environment containing a multiplicity of scavenging mechanisms? Is it a signal that activates preprogrammed molecular mechanism? Not all of these alternatives can be addressed post mortem. The chronic oxidant state can leave traces if its products are not eliminated, but is it necessarily lethal for the cell? An acute energy crisis may be lethal, but can it be detected in an infinitesimally small cell population? Discontinuous, punctual, short-lived radical emission can only be detected in living cells, and at the appropriate moment.

The hypothesis as it is generally applied to Parkinson's disease assimilates, perhaps somewhat abusively, oxidative stress and energy deficiency. High levels of free radicals that escape the normal detoxification mechanisms may cause damage to lipids, proteins, and DNA. Mitochondria, the major source of reactive oxygen species, are thought to be particularly prone to damage, both structural, through lipid oxidation, and functional, by direct inhibition of the electron transport chain, particularly complex I, as a result of excess superoxide production. A lethal decrease in ATP production is thought to ensue. There is a certain contradiction inherent in this hypothesis, since active complex I is needed for electron extraction that will be used for superoxide production. The process should therefore be self-inhibitory. There are some data, however, that support the theory.

Can Oxidative Stress be Detected In Vivo in Patients with Parkinson's Disease?

Indirect in vivo evidence of oxidative stress, through decreased glucose metabolism, has been reported in several brain structures such as the striatum, the thalamus, the cerebellum, or the cortex (40,41), but probably reflects the consequences of denervation of these structures after neuronal loss, rather than a primary metabolic defect. An increased lactate/*N*-acetylaspartate ratio was detected by nuclear magnetic resonance in the occipital cortex of demented patients with Parkinson's disease (42), and an increased lactate/creatine ratio in the striatum (43), suggesting altered aerobic metabolism in these brain structures, although the relationship to Parkinsonian neuropathology is not clear. Lactate and pyruvate levels in blood and cerebrospinal fluid are reported to be normal (44–46).

Oxidative Damage in Post-mortem Brain from Patients with Parkinson's Disease

Traces of oxidative damage in the substantia nigra have been reported in the brains of patients with Parkinson's disease: increased levels of malondialdehyde (47) and lipid hydroperoxides (48), particularly cholesterol hydroperoxide which is reported to increase 10 times compared to controls; but subsequent free radical degradation products could not be detected in electron spin resonance studies (48). Mitochondrial DNA is thought to be at risk for accumulated free radical damage, because of high oxygen consumption and limited mitochondrial DNA repair. There is no evidence, however, that DNA mutation attains pathogenic proportions (review in Ref. 5).

The observation that free radical-scavenging systems are intact and even elevated in patients with Parkinson's disease, sometimes cited as an adaptive response to oxidative stress, may also indicate that the neurons are, on the contrary, adequately protected. A decrease in catalase activity has been reported (49), and lower levels of reduced glutathione (24,50,51,52), but vitamin C (24) and vitamin E (53,54,55) levels are normal, and both cytosolic Cu/Zn superoxide dismutase, preferentially expressed in the vulnerable melanized dopaminergic neurons (56), and mitochondrial manganese-dependent superoxide dismutase activities are reported to increase (57,58). The number of glutathione peroxidase-positive glial cells surrounding surviving dopaminergc neurons also increases (34), although glutathione peroxidase levels seem to be unaltered (57,59) in the substantia nigra of patients.

The major evidence of oxidative stress in patients with Parkinson's disease, and it is indirect, is a decrease of approximately 35%, in the substantia nigra, in the activity of complex I of the mitochondrial electron transport chain (60–63) and in some of its subunits (64), as well as abnormal kyurenine metabolism (65), which could reduce available levels of nicotinamide–adenine dinucleotides needed to produce complex I substrates. Contradictory data is reported concerning mitochondrial deficiencies in muscle or platelets (review in Ref. 16). The evidence does not support a generalized mitochondrial disorder, although a local effect in the substantia nigra has been found consistently. The contribution of cell loss to this decrease is difficult to evaluate, however.

Recent evidence from differential screening of a cDNA library corresponding to the substantia nigra of a patient with Parkinson's disease, and cellular-level in situ hybridization on substantia nigra from a series of patients, showed that the role of complex I in neurodegeneration may involve more than a decrease in activity. Although a decrease in the expression of mRNA encoding a complex I subunit was observed in the less vulnerable dopaminergic neurons of the medial substantia nigra, an increase in the expression of this mRNA was observed in the dopaminergic neurons of the lateral part of the structure, those with the highest probability of degenerating (Figure 3; see also Ref. 66), suggesting that an increase rather than a decrease in complex I activity may be

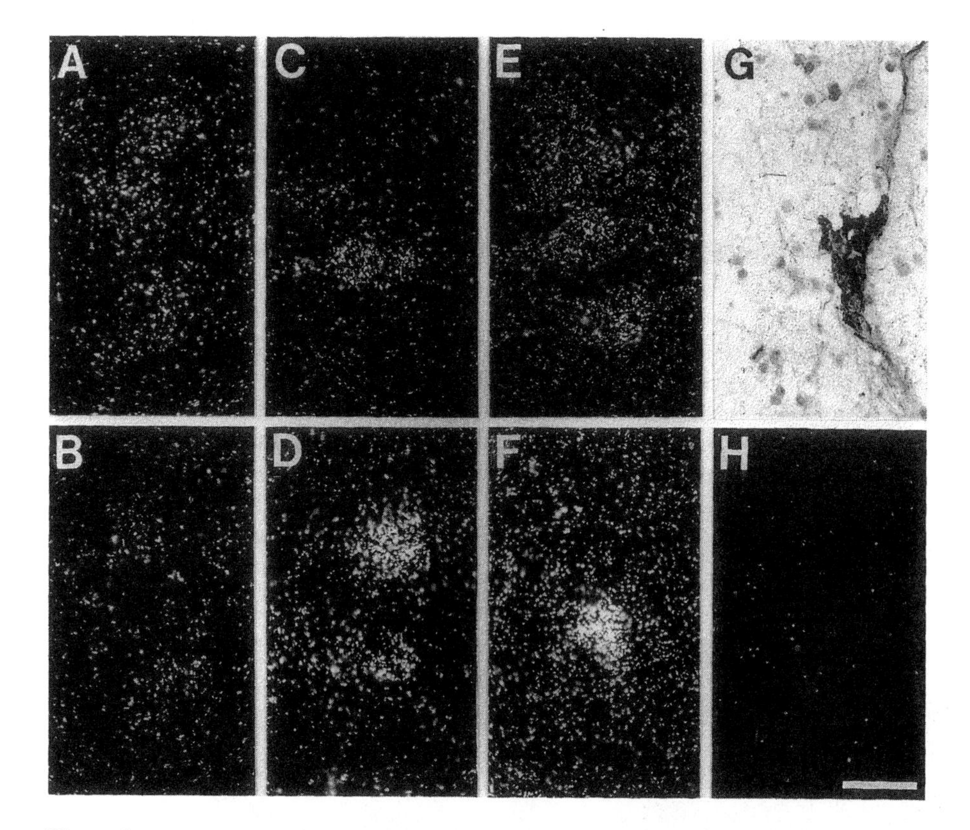

Figure 3. Upregulation in dopaminergic neurons of mitochondrially encoded subunits of complex I and complex IV respiratory enzymes in a patient with Parkinson's disease (PD), evidenced by in situ hybridization. (A–D) The density of silver grains corresponding to a ^{35}S-labeled antisense cRNA probe for subunit 1 of complex I (NADH dehydrogenase) in dopaminergic neurons in the median substantia nigra was decreased in the PD patient (B) compared to an age-matched control subject (A) with the same amount of mitochondrial DNA (not shown), but was increased in dopaminergic neurons in the lateral substantia nigra of the PD patient (D) compared to the control (C). (E,F). The density of silver grains corresponding to the probe for subunit 1 of complex IV (cytochrome *c* oxidase) was increased in dopaminergic neurons in the lateral substantia nigra of the PD patient (F) compared to the control (E), and also in the neurons of the medial substantia nigra (not shown). (G) Dopaminergic neurons were identified by colabeling with an antibody against tyrosine hydroxylase. (F) Nonspecific labeling with the complex I sense cRNA probe. A similar result was obtained with the complex IV probe (not shown). Bar = 100 μm.

associated with the initiation of apoptosis, although this remains to be confirmed. In addition, the expression of a mitochondrially encoded mRNA for a complex IV (cytochrome *c* oxidase) subunit is increased in all the dopaminergic neurons. These observations are not consistent with the notion of mitochondrial dysfunction in dopaminergic neurons in Parkinson's disease, and, indeed, no gross impairment of mitochondria in the dopaminergic neurons of patients with Parkinson's disease, even in those degenerating by apoptosis, is observable by electron microscopy (Figure 1), whereas radical attack is reported to result in mitochondrial swelling and other deformities (67).

It is difficult, on the basis of these data from living patients with Parkinson's disease and from post-mortem brain tissue, to conclude as to the existence of lethal oxidative stress or mitochondrial dysfunction in dopaminergic neurons of the substantia nigra. There is, however, another way in which mitochondrial free radical production might be implicated in the death of these neurons, as will be described below.

APOPTOSIS AND FREE RADICAL SIGNALING IN DOPAMINERGIC NEURONS IN VITRO

Two lines of evidence from recent studies on the cytotoxicity of TNF-α in nonneuronal cell systems have suggested a mechanism of apoptosis involving free radical production that may apply to neurodegeneration in Parkinson's disease. First of all, Schulze-Osthoff and his collaborators (68,69) clearly demonstrated on fibrosarcoma cell lines that free radical emission, localized at complex I of the electron transport chain, was necessary for the nuclear translocation of NFκB and cell death by apoptosis. Confirmed in TNF-α-intoxicated fibroblasts by Hennet and his collaborators (70), the association of free

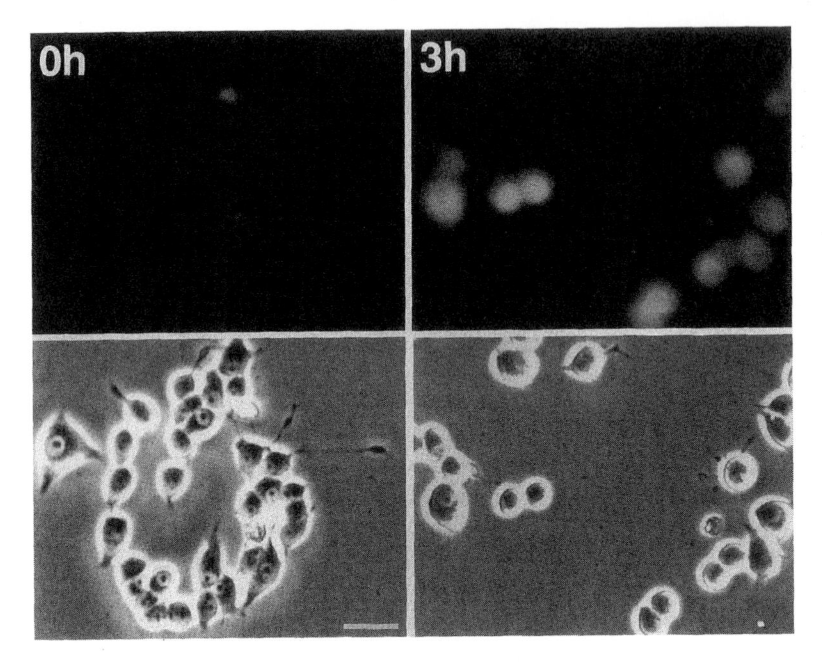

Figure 4. Activation of the apoptogenic ceramide-dependent transduction pathway in neuron-like PC12 cells induces superoxide production. Treatment of PC12 cells differentiated in the presence of nerve growth factor with a cell-permeant ceramide analog (c_2-ceramide: 25 μM) induces cell death by apoptosis. After a 12-h delay, the cells die rapidly over a period of 24 h. Apoptosis was characterized morphologically by cell body shrinkage, neurite retraction, chromatin condensation, and nuclear fragmentation (see Figure 6), and by internucleosomal DNA degradation (not shown). Superoxide production, detected with the fluorescent marker dihydrodichlorofluorescine diacetate (upper images), began to be observed about 2 h after the beginning of treatment, peaked between 3–5 h, and then declined. Phase-contrast micrographs (lower images) of the same field show neurite retraction after 3 h of treatment, compared to the cultures at 0 h, but no cell death. Bar = 50 μm.

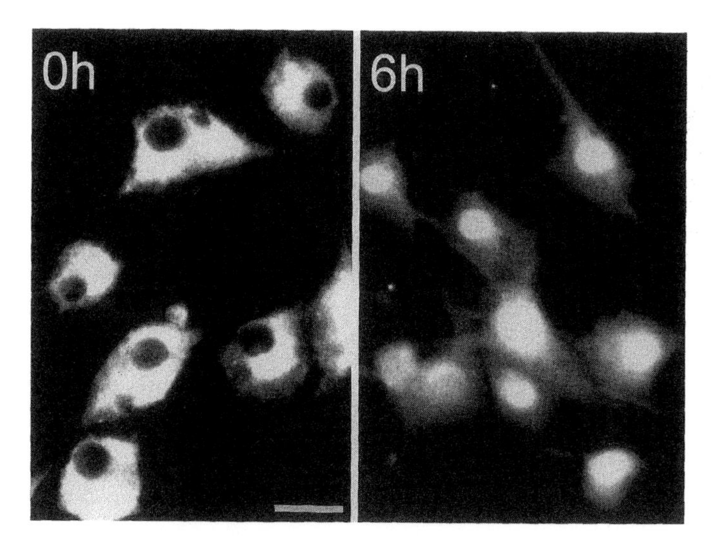

Figure 5. Translocation of activated transcription factor NFκB to the nucleus is an early event in ceramide-induced apoptosis. NFκB immunoreactivity was detected with a specific antibody revealed with a rhodamine-conjugated second antibody. In untreated cultures (0 h) of differentiated PC 12 cells, NFκB immunoreactivity was located in the cytoplasm. Translocation could be detected as early as 2 h after treatment with c_2-ceramide. At 6 h, most NFκB was found in cell nuclei. Bar = 30 μm.

radical emission and cell death has also been reported in other cell systems including neurons in culture (71–75). The second element, was the publication, at the same period, of a new signal transduction system, in which ceramide cleaved from membrane sphingomyelin activates a cascade of molecular events leading to apoptosis (review in Refs. 76,77). This signaling pathway, if it also functioned in neurons, seemed particularly interesting as a potential mechanism of neuronal cell death, and in particular the death of dopaminergic neurons in Parkinson's disease, because it can be activated by interleukin-1β, product of the human enzyme homologous to the nematode cell death gene *ced-3* (78), and by TNF-α, which, as indicated above, is found in the substantia nigra of parkinsonian patients. It has also been shown to be activated by the low-affinity nerve growth factor (NGF) receptor p75, which can also induce apoptosis, as well as by Fas, an immune system apoptogenic receptor, ionizing irradiation, and the *Drosophila* reaper protein (79), suggesting that it may represent a common cell death effector mechanism. Since the transcription factor NFκB is also activated by the ceramide-dependent signaling pathway, we hypothesized that free radical emission occurs within this transduction cascade, a plausible candidate mechanism for the death of dopaminergic neurons in Parkinson's disease.

We have been able to demonstrate in primary cultures of rat mesencephalon–the structure of origin of the neurons that die in patients with Parkinson's disease–that cell-permeant ceramide analogs can indeed activate the sphingomyelin-dependent signaling pathway in neurons in primary culture, leading to apoptosis (81), and in PC12 cells differentiated by nerve growth factor (Figures 4–6). Among the early events in the apoptotic process, 3–5 hours after the beginning of ceramide treatment, is the transient emission of superoxide ions from the mitochondria (Figure 4) and translocation of NFκB

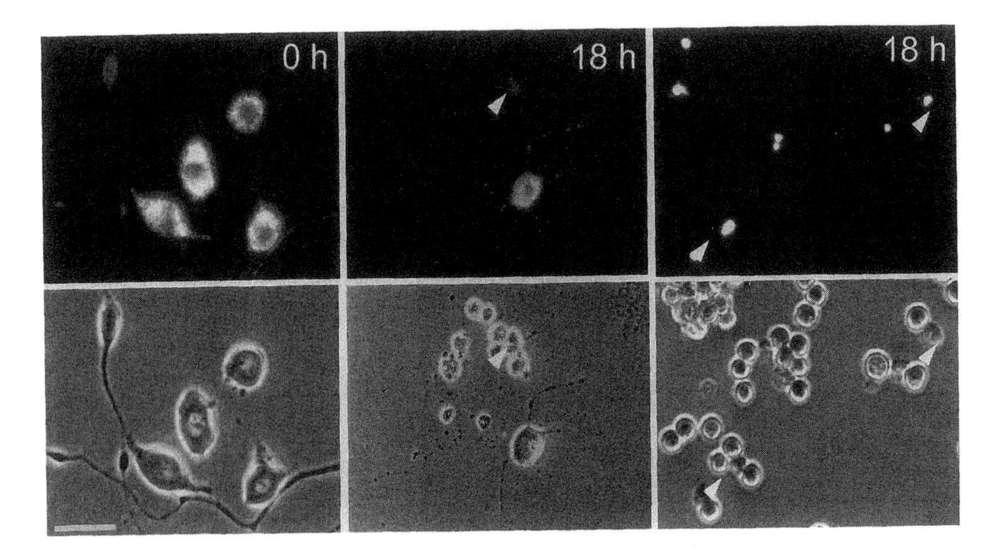

Figure 6. Dissipation of the mitochondrial membrane potential, a late event in ceramide-induced apoptosis, immediately precedes nuclear fragmentation and cell death. The fluorescent cation rhodamine 123 enters mitochondria with normal membrane potentials (positive outside, negative inside), along the charge gradient, as seen in untreated cultures of differentiated PC12 cells (phase-contrast image of the same field below). At 18 h (middle image), when cell death is very advanced, the remaining viable neurons still take up the fluorescent marker, which is excluded in a potential-dependent manner from those that are dying. The arrowheads indicate a dying neuron which still mantains a very limited membrane potential. The dead neurons exclude the marker. Condensed and fragmented nuclei (right image) can be detected at this time in the dead neurons using with propidium iodide, a fluorescent marker that penetrates nonviable cells and intercalates into the DNA. Arrowheads point to some of the dead cells and their nuclei. Fragmented nuclei are not found in neurons with normal mitochondrial membrane potentials. Bar = 50 μm.

into the cell nucleus (Figure 5). At this point in time, some morphological modifications are observed, such as neurite retraction and rounding of cell bodies, but the cells do not begin to die until much later, 12–14 hours after the beginning of treatment, concomitantly with dissipation of the mitochondrial membrane potential (Figure 6). Active regulated free radical production would then be an early event in the apoptotic program, whereas mitochondrial dysfunction, or at least dissipation of the membrane potential, is a late consequence.

CONCLUSION

In addition to the circumstantial evidence from post-mortem brain tissue that dopaminergic neurons in patients with Parkinson's disease may suffer from oxidative stress of mitochondrial origin, we have shown the existence of a transduction pathway that mediates the death by apoptosis of dopaminergic neurons in vitro, via the production of superoxide in mitochondria. We do not yet know what is the natural activator of this pathway in the dopaminergic neurons, although TNF-α remains an interesting candidate.

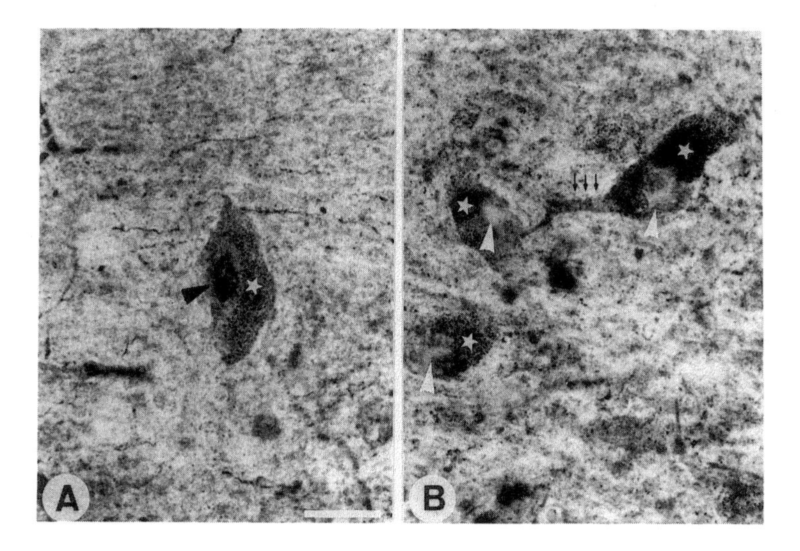

Figure 7. Immunocytochemical localization of NFκB in dopaminergic neurons. The dopaminergic neurons, identified by their neuromelanin content (stars), were labeled with an antibody against NFκB. In a patient with Parkinson's disease (A), activated NFκB was localized in the nucleus (black arrowhead). In a control subject (B), inactive NFκB was localized in the cytoplasm and nerve processes (small arrows), but the nuclei (white arrowheads) were unlabelled. Bar =100 μm.

Is this model of apoptosis relevant to the pathophysiology of Parkinson's disease? The only indication that we have for the moment that this is in fact the case is the presence of TNF-α in microglial cells in the substantia nigra of patients with Parkinson's disease (10), and preliminary data indicating that NFκB can be found post-mortem in the nuclei of some dopaminergic neurons in substantia nigra of patients with Parkinson's disease (Figure 7). This putative mechanism of neuronal death in Parkinson's disease presents the advantage that it may reconcile the presently disparate data concerning mitochondrial dysfunction, cytokine involvement, and free radical production. Within the context of the model, the term "oxidative stress" appears to be inappropriate, since a mitochondrially dependent signaling mechanism seems to be involved. The chronological sequence of the molecular events that occur after activation of the sphingomyelin transduction pathway indicates that superoxide production is not directly responsible for neuronal death. Its role is limited to signal transfer to a downstream element of the transduction cascade, and might possibly be one of the normal functions of this organelle. In parkinsonian brain, the observation that mitochondria are morphologically unaltered in apoptotic dopaminergic neurons also indicates that superoxide has not been dismutated to a more deleterious radical capable of inducing oxidative damage.

It should be stressed that this pathway does not pretend to explain the pathogenesis of Parkinson's disease. It may be the effector of cell death. TNF-α may or may not prove to be the activator. If so the essential question becomes what might activate microglial cells to mount a cytokine based cytotoxic attack. If the in vitro model presented here does prove to account for the way in which dopaminergic neurons die in Parkinson's disease, it constitutes a useful advance in our knowledge of the pathophysiology of parkinsonian neurodegeneration, because it defines a necessary criterion that must be met by any putative cause of the disease: it must be able to activate the transduction system. Even if the primary cause of neurodegeneration in this disease is not understood, identification of the molecular mechanism underlying apoptosis of the neurons opens new possibilities for therapeutic intervention, aimed not just at palliating the symptoms, but at stopping the degenerative process. In vitro this is already possible.

REFERENCES

1. Duvoisin RC. The cause of Parkinson's disease. In Marsden CD, Fahn S, eds. Movement Disorders. London: Butterworth Scientific; 1982:8–24.
2. Langston JW. Mechanisms underlying neuronal degeneration in Parkinson's disease: an experimental and theoretical treatise. Mov Disord 1989; 4 (supplement 1):S15–S25.
3. Agid Y, Ruberg M, Javoy-Agid F, et al. Are dopaminergic neurons selectively vulnerable to Parkinson's disease? In Narabayashi H, Nagatsu T, Yanagisawa N, Mizuno Y, eds. Advances in Neurology, vol. 60. New York: Raven Press; 1993:148–164.
4. Hirsch EC. Nerve cell death in Parkinson's disease. In Nappi G, Carceni T, Martignoni E, Giovannini P, eds. New Advances in Parkinson Disease and Other Extrapyramidal Disorders. London: Smith Gordon; 1994:7–14.
5. Schapira AHV. Nuclear and mitochondrial genetics in Parkinson's disease. J Med Genet 1995; 32:411–415.
6. Fearnley JM, Lees AJ. Ageing and Parkinson's disease: substantia nigra regional selectivity. Brain 1991; 114:2283–2301.
7. Hirsch EC, Graybiel AM, Agid, Y. Melanized dopaminergic neurons are differentially affected in Parkinson's disease. Nature 1988; 334:345–348.
8. Yuan J, Horwitz HR. 1990. The *Caenorhabditis elegans* genes *ced-3* and *ced-4* act cell autonomously to cause programmed cell death. Dev Biol 1990; 138:33–41.
9. McGeer PL, Itagaki A, Akiyama H, McGeer EG. Rate of cell death in parkinsonism indicates active neuropathological process. Ann Neurol 1988; 24:564–576.
10. Boka G, Anglade P, Wallach D, Javoy-Agid F, Agid Y, Hirsch EC. Immunocytochemical analysis of tumor necrosis factor and its receptors in Parkinson's disease. Neurosci Lett 1994; 172:151–154.
11. Javoy-Agid F. Dopaminergic cell death in Parkinson's disease. In Packer L, Prilipko L, Christen Y, eds. Free Radicals in the Brain. Aging, Neurological and Mental Disorders. Berlin: Springer-Verlag; 1992:99–108.
12. Hirsch E. 1992. Why are nigral catecholaminergic neurons more vulnerable than other cells in Parkinson's disease. Ann Neurol 1992; 32:S88–S93.
13. Hirsch EC. Does oxidative stress participate in nerve cell death in Parkinson's disease. Eur Neurol 1993; 33 (supplement 1):52–59.
14. Fahn S, Cohen G. The oxidant stress hypothesis in Parkinson's disease: evidence supporting it. Ann Neurol 1992; 32:804–812.
15. Jenner P. Oxidative damage in neurodegenerative disease. Lancet 1994; 344:796–798.
16. Beal MF. Mitochondrial function and oxidative damage in neurodegenerative diseases. Austin: RG Landes, 1995.
17. Calne D. The free radical hypothesis in Parkinson's disease: evidence against it. Ann Neurol 1992; 32:799–803.
18. Dexter DT, Wells FR, Agid F, Lees AJ, Jenner P, Marsden CD. Increased nigral iron content in postmortem parkinsonian brain. Lancet 1987; 2:1219–1220.
19. Dexter DT, Wells FR, Agid F, et al. Increased iron nigral content and alteration in other metals occurring in Parkinson's disease. J Neurochem 1989; 52:1830–1836.
20. Dexter DT, Carayon A, Javoy-Agid F, et al. Alterations in the levels of iron, ferritin and other trace metals in Parkinson's disease and other degenerative diseases affecting the basal ganglia. Brain 1991; 114:1953–1975.

21. Sofic E, Riederer P, Heinsen H, et al. Increased iron(III) and total iron content in post-mortem substantia nigra of parkinsonian brain. J Neural Transm 1988; 74:199–205.
22. Sofic E, Paulus W, Jellinger K, Riederer P, Youdim MBH. Selective increase of iron in substantia nigra zona compacta of parkinsonian brains. J Neurochem 1991; 56:978–982.
23. Hirsch EC, Brandel JP, Galle P, Javoy-Agid F, Agid Y. Iron and aluminium increase in the substantia nigra of patients with Parkinson's disease: an x-ray microanalysis. J Neurochem 1991; 56:446–451.
24. Riederer P, Sofic EM, Rausch WD, et al. Transition metals, ferritin, glutathione and ascorbic acid in parkinsonian brains. J. Neurochem 1989; 52:515–520.
25. Mochiuki H, Imai H, Endo K, et al. Iron accumulation in the substantia nigra of 1-methyl-4-phenyl-1,2,3,6-tetrahydropyridine (MPTP)-induced hemiparkinsonian monkeys. Neurosci Lett 1994; 168:251–253.
26. Temlett JA, Landsberg JP, Watt F, Grime GW. Increased iron in the substantia nigra compacta of the MPTP-lesioned hemi-parkinsonian African green monkey: evidence from proton microprobe elemental microanalysis. J Neurochem 1994; 62:134–146.
27. Michel PP, Vyas S, Agid Y. Toxic effects of iron for cultured mesencephalic dopaminergic neurons derived from rat embryonic brains. J Neurochem 1992; 59:118–127.
28. Sengstock GJ, Olanow CW, Menzies RA, Dunn AJ, Arendash GW. Infusion of iron into the rat substantia nigra: nigral pathology and dose-dependent loss of striatal dopaminergic markers. J Neurosci Res 1993; 35:67–82.
29. Kastner A, Hirsch EC, Lejeune O, Javoy-Agid F, Rascol O, Agid Y. Is the vulnerability of neurons in the substantia nigra of patients with Parkinson's disease related to their neuromelanin content? J Neurochem 1992; 59:1080–1089.
30. Graham DG. 1979. On the origin and significance of neuromelanin. Arch Pathol Lab Med 1979; 103:359–362.
31. Marsden CE. Neuromelanin and Parkinson's diesease. J Neural Transm 1983; 19 (supplement):121–141.
32. Michel PP, Hefti F. Toxicity of 6-hydroxydopamine and dopamine for dopaminergic neurons in culture. J Neurosci Res 1990; 26:428–435.
33. Herrero MT, Hirsch EC, Kastner A, et al. Does neuromelanin contribute to the vulnerability of catecholaminergic neurons in monkey intoxicated with MPTP? Neuroscience 1993; 56:499–511.
34. Damier P, Hirsch EC, Zhang P, Agid Y, Javoy-Agid F. Glutathione peroxidase, glial cells and Parkinson's disease. Neuroscience 1993; 52:1–6.
35. Hirsch EC, Mouatt A, Thomasset M, Javoy-Agid F, Agid Y, Graybiel AM. Expression of calbindin D28K-like immunoreactivity in catecholaminergic cell groups of the human midbrain: normal distribution and distribution in Parkinson's disease. Neurodegeneration 1992; 1:83–93.
36. Yamada T, McGeer PL, Baimbridge KG, McGeer P. Relative sparing in Parkinson's disease of substantia nigra dopamine neurons containing calbindin D28K. Brain Res 1990; 52:303–307.
37. German DC, Manaye KF, Sonsalla PK Brooks BA. Midbrain dopaminergic cell loss in Parkinson's disease and MPTP induced parkinsonism: sparing of calbindin-D28K containing cells. Ann NY Acad Sci 1992; 648:42–62.
38. Mattson MP, Rychlik B, Chu C, Christakos S. Evidence for calcium-reducing excito-protective roles for the calcium-binding protein calbindin-28K in cultured hippocampal neurons. Neuron 1991; 6:41–51.
39. Gaspar P, Ben Jalloun N, Febvret A. Sparing of the dopaminergic neurons containing calbindin D28K and of the dopaminergic mesocortical projections in weaver mutant mice. Neuroscience 1994; 61:293–305.
40. Peppard RF, Martin WR, Carr GD, et al. Cerebral glucose metabolism in Parkinson's disease with and without dementia. Arch Neurol 1992; 49:1262–1268.

41. Eberling JL, Richardson BC, Reed BR, Wolfe N, Jagust WJ. Cortical glucose metabolism in Parkinson's disease without dementia. Neurobiol Aging 1994; 15:329–335.
42. Bowen BC, Block RE, Sanchez-Ramos J, et al. Proton MR spectroscopy of the brain in 14 patients with Parkinson's disease. Am J Neuroradiol 1995; 16:61–68.
43. Chen YI, Jenkins BG, Rosen BR Evidence for impairment of energy metabolism in Parkinson's disease using in vivo localized MR spectroscopy. Proc Soc Magn Res 1994; 1:194.
44. DiMonte D, Tetrud JW, Langston JW. Blood lactate in Parkinson's disease. Ann Neurol 1991; 29:342–344.
45. Bravi D, Anderson JJ, Dagani F, et al. Effect of aging and dopaminomimetic therapy on mitochondrial respiratory function in Parkinson's disease. Mov Disord 1992; 7:228–231.
46. Nakagawa-Hattori Y, Yoshino H, Kondo T, Mizuno Y, Horai S. Is Parkinson's disease a mitochondrial disorder? J Neurol Sci 1992; 107:29–33.
47. Dexter DT, Carter CJ, Wells FR, et al. Basal lipid peroxidation is increased in Parkinson's disease. J Neurochem 1989; 52:381–389.
48. Dexter DT, Holley AE, Flitter WD, et al. Increased levels of lipid hydroperoxides in the parkinsonian substantia nigra: an HPLC and ESR study. Mov Disord 1994; 9:92–97.
49. Yoritaka A, Hattori N, Uchida K, Tanaka M, Standtman, Mizuno Y. Immunohistochemical detection of 4-hydroxynoneal protein adducts in Parkinson disease. Proc Natl Acad Sci USA 1996; 93:2696–2701.
50. Ambiani IM, VanWoert MH, Murphy S. Brain peroxidase and catalase in Parkinson's disease. Arch Neurol 1975; 32:114–118.
51. Perry TL, Godin DV, Hansen S. Parkinson's disease: A disorder due to nigral glutathione deficiency? Neurosci Lett 1982; 33:305–310.
52. Perry TL, Yong VW. Idiopathic Parkinson's disease, progressive supranuclear palsy and glutathione metabolism in the substantia nigra of patients. Neurosci Lett. 1986; 67:269–274.
53. Sian J, Dexter DT, Lees AJ, et al. Alternations in glutathione levels in Parkinson's disease and other neurodegenerative disorders affecting basal ganglia. Ann Neurol 1994; 36:348–355.
54. Dexter DT, Ward RJ, Wells FR, et al. Alpha-tocopherol levels in brain are not altered in Parkinson's disease. Ann Neurol 1992; 32:591–593.
55. Fernandez-Calle P, Molina JA, Jiménez-Jiménez FJ, et al. Serum levels of alpha-tocopherol (vitamin E) in Parkinson's disease. Neurology 1992; 42:1064–1066.
56. Jiménez-Jiménez FJ, Molina JA, Fernandez-Calle P, et al. Serum levels of β-carotene and other carotenoids in Parkinson's disease. Neurosci Lett 1993; 157:103–106.
57. Zhang P, Damier P, Hirsch EC, et al. Preferential expression of superoxide dismutase messenger RNA in melanized neurons in human mesencephalon. Neuroscience 1993; 55:167–175.
58. Martilla RJ, Lorentz H, Rinne UK. Oxygen toxicity protecting enzymes in Parkinson's disease: increase of superoxide dismutase-like activity in the substantia niga and basal nucleus. J Neurol Sci 1988; 86:321–331.
59. Saggu H, Cooksey J, Dexter D, et al. A selective increase in particulate superoxide dismutase activity in parkinsonian substantia nigra. J Neurochem 1989; 53:692–697.
60. Kish SJ, Morito C, Hornykiewicz O. Glutathione peroxidase activity in Parkinson's disease brain. Neurosci Lett 1985; 58:343–346.
61. Schapira AHV, Cooper JM, Dexter D, Clark JB, Jenner P, Marden CD. Mitochondrial complex I deficiency in Parkinson's disease. J Neurochem 1990; 54:823–827.
62. Schapira AHV, Mann VM, Cooper JM, et al. Anatomic and disease specificity of NADH CoQ reductase (complex I) deficiency in Parkinson's disease. J Neurochem 1990; 55:2142–2145.
63. Mann VM, Cooper JM, Krige D, Daniel SE, Schapira AH, Marsden CD. Brain, skeletal muscle and platelet homogenate mitochondrial function in Parkinson's disease. Brain 1992;

115:333–342.
64. Janetsky B, Hauck S, Youdim MBH, et al. Unaltered aconitase activity, but decreased complex I activity in substantia nigra pars compacta of patients with Parkinson's disease. Neurosci Lett 1994; 169:126–128.
65. Hattori N, Tanaka M, Ozawa T, Mizuno Y. Immunohistochemical studies on complexes I, II, III, and IV of mitochondria in Parkinson's disease. Ann Neurol 1991; 30:563–571.
66. Ogawa T, Matson WR, Beal MF, et al. Kyurenine pathway abnormalities in Parkinson's disease. Neurology 1992; 42:1702–1706.
67. Ruberg M, Brugg B, Prigent A, Hirsch E, Brice A, Agid Y. Apoptosis-related differential regulation of mitochondrial transcripts in Parkinson's disease. J Neurochem, in press.
68. Mehrotra S, Kakkar P, Viswanathan PN. Mitochondrial damage by active oxygen species in vitro. Free Radical Biol Med 1991; 10:277–285.
69. Schulze-Osthoff K, Bakker AC, Vanhaesebroeck B, Beyaert R, Jacobs WA, Fiers W. Cytotoxic activity of tumor necrosis factor is mediated by early damage of mitochondrial functions. J Biol Chem 1992; 267:5317–5323.
70. Schulze-Osthoff K, Beyaert R, Vandevoorde V, Haegeman, Fiers W. Depletion of the mitochondrial electron transport abrogates the cytotoxic and gene-inductive effects of TNF. EMBO J 1993; 12:3095–3104.
71. Hennet T, Richter C, Peterhans E. Tumor necrosis factor-α induces superoxide anion generation in mitochondria of L929 cells. Biochem J 1993; 289:587–592.
72. Vayssiere JL, Petit PX, Risler Y, Mignotte B. Commitment to apoptosis is associated with changes in mitochondrial biogenesis and activity in cell lines conditionally immortalized with simian virus 40. Proc Natl Acad Sci USA 1994; 91:11752–11756.
73. Greenlund LJS, Deckworth TL, Johnson EM Jr. Superoxide dimsutase delays neuronal apoptosis: a role for reactive oxygen species in programmed neuronal death. Neuron 1995; 14:303–315.
74. Reynolds IJ, Hastings TG. Glutamate induces the production of reactive oxygen species in cultured forebrain neurons following NMDA receptor activation. J Neurosci 1995; 15:3318–3327.
75. Mattson MP, Lovell MA, Furukawa K, Markesbery WR. Neurotrophic factors attenuate glutamate-induced accumulation of peroxides, elevation of intracellular Ca^{2+} concentration, and neurotoxicity and increase antioxidant enzyme activities in hippocampal neurons. J Neurochem 1995; 65:1740–1751.
76. Gunsekar PG, Kanthasamy AG, Borowitz JL, Isom GE. NMDA receptor activation produces concurrent generation of nitric oxide and reactive oxygen species: implication for cell death. J Neurochem 1995; 65:2016–2021.
77. Kolesnick R, Golde DW. The sphingomyelin pathway in tumor necrosis factor and interleukin-1 signaling. Cell 1994; 77:325–328.
78. Kolesnick RN, Haimowitz-Friedman A, Fuks Z. The sphinogomyelin signal transduction pathway mediates apoptosis for tumor necrosis factor, Fas, and ionizing radiation. Biochem Cell Biol 1994; 72:471–474.
79. Miura M, Zhu H, Rotello R, Hartwieg EA, Yuan J. Induction of apoptosis in fibroblasts by IL-1 beta-converting enzyme, a mammalian homolog of the *C. elegans* cell death gene *ced-3*. Cell 1993; 75:653–660.
80. Pronk GJ, Ramer K, Amiri P, Williams LT. Requirement of an ICE-like protease for induction of apoptosis and ceramide generation by REAPER. Science 1996; 271:808–810.
81. Brugg B, Michel PP, Agid Y, Ruberg M. Ceramide induces apoptosis in cultured mesencephalic neurons. J Neurochem 1996; 66:7

45

Protein Oxidation and Glycation in Neurodegenerative Diseases

Mark A. Smith, George Perry, and Lawrence M. Sayre
Case Western Reserve University, Cleveland, Ohio

OXIDATIVE STRESS, NEURODEGENERATIVE DISEASE, AND AGING

The etiology of neuronal death in neurodegenerative diseases exemplified by Alzheimer disease (AD), Parkinson disease (PD), and amyotrophic lateral sclerosis (ALS) remains elusive. However, recent advances in molecular genetics and neurochemistry have tied neuronal death to excitotoxicity and oxidative damage, both of which can arise, at least in part, from defects in energy metabolism (1–3). While there is substantial evidence for progressive oxidative modification of proteins in normal aging (4), a great deal of recent research has focused on modification of the neuronal cytoskeleton via oxidative stress mechanisms as a key aspect of irreversible cellular dysfunction, ultimately leading to cell death (reviewed in Refs. 5 and 6). Under normal conditions, damage by oxygen radicals is kept in check by an efficient antioxidant cascade. However, in pathological conditions, the equilibrium between oxidants and antioxidants is likely altered, and there is good reason to think that the central nervous system is particularly vulnerable to oxidative stress due to the high rate of oxygen utilization and high content of unsaturated lipids.

Two components of the neuronal cytoskeleton most implicated in age-related modifications are neurofilament (NF) "triplet" subunits, termed light (NF-L), medium (NF-M), and heavy (NF-H) (7), and the microtubule-associated protein, τ, of which six isoforms (comprising 352–441 amino acids) have been found in human brain (8). Both proteins have been immunolocalized to the aberrant filamentous protein inclusions associated with AD, PD, and ALS. Both NF and τ families contain conserved regions of tandem repeats that are responsible for supramolecular interactions with microtubules underlying axonal transport, in part regulated by phosphorylation/dephosphorylation of serine/threonine residues contained in these repeats. Phosphate levels in both NF and τ are abnormally high in neurodegenerative diseases and aging (9,10). Importantly, the Ser/Thr repeat regions are likely to be the most susceptible to oxidation and modification because of the high lysine content (7,11). It is of note that those diseases characterized by chronic, presumably long-term, oxidative damage are characterized by intraneuronal filamentous inclusions. There is evidence that oxidative stress-induced and related posttranslational modifications contribute to the formation of the inclusion bodies, this being at least in part a consequence of the high percentage of the oxidation-vulnerable lysine in NF and τ. Indeed, it is known that oxidatively stressed cells in culture often develop filamentous inclusions (e.g., 12,13) and, once formed, the inclusions can act to further accelerate oxidative events (e.g., 14).

We suspect that some of the modifications seen *in vivo* in aging and disease are analogous to those modifications found in experimental neuropathies caused by exposure to carbonyl reagents (15). The archetype of these agents is 2,5-hexanedione and its analogs, which, when administered systemically, induce axonal accumulations of NF and alterations of axonal transport (16). These carbonyl-derived neuropathies serve as a morphological model for the NF-containing axonal spheroids of motor neuron disease and ALS (17).

Protein damage which occurs under conditions of oxidative stress may represent (i) direct oxidation of protein side-chains (18); or (ii) adduction of products of lipid peroxidation and/or glycoxidation. Much immunocytochemical evidence exists for the presence of lipid or sugar-derived modifications in pathological cellular aberrations. We and others have determined that advanced glycation end-product (AGE) modifications are present on neurodegenerative structures, for example, the neurofibrillary tangles and amyloid plaques of AD (14,19,20). Moreover, we have also demonstrated direct protein side-chain oxidations and modification of proteins by 4-hydroxy-2-nonenal (HNE) and other lipid peroxidation products (14; M.A. Smith et al., unpublished data). Evidence continues to mount that bifunctional 4-hydroxy-2-alkenals, as opposed to other reactive lipid-derived aldehydes such as malondialdehyde (MDA) or 2-alkenals, are the major cytotoxic products of lipid peroxidation (21). A recent study concluded that protein adduction of HNE, but not of MDA, leads to intermolecular cross-linking and induces cytotoxicity (22).

NATURE OF DIRECT PROTEIN OXIDATION

Despite a substantial body of work on protein oxidation (reviewed in Refs. 18 and 23), there is still very little known about what actually occurs physiologically. Attempts to simulate oxidative stress in vitro have been confused in part by distinctive differences seen between the consequence of attack by authentic hydroxyl radical (HO·) generated by radiolysis, and exposure to hydroxyl-like reactivity such as peroxynitrite (24) or metal-catalyzed oxidation (MCO). The MCO process tends to represent "site-specific" oxidation at the sites of metal ion binding in the protein, and appears to result primarily in backbone cleavage events (23) as well as specific oxidations of side-chains with metal-coordinating capacity (e.g., His imidazole and Met thioether) (25,26). On the other hand, attack by HO· in the presence of O_2 appears to lead primarily to the generation of side-chain carbonyl groups for the aliphatic amino acids (27) and certain oxygenated derivatives of the aromatic amino acids such as Tyr (28) and Trp. This has led to the use of assays of protein-bound carbonyls as markers of protein oxidation (18) and modifications of these assays allows in situ detection of protein-bound carbonyls (29).

Direct evidence for protein cross-linking has so far been obtained mainly for coupling of aromatic ring-centered radicals of Tyr, Trp, and His (30,31), and from the reaction of HO·-oxidized His side-chains with Lys (32). Cross-linking arising from radical coupling of aliphatic amino acid-based radicals is negated in the presence of O_2 (23). However, the large amount of side-chain carbonyls ultimately generated in this case should engender substantial cross-linking through aldol condensation and/or Schiff base formation with free lysine amino groups, based on the biogenesis of collagen and elastin fibrils where lysine is enzymatically oxidized. However, there is as yet no direct evidence for nonenzymatic protein-based carbonyl cross-linking in vitro. Cross-linking may be limited to aldehydes generated from lysine (and arginine and proline), since most of the carbonyls generated from aliphatic side-chains are ketones, for which the Schiff base equilibrium is unfavorable.

Recent studies suggest that oxidative covalent cross-linking of proteins may occur predominantly when the proteins are already preassociated (31). This suggests that protein-derived reactive moieties have a limited lifetime and will preferentially react with solvent or other small molecules if no suitably reactive and/or suitably positioned protein group is nearby. Since both NF (33) and τ (34) can form noncovalent aggregates in the absence of oxidation, we hypothesize that the pathogenetic role of oxidative stress, involving direct oxidation and/or modification by reactive lipid-derived aldehydes, may be primarily in "irreversibly cementing", an altered but noncovalent–and therefore potentially reversible–interaction and protein aggregation initiated by other factors. Interestingly, a recent study showed that "artificial" enzymatic cross-linking of τ could be achieved only when the τ protein was preassociated (35). Since such cross-linking lowers susceptibility to proteolytic degradation, thereby inhibiting turnover, covalent cross-linking of cytoskeletal proteins may itself represent a cytotoxic event. At the same time, several oxidative modifications may involve chemical moieties which themselves support further oxidative stress through "redox cycling" or transition-metal sequestering roles.

CROSS-LINK VERSUS NON-CROSS-LINK MODIFICATIONS

Although covalent intermolecular cross-linking is most likely responsible for the irreversible insolubilization of aggregates of cytoskeletal proteins, it should be pointed out that cross-linking may *not* be an absolute requirement for insolubilization. First, oxidation of several neutral amino acid side-chains can lead to the generation of new side-chain anions (His to Asp, Pro to Glu, and Cys to cysteic acid), whereas oxidation of basic side-chains (Lys and Arg) can lead to their neutralization. Both types of oxidative change work in the direction of lowering the isolectric point (23). Second, oxidation/oxygenation of neutral side-chains results in a significantly decreased local hydrophobicity. It is not hard to imagine that bulk physicochemical shifts or even possibly relatively few alterations in selected protein domains can disrupt the normal supramolecular association of the protein and/or introduce new functionality which induces self-association. Insolubility may be a consequence of either change. It is of note that Li and colleagues (26) observed aggregation of human relaxin at neutral pH to result from non-cross-linked oxidative modification of three (two Met and one His) of the 53 amino acids in the sequence, though solubility could be achieved with SDS or by lowering the pH.

ALZHEIMER DISEASE PATHOLOGY

Almost a century since the first description of the characteristic extracellular senile plaques (SP) and the intraneuronal filamentous inclusions known as neurofibrillary tangles (NFT) that define AD (36), the mechanisms by which these lesions form are unknown. NFT correlate with dementia better than do SP (37,38) and the progression of NFT follows corticocortical connections (39). Thus, it is likely that elucidation of the mechanisms underlying NFT formation may lead to strategies that inhibit NFT deposition or hasten their removal, or, at the least, provide clues to the nature of toxic processes associated with the progression of AD.

The main protein component of NFT is τ (8,40), although NF subunits are also present (41). Both proteins are normally found as soluble components in the cell, but, in AD, have undergone modifications that lead to their deposition as highly insoluble filamentous aggregates (42). It has been shown that τ is "hyperphosphorylated" at the stage where it aggregates into paired helical filaments (PHF), the major constituent, along with straight filaments (SF), of NFT (43). The altered phosphate content has been suggested to be responsible for PHF formation (10,44), although the problem may be more one of defective dephosphorylation by phosphatases than of kinase hyperphosphorylation (45).

Although it seems reasonable that hyperphosphorylation of τ could initiate its diminished interaction with microtubules and consequent self-association, there are other theories of the cause of τ polymerization, including truncation of the N-terminal domain of τ outside the core repeat region (46). Moreover, there is much evidence that hyperphosphorylation alone cannot be the key determinant of insolubilization of mature NFT (47,48), including our recent demonstration that complete dephosphorylation of SDS-insoluble PHF with hydrofluoric acid does not effect solubility (49,50). On the other hand, although only a portion of PHF-τ can be solubilized by a variety of ionic detergents (e.g., SDS), chelators, denaturants, and chaotrophes, or by limited exposure to trifluoroacetic acid and 70% formic acid under conditions that solubilize the large bulk of β-amyloid plaques (51), NFT *can* be completely solubilized by hot aqueous glycine at pH 8 and by 1 N NaOH alone (49,50). These solubility properties of mature NFT summarized here appear inconsistent with a totally noncovalent association at this stage of pathology, but could be explained if NFT were covalently cross-linked by bonds that are unstable at high pH.

THE ROLE OF OXIDATIVE STRESS IN AD PATHOLOGY

Several compelling lines of evidence have implicated oxidative stress and free radical damage in the etiology and pathogenesis of AD, including an association of defects in energy metabolism (2,52) and compensatory upregulation of antioxidant enzymes (53,54), as well as free radical-like cytotoxicity associated with β-amyloid (55–57). Moreover, we and others reported immunocytochemical evidence for the close association of both simple and cross-linked AGE adducts and MDA adducts with the protein components of NFT and SP (14,19), and AGE modifications have been immunolocalized to τ isolated from AD brain (presumably in part coming from NFT) (14,20). That carbonyl-derived cross-links might be responsible for a significant fraction of NFT insolubility is consistent with our observation that NFT can be solubilized without substantial proteolysis by base treatment (49,50), since carbonyl-derived (e.g., aldol-like) cross-links are expected to be more base-labile than peptide bonds.

As we previously demonstrated for Alzheimer disease, induction of heme oxygenase is a robust indicator of the oxidative stress response in cells (58). Briefly, we demonstrated that heme oxygenase protein and mRNA were increased in brains of cases of Alzheimer disease (59) and that this increase was tightly correlated with regions of PHF pathology

(54). In further studies, we found that glycated τ (either glycated in vitro or isolated from AD brain) causes an oxidative response in neuroblastoma cells including induction of heme oxygenase, evidence of lipid peroxidation and induction of NFκB (14,60).

We demonstrated that an antibody recognizing a carbonyl-modified epitope of NF is able to strongly label NFT within the neuronal perikarya, neuropil threads, senile plaque neurites, and granulovacuolar degeneration in AD brain (61). In control patients, sporadic lesion-related immunoreactivity was seen. Also, carbonyl-modified NF could be isolated from homogenates of AD brain tissue but not from control tissue.

The NFT epitope recognized could be created in vitro by reaction of NF protein with aldehydes, including formaldehyde and glutaraldehyde, whose ability to fix tissue is a consequence of characteristic carbonyl condensation-dependent inter- and intramolecular cross-linking, for example, the lysine-derived pyridinium cross-links produced by glutaraldehyde (62). Further, antibody recognition was also effected by malondialdehyde, a known lysine-dependent cross-linking agent generated from lipid peroxidation (63). This latter observation is consistent with data showing a high amount of membrane disruption in degenerating neurites (64), as well as our previous demonstration that malondialdehyde adducts are associated with the pathological lesions of AD (14).

These data suggest that the common feature of epitope recognition is lysine-derived cross-links, rather than the particular carbonyl modification, since the cross-linking moieties are structurally diverse. It is probably the resultant altered secondary structure in the lysine-rich region of the neuronal proteins that is recognized. The recognition of PHF in non-aldehyde-fixed tissue (i.e., methacarn or cryostat sections) by these antisera strongly suggests that the NF modification already exists in vivo, that in PHF the NF component is cross-linked, and that these cross-links are formed by carbonyl reactions. In control patients, sporadic lesion-related immunoreactivity was seen, suggesting that this modification, while certainly disease-associated, is actually better correlated with the presence of abnormal intraneuronal inclusion bodies.

Two research groups have recently shown that PHF formation can be induced by disulfide bond formation between cysteine residues which become susceptible to oxidation due to a phosphorylation-induced detachment of τ from microtubules (65,66). On the other hand, under reducing conditions, τ appears to polymerize into SF-like rather than PHF-like structures (67). Unimpaired mitochondrial respiration normally maintains a reduced state in cells, where glutathione is mainly in the reduced as opposed to oxidized form. Intracellular sulfhydryl (SH) groups should thus remain mainly in their reduced form except under conditions of compromised mitochondrial respiration and/or oxidative stress. This suggests that the irreversible τ polymerization characterizing PHF likely represents disulfide-mediated dimerization which serves as a "nucleation step" (66), followed by additional oxidation-dependent cross-linking events which cement PHF assembly beyond the dimer stage. The recalcitrant solubility properties of NFT described above are consistent with the notion that insolubilization of mature NFT represents a more insidious level of oxidative stress-induced cross-linking than simple cystine-based dimerization.

IDENTIFICATION OF AN ADVANCED LIPID PEROXIDATION-DERIVED MODIFICATION IN NFT

We previously described a 2-pentylpyrrole modification of lysine as the only presently known "advanced" (stable end-product) adduct that forms from modification of proteins by the highly cytotoxic product of lipid peroxidation, 4-hydroxy-2-nonenal (HNE) (68). We raised two related rabbit polyclonal antibodies to this modification: one (KLH-ON) by direct derivatization of keyhole limpet hemocyanin (KLH, as carrier protein) with 4-oxononanal, which independently forms the HNE-derived pyrrole adduct in high yield; and the other (KLH-ACA-ON) by carbodiimide-mediated attachment of a *preformed* 2-pentylpyrrole derivative of 6-aminocaproic acid to KLH. Utilizing the corresponding BSA-derivatives (BSA-ON and BSA-ACA-ON) as coating agents in ELISA studies, these antibodies were shown to be highly specific for recognizing the 2-pentylpyrrole modification relative to "early" HNE adducts and other types of lysine modification. In preliminary studies with these antibodies on Alzheimer disease brain sections using the peroxidase–antiperoxidase technique, both the KLH-ACA-ON-elicited and KLH-ON-elicited antibodies specifically label neurofibrillary pathology and recognize relatively low levels of adduct in pyramidal neurons (88). These findings concur with the recent demonstration that HNE is cytotoxic to neurons (22), consistent with the notion that HNE might contribute to neurodegeneration.

THE ROLE OF OXIDATIVE STRESS IN PARKINSON DISEASE

Lewy bodies of Parkinson disease (PD) are eosinophilic intraneuronal filamentous inclusions found in magnocellular neurons of brain stem nuclei, predominantly substantia nigra and locus coeruleus. Lewy bodies consist of randomly oriented highly phosphorylated NF (69,70). Like PHF, Lewy bodies are insoluble in detergent and are resistant to removal from the extracellular space following neuronal death (71).

There is considerable evidence that oxidative stress is primarily involved in the pathogenesis of PD (reviewed in Ref. 5). For example, there is evidence of lipid peroxidative events (72,73) and of oxidative DNA damage (74). Additionally, we and others have demonstrated increased antioxidant defense mechanisms, including SOD in the substantia nigra of PD (75) and both SOD (76) and heme oxygenase-1 (77) in Lewy bodies.

THE ROLE OF OXIDATIVE STRESS IN AMYOTROPHIC LATERAL SCLEROSIS

ALS is a devastating neurodegenerative disease whose clinical parameters of paralysis, respiratory depression and death correlate closely with the degeneration and selective loss of motor neurons in the spinal cord and cortex. While a great many of the etiological aspects of ALS remain unclear, it is now apparent that specific alterations of Cu/Zn superoxide dismutase (SOD-1) can play a major pathological role in the disease. Recently, several familial, and apparently even some sporadic, ALS cases demonstrated mutations in the gene for SOD (78), and there are increased levels of SOD mRNA reported in ALS (79). Taken together, these findings suggest that oxidative free radical damage is an important etiological and pathogenic factor in ALS. Supporting this concept, an animal model of ALS was recently developed in mice transgenic for one of the mutated forms of SOD associated with familial ALS (80). Such transgenic mice develop many of the characteristic clinical and pathological correlates of ALS including motor deficits, motor neuron depletion, vacuolation, and formation of intraneuronal filamentous inclusions (81). We have demonstrated in preliminary experiments that these SOD-transgenic animals show an induction of the antioxidant enzyme heme oxygenase-1 in motor neurons, indicative of increased oxidative stress (82). We believe that these observations further indicate the merit of studying oxidative modification and stress in ALS and models of ALS.

THE EFFECT(S) OF PROTEIN MODIFICATION ON CELLULAR PROCESSES

Low levels of reactive oxygen species (ROS) affect signal transduction pathways. For example, the transcription factor NFκB (nuclear factor κB) is activated by ROS (83). Several heat shock proteins including HSP32, HSP70, and ubiquitin are induced during episodes of cellular stress (e.g., 84, 85). In previous studies, we demonstrated that the introduction of AGE-modified τ, but not unmodified τ, results in the activation of NFκB and induction of heme oxygenase antigen (14,60). Moreover, neurons exposed to AGE-modified τ exhibited evidence of oxidative stress including lipoperoxidative events and induction of malondialdehyde epitopes, and resulted in cytokine expression, an increased expression of β-protein precursor, and release of ~ 4 kDa β-amyloid protein (14,60).

CONCLUSION

The degree of cognitive and/or clinical impairment in neurodegenerative diseases is correlated to the degeneration and subsequent loss of specific neuronal populations. This fallout of neurons is in turn correlated with the pathological lesions (i.e., intraneuronal inclusions), comprising long-lived cytoskeletal proteins that are selectively vulnerable to oxidative stress, and it is tempting to speculate that free radical oxygen chemistry plays a pathogenetic role in all these neurodegenerative conditions. However, it is as yet undetermined how this oxidative stress will affect specific neuronal populations. For example, while cytoskeletal abnormalities are the commonlity between these neurodegenerative diseases, it is unlikely that the vulnerable neurons in Alzheimer and Parkinson disease (pyramidal neurons in the hippocampus/cortex and magnonuclear neurons in the substantia nigra, respectively) respond in identical ways to oxidative damage. Therefore, determining whether cytoskeletal inclusion body formation reflects mainly direct protein oxidation or adduction of lipid peroxidation products might give clues as to rational therapeutic protocols for each disease, for example, using water-soluble versus lipid-soluble antioxidants (86) or free radical versus carbonyl scavengers. We speculate that, while there will be a certain degree of overlap between these oxidative processes, each disease will show a predominance of specific oxidative modifications.

In conclusion, it is of considerable interest that preliminary epidemiological and clinical studies suggest that inhibitors of oxidative stress and glycation may prove highly effective in the treatment of neurodegenerative diseases (87).

ACKNOWLEDGMENTS

Work in the authors' laboratories is supported by grants from the National Institutes of Health, the American Health Assistance Foundation, and the American Federation of Aging Research. M.A. Smith is a Fellow of the American Philosophical Society.

REFERENCES

1. Beal MF. Aging, energy, and oxidative stress in neurodegenerative diseases. Ann Neurol 1995; 38:357–366.
2. Bowling AC, Beal MF. Bioenergetic and oxidative stress in neurodegenerative diseases. Life Sci 1995; 56:1151–1171.
3. Dawson R Jr, Beal MF, Bondy SC, Di Monte DA, Isom GE. Excitotoxins, aging, and environmental neurotoxins: implications for understanding human neurodegenerative diseases. Toxicol Appl Pharmacol 1995; 134:1–17.
4. Halliwell B, Gutteridge J. Oxygen radicals and the nervous system. Trends Neurosci 1985; 8:22–26.
5. Olanow CW. A radical hypothesis for neurodegeneration. Trends Neurosci 1993; 16:439–444.
6. Smith MA, Sayre LM, Monnier VM, Perry G. Radical AGEing in Alzheimer's disease. Trends Neurosci 1995; 18:172–176.
7. Geisler N, Kaufmann E, Fischer S, Plessmann U, Weber K. Neurofilament architecture combines structural principles of intermediate filaments with carboxy-terminal extensions increasing in size between triplet proteins. EMBO J 1983; 2:1295–1302.
8. Goedert M, Jakes R, Spillantini MG, et al. τ Protein in Alzheimer's disease. Biochem Soc Trans 1995; 23:80–85.
9. Gou JP, Eyer J, Leterrier JF. Progressive hyperphosphorylation of neurofilament heavy subunits with aging: possible involvement in the mechanism of neurofilament accumulation. Biochem Biophys Res Commun 1995; 215:368–376.
10. Hanger DP, Brion JP, Gallo JM, Cairns NJ, Luthert PJ, Anderton BH. τ in Alzheimer's disease and Down's syndrome is insoluble and abnormally phosphorylated. Biochem J 1991; 275:99–104.
11. Shaw G. Neurofilament protein. In Burgoyne RD, ed. The Neuronal Cytoskeleton. New York: Wiley-Liss; 1991:185–214.
12. Schipper HM. Mydlarski MB, Wang X. Cysteamine gliopathy *in situ*: a cellular stress model for the biogenesis of astrocytic inclusions. J Neuropathol Exp Neurol 1993; 52:399–410.
13. Manganaro F, Chopra VS, Mydlarski MB, Bernatchez G, Schipper HM. Redox perturbations in cysteamine-stressed astroglia: implications for inclusion formation and gliosis in the aging brain. Free Radical Biol Med 1995; 19:823–835.
14. Yan SD, Chen X, Schmidt AM, et al. Glycated τ protein in Alzheimer disease: a mechanism for induction of oxidant stress. Proc Natl Acad Sci USA 1994; 91:7787–7791.
15. Sayre LM, Autilio-Gambetti L, Gambetti P. Pathogenesis of experimental giant neurofilamentous axonopathies: a unified hypothesis based on chemical modification of neurofilaments. Brain Res 1985; 10:69–83.
16. Braendgaard H, Sidenius P. The retrograde fast component of axonal transport in motor and sensory nerves of the rat during administration of 2,5-hexanedione. Brain Res 1986; 378:1–7.
17. Griffin JW, Price DL, Hoffman PN. Neurotoxic probes of the axonal cytoskeleton. Trends Neurosci 1983; 6:490–495.
18. Stadtman ER. Protein oxidation and aging. Science 1992; 257:1220–1224.
19. Smith MA, Taneda S, Richey PL, et al. Advanced Maillard reaction products are associated with Alzheimer disease pathology. Proc Natl Acad Sci USA 1994; 91:5710–5714.
20. Ledesma MD, Bonay P, Colaco C, Avila J. Analysis of microtubule-associated protein τ glycation in paired helical filaments. J Biol Chem 1994; 269:21614–21619.
21. Esterbauer H, Schaur RJ, Zollner H. Chemistry and biochemistry of 4-hydroxynonenal, malonaldehyde and related aldehydes. Free Radical Biol Med 1991; 11:81–128.
22. Montine TJ, Amarnath V, Martin ME, Strittmatter WJ, Graham DG. *E*-4-hydroxy-2-nonenal is cytotoxic and cross-links cytoskeletal proteins in P19 neuroglial cultures. Am J Pathol 1996;

148:89–93.
23. Stadtman ER. Oxidation of free amino acids and amino acid residues in proteins by radiolysis and by metal-catalyzed reactions. Annu Rev Biochem 1993; 62:797–821.
24. King P, Antar S, Goshe M, Anderson VE. Carbonyl formation and protein cleavage: the potential for protein footprinting by oxidation using radiolysis or peroxynitrous acid decomposition. Free Radical Biol Med, submitted.
25. Uchida K, Kawakishi S. Reactions of a histidyl residue analogue with hydrogen peroxide in the presence of copper(II) ion. J Agric Food Chem 1990; 38:660–664.
26. Li S, Nguyen TH, Schoneich C, Borchardt RT. Aggregation and precipitation of human relaxin induced by metal-catalyzed oxidation. Biochemistry 1995; 34:5762–5772.
27. Liebster J, Kopoldova J. Radiation chemical reactions in aqueous oxygenated and oxygen-free solutions of aliphatic dipeptides and tripeptides. Radiat Res 1966; 27:162–173.
28. Gieseg SP, Simpson JA, Charlton TS, Duncan MW, Dean RT. Protein-bound 3,4-dihydroxyphenylalanine is a major reductant formed during hydroxyl radical damage to proteins. Biochemistry 1993; 32:4780–4786.
29. Smith MA, Perry G, Richey PL, Sayre LM, Anderson VE, Beal MF, Kowall N. Oxidative damage in Alzheimer's. Nature 1996; 382:120–121.
30. Huggins TG, Wells-Knecht MC, Detorie NA, Baynes JW, Thorpe SR. Formation of *o*-tyrosine and dityrosine in proteins during radiolytic and metal-catalyzed oxidation. J Biol Chem 1993; 268:12341–12347.
31. Brown KC, Yang SH, Kodadek T. Highly specific oxidative cross-linking of proteins mediated by a nickel–peptide complex. Biochemistry 1995; 34:4733–4739.
32. Guptasarma P, Balasubramanian D, Matsugo S, Saito I. Hydroxyl radical mediated damage to proteins, with special reference to the crystallins. Biochemistry 1992; 31: 4296–4303.
33. Lee MK, Marszalek JR, Cleveland DW. A mutant neurofilament subunit causes massive, selective motor neuron death: implications for the pathogenesis of human motor neuron disease. Neuron 1994; 13:975–988.
34. Crowther RA, Olesen OF, Smith MJ, Jakes R, Goedert M. Assembly of Alzheimer-like filaments from full-length τ protein. FEBS Lett 1994; 337:135–138.
35. Miller ML, Johnson GVW. Transglutaminase cross-linking of the τ protein. J Neurochem 1995; 65:1760–1770.
36. Khachaturian ZS. Diagnosis of Alzheimer's disease. Arch Neurol 1985; 42:1097–1105.
37. Blessed G, Tomlinson BE, Roth M. The association between quantitative measures of dementia and senile change in the cerebral grey matter of elderly subjects. Br J Psychol 1968; 114:797–811.
38. Delaere P, Duyckaerts C, Brion JP, Poulain V, Hauw JJ. Tau, paired helical filaments and amyloid in neocortex: a morphometric study of 15 cases with graded intellectual status of aging and senile dementia of Alzheimer type. Acta Neuropathol 1989; 77:645–653.
39. Pearson RC, Esiri MM, Hiorns RW, Wilcock GK, Powell T. Anatomical correlates of the distribution of the pathological changes in the neocortex in Alzheimer disease. Proc Natl Acad Sci USA 1985; 82:4531–4534.
40. Wischik CM, Novak M, Thogerson HC, et al. Isolation of a fragment of τ derived from the core of the paired helical filament of Alzheimer disease. Proc Natl Acad Sci USA 1988; 85:4506–4510.
41. Perry G, Rizzuto N, Autillio-Gambetti L, Gambetti P. Paired helical filaments from Alzheimer disease patients contain cytoskeletal components. Proc Natl Acad Sci USA 1985; 82:3916–3920.
42. Selkoe DJ, Ihara Y, Salazar FJ. Alzheimer's disease: insolubility of partially purified paired helical filaments in sodium dodycyl sulfate and urea. Science 1982; 215:1243–1245.
43. Wischik CM, Novak M, Edwards PC, Klug A, Tichelaar W, Crowther RA. Structural characterization of the core of the paired helical filament of Alzheimer disease. Proc Natl Acad Sci USA 1988; 85:4884–4888.

44. Iqbal K, Zaidi T, Bancher C, Grundke-Iqbal I. Alzheimer paired helical filaments. Restoration of the biological activity by dephosphorylation. FEBS Lett 1994; 349:104–108.
45. Trojanowski JQ, Lee VMY. Phosphorylation of paired helical filament τ in Alzheimer's disease neurofibrillary lesions: focusing on phosphatases. FASEB J 1995; 9:1570–1576.
46. Ksiezak-Reding H, Morgan K, Dickson DW. τ immunoreactivity and SDS solubility of two populations of paired helical filaments that differ in morphology. Brain Res 1994; 649:185–196.
47. Gustke N, Steiner B, Mandelkow EM, et al. The Alzheimer-like phosphorylation of τ protein reduces microtubule binding and involves Ser-Pro and Thr-Pro motifs. FEBS Lett 1992; 307:199–205.
48. Wang JZ, Gong CX, Zaidi T, Grundke-Iqbal I, Iqbal K. Dephosphorylation of Alzheimer paired helical filaments by protein phosphatase-2A and -2B. J Biol Chem 1995; 270: 4854–4860.
49. Smith MA, Siedlak SL, Richey PL, Nagaraj RH, Elhammer A, Perry G. Quantitative solubilization and analysis of insoluble paired helical filaments from Alzheimer disease. Brain Res 1996; 717:99–108.
50. Smith MA, Nagaraj RH, Perry G. Protocol for the quantitative analysis of paired helical filament solubilization: a method applicable to insoluble amyloids and inclusion bodies. Brain Res Protocols 1997; in press.
51. Masters CL, Simms G, Weinman NA, Multhaup G, McDonald BL, Beyreuther K. Amyloid plaque core protein in Alzheimer disease and Down syndrome. Proc Natl Acad Sci USA 1985; 82:4245–4249.
52. Blass JP, Baker AC, Ko L, Black RS. Induction of Alzheimer antigens by an uncoupler of oxidative phosphorylation. Arch Neurol 1990; 47:864–869.
53. Pappolla MA, Omar RA, Kim KS, Robakis NK. Immunohistochemical evidence of antioxidant stress in Alzheimer's disease. Am J Pathol 1992; 140:621–628.
54. Smith MA, Kutty RK, Richey PL, et al. Heme oxygenase-1 is associated with the neurofibrillary pathology of Alzheimer's disease. Am J Pathol 1994; 145:42–47.
55. Behl C, Davis J, Cole GM, Schubert D. Vitamin E protects nerve cells from amyloid-β protein toxicity. Biochem Biophys Res Commun 1992; 186:944–950.
56. Behl C, Davis JB, Lesley R, Schubert D. Hydrogen peroxide mediates amyloid β protein toxicity. Cell 1994; 77:817–827.
57. Hensley K, Carney JM, Mattson MP, et al. A model for β-amyloid aggregation and neurotoxicity based on free radical generation by the peptide: relevance to Alzheimer disease. Proc Natl Acad Sci USA 1994; 91:3270–3274.
58. Maines MD. Heme oxygenase: function, multiplicity, regulatory mechanisms, and clinical applications. FASEB J 1988; 2:2557–2568.
59. Premkumar DRD. Smith MA, Richey PL, et al. Induction of heme oxygenase-1 mRNA and protein in neocortex and cerebral vessels in Alzheimer's disease. J Neurochem 1995; 65:1399–1402.
60. Yan SD, Yan SF, Chen X, et al. Non-enzymatically glycated tau in Alzheimer's disease induces neuronal oxidant stress resulting in cytokine gene expression and release of amyloid-β peptide. Nature Medicine 1995; 1:693–699.
61. Smith MA, Rudnicka-Nawrot M, Richey PL, et al. Carbonyl-related posttranslational modification of neurofilament protein in the neurofibrillary pathology of Alzheimer's disease. J Neurochem 1995; 64:2660–2666.
62. Hardy PM, Hughes GJ, Rydon HN. The nature of the cross-linking of proteins by glutaraldehyde. Part 2. The formation of quaternary pyridinium compounds by the action of glutaraldehyde on proteins and the identification of a 3-(2-piperidyl)-pyridinium derivative, anabilysine, as a cross-linking entity. J Chem Soc Perkin Trans 1 1979; 2282–2285.
63. Chio KS, Tappel AL. Synthesis and characterization of the fluorescent products derived from malonaldehyde and amino acids. Biochemistry 1969; 8:2821–2826.

64. Praprotnik D, Smith MA, Richey PL, Vinters HV, Perry G. Plasma membrane fragility in dystrophic neurites in senile plaques of Alzheimer's disease: an index of oxidative stress. Acta Neuropathol 1996; 91:1–5.
65. Guttmann RP, Erickson AC, Johnson GVW. τ self-association: stabilization with a chemical cross-linker and modulation by phosphorylation and oxidation state. J Neurochem 1995; 64:1209–1215.
66. Schweers O, Mandelkow EM, Biernat J, Mandelkow E. Oxidation of cysteine-322 in the repeat domain of microtubule-associated protein τ controls the *in vitro* assembly of paired helical filaments. Proc Natl Acad Sci USA 1995; 92:8463–8467.
67. Wilson DM, Binder LI. Polymerization of microtubule-associated protein τ under near-physiological conditions. J Biol Chem 1995; 270:24306–24314.
68. Sayre LM, Arora PK, Iyer RS, Salomon RG. Pyrrole formation from 4-hydroxynonenal and primary amines. Chem Res Toxicol 1993; 6:19–22.
69. Galloway PG, Grundke-Iqbal I, Iqbal K, Perry G. Lewy bodies contain epitopes both shared and distinct from Alzheimer neurofibrillary tangles. J Neuropathol Exp Neurol 1988; 47:654–663.
70. Pollanen MS, Dickson DW, Bergeron C. Pathology and biology of the Lewy body. J Neuropathol Exp Neurol 1993; 52:183–191.
71. Galloway PG, Mulvihill P, Perry G. Filaments of Lewy bodies contain insoluble cytoskeletal elements. Am J Pathol 1992; 140:809–822.
72. Dexter DT, Carter CJ, Wells FR, et al. Basal lipid peroxidation in substantia nigra is increased in Parkinson's disease. J Neurochem 1989; 52:381–389.
73. Dexter DT, Holley AE, Flitter WD, et al. Increased levels of lipid hydroperoxides in the Parkinsonian substantia nigra: an HPLC and ESR study. Mov Disord 1994; 9:92–97.
74. Sanchez-Ramos J, Overvik E, Ames B. A marker of oxyradical-mediated DNA damage (8-hydroxy-2′-deoxyguanosine) is increased in nigro-striatum of Parkinson's disease brain. Neurodegeneration 1994; 3:197–204.
75. Saggu H, Cooksey J, Dexter D, et al. A selective increase in particulate superoxide dismutase activity in Parkinsonian substantia nigra. J Neurochem 1989; 53:692–697.
76. Nishiyama K, Murayama S, Shimizu J, et al. Cu/Zn superoxide dismutase-like immunoreactivity is present in Lewy bodies from Parkinson disease: a light and electron microscopic immunocytochemical study. Acta Neuropathol 1995; 89:474–474.
77. Castellani R, Smith MA, Richey PL, Perry G. Glycoxidation and oxidative stress in Parkinson disease and diffuse Lewy body disease. Brain Res 1996; 737:195–200.
78. Rosen DR, Siddique T, Patterson D, et al. Mutation in Cu/Zn superoxide dismutase gene are associated with familial amyotrophic lateral sclerosis. Nature 1993; 362:59–62.
79. Bergeron C, Muntasser S, Sommerville MJ, Weyer L, Percy ME. Copper/zinc superoxide dismutase mRNA levels in sporadic amyotrophic lateral sclerosis motor neurons. Brain Res 1994; 659:272–276.
80. Gurney ME, Pu H, Chiu AY, et al. Motor neuron degeneration in mice that express a human Cu,Zn superoxide dismutase mutation. Science 1994; 264:1772–1775.
81. Dal Canto MC, Gurney ME. Development of central nervous system pathology in a murine transgenic model of human amyotrophic lateral sclerosis. Am J Pathol 1994; 145:1271–1279.
82. Smith MA, Dal Canto M, Richey PL, Perry G. The transgenic SOD-1 mouse model of ALS: selective oxidative stress in motor neurons. Soc Neurosci Abst 1996; 22, 712.
83. Schreck R, Rieber P, Baeuerle P. Reactive oxygen intermediates as apparently widely used messengers in the action of NF-κB transcription factor and HIV-1. EMBO J 1991; 10:2247–2258.
84. Lindquist S. The heat-shock response. Annu Rev Biochem 1986; 55:1151–1191.
85. Shibahara S, Müller RM, Taguchi H. Transcriptional control of rat heme oxygenase by heat shock. J Biol Chem 1987; 262:12889–12892.

86. Dean RT, Hunt JV, Grant AJ, Yamamoto Y, Niki E. Free radical damage to proteins: The influence of the relative localization of radical generation, antioxidants, and target proteins. Free Radical Biol Med 1991; 11:161–168.
87. Münch G, Taneli Y, Schraven E, et al. The cognition-enhancing drug tenilsetam is an inhibitor of protein crosslinking by advanced glycosylation. J Neural Trans–Parkinsons Dis Dementia Sect 1994; 8:193–208.
88. Sayre LM, Zelasko DA, Harris PLR, Perry G, Salomon RG, Smith MA. 4-Hydroxynonenal-derived advanced lipid peroxidation end products are increased in Alzheimer's disease. J Neurochem 1997; 68:2092–2097.

46

Glycation as Oxidative Stress and Redox Regulation: Implications for Aging, Diabetes, and Familial Amyotrophic Lateral Sclerosis

Naoyuki Taniguchi, Junichi Fujii, Hideaki Kaneto, Michio Asahi, Theingi Myint, Nobuko Miyazawa, Keiichiro Suzuki, and Kazi Nazrul Islam
Osaka University Medical School, Osaka, Japan

INTRODUCTION

Oxidative stress due to production of reactive oxygen species occurs in a variety of pathological and physiological conditions including diabetes and inflammation (1). Under the normal aging process or diabetic conditions, nonenzymatic glycosylation, designated as the glycation reaction (the Maillard reaction), occurs and this reaction also produces reactive oxygen species (2). On the other hand, in inflammatory processes, reactive oxygen species such as NO are produced by stimulation by cytokines such as interleukin-1 (IL-1) and tumor necrosis factor (TNF)-α. The antioxidative enzymes such as Cu,Zn-superoxide dismutase (SOD), Mn-SOD, glutathione peroxidase (GPx), and catalase play a pivotal role in scavenging the reactive oxygen species. However, under hyperglycemic conditions or in inflammatory processes, the major scavenging enzymes, Cu,Zn-SOD and GPx, are inactivated by the reactive oxygen species generated in the cells. Moreover, gene expression of other antioxidative enzymes including Mn-SOD, catalase and glutathione *S*-transferases 1 and 2 are downregulated by transforming growth factor β1 (3). These events may suggest that oxidative stress is enhanced by downregulation of antioxidative enzymes at transcriptional and posttranslational levels. In this review we focus on the role of antioxidative enzymes in inflammation and diabetes as well as in a typical neurodegenerative disease, familial amyotrophic lateral sclerosis.

GLYCATION REACTION OF Cu,Zn-SOD IN VIVO AND IN VITRO

Our previous studies indicated that Cu,Zn-SOD, which is fairly abundant in most tissues, undergoes inactivation by the glycation reaction in vivo (4). Subsequent work indicated that the glycated forms of Cu,Zn-SOD are increased in the erythrocytes of patients with

Werner's syndrome, an accelerated aging disease (5) as well as in patients with diabetes (6). The in vitro-glycated sites were identified and it was found that Lys-122 and Lys-128 residues are likely candidates for glycation reaction in vitro (7). Moreover, Cu,Zn-SOD is site-specifically cleaved by glycation reaction and finally undergoes random fragmentations after incubation with glucose (8). In these processes, hydroxy radicals are produces and then the Fenton-type reaction plays a key role in site-specific and random fragmentations. These findings indicated that Cu,Zn-SOD incubated with reducing sugars such as glucose and fructose produces superoxide anion in the presence of transition metals via ene–diol structures as an intermediate product of the Maillard reaction, and then superoxide anion is converted to hydroxy radicals via hydrogen peroxide (Figure 1). The hydroxy radical is a potent reactive oxygen species that cleaves proteins, DNA (9), and lipids.

Very recently we have found that glycated Cu,Zn-SOD is abundant in rat lens as compared to other tissues, as judged by use of a boronate affinity column to estimate the glycated form of the Cu,Zn-SOD. In normal rats lens levels of glycated Cu,Zn-SOD showed a gradual increase with age, whereas in diabetic rats substantial increases were observed (10). Immunoblotting analyses using anti-hexitollysine IgG indicated that glycated Cu,Zn-SOD contains Amadori products. Moreover, Cu,Zn-SOD in lenses was site-specifically fragmented, probably because of glycation (11), as was observed in studies in vitro. Several reports concerning the role of glycation in cataractogenesis have already been published. Current opinion favors the postulate that the etiology of cataract formation is related to the progressive aggregation of lens crystalline. In strepotozocin-diabetic rats, the glycation of crystallins increased time-dependently and glycation has been observed to lead to inner-crystallins cross-links. These studies support the protein glycation theory and suggest that the primary cause of cataractogenesis is related to the glycation of lens proteins including lens crystallins and Cu,Zn-SOD. Our study shows that the lens of diabetic and aging rats accumulates large amounts of glycated Cu,Zn-SOD that are mainly localized in epithelial cells. The glycation of Cu,Zn-SOD eventually brings

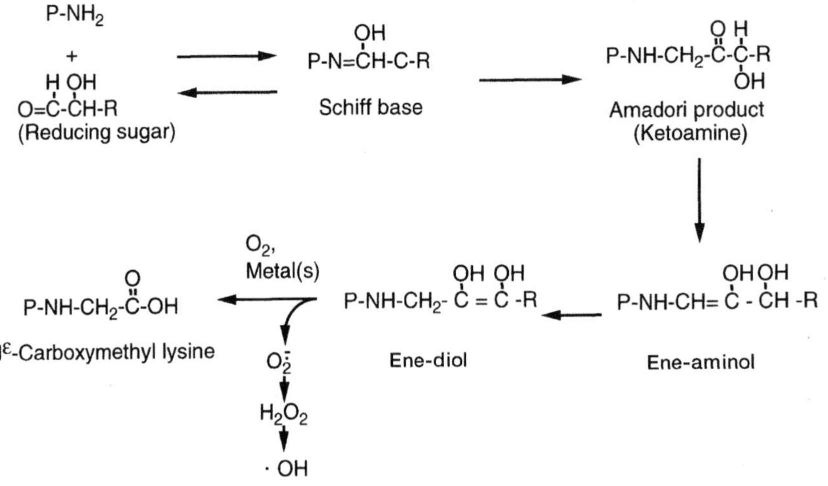

Figure 1. Generation of superoxide anion at the initial stage of glycation reaction.

about fragmentation and leads to the formation of superoxide and hydroxy radicals. These data suggest that glycation and fragmentation of Cu,Zn-SOD in lens is strongly associated with diabetic complication in the lens.

APOPTOTIC CELL DEATH IS TRIGGERED BY NO AND GLYCATION IN PANCREATIC β-CELLS

Insulin-dependent diabetes mellitus (IDDM) is mediated by an autoimmune mechanism of the inflammatory process that is characterized by destruction of pancreatic β-cells. Interleukin (IL)-1β has been proposed to play an important role in mediating both destruction and dysfunction of pancreatic β-cells. Incubation of the β-cells with IL-1β results in inhibition of glucose-induced insulin secretion. The deleterious effects of IL-1β have been proposed to involve generation of reactive oxygen species, including NO, and mitochondrial function. IL-1β was reported to induce the expression of inducible NOS, which was preceded by expression of c-*fos* mRNA. We have demonstrated that both exogenous NO and NO generated endogenously by IL-1β brought about apoptosis of isolated rat pancreatic islet cells as well as HIT cells, a pancreatic β-cell tumor cell line (12). This apoptosis was characterized by cleavage of DNA into nucleosomal fragments of 180–200 bp and morphologically by nuclear shrinkage, chromatin condensation, and apoptotic body formation. The IL-β-induced DNA cleavage of β-cells occurred in a time- and dose-dependent manner. Actinomycin D, cycloheximide, and NOS inhibitors inhibited the DNA cleavage, which was correlated with the amount of NO produced, indicating that NO produced by HIT cells themselves could mediate the apoptosis. Furthermore, in the presence of TNF-α, internucleosomal DNA cleavage is an important initial step in the destruction and dysfunction of pancreatic β-cells induced by inflammatory stimulation. Reactive oxygen species have been demonstrated to cause apoptotic cell death. Recent reports have shown that NO also causes apotosis in macrophage cells. The cells generate reactive oxygen species including NO. The similar apoptotic cell death in pancreatic β-cells is observed with reducing sugars via glycation reaction as shown below.

NO SELECTIVELY INACTIVATES GLUTATHIONE PEROXIDASE ACTIVITY

GPx is a tetrameric selenoenzyme carrying one essential selenocysteine residue per subunit for its activity. Three isozymes designated as cellular GPx, extracellular GPx, and phospholipid hydroperoxide GPx are known. As described above, oxidative stress is one of the direct causes of apoptotic cell death and GPx as well as bcl-2, a proto-oncogene that blocks apoptotic death in multiple contexts, can prevent apoptosis (13). *S*-Nitro-*N*-acetyl-DL-penicillamine (SNAP), an NO donor, inactivated bovine GPx in a dose- and time-dependent manner (14). The IC_{50} of SNAP for GPx was 2 μM at 1 h of incubation and was 20% of the IC_{50} for another thiol enzyme, glyceraldehyde-3-phosphate dehydrogenase, in which a specific cysteine residue is known to be nitrosylated. Incubation of the inactivated GPx with 5 mM dithiothreitol within 1 h restored about 50% of activity of the start of the SNAP incubation. However, longer exposure to NO donors irreversibly inactivated the enzyme. The similarity of the inactivation with SNAP and reaction of GPx with dithiothreitol to that of glyceraldehyde-3-phosphate dehydrogenase suggested that NO released from SNAP modified a cysteine-like essential residue on GPx. When U937 cells were incubated with 100 μM SNAP for 1 h, a significant decrease in GPx activity was observed, although the change was less dramatic than that with the purified enzyme, and intracellular peroxide levels increased as judged by FACS using a peroxide-sensitive dye. Other major antioxidative enzymes, Cu,Zn-SOD, Mn-SOD, and catalase, were not affected by SNAP, which suggested that the increased accumulation of peroxides in SNAP-treated cells was due to inhibition of GPx activity by NO. Moreover, stimulation with lipopolysaccharide significantly decreased intracellular GPx activity in RAW 264.7 cells and this effect was blocked by NO inhibitor N^{ω}-methyl-L-arginine. This indicated that GPx was also inactivated by endogenous NO. This mechanism may at least in part explain the cytotoxic effects of NO on cells and NO-induced apoptotic cell death.

MUTANT Cu,Zn-SODs RELATED TO FAMILIAL AMYOTROPHIC LATERAL SCLEROSIS

Defects in the gene encoding Cu,Zn-SOD were demonstrated in many cases of familial amyotrophic lateral sclerosis (FALS), a motor neuron disease (15). Over 40 mutations were found in patients with FALS. Some were in a domain that shows little conservation of amino acid residues among different species. These mutations were assumed to affect the subunit interaction or folding of the enzyme. It is not clear, however, whether these mutations actually decrease SOD activity by reducing protein stability or increase susceptibilities of the enzymes to protease due to missfolding. The mechanism by which FALS is caused seemed to be more complex because both suppression and overexpression of SOD activity induced neuronal cell death. Several groups hypothesize that a gain of function of these mutant enzymes is a cause of the disease. We have produced wild-type and six mutant Cu,Zn-SODs related to FALS in a baculovirus/insect cell expression system (16). Incubation of the mutant enzyme with 100 mM glucose reduced the activity more prominently than the wild-type enzyme by the glycation reaction. Moreover, the formation of Amadori products was remarkable as judged by ELISA using anti-hexitollysine IgG. This suggests that the mutant enzymes are vulnerable to the reaction. Decreased scavenging ability, vulnerability to glycation, and free copper release from the inactivated enzymes are all likely reasons why the mutant enzymes might produce hydroxy radicals which may be a potent causative factor of FALS via the Fenton chemistry.

FRUCTOSE IS A STRONG REDUCING SUGAR IN TERMS OF GLYCATION REACTION

Reducing sugars are known to produce reactive oxygen species mainly through the glycation reaction. D-Ribose and 2-deoxy-D-ribose, which rank at the top among reducing sugars, have been known to induce cell death in mononuclear cells. Under diabetic conditions, glucose is converted to fructose through the polyol pathway and fructose level is actually increased. Fructose has a stronger reducing capacity than glucose and the glycation reaction is easily induced by fructose. Pancreatic β-cells are also exposed to a large amount of fructose as well as glucose under diabetic conditions. Fructose produced outside is also thought to enter into β-cells, because the cells have a glucose transporter, GLUT2, which has the ability to transport fructose as well as glucose. Fructose and ribose brought about apoptosis in isolated rat pancreatic islet cells as well as in HIT cells (17). This apoptosis was characterized biochemically by cleavage of DNA into nucleosomal fragments of 180–200 bp and morphologically by nuclear shrinkage, chromatin condensation, and apoptotic body formation. The DNA cleavage occurred in a time- and dose-dependent manner and the extent of DNA damage corresponded to the reducing capacity of the sugars. *N*-Acetyl-L-cysteine, an antioxidant which elevates intracellular GSH levels, and aminoguanidine, an inhibitor of glycation reaction, inhibited the DNA cleavage. The extent of DNA cleavage was more remarkable after preincubation with buthionine sulfoximine, which depletes intracellular GSH levels. These results clearly indicate that reactive oxygen species which are thought to be produced from peroxidation products were also increased. We also demonstrated that proteins in β-cells were actually glycated by using an antibody which can specifically recognize the protein glycated by fructose but not by glucose or ribose. Reducing sugars also increased intracellular peroxide levels preceding the induction of apoptosis as judged by FACS analysis using dichlorofluorescin diacetate. Levels of carbonyls, an index of oxidative modification, and malondialdehyde, a lipid peroxidation product, were also increased. These data suggest that reducing sugars trigger oxidative modification and apoptosis in pancreatic β-cells by provoking oxidative stress mainly through a glycation reaction, which may explain the deterioration of β-cells under diabetic conditions.

PERSPECTIVES

Oxidative stress causes a variety of pathophysiological changes in tissues. Antioxidative enzymes play a key role in scavenging reactive oxygen species generated in the inflammatory process and in diabetes. However, major enzymes such as Cu,Zn-SOD and GPx are posttranslationally inactivated by the glycation reaction under diabetic conditions and NO in inflammatory processes, respectively. The prevention and inhibition of the inactivation of the antioxidative enzymes may provide new vistas in the research on reactive oxygen species.

REFERENCES

1. Taniguchi N. Clinical significances of superoxide dismutases: Changes in aging, diabetes, ischemia, and cancer. Adv Clin Chem 1992; 29:1–59.
2. Sakurai T, Tsuchiya S. Superoxide production from nonenzymatically glycated protein. FEBS Lett 1988; 236:406–410.
3. Kayanoki Y, Fujii J, Suzuki K, Kawata S, Matsuzawa Y, Taniguchi N. Suppression of antioxidative enzyme expression by transforming growth factor-β1 in rat hepatocytes. J Biol Chem 1994; 269:15488–15492.
4. Arai K, Iizuka S, Tada Y, Oikawa K, Taniguchi N. Increase in the glucosylated form of erythrocyte Cu,Zn-superoxide dismutase in diabetes and close association of the nonenzymatic glucosylation with the enzyme activity. Biochim Biophys Acta 1987; 924:292–296.
5. Taniguchi N, Kinoshita N, Arai K, Iizuka S, Usui M, Naito T. Inactivation of erythrocyte Cu-Zn-superoxide dismutase through nonenzymatic glycosylation. In Baynes JW, Monnier VM, eds. The Maillard Reaction in Aging, Diabetes and Nutrition. New York: A.R. Liss; 1989:

277–290.
6. Kawamura N, Ookawara T, Suzuki K, Konishi K, Mino M, Taniguchi N. Increased glycated Cu,Zn-superoxide dismutase levels in erythrocytes of patients with insulin-dependent diabetes mellitus. J Clin Endocrinol Metab 1992; 74:1352–1354.
7. Arai K, Maguchi S, Fujii S, Ishibashi H, Oikawa K, Taniguchi N. Glycation and inactivation of human Cu-Zn-superoxide dismutase. Identification of the in vitro glycation sites. J Biol Chem 1987; 262: 16969–16972.
8. Ookawara T, Kawamura N, Kitagawa Y, Taniguchi N. Site-specific and random fragmentation of Cu,Zn-superoxide dismutase by glycation reaction. Implication of reactive oxygen species. J Biol Chem 1992; 267:18505–18510.
9. Kaneto H, Fujii J, Suzuki K, et al. DNA cleavage induced by glycation of Cu,Zn-superoxide dismutase. Biochem J 1994; 304:219–225.
10. Myint T, Hoshi S, Ookawara T, Miyazawa N, Suzuki K, Taniguchi N. Immunological detection of glycated proteins in normal and streptozotocin-induced diabetic rats using anti hexitol-lysine IgG. Biochim Biophys Acta 1995; 1272:73–79.
11. Takata I, Kawamura N, Myint T, Glycated Cu,Zn-superoxide dismutase in rat lenses: evidence for the presence of fragmentation in vivo. Biochem Biophys Res Commun 1996; 219:243–248.
12. Kaneto H, Fujii J, Seo HG, Apoptotic cell death triggered by nitric oxide in pancreatic β-cells. Diabetes 1995; 44:733–738.
13. Hockenbery DM, Oltvai ZN, Yin X-M, Milliman CL, Korsmeyer SJ. Bcl-2 functions in an antioxidant pathway to prevent apoptosis. Cell 1993; 75:241–251.
14. Asahi M, Fujii J, Suzuki K, et al. Inactivation of glutathione peroxidase by nitric oxide. Implication for cytotoxicity. J Biol Chem 1995; 270:21035–21039.
15. Brown RH Jr. Amyotrophic lateral sclerosis: recent insights from genetics and transgenic mice. Cell 1995; 80:687–692.
16. Fujii J, Myint T, Seo HG, Kayanoki Y, Ikeda Y, Taniguchi N. Characterization of wild-type and amyotrophic lateral sclerosis-related mutant Cu,Zn-superoxide dismutases overproduced in baculovirus-infected insect cells. J Neurochem 1995; 64:1456–1461.
17. Kaneto H, Fujii J, Myint T, Reducing sugar triggers oxidative modification and apoptosis in pancreatic β-cells by provoking oxidative stress through glycation reaction. Biochem J, 1996; 320:855–863.

47

Oxidative Stress in Ischemia–Reperfusion: Reilly's Irritation Syndrome Revisited

Makoto Suematsu and Masaharu Tsuchiya
Keio University School of Medicine, Shinjuku-ku, Tokyo, Japan

WHAT IS STRESS? REILLY'S IRRITATION SYNDROME REVISITED

When exposed to a series of noxious stimuli such as surgical insults and bacterial infection, the autonomic nervous system exhibits supraphysiological responses and thereby causes nonspecific vasomotor derangements in multiple organ systems. This phenomenon, known as irritation syndrome, was proposed in the late 1930s by Dr. James Reilly, a French experimental pathologist in Claude Bernard Hospital. When he studied the pathophysiology of typhus, he recognized that, independently of the nature of the stimulus (e.g., bacterial toxin, surgical insult, temperature changes), application of excessive levels of the stimulation (irritation) in a particular local organ evoked autonomic nervous perturbation which in turn caused hemorrhagic changes in a variety of organs including lung, stomach, and adrenal glands (1). Prior to the occurrence of the hemorrhagic changes, the organ hemodynamics exhibit a biphasic fluctuation similar to ischemia–reperfusion that results from alternate excitation of sympathetic and parasympathetic nervous systems.

Reilly and his colleagues applied a series of repeated electrical stimuli to the sympathetic nerve fibers or carried out croton oil immersion into the autonomic nervous ganglia in order to reproduce nonspecifically such hemodynamic perturbation (1). When the animals exposed to the irritation were pretreated with chlorpromazine, a major tranquilizer blocking the autonomic nervous systems, organ vascular systems became insensitive to the irritation and the irritation-induced hemorrhagic changes were thereby prevented. Later, Dr. Hans Selye, who had learned of Reilly's concept during his stay in Claude Bernard Hospital, proposed "general adaptation syndrome" in which the hormonal responses, including ACTH and adrenocortical steroids, play a crucial role in the pathological changes among systemic organs. This concept was widely accepted across the world. However, it should be noted that Reilly's irritation syndrome first provided the fundamental concept of "stress" from the viewpoint of autonomic nervous abberations.

LEUKOCYTE ADHESION IN MICROCIRCULATION: CAUSE OR RESULT OF THE INJURY?

Although it was known that ischemia–reperfusion evokes organ vascular damage leading to hemorrhagic changes, the mechanisms were quite unknown until oxygen free radicals emerged as a primary mediator causing cell injury. Since microvascular endothelial cells are known to possess in abundance xanthine oxidase, an oxyradical-producing enzyme, these cells have been considered to constitute a target primarily exposed to oxidant stress-induced cell injury. On the other hand, adhesion of circulating neutrophils to microvasculature have been postulated as an important cellular source of oxyradical generation. In fact, depletion of neutrophils or immunoneutralization of adhesion molecules using monoclonal antibodies which interfere with the interaction between the endothelium and neutrophils successfully attenuated the reperfusion – induced tissue injury.

Recently, reactive oxygen species have attracted great interest as triggering molecules leading to expression of adhesion molecules in endothelial cells (2). Since neutrophils produce large amounts of oxidants upon adhesion to endothelial cells in vitro or in vivo (3), neutrophil-dependent oxidative stress can play an important role in further recruitment of inflammatory cells. However, there are still unresolved questions: Which cells can initially produce oxidants – endothelial cells, neutrophils or parenchymal cells? Which adhesion molecules play a critical role in the initial recruitment of neutrophils in microvessels?

Based on observations in the mesenteric postcapillary venules in vivo or using the in vitro flow chamber technique, sequential multistep pathways involving selectin-dependent rolling and integrin-dependent firm adhesion are now believed to constitute fundamental mechanisms for neutrophil adhesion in microvessels. However, is such a central dogma for tissue neutrophil accumulation applicable to other organs such as lung or liver? Although the signal transducing mechanisms for inducing expression of adhesion molecules have been studied extensively using cultured endothelial cells derived from larger sizes of vessels (e.g., human umbilical vein endothelial cells), can this information generally be applied to explain the microvascular events?

TEMPORAL AND SPATIAL INFORMATION ABOUT MICROVASCULAR EVENTS DURING REPERFUSION INJURY

Intravital digital microfluorography is a powerful tool for revealing critical cellular events in microcirculation as a function of time and space simultaneously and shedding light on several answers to some of these unresolved questions. In the experimental model of ischemia–reperfusion in the rat microcirculation in spinotrapezius muscle (4), we observed spatial and temporal alterations in leukocyte adhesion and irreversible cell injury with dual-color digital microfluorography using intravascular administration of carboxy-fluorescein succinimidyl ester (excitation maximum at 480 nm for leukocyte staining) and propidium iodide (PI) (excitation maximum at 530 nm for staining of nuclei in damaged cells), respectively. According to these data, during the 1-h ischemic period, leukocytes exhibited low shear-dependent adhesion in postcapillary venules as well as intracapillary entrapment, while the muscle tissue did not display any evidence for cell injury. However, such leukocyte entrapment is a quite transient event: In response to reperfusion, a majority of these cells were detatched from microvessels immediately as a result of the increasing disperse force yielded by re-establishment of blood flow. On the other hand, within 15 min after the start of reperfusion, the muscle tissue exhibited positive staining with PI, suggesting the initial cell damage. Most of the initially stained nuclei were those in myocytes, but not in capillary endothelial cells as judged from their round and large shapes. In addition, during the early period after reperfusion, there was no significant increase in the density of leukocytes either in venules or in capillaries. Histochemical analysis revealed that a particular type of muscle fiber, such as fast-twitch glycolytic fiber (white fiber) rather than oxidative fiber (red fiber), is more sensitive to reoxygenation insult. Interestingly, administration of a xanthine oxidase inhibitor, BOF-4272, which specifically blocks the enzyme activity without scavenging oxidants, was able to block the reperfusion-induced myocyte injury (4). Since xanthine oxidase is known to be localized in microvascular endothelium but not in the muscle parenchyma, these results collectively suggest that endothelial cells serve as a primary source of oxidative stress which evokes the initial myocyte damage after reperfusion independently of the leukocyte adherence. At the same time, our results suggest that microvascular endothelium can hardly be damaged but is functionally activated and thereby induces subsequent leukocyte sequestration in the postischemic tissue.

We are now investigating the initial functional changes of microvascular endothelium by visually imaging the luminal expression of adhesion molecules involved in the leukocyte adherence (e.g., ICAM-1, P-selectin, SLeX (5)) using laser confocal video microscopy. Such a technique allows us to trace the time history of the molecular expression in situ in the same microvascular beds and may thus provide important information showing the rapidity of the endothelial responses after exposure to ischemia–reperfusion.

REFERENCES

1. Tsuchiya M, Suematsu M, Miura S, Suzuki M, Yamakawa H. Les aspects microcirculatoires du phenomene de Reilly. Bull Acad Natle Med 1988; 172:409–419.
2. Patel KD, Zimmerman GA, Prescott SM. Oxygen radicals induce human endothelial cells to express GMP-140 and bind neutrophils. J Cell Biol 1991; 112:749–759.
3. Suematsu M, Kurose I, Asako H, Miura S, Tsuchiya M. Oxyradical-dependent photoemission during endothelium-granulocyte interactions in the mesenteric microvessels treated with platelet-activating factor. J Biochem 1989; 103:355–360.
4. Suematsu M, DeLano FA, Poole D, et al. Spatial and temporal correlation between leukocyte behaviour and cell injury in post-ischemic rat skeletal muscle microcirculation. Lab. Invest 1994; 70: 684–695.
5. Tamatani T, Suematsu M, Tezuka K, et al. Recognition of consensus carbohydrate structure in ligands for selectins by novel antibody against sialyl Lewis X. Am J Physiol 1995; 269:H1282–H1287.

48

Evaluation of the Antiperoxidase Defense System of the Lens Utilizing GSH Peroxidase Transgenics and Knockouts

Abraham Spector, Ren-Rong Wang, Wanchao Ma, Yinqing Yang, and Wan-Cheng Li
Columbia University, New York, New York

Ye-Shih Ho
Wayne State University, Detroit, Michigan

The lens of the eye is a deceptively simple tissue. It is a completely transparent, avascular body stationed behind the cornea and held in correct position by the zonular fibers (1,2). The function of the lens is to focus light onto the retina, changing shape to allow clear delineation of objects at varying distances from the eye. The lens is enclosed by an extracellular matrix, the capsule. It contains only a single layer of epithelial cells on the anterior side of the tissue which, in the equatorial region, terminally differentiate into fiber cells. These long hexagonal cells, extending both to the anterior and posterior of the tissue, gradually lose their nuclei and other intracellular organelles and finally become relatively inert bags of protein incapable of repairing damage. The newly formed fibers displace the previously formed fibers inward toward the center of the tissue. This process continues throughout life. Thus, the youngest and metabolically most active section of the lens is in the periphery and the oldest and least active section is in the center of the lens, the nucleus. The metabolically active periphery, consisting of the epithelium and the newly formed fibers, is primarily responsible for maintaining the homeostasis of the tissue and protecting the relative inert inner region from stress.

All insults to the lens are believed to result in the loss of transparency, that is, cataract. Thus, it is difficult to ascertain the basis of the pathology by inspection of the tissue. A large number of apparently unrelated factors appear capable of causing cataract (2) including diseases such as diabetes and glaucoma, corticosteroids, radiation (X-ray, microwave, ultraviolet), mutation of the lens crystallins (the major lens proteins), hyperbaric oxygen, and oxidative stress. The impact of these and possibly other factors results in an increasing prevalence of cataract with aging. Cataract is the leading cause of morbidity and functional impairment among the elderly with more than one million operations per year to remove the opaque lens in the United States alone (3,4).

Examination of human cataract indicates that in most cases there is extensive oxidation of the lens proteins, which are present in remarkably high concentration (5). It was found that in a significant proportion of cataract patients, H_2O_2 is elevated in the aqueous fluid that is present in front of the lens and in the lens itself (6,7). Furthermore, H_2O_2 can cause cataract and produces similar changes to protein as one observes in cataract (8–11). These changes can lead to protein aggregation, light scattering, and eventually opacification. Such observations have led to the view that H_2O_2 may be an initiating or contributing factor for the development of a significant fraction of cataract, particularly cataract presenting in the older population. There is some evidence that in this population, the enzymes which protect the lens from oxidative stress have decreased in activity (12,13).

The possibility that oxidative stress and H_2O_2 in particular may be important factors in the development of cataract has led our laboratory to investigate the impact of oxidative stress upon the development of the disease. Utilizing murine lenses in cultures, it was found that a photochemical stress generated with daylight radiation in the presence of low levels of riboflavin and 4% oxygen (to mimic the environment of the lens) produces H_2O_2, superoxide, and hydroxyl radicals (9). This stress resulted in the development of cataract, but only if H_2O_2 was present (9,14). Adding catalase, an H_2O_2-degrading enzyme, to the medium eliminated the loss of transparency (9). This was true even though riboflavin had entered the lens and presumably generated H_2O_2 in the tissue. Since H_2O_2 rapidly diffuses through cell walls, a significant increase in H_2O_2 concentration in the lens cannot occur with catalase in the medium. Thus, at least under these conditions, it is H_2O_2 which is the major oxidative stressing factor.

Examination of the sequence of events leading to cataract in the rat lens subjected to photochemical insult indicated that a significant change in the redox set point of the cell is the first indication that the lens is being stressed (15,16). The redox set point can be defined as the ratio of reduced to 1/2 oxidized glutathione, GSH/GS-. The lens normally has only trace amounts of GSSG and hence a strongly reducing environment, but within an hour of photochemical stress utilizing 4 μM riboflavin and 4% O_2 the redox set point has dropped to 0.5, reflecting an oxidizing environment. Even after 15 min of stress, a significant decrease in the reducing environment is observed. It is interesting that this large change is observed primarily in the epithelium but not in the bulk of the lens. During the early stages of stress, the epithelial cells are capable of returning to a normal reducing

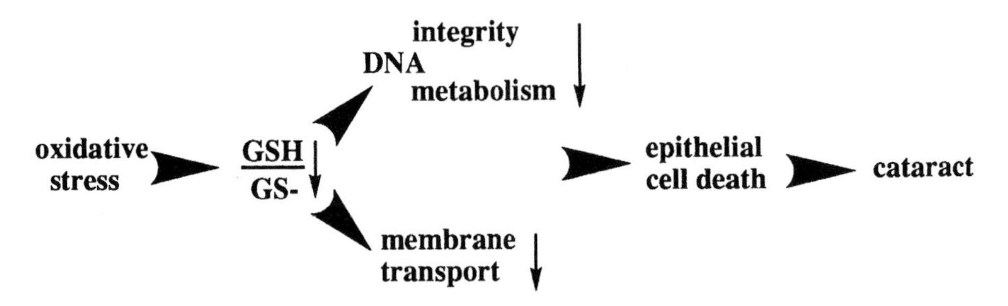

Figure 1. Sequence of events initiated by oxidative stress leading to cataract. GSH/GS-indicates the ratio of reduced to oxidized glutathione, which is a measure of the redox set point of the system. ↓ indicates a decrease in a given parameter. For a more complete description of the sequence of events leading to cataract, see Ref. 16.

environment if the insulting environment is removed, but this is true only with a stress period of a few hours. After that time there is a slow but steady irreversible decline in the ability of the system to reduce GSSG.

Two other changes occur almost at the same time: damage to the DNA system measured by thymidine incorporation or single strand breaks, and loss of transport function. This then leads to cell death, culminating in opacification of the lens (Figure 1) (16). The delineation of the sequence of events leading to cataract when the murine lens is subjected to photochemical stress now allows the investigator to evaluate the effect of manipulating the antioxidative defenses of the tissue.

It can be argued that if the defenses of the lens against H_2O_2 stress can be strengthened, then perhaps cataract initiated by such stress can be prevented. While the major systems involved in the antioxidative metabolism are well known, there is only a rudimentary understanding of how they interact with each other, not only in the lens but in all tissues. We have, therefore, attempted to define more precisely the contribution of glutathione peroxidase-1 (GSHPx-1), GSSG reductase (GSSG Red), catalase (CAT), and GSH to the degradation of H_2O_2 at the tissue level. The work was initiated by examination of GSHPx-1 (17). Genomic clones for GSHPx-1 were isolated from a bacteriophage FIX II genomic library prepared with mouse DNA by hybridization selection with a corresponding rat cDNA clone. A 5.3 kb genomic fragment was isolated from the library by *Sac*I digestion and subcloned into pBluescript SK (pSK). The fragment was shown to contain the entire 1.4 kb mouse GSHPx-1 gene. Transgenics were produced utilizing this fragment. Knockouts were made by introducing a neomycin resistance gene cassette into exon 2 at an *Eco*RI site and a herpes thymidine kinase gene expression cassette 3′ to the targeting sequence. Animals 8 to 12 weeks of age were used for experimentation. Both transgenics and knockouts appeared normal on the basis of general appearance, weight, color, and activity. Gross inspection of the lens showed no pathology. Examination of a number of tissues for GSHPx-1 activity gave the results shown in Table 1. In all tissues examined, there was no significant GSHPx-1 activity in the knockouts in comparison to normal tissue levels. The tissues from transgenics showed varied augmentation of activity. In the lens the activity was increased about 4.5-fold, whereas in the liver little change was observed. Examination of the CAT activity showed no increase as a result of the knockout of the GSHPx-1 activity nor any significant change with augmentation of the GSHPx-1 activity.

Surprisingly, the 4.5-fold increase in GSHPx-1 activity in the transgenic lens resulted in no change in the ability of the lens to metabolize H_2O_2. This was observed both

Table 1. Glutathione Peroxidase-1 Activity in Various Tissues of Normal, Transganic, and Knockout Mice

	GSHPx-1 activity[a] (mU/mg protein)			
Genotype	Lens	Brain	Heart	Liver
normal	16.9 ± 1.2	69.6 ± 3.8	178 ± 49	1667 ± 29
transgenic	76.0 ± 4.4	235 ± 21	430 ± 93	1999 ± 57
knockout	0.4 ± 0.2	ND	3.9 ± 0.5	8.4 ± 2

[a]GSHPx-1 activity was measured with 2 mM. GSH and corrected to theoretical values with 4 mM GSH. Values represent the average of two or more experiments ± SD. ND = not detected. See Ref. 17 for further information.

when the lens was confronted initially with a high concentration of H_2O_2 and when the photochemical system was used and the H_2O_2 gradually accumulated. In the absence of GSHPx-1 little change was found in H_2O_2 decay compared to the normal lens, but with the photochemical system a significantly faster rate of accumulation of H_2O_2 was observed in cultures containing the knockout lens (Figure 2).

These observations raise a number of questions, such as why the elevation of GSHPx-1 activity has little effect on the degradation of H_2O_2. Since GSHPx-1 is dependent upon GSSG Red to maintain the GSH which is used as a cofactor by the enzyme, a comparison of the relative distribution and activity of the enzymes was made (17). It was found that in the normal lens there is approximately 12-fold more GSHPx-1 activity than GSSG Red activity under optimal conditions. However, while the GSHPx-1 is distributed relatively evenly throughout the lens, the GSSG Red is concentrated in the epithelium, being 8-fold more active in this region than in the remainder of the lens. Thus, in the epithelium, GSHPx-1 is only 1.5-fold more active than GSSG Red under normal conditions. However, in the transgenic, GSHPx-1 is not only markedly increased in the lens, giving a 50-fold excess over GSSG Red, but it is concentrated in the epithelium so that even in this region there is about a 12-fold excess of GSHPx-1 activity. Thus, the increased GSHPx-1 activity would be

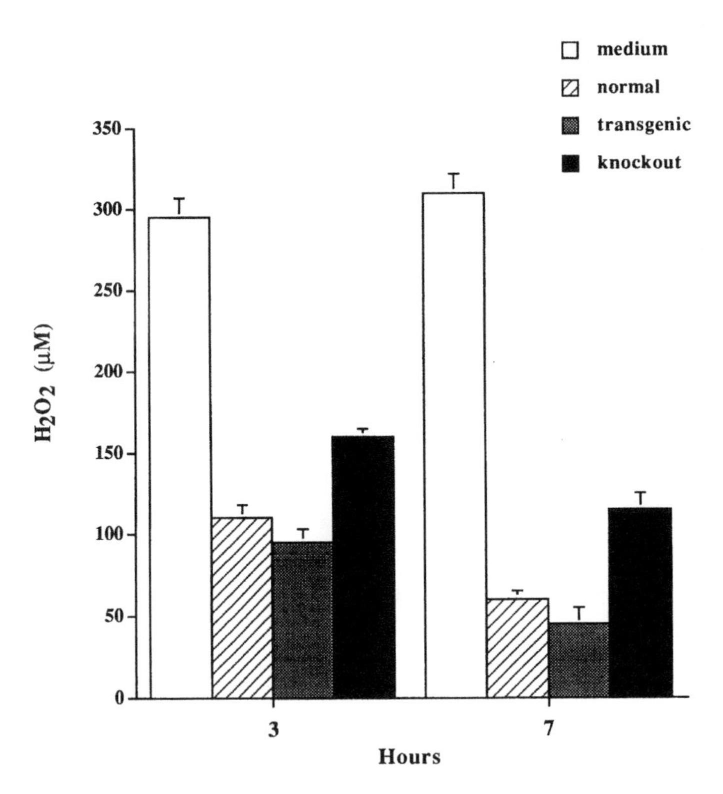

Figure 2. H_2O_2 concentrations generated by photochemical stress (1.7 μM riboflavin, 4% O_2, daylight radiation with two 15 W lamps with a cutoff of approximately 350 nm). The lenses were placed in 300 μl of Sigma M3769 medium under the above conditions. The H_2O_2 levels at 3 and 7 h after initiating the photochemical stress are shown. The results are the mean ± SD from 5 experiments. See Ref. 17 for additional information.

expected to quickly exhaust the available GSH since the GSSG Red could not keep up with the rate at which peroxidase was oxidizing GSH.

To get a better insight into the problem, nonprotein thiol (NP-SH) levels were measured in the epithelium and in the lens minus the epithelium (17,18). (Since in the lens, GSH represents almost all of the NP thiol, this is a reasonable estimate of GSH concentration.) It was found that when lenses from either normal or transgenic animals were subjected to a H_2O_2 stress of 300 μM, the NP-SH changed markedly only in the epithelial layer, indeed, the changes found in the whole lens could be accounted for by the drop in NP-SH in the epithelial cell fraction. Thus, it would appear that even in the absence of additional GSHPx-1, GSSG Red was already operating at a maximum rate and could not keep up with the demand in the epithelial cell layer. The NP-SH decreased about 2-fold in 30 min throughout the normal lens and at about the same rate in the transgenics when GSSG Red was inhibited by BCNU (1,3-bis(2-chloroethyl)-1-nitrosourea). Now if, in the absence of GSSG Red inhibition, the drop in NP-SH was due to GSHPx-1 activity, then in the knockout there should be a higher level of NP-SH. But this is not found; the level of NP-SH decreases at about the same rate regardless of the presence or absence of GSHPx-1. Thus, GSSG Red appears to be responsible for maintaining the NP-SH in the lens but it cannot maintain NP-SH levels in the epithelium under oxidative stress. The results suggest that other reactions are involved in oxidizing the NP-SH.

To assess the contributions of each of the presumed major components contributing to lens H_2O_2 degradation, normal and knockout lenses were examined. 3-AT (3-amino-triazole) was used to inhibit CAT and BCNU to inhibit GSSG Red. If it is assumed that H_2O_2 consumption is primarily dependent on GSHPx-1, GSSG Red and CAT, and the concentration of NP-SH, then it should be possible to define the contribution of each component. This can be done by comparing the H_2O_2 consumption per mg lens protein from normal and knockout animals of comparable age under various conditions
Thus:

(1) N	=	GSHPx-1	+ GSSG Red	+ CAT	+ NP-SH
(2) N + BCNU	=	GSHPx-1		+ CAT	+ NP-SH
(3) N + 3-AT	=	GSHPx-1	+ GSSG Red		+ NP-SH
(4) N + 3-AT + BCNU	=	GSHPx-1			+ NP-SH
(5) K	=		GSSG Red	+ CAT	+ NP-SH
(6) K + BCNU	=			CAT	+ NP-SH
(7) K + 3-AT	=		GSSG Red		+ NP-SH
(8) K + 3-AT + BCNU	=				NP-SH

where N = normal lens and K = knockout lens.

By appropriate subtraction, the contribution of each component can now be determined.

GSHPx-1	= 1–5
CAT	= 1–3 or 5–7
GSSH Red	= 1–2 or 5–6
NP-SH	= 8
NP-SH +GSSG Red	= 7

Table 2. Percentage of H_2O_2 Degradation Attributel to Constituents of the Anti-H_2O_2. Defense of the Lens

	Percentage of total degradation			
	300 μM initial H_2O_2		80 μMinitial H_2O_2	
	30 min	60 min	30 min	60 min
GSHPx-1	15.7	13.6	16.3	14.7
CAT	27.0	30.9	8.2	14.7
GSSG Red	17.0	26.5	12.2	13.7
NP-SH	34.0	31.6	57.1	50.9
NP–SH+GSSG Red	54.1	55.4	73.5	72.5

See Refs. 17 and 18 for further information.

In most cases, the analyses considered above are obtained by inhibiting or eliminating the enzyme in question, although data derived from inhibition of two enzymes gave comparable results. The normal and knockout systems gave similar results for the contribution of CAT and GSSG Red.

A summary of some of the results is shown in Table 2. [See Spector et al. (18) for further information]. Two different H_2O_2 concentrations were used to explore the range of elevated H_2O_2 concentrations found in the lens and aqueous of cataract patients. Also, since CAT activity is linearly related to H_2O_2 concentration, it was of interest to examine the effect of varying H_2O_2 concentrations on this enzyme. Surprisingly, GSHPx-1 contributed only a small fraction to the total H_2O_2 degradation. From about 14% to 16% can be attributed to GSHPx-1 activity. Since the enzyme has a high affinity for H_2O_2, there is little change in the GSHPx-1 contribution observed with initial H_2O_2 concentrations of 300 μM and 80 μM H_2O_2. CAT contributed about 27–31% of the total H_2O_2 degradation at 300 μM H_2O_2 and about 8–15% at 80 μM, reflecting the concentration dependence of the enzyme. Thus, the CAT contribution is greater than that of GSHPx-1 at higher H_2O_2 concentrations and equal to or less than that of GSHPx-1 at lower H_2O_2 levels. What is striking is that even under the best conditions the contribution of the two enzymes, previously believed to be the major H_2O_2-degrading enzymes, accounts for less than 50% of the H_2O_2 degradation. The results suggest that lens NP-SH, which is essentially GSH, is a major factor involved in eliminating H_2O_2. In the absence of GSHPx-1 and CAT, approximately 31–34% is degraded at 300 μM and 51–57% at 80 μM. And if GSSG Red was present to reduce the GSSG formed, then from 54% to 75% of the H_2O_2 was found to be consumed.

To get an assessment of the role of NP-SH from another perspective, NP-SH was determined in both the epithelial cell layer and the whole lens under conditions identical to those used to assay the H_2O_2 degradation (18). Analyses of normal lenses gave values of approximately 30 nmol NP-SH for the whole lens and about 4 nmol for the epithelium. Following exposure of the normal lens to an initial concentration of 300 μM H_2O_2, at the end of an hour the only change in NP-SH is observed in the epithelial cell layer (Table 3). As mentioned earlier, it appears that the most active metabolic region cannot maintain its redox set point but prevents oxidation from occurring in the remainder of the lens. (It

Table 3. Change in Lens NP-SH When Subjected to 300 μM H_2O_2 for 1 h

Antioxidative system				ΔNP-SH (nmol) Capsule epithelium	Lens
GSHPx-1	CAT	GSSG Red	NP-SH	2.9	2.9
GSHPx-1	CAT	–	NP-SH	4.2	17.2
GSHPx-1	–	GSSG Red	NP-SH	–	11.6
GSHPx-1	–	–	NP-SH	–	18.1
–	CAT	GSSG Red	NP-SH	3.2	2.3
–	CAT	–	NP-SH	4.4	14.2
–	–	GSSG Red	NP-SH	–	2
–	–	–	NP-SH	4.3	15.5

should be noted that following similar experiments, assays with the lens stripped of the capsule epithelium confirmed this conclusion.) If GSSG Red is removed, the epithelium can no longer protect the remainder of the lens and extensive oxidation is observed. Elimination of GSHPx-1 causes little change in the NP-SH status. In contrast, inhibition of CAT in the presence of GSHPx-1 causes a large decrease in NP-SH due to the utilization of GSH by GSHPx-1 but if GSHPx-1 is absent, the change is small, and unpublished work from this laboratory suggests it is confined to the epithelium. Thus, the GSSG Red can maintain NP-SH in the absence of GSHPx-1 in the presence of a substantial oxidative stress except in the epithelial cell layer.

Similar results were obtained with a lesser stress of 80 μM H_2O_2. However, in the normal lens or lenses lacking either GSHPx-1 or CAT, hardly any change in NP-SH was observed even in the epithelium. Only when GSSG Red was inhibited was a substantial oxidation of NP-SH found.

If it is accepted that the reaction of thiol groups with H_2O_2 is similar to the reaction with selenium, then it requires two thiols to reduce one H_2O_2. Then it can be assumed that nonenzymatically 0.5 ΔGSH = H_2O_2 consumed if ΔGSH = ΔNP-SH and in the GSHPx-1 knockout in the presence of BCNU and 3-AT only a nonenzymatic reaction with GSH

Table 4. A Comparison of the ΔH_2O_2 and 0.5 ΔNP-SH in Lenses from GSHPx-1 Knockouts in the Presence of BCNU (0.12 mM) + 3-AT (20 mM)

	H_2O_2 degraded (nmol)	0.5 NP-SH(nmol)
Insult at 30 min		
Initial insult 80 μM	2.8 ± 0.2	2.9 ± 0.3
Initial insult 300 μM	5.4 ± 0.2	6.3 ± 0.2
Insult at 60 min		
Initial insult 80 μM	5.2 ± 0.1	3.7 ± 0.3
Initial insult 300 μM	9.3 ± 0.2	7.8 ± 0.2

See Ref. 18 far further information.

occurs. When a comparison is then made between the loss of 0.5 NP-SH and the H_2O_2 degraded, a reasonable agreement is observed. At the 80 μM H_2O_2 level from 70% to 100% of the 0.5 ΔNP-SH equals the H_2O_2 degraded, and with 300 μM H_2O_2 from about 83% to 117% (Table 4). Thus, these results suggest that the nonenzymatic reaction with GSH accounts for much of the degradation of H_2O_2 in the lens. However, it should be noted that no other enzymes that may require GSH to degrade H_2O_2 are considered in this analysis.

Now if the degradation of H_2O_2 by the lens in the absence of CAT and GSHPx-1 activity is nonenzymatic and is occurring exclusively through reaction with GSH (NP-SH), then the rate of degradation of the H_2O_2 should correspond to the rate observed when GSH and H_2O_2 interact in the test tube. Since the average GSH concentration in the lens is approximately 6 mM (being much higher in the periphery and much lower in the center), an experiment was performed to determine H_2O_2 decay rates in the presence of 6 mM GSH. It was found that the initial rate of H_2O_2 degradation was only 38% of the lens rate at an initial concentration of 300 μM and 19% with 80 μM H_2O_2. The overall results suggest that while GSH may be involved in up to about 70% of the reactions involving H_2O_2 degradation in the absence of GSHPx-1, it is probable that much of this reactivity is due to reactions involving other enzymes that require GSH as a cofactor.

ACKNOWLEDGMENTS

This work was supported by grants to A.S. from the National Eye Institute and Research to Prevent Blindness. The assistance of Elaine Bluberg in the preparation of this chapter is gratefully acknowledged.

REFERENCES

1. Horwitz J, Jaffe N. Anatomy and embryology. In Podos SM, Yanoff M, eds. Textbook of Ophthalmology 3, Lens and Cataract. New York: Gower Medical Publishing; 1992:2–8.
2. Harding JJ, Crabbe MJ. The lens: development, proteins, metabolism and cataract. In Davson H, ed. The Eye, vol. 1B. Orlando: Academic Press; 1984:207–492.
3. Kupfer C, Underwood B, Gillen T. Leading causes of visual impairment world wide. In Albert DM, Jakobiec FA, eds. Principles and Practice of Ophthalmology, Basic Science. Philadelphia: W.B. Saunders; 1994:1249–1255.
4. Pizzarello LD. The dimensions of the problems of eye disease among the elderly. Ophthalmology 1987; 94:1191–1195.
5. Spector A. Oxidative stress induced cataract: mechanism of action. FASEB J 1995; 9:1173–1182.
6. Spector A, Garner WH. Hydrogen peroxide and human cataract. Exp Eye Res 1981; 33:673–681.
7. Ramachandran S, Morris SM, Devamanoharan PS, Henein M, Varma SD. Radio-isotopic determination of hydrogen peroxide in aqueous humor and urine. Exp Eye Res 1991; 53:503–506.
8. Garner MH, Garner WH, Spector A. The effect of H_2O_2 on Na/K-ATPase. Invest Ophthalmol Vis Sci 1982; 22(supplement):34.
9. Spector A, Wang G-M, Wang R-R, Garner WH, Moll H. The prevention of cataract caused by oxidative stress in cultured rat lenses. I. H_2O_2 and photochemically induced cataract. Curr Eye Res 1993; 12:163–179.

10. Giblin F, McCready J, Schrimscher L, Reddy V. The role of glutathione metabolism in the detoxification of H_2O_2 in rabbit lens. Invest Ophthalmol Vis Sci 1987; 22:330–335.
11. Zigler JS, Huang Q-L, Du X-Y. Oxidative modification of lens crystallins by H_2O_2 and chelated iron. Free Radical Biol Med 1989; 7:499–505.
12. Hockwin O, Ohrloff C. The eye in the elderly: lens. In Platt D, ed. Geriatrics. Berlin: Springer-Verlag, 1984:373–424.
13. Rathbun WB, Bovis MG. Activity of glutathione peroxidase and glutathione reductase in the human lens related to age. Curr Eye Res 1986; 5:381–385.
14. Spector A, Wang G-M, Wang R-R. Photochemically induced cataracts in rat lenses can be prevented by AL-3823A, a glutathione peroxidase mimic. Proc Natl Acad Sci USA 1993; 90:7485–7489.
15. Spector A, Wang G-M, Wang R-R, Li W-C, Kuszak JR. A brief photochemically induced oxidative insult causes irreversible lens damage and cataract. I. Transparency and epithelial cell layer. Exp Eye Res 1995; 60:472–481.
16. Spector A, Wang G-M, Wang R-R, Li W-C, Kleiman NJ. A brief photochemically induced oxidative insult causes irreversible lens damage and cataract. II. Mechanism of action. Exp Eye Res 1995; 60:483–492.
17. Spector A, Yang Y, Ho Y-S, et al. Variation in cellular glutathione peroxidase activity in lens epithelial cells transgenics and knockouts does not significantly change the response to H_2O_2 stress. Exp Eye Res 1996; 62:521–539.
18. Spector A, Ma W, Wang R-R, Yang Y, Ho Y-S. The contribution of GSH peroxidase-1 catalase and GSH to the degradation of H_2O_2 by the mouse lens. Exp Eye Res 1997, in press.

49

The Rheumatoid Joint: Redox-Paradox?

Vanessa Gilston, David R. Blake and Paul G. Winyard
St. Bartholomew's and The Royal London School of Medicine and Dentistry, Queen Mary and Westfield College, University of London, London, UK†

INTRODUCTION

Rheumatoid arthritis is a chronic inflammatory disease, some characteristic features of which are inflammation within the joints, the proliferation of synovial tissue adjacent to cartilage/bone, and the proteinase-mediated destruction of cartilage/bone at these sites. In several senses the redox-related biochemistry within the rheumatoid joint is paradoxical. Within the synovial tissue, a reducing environment exists. For example, the ratio of protein -SH to -S-S- is increased compared with control synovium (1). Measurements of pO_2 and pH within both synovial fluid and synovial tissue from rheumatoid joints are consistent with a state of chronic hypoxia and acidosis (2,3) (G.A. Ellis, S.E. Edmonds and D.R. Blake, unpublished data). Yet within the synovial fluid of the knee joint space there is evidence of oxidizing conditions, such as a decreased ratio of ascorbate to dehydroascorbate (4) and increased concentrations of markers of free radical damage to lipids and proteins. In an additional paradox, iron-loaded ferritin accumulates within the macrophage-like cells of the inflamed tissue (5), while high concentrations of ferritin molecules within the adjacent synovial fluid have low iron loading (6).

Molecular oxygen (O_2) may be reduced by cellular enzymes to form reactive oxygen intermediates (ROI). The production of ROI is a critical event in the cascades that comprise the inflammatory reaction, both to limit tissue damage and to prevent or inhibit infection. However, overproduction of ROI has been implicated in the continuous inflammatory reaction in patients with rheumatoid arthritis. Until recently, most research into the role of free radicals in inflammation was directed toward the putative role of ROI as damaging agents. Oxidative modification has been studied in biomolecules such as α_1-antitrypsin, cartilage proteoglycans, collagen, IgG, and so on (3, 7). Thus, ROI were studied within the context of a direct, destructive role in pathophysiology. This may be true for high concentrations of certain ROI, but recent evidence suggests that some ROI play important roles as cellular messengers at relatively low concentrations.

---

†Please Note: This is *not* the correspondence address but Institute only.

Correspondence address: Bone & Joint Research Unit, 25–29 Ashfield Street, London, E1 2AD, UK.

In the course of the present decade, researchers have increasingly recognized this cellular control aspect of free radical action. Endothelium-derived relaxing factor (EDRF) has been identified as ·NO, a reactive nitrogen intermediate [RNI; (8)]. Moreover, ROI can activate apoptosis, a *programmed* form of cell death, at least in some circumstances (9). ROI appear to activate certain protein kinases involved in cellular signal transduction, such as protein kinase C (see, for example Refs. 10 and 11) and mitogen-activated protein (MAP) kinase (12). ROI may also activate, or suppress, various protein phosphatases (11, 13).

A key feature of inflammatory diseases such as rheumatoid arthritis is the increased expression of certain genes which encode "inflammatory" proteins (such as certain cytokines) and proteinases involved in tissue destruction, such as collagenases, gelatinases, and stromelysins. In turn, an important characteristic of gene expression is the control of gene transcription by specific proteins, transcription factors, which bind to short DNA sequence elements located adjacent to the promoter or in enhancer regions of genes. Once bound to DNA, transcription factors interact with each other and with the proteins of the transcriptional apparatus itself (e.g., RNA polymerase) to regulate gene expression. Recently, there has been considerable interest in the idea that transcription factors may be useful targets for novel therapeutic strategies in the treatment of human diseases, including inflammatory diseases (14).

The two transcription factors, activator protein-1 (AP-1) and NFκB, can both be regulated by intracellular ROI (15, 16). They have been implicated in the transcriptional regulation of a wide range of genes involved in cellular inflammatory responses and tissue destruction. Inappropriate activation, such as the overexpression of proinflammatory genes, may be involved in the progression of inflammatory diseases such as rheumatoid arthritis. The overproduction of ROI may, in part, be counteracted through the use of antioxidants. Nevertheless, it should be remembered that these processes, when properly controlled, are physiological. Therefore, both ROI and RNI can no longer be regarded solely as damaging species whose complete elimination by antioxidant therapy is bound to have beneficial effects on human health. Thus far there is still no convincing evidence from the clinical studies conducted that supplementation with antioxidant nutrients can influence the process of on-going joint inflammation (17).

An explanation for the paradox of oxidative damage to biomolecules within the chronically hypoxic joint lies in the hypothesis that the inflamed joint is susceptible to hypoxia–reperfusion injury (16). In this review we will consider the potential repercussions of this hypothesis for redox-regulated transcription factor activation and related therapeutic stategies.

HYPOXIA–REPERFUSION INJURY

Hypoxic tissue damage has long been recognized as a factor in the pathology related to human disease states such as transient coronary and cerebral ischemia. However, it has now become clear that in many clinical situations a substantial part of the injury is due to the reintroduction of oxygen to the tissue by the restoration of the blood supply, an event known as reperfusion. This hypoxia–reperfusion cycle generates ROI owing to the uncoupling of a variety of intracellular redox systems, thereby disturbing the delicate redox balance of the cell.

The chronic state of the rheumatoid joint is both hypoxic and acidotic (3). In cells of the synovial membrane the intracellular redox potential is lowered relative to normal cells. Thus, in rheumatoid synovitis 88% of the sulfur-containing amino acids are reduced, in contrast to 63% in control synoviocytes (1). However, we have previously suggested (3, 16) that cycles of ROI generation may be superimposed on this apparently more reduced environment. This may occur by a hypoxia–reperfusion mechanism, summarized as follows.

A moving limb will create pressure fluxes within the joint. Patients with rheumatoid arthritis have significantly greater resting pressures within the joint, which will rise steadily during movement to levels well above the capillary perfusion pressure. This is in complete contrast to the normal knee joint whose pressure will remain subatmospheric during exercise. The increase in capillary perfusion pressure in the rheumatoid joint impedes blood flow to the area and thus creates temporary synovial ischemia. A reactive hyperemia will then occur within the synovial membrane on the cessation of exercise. Thus, movement of the inflamed joint will provide the potential pathophysiological environment for the production of ROI, by way of a hypoxia–reperfusion mechanism. By this mechanism the chronic reducing environment of the rheumatoid synovium will suffer intermittent reperfusion events which permit the generation of ROI. This may explain the "redox-paradox" of the extracellular presence of oxidatively modified biomolecules within the inflamed joint, in conjunction with the chronic reducing environment of the synovium.

Two redox-controlled transcription factors which are modulated as a result of hypoxia and reoxygenation are AP-1 and NFκB (18). It seems feasible that the AP-1–NFκB "yin-yang" relationship (18) that occurs in tissue hypoxia–reperfusion may be relevant to the dual proliferative/inflammatory response characteristic of rheumatoid synovitis, which has sometimes been referred to as "tumor-like proliferation" (19).

REDOX CONTROL OF AP-1 AND NFκB

Activator protein-1 (AP-1) is a protein dimer composed of the proto-oncogene products, Fos and Jun. mRNA levels for c-*fos* and c-*jun* are strongly induced in response to hydrogen peroxide and other oxidative stresses, such as ultraviolet light and ionizing radiation, in both fibroblasts and T cells (20). In contrast, AP-1 binding activity is only weakly induced by hydrogen peroxide (21). Treatment with antioxidants such as pyrrolidine dithiocarbamate activates AP-1 and it has been suggested that the AP-1 DNA binding site is an antioxidant response element (21). The fact that AP-1 can be upregulated by both pro-oxidant and antioxidant conditions will be discussed later.

The AP-1 site, also referred to as the tetradeconyl phorbol acetate-responsive element (TRE) is found in various genes, including those encoding human collagenase, stromelysin, transforming growth factors (TGFs) α and β, interleukin-2, and tissue inhibitor of metalloproteinases-1 (TIMP-1) (22). The matrix metalloproteinases, collagenase and stromelysin, are thought to play a key part in cartilage destruction and inactivation of proteinase inhibitors in inflammatory joint disease (23, 24) and possibly in atherosclerosis (25). TGF-β induces the expression of collagen type I and type III genes, downregulates collagenase production, and induces TIMP-1, an inhibitor of stromelysin and collagenase (26).

The target genes for NFκB comprise a growing list of genes intrinsically linked to a coordinated inflammatory response. These include genes encoding tumor necrosis factor (TNF-α), interleukin (IL)-1, IL-6, IL-8, the IL-2 receptor β chain, inducible nitric oxide synthase (iNOS), MHC class I antigens, E-selectin, vascular cell adhesion molecule-1,

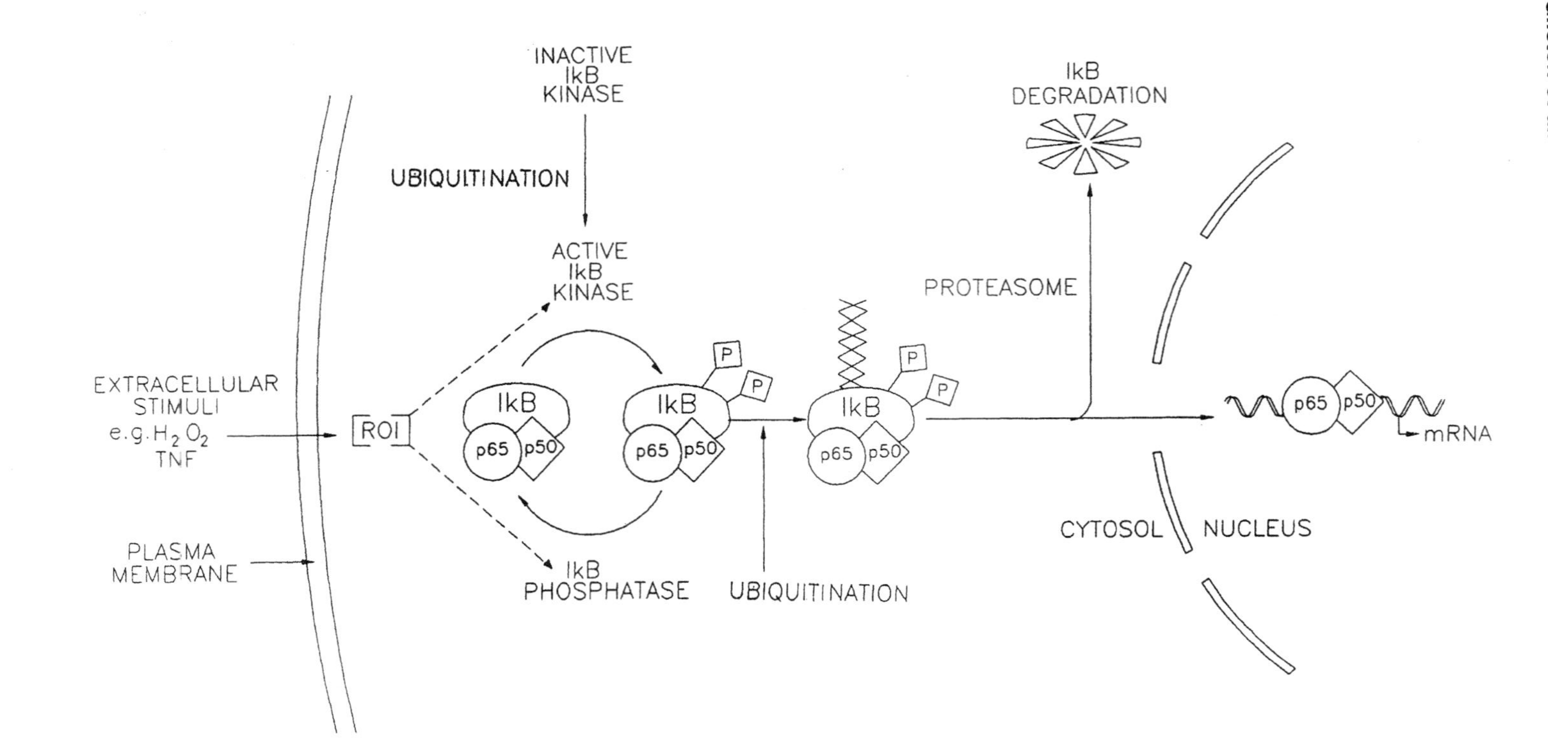

Figure 1. The activation of NFκB. ROI participate in either the activation of a kinase or the suppression of a phosphatase involved in regulating the phosphorylation status of IκB. Ubiquitination is necessary for both IκB kinase activation and the recognition of IκB by the 26S proteasome complex. Once phosphorylated and ubiquitinated, the IκB subunit is degraded by the 26S proteasome. Nuclear location sequences on both subunits of NFκB are unmasked and free NFκB dimers can translocate to the nucleus and thus regulate gene transcription.

serum amyloid A precursor, c-Myc (15), and the H-chain of ferritin (27). The DNA binding, nuclear form of NFκB is a protein heterodimer made up of one Rel-A (p65) subunit and one p50 subunit. In nonstimulated cells, NFκB exists in an inactive, cytosolic form bound to its inhibitor, IκB. Activators of NFκB (such as TNF-α, IL-1, phorbol esters, viruses, lipopolysaccharide, calcium ionophores, cycloheximide, and ionizing radiation) induce the dissociation of IκB from the NFκB–IκB complex, and positively charged nuclear location sequences (NLS) in Rel-A and p50 are unmasked. NFκB is then translocated to the nucleus, where it controls gene expression. The events of NFκB activation are summarized in Figure 1. The importance of ROI in the expression of the genes coding for these proteins was highlighted by Baeuerle and colleagues (28). They showed that NFκB activity was induced by hydrogen peroxide in a human T cell line. This effect was blocked by the antioxidant *N*-acetylcysteine. Other more recently reported activators include oxidized LDL (29) and nitric oxide (endothelial-derived relaxing factor) (30), although the latter effect is disputed (31).

Not only *N*-acetylcysteine but also other antioxidants such as pyrrolidine dithiocarbamate (PDTC), diethyl dithiocarbamate, 2-mercaptoethanol, *o*-phenanthroline and deferoxamine inhibited the activation of NFκB by the recognized stimuli of this transcription factor mentioned above. In addition, α-tocopherol was recently reported to suppress NFκB activation (31). However, some of the effects of α-tocopherol on cellular signaling may not be due solely, or at all, to an antioxidant effect, since α-tocopherol has been shown to inhibit protein kinase C activity via a mechanism that does not involve the antioxidant action of α-tocopherol (32). Many clinically useful antirheumatic compounds contain thiol groups (e.g., D-penicillamine and gold compounds such as aurothioglucose and aurothiomalate) (33) and some of these have been shown to inhibit NFκB activation. The action of these compounds will be discussed further later in this chapter.

IκB dissociation from NFκB involves its phosphorylation-controlled proteolytic degradation (34–36), while ROI appear to control this IκB phosphorylation. This explains why NFκB activation is blocked by a host of antioxidants. The precise mechanism by which ROI control the phosphorylation of IκB is unclear (34). ROI might either activate an IκB kinase or suppress a phosphatase activity. Experiments using specific inhibitors of different protease classes indicated that the intracellular proteinase activity responsible for the degradation of the phosphorylated form of IκB was the 26S proteosome (15, 34, 36).

Chen and coworkers (37) have recently identified a large multisubunit kinase that phosphorylates the IκB subunit at serines 32 and 36. They also found that ubiquitination was essential for IκB degradation and that phosphorylation alone was not sufficient for recognition by the 26S proteasome. As shown in Figure 1, ubiquitination was also required for the activation of the IκB kinase. Thus, ubiquitination appears to be a prerequiste for both the activation of the IκB kinase and the recognition of the IκB subunit by the 26S protease.

A further important control step in the activation of both AP-1 and NFκB appears to be the signal responsible for the DNA binding of these two transcription factors to their responsive element. In both cases it appears that the reduction of specific cysteine residues within the DNA recognition sites of AP-1 and NFκB is an important regulatory signal for the induction of DNA binding. Two enzymes involved in this reduction are thioredoxin and redox factor-1, which are discussed below.

In the case of the Fos and Jun heterodimer (AP-1), binding has been shown to be modulated by the reduction–oxidation of a single conserved cysteine residue in the DNA recognition site of the two subunits. Since the reduced state of the conserved cysteines is critical for DNA binding, it has been speculated (21) that redox modification of this domain may be part of the mechanism controlling transcriptional activity.

Furthermore, a nuclear protein has been identified that was able to reduce the Fos and Jun heterodimer, thus stimulating DNA binding in vitro (38). This ubiquitous nuclear protein, known as redox factor-1 (Ref-1), is a bifunctional protein which also possesses apurinic/apyrimidinic endonuclease DNA repair activity (39). However, both the redox and DNA repair activities of Ref-1 can be distinguished biochemically, which suggests that a link may exist between transcription factor regulation, oxidative signaling and DNA repair processes.

Abate and colleagues (38) showed that the oxidation of Ref-1 significantly diminished its ability to stimulate the DNA binding activity of AP-1. However, upon the addition of thioredoxin, an enzyme that catalyzes the reduction of cysteine residues, the stimulatory activity of Ref-1 was restored and AP-1 binding was resumed. Thioredoxin alone was unable to enhance AP-1/DNA binding, suggesting that it increases the reducing efficiency of Ref-1, rather than acting directly on the Fos and Jun subunits of AP-1 (39).

The same group has now shown that both Ref-1 and AP-1 can be activated in the response of HT29 colon cancer cells to hypoxia (40). Elevation of the Ref-1 gene steady-state mRNA levels occurs as an early event following induction of hypoxia and persists when cells are restored to a normally oxygenated environment. This is supportive of the data recently published by Rupec and Baeuerle, who showed the activation of AP-1 in hypoxic conditions (18). AP-1 activation during hypoxia appeared to rely on the activation of so-called "primary transcription factors," such as the serum-responsive factor (SRF), which has also been shown to be activated by antioxidants (21). These newly synthesized "primary transcription factors" are then able to upregulate c-*fos* and c-*jun* gene transcription and therefore AP-1 activation. Earlier we mentioned how c-*fos* and c-*jun* could be upregulated by oxidative stress (20), yet AP-1 binding to DNA is only weakly induced by hydrogen peroxide (21). Thus, it appears that both antioxidant and pro-oxidant conditions increase the expression of the genes encoding the components of AP-1, but AP-1–DNA binding occurs preferentially during hypoxia. Meyer and colleagues (21) showed that the induction of c-*fos* and c-*jun* mRNAs by both hydrogen peroxide and the antioxidant PDTC occurred with very similar kinetics, suggesting that antioxidant and oxidant conditions in the cell may funnel into the same pathway. On the other hand, AP-1 may exist in a latent form which is only fully activated when cells regain a "normoxic" or hypoxic state. Figure 2 shows the contrasts in AP-1 and NFκB activation during hypoxia–reperfusion when c-*fos* and c-*jun* are upregulated in response to reoxygenation.

The inducible transcription factor NFκB has also been shown to require thioredoxin for DNA binding (41). In vitro, human thioredoxin regulates the DNA binding activity of NFκB, apparently via the chemical reduction of a cysteine-62 residue that is critical for DNA binding (42, 43). Alkylating agents such as *N*-ethylmaleimide and oxidizing agents such as diamide have both been shown to modify free sulfydryls and inhibit NFκB–DNA binding (44).

Among the most important antioxidants is reduced glutathione (GSH), which is present at millimolar concentrations within the cell. However, as discussed above, thioredoxin has been shown to be involved in the DNA binding of NFκB (41). The fact that thioredoxin is required, even when the GSH to GGSG ratio in the nucleus is very

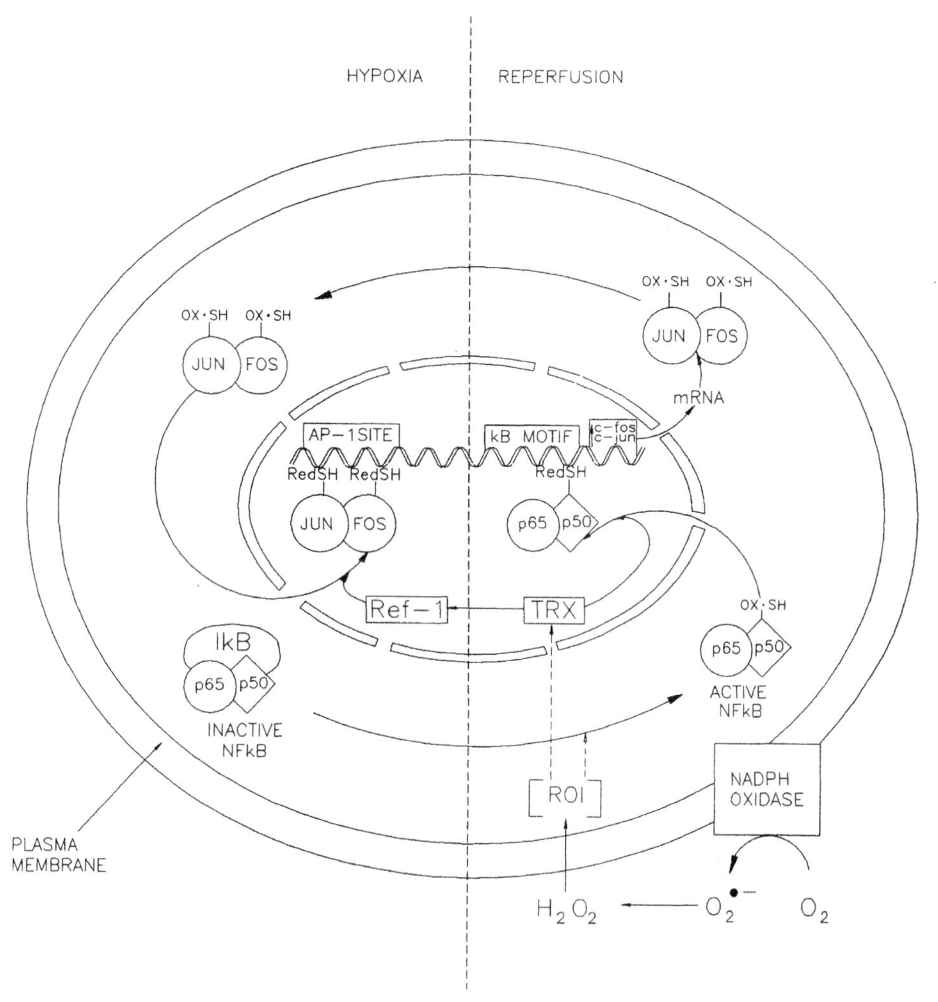

Figure 2. Contrasts in AP-1 and NFκB activation during hypoxia–reperfusion. Upon reperfusion, ROI are produced and NFκB may be activated. ROI upregulate the expression of the enzyme thioredoxin. Thioredoxin catalyzes the reduction of a specific cysteine residue on the p50 subunit of NFκB enabling it to bind to DNA. The proto-oncogenes c-*fos* and c-*jun* are upregulated in response to both oxidative stress and hypoxic conditions, possibly through signals that funnel into the same pathway. The gene products, namely Fos and Jun, dimerize to form AP-1 that is unable to bind DNA without specific reduction of its subunits. Hypoxia upregulates the expression of another reducing enzyme, redox factor-1 (Ref-1) that is able to reduce AP-1 and promote DNA binding. The efficiency of Ref-1 can be increased by thioredoxin.

high, is not fully understood. However, thioredoxin is more effective at reducing (poly)peptides with exposed disulfides, whereas GSH is more effective in reducing small disulfides (45). The upregulation of thioredoxin during cellular oxidative stress may serve as a protective mechanism when levels of GSH decrease by maintaining a system able to catalyze the reduction of critical protein disulfides.

To explain the observation that activation of NFκB can occur in response to both ROI and the reducing enzyme thioredoxin, Hayashi and coworkers (46) pointed out that ROI induce the expression of thioredoxin, which may in turn activate NFκB. Perhaps a *transient* flux of ROI is required for IκB phosphorylation (the step which is inhibited by antioxidants), while thioredoxin is required (at a later time-point after the stimulus) for the efficient binding of NFκB to nuclear DNA. In the case of AP-1, DNA binding may be enhanced as a result of thioredoxin still present when hypoxia resumes. Thus, it appears that two cellular redox systems, namely, the thioredoxin and Ref-1 enzymes, can act synergistically to stimulate the DNA binding of AP-1. Recently, Ref-1 has been implicated in NFκB activation, exerting its effects on the p50 subunit in much the same manner as for AP-1 (39).

AP-1 AND NFκB IN THE RHEUMATOID SYNOVIUM

A recent immunohistochemical and in situ hybridization study of synovial tissue from rheumatoid arthritis patients showed numerous cells expressing *jun*-B and c-*fos* (the genes encoding AP-1 subunits) (47). These positive cells were within the lining layer and diffuse infiltrates of the synovial membrane. They were identified as fibroblast-like, using cell-specific markers. Expression of *jun*-B/c-*fos* was not detected in lymphocytes or macrophage-like cells, while the number of cells which were positive was considerably smaller in osteoarthritis and healthy control synovia. The spontaneous expression of c-*fos* and other proto-oncogenes in rheumatoid synovial cells has been reviewed (48). The cellular distribution of Jun/Fos within the rheumatoid synovium contrasts with that seen for the p50 and p65 subunits of NFκB in a recent immunohistochemical study by Handel and colleagues (49). These authors found that p50/p65 was present in the nuclei of macrophage-like cells of the synovial lining and sublining areas, as well as in endothelial cells. There was little staining in normal control synovium samples. The contrast in subunit distribution highlights the importance of cell type-specific signaling and gene expression.

We have used antibodies specific for the DNA binding, nuclear form, of NFκB to perform immunohistochemical studies in the synovium of rheumatoid arthritis and osteoarthritis patients (50). The basis of such "activity-specific" antibodies is that they were raised against the NLS of Rel-A. Binding of IκB to the p50–Rel-A heterodimer sterically masks the NLS [(51); see also above]. In agreement with Handel et al. (49), we found that in the synovium of rheumatoid arthritis patients both vascular endothelial cells and macrophage-like cells showed the presence of the activated form of NFκB (see Figures 3a and 3b). Again, staining was nuclear–as would be expected–and synovial tissue from controls exhibited either no staining or only weak staining (see Figure 3c). We also found that the relative amounts of staining in these two cell types was dependent on the chronicity of the disease (see Figures 3a and 3b).

Staining for activated NFκB was not associated with synovial lymphocyte aggregates. This result is consistent with previous observations of an apparent lack of recent activation of T lymphocytes within the rheumatoid synovium. Thus, while synovial T cells have been implicated in the pathology of rheumatoid arthritis, T cell-derived cytokines and other markers of recent T cell activation such as the IL-2 receptor have only been demonstrated at low levels in the rheumatoid joint (52). As mentioned earlier, induced expression of ferritin appears to occur in the rheumatoid synovial fluid, but these

(a)

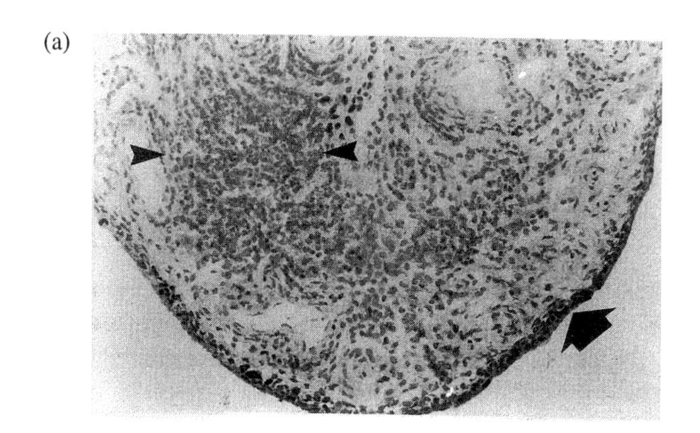

(b)

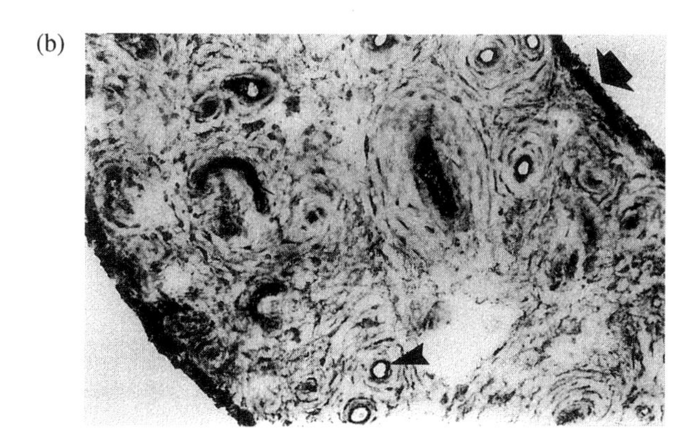

(c)

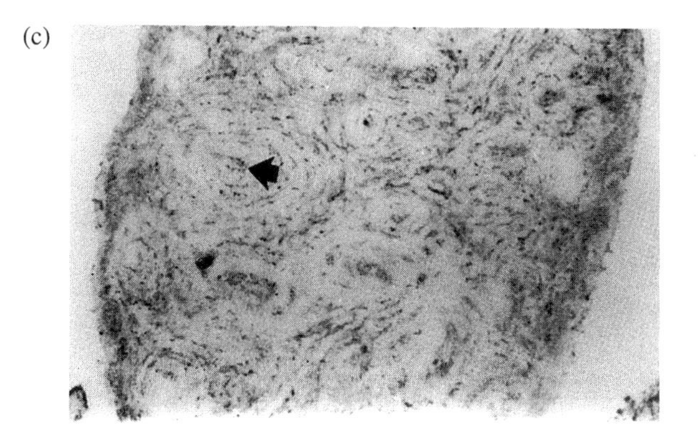

Figure 3. Immunohistochemical staining of the active form of NFκB within human synovial membrane sections. (a) Villous projection of synovium from a patient with chronic rheumatoid arthritis, showing positive staining of synovial lining cells (arrow) in the absence of staining of a lymphoid aggregate (area between arrowheads). (b) Villous projection of synovium from a patient with an acute flare of rheumatoid arthritis. There is staining of surface lining cells (arrow) as well as strong staining of subsynovial vessels, seen both in transverse and longitudinal sections (arrowheads). (c) Weak background staining only, in synovium obtained at autopsy from a subject with no history of arthritis. Negative staining of blood vessels in the subsynovium (arrow). × 200. (From Ref. 50.)

ferritin molecules have low iron loading (6). This observation might be explained by the induced expression of ferritin subunits, related to the activation of NFκB (27).

WHAT IS THE CELLULAR SOURCE OF THE ROI INVOLVED IN NFκB ACTIVATION?

There are many potential sources of ROI/RNI that may be involved in inflammatory cell signaling. These include NADPH oxidase, cytochrome P450, nitric oxide synthases, cyclo-oxygenases and lipoxygenases, the mitochondrial respiratory chain, and xanthine oxidase (see Figure 4). Since these systems have been discussed extensively elsewhere, in the context of their radical-generating properties (see some of the references given in the following paragraphs), they will be described only briefly here, in relation to their possible roles in cell signaling.

The plasma membrane-bound NADPH oxidase of polymorphonuclear leukocytes (PMNs) contains cytochrome b_{245} and catalyses the univalent reduction of molecular oxygen to generate the superoxide anion radical, $O_2\cdot^-$ (53, 54). In addition to phagocytic cells, other cell types such as lymphocytes (55) and glomerular mesangial cells (56) also

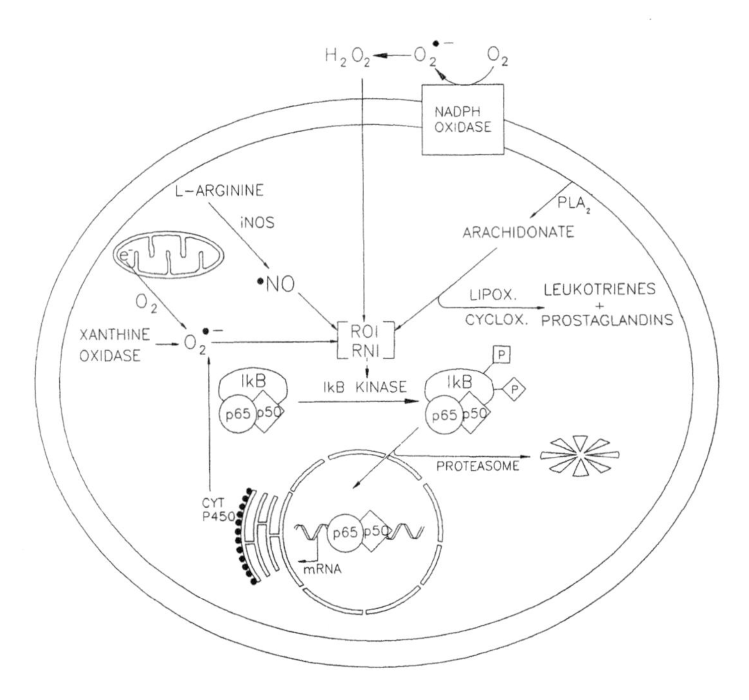

Figure 4. ROI-generating systems potentially involved in NFκB activation. ROI are generated by a diverse range of systems operating within the cell. The importance of each system may vary between different cell lines and cell types. It has been postulated that the signals from a diverse range of cell stimuli funnel into a common pathway involving ROI generation and thereby regulate NFκB activation.

possess a plasma membrane; NADPH oxidase. Indeed, in glomerular mesangial cells, the NADPH oxidase inhibitor 4′-hydroxy-3′-methoxyacetophenone (apocynin) inhibited the activation of NFκB in response to TNF-α or aggregated IgG (56).

The endoplasmic reticulum contains an NADPH-dependent electron transport complex where the family of heme proteins, termed cytochromes P450, serve as catalysts for the hydroxylation of a variety of different organic compounds. However, in addition to this activity, some cytochrome P450 systems exhibit a concurrent oxidase activity and can catalyze the reaction between NADPH and oxygen to produce superoxide (57). Thus, cytochromes P450 may be important ROI-generating systems in certain cell types.

The free radical nitric oxide (·NO) has been shown to be an important factor in bioregulation, initially by the demonstration of its role as endothelium-derived relaxing factor (EDRF) and subsequently by the observation of constitutive NO-synthase enzymes in numerous other cell types and tissues (8). In addition to the generation of ·NO by the constitutive enzymes, ·NO is generated by macrophages, neutrophils, and lymphocytes via the action of a calcium-independent, cytokine-inducible-·NO-synthase, the expression of which results in the release of ·NO in amounts greatly exceeding those from the constitutive pathway. Lander and colleagues (30) showed that ·NO-generating compounds such as *S*-nitroso-*N*-acetylpenicillamine (SNAP) induced the activation of NFκB binding in vitro in human peripheral blood mononuclear cells. Furthermore, the transcription of the iNOS gene is itself induced by NFκB (15, 58).

The arachidonic acid cascade is a central process in generating the important inflammatory mediators prostaglandins and leukotrienes (59). Arachidonic acid is released from the plasma membrane of the cell by the action of a phospholipase A_2. The released lipid is then metabolized in a process which can give rise to intracellular ROI (60). Thus, arachidonic acid can be acted on by either cyclo-oxygenases (also known as prostaglandin H synthases) which catalyze conversion to prostaglandins, or by lipoxygenases which catalyze the formation of leukotrienes and hydroxyeicosatetraenoic acids. Both a constitutive form and an inducible form of cyclo-oxygenase have been identified (59, 61), known as COX-1 (prostaglandin H_2 synthase-1; PGHS-1) and COX-2 (PGHS-2), respectively. The mode of action of nonsteroidal anti-inflammatory drugs involves the inhibition of cyclo-oxygenase activity and an area of current interest is the development of new drugs, with fewer side-effects, which selectively inhibit COX-2, but not COX-1. Aspirin (acetylsalicylic acid), a potent cyclo-oxygenase inhibitor, has been shown to block the activation of NFκB in the human Jurkat T cell line, albeit at high concentrations (64). However, the related compound sodium salicylate also inhibited NFκB activation. Sodium salicylate has anti-inflammatory activity, but does not inhibit cyclo-oxygenase. This is consistent with the suggestion (64) that the anti-inflammatory activity of aspirin may not be wholly attributable to its ability to inhibit cyclo-oxygenase, and that the inhibition of NFκB activation may play a role. A number of new anti-inflammatory drugs have also been targeted toward the selective inhibition of 5-lipoxygenase, an enzyme which catalyzes the formation of the proinflammatory molecule leukotriene B4 (59). On the other hand, 15-lipoxygenase has been implicated in the oxidative modification of LDL in atherogenesis (62, 63), a process in which NFκB activation has been implicated (29).

Most of the oxygen consumed by mammalian cells is converted to water via the mitochondrial electron transport system, in which electrons flow from NADH to sequentially reduce flavoproteins, ubiquinone, mitochondrial cytochromes, and finally molecular oxygen. The last reaction of this respiratory chain, catalyzed by cytochrome oxidase, is the donation of four electrons to each O_2 molecule to form water. However, up to 5% of the electrons entering the mitochondrial electron transport chain can become uncoupled from it and singly leak out onto O_2 to form $O_2\cdot^-$ (65). This source of free radicals is thought to be involved in TNF-mediated cytotoxicity (66), and will be returned to later.

The enzyme xanthine dehydrogenase is present within the cell cytosol but may also be bound on the endothelial cell surface. It catalyzes the oxidation of hypoxanthine and xanthine to uric acid. Xanthine dehydrogenase is thought to be located predominantly in the liver, small intestine, and capillary endothelium in man (67). However, the distribution is different in other species. In healthy tissue, most of the enzyme is present as the "D form," which transfers electrons to NAD^+:

$$\text{xanthine} + H_2O + NAD^+ \rightarrow \text{uric acid} + NADH + H^+$$

However, about 10% of the enzyme is present as an oxidase ("type O") form, which transfers electrons to molecular oxygen to form $O_2\cdot^-$:

$$\text{xanthine} + H_2O + 2O_2 \rightarrow \text{uric acid} + 2O_2\cdot^- + 2H^+$$

Both reactions are inhibited by oxypurinol, the principal metabolite of allopurinol. The original mechanism proposed by McCord for the production of $O_2\cdot^-$ in ischemic tissues involves changes in purine metabolism within ischemic cells (68). During temporary ischemia, low oxygen concentrations cause a decline in mitochondrial oxidative phosphorylation by decreasing the capacity of the respiratory chain for coupled electron transport. This increases the dependence of the cell on ATP production via anaerobic glycolysis. Anaerobic glycolysis is an inefficient means of ATP production from glucose and leads to raised concentrations of adenosine and of its breakdown products, including hypoxanthine and xanthine, which are substrates for the xanthine dehydrogenase enzyme system. Cellular levels of ATP fall. Cells are no longer able to maintain proper ion gradients across their membranes, and this precipitates a redistribution of Ca^{2+} ions. The elevated cytosolic Ca^{2+} concentration activates a protease, possibly a calpain, which catalyzes the conversion of xanthine dehydrogenase to xanthine oxidase. Reperfusion of the temporarily ischemic organ restores a supply of the remaining substrate required for xanthine oxidase activity (i.e., O_2) and $O_2\cdot^-$ is generated.

The conversion of xanthine dehydrogenase to its oxidase form may also be induced by exposure of cells to TNF (69). Thus, some studies have demonstrated a protective effect of xanthine oxidase inhibitors such as allopurinol toward TNF-mediated cytotoxicity (70, 71). In contrast, other studies have shown that the cytotoxicity of TNF was mainly related to uncoupling of the mitochondrial respiratory chain [(66), and see below]. Thus, it appears that the relative importance of different sources of ROI, and therefore the potential source of ROI involved in activating NFκB, may be dependent on the type of cell. It also seems plausible that the environment of the cell, for example, oxygen tension or prior exposure to cytokines, as well as the nature of the NFκB-activating stimulus, may play a role. This suggests that ROI-mediated NFκB activation may not be important in every cell type, in all environments (72).

Some studies have begun to address these issues. Recently, Los and coworkers (73) studied the activation of the CD28-responsive complex, which has been shown to consist of protein subunits of the NFκB family, as a consequence of the triggering of the CD28 surface receptor of isolated human peripheral blood T lymphocytes. It was shown that the intracellular formation of ROI was a required step in this process. To identify the source of the ROI, the effects of inhibitors of the ROI-generating enzymes discussed above were tested. No significant effects were seen with the NADPH oxidase inhibitor diphenylene iodonium, or the respiratory chain inhibitor rotenone, or the xanthine oxidase inhibitor allopurinol. However, specific inhibitors of either 5-lipoxygenase (such as ICI 230487) or phospholipase A_2 (such as *p*-bromophenacyl bromide) prevented the activation of the NFκB/CD28-responsive complex. These results are of particular interest because the experiments were carried out in primary human cells. Many experiments which have been carried out in relation to the role of ROI/RNI as cellular messengers that induce gene expression have been done in cell lines. In such cells, some signaling pathways may be redundant.

In contrast to the study by Los et al. (73), experiments in the TNF-α-sensitive L929 fibrosarcoma cell line (74) showed that the activation of NFκB by TNF-α was blocked by rotenone, a specific inhibitor of the electron flow within the mitochondrial electron transport chain, which inhibits ROI generation. L929 subclones that lacked a functional respiratory chain were resistant to NFκB activation by TNF-α. These subclones were likewise resistant to the cytotoxicity of TNF-α, which also appears to involve ROI production. In addition, while rotenone reduced TNF cytotoxicity, inhibitors of two of the other potential ROI sources (xanthine oxidase and NADPH oxidase) had no effect (66). Similarly, mitochondrial respiratory chain uncoupling in isolated rat hepatocytes was associated with NFκB activation (75).

NFκB AND ANTI-INFLAMMATORY DRUGS

Owing to the range of inflammatory genes that are induced by activated NFκB, it has been suggested that the activation of this transcription factor by ROI, produced as an early event, could play a critical role in inflammatory reactions (76). As indicated above, some of the cytokines whose genes are switched on by activated NFκB, such as TNF-α and IL-1, are themselves activators of NFκB, giving the potential for a positive feedback cycle within the inflammatory response. Even if the initial cellular stimulus for NFκB activation is not ROI/RNI, the apparent merging of all the pathways of signal transduction involving NFκB on a ROI-dependent step suggests a possible therapeutic target.

The anti-inflammatory action of several well-known drugs, including sodium salicylate (64) and gold(I) thiolate compounds such as aurothioglucose (77), has been suggested to be due to the inhibition of NFκB activation. The DNA binding activity of NFκB is thought to be Zn^{2+}-dependent. Yang and colleagues (77) suggested that cysteine residues may be involved in the binding of Zn^{2+}. Thus, in the case of aurothioglucose, it was proposed that Au(I) oxidizes the Zn^{2+}-associated thiolate anions to disulfides, thereby preventing the binding of NFκB to DNA. It has also recently been suggested that glucocorticoids exert their immunosuppressive activity by inducing the synthesis of IκB (78, 79), which blocks the NFκB-mediated expression of the genes described above. Another mechanism which may contribute to immunosuppression is the direct interaction of the glucocorticoid-receptor complex with NFκB, thereby preventing its association with DNA (see Refs. 78 and 79). A similar situation has been shown for the estrogen receptor (80). This receptor physically interacts with NFκB and another transcription factor, C/EBPβ (NF-IL6). Both NFκB and C/EBPβ regulate IL-6 gene expression in human osteoblasts. Their interaction with the estrogen receptor results in inhibition of the IL-6 promoter. This estrogen effect may be involved in the bone resorption associated with osteoporosis (the most common form is postmenopausal in women) and, possibly, in the marked preponderance of rheumatoid arthritis in females.

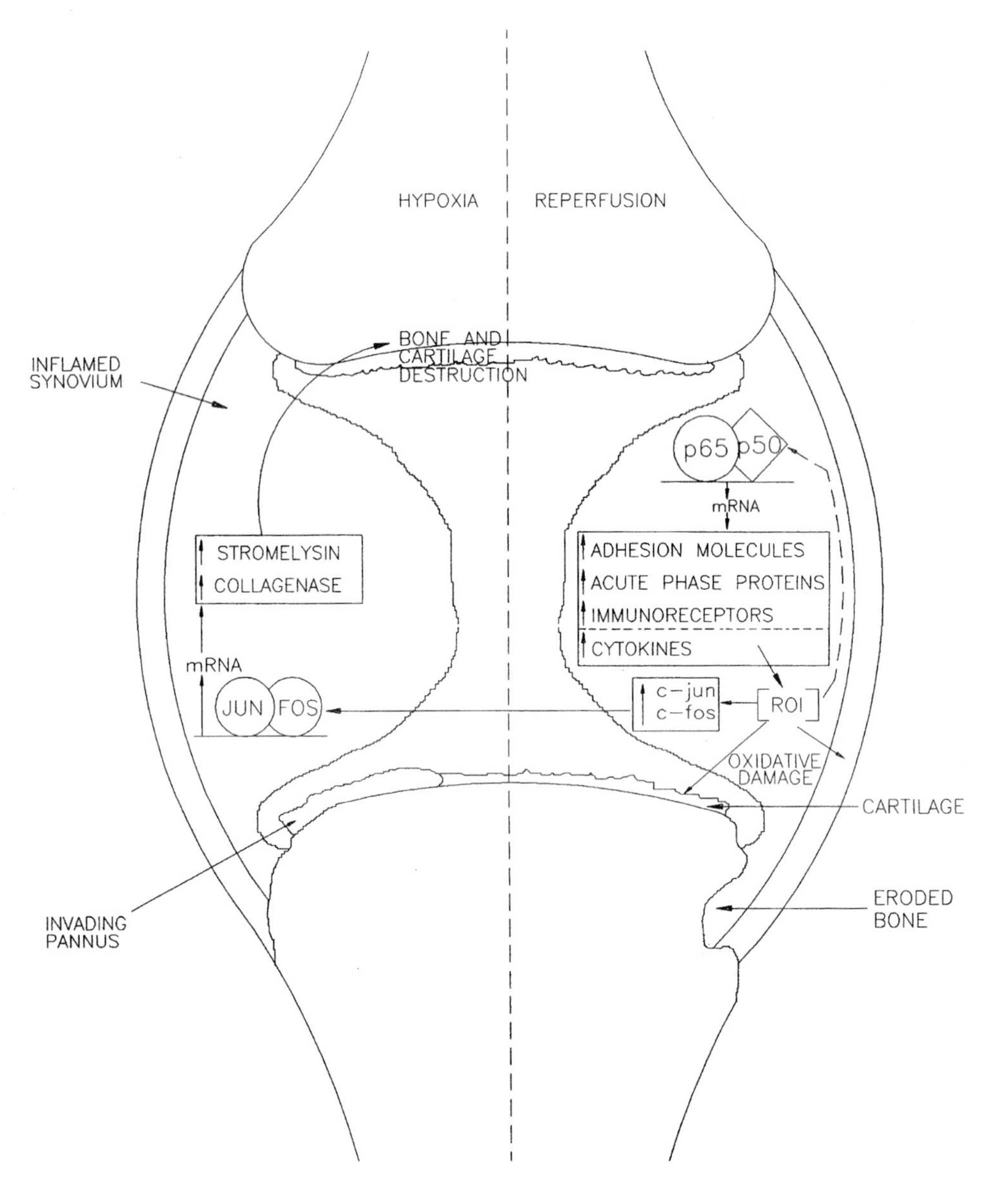

Figure 5 The consequences of hypoxia–reperfusion in the rheumatoid joint. Transcription factor activation may be regulated by hypoxia–reperfusion within the rheumatoid joint. Upregulation of AP-1 during hypoxia may be involved in the transcription of enzymes that promote cartilage destruction. The increase of ROI upon reperfusion may cause both the direct oxidative modification of important biomolecules and the indirect production of inflammatory proteins via the activation of NFκB. Hypoxia and reperfusion within the rheumatoid joint may be involved in tissue proliferation, joint inflammation, and cartilage/bone destruction.

We discussed above that one important NFκB-regulated gene is that encoding the cytokine TNF-α. It was also mentioned that ROI are involved in TNF-induced activation of NFκB, while ROI have also been shown to play a role in TNF-induced apoptosis (69). TNF appears to play a central role in several inflammatory conditions, including rheumatoid arthritis (81) and septic shock (82). Indeed, a marked feature of the pathology of the TNF transgenic mouse is a synovitis (83). Recent clinical studies have shown that intravenous administration of anti-TNF antibodies produced dramatic anti-inflammatory effects in rheumatoid arthritis patients (81).

CONCLUSION

The diagram of the joint shown in Figure 5 attempts to summarize the apparently paradoxical, redox-related events which can be explained by cycles of hypoxia–reperfusion. This is, of course, a highly schematic overview emphasizing the upregulation of genes by AP-1 and NFκB in the contrasting environments of hypoxia and reperfusion.

An improved understanding of the transcriptional mechanisms which regulate the expression of genes encoding important inflammatory proteins, such as TNF, will hopefully lead to the identification of regulatory steps that can be successfully targeted by novel anti-inflammatory compounds. Among the pathways thus far identified, it appears that ROI constitute a common intracellular messenger in the regulation of both AP-1 and NFκB by a diverse range of extracellular stimuli. As such, these redox-related transcription factors may represent important targets for both existing and new antioxidant drugs.

ACKNOWLEDGMENTS

We thank the British Technology Group and the Arthritis and Rheumatism Council for financial support.

REFERENCES

1. Butcher RG, Bitensky L, Cashman B, Chayen J. Differences in the redox balance in human rheumatoid and non-rheumatoid synovial lining cells. Beitr Path Bd 1973; 148:265–274.
2. Poulter LW, Bitensky L, Cashman B, Chayen J. The maintenance of human synovial tissue in vitro. Virchows Arch B Zellpath 1970; 4:303–309.
3. Merry P, Winyard PG, Morris CJ, Grootveld M, Blake DR. Free radicals, inflammation, and synovitis: the current status. Ann-Rheum Dis 1989; 48:864–870.
4. Lunec J, Blake DR. The determination of dehydroascorbic acid and ascorbic acid in the serum and synovial fluid of patients with rheumatoid arthritis. Free Radical Res Commun 1985; 1:31–39.
5. Trenam CW, Winyard PG, Morris CJ, Blake DR. Iron-promoted oxidative damage in rheumatic diseases. In Lauffer RB, ed. Iron and Human Diseases. Boca Raton, Florida: CRC Press; 1992: 396–412.
6. Brailsford S, Lunec J, Winyard PG, Blake DR. A possible role for ferritin during inflammation. Free Radical Res Commun 1985; 1:101–109.

7. Halliwell B, Gutteridge JMC. The importance of free radicals and catalytic metal ions in human disease. Mol Aspects Med 1985; 8:89–193.
8. Snyder SH. No endothelial NO. Nature 1995; 377:196–197.
9. Jacobson MD. Reactive oxygen species and programmed cell death. TIBS 21: 83–86.
10. Brawn MK, Chiou WJ, Leach KL. Oxidant-induced activation of protein-kinase-C in UC11MG cells. Free Radical Res 1995; 22:23–37.
11. Whisler RL, Goyette MA, Grants IS, Newhouse YG. Sublethal levels of oxidant stress stimulate multiple serine threonine kinases and suppress protein phosphatases in Jurkat T-cells. Arch Biochem Biophys 1995; 319:23–35.
12. Fialkow L, Chan CK, Rotin D, Grinstein S, Downey GP. Activation of the mitogen-activated protein-kinase signaling pathway in neutrophils–role of oxidants. J Biol Chem 1994; 269:31234–31242.
13. Keyse SM, Emslie EA. Oxidative stress and heat shock induce a human gene encoding a protein-tyrosine phosphatase. Nature 1992; 359:644–647.
14. McKay IA, Winyard PG, Leigh IM, Bustin SA. Nuclear transcription factors: potential targets for new modes of intervention in skin disease. Br J Dermatol 1994; 131:591–597.
15. Baeuerle PA, Henkel T. Function and activation of NF-κB in the immune system. Annu Rev Immunol 1994; 12:141–179.
16. Blake DR, Winyard PG, Marok R. The contribution of hypoxia-reperfusion injury to inflammatory synovitis: the influence of reactive oxygen intermediates on the transcriptional control of inflammation. Ann NY Acad Sci 1994; 723:308–317.
17. Kus ML, Fairburn K, Blake DR, Winyard PG. A vascular basis for free radical involvement in inflammatory joint disease. In Blake DR and Winyard PG, eds. Immunopharmacology of Free Radical Species. London: Academic Press; 1995: 97–112.
18. Rupec RA, Baeuerle PA. The genomic response of tumor cells to hypoxia and reoxygenation–differential activation of transcription factors AP-1 and NF-κB. Eur J Biochem 1995; 234:632–640.
19. Fassbender HG. Current understanding of rheumatoid arthritis. Inflammation 1984; 8 (supplement):S27–S42.
20. Amstad PA, Krupitza G, Cerutti PA. Mechanism of *c-fos* induction by active oxygen. Cancer Res 1992; 52:3952–3960.
21. Meyer MR, Schreck R, Baeuerle PA. Hydrogen peroxide and antioxidants have opposite effects on activation of NFκB and AP-1 in intact cells: AP-1 as secondary antioxidant-responsive factor. EMBO J 1993; 12:2005–2015.
22. Karin M. The AP-1 complex and its role in transcriptional control by protein kinase C. In Cohen P and Foulkes JG, eds. Molecular Aspects of Cellular Regulation, vol. 6. The Hormonal Control of Gene Transcription. Amsterdam: Elsevier; 1991: 235–253.
23. Gravallese EM, Darling JM, Ladd AL, Katz JN, Glimcher LH. In situ hybridization studies of stromelysin and collagenase messenger RNA expression in rheumatoid synovium. Arthritis Rheum 1991; 34:1076–1084.
24. Winyard PG, Zhang Z, Chidwick K, Blake DR, Carrell RW, Murphy G. Proteolytic inactivation of human α_1-antitrypsin by human stromelysin. FEBS Lett 1991; 279:91–94.
25. Henney AM, Wakeley PR, Davies MJ, et al. Localization of stromelysin gene expression in atherosclerotic plaques by *in situ* hybridization. Proc Natl Acad Sci USA 1991; 88:8154–8158.
26. Edwards DR, Murphy G, Reynolds JJ, et al. Transforming growth factor beta modulates the expression of collagenase and metalloproteinase inhibitor. EMBO J 1987; 6:1899–1904.
27. Kwak EL, Larochelles DA, Beaumont C, Torti SV, Torti FM. Role for NF-κB in the regulation of ferritin H by tumour necrosis factor-α. J Biol Chem 1995; 270:15285–15293.
28. Schreck R, Rieber P, Baeuerle PA. Reactive oxygen intermediates as apparently widely used messengers in the activation of the NF-κB transcription factor and HIV-1. EMBO J 1991; 10:2247–2258.

29. Liao F, Andalibi A, deBeer FC, Fogelman AM, Lusis AJ. Genetic control of inflammatory gene induction and NF-κB like transcription factor activation in response to an atherogenic diet in mice. J Clin Invest 1993; 91:2572–2579.
30. Lander HM, Sehajpal P, Levine DM, Novogrodsky A. Activation of human peripheral blood mononuclear cells by nitric oxide-generating compounds. J Immunol 1993; 150:1509–1516.
31. Schreck R, Albermann K, Baeuerle PA. Nuclear factor κB: an oxidative stress-responsive transcription factor of eukaryotic cells (a review). Free Radical Res Commun 1992; 17:221–237.
32. Tasinato A, Boscoboinik D, Bartoli GM, Maroni P, Azzi A. *d*-α-Tocopherol inhibition of vascular smooth muscle cell proliferation occurs at physiological concentrations, correlates with protein kinase C inhibition, and is independent of its antioxidant properties. Proc Natl Acad Sci USA 1995; 92:12190–12194.
33. Drury PL, Rudge SR, Perrett D. Structural requirements for activity of certain "specific" antirheumatic drugs: more than a simple thiol group? Br J Rheumatol 1984; 23:100–106.
34. Traenckner EB-M, Wilk S, Baeuerle PA. A proteasome inhibitor prevents activation of NF-κB and stabilises a newly phosphorylated form of IκB-α that is still bound to NF-κB. EMBO J 1994; 13:5433–5441.
35. Brown K, Gerstberger S, Carlson L, Franzoso G, Siebenlist U. Control of IκB-α proteolysis by site-specific, signal-induced phosphorylation. Science 1995; 267:1485–1488.
36. DiDonato JA, Mercurio F, Karin M. Phosphorylation of IκB-α precedes but is not sufficient for its dissociation from NF-κB. Mol Cell Biol 1995; 15:1302–1311.
37. Chen JZ, Parent L, Maniatis T. Site-specific phosphorylation of IκB-α by a novel ubiquination-dependent protein kinase activity. Cell 1996; 84:853–862.
38. Abate C, Patel L, Rauscher FJ, Curran T. Redox regulation of Fos and Jun DNA-binding activity in vitro. Science 1990; 249:1157–1161.
39. Xanthoudakis S, Miao G, Wang F, Yu ching E. Pan, Curran T. Redox activation of Fos and Jun DNA binding activity is mediated by a DNA repair enzyme. EMBO J 1992; 11:3323–3335.
40. Yao KS, Xanthoudakis S, Curran T, O'Dwyer PJ. Activation of AP-1 and of a nuclear redox factor, Ref-1, in the response of HT29 colon cancer cells to hypoxia. Mol Cell Biol. 1994; 14:5997–6003.
41. Matthews JR, Wakasugi N, Virelizier JL, Yodoi J, Hay RT. Thioredoxin regulates the DNA binding activity of NFκB by reduction of a disulphide bond involving cysteine 62. Nucleic Acids Res 1992; 20:3821–3830.
42. Ghosh G, van Duyne G, Ghosh S, Sigler PB. Structure of NF-κB p50 homodimer bound to a κB site. Nature 1993; 373:303–310.
43. Muller CW, Rey FA, Sodeoka M, Verdine GL, Harrison SC. Structure of the NF-κB p50 homodimer bound to DNA. Nature 1995; 373:311–317.
44. Toledano MB, Leonard WJ. Modulation of transcription factor NF-κB binding activity by oxidation–reduction in vitro. Proc Natl Acad Sci USA 1991; 88:4328–4332.
45. Schulze-Ostoff K, Los M, Baeuerle PA, Redox signalling by transcription factors NFκB and AP-1 in lymphocytes. Biochem Pharmacol 1995; 50:735–741.
46. Hayashi T, Ueno Y, Okamato T. Oxidoreductive regulation of nuclear factor κB: involvement of a cellular reducing catalyst thioredoxin. J Biol Chem 1993; 268:11380–11388.
47. Kinne RW, Boehm S, Iftner T, et al. Synovial fibroblast-like cells strongly express *jun*-B and c-*fos* proto-oncogenes in rheumatoid- and osteoarthritis. Scand J Rheumatol 1995; 24 (supplement 101):121–125.
48. Gay S, Gay RE, Koopman WJ. Molecular and cellular mechanisms of joint destruction in rheumatoid arthritis: two cellular mechanisms explain joint destruction? Ann Rheum Dis 1993; 52:S39–S47.
49. Handel ML, McMorrow LB, Gravallese EM. Nuclear factor-κB in rheumatoid synovium.

Localisation of p50 and p65. Arthritis Rheum 1995; 38:1762–1770.
50. Marok R, Winyard PG, Coumbe A, et al. Activation of the transcription factor NF-κB in the inflamed human synovium. Arthritis Rheum 1996; 39:583–591.
51. Kaltschmidt C, Kaltschmidt B, Henkel T, Stockinger H, Baeuerle PA. Selective recognition of the activated form of transcription factor NF-κB by a monoclonal antibody. Biol Chem Hoppe-Seyler 1995; 376:9–16.
52. Salmon M, Gaston JSH. The role of T-lymphocytes in rheumatoid arthritis. Br Med Bull 1995; 51:332–345.
53. Forman HJ, Thomas MJ. Oxidant production and bactericidal activity of phagocytes. Annu Rev Physiol 1986; 48:669–680.
54. Segal AW. Components of the microbicidal oxidase of phagocytes. Biochem Soc Trans 1991; 49:49–50.
55. Hancock JT, Maly FE, Jones OTG. Properties of the superoxide-generating oxidase of B-lymphocyte cell lines. Determination of Michaelis parameters. Biochem J 1989; 262:373–375.
56. Satriano J, Schlondroff D. Activation and attenuation of transcription factor NF-κB in mouse glomerular mesangial cells in response to tumor-necrosis-factor-alpha, immunoglobulin G, and adenosine 3′/5′ – cyclic monophosphate–evidence for involvement of reactive oxygen species. J Clin Invest 1994; 94:1629–1636.
57. White RE, Coon MJ. Oxygen activation by cytochrome P-450. Annu Rev Biochem 1980; 49:315–356.
58. Barnes PJ. Anti-inflammatory mechanisms of glucocorticoids. Biochem Soc Trans 1995; 23:940–945.
59. Lewis AJ, Keft AF. A review on the strategies for the development and application of new anti-arthritic agents. Immunopharmacol Immunotoxicol 1995; 17:607–663.
60. Cadenas E. Biochemistry of oxygen toxicity. Annu Rev Biochem 1989; 58:79–110.
61. Vane J. Towards a better aspirin. Nature 1994; 367:215–216.
62. Yla-Herttuala S, Rosenfeld ME, Parthasarathy S, et al. Gene expression in macrophage-rich human atherosclerotic lesions. 15-Lipoxygenase and acetyl low density lipoprotein receptor messenger RNA colocalize with oxidation specific lipid–protein adducts. J Clin Invest 1991;87:1146–1152.
63. Jessup W, Darley-Usmar V, O'Leary V, Bedwell S. 5-Lipoxygenase is not essential inmacrophage-mediated oxidation of low-density lipoprotein. Biochem J 1991; 278:163–169.
64. Kopp E, Ghosh S. Inhibition of NF-κB by sodium salicylate and aspirin. Science 1994; 265:956–959.
65. Fridovich I. Hypoxia and oxygen toxicity. Adv Neurol 1979; 26:255–259.
66. Schulze-Osthoff K, Bakker AC, Vanhaesebroeck B, Beyaert R, Jacob WA, Fiers W. Cytotoxic activity of tumor necrosis factor is mediated by early damage of mitochondrial functions. Evidence for the involvement of mitochondrial radical generation. J Biol Chem 1992; 267:5317–5323.
67. Jarasch E-D, Bruder G, Heid HW. Significance of xanthine oxidase in capillary endothelial cells. Acta Physiol Scand 1986; (supplement) 548:39–46.
68. McCord JM. Oxygen derived free radicals in postischemic tissue injury. N Engl J Med 1985; 312:159–163.
69. Larrick JW, Wright SC. Cytotoxic mechanism of tumor necrosis factor-α. FASEB J 1990; 4:3215–3223.
70. Adamson GM, Billings RE. The role of xanthine oxidase in oxidative damage caused by cytokines in cultured mouse hepatocytes. Life Sci 1994; 55:1701–1709.
71. Olah T, Regely K, Mandi Y. The inhibitory effects of allopurinol on the production and cytotoxicity of tumor necrosis factor. Naunyn–Schmiedebergs Arch Pharmacol 1994; 350:96–99.

72. Brennan P, O'Neil LAJ. Effects of oxidants and anti-oxidants on nuclear factor κB activation in three different cell lines: evidence against a universal hypothesis involving oxygen radicals. Biochim Biophys Acta 1995; 1260:167–175.
73. Los M, Schenk H, Hexel K, Baeuerle PA, Droge W, Schulze-Osthoff, K. IL-2 gene expression and NF-κB activation through CD28 requires reactive oxygen production by 5-lipoxygenase. EMBO J 1995; 14:3731–3740.
74. Schulze-Osthoff K, Beyaert R, Vandevoorde V, Haegeman G, Fiers W. Depletion of the mitochondrial electron transport abrogates the cytotoxic and gene-inductive effects of TNF. EMBO J 1993; 12:3095–3104.
75. Garciaruiz C, Colell A, Morales A, Kaplowitz N. Role of oxidative stress generated from the mitochondrial electron-transport chain and mitochondrial glutathione status in loss of mitochondrial-function and activation of transcription factor nuclear factor-kappa-B–studies with isolated mitochondria and rat hepatocytes. Mol Pharmacol 1993; 48:825–834.
76. Kaltschmidt C, Kaltschmidt B, Lannes-Vieira J, et al. Transcription factor NF-κB is activated in microglia during experimental autoimmune encephalomyelitis. J Neuroimmunol. 1994; 55:99–106.
77. Yang JP, Merin JP, Nakano T, Kato T, Kitade Y, Okamoto T. Inhibition of the DNA-binding activity of NF-κB by gold compounds in vitro. FEBS Lett 1995; 361:89–96.
78. Scheinman RI, Cogswell PC, Lofquist AK, Baldwin AS. Role of transcriptional activation of IκB-α in mediation of immunosuppression by glucocorticoids. Science 1995; 270:283–286.
79. Auphan N, DiDonato JA, Rosette C, Helmberg A, Karin M. Immunosuppression by glucocorticoids: inhibition of NF-κB activity through induction of IκB synthesis. Science 1995; 270:286–290.
80. Stein B, Yang MX. Repression of the interleukin-6 promoter by estrogen receptor is mediated by NFκB and C/EBPβ. Mol Cell Biol 1995; 15:4971–4979.
81. Chernajovsky Y, Feldmann M, Maini RN. Gene therapy of rheumatoid arthritis via cytokine regulation: future perspectives. Br Med Bull 1995; 51:503–51.
82. Suitters AJ, Foulkes R, Opal SM, et al. Differential effect of isotype on efficacy of anti-tumour necrosis factor α chimeric antibodies in experimental septic shock. J Exp Med 1994; 179:849–856.
83. Keffer J, Probert L, Cazlaris H, et al. Transgenic mice expressing human tumour necrosis factor: a predictive genetic model of arthritis. EMBO J 1991; 10:4025–4031.

50

Chronic Oxidative Stress in Rheumatoid Arthritis: Implications for T Cell Function

Madelon M. Maurice, Ellen A. M. van der Voort, Anita I. van Vilet, Paul-Peter Tak, Ferdinand C. Breedveld, and Cornelis L. Verweij
University Hospital, Leiden, The Netherlands

Hajime Nakamura and Sussanne Thorell
Karolinska Institutet, Stockholm, Sweden

INTRODUCTION

General Background

Rheumatoid arthritis (RA) is a common, chronic inflammatory joint disease of which the etiology is still incompletely understood. In the joint of an RA patient the synovial membrane, which covers the joint cavity, is inflamed and becomes infiltrated with predominantly $CD4^+$ T cells and macrophages and to a lesser extent with B cells, fibroblasts, and dendritic cells (Figure 1). Neutrophils are the major cell type in the synovial fluid in addition to the above-mentioned mononuclear cells. When the inflammation progresses, the synovial membrane becomes rampant and through the action of destructive enzymes such as collagenase and matrix metalloproteinases the adjacent cartilage and bone are destroyed, with disability as a consequence.

Role of T Cells in RA

Most of the synovial tissue infiltrating T cells display an activated phenotype consisting predominantly of the "primed" CD45RO subset and expressing high levels of the IL-2 receptor, adhesion molecules, and MHC class II (1–12). A strong indication for a role for T cells in the pathogenesis of RA is found in the association of the more severe types of disease with HLA-DR1/DR4, suggesting an antigen-driven mechanism of disease. Therefore, it is generally believed that T cells play an important role in the initiation and/or perpetuation of RA, and several assignments point to an involvement of T cells in the pathogenesis of RA.

A number of "rheumatoid" self-antigens have been proposed, including type II collagen (13,14), cartilage proteoglycan (15), and chondrocyte antigens (16,17), but definitive evidence in favor of any of them is still incomplete. Furthermore, despite

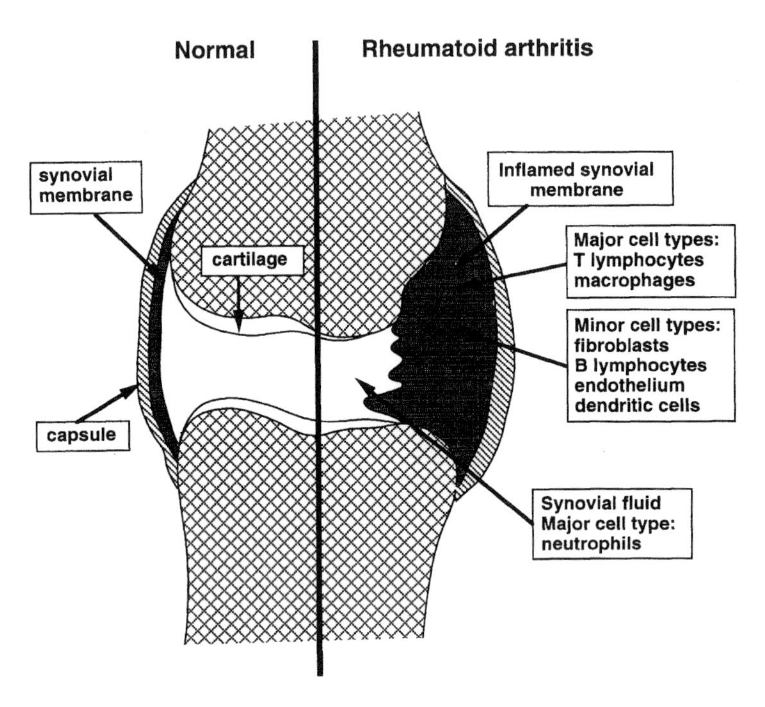

Figure 1. Schematic representation of a normal and a rheumatoid joint.

extensive studies, a consensus in T cell receptor (TCR) Vα/β usage in synovial tissue and fluid could not be demonstrated (18–20).

In contrast to the activated phenotype, RA joint T cells reveal impaired Ca^{2+} and proliferative responses, and low levels of T cell-derived cytokines are detected in RA joints (21–30). On the basis of these findings, it has been suggested that T cell are not important in perpetuating disease in the later stages (31). This postulate is supported by disappointing results of therapeutic trials affecting T cell function in late-stage RA patients. The depletion of CD4, CDW52, CD7, CD5, and IL-2 receptor-bearing cells (32–35) has shown only moderate or no benefit in placebo-controlled studies. Since there is a subset of T cells (Th1 cells) producing the proinflammatory mediators IL-2 and IFN-γ, and a second subset (Th2 cells) producing the anti-inflammatory cytokine IL-4 (36), it is possible that therapies that affect overall T cell function may be ineffective because both pathogenic and protective T cell functions would be downregulated.

Possible Role of Oxidative Stress in T Cell Function in Established RA

In RA, much evidence has been found for increased production of reactive oxygen species (ROS) in vivo in patients with active rheumatoid disease. Increased lipid peroxidation products (37,38), degradation of hyaluronic acid by free radical mechanisms (39), and depletion of ascorbate (40) are reported in serum and synovial fluid, together with increased exhalation of pentane (41).

Lymphocytes are known to be highly sensitive to oxidative stress, and they are exposed to oxidative conditions during the course of their normal function at sites of inflammation. On the basis of findings in HIV patients (42–44), we hypothesized that a chronic increase in the production of ROS in RA might lead to a depletion of the antioxidative capacity of the T cells, which leads to a dysfunctioning of the T cells at the site of inflammation.

To study the antioxidant capacity of T cells in RA at the site of inflammation, we measured the levels of the two key redox mediators: intracellular glutathione (GSH) and extracellular thioredoxin (TRX). GSH is an intracellular nonprotein thiol with both oxidant-scavenging and redox-regulating capacities. Modulation of intracellular GSH levels has previously been shown to influence T cells function both in vitro (45) and in AIDS patients in vivo (42–44). TRX, a 13 κDa protein, is another major antioxidant which catalyzes dithiol/disulfide exchange reactions in a complex with TRX reductase (46,47). TRX has previously been reported to be induced under conditions of oxidative stress as determined in vitro (47,48) and in AIDS patients in vivo (49).

RESULTS

Intracellular GSH Levels of RA Synovial Fluid T Cells Are Decreased Compared with Peripheral Blood T Cells

Intracellular levels of GSH within CD4-, CD8-, CD14-, and CD20-positive RA peripheral blood (PB) and synovial fluid (SF) cell subsets (Figure 2) of 12 RA patients were determined by means of a flow cytometric assay as previously described (42). Significantly decreased levels of intracellular GSH were found within SF $CD4^+$ and $CD8^+$ T cell subsets when compared to GSH levels in PB $CD4^+$ ($p < 0.005$) and

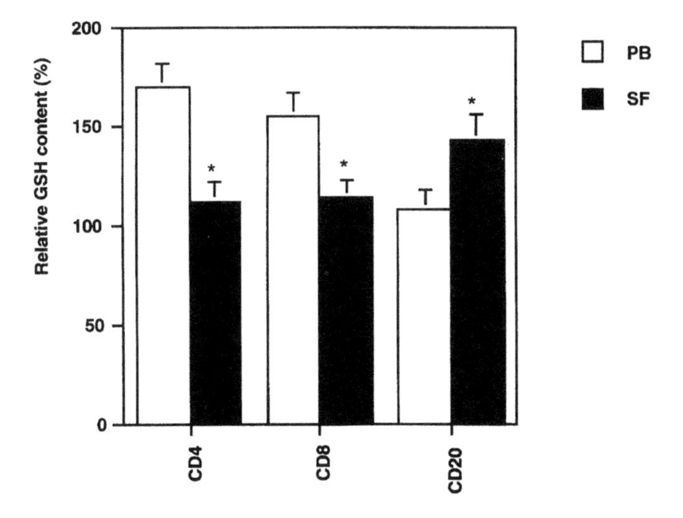

Figure 2. Relative GSH contents of $CD4^+$, $CD8^+$ and $CD20^+$ peripheral blood (PB) and synovial fluid (SF) lymphocytes of 12 RA patients as determined in a flow cytometric assay. SF $CD4^+$ and $CD8^+$ T cells have significantly decreased intracellular GSH levels when compared with PB $CD4^+$ ($p < 0.005$) and $CD8^+$ ($p < 0.005$) T cells. In contrast, $CD20^+$ SF B cells were found to have significantly increased levels of GSH ($p < 0.005$). Statistical differences were determined in a paired Wilcoxon signed-rank test.

$CD8^+$ ($p < 0.005$) T cells. These findings were confirmed by use of a modified enzymic recycling assay according to Tietze (50,51) to determine intracellular GSH in lysates of purified PB and SF T cells. We could not detect major differences in GSH levels of $CD45RA^+$ compared to $CD45RO^+$ lymphocytes within the blood of 10 RA patients, excluding the enrichment of $CD45RO^+$ T cells in the SF as an explanation of the above findings. Within PB and SF $CD14^+$ monocytes, no differences were found with respect to intracellular GSH levels ($p > 0.05$), whereas, in contrast to the findings for T cells, a significant increase of intracellular GSH was found in SF $CD20^+$ B cells when compared to PB $CD20^+$ B cells ($p < 0.005$, Figure 1). However, the latter observation could possibly be explained by an increase in cell size of the SF B cell subset when compared to PB B cells as determined by increase in forward and side scatter.

TRX Levels Are Significantly Elevated in SF of RA Patients When Compared with Plasma Levels: A Specific Feature of RA?

As a second marker of an altered redox balance in the joints of RA patients, levels of TRX were measured in paired plasma and SF samples of 12 RA patients by means of a sandwich ELISA as previously described (49). The levels of TRX were significantly elevated in the SF when compared to plasma ($p < 0.01$). Mean TRX levels were 122 ± 74 and 282 ± 147 ng/ml for plasma and SF samples, respectively.

Furthermore, preliminary studies have shown that TRX levels in the SF of RA patients were significantly increased when compared with a group of 9 patients with other inflammatory arthritides. These findings suggest a local increase in the secretion of TRX at the site of inflammation that is rather specific for RA patients.

However, plasma TRX levels of RA patients were also elevated when compared with the plasma TRX levels of the non-RA group ($p < 0.05$). This finding could reflect either an additional systemic effect of the chronic inflammation or, alternatively, a leakage of TRX from the inflamed joint to the plasma of the RA patients.

Effects of In Vitro GSH Modulation of SF T Cell Function In Vitro

Intracellular GSH levels can be elevated in vitro by supplementation of cells with *N*-acetylcysteine (NAC), a thiol-containing compound that can serve as a precursor for GSH synthesis besides acting as an oxidant scavenger itself. A decrease of the intracellular GSH level can be obtained by culturing cells in the presence of L-buthionine (*S*,*R*)-sulfoximine (BSO), a specific inhibitor of the enzyme γ-glutamylcysteine synthetase which catalyzes the rate-limiting step in the de novo GSH synthesis.

To determine the implication of a change in intracellular GSH levels on the function of PB or SF T cells, we first tested T cell proliferation after in vitro modulation of intracellular GSH levels. T cells were incubated for 24 h with NAC or BSO, subsequently stimulated with anti-CD3 and anti-CD28 antibodies, and tested for proliferative capacity as determined by $[^3H]$thymidine incorporation during the last 16 h of a 72 h culture period. In Figures 3A and 3B the results are shown for T cells from two healthy individuals. Normal T cells supplemented with NAC showed an optimal increase in proliferation in the presence of 1–2.5 mM NAC, whereas GSH depletion with 200 μM BSO led to clearly decreased proliferative responses (Figures 3A and 3B). T cells from the peripheral blood of two RA patients showed a similar pattern of proliferative responses in the presence of

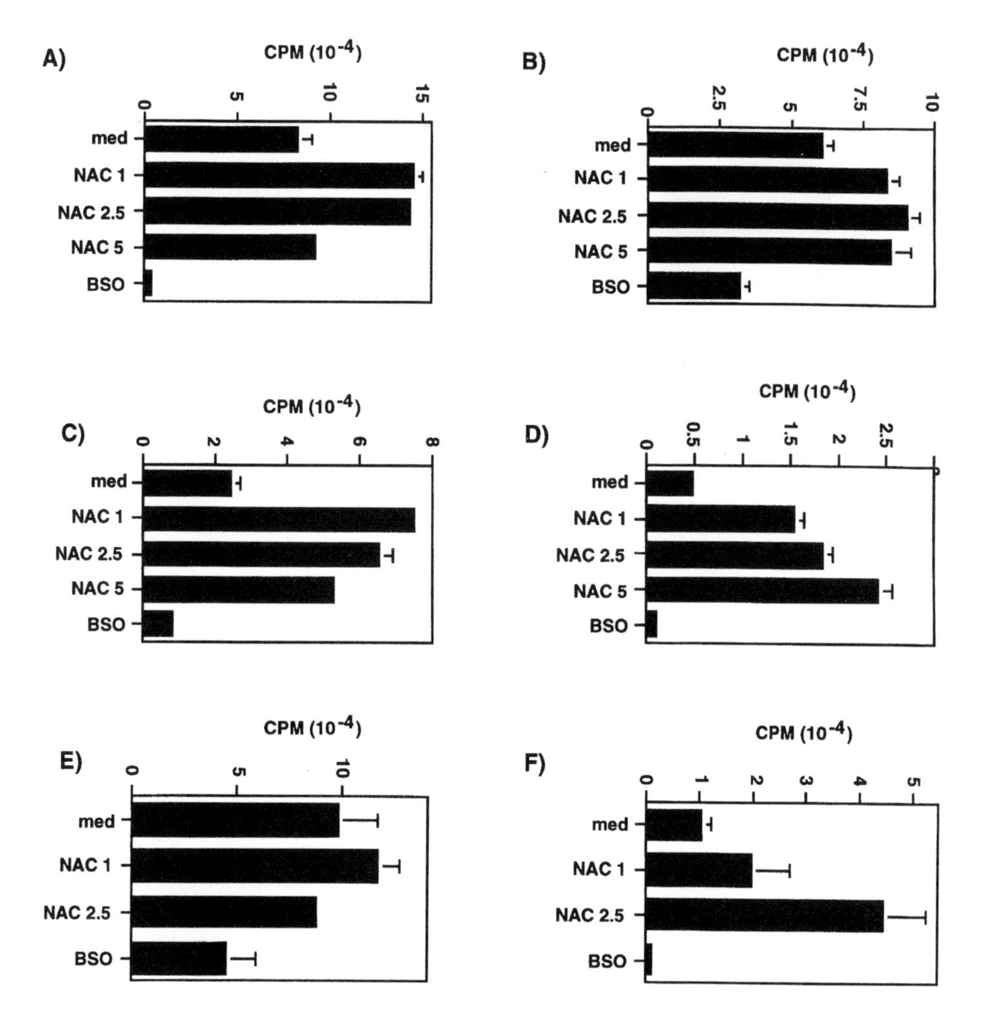

Figure 3. Modulation of GSH by NAC and BSO influences proliferation of purified PB and SF T cells. Representative studies are shown for PB T cells from two healthy individuals stimulated with anti-CD3 and anti-CD28 mAb (1 μg/ml) in the presence of graded amounts of NAC or 200 μM BSO (A and B) and T cells from PB (C and E) and SF (D and F) from two RA patients.

graded amounts of NAC. Optimal responses could be found with supplementation of cells with 1–2.5 mM NAC and suppression of response was obtained with 200 μM BSO (Figures 3C and 3D). As anticipated from GSH measurements in SF T cells, the SF T cells revealed optimal proliferation at higher concentrations of NAC (2.5–5 mM) which was also found to be more pronounced in terms of stimulation index, mostly 4- to 5-fold higher than basal responses (Figures 3E and 3F).

To confirm modulation of intracellular GSH levels in vitro, SF T cells were incubated for 24 h with NAC or BSO, after which intracellular GSH levels were determined. As shown in Figure 4, NAC could upregulate GSH levels within SF T cells, whereas a decrease in GSH content was observed in the presence of BSO.

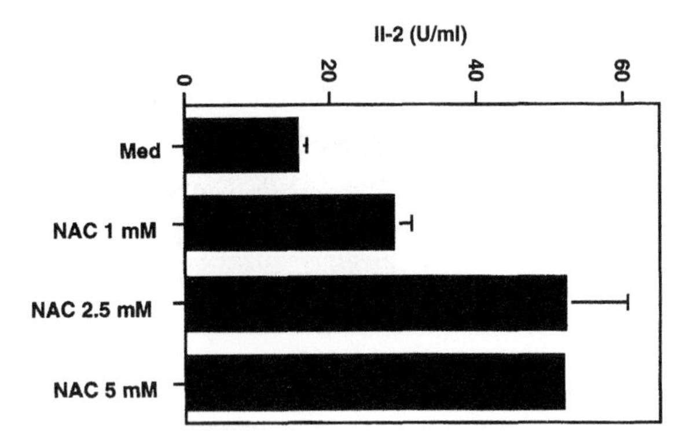

Figure 4. L-2 production of SF T cells after increase of intracellular GSH levels by NAC. T cells were stimulated for 24 h with anti-CD3 and anti-CD28 mAb (1 μg/ml) in the presence of graded amounts of NAC.

As a second measure of T cell function, we determined the IL-2 production of SF T cells after supplementation of the cultures with NAC (Figure 4). In addition to proliferative responses, IL-2 production of SF T cells upon stimulation with anti-CD3 and anti-CD28 was increased when compared with basal levels after treatment of cells with graded amounts of NAC.

CONCLUSIONS

Summary

In summary, the results presented here provide evidence for an altered redox state of T cells in the SF of RA patients in terms of decreased intracellular GSH levels and increased extracellular levels of TRX. These findings are likely to be a consequence of chronic exposure to oxidative stress. The decrease in intracellular GSH levels was found to be specific for T cells in the SF, underscoring the sensitivity of T cells for oxidative conditions. TRX levels in the SF of RA patients were significantly higher than the levels determined in SF of patients with other inflammatory joint diseases, indicating that secretion of high levels of TRX is a specific feature of the chronic pro-oxidant conditions in RA. Support for functional consequences of an altered redox balance in SF T cells was provided by the observation of much higher SF T cell proliferative responses and higher amounts of IL-2 production in the presence of NAC, a replenisher of GSH and antioxidant by itself.

Oxidative Stress: Possible Role in Pathogenesis of RA?

Under chronic inflammatory conditions, long-term exposure to oxidants might lead to chronic changes in the delicate balance of the pro-oxidant and antioxidant status of the T cell and consequently counteract the initially stimulatory effects of oxidative stress. In this respect it is of interest to mention that depletion of antioxidants predisposes to the development of human autoimmune diseases (52,53). In vivo evidence for a role of an altered redox balance in RA is demonstrated upon treatment of RA patients with potential aliphatic thiols with strong reducing properties, which has proved to be beneficial (54–56).

In RA, TNF-α has been proposed to act as a suppressor of T cell function with respect to proliferation and cytokine production when chronically present in T cell cultures (57), or to induce a permanent arrest of SF T cells in the S phase of the cell cycle (58). It is of interest that TNF-α was previously shown to induce oxidative stress and can also lower intracellular GSH levels (59). Furthermore, cells with decreased GSH levels were observed to become arrested in the S and G2 phases of the cell cycle (60).

Treatment of patients with monoclonal antibodies directed against TNF-α has previously been shown to be beneficial in RA. In association with an improved clinical status, in vitro proliferative responses of T cells from RA patients responding to anti-TNF-α were found to be improved. It is therefore tempting to speculate that a relief of chronic oxidative stress may be held responsible for the restoration of T cell function upon anti-TNF-α treatment in RA patients.

In accordance with the disappointing results of T cell-directed therapies in RA, improvement of T cell function in RA might possibly turn on regulatory, protective T cell subsets which normally regulate the course of an inflammation. The precise role of downregulated T cells in established RA remains to be resolved.

REFERENCES

1. Forre O, Thoen J, Lea T, et al. In situ characterization of mononuclear cells in rheumatoid arthritis, using monoclonal antibodies. Scand J Immunol 1982; 16:315–319.
2. Poulter LW, Duke O, Panayi GS, Hobbe S, Raftery MJ, Janossy G. Activated T lymphocytes of the synovial membrane in rheumatoid arthritis and other arthropathies. Scand J Immunol 1988; 22:683–690.
3. Keystone EC, Snow KM, Bombardier C, Chang C, Nelson DL, Rubin LA. Elevated soluble interleukin-2 receptor levels in the sera and synovial fluids of patients with rheumatoid arthritis. Arthritis Rheum 1988; 31:844–849.
4. Nakao H, Eguchi K, Kawakami A, et al. Phenotypic characterization of lymphocytes infiltrating synovial tissue from patients with rheumatoid arthritis. J Rheumatol 1990; 17:142–148.
5. Pincus SH, Clegg DO, Ward JR. Characterization of T cells bearing HLA-DR antigens in rheumatoid arthritis. Arthritis Rheum 1985; 28:8–15.
6. Pitzalis C, Kingsley G, Haskard D, Panayi G. The preferential accumulation of helper-inducer T lymphocytes in inflammatory lesions: evidence for regulation by selective endothelial and homotypic adhesion. Eur J Immunol 1988; 18:1397–1404.
7. Postigo AA, Garcia-Vicuna R, Laffon A, Sanchez-Madrid F. The role of adhesion molecules in the pathogenesis of rheumatoid arthritis. Autoimmunity 1993; 16:69–76.
8. Potocnik AJ, Kinne R, Menninger H, Zacher J, Emmerich F, Kroczek RA. Expression of activation antigens on T cells in rheumatoid arthritis patients. Scand J Immunol 1990; 31:213–224.
9. Rodriguez RM, Pitzalis C, Kingsley GH, Henderson E, Humphries MJ, Panayi GS. T lymphocyte adhesion to fibronectin (FN): a possible mechanism for T cell accumulation in the rheumatoid joint. Clin Exp Immunol 1992; 89:439–445.
10. Smith MD, Roberts-Thomson PJ. Lymphocyte surface marker expression in rheumatoid diseases: evidence for prior activation of lymphocytes in vivo. Ann Rheum Dis 1990; 49:81–87.

11. Takahashi H, Söderström K, Nilsson E, Kiessling R, Patarroyo M. Integrins and other adhesion molecules on lymphocytes from synovial fluid and peripheral blood of rheumatoid arthritis patients. Eur J Immunol 1992; 22:2879–2885.
12. Wood NC, Symons JA, Duff GW. Serum interleukin-2-receptor in rheumatoid arthrits: a prognostic indicator of disease activity? J Autoimmun 1988; 1:353–361.
13. Londei M, Savill CM, Verhoef A, et al. Persistence of collagen type II specific T cell clones in the synovial membrane of a patient with rheumatoid arthritis. Proc Natl Acad Sci USA 1989; 86:636–640.
14. Klimiuk PS, Clague RB, Grennan DM, Dyer PA, Smeaton I, Harris R. Autoimmunity to native type II collagen: a distinct genetic subset of rheumatoid arthritis. J Rheumatol 1985; 12:865–870.
15. Golds EE, Stephen IBM, Esdaile JM, Strawczynski H, Poole AR. Lymphocyte transformation to connective tissue antigens in adult and juvenile rheumatoid arthritis, osteoarthritis, ankylosing spondylitis, systemic lupus erythematosis, and a non-arthritic control population. Cell Immunol 1983; 82:196–209.
16. Alsalameh S, Mollenhauer J, Hain N, Stock K-P, Kalden JR, Burmester GR. Cellular immune response toward human articular chondrocytes. Arthritis Rheum 1990; 33:1477–1486.
17. Mollenhauer J, von der Mark K, Burmester GR, Gluckert K, Lutjen-Drecoll E, Brune K. Serum antibodies against chondrocyte cell surface proteins in osteoarthritis and rheumatoid arthrits. J Rheumatol 1988; 15:1811–1817.
18. Dedeoglu F, Kaymaz H, Seaver N, Schluter SF, Yocum DE, Marchalonis JJ. Lack of preferential Vb usage in synovial T cells of rheumatoid arthritis patients. Immunol Res 1993; 12:12–20.
19. Olive C, Gatenby PA, Serjeantson SW. Analysis of T cell receptor Va and Vb gene usage in synovia of patients with rheumatoid arthritis. Immunol Cell Biol 1991; 69:349–354.
20. Struyk L, Kurnick JT, Hawes GE, et al. T-cell receptor V-gene usage in synovial fluid of patients with chronic arthritis. Hum Immunol 1993; 37:237–251.
21. Nykanen P, Bergroth V, Raunio P, Nordstrom D, Kontinen YT. Phenotypic characterizatin of ^{3}H-thymidine incorporating cells in rheumatoid arthritis synovial membrane. Rheumatol Int 1986; 6:269–271.
22. Haraoui B, Wilder RL, Malone DG, Allen JB, Katona IM, Wahl SM. Immune function in severe, active rheumatoid arthritis: a relationship between peripheral blood mononuclear cell proliferation to soluble antigens and mononuclear cell subset profiles. J Immunol 1984; 133:697–701.
23. Kingsley GH, Pitzalis C, Panayi GS. Abnormal lymphocyte reactivity to self-major histocompatibility antigens in rheumatoid arthritis. J Rheumatol 1987; 14(4):677–683.
24. Verwilghen J, Vertessen S, Stevens EAM, Dequeker J, Ceuppens JL. Depressed T cell reactivity to recall antigens in rheumatoid arthritis. Clin Immunol 1990; 10:90.
25. Keystone EC, Poplonski L, Miller RG, Gorczynski R, Gladman D, Snow K. Reactivity of T cells from patients with rheumatoid arthritis to anti-CD3 antibody. Clin Immunol Immunopathol 1988; 48(3):325–337.
26. Pope RM, McChesney L, Talal N, Fischbach M. Characterization of the defective autologous mixed lymphocyte response in rheumatoid arthritis patients. Arthritis Rheum 1984; 27(11):1234–1244.
27. Mirza NM, Relias V, Yunis EJ, Pachas WN, Dasgupta JD. Defective signal transduction via T-cell receptor–CD3 structure in T cells from rheumatoid arthritis patients Hum Immunol 1993; 36:91–98.
28. Allen ME, Young SP, Michell RH, Bacon PA. Altered T lymphocyte signaling in rheumatoid arthritis. Eur J Immunol 1995; 25:1547–1554.
29. Firestein GS, Xu WD, Townsend K, et al. Cytokines in chronic inflammatory arthritis. I. Failure to detect T cell lymphokines (Il-2 and I1–3) and presence of macrophage colony-stimulating factor (CSF-1) and novel mast cell growth factor in rheumatoid synovitis. J Exp

Med 1988; 168:1573–1586.

30. Firestein GS, Zvaifler NJ. Peripheral blood and synovial fluid monocyte activation in inflammatory arthritis. II. Low levels of synovial fluid and synovial tissue interferon suggest that gamma-interferon is not the primary macrophage activating factor. Arthritis Rheum 1987; 30:864–871.
31. Firestein GS, Zvaifler NJ. How important are T cells in chronic rheumatoid arthritis? Arthritis Rheum 1990; 33(6):768–773.
32. van der Lubbe PA, Dijkmans BAC, Markusse HM, Nassander U, Breedveld FC. A randomized, double-blind, placebo-controlled study of CD4 monoclonal antibody therapy in early rheumatoid arthritis. Arthritis Rheum 1995; 38:1097–1106.
33. Kirkham BW, Thien F, Pelton BK, et al. Chimeric CD7 monoclonal antibody therapy in rheumatoid arthritis. J Rheumatol 1992; 19:1348–1352.
34. Moreland LW, Sewell KL, Trentham DE, et al. Interleukin-2 diphtheria fusion protein (DAB486IL-2) in refractory rheumatoid arthritis: a double-blind, placebo-controlled trial with open-label extension. Arthritis Rheum 1995; 38:1177–1186.
35. Olsen NJ, Cush JJ, Lipsky PE, et al. Multicenter trial of an anti-CD5 immunoconjugate in rheumatoid arthritis (RA). Arthritis Rheum 1994; 37:S295.
36. Mosmann TR, Coffman RL. Th1 and Th2 cells: different patterns of lymphokine secretion lead to different functional properties. Annu Rev Immunol 1989; 7:145–173.
37. Merry P, Grootveld M, Lunec J, Blake DR. Oxidative damage to lipids within the inflamed human joint provides evidence of radical-mediated hypoxic-reperfusion injury. Am J Clin Nutr 1991; 57:362S–369S.
38. Rowley DA, Gutteridge JMC, Blake DR, Farr M, Halliwell B. Lipid peroxidation in rheumatoid arthritis: thiobarbituric acid-reactive material and catalytic iron salts in synovial fluid from rheumatoid patients. Clin Sci 1984; 66:691–695.
39. Grootveld M, Henderson EB, Farell A, Blake DR, Parkes HG, Haycock P. Oxidative damage to hyaluronate and glucose in synovial fluid during exercise of the inflamed rheumatoid joint. Biochem J 1991; 273:459–467.
40. Lunec J, Blake DR. The determination of dehydroascorbic acid and ascorbic acid in serum and synovial fluid of patients with rheumatoid arthritis. Free Radical Res Commun 1985; 1:31–39.
41. Humad S, Zarling E, Clapper M, Skosey JK. Breath pentane excretion as marker of disease activity in rheumatoid arthritis. Free Radical Res Commun 1988; 5:101–106.
42. Roederer M, Staal FJT, Osada H, Herzenberg LA. CD4 and CD8 T cells with high intracellular glutathione levels are selectively lost as the HIV infection progresses. Int Immunol 1991; 3:933–937.
43. Staal FJT, Ela SW, Roederer M, Anderson MT, Herzenberg LA. Glutathione deficiency and human immunodeficiency virus infection. Lancet 1992; 339:909–912.
44. Staal FJT, Anderson MT, Staal GEJ, Herzenberg LA, Gitler C. Redox regulation of signal transduction: tyrosine phosphorylation and calcium influx. Proc Natl Acad Sci 1994; 91(9):3619–3622.
45. Suthanthiran M, Anderson ME, Sharma VK, Meister A. Glutathione regulates activation-dependent DNA synthesis in highly purified normal human T lymphocytes stimulated via the CD2 and CD3 antigens. Proc Natl Acad Sci USA 1990; 87:3343–3347.
46. Holmgren A. Thioredoxin. Annu Rev Biochem 1985; 54:237–271.
47. Holmgren A. Thioredoxin and glutaredoxin systems. J Biol Chem 1989; 264(24):13963–13966.
48. Yodoi J, Tursz T. ADF, a growth-promoting factor derived from an adult T cell leukemia and homologous to thioredoxin: involvement in lymphocyte immortalization by HTLV-1 and EBV. Adv Cancer Res 1991; 57:381–411.
49. Nakamura H, De Rosa S, Roederer M, et al. Elevation of plasma thioredoxin levels in HIV-infected individuals. Intl Immunol 1996; 8(4):603–611.

50. Tietze F. Enzymic method for quantitative determination of nanogram amounts of total and oxidized glutathione: applications to mammalian blood and other tissues. Anal Biochem 1969; 27:502–522.
51. Griffith OW. Determination of glutathione and glutathione disulfide using glutathione reductase and 2-vinylpyridine. Anal Biochem 1980; 106:207–212.
52. Heliövaara M, Knekt P, Aho K, Alfthan G, Aromaa A. Serum antioxidants and risk of rheumatoid arthritis. Ann Rheum Dis 1994; 53:51–53.
53. Salonen J, Nyyssönen K, Tuomainen T-P, et al. Increased risk of non-insulin dependent diabetes mellitus at low plasma vitamin E concentrations: a four year follow up study in men. Br Med J 1995; 311:1124–1127.
54. Huck F, Médicis R, Lussier A, Dupuis G, Federlin P. Reducing properties of some slow acting antirheumatic drugs. J Rheum 1984; 11(5):605–609.
55. Halliwell B, Hoult JR, Blake DR. Oxidants, inflammation and anti-inflammatory drugs. FASEB J 1988; 2:2867–2873.
56. Munthe E, Kass E, Jelum E. D-Penicillamine induced increase in intracellular glutathione correlating to response in rheumatoid arthritis. J Rheumatol 1981; 8(supplement 7):14–19.
57. Cope AP, Londei M, Chu NR, et al. Chronic exposure to tumor necrosis factor (TNF) in vitro impairs the activation of T cells through the T cell receptor/CD3 complex; reversal in vivo by anti-TNF antibodies in patients with rheumatoid arthritis. J Clin Invest 1994; 94:749–760.
58. Lai N-S, Lan J-L, Yu C-L, Lin R-H. Role of tumor necrosis factor-α in the regulation of activated synovial T cell growth: downregulation of synovial T cells in rheumatoid arthritis patients. Eur J Immunol 1995; 25:3243–3248.
59. Ishii Y, Partridge CA, Del Vecchio PJ, Malik AB. Tumor-necrosis factor-α-mediated decrease in glutathione increases the sensitivity of pulmonary vascular endothelial cells to H_2O_2. J Clin Invest 1992; 89:794–802.
60. Poot M, Teubert H, Rabinovitch S, Kavanagh TJ. De novo synthesis of glutathione is required for both entry into and progression through the cell cycle. J Cell Physiol 1995; 163:5

Index

AAPH
  apoptosis, 13
  cell growth, 11-12
Acquired immunodeficiency syndrome. *see* AIDS
Actinomycin D, NFκB, 68
Activator protein-1 (AP-1), 518-24
  rheumatoid synovium, 524-526
Adenine nucleotide translocator (ANT), 323
Adriamycin, reactive oxygen species (ROS), 370
Advanced glycation end-product (AGE), 486
Aging
  heat stress protein (HSP), 121-122
  protein oxidation, 491-492
AH109A carcinoma cells, 369-370
AIDS. *see also* HIV; HIV-1
  antioxidants, 107
  atherosclerosis, 409-423
  β-carotene, 413-423
  glutathione (GSH), 379-386, 429-434
  Immunocal, 452-455
  L-carnitine, 437-444
  lipoate, 262-263
  N-acetylcysteine (NAC), 379-386
  NFκB, 76, 83
  precautions, 386
  selenium, 413-423
α-lipoic acid. *see* Lipoate
Alzheimer's disease
  pathology, 487-488
    oxidative stress, 488-489
Aminotriazole, NFκB, 84-85
Amsacrine, NFκB, 68
Amyotrophic lateral sclerosis (ALS), oxidative stress, 490-491
Antigen receptor signals, lymphocytes, 35-37, 40
Anti-inflammatory drugs, NFκB, 529-531
Antimycin, 274, 277, 280, 339, 345
Anti-NFκB compounds, screening, 80
Antioxidant defense, HIV, 413
Antioxidant enzymes
  assays, 324
  perinatal development, 323-330
Antioxidant enzyme studies
  rat-brain mitochondria, methods, 324-325
  rat brain mitochondria, results, 325-330
Antioxidants, 113-114, 224-225
  defined, 224
  HIV, 105
  hydrogen peroxide (H_2O_2), 98
  iron, 225-226
  NFκB, 92-94, 258
Apoliprotein A1, CD4 T cells, 411
Apoptosis, 120
  ceramide, 477
  defined, 179

[Apoptosis]
deoxy-D-ribose (dRib), 199-203
desferal, 202
Fas-Fas ligand (FasL) system, 438-439
flow cytometry, 200
free radicals, 12-14
glutathione (GSH), 180-187
hydrogen peroxide (H_2O_2), 194-195, 205-210
IMMUNOCAL™, 451-452
L-Buthionine-(S, R)-sulfoxine (BSO), 202
L-carnitine, 440-444
lipoic acid, 209
mitochondria, 213-219
mitochondrial inner transmembrane potential, 214-215
nitric oxide (NO), 13-14, 205-210, 499
nonoxidative stimuli, 179-180
Parkinson's disease, 469-480
permeability transition, 214-215
phases, 213
phosphotyrosine phosphatases (PTP), 38
quercetin, 292
radiation-induced, 15-26
redox regulation, 195-196, 518
signal transduction pathways, 17
SOD, 208-210
tumor necrosis factor-α (TNF-α), 476-478
Arachidonic acid (AA), 389
bombesin/gastrin-releasing peptides (GRP), 167-168
flavonoids, 292
HIV-1 long terminal repeat (LTR), 389-396
NFκB, 527
vasopressin, 167-168
Arachidonic acid (AA) metabolites, release of, 46-47
Arginine, *N*-acetylcysteine (NAC), 376
Ascorbate, dihydrolipoate (DHLA), 256
Atherosclerosis, AIDS, 409-423
ATP hydrolysis, mitochondrial matrix, 344-345
Aurothioglucose, NFκB, 529
Azide, NFκB, 84-85

Bacterial infection, heat stress protein (HSP), 121-122
Bathocuproine disulfonate (BCPS), glutathione (GSH), 183
Bathophenanthroline disulfonate (BPS), glutathione (GSH), 183
β-carotene, AIDS, 413-423
B cell activation, BCR-mediated, FcγRIIB, 149-150
B cell receptors (BCR), 147-149, 153-154
B cells
antigen receptor signals, 37
ultraviolet radiation, 38
Bcl-2
mitochondria, 216-218
oxidative stress, 120
bis(maltolato)oxovanadium(IV) (BMOV), 40-42
Bleomycin, NFκB, 69
Bombesin, mitogen-activated protein (MAP) kinases, 167
Bombesin/gastrin-releasing peptides (GRP)
arachidonic acid, 167-168
Ca^{2+}, 166-167
DAG, 166
epidermal growth factor (EGF) receptor, 167
function, 163-164
inositol phosphatidyl, 166-167
insulin, 164
PLC, 170-172
prostaglandin, 167-168
protein kinase C, 166-167
proto-oncogenes, 168
rat-1 cells, 170-172
receptors, 165-166
subfamilies, 164
Swiss 3T3 cells, 164-172
tumoregenesis, 164
tyrosine kinase pathway, 170-172
tyrosine phosphorylation, 168-170
Bovine pulmonary artery endothelial cells (BPAECs), phospholipase modulation, 45-58
Brain, manganese superoxide dismutase (Mn-SOD), 308
Breast cancer cells, IMMUNOCAL™, 457
BRLP-42, 122
Bromosulfophthalein (BSP), glutathione (GSH), 183
Buthionine sulfoximine (BSO)
glutathione (GSH), 457
IMMUNOCAL™, 449
BXT-51072k, 314-320

Ca^{2+}, bombesin/gastrin-releasing peptides (GRP), 166-167
Calcium flux
flow cytometry, 258
hydrogen peroxide (H_2O_2), 258

[Calcium flux]
Wurzburg vs. Jurkat cells, 258
Camptothecin, NFκB, 68
Cancer
flavonoids, 290-291
heat stress protein (HSP), 121-122
Cancer cells, proliferation, 163
Cancer clinical trials, IMMUNOCAL™, 457-458
Cancer treatments, free radicals, 369-371
Catalase
enzyme activities, perinatal brain development, 323-328
erythrocyte hemolysis, 11
hydrogen peroxide (H_2O_2), 194
ischemia-reperfusion injury, 370
superoxide, 194
Cataract, etiology, 507-509
Catechin, 289
CD4 T cells, 409
apoliprotein A1, 411
fatty acids, 412
GSH, 379-386
HDL cholesterol, 411
lipidic anomalies, 412
lipids, 410-413
liprotein A, 412
triglycerides, 411
Cell activation
IgG antibodies, 147-157
ITAM-mediated, FcγRIIB, 151
Cell death. *see* Apoptosis
Cell growth, 163
flavonoids, 291
free radicals, 11-14
hydrogen peroxide (H_2O_2), 191
redox regulation, 192, 195
superoxide, 191
Cell regulation, hydrogen peroxide (H_2O_2), 99
Cell shrinkage, glutathione (GSH), 184, 186
Ceramide
apoptosis, 477
Fas-Fas ligand (FasL) system, 437
HIV infection, 439-440
L-carnitine, 440-444
tumor necrosis factor-α (TNF-α), 438-439
Ceruloplasmin, 226
c-fos, bombesin/gastrin-releasing peptides (GRP), 168
CGMP, low-density lipoproteins (LDL), 130-132
Chronic fatigue syndrome
cysteine, 375
glutamine, 375
Chronic gastritis, thioredoxin (TRX), 363-364, 366
Chronic hepatitis B-related cirrhosis, HIV-1, 463-466
Chronic hepatitis (CH), thioredoxin (TRX), 360, 362
Chronic inflammatory diseases, thioredoxin (TRX) serum levels, 359-366
c-myc, bombesin/gastrin-releasing peptides (GRP), 168
Colon cancer cells, glutathione (GSH), 455-456
Copper, HIV, 419
Cu, An-SOD, familial amyotrophic lateral sclerosis, 500
Cu, Zn-SOD
diabetes, 498-499
glycation reaction, 497-499
HIV-1, 307, 310, 329-330, 490
Werner's syndrome, 498
Cyanidanol
hepatoprotective effect, 287
immunomodulation, 292
Cyanide, NFκB, 84-85
Cyclosporin A, permeability transition (PT), 269
Cysteine
chronic fatigue syndrome, 375
DHLA, 263
HIV, 262, 373-376
milk, 458
N-acetylcysteine (NAC), 376
Cysteine delivery system, IMMUNOCAL™, 448-449, 452
Cytochrome oxidase, 329-330

DAG
bombesin/gastrin-releasing peptides (GRP), 166
hydrogen peroxide (H_2O_2), 49-50
Daunomycin, NFκB, 68-70
Deferoxamine
NFκB, 258
tumor necrosis factor-α (TNF-α) 102, 104
Deoxy-D-ribose (dRib)
apoptosis, 199-203
GSH, 199-203
Desferal, apoptosis, 202
Diabetes, Cu, Zn-SOD, 498-499
Dihydrolipoamide dehydrogenase, 253
Dihydrolipoate (DHLA), 253-254
antioxidant properties, 254-256

[Dihydrolipoate (DHLA)]
ascorbate, 256
cysteine, 263
NFκB, 258, 261
recycling oxidized antioxidants, 256-257
superoxides, 255-256
vitamin E, 256-257
Dimethylthiourea (DMTU), tumor necrosis factor-α (TNF-α)100-101
DMSO
adriamycin, 370-371
hyperthermia, 370
DNA, NFκB, 68-70

E. coli
8-OxoG, 353-354
glutaredoxin (GRX), 233-234
hydrogen peroxide (H2O2), 29-32
oxidative regulation, 104
oxidative stress response, 29-32
thioredoxin (TRX), 230-231, 247, 359
Ebselen, 314
Eicosanoid cascade, HIV, 395
Eicosapentaenoic acid (EPA), 389
8-oxo-7,8-dihycho.2'-deoxyguanosine (Oxo8dG), 430, 432
8-oxoguanine (8-OxoG)
formation, 351-352
mutagenesis, 352-353
repair, 353-355
Electron leak pathway, superoxide, 334
Electron univalent leak, 333
Electrophoresis, 324-325
Electrophoretic mobility shift assay (EMSA), 391, 393
Endothelial cells
hydrogen peroxide (H_2O_2), 313-321
low density lipoproteins (LDL), 129, 134-135
nitric oxide (NO), 129-130, 134-135
tumor necrosis factor-α (TNF-α), 313-321
Enzyme-linked immunosorbent assay (ELISA), thioredoxin (TRX), 360-361, 365
Epidermal growth factor (EGF) receptor, bombesin/gastrin-releasing peptides (GRP), 167
Epstein-Barr virus (EBV), gastric cancer, 366
Erythrocytes
antioxidant supplementation, 413-423
hemolysis, 10-11
ESR spin probes, free radicals, 13
Etoposide, NFκB, 68-70
Eukaryotes, 8-OxoG repair, 354-355

Familial amyotrophic laterial sclerosis, CU, An-SOD, 500
Fas-Fas ligand (FasL) system
apoptosis, 438-439
ceramide, 437
HIV infection, 439
Fatty acids
CD4 T cells, 412
erythrocyte membrane, HIV, 421-422
Fc cell receptors (FCR), 147-149, 154-156
FcγRIIB, 148-149, 156-157
BCR-mediated B cell activation, 149-150
FCR-mediated mast cell activation, 150
function, 148-149
inhibitory biochemical mechanisms, 153-154
inhibitory biochemical significance, 154-156
ITAM-based immunoreceptor cell activation, 151
ITIM, 151-153
TCR-mediated T cell activation, 150
Flavonoids, 285-293
enzymes, 288-290
inhibition, 288-289
stimulation, 289-290
functions, 285-286
gene expression, 292
metabolism, 286-287
physiological effect
anticarcinogenicity, 290-291
immunomodulation, 292
inflammatory effects, 291-292
signal transduction, 292
physiological effects, 287, 290-293
protein interactions, 287-288
receptors, 288
Flow cytometry
apoptosis, 200
calcium flux, 258
Focal adhesion kinase ($p125^{fak}$), tyrosine phosphorylation, 168-170
Fpg protein, 354-355
Free radicals
apoptosis, 12-14
radiation-induced, 15-26
cancer treatments, 369-371
cell growth, 11-12
reactivity, 9-10
Fructose, glycation, 500-501
Gastric cancer, etiology, 366

General adaptation syndrome, 503
Genistein, 289
phospholipase D (PLD) activation, 54-57
Glutamine
chronic fatigue syndrome, 375
N-acetylcysteine (NAC), 376
Glutaredoxin (GRX), 233-234
oxidative stress, 234-235
Glutathione (GSH)
AIDS, 379-386, 429-434
apoptosis, 180-187, 199-203
buthionine sulfoximine (BSO), 457
CD4 T cells, 379-386
colon cancer cells, 455-456
deoxy-D-ribose (dRib), 199-203
HIV, 262-263, 373-376
HPLC, 200-201
hydrogen peroxide (H_2O_2), 195
IMMUNOCAL™, 447-448, 450-452
inhibitors, 182-185
L-buthionine-(S, R)-sulfoxine (BSO), 202
lipoate, 257, 263
lung cancer cells, 455
N-acetylcysteine (NAC), 202
NFκB, 78, 86, 103
rheumatoid arthritis (RA), 539-540
transporters, 181, 183
Glutathione (GSH) compartmentation, oxidative stress, 269-281
Glutathione peroxidase-1 (GSHPx-1), 509-514
Glutathione peroxidase (GPX)
enzyme activities, perinatal brain development, 323-332
hydrogen peroxide (H_2O_2), 194, 314
nitric oxide (NO), 499-500
Glutathione reductase (GR), enzyme activities, perinatal brain development, 323-332
Glutathione S-transferase (GT), enzyme activities, perinatal brain development, 323-332
Glycation, fructose, 500-501
Glycation reaction, 497
Gold ions, NFκB, 79

$H^+/2e^-$, superoxide radiacls, 345
H^+-cycle, 333-347
HDL cholesterol, CD4 T cells, 411
Heat production, superoxide radicals, 346-347
Heat shock, 115, 117
Heat shock protein (HSP)
composition, 113
[Heat shock protein (HSP)]
protective function, 114, 119-122
redox regulation, 113-122
Heat shock response, quercetin, 292
Helicobacter pylori infection, 360
gastric cancer, 366
thioredoxin (TRX), 364
Hematogenic cancer cell metastasis, NFκB, 76, 80
Heme oxygenase-1
low-density lipoproteins (LDL), 132-134
western blot analysis, 129-130
Heme oxygenase, Alzheimer's disease and, 488-489
Hepatocellular carcinoma (HCC), thioredoxin (TRX), 360, 363
Herpes simplex virus, hydrogen peroxide (H_2O_2), 196
High-performance liquid chromatography (HPLC), GSH, 200-201, 380
HIV
antioxidant defense, 413
ceramide, 439-440
cysteine, 273, 373-376
eicosanoid cascade, 395
erythrocyte membranes
fatty acids, 421-422
malonic dialdehyde (MDA), 421
vitamin C, 420-421
vitamin E, 420-421
zinc, 419-420
Fas-Fas ligand (FasL) system, 439
glutathione (GS), 373-376
heat stress protein (HSP), 121-122
IMMUNOCAL™, 451-452
lipidic anomalies, 412-413
N-acetylcysteine (NAC), 105, 262-263, 373-376
NFκB, 79-80, 103-104, 262, 394-395
peripheral blood lymphocytes, protein degradation, 399-406
plasma
copper, 419
selenium, 413, 415-417
zinc, 417-419
replication, oxidant-induced, 104-107
vitamin A, 413
vitamin C, 413
vitamin E, 413
HIV-1
chronic hepatitis B-related cirrhosis, interferon-α, 463-466
Cu, Zn-SOD, 307
lipoate, 262

HIV-1
  oxidative stress, 306-307
  radical oxygen intermediates (ROI), 196
  superoxide dismutase (SOD), 463-466
HIV-1 long terminal repeat (LTR), arachidonic acid (AA), 389-396
Human promonocytic cells (U937), 390
Human umbilical vein endothelial cells (HUVEC), 314
Humanized native milk serum protein isolate (HNMPI) *see* IMMUNOCAL™
Hydroethydine oxidation, peripheral blood lymphocytes, 399-402
Hydrogen peroxide (H_2O_2)
  apoptosis, 25-26, 194-195
  bacterial response, 29-32
  calcium flux, 258
  cell growth, 191-192, 194-195
  DAG, 49-50
  DNA, 17
  endothelial cells, 313-321
  generation, 192-193
  heat stress protein (HSP), 116
  herpes simplex virus, 196
  HIV, 104, 106-07
  intracellular levels, 194
  iron, 225
  irradiation, 17-25
  lens, 507-504
  messenger functions, 97-99
  NFκB, 78, 83-86, 94, 103-104, 258-260
  permeability transition (PT), 269, 280
  phospholipase D (PLD) activation, 53-58
  reactivity, 10
  regulation, 194
  tyrosine kinase signal pathways, 40-41
Hydroperoxides, reactivity, 10
Hyperthermia, reactive oxygen species (ROS), 370
Hypochlorite. *see* Hypochlorous acid (HOCI)
Hypochlorous acid (HOCI)
  apoptosis, 25-26
  iron, 226
  irradiation, 17-25
  lipoate, 255
Hypoxia-reperfusion injury, 518-519

ICAM-1, TNF-α, 314, 317, 320
IgG antibodies, cell activation, 147-157
IκB, 83, 89-92, 102, 104, 521
IκB kinase, 76, 94
IκB protein family, NFκB activation, 65-66, 258
IL-1B, NFκB, 70-72, 90-92
Immunoblotting analysis, thioredoxin (TRX), 361-362
IMMUNOCAL™
  AIDS, 452-455
  apoptosis, 451-452
  breast cancer cells, 457
  buthionine sulfoximine, 449
  cancer clinical trials, 457-458
  cysteine delivery system, 448-449, 452
  glutathione (GSH), 447-448, 450-452
  HIV infection, 451-452
  human milk, 458-459
  liver cancer cells, 457
  pancreas cancer cells, 457
  wasting syndrome, 452-455
Immunomodulation
  flavonoids, 292
  superoxide dismutases (SOD), 305-310
Immunoreceptor family, 147-149
Immunoreceptor tyrosine-based activation motif (ITAM), 147-149, 154-155
Immunoreceptor tyrosine-based inhibition motif (ITIM), 150-153
Inflammation
  flavonoids, 291-292
  heat stress protein (HSP), 119, 121-122
  hydrogen peroxide (H_2O_2), 98-99, 196
  superoxide, 196, 310
Inositol phosphatidyl, bombesin/gastrin-releasing peptides (GRP), 166-167
Insulin, bombesin/gastrin-releasing peptides (GRP), 164
Insulin-dependent diabetes mellitus (IDDM), 499
Insulin receptor B-chain, regulation, 373-374
Interferon, manganese superoxide dismutase (Mn-SOD), 307
Interferon-α, HIV-1, 463-466
Interleukin-8 (IL-8), hydrogen peroxide (H_2O_2), 100
Intravita digital microfluorography, 504-505
Ionizing radiation, lymphocytes, 38-39
Iron, 223
  antioxidants, 225-226
  HIV, 104-107
Irritation syndrome, 503-505
Ischemia, heat stress protein (HSP), 119
Ischemia-reperfusion, 503-505
  flavonoids, 292
  leukocyte adhesion, 504
  microvascular events, 504-505
  reactive oxygen species (ROS), 370

Jurkat JR (Wurzburg) cells, NFκB, 84-86
Jurkat T lymphocytes
glutathione (GSH), 180-184, 456
lipoate, 258-260, 262

L-arginine transport
low-density lipoproteins (LDL), 129-130, 134-135
endothelial cells, 129
Laser confocal video microscopy, 505
L-buthionine-(S, R)-sulfoxine (BSO)
apoptosis, 202
GSH, 202
L-carnitine
AIDS, 437-444
apoptosis, 440-444
ceramide, 440-444
Lens (eye), antiperoxidase defense system, 507-514
Leukocyte adhesion, ischemia-reperfusion, 504
Lewy bodies, 490
Lipid peroxide (LOOH), iron, 226
Lipid peroxidation
heat stress protein (HSP), 119
iron, 226
peroxynitrite, 14
superoxide dismutases (SOD), 3-5
vitamin E, 256
Lipoate
acquired immunodeficiency syndrome (AIDS), 262-263
antioxidant properties, 254-256
glutathione (GSH)257, 263
hydrochlorous acid (HOCl), 255
NFκB, 260-261
nitric oxide (NO), 255
therapeutic potential, 251-263
vitamin C deficiency, 257
vitamin E deficiency, 257
Lipoic acid, apoptosis, 209
Lipoyl residues, oxidative metabolism, 252-253
Liprotein A, CD4 T cells, 412
Liver
cyanidanol, 287
silybinin, 287
Liver cancer cells, IMMUNOCAL™, 457
Liver cirrhosis (LC), thioredoxin (TRX), 360, 362
Long terminal repeat (LTR)-chloramphenicol acyltransferase (CAT) assay, 390
Low-density lipoproteins (LDL)
[Low-density lipoproteins (LDL)]
cGMP, 130-132
heme oxygenase-1, 132-134
L-arginine transport, 129-130, 134-135
nitric oxide (NO)127, 134
PGI2, 130-132, 134-135
Lung cancer cells, glutathione (GSH), 455
Lymphocytes
antigen receptor signaling, 35-37, 40
hydrogen peroxide (H_2O_2), 39-40
ionizing radiation, 38-39
phosphotyrosine phosphatases, 40-41
ultraviolet radiation, 38-40

Maillard reaction, *see* Glycation reaction
Malignant disease, thioredoxin (TRX) serum levels, 359-366
Malondialdehyde (MDA), 430
erythrocyte membrane, HIV, 421
Manganese superoxide dismutase (Mn-SOD), 305-306, 329-330
interferon, 307
nitric oxide (NO), 310
peripheral blood lymphocytes, 399-402
TNF-α, 310
Mast cell activation, FcR-mediated, FcγRIIB, 150, 155
Metal-catalyzed oxidation (MCO), 486
Methylene blue (MB), NFκB, 66-67, 72
Milk, IMMUNOCAL™, 458-459
Mitochondria
apoptosis, 213-219
Bcl-2, 216-217
brain
catalase, 325
SOD, 323-330
heart, 325
heat stress protein (HSP), 119-120
liver, 325, 346-347
oxidative stress, 215-218
reactive oxygen species (ROS), 217-218
superoxide, 333-347
Mitochondrial inner transmembrane potential, apoptosis, 214-215
Mitochondrial matrix
ATP hydrolysis, 344-345
succinate oxidation, 344-345
Mitochondrial proton leakage, superoxide radicals, 341-345
Mitochondrial respiratory chain, 333-347
Mitogen-activated protein (MAP) kinases, bombesin, 167

Mitogenic neuropeptides, signal transduction pathways, 163-173
Multiparameter fluorescence activated cell sorter (FACS), GSH, 380
MutY protein, 353-354

N-acetylcysteine (NAC)
  AIDS, 379-386
  apoptosis, 14, 202
  arginine, 376
  cysteine, 376
  glutamine, 376
  GSH, 202
  HIV, 105, 262-263, 373-376
  NFκB, 78-79, 103, 258, 260-262
  tumor necrosis factor-α, 100-101
NADH oxidation, 271-281
NADPH oxidase, NFκB, 526-527
NADPH oxidation, 271-281
Necrosis, 120
Neurofibrillary tangles (NFT), 487-490
  advanced lipid peroxidation-derived modification, 489-490
Neuromelanin, 472
Neutrophils, apoptosis, 205-210
NFκB, 518-24
  activation
    antioxidants, 66-68, 92-94, 257-261, 520-521
    cellular source, 526-529
    cytokines, 70-71, 258
    DNA, 68-70, 373
    hydrogen peroxide (H_2O_2), 78, 83-86, 94, 258-260
    IκB family, 65-66
    inhibition, 258
    kinase pathways, 76, 94
    nitric oxide (NO), 92, 527
    nuclear translocation, 63-66, 76, 90-92
    oxidation-induced, 102-104
    photosensitization, 66-68
    redox regulation, 76-78, 89-95, 103, 118-119, 518
    Rel family, 64-66, 258
  anti-inflammatory drugs, 529-531
  apoptosis, 17
  arachidonic acid (AA), 527
  flavonoids, 305-306
  HIV, 394-395
  NADPH oxidase, 526-527
  pathophysiology, 75-76, 79-80
  phosphotyrosine phosphatases (PTP), 41
  radiation, 38
[NFκB]
  regulatory effects, 257-261
  rheumatoid synovium, 524-526
  structure, 76, 102
  substantia nigra, 479
  therapeutic relevance, 257-263
NFκB/IκBa system, 89-92
NFκB kinase, 76
NFκB signaling pathway, redox regulation, 75-81, 83, 89-95, 103, 118-119
Nitric oxide (NO)
  apoptosis, 13-14, 499
  glutathione peroxidase (GPX)499-500
  iron, 226
  lipoate, 255
  low-density lipoproteins (LDL), 127, 134
  manganese superoxide dismutase (Mn-SOD), 310
  NFκB, 92, 527
  prostacyclin (PCGI2) synthesis, 129
  reactivity, 9-10
  superoxide dismutases (SOD), 308-310
  thioredoxin system, 239-240
  tyrosine phosphorylation signaling pathways, 139-145
Nitrogen dioxide, reactivity, 9-10
Nuclear factor kappa B (NFκB) *see* NFκB

OGG1 gene, 354-355
OGG1 protein, 354-355
Oligomycin, 274, 277, 280
Oxidant-antioxidant balance, 1-5
Oxidative metabolism, lipoyl residues, 252-253
Oxidative stress
  amyotrophic lateral sclerosis (ALS), 490-491
  glutaredoxin (GRX), 234-235
  glutathione (GSH) compartmentation, 269-281
  heat shock protein (HSP), 113-114, 119-122
  HIV-1, 306-307
  mitochondria, 215-218
  neurodegenerative disease, 308
  Parkinson's disease, 472-474, 490
  spontaneous mutations, 351-356
  thioredoxin (TRX), 231-232
  thioredoxin (TRX)/adult t cell leukemia-derived factor (ADF), 248
Oxygen, 223-224
OxyR, 30-31, 104
Ozone, reactivity, 10

Pancreas cancer cells, IMMUNOCAL™, 457
Pancreatic β-cells, apoptosis, 499
Paraquat, superoxide dismutases (SOD) synthesis, 2
Parkinson's disease
apoptosis, 469-480
oxidative damage, 474-476
oxidative stress, 472-474, 490
Paxillan, tyrosine phosphorylation, 170
Peripheral blood lymphocytes
HIV, 399-406, 437-444
hydroethydine oxidation, 399-402
Mn-SOD, 399-402
protein synthesis and degradation, 402-406
Peripheral blood mononuclear cells (PBMC)
deoxy-D-ribose (dRib), 199-203
HIV, GSH, 451
Permeability transition (PT), 269-281
apoptosis, 214-215
cyclosporin A, 269
hydrogen peroxide (H_2O_2), 269
reactive oxygen species (ROS), 215-218
t-butylhydroperoxide, 269
Permeability transition studies, 270-281
methods, 270-271
results, 271-281
Phagocytosis
glutathione (GSH), 186
heat stress protein (HSP), 121-122
Phenol-3, 6-dibromosulfophthalein disulfonate (diBSP), glutathione (GSH), 183
Phorbol esters (PMA), NFκB, 94
Phospholipase A_2, activation, 46-47
Phospholipase C (PLC)
activation, 47-48
bombesin/gastrin-releasing peptides (GRP), 170-172
bovine pulmonary endothelium cells (BPECs), 47-48
Phospholipase D (PLD)
activation, 51-58
mechanism, 54-57
phosphatase inhibitors, 57-58
physiological significance, 58-59
tyrosine phosphorylation, 54-57
bovine pulmonary endothelium cells (BPECs), 51-53
Phosphotyrosine phosphatases (PTP)
apoptosis, 38
NFκB, 41
tyrosine phosphorylation, 36-37, 40-41
Phytochemicals, 285
Ploymorphonuclear leukocytes (PMN), apoptosis, 205-210
Programmed cell death *see* Apoptosis
Prokaryotes, 8-OxoG repair, 353-354
Prostacyclin (PGI2), low-density lipoproteins (LDL), 130-132, 134-135
Prostacyclin (PGI2) synthesis, nitric oxide (NO), 129
Prostaglandin
bombesin/gastrin-releasing peptides (GRP), 167-168
vasopressin, 167-168
Protein degradation, 402-405
ubiquitin-dependent proteolytic system, 404-405
Protein glycation theory, 498-499
Protein kinase C cascade
apoptosis, 17
NFκB, 102
radiation, 38
Protein kinase C (PKC), bombesin/gastrin-releasing peptides (GRP), 166-167
Protein oxidation, 486-487
aging, 491-492
Protein tyrosine kinases (PTK), tyrosine phosphorylation, 35-37
Protein tyrosine phosphatases (PTP), tyrosine phosphorylation, 57-58
Protein X, 253
Proton leakage, superoxide radicals, 346-347
Pyrrolidine dithiocarbamate (PDTC), NFκB, 69-70, 72, 258

Q-cycle, 333-347
Quercetin, 289-290
apoptosis, 292
heat shock response, 292

Radical oxygen intermediates (ROI), 75
HIV-1, 196
NFκB, 78-79, 92-94
Rat-1 cells, bombesin/gastrin-releasing peptides (GRP), 170-172
Reactive oxygen cycle, 334-347
Reactive oxygen intermediates (ROI), rheumatoid arthritis, 517-518
Reactive oxygen species (ROS)
adriamycin, 370
apoptosis, 195-196, 217-218
heat shock protein (HSP), 113, 115
HIV, 104

[Reactive oxygen species (ROS)]
hyperthermia, 370
ischemia/reperfusion injury, 370
mitochondria, 217-218
NFκB, 66-68, 72
permeability transition (PT), 215-218
reactivity, 9-10
Redox factor-1 (Ref-1), 522
Redox index, 326
Redox regulation
apoptosis, 195-196
cell growth, 192, 195
defined, 75
heat shock protein (HSP), 113-122
heat shock protein (HSP) vs. NFκB, 118-119
NFκB signaling pathway, 75-81, 83, 89-95, 103
thioredoxin (TRX)/adult t cell leukemia-derived factor (ADF), 247-249
Reilly's irritation syndrome, 503-505
Rel protein family, NFκB activation, 64-65
Reperfusion injury, superoxide dismutases (SOD), 3-4
Rheumatoid arthritis (RA)
glutathione (GSH), 539-540
NFκB, 76, 524-526, 531
oxidative stress, 542-543
pathology, oxidative stress, 542-543
reactive oxygen intermediates (ROI), 517-518
T cells, 537-539
GSH, 540-542
thioredoxin (TRX), 540
Rheumatoid synovitis, 519
Rheumatoid synovium
activator protein-1 (AP-1), 524-526
NFκB, 524-526
Rose bengal, NFκB, 68, 72
rpoS-encoded σs subunit, 31-32

Saccharomyces cerevisiae, 8-OxoG, 354-355
Salmonella typhimurium
hydrogen peroxide (H_2O_2), 30-31
oxidative regulation, 104
Scurvy, lipoate, 254
Selenium
AIDS, 413-423
HIV, 413, 415-417
thioredoxin system, 236-239
Selenocysteine, thioredoxin reductase (TR), 240-242
Sepsis, heat stress protein (HSP), 119
Serum-responsive factor (SRF), 522
Signal transduction pathways, mitogenic neuropeptides, 163-173
Silybinin, hepatoprotective effect, 287
Singlet oxygen ($^1O_2$)
NFκB, 66-67, 72
reactivity, 10
Smooth-muscle cells
heme oxygenase-1, 129-130, 132-134
low density lipoproteins (LDL), 129-135
S-nitro-N-acetyl-DL-penicillamine (SNAP), 499-500
Sodium oxodiperoxo(1, 10-phenathroline) vanadate(V) [pV(phen)], 40-41
Sodium salicylate, NFκB, 529
SoxR, 29-30
soxRS regulon, regulation, 29-30
SoxS, 29-30
Spontaneous mutations, oxidative stress, 351-356
Substantia nigra
NFκB, 479
oxidative damage, 474-476
tumor necrosis factor-α (TNF-α), 478-479
Succinate oxidation, mitochondrial matrix, 344-345
Sulfide, 274, 277, 280
Superoxide
cell growth, 191
electron leak pathway, 334
generation, 192-193, 337-341
mitochondria, 333-347
regulation, 194
superoxide dismutases (SOD), 340
Superoxide dismutases (SOD), 1-5
adriamycin, 370-371
apoptosis, 208-210
brain, 308
enzyme activities, perinatal brain development, 323-332
erythrocyte hemolysis, 11
evolution, 97
HIV-1, 306-307, 463-466
hydrogen peroxide (H_2O_2), 194
hyperthermia, 370
immunomodulation, 305-310
ischemia-reperfusion injury, 370
nitric oxide (NO), 308-310
oral administration, 310
protective functions, 114
reactivity, 9-10
reperfusion injury, 3-4, 371
superoxide, 340
[Superoxide dismutases (SOD)]

toxicity, 1-3
Superoxide radicals
$H^+/2e^-$, 345
heat production, 346-347
proton leakage, 346-347
Superoxides, dihydrolipoate (DHLA), 255-256
Swiss 3T3 cells, bombesin/gastrin-releasing peptides (GRP), 164-172

T-butylhydroperoxide, permeability transition (PT), 269, 280
T cell activation, TCR-mediated, FcγRIIB, 150, 155-156
T cell receptors (TCR), 147-149, 155-156
T cells
antigen receptor signals, 37
rheumatoid arthritis (RA), 537-539
GSH, 540-542
ultraviolet radiation, 38-39
Tetradeconyll phorbol acetate-responsive elemenet (TRE) *see* Activator protein-1 (AP-1)
Thapsigargin, 262
Thiobarbituric acid-reactive substances (TBARS), 369
Thiols, 251-252, 257
HIV, 262
Thioredoxin reductase (TR), 253-254
mammalian structure, 240-242
mammalian vs. *E. coli*, 235-236
selenocysteine, 240-242
Thioredoxin system, 229-230
nitric oxide (NO), 239-240
selenium, 236-239
Thioredoxin (TRX)
biological activities, 359-360
heat stress protein (HSP), 115-118
NFκB, 76-78, 92, 522-524
oxidative stress, 231-232
protective function, 114
rheumatoid arthritis (RA), 540
serum levels, gender differences, 365
serum level studies
in chronic gastritis patients, 363-364
disucssion, 364-366
in HCC patients, 363
in healthy volunteers, 362
immunoblotting analysis, 362
in LC/CH patients, 362
methods, 360-361
structure, 230-231, 247, 359
Thioredoxin (TRX)/adult t cell leukemia-derived factor (ADF)
intracellular signaling, 248
oxidative stress, 248
redox regulation, 247-249, 248
ultraviolet irradiation, 248
Topoisomerase poisons, NFκB, 68-70
trans-acting transcriptional activator (Tat), 306-307
Transferrin, 226
Triglycerides, CD4 T cells, 411
Tumoregenesis, bombesin/gastrin-releasing peptides (GRP), 164
Tumor necrosis factor-α (TNF-α)
apoptosis, 476-478
ceramide, 438-439
endothelial cells 313-321
heat stress protein (HSP), 119
HIV, 104, 106
manganese superoxide dismutase (Mn-SOD), 310
NFκB, 70-72, 83, 90-92, 258
oxidants, 100-102, 154
substantia nigra, 478-479
Tyrosine kinase pathway, bombesin/gastrin-releasing peptides (GRP), 170-172
Tyrosine kinase signal pathways
hydrogen peroxide (H_2O_2), 40-41
NFκB, 102
Tyrosine phosphorylation
bombesin/gastrin-releasing peptides (GRP), 168-172
focal adhesion kinase ($p125^{fak}$), 168-170
lymphocytes, 35-41
paxillan, 170
phospholipase D (PLD) activation, 54-57
vascular endothelial cells (ECs), 45-46
Tyrosine phosphorylation signaling pathways
endogenous nitric oxide (NO),142-144
exogenous nitric oxide (NO),141-142

Ubiquitin-dependent proteolytic system, protein degradation, 404-405
Ultraviolet radiation
lymphocytes, 38-39
NFκB, 72
thioredoxin (TRX)/adult t cell leukemia-derived factor (ADF), 248
Uric acid, erythrocyte hemolysis, 11

Vascular endothelial cells (ECs)
phospholipase modulation, 45-60

[Vascular endothelial cells (ECs)]
tyrosine phosphorylation, 45-46
Vasopressin
arachidonic acid, 167-168
prostaglandin, 167-168
Vitamin A, and HIV, 413
erythrocyte membrane, 420-421
Vitamin C
erythrocyte hemolysis, 11
HIV, 413
NFκB, 103
radical-scavenging, 9
Vitamin C deficiency, lipoate, 257
Vitamin E
dihydrolipoate (DHLA), 256-257
erythrocyte hemolysis, 11
HIV, 413
erythrocyte membrane, 420-421
lipid peroxidation, 256
radical-scavenging, 9
Vitamin E deficiency, lipoate, 254, 257
VX2 carcinoma cells, 369-370

Wasting syndrome, IMMUNOCAL™, 452-455
Werner's syndrome, Cu, Zn-SOD, 498
Western blot analysis, heme oxygenase-1, 129-130
Whey proteins, milk, 458
Wurzburg cells, NFκB, 84-86, 258-259

X-rays, NFκB, 72

Zidovudine (AZT), 429-434
GSH, 432-433
malondialdehyde (MDA), 433
mitochondrial DNA, 434
Zinc
erythrocyte membrane, HIV, 419
HIV, 417-419